Springer Proceedings in Mathematics & Statistics

Volume 527

This book series features volumes composed of selected contributions from workshops and conferences in all areas of current research in mathematics and statistics, including data science, operations research and optimization. In addition to an overall evaluation of the interest, scientific quality, and timeliness of each proposal at the hands of the publisher, individual contributions are all refereed to the high quality standards of leading journals in the field. Thus, this series provides the research community with well-edited, authoritative reports on developments in the most exciting areas of mathematical and statistical research today.

Preface

In September 2022, the four of us organized the conference "*Elliptic Curves and Modular Forms in Arithmetic Geometry*" at the University of Milano; the goal of the conference, which was preceded by the School on Arithmetic Geometry that took place at Universität Duisburg-Essen, was to celebrate Massimo Bertolini, and the impact of his mathematics on number theory, on the occasion of his 60th birthday.

The two events in Essen and Milano turned out to be successful both scientifically and in terms of attendance: it is remarkable that a significant number of participants in Milano were Ph.D. students or early postdocs. The present volume mainly consists of (but is not limited to) articles that grew out of talks delivered at the conference.

It is therefore our pleasure to dedicate this volume to Massimo, in admiration for his fundamental contributions to arithmetic geometry and as a sign of our heartfelt gratitude for the intellectual influence he had and still has on us and on many other students and collaborators.

Padova, Italy — Matteo Longo
Milan, Italy — Marco Adamo Seveso
Milan, Italy — Rodolfo Venerucci
Genova, Italy — Stefano Vigni
October 2025

Contents

On the Tate-Shafarevich Groups of CM Elliptic Curves Over Anticyclotomic $\mathbb{Z}_p$-Extensions at Inert Primes 1
Ashay A. Burungale, Shinichi Kobayashi, and Kazuto Ota

Nonvanishing of Generalised Kato Classes and Iwasawa Main Conjectures 23
Francesc Castella

Flach Classes and Generalised Hecke Eigenvalues 51
Henri Darmon and Alice Pozzi

On Constructing Extensions of Residually Isomorphic Characters 75
Samit Dasgupta

On the Proalgebraic Fundamental Group of Topological Spaces and Amalgamated Products of Affine Group Schemes 101
Christopher Deninger and Michael Wibmer

Non-Archimedean Plectic Jacobians 153
Michele Fornea and Lennart Gehrmann

A p-Adic Gross-Zagier Formula for the Triple p-Adic L-Function at Non-crystalline Points 183
Francesca Gatti and Victor Rotger

On the Anticyclotomic Mazur–Tate Conjecture for Elliptic Curves with Supersingular Reduction 229
Chan-Ho Kim

The Bertolini-Darmon-Prasanna p-Adic L-Function via q_{dR}-Expansions 241
Daniel Kriz

Anticyclotomic p-Adic L-Functions for Rankin–Selberg Product 283
Yifeng Liu

Spherical Varieties and p-Adic Families of Cohomology Classes 321
David Loeffler, Robert Rockwood, and Sarah Livia Zerbes

Reciprocity Laws for Generalized Heegner Classes 349
Marco Adamo Seveso

Congruences of Modular Forms and Modularity of Tate–Shafarevich Classes 377
Matteo Tamiozzo

On the Tate-Shafarevich Groups of CM Elliptic Curves Over Anticyclotomic $\mathbb{Z}_p$-Extensions at Inert Primes

Ashay A. Burungale, Shinichi Kobayashi, and Kazuto Ota

Abstract We investigate the asymptotic behavior of the p-part of the Tate-Shafarevich group of a CM elliptic curve over the anticyclotomic $\mathbb{Z}_p$-extension of the CM field at inert primes $p \geq 5$. A key tool of the proof is our recent work on anticyclotomic CM Iwasawa theory at inert primes.

Keywords CM elliptic curves · Iwasawa theory · Tate-Shafarevich groups

1 Introduction

Let E be an elliptic curve defined over a number field K. A mysterious arithmetic invariant is the conjecturally finite Tate-Shafarevich group $Ш(E/F)$ for a finite extension F/K. One may seek to study the variation of $Ш(E/F)$ as F-varies.

Let p be a prime, K_∞ a $\mathbb{Z}_p$-extension of K and K_n its n-th layer. In light of Iwasawa's seminal formula [17] for the p-part of the class number of K_n, one may seek to explore the asymptotic behavior of the p-part of the Tate-Shafarevich groups $Ш(E/K_n)$ as n varies. Mazur initiated the study of this problem in the early 70's [24], essentially treating the ordinary case: if E has good ordinary reduction at the primes above p, then there exist integers $\lambda, \mu \in \mathbb{Z}_{\geq 0}$ such that for any sufficiently large n one has

$$\mathrm{length}_{\mathbb{Z}_p} Ш(E/K_n)[p^\infty] - \mathrm{length}_{\mathbb{Z}_p} Ш(E/K_{n-1})[p^\infty] = \lambda + \mu\phi(p^n). \quad (1)$$

A. A. Burungale
The University of Texas at Austin, Texas, USA
e-mail: ashayburungale@gmail.com

S. Kobayashi (✉)
Faculty of Mathematics, Kyushu University, Fukuoka, Japan
e-mail: kobayashi@math.kyushu-u.ac.jp

K. Ota
Department of Mathematics, Graduate School of Science, Osaka University Toyonaka, Osaka, Japan
e-mail: kazutoota@math.sci.osaka-u.ac.jp

M. Longo et al. (eds.), *Elliptic Curves and Modular Forms in Arithmetic Geometry*, Springer Proceedings in Mathematics & Statistics 527,
https://doi.org/10.1007/978-3-032-13123-2_1

Here it is assumed that $Ш(E/K_m)[p^\infty]$ is finite for any $m \geq 0$, the integers λ and μ are the Iwasawa invariants of an Iwasawa module associated to certain Selmer groups, and $\phi(p^n) = p^{n-1}(p-1)$ (see [12, 24] and [21, Theorem C]).

In contrast the supersingular case exhibits new phenomena and is still not well-understood. For the rest of this section, suppose that E is defined[1] over $\mathbb{Q}$ and has good supersingular reduction at p. The case of the cyclotomic $\mathbb{Z}_p$-extension of $\mathbb{Q}$ has been studied by Nasybullin [26], Kurihara [20], Perrin-Riou [28], the second-named author [19] and [35]. For example, if $p \geq 5$, for any sufficiently large n, one has

$$\text{length}_{\mathbb{Z}_p} Ш(E/K_n)[p^\infty] - \text{length}_{\mathbb{Z}_p} Ш(E/K_{n-1})[p^\infty]$$
$$= \begin{cases} \sum_{m=0}^{\frac{n-2}{2}} \phi(p^{n-1-2m}) + \lambda_+ + \phi(p^n)\mu_+ & (n\text{: even}) \\ \sum_{m=0}^{\frac{n-3}{2}} \phi(p^{n-1-2m}) + \lambda_- + \phi(p^n)\mu_- & (n\text{: odd}) \end{cases} \quad (2)$$

where the integers $\lambda_\pm, \mu_\pm \in \mathbb{Z}_{\geq 0}$ are Iwasawa invariants of the associated $\pm$-Selmer groups (cf. [19, Theorem 1.4]). Note that the above formula (2) always differs from the ordinary case (1), even if $\lambda_\pm = \mu_\pm = 0$ (for example if the algebraic part of the L-value $L(E, 1)$ is p-indivisible). In this cyclotomic setting Kato and Rohrlich proved that $\text{rank}_{\mathbb{Z}} E(K_\infty) < +\infty$, which is an ingredient in the proof of (2).

If K differs from $\mathbb{Q}$ and K_∞ differs from the cyclotomic $\mathbb{Z}_p$-extension, the prior results consider the case of imaginary quadratic fields K with *p split* (cf. [9, 16, 22, 23]). Under the Heegner hypothesis[2] for E over K, Lei-Lim-Muller recently found an asymptotic formula for $\text{length}_{\mathbb{Z}_p} Ш(E/K_n)$ along the anticyclotomic $\mathbb{Z}_p$-extension of K (see [22, Theorem A]). In this case $\text{rank}_{\mathbb{Z}} E(K_n)$ grows unlike the cyclotomic case, and the asymptotic formula also differs from (2). Other $\mathbb{Z}_p$-extensions of K have also been considered by Iovita and Pollack if the algebraic part of the L-value $L(E/K, 1)$ is p-indivisible (see [16, Theorem 5.1 and Corollary 5.11]).

The aim of this paper is to initiate the study of the variation of $\text{length}_{\mathbb{Z}_p} Ш(E/K_n)$ over imaginary quadratic fields K with *p inert*, a setting complementing[3] the conjectural backdrop. We consider the variation for CM elliptic curves E with CM by an order of K along the anticyclotomic $\mathbb{Z}_p$-extension of K (see Theorems 1.1 and 1.3). In this case E has supersingular reduction at p. Moreover, the variation of $\text{rank}_{\mathbb{Z}} E(K_n)$ notably differs from the above cases (cf. [6, Corollary 3.11] or (5)). The pertinent CM Iwasawa theory was envisioned by Rubin in the late 80's [31]. The recent proof of Rubin's conjecture [5] on local units along the anticyclotomic $\mathbb{Z}_p$-extension of the unramified quadratic extension of $\mathbb{Q}_p$ has led to development of Rubin's theory [6–8], and underlies this paper. see also [7].

[1] Even if K differs from $\mathbb{Q}$.

[2] In the late 80's Perrin-Riou [27] initiated Iwasawa theory of Heegner points, based on which, the anticyclotomic Tate-Shafarevich groups have been studied by Bertolini [2] and Howard [15], among others.

[3] This setting complements the conjectural framework of Iwasawa theory due to Coates, Perrin-Riou and Greenberg. In the existent framework, one can often make an educated guess regarding the variation of Tate–Shafarevich groups.

1.1 Results

We first introduce the setting.

Let E be a CM elliptic curve defined over $\mathbb{Q}$ and K the CM field. Let $p \geq 5$ be a prime of good reduction for E which remains inert in K. Let K_∞ denote the anticyclotomic $\mathbb{Z}_p$-extension of K and K_n its n-th layer. A basic problem is to study the behaviour of $\text{length}_{\mathbb{Z}_p} Ш(E/K_n)$ as n varies.

Let φ be the Hecke character over K associated to E, and so $L(E/\mathbb{Q}, s) = L(\varphi, s)$. In the early 80's Greenberg found the interesting root number formula

$$\varepsilon(\varphi\chi) = (-1)^{n-1}\varepsilon(\varphi) \tag{3}$$

for χ a character of $\text{Gal}(K_\infty/K)$ of order $p^n > 1$. In particular the root numbers vary along the alternate anticyclotomic layers, and non-torsion rational points appear in one of the layers. This perhaps suggests that the variation of $\text{length}_{\mathbb{Z}_p} Ш(E/K_n)$ depends on the parity of n.

1.1.1. Our main result considers one of the anticyclotomic layers:

Theorem 1.1 *Let E be a CM elliptic curve defined over $\mathbb{Q}$ and $\varepsilon \in \{\pm 1\}$ its root number. Let K denote the associated CM field. Let $p \geq 5$ be a prime of good supersingular reduction for E, and K_∞ the anticyclotomic $\mathbb{Z}_p$-extension of K. Then there exist integers $\mu, \lambda \in \mathbb{Z}_{\geq 0}$ such that for any sufficiently large $n \geq 0$ with $(-1)^{n+1} = \varepsilon$, the cokernel of the restriction map*

$$Ш(E/K_{n-1})[p^\infty] \to Ш(E/K_n)[p^\infty]$$

is finite, and

$$\text{length}_{\mathbb{Z}_p} \left(\text{Coker}\left(Ш(E/K_{n-1})[p^\infty] \to Ш(E/K_n)[p^\infty]\right)\right) = \lambda + \phi(p^n)\mu. \tag{4}$$

The main text treats the case of a certain CM abelian varieties over K, namely the Weil restriction of Gross curves as in Sect. 2.1 (cf. Theorem 4.11).

Remark 1.2

(a) For any n sufficiently large, one has

$$\text{rank}_{\mathbb{Z}} E(K_n) - \text{rank}_{\mathbb{Z}} E(K_{n-1}) = c_n\phi(p^n), \tag{5}$$

where $c_n = 2$ if $(-1)^{n+1} = -\varepsilon$ and $c_n = 0$ if $(-1)^{n+1} = \varepsilon$. Since the variation of $\text{rank}_{\mathbb{Z}} E(K_n)$ appears to be intertwined with the variation of $\text{length}_{\mathbb{Z}_p} Ш(E/K_n)$, we expect the phenomenon to differ from (4) if $(-1)^{n+1} = -\varepsilon$.

(b) Suppose that $\text{length}_{\mathbb{Z}_p} Ш(E/K_m)$ is finite for all m. Then for $n \gg 0$ with $(-1)^{n+1} = \varepsilon$, the left hand side of (4) coincides with the quantity $\text{length}_{\mathbb{Z}_p}(Ш(E/K_n))[p^\infty] - \text{length}_{\mathbb{Z}_p}(Ш(E/K_{n-1})[p^\infty])$ (cf. Lemma 4.5).

(c) The integers λ and μ are given explicitly in terms of Iwasawa invariants associated to strict Selmer groups and certain localization maps. Alternatively, they may be described in terms of Rubin's p-adic L-function (cf. Theorem 4.11 and Remark 4.12).

(d) For χ as in Theorem 1.1, we recently [7] determined the asymptotic p-adic valuation of the algebraic part of L-values $L(\varphi\chi, 1)$. The formula has the same shape as Pollack's formula [29] for the cyclotomic $\mathbb{Z}_p$-deformation, yet the variation of the Tate-Shafarevich groups differs from (2). This is due to the regulators $\mathrm{reg}(E/K_n)$: while in the cyclotomic

$$\mathrm{reg}(E/K_n) = p^{\mathrm{rank}_{\mathbb{Z}} E(K_\infty)} \mathrm{reg}(E/K_{n-1})$$

for $n \gg 0$, in the present case $\mathrm{reg}(E/K_n) = p^{\mathrm{rank}_{\mathbb{Z}} E(K_n)} \mathrm{reg}(E/K_{n-1})$ for $n \gg 0$ with $(-1)^{n+1} = \varepsilon$. In view of the variation of $\mathrm{rank}_{\mathbb{Z}} E(K_n)$ as in (5) the results of [7] are thus compatible with Theorem 1.1. In fact, employing the Birch and Swinnerton-Dyer formula for CM abelian varieties [3, 32], this argument may lead to a (different) proof of Theorem 1.1.

(e) In the inert case, Çiperiani [10] investigates the cotorsionness of $\varinjlim_n Ш(E/K_n)[p^\infty]$ under Heegner hypothesis. Although it is not satisfied in our case, the cotorsionness follows from [1, Proposition 3.3 (i)] and (5).

(f) The vanishing of the μ-invariant appearing in (4) is established in the sequel [4].

1.1.2. In primitive settings the asymptotic behavior of $Ш(E/K_n)$, without fixing the parity of n, is given by the following.

Theorem 1.3 *Let E be a CM elliptic curve over $\mathbb{Q}$, and $\varepsilon \in \{\pm 1\}$ the root number. Let K be the CM field. Let $p \geq 5$ be a prime of good supersingular reduction for E, and K_∞ the anticyclotomic $\mathbb{Z}_p$-extension of K.*

(1) Suppose that the Iwasawa module $\mathscr{X}^0$ associated to strict Selmer groups as in Sect. 3.3 is finite. Then for any sufficiently large $n \geq 0$, the restriction map

$$Ш(E/K_{n-1})[p^\infty] \to Ш(E/K_n)[p^\infty]$$

is surjective. In particular, if $Ш(E/K_n)[p^\infty]$ is also finite for sufficiently large n, then there exists an integer $m \geq 0$ such that for any $n \geq m$ we have

$$Ш(E/K_n)[p^\infty] \cong Ш(E/K_m)[p^\infty].$$

(2) Suppose that $L(E, 1)/\Omega$ is p-indivisible for Ω the Néron period if $\varepsilon = 1$, and that $L'(E, 1)/\Omega \cdot \mathrm{reg}(E)$ is p-indivisible and $E(\mathbb{Q}) \otimes \mathbb{Z}_{(p)}$ is dense[4] in $E(\mathbb{Q}_p) \otimes \mathbb{Z}_{(p)}$ with respect to the p-adic topology if $\varepsilon = -1$. Then

[4] That is $E(\mathbb{Q}) \not\subset pE(\mathbb{Q}_p)$.

$$Ш(E/K_n)[p^\infty] = \{0\}$$

for any integer $n \geq 0$.

The main text treats the case of a certain CM abelian varieties (see Theorems 4.15 and 4.17).

Remark 1.4 Curiously, the variation of Tate-Shafarevich groups as in Theorems 1.1 and 1.3 resembles the case of anticyclotomic $\mathbb{Z}_p$-deformation of an elliptic curve over an imaginary quadratic field with p split, under Heegner hypothesis, due to Lei-Lim-Muller [22] (for example, the quantity of [22, (1.3)] does not appear in [22] or in our cases). In the latter case the root number is -1 for any finite order twist, unlike our setting (cf. (3)). See also §1.2.1 for the difference of the proof. It is instructive to extensively compare and contrast our results and strategy with that of [22], which we leave to the interested reader.

1.2 Remarks on the Proof

One typically expects Iwasawa theory to encode the variation of the Tate-Shafarevich groups. In the supersingular settings of [19, 22] plus/minus Iwasawa theory underlies the proofs, and so it does in our setting. Such a theory was suggested by Rubin [31] and recently developed in [5–7] (see Sect. 3).

If $(-1)^{n+1} = \varepsilon$, then $\mathrm{rank}_{\mathbb{Z}}\, E(K_n) = \mathrm{rank}_{\mathbb{Z}}\, E(K_{n-1})$ for $n \gg 0$, and the situation is similar to that of the cyclotomic deformation [19] where $\mathrm{rank}_{\mathbb{Z}}\, E(K_\infty) < +\infty$. Our proof of Theorem 1.1 is based on a comparison between strict Selmer groups and usual Selmer groups and a control theorem (see Sect. 4.2), the argument being a variant of the cyclotomic case [19]. Under the assumptions of Theorem 1.3, a similar argument applies (see Sect. 4.3). In the root number -1 case of its part (2) we rely on the p-adic Beilinson formula established in [6]. We would like to emphasise that Rubin's conjecture is central to the proofs.

1.2.1. If $(-1)^{n+1} = -\varepsilon$, then $\mathrm{rank}_{\mathbb{Z}}\, E(K_{n+1}) - \mathrm{rank}_{\mathbb{Z}}\, E(K_n) = 2\phi(p^{n+1})$ for $n \gg 0$, and the situation appears to be similar to the anticyclotomic deformation [22], where a system of Heegner points controls the growth of $\mathrm{rank}_{\mathbb{Z}} E(K_n)$. Moreover, the associated BDP Selmer group satisfies a control theorem, leading to the main result of [22]. However, neither the Heegner hypothesis nor the BDP Selmer group are available in our setting, and the strategy of [22] does not seem to apply. We hope to consider this problem in the near future.

2 Backdrop

This section introduces the notation, and presents some basic properties of Galois cohomology groups and Selmer groups.

2.1 Set Up

Let $p \geq 5$ be a prime. Fix embeddings $\iota : \overline{\mathbb{Q}} \hookrightarrow \mathbb{C}$ and $\iota_p : \overline{\mathbb{Q}} \hookrightarrow \overline{\mathbb{Q}}_p$.

Let K be an imaginary quadratic field and h_K the class number. Suppose that p remains inert in K and that

$$p \nmid h_K.$$

Denote by H the Hilbert class field of K and by $N_{H/K}$ the norm map.

Let φ be a Hecke character over K of infinity type $(1, 0)$ and conductor $\mathfrak{f}$. We assume that the Hecke character

$$\psi := \varphi \circ N_{H/K}$$

over H is attached to a $\mathbb{Q}$-curve E in the sense[5] of Gross [14] such that E has good (supersingular) reduction at all primes of H dividing p. Denote by j_E the j-invariant of E. In view of [14, Theorem 10.1.3 and Lemma 11.1.1], we fix a model of E over $\mathbb{Q}(j_E)$. Let

$$A = \mathrm{Res}_{\mathbb{Q}(j_E)/\mathbb{Q}}(E)$$

be the Weil restriction of $E/\mathbb{Q}(j_E)$, which is an h_K-dimensional abelian variety of GL_2-type over $\mathbb{Q}$ associated to the theta series (of weight two) associated to φ. Note that the base change A/K of A to K is a simple CM abelian variety (cf. [25, Theorem 2]). Put $L := \mathrm{End}_K(A) \otimes \mathbb{Q}$, which is a CM-field containing K such that $[L : K] = h_K$ and $\mathrm{End}_{\mathbb{Q}}(A) \otimes \mathbb{Q}$ is a maximal totally real subfield of L. Let $\mathcal{O}_{K,p}$ denote the p-adic completion of $\mathcal{O}_K$.

Lemma 2.1 *The $\mathcal{O}_{K,p}$-algebra $\mathrm{End}_K(A) \otimes_{\mathcal{O}_K} \mathcal{O}_{K,p}$ is finite and étale.*

Proof This lemma is due to Gross: [14, Theorem 15.2.5] proves the assertion at least if $2 \nmid h_K$, and the same argument applies if $2 \mid h_K$. □

By Lemma 2.1, we have $\mathcal{O}_L \otimes_{\mathbb{Z}} \mathbb{Z}_p = \mathrm{End}_K(A) \otimes_{\mathcal{O}_K} \mathcal{O}_{K,p}$. In turn for $N \geq 1$, the set $A[p^N]$ of p^N-torsion of A is naturally an $\mathcal{O}_L \otimes_{\mathbb{Z}} \mathbb{Z}_p$-module equipped with continuous G_K-action. We fix a prime $\mathfrak{p}$ of L above $p\mathcal{O}_K$, and denote by $\mathcal{O}$ the integer ring of the completion of L at $\mathfrak{p}$. Put

$$T = T_p(A) \otimes_{\mathcal{O}_L \otimes \mathbb{Z}_p} \mathcal{O}, \quad W = A[p^\infty] \otimes_{\mathcal{O}_L \otimes \mathbb{Z}_p} \mathcal{O}.$$

Then T is naturally an $\mathcal{O}$-module free of rank one equipped with continuous G_K-action. By [31, (3)], we have an isomorphism of $\mathcal{O}[G_K]$-modules

$$T^\tau \cong \mathrm{Hom}_{\mathcal{O}}(W, L_{\mathfrak{p}}/\mathcal{O}(1)), \tag{6}$$

[5] That is E is an elliptic curve defined over H with complex multiplication by $\mathcal{O}_K$ such that E is isogenous to all of its $\mathrm{Gal}(H/\mathbb{Q})$-conjugates.

where T^τ denotes the representation induced by twisting the action of G_K by the complex conjugation $\tau \in \mathrm{Gal}(K/\mathbb{Q})$, and for a $\mathbb{Z}_p[G_K]$-module M we put $M(1) = M \otimes_{\mathbb{Z}_p} \varprojlim_n \mu_{p^n}$ with μ_{p^n} being the group of the p^n-th roots of unity.

For $n \geq 0$, let K_n be the n-th layer of the anticyclotomic $\mathbb{Z}_p$-extension K_∞ of K. We put $\Gamma = \mathrm{Gal}(K_\infty/K)$ and $\Lambda = \mathscr{O}[\![\Gamma]\!]$. Denote by $Ш_n$ the $\mathfrak{p}$-primary part of the Tate-Shafarevich group of A/K_n, which fits into the short exact sequence (7) below.

2.2 Selmer Groups

We introduce Selmer groups with varying local conditions.

For $n \geq 0$ and a prime v of K_n, we denote by $K_{n,v}$ the completion. Denote by $H^1_{\mathrm{f}}(K_{n,v}, W)$ the image of

$$A(K_{n,v}) \otimes_{\mathbb{Z}} \mathbb{Q}_p/\mathbb{Z}_p \to H^1(K_{n,v}, A[p^\infty]) \to H^1(K_{n,v}, W),$$

where the first map is the Kummer map. Define $H^1_{\mathrm{f}}(K_{n,v}, T^\tau)$ to be the orthogonal complement of $H^1_{\mathrm{f}}(K_{n,v}, W)$ under the perfect pairing

$$H^1(K_{n,v}, W) \times H^1(K_{n,v}, T^\tau) \to L_{\mathfrak{p}}/\mathscr{O}$$

induced by (6) and the local duality.

For $M \in \{W, T^\tau\}$, define

$$\mathrm{Sel}(K_n, M) = \mathrm{Ker}\left(H^1(K_n, M) \to \prod_v \frac{H^1(K_{n,v}, M)}{H^1_{\mathrm{f}}(K_{n,v}, M)}\right),$$

where v ranges over all primes of K_n. Then we have a short exact sequence

$$0 \to A(K_n) \otimes L_{\mathfrak{p}}/\mathscr{O} \to \mathrm{Sel}(K_n, W) \to Ш_n \to 0. \tag{7}$$

Also define

$$\mathrm{Sel}_{\mathrm{rel}}(K_n, M) = \mathrm{Ker}\left(H^1(K_n, M) \to \prod_{v \nmid p} \frac{H^1(K_{n,v}, M)}{H^1_{\mathrm{f}}(K_{n,v}, M)}\right),$$
$$\mathrm{Sel}_0(K_n, M) = \mathrm{Ker}\left(\mathrm{Sel}_{\mathrm{rel}}(K_n, M) \to H^1(K_{n,p}, M)\right).$$

We have the following preliminary.

Lemma 2.2 *For $1 \leq N \leq \infty$, let $M \in \{W[p^N], T^\tau/p^N\}$, where we regard T^τ/p^N as T^τ for $N = \infty$.*

(1) For any prime v of K dividing the conductor $\mathfrak{f}$ of φ and for $0 \leq i \leq 2$, we have $H^i(K_{n,v}, M) = \{0\}$.

(2) For $i \in \{0, 2\}$, we have $H^i(K_{n,p}, M) = \{0\}$.

Proof (1) Recall that $p \nmid \mathfrak{f}$. Since K_n/K is unramified at v, the elliptic curve $E \otimes_H H_n$ has bad, potentially good reduction at the primes w of $H_n := HK_n$ dividing v. Then the local Tamagawa number of E at w is less than 5 (cf. [34, Appendix C, Theorem 15.1]). Hence, $H^0(H_{n,w}, E[p^N]) = \{0\}$. By the local duality and Euler characteristic formula, it follows that $H^1(H_{n,w}, E[p^N]) = H^2(H_{n,w}, E[p^N]) = \{0\}$.

Since $A = \mathrm{Res}_{H/K}(E)$ over K, by Shapiro's lemma, $H^i(K_{n,v}, W[p^N])$ may be regarded as a summand of $H^i(H_{n,w}, E[p^N])$, and so we obtain the assertion for $W[p^N]$. In view of (6) we have $H^i(K_{n,w}, T^\tau/p^N) \cong H^i(K_{n,\overline{w}}, T/p^N) \cong H^i(K_{n,\overline{w}}, W[p^N])$, where $\overline{w}$ denotes the complex conjugate of a place w of K_n. Since $\overline{\mathfrak{f}} = \mathfrak{f}$, the assertion for T^τ/p^N then follows from that for $W[p^N]$.

(2) Note that p splits completely in H/K as p is inert in K, and that the formal group of E over the completion of $\mathscr{O}_H$ at each prime above $p\mathscr{O}_K$ is isomorphic to the Lubin-Tate formal group over $\mathscr{O}_{K,p}$ with parameter $-p$. Since $H^0(K_{n,p}, E[p^\infty]) = \{0\}$ (cf. [31, Proposition 4.5]), the same argument as in the proof (1) implies the lemma. □

Corollary 1 *For $M \in \{W, T^\tau\}$, we have*

$$\mathrm{Sel}_{\mathrm{rel}}(K_n, M) = H^1(\mathscr{K}_{p\mathfrak{f}}/K_n, M), \tag{8}$$

where $\mathscr{K}_{p\mathfrak{f}}$ denotes the maximal algebraic extension of K unramified outside $p\mathfrak{f}$.

Proof This is a simple consequence of Lemma 2.2 (1). □

3 Supersingular Iwasawa Theory

The section describes the underlying plus/minus Iwasawa theory following [5–7].

3.1 Plus/Minus Selmer Groups

Denote by Ψ_∞/Ψ the anticyclotomic $\mathbb{Z}_p$-extension of

$$\Psi := K \otimes_{\mathbb{Q}} \mathbb{Q}_p$$

and by Ξ the set of finite characters of $\mathrm{Gal}(K_\infty/K) = \mathrm{Gal}(\Psi_\infty/\Psi)$. (Note that $p \nmid h_K$.) Put

$$\Xi^+ = \{\chi \in \Xi \mid \mathrm{cond}^{\mathrm{r}}\chi \text{ is an even power of } p\},$$
$$\Xi^- = \{\chi \in \Xi \mid \mathrm{cond}^{\mathrm{r}}\chi \text{ is an odd power of } p\}.$$

Fixing a Weierstrass model of E over $\mathscr{O}_H$ which is minimal at the prime above $p\mathscr{O}_K$ compatible with ι_p, denote by λ the logarithm of the formal group $\hat{E}$ of E. For $\chi \in \Xi$ and $x \in \hat{E}(\Psi_\infty)$, we define

$$\lambda_\chi(x) = p^{-n} \sum_{\sigma \in \mathrm{Gal}(\Psi_n/\Psi)} \lambda(x^\sigma)\chi^{-1}(\sigma),$$

where $n \geq 0$ is any integer such that $x \in \hat{E}(\Psi_n)$ and χ factors through $\mathrm{Gal}(\Psi_n/\Psi)$. As in [31], we define

$$\hat{E}(\Psi_n)^\pm = \left\{ x \in \hat{E}(\Psi_n) \middle| \lambda_\chi(x) = 0 \text{ for all } \chi \in \Xi^\pm \right\},$$

and denote by $H^1_\pm(\Psi_n, E[p^\infty])$ the image of $\hat{E}(\Psi_n)^\pm \otimes \mathbb{Q}_p/\mathbb{Z}_p$ under the Kummer map $E(\Psi_n) \otimes \mathbb{Q}_p/\mathbb{Z}_p \hookrightarrow H^1(\Psi_n, E[p^\infty])$.

Since $A = \mathrm{Res}_{H/K}(E)$ and $p\mathscr{O}_K$ splits completely in H, we have

$$T|_{G_\Psi} \cong T_p(E)|_{G_\Psi} \otimes_{\mathscr{O}_{K,p}} \mathscr{O}, \qquad W|_{G_\Psi} \cong E[p^\infty]|_{G_\Psi} \otimes_{\mathscr{O}_{K,p}} \mathscr{O}. \tag{9}$$

Denote by $H^1_\pm(\Psi_n, W)$ the subgroup of $H^1_\pm(\Psi_n, W)$ which corresponds to that for $E[p^\infty]$ via the isomorphisms[6] above. Define $H^1_\pm(\Psi_n, T^\tau)$ as the orthogonal complement of $H^1_\pm(\Psi_n, W)$ under the perfect pairing

$$H^1(\Psi_n, W) \times H^1(\Psi_n, T^\tau) \to L_\mathfrak{p}/\mathscr{O} \tag{10}$$

induced by the local duality and (6).

Definition 3.1 For $M \in \{W, T^\tau\}$, define

$$\mathrm{Sel}_\pm(K_n, M) = \mathrm{Ker}\left(\mathrm{Sel}_{\mathrm{rel}}(K_n, M) \to \frac{H^1(\Psi_n, M)}{H^1_\pm(\Psi_n, M)}\right).$$

For a $\mathbb{Z}_p$-module L, put $L^\vee = \mathrm{Hom}(L, \mathbb{Q}_p/\mathbb{Z}_p)$. For $* \in \left\{\emptyset, 0, \pm, \mathrm{rel}\right\}$, define

$$\mathscr{X}_n^* = \mathrm{Sel}_*(K_n, W)^\vee.$$

By the Poitou-Tate duality, we have a short exact sequence

$$0 \to \mathrm{Sel}(K_n, T^\tau) \to \mathrm{Sel}_{\mathrm{rel}}(K_n, T^\tau) \to \frac{H^1(\Psi_n, T^\tau)}{H^1_\mathrm{f}(\Psi_n, T^\tau)} \to \mathscr{X}_n \to \mathscr{X}_n^0 \to 0, \tag{11}$$

[6] Note that $H^1_\pm(\Psi_n, W)$ is independent of the choice of the isomorphism.

where the local conditions defining $\mathrm{Sel}_{\mathrm{rel}}(K_n, T^\tau)$ and $\mathrm{Sel}_0(K_n, W)$ are the orthogonal complements of each other under the pairing (10) by Lemma 2.2.

3.2 Coleman Maps

Following [6], we introduce plus/minus Coleman maps, and study the structure of certain Λ-modules.

Fixing a topological generator γ of Γ, for $n \geq 1$ put

$$\omega_n^+ = \prod_{1\leq k\leq n,\ (-1)^k=1} \Phi_k(\gamma),$$

$$\omega_n^- = (\gamma - 1) \prod_{1\leq k\leq n,\ (-1)^k=-1} \Phi_k(\gamma),$$

$$\omega_n = \omega_n^+\omega_n^- = \gamma^{p^n} - 1 \in \mathbb{Z}[\gamma],$$

where $\Phi_k(X)$ denotes the p^k-th cyclotomic polynomial. Also put $\omega_0^+ = 1$ and $\omega_0^- = \gamma - 1$.

By [31, Corollary 5.7 and Proposition 8.1] and [6, (2.7)], the Λ-modules

$$\varprojlim_m H^1_\pm(\Psi_m, T^\tau)$$

are free of rank one, and we fix a generator $v_\pm = (v_{\pm,m})$ of $\varprojlim_m H^1_\pm(\Psi_m, T^\tau)$. Denote by $(v_{+,n}^\perp, v_{-,n}^\perp)$ the dual basis of $(v_{-,n}, v_{+,n})$ under the perfect pairing

$$H^1(\Psi_n, T) \times H^1(\Psi_n, T^\tau) \to \Lambda_n := \mathcal{O}[\mathrm{Gal}(\Psi_n/\Psi)],\quad (a, b) \mapsto \sum_{\sigma\in\mathrm{Gal}(\Psi_n/\Psi)} (a, b^\sigma)_n\sigma, \qquad (12)$$

where $(-, -)_n$ denotes the perfect pairing

$$(-, -)_n : H^1(\Psi_n, T) \times H^1(\Psi_n, T^\tau) \to \mathcal{O}$$

induced by (6) (cf. [6, Lemma 2.3]). Note that (12) is sesquilinear with respect to the involution of Λ_n induced by $\sigma \mapsto \sigma^{-1}$ for $\sigma \in \mathrm{Gal}(\Psi_n/\Psi)$,

Define

$$c_n^\pm = \omega_n^\mp v_{\pm,n}^\perp \in H^1(\Psi_n, T).$$

By [6, Lemma 2.5 and Theorem 2.7], we have the following.

Theorem 3.2 *(1) The point $c_n^\pm$ generates the image of $\hat{E}(\Psi_n)^\pm \otimes_{\mathcal{O}_\Psi} \mathcal{O}$ in $H^1_{\mathrm{f}}(\Psi_n, T)$ over Λ, where $H^1_{\mathrm{f}}(\Psi_n, T)$ is the Bloch-Kato subgroup of $H^1(\Psi_n, T)$ identified with the image of $\hat{E}(\Psi_n) \otimes_{\mathcal{O}_\Psi} \mathcal{O}$ under the Kummer map via (9).*

(2) We have

$$H^1_{\mathrm{f}}(\Psi_n, T) = \left(\hat{E}(\Psi_n)^+ \otimes \mathcal{O}\right) \oplus \left(\hat{E}(\Psi_n)^- \otimes \mathcal{O}\right) = \Lambda_n c_n^+ \oplus \Lambda_n c_n^-. \qquad (13)$$

Define a homomorphism of Λ-modules

$$\widetilde{\mathrm{Col}}_n^{\pm} : H^1(\Psi_n, T^\tau) \to \Lambda_n, \quad z \mapsto \sum_{\sigma \in \mathrm{Gal}(\Psi_n/\Psi)} ((c_n^{\pm})^\sigma, z)_n \sigma.$$

Lemma 3.3 *We have* $\widetilde{\mathrm{Col}_n^{\pm}}) = \omega_n^{\mp}\Lambda_n$ *and* $\mathrm{Ker}\left(\widetilde{\mathrm{Col}}_n^{\pm}\right) = H^1_{\pm}(\Psi_n, T^\tau)$.

Proof The assertion regarding the image just follows from the perfectness of (12) and the definition of $c_n^{\pm}$. As for the description of the kernel, it follows from Theorem 3.2 (1), and the fact that $\hat{E}(\Psi_n)^{\pm} \otimes \mathscr{O}$ and $H^1_{\pm}(\Psi_n, T^\tau)$ are the orthogonal complements of each other under the pairing $(-,-)_n$, the latter being a consequence of [31, Proposition 5.6] and Lemma 2.2 (2). □

In view of Lemma 3.3 the map $\widetilde{\mathrm{Col}}_n^{\pm}$ induces an isomorphism

$$\widetilde{\mathrm{Col}}_n^{\pm} : \frac{H^1(\Psi_n, T^\tau)}{H^1_{\pm}(\Psi_n, T^\tau)} \cong \omega_n^{\mp}\Lambda_n.$$

In turn the latter induces

$$\mathrm{Col}_n^{\pm} : \frac{H^1(\Psi_n, T^\tau)}{H^1_{\pm}(\Psi_n, T^\tau)} \cong \Lambda_n^{\pm} := \mathscr{O}[\gamma - 1]/(\omega_n^{\pm}), \tag{14}$$

via

$$z \bmod H^1_{\pm}(\Psi_n, T^\tau) \mapsto \sum_{\sigma \in \mathrm{Gal}(\Psi_n/\Psi)} ((v_n^{\pm})^\sigma, z)_n \sigma \bmod (\omega_n^{\pm}).$$

3.3 Iwasawa Invariants

We introduce and study Iwasawa invariants which appear in the statement of our main result.

Let $\varepsilon \in \{\pm\}$ be the sign of the root number of the Hecke character φ.

For $* \in \{\mathrm{rel}, 0, \pm\}$, put

$$\mathscr{S}_* = \varprojlim_m \mathrm{Sel}_*(K_m, T^\tau), \quad \mathscr{X}^* = \left(\varinjlim_m \mathrm{Sel}_*(K_m, W)\right)^\vee.$$

Using the p-adic Weierstrass preparation theorem, let $\mathscr{L}_0$ and $\mathscr{L}_1$ be elements in $\mathscr{O}[\gamma - 1]$ such that

$$\mathrm{Char}_\Lambda(\mathscr{X}^0) = \mathscr{L}_0\Lambda, \quad \mathrm{Char}_\Lambda\left(\mathrm{Coker}\left(\mathscr{S}_\varepsilon \to \varprojlim_m H^1_\varepsilon(\Psi_m, T^\tau)\right)\right) = \mathscr{L}_1\Lambda.$$

Proposition 3.4 *The elements $\mathscr{L}_0$ and $\mathscr{L}_1$ are non-zero.*

Proof Fixing a basis of T^τ, let

$$z_A \in \varprojlim_m H^1(K_m, T^\tau)$$

be the element constructed from elliptic units as in [31, Theorem 3.2] (see also [6, §3.1.2]). It interpolates algebraic parts of the L-values $L(\overline{\varphi}\chi, 1)$ as χ varies over finite characters of $\mathrm{Gal}(K_\infty/K)$, and we have $\mathrm{loc}_p(z_A) \in \varprojlim_m H^1_\varepsilon(\Psi_m, T^\tau)$ (cf. [31, Corollary 3.3]).

By [11, Theorem 1], for $k \gg 0$, we have

$$L(\overline{\varphi}^{1+2k}, 1+k) \neq 0.$$

In view of [18, Theorem 12.5, Proposition 15.9 and (15.16.1)], observe that $\mathrm{loc}_p(z_A)$ interpolates those Hecke L-values as well. Therefore, it follows that $\mathrm{loc}_p(z_A) \in \varprojlim_m H^1_\varepsilon(\Psi_m, T^\tau)$ is non-torsion, and we deduce that $\mathscr{L}_1 \neq 0$.

By the usual Euler system argument (cf. [1, Proposition 3.3 (ii)]), the fact that z_A is non-torsion implies that $\mathscr{L}_0 \neq 0$. □

Put

$$\mathscr{Y}_n = \mathrm{Coker}\left(\mathrm{Sel}_{\mathrm{rel}}(K_n, T^\tau) \to \frac{H^1(\Psi_n, T^\tau)}{H^1_{\mathrm{f}}(\Psi_n, T^\tau)}\right), \quad \mathscr{Y}_n' = \mathrm{Coker}\left(\mathscr{S}_{\mathrm{rel}} \to \frac{H^1(\Psi_n, T^\tau)}{H^1_{\mathrm{f}}(\Psi_n, T^\tau)}\right).$$

Lemma 3.5 *For any integer $n \geq 0$, there exists an isomorphism*

$$a_n : \mathscr{Y}_n' \cong \Lambda_n^{-\varepsilon}/(\mathscr{L}_1) \oplus \Lambda_n^{\varepsilon}$$

of Λ-modules such that the diagram

$$\begin{array}{ccc} \mathscr{Y}_{n+1}' & \xrightarrow{a_{n+1}} & \Lambda_{n+1}^{-\varepsilon}/(\mathscr{L}_1) \oplus \Lambda_{n+1}^{\varepsilon} \\ \downarrow & & \downarrow \\ \mathscr{Y}_n' & \xrightarrow{a_n} & \Lambda_n^{-\varepsilon}/(\mathscr{L}_1) \oplus \Lambda_n^{\varepsilon} \end{array}$$

commutes, where the vertical arrows are the natural projections.

Proof By the proof of Rubin's conjecture [5], we have

$$\varprojlim_m H^1(\Psi_m, T^\tau) = \varprojlim_m H^1_+(\Psi_m, T^\tau) \oplus \varprojlim_m H^1_-(\Psi_m, T^\tau). \tag{15}$$

Note that [1, Theorem 4.2] and (15) imply that

$$\mathscr{S}_\varepsilon = \mathscr{S}_{\mathrm{rel}}.$$

In view of (13) and the fact that $H^1_\pm(\Psi_n, T^\tau)$ and $\hat{E}^\pm(\Psi_n) \otimes \mathscr{O}$ are the orthogonal complements of each other, it follows that

$$\frac{H^1(\Psi_n, T^\tau)}{H^1_{\mathrm{f}}(\Psi_n, T^\tau)} = \frac{H^1(\Psi_n, T^\tau)}{H^1_+(\Psi_n, T^\tau)} \oplus \frac{H^1(\Psi_n, T^\tau)}{H^1_-(\Psi_n, T^\tau)},$$

and so

$$\mathscr{Y}_n' = \mathrm{Coker}\left(\mathscr{S}_\varepsilon \to \frac{H^1(\Psi_n, T^\tau)}{H^1_{-\varepsilon}(\Psi_n, T^\tau)}\right) \oplus \frac{H^1(\Psi_n, T^\tau)}{H^1_\varepsilon(\Psi_n, T^\tau)}.$$

Since $\varprojlim_m H^1_\varepsilon(\Psi_m, T^\tau) \cong \Lambda$, by (14) and the definition of $\mathscr{L}_1$, it suffices to show surjectivity of the map

$$\varprojlim_m H^1_\varepsilon(\Psi_m, T^\tau) \to \frac{H^1(\Psi_n, T^\tau)}{H^1_{-\varepsilon}(\Psi_n, T^\tau)}.$$

We first note that the natural map

$$\mathrm{pr}_n : \varprojlim_m H^1(\Psi_m, T^\tau) \to H^1(\Psi_n, T^\tau)$$

is surjective by Lemma 2.2 (2). Hence, denoting by $\mathscr{I}_\pm$ the image of $\varprojlim_m H^1_\pm(\Psi_m, T^\tau)$ under pr_n, the natural map

$$\varprojlim_m H^1_\varepsilon(\Psi_m, T^\tau) \to H^1(\Psi_n, T^\tau)/\mathscr{I}_{-\varepsilon}$$

is surjective. Since $\mathscr{I}_{-\varepsilon} \subseteq H^1_{-\varepsilon}(\Psi_n, T^\tau)$, the existence of a_n follows. Note that the asserted commutativity of a_n is just a consequence of the commutativity of (14). □

4 Proofs of the Main Results

Based on the supersingular CM Iwasawa theory, the section proves the main results.

Notation 4.1 For an element $\mathscr{L} \in \Lambda \setminus \{0\}$, denote by $\mu(\mathscr{L})$ its μ-invariant and $\lambda(\mathscr{L})$ the λ-invariant. Fix an element $\widetilde{\mathscr{L}} \in \Lambda$ dividing $\mathscr{L}$ such that

(i) $\mu(\mathscr{L}) = \mu(\widetilde{\mathscr{L}})$,
(ii) For sufficiently large $m \geq 1$, $\mathscr{L}/\widetilde{\mathscr{L}}$ divides a power of ω_m, and $\widetilde{\mathscr{L}}$ is coprime to ω_m.

The element $\widetilde{\mathscr{L}}$ is unique up to a unit of Λ, and put $\lambda'(\mathscr{L}) = \lambda(\widetilde{\mathscr{L}})$.

4.1 Preliminaries

The subsection relates certain Iwasawa modules, and presents a control theorem for strict Selmer groups (cf. Remark 4.8).

Lemma 4.2 *There exist a finitely generated $\mathbb{Z}_p$-module $\mathscr{A}$ and an integer $k \geq 0$ such that for any integer $n \geq k$, we have a short exact sequence*

$$0 \to \mathscr{A} \to \mathscr{Y}_n' \to \mathscr{Y}_n \to 0 \tag{16}$$

for $\mathscr{Y}_n' \to \mathscr{Y}_n$ being the natural projection, for which the following diagram

$$\begin{array}{ccccccccc} 0 & \longrightarrow & \mathscr{A} & \longrightarrow & \mathscr{Y}_{n+1}' & \longrightarrow & \mathscr{Y}_{n+1} & \longrightarrow & 0 \\ & & \downarrow{\scriptstyle \times p} & & \downarrow & & \downarrow & & \\ 0 & \longrightarrow & \mathscr{A} & \longrightarrow & \mathscr{Y}_n' & \longrightarrow & \mathscr{Y}_n & \longrightarrow & 0 \end{array} \tag{17}$$

commutes.

Proof In view of (8) and the short exact sequence

$$\varprojlim_m H^1(\mathscr{K}_{\mathfrak{pf}}/K_m, T^\tau) \to H^1(\mathscr{K}_{\mathfrak{pf}}/K_n, T^\tau) \to \varprojlim_m H^2(\mathscr{K}_{\mathfrak{pf}}/K_m, T^\tau)[\omega_n] \to 0,$$

we have a short exact sequence

$$\mathscr{S}_{\mathrm{rel}} \to \mathrm{Sel}_{\mathrm{rel}}(K_n, T^\tau) \to \varprojlim_m H^2(\mathscr{K}_{\mathfrak{pf}}/K_m, T^\tau)[\omega_n] \to 0.$$

By applying the snake lemma to the commutative diagram with exact rows

$$\begin{array}{ccccccccc} & & \mathscr{S}_{\mathrm{rel}} & \longrightarrow & \mathrm{Sel}_{\mathrm{rel}}(K_n, T^\tau) & \longrightarrow & \varprojlim_m H^2(\mathscr{K}_{\mathfrak{pf}}/K_m, T^\tau)[\omega_n] & \longrightarrow & 0 \\ & & \downarrow & & \downarrow & & \downarrow & & \\ 0 & \longrightarrow & \dfrac{H^1(\Psi_n, T^\tau)}{H^1_{\mathrm{f}}(\Psi_n, T^\tau)} & \xrightarrow{=} & \dfrac{H^1(\Psi_n, T^\tau)}{H^1_{\mathrm{f}}(\Psi_n, T^\tau)} & \longrightarrow & 0 & \longrightarrow & 0, \end{array}$$

we have a natural short exact sequence

$$\mathrm{Sel}(K_n, T^\tau) \to \varprojlim_m H^2(\mathscr{K}_{\mathfrak{pf}}/K_m, T^\tau)[\omega_n] \to \mathscr{Y}_n' \to \mathscr{Y}_n \to 0. \tag{18}$$

Note that there exists $k \geq 0$ such that for $n \geq k$ we have

$$\varprojlim_m H^2(\mathscr{K}_{\mathfrak{pf}}/K_m, T^\tau)[\omega_n] = \varprojlim_m H^2(\mathscr{K}_{\mathfrak{pf}}/K_m, T^\tau)[\omega_k].$$

Hence the assertion follows from the commutative diagram

$$\begin{array}{ccc} \mathrm{Sel}(K_n, T^\tau) & \longrightarrow & \varprojlim_m H^2(\mathscr{K}_{\mathfrak{pf}}/K_m, T^\tau)[\omega_k] \\ \Big\downarrow \mathrm{cores}_{n/n-1} & & \Big\downarrow \times p \\ \mathrm{Sel}(K_{n-1}, T^\tau) & \longrightarrow & \varprojlim_m H^2(\mathscr{K}_{\mathfrak{pf}}/K_m, T^\tau)[\omega_k], \end{array}$$

where $\mathrm{cores}_{n/n-1}$ denotes the map induced by the corestriction map. □

Remark 4.3 In view of the proof we have

$$\mathscr{A} = \mathrm{Coker}\left(\mathrm{Sel}(K_n, T^\tau) \to \varprojlim_m H^2(\mathscr{K}_{\mathfrak{pf}}/K_m, T^\tau)[\omega_n]\right)$$

for sufficiently large n (cf. (18)).

For $n \geq 1$, put

$$\mathscr{Y}_{n/n-1} = \mathrm{Ker}(\mathscr{Y}_n \to \mathscr{Y}_{n-1}), \quad \mathscr{Y}'_{n/n-1} = \mathrm{Ker}(\mathscr{Y}'_n \to \mathscr{Y}'_{n-1}),$$
$$\mathscr{X}_{n/n-1} = \mathrm{Ker}(\mathscr{X}_n \to \mathscr{X}_{n-1}), \quad \mathscr{X}^0_{n/n-1} = \mathrm{Ker}(\mathscr{X}^0_n \to \mathscr{X}^0_{n-1}).$$

By applying the snake lemma to (17), for $n \geq k$ we have

$$0 \to \mathscr{A}[p] \to \mathscr{Y}'_{n/n-1} \to \mathscr{Y}_{n/n-1} \to \mathscr{A}/p \to 0. \tag{19}$$

Lemma 4.4 *For $n \geq 0$, there exists a natural exact sequence*

$$0 \to \mathscr{Y}_{n/n-1} \to \mathscr{X}_{n/n-1} \to \mathscr{X}^0_{n/n-1} \to 0. \tag{20}$$

Proof By Lemma 2.2 (2), the natural maps $\mathscr{Y}_n \to \mathscr{Y}_{n-1}$, $\mathscr{X}_n \to \mathscr{X}_{n-1}$ and $\mathscr{X}^0_n \to \mathscr{X}_{n-1}$ are surjective. Hence the lemma follows from the snake lemma and the short exact sequence

$$0 \to \mathscr{Y}_n \to \mathscr{X}_n \to \mathscr{X}^0_n \to 0,$$

the latter being a simple consequence of (11). □

Lemma 4.5 *For $n \geq 0$, we have a natural exact sequence*

$$\begin{aligned} 0 \to \mathrm{Coker}(\text{Ш}_{n-1} \to \text{Ш}_n)^\vee &\to \mathscr{X}_{n/n-1} \\ &\to (A(K_n)/A(K_{n-1}) \otimes_{\mathrm{End}_K(A)} L_\mathfrak{p}/\mathscr{O})^\vee \to \mathrm{Ker}(\text{Ш}_{n-1} \to \text{Ш}_n)^\vee \to 0. \end{aligned}$$

In particular, if $\mathrm{Coker}(\text{Ш}_{n-1} \to \text{Ш}_n)$ *is finite, then*

$$(\mathscr{X}_{n/n-1})_{\mathrm{tor}} \cong \mathrm{Coker}(\text{Ш}_{n-1} \to \text{Ш}_n)^\vee.$$

Proof We simply apply the the snake lemma to the commutative diagram

$$\begin{array}{ccccccccc} 0 & \longrightarrow & \text{Ш}_n^\vee & \longrightarrow & \mathscr{X}_n & \longrightarrow & (A(K_n)\otimes L_{\mathfrak{p}}/\mathscr{O})^\vee & \longrightarrow & 0 \\ & & \downarrow & & \downarrow & & \downarrow & & \\ 0 & \longrightarrow & \text{Ш}_{n-1}^\vee & \longrightarrow & \mathscr{X}_{n-1} & \longrightarrow & (A(K_{n-1})\otimes L_{\mathfrak{p}}/\mathscr{O})^\vee & \longrightarrow & 0. \end{array}$$

□

Remark 4.6 For any sufficiently large $n \geq 0$ such that $(-1)^{n+1} = \varepsilon$, $\mathrm{Coker}(\text{Ш}_{n-1} \to \text{Ш}_n)$ is finite and $A(K_n) = A(K_{n-1})$. Indeed, as mentioned in the proof of Lemma 3.4, the elliptic unit $z_A \in \varprojlim_m H^1_\varepsilon(\Psi_m, T^\tau)$ is non-torsion, and Rubin's p-adic L-function [31, §10] is non-zero. The latter interpolates algebraic part of the L-values $L(\overline{\varphi}\chi, 1)$ for χ of order $p^n > 1$ with $(-1)^{n+1} = \varepsilon$. Hence, for such a sufficiently large n, we have $L(\overline{\varphi}\chi, 1) \neq 0$. Then [6, Theorem 3.9] implies that $\mathrm{Coker}(\text{Ш}_{n-1} \to \text{Ш}_n)$ is finite and $A(K_n) = A(K_{n-1})$. In turn we have $\mathscr{X}_{n/n-1} \cong \mathrm{Coker}(\text{Ш}_{n-1} \to \text{Ш}_n)^\vee$, which is finite.

A principal result of this subsection is the following.

Proposition 4.7 *For any sufficiently large* $n \geq 0$, *we have*

$$\mathrm{length}_{\mathscr{O}}(\mathscr{X}^0_{n/n-1}) = \phi(p^n)\mu(\mathscr{L}_0) + \lambda'(\mathscr{L}_0).$$

Proof By the same argument[7] as in the proof of [19, Lemma 10.5 ii)], it suffices to show that for $n \geq 0$ the $\mathscr{O}$-module $\mathscr{X}^0_{n+1/n}$ is isomorphic to the kernel of the natural projection $\mathscr{X}^0/\omega_{n+1} \to \mathscr{X}^0/\omega_n$. The latter is equivalent to proving that the map $\mathscr{X}^0/\omega_n \to \mathscr{X}^0_n$ is an isomorphism.

By Lemma 2.2 (1), we have the following commutative diagram with exact rows

$$\begin{array}{ccccccc} 0 & \longrightarrow & \mathrm{Sel}_0(K_n, W) & \longrightarrow & H^1(\mathscr{K}_{\mathrm{p}\mathfrak{f}}/K_n, W) & \longrightarrow & H^1(\Psi_n, W) \\ & & \downarrow & & \downarrow h_n & & \downarrow \\ 0 & \longrightarrow & \mathrm{Sel}_0(K_\infty, W)[\omega_n] & \longrightarrow & H^1(\mathscr{K}_{\mathrm{p}\mathfrak{f}}/K_\infty, W)[\omega_n] & \longrightarrow & H^1(\Psi_\infty, W). \end{array}$$

By Lemma 2.2 (2), the restriction maps $H^1(\Psi_n, W) \to H^1(\Psi_\infty, W)$ and h_n are injective. Since $\mathrm{Gal}(K_\infty/K_n) \cong \mathbb{Z}_p$ has cohomological dimension one, h_n is surjective. Hence, by applying the snake lemma to the diagram above, the homomorphism $\mathscr{X}^0/\omega_n \to \mathscr{X}^0_n$ is bijective. □

[7] Note that $\lambda(\mathscr{L}_0/\widetilde{\mathscr{L}_0}) = \mathrm{rank}_{\mathscr{O}}(\mathscr{X}^0/\omega_m)$.

Remark 4.8 As seen in the proof, the restriction map $\mathrm{Sel}_0(K_n, W) \to \mathrm{Sel}_0(K_\infty, W)[\omega_n]$ is an isomorphism for $n \geq 0$.

4.2 Anticyclotomic Layers with Root Number +1

The aim of this subsection is to prove Theorem 4.11.

We begin with a couple of preparatory lemmas.

Lemma 4.9 *For any sufficiently large $n \geq 1$ with $(-1)^{n+1} = \varepsilon$, we have*

$$\mathrm{length}_{\mathcal{O}}(\mathscr{Y}_{n/n-1}) = \mathrm{rank}_{\mathcal{O}}(\mathscr{A}) + \mathrm{length}_{\mathcal{O}}(\mathscr{Y}'_{n/n-1}), \tag{21}$$

where every term is finite.

Proof By Remark 4.6, $\mathscr{X}_{n/n-1}$ is finite. Hence, the modules appearing in (20) and (19) are finite. By (19) we then have

$$\begin{aligned}\mathrm{length}_{\mathcal{O}}(\mathscr{Y}_{n/n-1}) &= \mathrm{length}_{\mathcal{O}}(\mathscr{Y}'_{n/n-1}) + \mathrm{length}_{\mathcal{O}}(\mathscr{A}/p) - \mathrm{length}_{\mathcal{O}}(\mathscr{A}[p]) \\ &= \mathrm{rank}_{\mathcal{O}}(\mathscr{A}) + \mathrm{length}_{\mathcal{O}}(\mathscr{Y}'_{n/n-1}).\end{aligned}$$

□

Lemma 4.10 *For any sufficiently large $n \geq 1$ with $(-1)^{n+1} = \varepsilon$, we have*

$$\mathrm{length}_{\mathcal{O}}(\mathscr{Y}'_{n/n-1}) = \lambda'(\mathscr{L}_1) + \phi(p^n)\mu(\mathscr{L}_1).$$

Proof If $(-1)^{n+1} = \varepsilon$, then $\omega_n^{-\varepsilon} = \Phi_n \omega_{n-1}^{-\varepsilon}$ and $\omega_n^{\varepsilon} = \omega_{n-1}^{\varepsilon}$. In view of the short exact sequence

$$0 \to \mathscr{Y}'_{n/n-1} \to \mathscr{Y}'_n \to \mathscr{Y}'_{n-1} \to 0,$$

Lemma 3.5 implies that

$$\mathscr{Y}'_{n/n-1} \cong \mathrm{Ker}(\Lambda/(\Phi_n \omega_{n-1}^{-\varepsilon}, \mathscr{L}_1) \to \Lambda/(\omega_{n-1}^{-\varepsilon}, \mathscr{L}_1)),$$

where $\Lambda/(\Phi_n \omega_{n-1}^{-\varepsilon}, \mathscr{L}_1) \to \Lambda/(\omega_{n-1}^{-\varepsilon}, \mathscr{L}_1)$ is the natural projection.

Put $R = \mathcal{O}[\gamma - 1]/(\mathscr{L}_1)$. We then have

$$\mathscr{Y}'_{n/n-1} \cong \omega_{n-1}^{-\varepsilon} R / \Phi_n \omega_{n-1}^{-\varepsilon} R. \tag{22}$$

Consider the surjective map

$$R/\Phi_n R \longrightarrow \omega_{n-1}^{-\varepsilon} R / \Phi_n \omega_{n-1}^{-\varepsilon} R \tag{23}$$

given by the multiplication by $\omega_{n-1}^{-\varepsilon}$. Suppose that the image of an element $g \in \mathcal{O}[\gamma - 1]$ is in the kernel of (23). Then

$$\omega_{n-1}^{-\varepsilon} g = \Phi_n \omega_{n-1}^{-\varepsilon} h_1 + h_2 \mathscr{L}_1$$

for some $h_i \in \mathscr{O}[\gamma - 1]$. Hence $g = \Phi_n h_1 + \tilde{h}_2 \widetilde{\mathscr{L}_1}$ for some $\tilde{h}_2 \in \mathscr{O}[\gamma - 1]$, where $\widetilde{\mathscr{L}_1}$ is as in Notation 4.1. Hence, the kernel of (23) is $(\Phi_n, \widetilde{\mathscr{L}_1})/(\Phi_n, \mathscr{L}_1)$. By (22), we then have

$$\mathscr{Y}'_{n/n-1} \cong \mathscr{O}[\gamma - 1]/(\Phi_n, \widetilde{\mathscr{L}_1}), \tag{24}$$

whose $\mathscr{O}$-length is $\lambda'(\mathscr{L}_1) + \phi(p^n)\mu(\mathscr{L}_1)$ for a sufficiently large n. □

Theorem 4.11 *For any sufficiently large $n \geq 0$ with $(-1)^{n+1} = \varepsilon$, we have*

$$\mathrm{length}_{\mathscr{O}}(\mathrm{Coker}(\mathrm{III}_{n-1} \to \mathrm{III}_n)) = \phi(p^n)\mu(\mathscr{L}_0\mathscr{L}_1) + \lambda'(\mathscr{L}_0\mathscr{L}_1) + \mathrm{rank}_{\mathscr{O}}(\mathscr{A}).$$

Proof We may assume that $A(K_n) = A(K_{n-1})$ and $\mathrm{Coker}(\mathrm{III}_{n-1} \to \mathrm{III}_n)$ is finite (cf. Remark 4.6). Then Lemma 4.5 implies that $\mathscr{X}_{n/n-1}$ is also finite.

By (20) and (21), we have

$$\begin{aligned}\mathrm{length}_{\mathscr{O}}(\mathscr{X}_{n/n-1}) &= \mathrm{length}_{\mathscr{O}}(\mathscr{X}^0_{n/n-1}) + \mathrm{length}_{\mathscr{O}}(\mathscr{Y}_{n/n-1})\\ &= \mathrm{rank}_{\mathscr{O}}(\mathscr{A}) + \mathrm{length}_{\mathscr{O}}(\mathscr{X}^0_{n/n-1}) + \mathrm{length}_{\mathscr{O}}(\mathscr{Y}'_{n/n-1}).\end{aligned}$$

Hence,

$$\mathrm{length}_{\mathscr{O}}(\mathrm{Coker}(\mathrm{III}_{n-1} \to \mathrm{III}_n)) = \mathrm{rank}_{\mathscr{O}}(\mathscr{A}) + \mathrm{length}_{\mathscr{O}}(\mathscr{X}^0_{n/n-1}) + \mathrm{length}_{\mathscr{O}}(\mathscr{Y}'_{n/n-1})$$

by Lemma 4.5. Therefore, Proposition 4.7 and Lemma 4.10 conclude the proof. □

Remark 4.12 By [1, Proposition 3.3(ii) and Theorem 4.2] and (15), the product $\mathscr{L}_0\mathscr{L}_1$ coincides with Rubin's p-adic L-function [31, §10], alternatively a generator of the characteristic ideal of $\mathscr{X}^{-\varepsilon}$, up to an element of $\Lambda^\times$.

4.3 Primitive Case

The aim of this subsection is to establish asymptotic variation of the Tate-Shafarevich groups, without fixing parity of the anticyclotomic layers, under primitivity assumptions implying the vanishing of certain Iwasawa invariants.

We begin with a preliminary.

Lemma 4.13 *Suppose that $\mathscr{L}_1 \in \Lambda^\times$. For $(-1)^{n+1} = -\varepsilon$, we have*

$$\mathscr{Y}'_{n/n-1} \cong \mathrm{Ker}(\mathscr{Y}'_n \to \mathscr{Y}'_{n-1}) \cong \mathscr{O}[\gamma - 1]/(\Phi_n).$$

Proof Since $\omega_n^{\varepsilon} = \Phi_n \omega_{n-1}^{\varepsilon}$ and $\omega_n^{-\varepsilon} = \omega_{n-1}^{-\varepsilon}$, the short exact sequence

$$0 \to \mathscr{Y}'_{n/n-1} \to \mathscr{Y}'_n \to \mathscr{Y}'_{n-1} \to 0$$

may be identified with

$$0 \to \omega_{n-1}^{\varepsilon}\Lambda/(\Phi_n\omega_{n-1}^{\varepsilon})\Lambda \to \Lambda/(\Phi_n\omega_{n-1}^{\varepsilon}) \to \Lambda/(\omega_{n-1}^{\varepsilon}) \to 0$$

by Lemma 3.5. The latter may be identified with

$$0 \to \Lambda/(\Phi_n) \xrightarrow{\times\omega_{n-1}^{\varepsilon}} \Lambda/(\Phi_n\omega_{n-1}^{\varepsilon}) \to \Lambda/(\omega_{n-1}^{\varepsilon}) \to 0,$$

concluding the proof. □

Remark 4.14 By (19) and Lemma 4.13, we have[8] $\mathscr{A}[p] = \{0\}$.

Theorem 4.15 *Suppose that the strict Selmer group $\mathscr{X}^0$ is finite. Then for sufficiently large $n \geq 0$ such that* $\mathrm{Coker}(\text{Ш}_{n-1} \to \text{Ш}_n)$ *is finite, the restriction map $\text{Ш}_{n-1} \to \text{Ш}_n$ is surjective. In particular, if Ш_n is also finite for sufficiently large n, then there exists an integer $m \geq 0$ such that for any $n \geq m$ we have*

$$\text{Ш}_n \cong \text{Ш}_m.$$

Proof By the assumption, note that $\mathscr{L}_0 \in \Lambda^\times$. In view of [1, Proposition 3.3 (ii) and Theorem 4.2], we also have $\mathscr{L}_1 \in \Lambda^\times$.

Since $\mathscr{X}^0$ is finite, by Remark 4.3 and the global duality, $\mathscr{A}$ is also finite. Hence, in view of Theorem 4.11 it suffices to consider the case of $n \gg 0$ with $(-1)^{n+1} = -\varepsilon$. By (20) and Proposition 4.7, we have

$$\mathscr{Y}_{n/n-1} \cong \mathscr{X}_{n/n-1}. \tag{25}$$

In turn Remark 4.14 implies that $\mathscr{A} = \{0\}$. Hence, $\mathscr{Y}_{n/n-1}$ is p-torsion free by (19) and Lemma 4.13. Therefore, Lemma 4.5 and (25) conclude the proof. □

Remark 4.16 The assumption on the finiteness of $\mathrm{Coker}(\text{Ш}_{n-1} \to \text{Ш}_n)$ is satisfied for any sufficiently large n. If $(-1)^{n+1} = \varepsilon$, this is the content of Remark 4.6. In the case $(-1)^{n+1} = -\varepsilon$, we have $\mathrm{ord}_{s=1} L(\varphi\overline{\chi}, s) \leq 1$ for any finite character χ of $\mathrm{Gal}(K_n/K)$ of conductor p^{n+1} by Rohrlich[9] [30], and we may apply [6, Theorem 3.9].

We now give a sufficient condition for the triviality of Ш_n for all $n \geq 0$. Let $\Omega \in \mathbb{C}^\times$ be a CM period as in [6, §3.1.1].

Theorem 4.17 *Suppose that* $\mathrm{Coker}(\text{Ш}_{n-1} \to \text{Ш}_n)$ *is finite for all $n \geq 1$ such that $(-1)^{n+1} = -\varepsilon$. Suppose also that $p \nmid h_K$ and one of the following conditions holds.*

(1) The L-value $L_{p\mathfrak{f}}(\varphi, 1)/\Omega$ is p-indivisible, where $L_{p\mathfrak{f}}(\varphi, s)$ denotes the Hecke L-function associated to φ with Euler factor at the primes dividing $p\mathfrak{f}$ removed.

---

[8] Note that this assertion is independent of the parity of n.

[9] In [30] it is assumed that $h_K = 1$, however the hypothesis is inessential.

(2) *We have* $\mathrm{ord}_{s=1} L(\varphi, s) = 1$, *and* $L'_{\mathrm{pf}}(\varphi, 1)/\Omega \cdot \mathrm{reg}_A$ *is p-indivisible, where* reg_A *denotes the regulator, and the localization map* $A(\mathbb{Q}) \otimes \mathcal{O} \to A(\mathbb{Q}_p)/p$ *is surjective.*

Then, for any $n \geq 0$ *we have*

$$\text{Ш}_n = \{0\}.$$

Proof (1) In this case, we have $\varepsilon(\varphi) = +1$. By the assumption, the image of elliptic unit z_A (as in the proof of Proposition 3.4) in $H^1(\Psi, T^\tau)/H^1_{\mathrm{f}}(\Psi, T^\tau)$ is p-indivisible. So the Euler system argument implies that $\mathscr{X}_0 = \mathscr{X}_0^0 = \text{Ш}_0 = \{0\}$ (cf. [33, Theorem 2.2.10]). It then suffices to show that for $n \geq 1$ the map $\text{Ш}_{n-1} \to \text{Ш}_n$ is surjective.

Note that by Remark 4.8 and Nakayama's lemma, $\mathscr{X}^0 = \{0\}$, and hence $\mathscr{X}_n^0 = 0$ for $n \geq 0$. In particular, (25) holds for arbitrary $n \geq 0$, and by (19) we also have $\mathscr{Y}'_{n/n-1} \cong \mathscr{Y}_{n/n-1}$ as $\mathscr{A} = \{0\}$.

Suppose that $(-1)^{n+1} = \varepsilon$. Then, $\mathscr{Y}'_{n/n-1} = \{0\}$ by (24), and hence $\mathscr{Y}_{n/n-1} = \{0\}$. In turn Lemma 4.5 and (25) yield the surjectivity of $\text{Ш}_{n-1} \to \text{Ш}_n$.

Suppose that $(-1)^{n+1} = -\varepsilon$. By (25) and Lemma 4.13, $\mathscr{X}_{n/n-1}$ is torsion-free. Hence, by the finiteness of $\mathrm{Coker}(\text{Ш}_{n-1} \to \text{Ш}_n)$, the surjectivity of $\text{Ш}_{n-1} \to \text{Ш}_n$ follows.

(2) By the assumption and the Gross-Zagier and Kolyvagin theorem, note that $\mathrm{rank}_{\mathcal{O}} \mathscr{X}_0 = 1$ and Ш_0 is finite. In view of the assumption and [6, Theorem 4.8], the special value of Rubin's p-adic L-function $\mathscr{L}$, as defined in [6, (3.14)], at the trivial character is p-indivisible: the assumption implies that the rational point $P \in A(\mathbb{Q})$ appearing in [6, Theorem 4.8] is locally p-indivisible. Therefore, $\mathscr{L}$ is a p-adic unit.

Since the ideal of Λ generated by $\mathscr{L}$ coincides with the characteristic ideal of the quotient of $\varprojlim_m H^1_-(\Psi_m, T^\tau)$ by the submodule generated by the image of the elliptic unit z_A (cf. [5, 6.0.1]), we conclude that the image of z_A in $H^1(\Psi, T^\tau)$ is p-indivisible. Hence, by the usual Euler system argument, we have $\mathscr{X}_0^0 = \{0\}$.

In turn, (11) leads to the short exact sequence

$$0 \to \mathrm{Sel}(K, T^\tau) \to \mathrm{Sel}_{\mathrm{rel}}(K, T^\tau) \to H^1(\Psi, T^\tau)/H^1_{\mathrm{f}}(\Psi, T^\tau) \to \mathscr{X}_0 \to 0. \quad (26)$$

Since $H^1(\Psi, T^\tau)/H^1_{\mathrm{f}}(\Psi, T^\tau) \cong \mathcal{O}$, the middle map is just the zero map and so $\mathscr{X}_0 \cong \mathcal{O}$. The latter implies that $\text{Ш}_0 = \{0\}$. We now proceed just as in the proof of part (1). □

Acknowledgements We thank Shilin Lai and Antonio Lei for helpful comments. We also thank the referee for valuable suggestions. This work was partially supported by the NSF grants DMS 2303864 and DMS 2302064, and the JSPS KAKENHI grants JP17H02836, JP18H05233, JP22H00096, JP21K13774 and JP21H00969.

References

1. A. Agboola and B. Howard, *Anticyclotomic Iwasawa theory of CM elliptic curves. II*, Math. Res. Lett. 12 (2005), no. 5-6, 611–621
2. Bertolini, M.: Selmer groups and Heegner points in anticyclotomic $\mathbb{Z}_p$-extensions. Compositio Math. **99**(2), 153–182 (1995)
3. Burungale, A., Flach, M.: The Birch and Swinnerton-Dyer conjecture for certain elliptic curves with complex multiplication. Camb. J. Math. **12**(2), 357–415 (2024)
4. A. Burungale, W. He, S. Kobayashi and K. Ota, *Hecke L-values, definite Shimura sets and mod ℓ non-vanishing*, preprint, arXiv:2408.13932
5. A. Burungale, S. Kobayashi and K. Ota, *Rubin's conjecture on local units in the anticyclotomic tower at inert primes,* Ann. of Math. (2) 194 (2021), no. 3, 943-966
6. Burungale, A., Kobayashi, S., Ota, K.: p-adic L-functions and rational points on CM elliptic curves at inert primes. J. Inst. Math. Jussieu **23**(3), 1417–1460 (2024)
7. A. Burungale, S. Kobayashi and K. Ota, *The p-adic valuation of local resolvents, generalized Gauss sums and anticyclotomic Hecke L-values of imaginary quadratic fields at inert primes*, arXiv:2508.17776
8. A. Burungale, S. Kobayashi, K. Ota and S. Yasuda, *Kato's epsilon conjecture for anticyclotomic CM deformations at inert primes*, J. Number Theory **270**, 17–67 (2025)
9. Çiperiani, M.: Tate-Shafarevich groups in anticyclotomic $\mathbb{Z}_p$-extensions at supersingular primes. Compos. Math. **145**(2), 293–308 (2009)
10. Çiperiani, M.: Supersingular elliptic curves over $\mathbb{Z}_p$-extensions. J. Reine Angew. Math. **801**, 45–56 (2023)
11. Greenberg, R.: On the critical values of Hecke L-functions for imaginary quadratic fields. Invent. Math. **79**(1), 79–94 (1985)
12. R. Greenberg, *Iwasawa theory for elliptic curves, Arithmetic theory of elliptic curves* (Cetraro, 1997), Lecture Notes in Math. 1716, 51–144, Springer, Berlin, 1999
13. R. Greenberg, *Introduction to Iwasawa theory for elliptic curves, Arithmetic algebraic geometry* (Park City, UT, 1999), 407–464, IAS/Park City Math. Ser., 9, Amer. Math. Soc., Providence, RI, 2001
14. Gross, B.: Arithmetic on elliptic curves with complex multiplication. Lecture Notes in Mathematics, vol. 776. With an appendix by B. Mazur, Springer, Berlin (1980)
15. Howard, B.: The Heegner point Kolyvagin system. Compos. Math. **140**(6), 1439–1472 (2004)
16. Iovita, A., Pollack, R.: Iwasawa theory of elliptic curves at supersingular primes over $\mathbb{Z}_p$-extensions of number fields. J. Reine Angew. Math. **598**, 71–103 (2006)
17. K. Iwasawa, *On $\mathbb{Z}_l$-extensions of algebraic number fields*, Ann. of Math. (2) 98 (1973), 246–326
18. K. Kato, *p-adic Hodge theory and values of zeta functions of modular forms*, Cohomologies p-adiques et applications arithmétiques. III. Astérisque No. 295 (2004), ix, 117–290
19. Kobayashi, S.: Iwasawa theory for elliptic curves at supersingular primes. Invent. Math. **152**(1), 1–36 (2003)
20. M. Kurihara, *On the Tate Shafarevich groups over cyclotomic fields of an elliptic curve with supersingular reduction. I*, Invent. Math. 149 (2002), no. 1, 195–224
21. Lee, J.: Structure of the Mordell-Weil group over the $\mathbb{Z}_p$-extensions. Trans. Amer. Math. Soc. **373**(4), 2399–2425 (2020)
22. A. Lei, M.F. Lim and K. Müller, *Asymptotic formula for Tate–Shafarevich groups of p-supersingular elliptic curves over anticyclotomic extensions*, Adv. Math. 434 (2023), Paper No. 109320, 28 pp
23. Matar, A.: Kolyvagin's work and anticyclotomic tower fields: the supersingular case. Acta Arith **201**(2), 131–147 (2021)
24. Mazur, B.: Rational points of abelian varieties with values in towers of number fields. Invent. Math. **18**, 183–266 (1972)
25. Nakamura, T.: On abelian varieties associated with elliptic curves with complex multiplication. Acta Arith **97**(4), 379–385 (2001)

26. A. G. Nasybullin, *Elliptic curves with supersingular reduction over* Γ*-extensions*, Uspehi Mat. Nauk 32 (1977), no. 2(194), 221–222
27. Perrin-Riou, B.: Fonctions L p-adiques, théorie d'Iwasawa et points de Heegner. Bull. Soc. Math. France **115**(4), 399–456 (1987)
28. Perrin-Riou, B.: Arithmétique des courbes elliptiques à réduction supersingulière en p. Experiment. Math. **12**(2), 155–186 (2003)
29. R. Pollack, *On the p-adic L-function of a modular form at a supersingular prime*, Duke Math. J. 118, 2003, no. 3, 523–558
30. Rohrlich, D.: On L-functions of elliptic curves and anticyclotomic towers. Invent. Math. **75**(3), 383–408 (1984)
31. Rubin, K.: Local units, elliptic units, Heegner points and elliptic curves. Invent. Math. **88**(2), 405–422 (1987)
32. Rubin, K.: The main conjectures of Iwasawa theory for imaginary quadratic fields. Invent. Math. **103**(1), 25–68 (1991)
33. K. Rubin, *Euler systems*, Annals of Mathematics Studies 147, Princeton University Press, Princeton, NJ, 2000, xii+227
34. J. Silverman, *The arithmetic of elliptic curves*, Second edition, Graduate Texts in Mathematics, 106. Springer, Dordrecht, 2009. xx+513 pp
35. Sprung, F.: The Šafarevič-Tate group in cyclotomic $\mathbb{Z}_p$-extensions at supersingular primes. J. Reine Angew. Math. **681**, 199–218 (2013)

Nonvanishing of Generalised Kato Classes and Iwasawa Main Conjectures

Francesc Castella

Abstract A construction due to Darmon-Rotger gives rise to generalised Kato classes $\kappa_p(E)$ in the p-adic Selmer group $\mathrm{Sel}(\mathbf{Q}, V_pE)$ of elliptic curves $E/\mathbf{Q}$ of positive even analytic rank, where $p > 3$ is any prime of good ordinary reduction for E. Darmon and Rotger conjectured that $\kappa_p(E) \neq 0$ precisely when $\mathrm{Sel}(\mathbf{Q}, V_pE)$ is two-dimensional. The first cases of this conjecture were obtained by the author with M.-L. Hsieh. In this note we give a new proof of the implication $\kappa_p(E) \neq 0 \Rightarrow \dim_{\mathbf{Q}_p} \mathrm{Sel}(\mathbf{Q}, V_pE) = 2$ established in op. cit., and show that the converse implication holds if and only if the restriction map $\mathrm{loc}_p : \mathrm{Sel}(\mathbf{Q}, V_pE) \to E(\mathbf{Q}_p)\widehat{\otimes}\mathbf{Q}_p$ is nonzero. The present approach is an adaptation of the method introduced by the author, where the case of CM elliptic curves $E/\mathbf{Q}$ is studied.

Keywords Elliptic curves · Iwasawa theory · P-adic families of molecular forms

2020 Mathematics Subject Classification Primary 11G05 · Secondary 11G40

1 Introduction

After the groundbreaking works of Gross–Zagier and Kolyvagin in the 1980s, the construction of non-torsion rational points (or more generally, Selmer classes) on elliptic curves of $E/\mathbf{Q}$ with $\mathrm{ord}_{s=1} L(E, s) \geq 2$, akin to the construction of Heegner points in the cases of analytic rank 0 or 1, is widely regarded as one of the central open problems in number theory.

In a recent series of spectacular works [9, 11, 21, 23, 24] (culminating in the collective volume [4] with applications to the theory of Stark–Heegner points), Darmon–Rotger and Bertolini–Seveso–Venerucci revisited the construction of diagonal cycle classes due to Gross–Kudla [26] and Gross–Schoen [28], obtaining in particular an interpolation of these classes in p-adic families. Directly from their geometric construction, one obtains classes attached to triples (f, g, h) of cuspidal eigenforms of

F. Castella (✉)
Department of Mathematics, University of California, Santa Barbara, USA
e-mail: castella@ucsb.edu

M. Longo et al. (eds.), *Elliptic Curves and Modular Forms in Arithmetic Geometry*, Springer Proceedings in Mathematics & Statistics 527,
https://doi.org/10.1007/978-3-032-13123-2_2

"balanced" weights (k, l, m) (meaning that none of k, l, or m is larger than the sum of the other two), but after p-adic interpolation one also gets classes for weights beyond this range, such as the "unbalanced" weight $(2, 1, 1)$.

Of special relevance to the Birch–Swinnerton-Dyer conjecture and its equivariant refinements is the case where f is the newform of weight 2 associated to an elliptic curve $E/\mathbf{Q}$, and g and h are the weight 1 cuspidal eigenforms associated (as a consequence of the proof of Serre's conjecture [37]) to degree 2 odd irreducible Artin representations ϱ_1 and ϱ_2, respectively. In this case, one of the main results of [23] and [11] relates the resulting *generalised Kato classes*

$$\kappa_p(f, g, h) \in \mathrm{H}^1(\mathbf{Q}, V_pE \otimes \varrho), \quad \varrho := \varrho_1 \otimes \varrho_2,$$

obtained from a specialisation in weights $(2, 1, 1)$ of a p-adic family of diagonal cycles classes, to the value at $s = 1$ of the twisted Hasse–Weil L-funcion $L(E \otimes \varrho, s)$. Together with global duality, this relation leads to a proof of the implication

$$L(E \otimes \varrho, 1) \neq 0 \quad \Longrightarrow \quad \mathrm{Hom}_{G_{\mathbf{Q}}}(V_\varrho^\vee, E(H) \otimes L) = 0,$$

where L is any number field over which ϱ is defined (assumed to admit an embedding into $\mathbf{Q}_p$ for simplicity), $V_\varrho^\vee$ is the linear dual of an L-vector space affording ϱ, and H is the fixed field of $\ker(\varrho)$. More precisely, by virtue of the explicit reciprocity law obtained in *loc. cit.*, the nonvanishing of $L(E \otimes \varrho, 1)$ implies that the class $\kappa_p(f, g, h)$ is *non-crystalline* at p, from where the bound on the ϱ-isotypical component of $E(H)$ follows by global duality.

Interestingly, the same explicit reciprocity law shows that the generalised Kato classes $\kappa_p(f, g, h)$ are Selmer whenever $L(E \otimes \varrho, 1) = 0$. Moreover, the representations ϱ for which the construction applies necessarily have real traces, and in many cases the sign in the functional of the self-dual L-function $L(E \otimes \varrho, s)$ is $+1$. Thus the results of [23] and [11] provide a construction of (possibly zero *a priori*!) Selmer classes in situations where $\mathrm{ord}_{s=1} L(E \otimes \varrho, s) \geq 2$.

In [22], Darmon–Rotger carried out a systematic study of their construction in relation with the Birch–Swinnerton-Dyer conjecture and the elliptic Stark conjecture of [20]. (More recently, a similar study was carried out by Bertolini–Seveso–Venerucci [10], which in particular allows one to interpret and refine some of the rationality conjectures in [22] in terms of a p-adic Birch–Swinnerton-Dyer conjecture for p-adic Garrett–Rankin L-functions.) In particular, when $L(E \otimes \varrho, 1) = 0$ with sign $+1$ they conjectured the equivalence

$$\kappa_p(f, g, h) \neq 0 \quad \overset{?}{\Longleftrightarrow} \quad \dim_{\mathbf{Q}_p} \mathrm{Sel}(\mathbf{Q}, V_pE \otimes \varrho) = 2. \tag{1}$$

This leads to the expectation, when combined with the equivariant Birch–Swinnerton-Dyer conjecture, that $\kappa_p(f, g, h)$ is a nonzero class in $\mathrm{Sel}(\mathbf{Q}, V_pE \otimes V_\varrho)$ if and only if $\mathrm{ord}_{s=1} L(E \otimes \varrho, s) = 2$.

Of special interest to the Birch–Swinnerton-Dyer conjecture is the case in which ϱ contains the trivial representation. This occurs precisely when $\varrho_2 \simeq \varrho_1^\vee$, so that

$$V_\varrho \simeq L \oplus \mathrm{ad}^0(V_{\varrho_1}),$$

where $\mathrm{ad}^0(V_{\varrho_1})$ is the 3-dimensional representation of $G_{\mathbf{Q}}$ acting on the trace zero endomorphisms of V_{ϱ_1}. By the Artin formalism, we then have

$$L(E \otimes \varrho, s) = L(E, s) \cdot L(E \otimes \mathrm{ad}^0(\varrho_1)).$$

In particular, if $L(E \otimes \mathrm{ad}^0(\varrho_1)) \neq 0$, denoting by $\kappa_p(E)$ the image of $\kappa_p(f, g, h)$ under the resulting natural projection $\mathrm{H}^1(\mathbf{Q}, V_p E \otimes \varrho) \to \mathrm{H}^1(\mathbf{Q}, V_p E)$, by the Bloch–Kato conjecture [7] (which in this case predicts that the dimensions of the Selmer groups for $V_p E$ and $V_p E \otimes \varrho$ are the same) the equivalence (1) amounts to the conjectural equivalence

$$\kappa_p(E) \neq 0 \quad \overset{?}{\iff} \quad \dim_{\mathbf{Q}_p} \mathrm{Sel}(\mathbf{Q}, V_p E) = 2. \tag{2}$$

In [17], the author and M.-L. Hsieh proved the first cases of conjecture (2), in situations where ϱ_1 and ϱ_2 are induced from inverse finite order characters of an imaginary quadratic field in which p splits. One of the key ingredients in the proof was a leading coefficient formula for the p-adic L-function introduced in [2] in terms of Howard's derived p-adic heights [30].

The main result of this note is the following, which includes a new proof of [17, Theorem A]. Let $w \in \{\pm 1\}$ denote the sign in the functional equation of $L(E, s)$.

Theorem A *Let $E/\mathbf{Q}$ be an elliptic curve of conductor N with good ordinary reduction at $p > 3$. Assume that:*

(i) $E[p]$ is absolutely irreducible,
(ii) there exists a prime $q \| N$, and if $q \equiv \pm 1 \pmod{p}$ then $E[p]$ is ramified at q,
(iii) either $p > 5$ or $\mathrm{Gal}(\mathbf{Q}(E[p])/\mathbf{Q})$ contains a conjugate of $\mathrm{SL}_2(\mathbf{F}_p)$.

Let $\kappa_p(E) \in \mathrm{H}^1(K, V_p E)$ be the generalised Kato class attached to any auxiliary pair (K, χ) as in Choice 5.2.1. If $w = +1$ and $L(E, 1) = 0$ (so $\mathrm{ord}_{s=1} L(E, s) \geq 2$), then $\kappa_p(E) \in \mathrm{Sel}(\mathbf{Q}, V_p E)$ and the following hold:

(1) If $\kappa_p(E)$ is nonzero, then $\dim_{\mathbf{Q}_p} \mathrm{Sel}(\mathbf{Q}, V_p E) = 2$.
(2) If $\dim_{\mathbf{Q}_p} \mathrm{Sel}(\mathbf{Q}, V_p E) = 2$, then $\kappa_p(E)$ is non-zero if and only if the restriction map

$$\mathrm{loc}_p : \mathrm{Sel}(\mathbf{Q}, V_p E) \to E(\mathbf{Q}_p) \hat{\otimes} \mathbf{Q}_p$$

is nonzero, and in that case $\kappa_p(E)$ spans the one-dimensional subspace $\ker(\mathrm{loc}_p)$.

Remark

(1) Theorem A recovers [17, Theorem A] under slightly weaker hypotheses on $E[p]$, and moreover, shows that the condition $\mathrm{Sel}(\mathbf{Q}, V_pE) \neq \ker(\mathrm{loc}_p)$ is *necessary* for the nonvanishing of $\kappa_p(E)$ when $\mathrm{Sel}(\mathbf{Q}, V_pE)$ is 2-dimensional.

(2) When $\mathrm{Sel}(\mathbf{Q}, V_pE)$ is 2-dimensional and different from $\ker(\mathrm{loc}_p)$, Theorem A shows that

$$\kappa_p(E) = \lambda \cdot \big(\log_p(Q)P - \log_p(P)Q\big)$$

for some $\lambda \in \mathbf{Q}_p^\times$, where (P, Q) is any basis for $\mathrm{Sel}(\mathbf{Q}, V_pE)$ and $\log_p$ is the composition of loc_p with the formal group logarithm on $E(\mathbf{Q}_p)\hat{\otimes}\mathbf{Q}_p$. This is consistent with the refined conjecture in [22, Conj. 3.12] specialised to the "adjoint rank (2, 0) setting" (see [22, Sect. 4.5.3]), which further predicts a rationality statement for λ (see also [10, Conj. 3.4, Remark 3.5.2]).

To prove Theorem A, in this paper we adapt the method introduced in [13], where a similar result is proved for elliptic curves $E/\mathbf{Q}$ with complex multiplication (note that the two results are complementary: the existence of a prime $q \,\|\, N$ of multiplicative reduction for E excludes the CM case). More precisely, the approach is based on a study of the relation between two different formulations of the Iwasawa–Greenberg main conjecture for triple products:

(IMC-1) one in terms of Hsieh's triple product p-adic L-functions [31];

(IMC-2) another without reference to p-adic L-functions, phrased in terms of the p-adic family of diagonal cycles used in the construction of $\kappa_p(f, g, h)$ and $\kappa_p(E)$.

In the case where $\varrho_1 = \varrho_2^\vee$ is dual to ϱ_2, and is induced from a finite order Hecke character of an imaginary quadratic field K, conjecture (IMC-1) can be related to the Iwasawa main conjecture for E/K in the anticyclotomic setting, and using global duality and the explicit reciprocity laws of [24] and [11], one can easily show that (in general) conjectures (IMC-1) and (IMC-2) are equivalent. Thus, from the results on the anticyclotomic main conjecture stemming from the works of Bertolini–Darmon [3] and Skinner–Urban [47], we deduce under mild hypotheses a proof of conjecture (IMC-2), from where the proof of our main result follow easily from a variant of Mazur's control theorem and global duality.

We conclude this introduction by noting that even though throughout the paper p is assumed to be a prime of *good* ordinary reduction, it seems possible to extend our results to the multiplicative case; it would then be interesting to compare the resulting $\kappa_p(E)$ with the p-adic limits of Heegner points studied in [19, 25] (see also [12] for a related construction).

1.1 Acknowledgements

I heartily thank Matteo Longo, Marco Adamo Seveso, Rodolfo Venerucci, and Stefano Vigni for the opportunity to contribute this note to the proceedings in celebration of Massimo Bertolini's 60th birthday. The chief insight exploited in this paper is that the nonvanishing of $\kappa_p(E)$ can be related to an Iwasawa main conjecture in the spirit of the Heegner point main conjecture studied by Massimo about 30 years ago [5]. Some of Massimo's later results also play a role in our proof. It is a pleasure to dedicate this note to him as a humble attestation of his lasting and growing impact on the subject. We also thank the referee for their careful reading and the helpful suggestions. During the preparation of this note, the author was partially supported by the NSF grants DMS-1946136 and DMS-2101458.

2 p-Adic L-Functions

Fix a prime $p > 3$. Let K be an imaginary quadratic field of discriminant D_K and assume that

$$(p) = \mathfrak{p}\overline{\mathfrak{p}} \text{ splits in } K,$$

with $\mathfrak{p}$ the prime of K above p induced by a fixed embedding $\iota_p : \overline{\mathbf{Q}} \hookrightarrow \overline{\mathbf{Q}}_p$. In this section we recall from [31] the construction of the triple product p-adic L-function for Hida families, and its relation with the anticyclotomic p-adic L-functions of Bertolini–Darmon [2] in the case where two of the Hida families are CM Hida families attached to dual ray class characters of K.

2.1 Triple Product p-Adic L-Function

Let $\mathbb{I}$ be a normal domain finite flat over

$$\Lambda := \mathscr{O}[\![1 + p\mathbb{Z}_p]\!],$$

where $\mathscr{O}$ is the ring of integers of a finite extension of $\mathbf{Q}_p$. For a positive integer N prime to p and a Dirichlet character $\chi : (\mathbb{Z}/Np\mathbb{Z})^\times \to \overline{\mathbf{Q}}_p^\times$, assumed to be valued in $\mathscr{O}$ after extending scalars in Λ if necessary, we denote by $S^o(N, \chi, \mathbb{I}) \subset \mathbb{I}[\![q]\!]$ the space of ordinary $\mathbb{I}$-adic cusp forms of tame level N and branch character χ as defined in [31, Sect. 3.1].

Denote by $\mathfrak{X}_{\mathbb{I}}^+ \subset \operatorname{Spec} \mathbb{I}(\overline{\mathbf{Q}}_p)$ the set of *arithmetic points* of $\mathbb{I}$, consisting of the ring homomorphisms $Q : \mathbb{I} \to \overline{\mathbf{Q}}_p$ such that $Q|_{1+p\mathbb{Z}_p}$ is given by $z \mapsto z^{k_Q}\epsilon_Q(z)$ for some $k_Q \in \mathbb{Z}_{\geq 2}$ called the *weight of* Q and $\epsilon_Q(z) \in \mu_{p^\infty}$. As in [31, Sect. 3.1], we say that $f = \sum_{n=1}^{\infty} a_n(f)q^n \in S^o(N, \chi, \mathbb{I})$ is a *primitive Hida family* if for every $Q \in$

$\mathfrak{X}_{\mathbb{I}}^{+}$ the specialization $\boldsymbol{f}_Q$ gives the q-expansion of an ordinary p-stabilised newform of weight k_Q and tame conductor N. Attached to such $\boldsymbol{f}$ we let $\mathfrak{X}_{\mathbb{I}}^{\mathrm{cls}}$ be the set of ring homomorphisms Q as above with $k_Q \in \mathbb{Z}$ such that $\boldsymbol{f}_Q$ is the q-expansion of a classical modular form; so $\mathcal{X}_{\mathbb{I}}^{\mathrm{cls}} \supset \mathfrak{X}_{\mathbb{I}}^{+}$ by Hida theory.

For $\boldsymbol{f}$ a primitive Hida family of tame level N, we let

$$\rho_{\boldsymbol{f}} : G_{\mathbf{Q}} \to \mathrm{GL}_2(\mathrm{Frac}\,\mathbb{I})$$

denote the associated Galois representation, where Frac $\mathbb{I}$ is the field of fractions of $\mathbb{I}$. It will be convenient for us to take $\rho_{\boldsymbol{f}}$ to be the *dual* of the Galois representation considered in [31, Sect. 3.2]; in particular, det $\rho_{\boldsymbol{f}} = \chi_{\mathbb{I}} \cdot \varepsilon_{\mathrm{cyc}}$ in the notations of *loc. cit.*, where $\varepsilon_{\mathrm{cyc}}$ is the p-adic cyclotomic character. We assume that the associated the residual representation $\bar{\rho}_{\boldsymbol{f}}$ is absolutely irreducible, and also denote by

$$\rho_{\boldsymbol{f}} : G_{\mathbf{Q}} \to \mathrm{Aut}_{\mathbb{I}}(V_{\boldsymbol{f}}) \simeq \mathrm{GL}_2(\mathbb{I})$$

the realisation of $\rho_{\boldsymbol{f}}$ on a module $V_{\boldsymbol{f}} \simeq \mathbb{I}^{\oplus 2}$. By [51, Theorem 2.2.2], restricted to $G_{\mathbf{Q}_p}$ the Galois representation $V_{\boldsymbol{f}}$ fits into a short exact sequence

$$0 \to V_{\boldsymbol{f}}^{+} \to V_{\boldsymbol{f}} \to V_{\boldsymbol{f}}^{-} \to 0,$$

where $V_{\boldsymbol{f}}^{-}$ is free of rank 1 over $\mathbb{I}$, with the $G_{\mathbf{Q}_p}$-action given by the unramified character sending an arithmetic Frobenius Fr_p to $a_p(\boldsymbol{f})$.

Associated with $\boldsymbol{f}$ there is a $\mathbb{I}$-algebra homomorphism

$$\lambda_{\boldsymbol{f}} : \mathbb{T}(N, \mathbb{I}) \to \mathbb{I},$$

where $\mathbb{T}(N, \mathbb{I})$ is the Hecke algebra acting on $\oplus_{\chi} S^{o}(N, \chi, \mathbb{I})$, where χ runs over the characters of $(\mathbb{Z}/pN\mathbb{Z})^{\times}$. Let $\mathbb{T}_{\mathfrak{m}}$ be the local component of $\mathbb{T}(N, \mathbb{I})$ through which $\lambda_{\boldsymbol{f}}$ factors, and following [29] define the *congruence ideal* $C(\boldsymbol{f})$ of $\boldsymbol{f}$ by

$$C(\boldsymbol{f}) := \lambda_{\boldsymbol{f}}(\mathrm{Ann}_{\mathbb{T}_{\mathfrak{m}}}(\ker \lambda_{\boldsymbol{f}})) \subset \mathbb{I}.$$

When, in addition to absolutely irreducible, $\bar{\rho}_{\boldsymbol{f}}$ is also p-distinguished (i.e., the semi-simplification of $\bar{\rho}_{\boldsymbol{f}}|_{G_{\mathbf{Q}_p}}$ is non-scalar), it follows from the results of [52] and [29] that $C(\boldsymbol{f})$ is generated by a nonzero element $\eta_{\boldsymbol{f}} \in \mathbb{I}$.

2.1.1 Triple Products of Hida Families

Let

$$(\boldsymbol{f}, \boldsymbol{g}, \boldsymbol{h}) \in S^{o}(N_f, \chi_f, \mathbb{I}_f) \times S^{o}(N_g, \chi_g, \mathbb{I}_g) \times S^{o}(N_h, \chi_h, \mathbb{I}_h)$$

be a triple of primitive Hida families with

$$\chi_f \chi_g \chi_h = \omega^{2a} \text{ for some } a \in \mathbb{Z}, \tag{3}$$

where ω is the Teichmüller character. Note that (3) does not uniquely determine a; we fix once and for all one of the two possible choices for a modulo $(p-1)/2$. Put

$$\mathcal{R} = \mathbb{I}_f \hat{\otimes}_{\mathcal{O}} \mathbb{I}_g \hat{\otimes}_{\mathcal{O}} \mathbb{I}_h,$$

which is a finite extension of the three-variable Iwasawa algebra $\Lambda \hat{\otimes}_{\mathcal{O}} \Lambda \hat{\otimes}_{\mathcal{O}} \Lambda$, and let

$$\mathfrak{X}^f_{\mathcal{R}} := \left\{ (Q_1, Q_2, Q_3) \in \mathfrak{X}^+_{\mathbb{I}_\varphi} \times \mathfrak{X}^{\mathrm{cls}}_{\mathbb{I}_g} \times \mathfrak{X}^{\mathrm{cls}}_{\mathbb{I}_h} \ : \ k_{Q_1} \geq k_{Q_2} + k_{Q_3} \text{ and } k_{Q_1} \equiv k_{Q_2} + k_{Q_3} \pmod 2 \right\}$$

be the weight space for $\mathcal{R}$ in the so-called $\boldsymbol{f}$*-unbalanced range.*

Let $\mathbf{V} = V_{\boldsymbol{f}} \hat{\otimes}_{\mathcal{O}} V_{\boldsymbol{g}} \hat{\otimes}_{\mathcal{O}} V_{\boldsymbol{h}}$ be the triple tensor product Galois representation attached to $(\boldsymbol{f}, \boldsymbol{g}, \boldsymbol{h})$, and writing $\det \mathbf{V} = \mathcal{X}^2 \varepsilon_{\mathrm{cyc}}$ with $\mathcal{X}(\mathbf{c}) = (-1)^a$ for the complex conjugation $\mathbf{c}$ (as is possible by (3)) define

$$\mathbf{V}^\dagger := \mathbf{V} \otimes \mathcal{X}^{-1}, \tag{4}$$

which is a self-dual twist of $\mathbf{V}$. Define the rank four $G_{\mathbf{Q}_p}$-invariant subspace $\mathscr{F}^f_p(\mathbf{V}^\dagger) \subset \mathbf{V}^\dagger$ by

$$\mathscr{F}^f_p(\mathbf{V}^\dagger) := V^+_{\boldsymbol{f}} \hat{\otimes}_{\mathcal{O}} V_{\boldsymbol{g}} \hat{\otimes}_{\mathcal{O}} V_{\boldsymbol{h}} \otimes \mathcal{X}^{-1}. \tag{5}$$

For every $\underline{Q} = (Q_1, Q_2, Q_3) \in \mathfrak{X}^f_{\mathcal{R}}$ we denote by $\mathscr{F}^f_p(\mathbf{V}^\dagger_{\underline{Q}}) \subset \mathbf{V}^\dagger_{\underline{Q}}$ the corresponding specialisations. Finally, for every rational prime ℓ denote by $\varepsilon_\ell(\mathbf{V}^\dagger_{\underline{Q}})$ the epsilon factor attached to the restriction of $\mathbf{V}^\dagger_{\underline{Q}}$ to $G_{\mathbf{Q}_\ell}$ as in [48, p. 21], and assume that

$$\text{for some } \underline{Q} \in \mathfrak{X}^f_{\mathcal{R}}, \text{ we have } \varepsilon_\ell(\mathbf{V}^\dagger_{\underline{Q}}) = +1 \text{ for all prime factors } \ell \text{ of } N_f N_g N_h. \tag{6}$$

As explained in [31, Sect. 1.2], condition (6) is independent of $\underline{Q}$, and it implies that the sign in the functional equation for the triple product L-function

$$L(\mathbf{V}^\dagger_{\underline{Q}}, s)$$

(with center at $s = 0$) is $+1$ for all $\underline{Q} \in \mathfrak{X}^f_{\mathcal{R}}$.

Theorem 2.1.1 *Let* $(\boldsymbol{f}, \boldsymbol{g}, \boldsymbol{h})$ *be a triple of primitive Hida families as above satisfying conditions (3) and (6). Assume in addition that:*

- $\gcd(N_f, N_g, N_h)$ *is square-free,*
- *the residual representation* $\bar{\rho}_f$ *is absolutely irreducible and p-distinguished,*

and fix a generator η_f *of the congruence ideal of* $\boldsymbol{f}$*. Then there exists a unique element*

$$\mathscr{L}_p^f(\boldsymbol{f}, \boldsymbol{g}, \boldsymbol{h}) \in \mathcal{R}$$

such that for all $\underline{Q} = (Q_0, Q_1, Q_2) \in \mathfrak{X}_{\mathcal{R}}^f$ *of weight* (k_0, k_1, k_2) *with* $\epsilon_{Q_0} = 1$ *we have*

$$(\mathscr{L}_p^f(\boldsymbol{f}, \boldsymbol{g}, \boldsymbol{h})(\underline{Q}))^2 = \Gamma_{\mathbf{V}_{\underline{Q}}^\dagger}(0) \cdot \frac{L(\mathbf{V}_{\underline{Q}}^\dagger, 0)}{(\sqrt{-1})^{2k_0} \cdot \Omega_{f_{Q_0}}^2} \cdot \mathcal{E}_p(\mathscr{F}_p^f(\mathbf{V}_{\underline{Q}}^\dagger)) \cdot \prod_{\ell \in \Sigma_{\mathrm{exc}}} (1 + \ell^{-1})^2,$$

where:

- $\Gamma_{\mathbf{V}_{\underline{Q}}^\dagger}(0) = \Gamma_{\mathbf{C}}(c_{\underline{Q}})\Gamma_{\mathbf{C}}(c_{\underline{Q}} + 2 - k_1 - k_2)\Gamma_{\mathbf{C}}(c_{\underline{Q}} + 1 - k_1)\Gamma_{\mathbf{C}}(c_{\underline{Q}} + 1 - k_2)$, *with*

$$c_{\underline{Q}} = (k_0 + k_1 + k_2 - 2)/2$$

  and $\Gamma_{\mathbf{C}}(s) = 2(2\pi)^{-s}\Gamma(s)$*;*
- $\Omega_{f_{Q_0}}$ *is the canonical period*

$$\Omega_{f_{Q_0}} := (-2\sqrt{-1})^{k_0+1} \cdot \frac{\|f_{Q_0}^\circ\|_{\Gamma_0(N_f)}^2}{\iota_p(\eta_{f_{Q_0}})} \cdot \left(1 - \frac{\chi_f'(p)p^{k_0-1}}{\alpha_{Q_0}^2}\right)\left(1 - \frac{\chi_f'(p)p^{k_0-2}}{\alpha_{Q_0}^2}\right),$$

  with $f_{Q_0}^\circ \in S_{k_0}(N_f)$ *the newform of conductor* N_f *associated with* f_{Q_0}, χ_f' *the prime-to-p part of* χ_f, *and* α_{Q_0} *the specialization of* $a_p(\boldsymbol{f}) \in \mathbb{I}_f^\times$ *at* Q_0*;*
- $\mathcal{E}_p(\mathscr{F}_p^f(\mathbf{V}_{\underline{Q}}^\dagger))$ *is the modified p-Euler factor*

$$\mathcal{E}_p(\mathscr{F}_p^f(\mathbf{V}_{\underline{Q}}^\dagger)) := \frac{L_p(\mathscr{F}_p^f(\mathbf{V}_{\underline{Q}}^\dagger), 0)}{\varepsilon_p(\mathscr{F}_p^f(\mathbf{V}_{\underline{Q}}^\dagger)) \cdot L_p(\mathbf{V}_{\underline{Q}}^\dagger/\mathscr{F}_p^f(\mathbf{V}_{\underline{Q}}^\dagger), 0)} \cdot \frac{1}{L_p(\mathbf{V}_{\underline{Q}}^\dagger, 0)},$$

and Σ_{exc} *is an explicitly defined subset of the prime factors of* $N_f N_g N_h$, *[31, p. 416].*

Proof This is [31, Theorem A], which also includes an interpolation formula in the cases where ϵ_{Q_0} is not necessarily trivial. □

2.2 CM Hida Families

Let K_∞ be the unique $\mathbb{Z}_p^2$-extension of K, and let $K_{\mathfrak{p}^\infty}$ be the maximal subfield of K_∞ unramified outside $\mathfrak{p}$. Put

$$\Gamma_\infty := \mathrm{Gal}(K_\infty/K) \simeq \mathbb{Z}_p^2, \qquad \Gamma_{\mathfrak{p}^\infty} := \mathrm{Gal}(K_{\mathfrak{p}^\infty}/K) \simeq \mathbb{Z}_p.$$

For every ideal $\mathfrak{C} \subset \mathcal{O}_K$ we let $K(\mathfrak{C})$ be the ray class field of K of conductor $\mathfrak{C}$ (so in particular $K_{\mathfrak{p}^\infty}$ is the maximal $\mathbb{Z}_p$-extension of K inside $K(\mathfrak{p}^\infty)$). Let $\mathrm{Art}_\mathfrak{p}$ be the restriction of the global Artin map to $K_\mathfrak{p}^\times$, with geometric normalisation. Identifying $\mathbb{Z}_p^\times$ and $\mathcal{O}_{K_\mathfrak{p}}^\times$ via $\iota_p : \overline{\mathbf{Q}} \hookrightarrow \overline{\mathbf{Q}}_p$, the map $\mathrm{Art}_\mathfrak{p}$ induces an embedding $1 + p\mathbb{Z}_p \to \Gamma_{\mathfrak{p}^\infty}$. Write $I_\mathfrak{p}^\mathrm{w} = \mathrm{Art}_\mathfrak{p}(1 + p\mathbb{Z}_p)|_{K_{\mathfrak{p}^\infty}}$ and let $b \geq 0$ be such that $[\Gamma_{\mathfrak{p}^\infty} : I_\mathfrak{p}^\mathrm{w}] = p^b$. (Note that $b = 0$ if the class number of K is coprime to p).

Fix a topological generator $\gamma_\mathfrak{p} \in \Gamma_{\mathfrak{p}^\infty}$ with $\gamma_\mathfrak{p}^{p^b} = \mathrm{Art}_\mathfrak{p}(1 + p)|_{K_{\mathfrak{p}^\infty}}$, and for each variable S let $\Psi_S : G_K \to \mathcal{O}[\![S]\!]^\times$ be the universal character given by

$$\Psi_S(\sigma) = (1 + S)^{l(\sigma)}, \tag{7}$$

where $l(\sigma) \in \mathbb{Z}_p$ is such that $\sigma|_{K_{\mathfrak{p}^\infty}} = \gamma_\mathfrak{p}^{l(\sigma)}$. Let $\mathbf{v} \in \mathcal{O}$ be such that $\mathbf{v}^{p^b} = 1 + p$ (after enlarging $\mathcal{O}$ if necessary). For any finite order character $\psi : G_K \to \mathcal{O}^\times$ of conductor $\mathfrak{C}$ put

$$\boldsymbol{\theta}_\psi(S)(q) = \sum_{(\mathfrak{a},\mathfrak{p}\mathfrak{C})=1} \psi(\sigma_\mathfrak{a})\Psi^{-1}_{\mathbf{v}^{-1}(1+S)-1}(\sigma_\mathfrak{a})q^{\mathbf{N}(\mathfrak{a})} \in \mathcal{O}[\![S]\!][\![q]\!],$$

where $\sigma_\mathfrak{a} \in \mathrm{Gal}(K(\mathfrak{C}\mathfrak{p}^\infty)/K)$ is the Artin symbol of $\mathfrak{a}$. Then $\boldsymbol{\theta}_\psi(S)$ is a primitive Hida family defined over $\mathcal{O}[\![S]\!]$ of tame level $D_K\mathbf{N}(\mathfrak{C})$ and tame character $(\psi \circ \mathscr{V})\eta_{K/\mathbf{Q}}\omega^{-1}$, where

$$\mathscr{V} : G_\mathbf{Q}^{\mathrm{ab}} \to G_K^{\mathrm{ab}}$$

is the transfer map and $\eta_{K/\mathbf{Q}}$ is the quadratic character associated to $K/\mathbf{Q}$.

2.3 Anticyclotomic p-Adic L-Functions

Let $f \in S_2(\Gamma_0(pN_f))$, with $p \nmid N_f$, be a p-ordinary p-stabilised newform of tame level N_f defined over $\mathcal{O}$. Assume that f is the ordinary p-stabilisation of the newform $f^\circ \in S_2(\Gamma_0(N_f))$, and let $\alpha_p \in \mathcal{O}^\times$ be the U_p-eigenvalue of f. Write

$$N_f = N^+N^- \tag{8}$$

with N^+ (resp. N^-) divisible only by primes which are split (resp. inert) in K, and fix an ideal $\mathfrak{N}^+ \subset \mathcal{O}_K$ with $\mathcal{O}_K/\mathfrak{N}^+ \simeq \mathbf{Z}/N^+\mathbf{Z}$.

Let Γ^- be the Galois group of the anticyclotomic $\mathbf{Z}_p$-extension K_∞^-/K; this corresponds to the maximal quotient of Γ_∞ where complex conjugation $\mathbf{c}$ acts as inversion. Denote by $\pi_- : \Gamma_\infty \to \Gamma^-$ the natural projection. Then π_- yields a canonical isomorphism $\Gamma_\infty^{1-\mathbf{c}} := \{\gamma^{1-\mathbf{c}} \mid \gamma \in \Gamma_\infty\} \simeq \Gamma^-$. Similarly, restriction to $K_{\mathfrak{p}^\infty}$ gives an isomorphism $\Gamma_\infty^{1-\mathbf{c}} \simeq \Gamma_{\mathfrak{p}^\infty}$. Let $\gamma^- \in \Gamma^-$ be the topological generator mapping to $\gamma_\mathfrak{p}$ under the composite isomorphism $\Gamma^- \simeq \Gamma_{\mathfrak{p}^\infty}$. We then identity $\mathcal{O}[\![\Gamma^-]\!]$ with the one variable power series ring $\mathcal{O}[\![T]\!]$ via $\gamma^- \mapsto 1 + T$.

Remark 2.3.1 Note that letting $\Psi : G_K \to \mathscr{O}[\![T]\!]^\times$ be the composition of the character $G_K \to \Gamma^- \hookrightarrow \mathscr{O}[\![\Gamma^-]\!]^\times$ with the identification $\mathscr{O}[\![\Gamma^-]\!]^\times \simeq \mathscr{O}[\![T]\!]^\times$ defined by γ^-, it follows from the above discussion that

$$\Psi = \Psi_T^{1-\mathbf{c}},$$

where $\Psi_T : G_K \to \mathscr{O}[\![T]\!]^\times$ is the universal character in (7) for the variable T. This equality will be often implicitly used later in the paper.

Theorem 2.3.2 *Let χ be a ring class character of K of conductor $c\mathcal{O}_K$ with values in $\mathscr{O}$, and assume that:*

(i) $(pN_f, cD_K) = 1$,
(ii) *N^- is the squarefree product of an odd number of primes,*
(iii) *if $q \mid N^-$ is a prime with $q \equiv 1 \pmod p$, then $\bar{\rho}_f$ is ramified at q.*

Then there exists a unique $\Theta_{f/K,\chi}(T) \in \mathscr{O}[\![T]\!]$ such that for every p-power root of unity ζ:

$$\Theta_{f/K,\chi}(\zeta-1)^2 = \frac{p^n}{\alpha_p^{2n}} \cdot \mathcal{E}_p(f,\chi,\zeta)^2 \cdot \frac{L(f^\circ/K \otimes \chi\epsilon_\zeta, 1)}{(2\pi)^2 \cdot \Omega_{f^\circ,N^-}} \cdot u_K^2 \sqrt{D_K}\, \chi\epsilon_\zeta(\sigma_{\mathfrak{N}^+}) \cdot \epsilon_p,$$

where:

- $n \geq 0$ *is such that ζ has exact order p^n,*
- $\epsilon_\zeta : \Gamma_\infty^- \to \mu_{p^\infty}$ *be the character defined by $\epsilon_\zeta(\gamma^-) = \zeta$,*
- $\mathcal{E}_p(f,\chi,\zeta) = \begin{cases} (1-\alpha_p^{-1}\chi(\mathfrak{p}))(1-\alpha_p\chi(\bar{\mathfrak{p}})) & \text{if } n = 0, \\ 1 & \text{if } n > 0, \end{cases}$
- $\Omega_{f^\circ,N^-} = 4\|f^\circ\|^2_{\Gamma_0(N_f)} \cdot \eta_{f,N^-}^{-1}$ *is the* Gross period *of f° (see [31, p. 524]),*
- $u_K = |\mathcal{O}_K^\times|/2$,
- $\epsilon_p \in \{\pm 1\}$ *is the local root number of f° at p.*

Proof See [2] for the original construction, and [16, Theorem A] for the stated interpolation property. □

When χ is the trivial character, we write $\Theta_{f/K,\chi}(T)$ simply as $\Theta_{f/K}(T)$.

2.4 *Factorisation of Triple Product p-Adic L-Functions*

Let $f \in S_2(pN_f)$ be a p-stabilised newform as in the preceding section. By Hida theory, f is the specialisation of a unique primitive Hida family $\boldsymbol{f} \in S^o(N_f, \mathbb{I})$ at an arithmetic point $Q_0 \in \mathfrak{X}_{\mathbb{I}}^+$ of weight 2. Let $\ell \nmid pN_f$ be a prime split in K, and let χ be a ring class character of K of conductor $\ell^m\mathcal{O}_K$ for some $m > 0$. Denoting by $\mathbf{c}$ the non-trivial automorphism of $K/\mathbf{Q}$, write $\chi = \psi/\psi^{\mathbf{c}}$ with ψ a ray class character modulo $\ell^m \mathcal{O}_K$ with $\psi^{\mathbf{c}}(\sigma) = \psi(\mathbf{c}\sigma\mathbf{c}^{-1})$. Let

$$\boldsymbol{g} = \boldsymbol{\theta}_{\psi}(S_1) \in \mathcal{O}[\![S_1]\!][\![q]\!], \quad \boldsymbol{g}^* = \boldsymbol{\theta}_{\psi^{-1}}(S_2) \in \mathcal{O}[\![S_2]\!][\![q]\!] \tag{9}$$

be the primitive CM Hida families (of tame level $C = D_K\ell^{2m}$) attached to ψ and ψ^{-1}, respectively.

The triple $(\boldsymbol{f}, \boldsymbol{g}, \boldsymbol{g}^*)$ satisfies conditions (3) and (6) and the associated triple product p-adic L-function $\mathscr{L}_p^f(\boldsymbol{f}, \boldsymbol{g}, \boldsymbol{g}^*)$ of Theorem 2.1.1 is an element in $\mathcal{R} = \mathbb{I}\hat{\otimes}_{\mathcal{O}}\mathcal{O}[\![S_1]\!]\hat{\otimes}_{\mathcal{O}}\mathcal{O}[\![S_2]\!] \simeq \mathbb{I}[\![S_1, S_2]\!]$. Put $S = S_1$. In the following, we let

$$\mathscr{L}_p^f(f, \underline{\boldsymbol{g}\boldsymbol{g}^*}) \in \mathcal{O}[\![S]\!] \tag{10}$$

be the image of $\mathscr{L}_p^f(\boldsymbol{f}, \boldsymbol{g}, \boldsymbol{g}^*)$ under the map $\mathbb{I}[\![S_1, S_2]\!] \to \mathcal{O}[\![S_1, S_2]\!]$ given by the specialisation $Q_0 : \mathbb{I} \to \mathcal{O}$ composed with the quotient $\mathcal{O}[\![S_1, S_2]\!] \to \mathcal{O}[\![S_1, S_2]\!]/(S_1 - S_2)$.

Let $K(\chi)$ be the field obtained by adjoining to K the values of χ, and put $K(\chi, \alpha_p) = K(\chi)(\alpha_p)$.

Proposition 2.4.1 *Assume that:*

(i) N^- *is the squarefree product of an odd number of primes,*
(ii) if $q \mid N^-$ *is a prime with* $q \equiv 1 \pmod{p}$*, then* $\bar{\rho}_f$ *is ramified at* q.

Then

$$\mathscr{L}_p^f(f, \underline{\boldsymbol{g}\boldsymbol{g}^*})(S) = \pm\mathbf{w}^{-1} \cdot \Theta_{f/K}(T) \cdot C_{f,\chi} \cdot \sqrt{L^{\mathrm{alg}}(f/K \otimes \chi, 1)} \cdot \frac{\eta_{f^\circ}}{\eta_{f^\circ, N^-}},$$

where $T = \mathbf{v}^{-1}(1 + S) - 1$, $\mathbf{w}$ *is a unit in* $\mathcal{O}[\![T]\!]$, $C_{f,\chi} \in K(\chi, \alpha_p)^\times$, *and*

$$L^{\mathrm{alg}}(f/K \otimes \chi, 1) := \frac{L(f/K \otimes \chi, 1)}{4\pi^2 \| f^\circ \|^2_{\Gamma_0(N_f)}} \in K(\chi).$$

Proof This is immediate from [31, Proposition 8.1] and the interpolation property of $\Theta_{f/K,\chi}(0)$. □

3 Selmer Group Decompositions

In this section we introduce two different Selmer groups associated to triple products of modular forms, and relate them to anticyclotomic Selmer groups attached to a single modular form.

3.1 Selmer Groups for Triple Products of Modular Form

Let $(\boldsymbol{f}, \boldsymbol{g}, \boldsymbol{h})$ be a triple of primitive Hida families as in Sect. 2.1.1 satisfying (3), and let $\mathbf{V}^{\dagger} = \mathbf{V} \otimes \mathcal{X}^{-1}$ be the self-dual twist of the associated big Galois representation.

Definition 3.1.1 Put

$$\mathscr{F}_p^{\mathrm{bal}}(\mathbf{V}^{\dagger}) = \mathscr{F}_p^2(\mathbf{V}^{\dagger}) := \left(V_{\boldsymbol{f}}^+ \otimes V_{\boldsymbol{g}}^+ \otimes V_{\boldsymbol{h}} + V_{\boldsymbol{f}}^+ \otimes V_{\boldsymbol{g}} \otimes V_{\boldsymbol{h}}^+ + V_{\boldsymbol{f}} \otimes V_{\boldsymbol{g}}^+ \otimes V_{\boldsymbol{h}}^+\right) \otimes \mathcal{X}^{-1},$$

and define the *balanced local condition* $\mathrm{H}^1_{\mathrm{bal}}(\mathbf{Q}_p, \mathbf{V}^{\dagger})$ by

$$\mathrm{H}^1_{\mathrm{bal}}(\mathbf{Q}_p, \mathbf{V}^{\dagger}) := \mathrm{im}\left\{\mathrm{H}^1(\mathbf{Q}_p, \mathscr{F}_p^{\mathrm{bal}}(\mathbf{V}^{\dagger})) \rightarrow \mathrm{H}^1(\mathbf{Q}_p, \mathbf{V}^{\dagger})\right\}.$$

Similarly, put

$$\mathscr{F}_p^{\boldsymbol{f}}(\mathbf{V}^{\dagger}) := \left(V_{\boldsymbol{f}}^+ \otimes V_{\boldsymbol{g}} \otimes V_{\boldsymbol{h}}\right) \otimes \mathcal{X}^{-1},$$

and define the *$\boldsymbol{f}$-unbalanced local condition* $\mathrm{H}^1_{\boldsymbol{f}}(\mathbf{Q}_p, \mathbf{V}^{\dagger})$ by

$$\mathrm{H}^1_{\boldsymbol{f}}(\mathbf{Q}_p, \mathbf{V}^{\dagger}) := \mathrm{im}\left\{\mathrm{H}^1(\mathbf{Q}_p, \mathscr{F}_p^{\boldsymbol{f}}(\mathbf{V}^{\dagger})) \rightarrow \mathrm{H}^1(\mathbf{Q}_p, \mathbf{V}^{\dagger})\right\}.$$

It is easy to see that the maps appearing in these definitions are injective, and in the following we shall use this to identify $\mathrm{H}^1_?(\mathbf{Q}_p, \mathbf{V}^{\dagger})$ with $\mathrm{H}^1(\mathbf{Q}_p, \mathscr{F}_p^?(\mathbf{V}^{\dagger}))$ for $? \in \{\mathrm{bal}, \boldsymbol{f}\}$. Fix a finite set of primes S containing ∞ and the primes dividing $N_f N_g N_g$, and let $G_{\mathbf{Q},S}$ be the Galois group of the maximal extension of $\mathbf{Q}$ unramified outside S.

Definition 3.1.2 Let $? \in \{\mathrm{bal}, \boldsymbol{f}\}$, and define the Selmer group $\mathrm{Sel}^?(\mathbf{Q}, \mathbf{V}^{\dagger})$ by

$$\mathrm{Sel}^?(\mathbf{Q}, \mathbf{V}^{\dagger}) := \ker\left\{\mathrm{H}^1(G_{\mathbf{Q},S}, \mathbf{V}^{\dagger}) \rightarrow \frac{\mathrm{H}^1(\mathbf{Q}_p, \mathbf{V}^{\dagger})}{\mathrm{H}^1_?(\mathbf{Q}_p, \mathbf{V}^{\dagger})}\right\}.$$

We call $\mathrm{Sel}^{\mathrm{bal}}(\mathbf{Q}, \mathbf{V}^{\dagger})$ (resp. $\mathrm{Sel}^{\boldsymbol{f}}(\mathbf{Q}, \mathbf{V}^{\dagger})$) the *balanced* (resp. *$\boldsymbol{f}$-unbalanced*) Selmer group.

Let $\mathbf{A}^{\dagger} = \mathrm{Hom}_{\mathbb{Z}_p}(\mathbf{V}^{\dagger}, \mu_{p^\infty})$ be the arithmetic dual of $\mathbf{V}^{\dagger}$, and for $? \in \{\mathrm{bal}, \boldsymbol{f}\}$ define $\mathrm{H}^1_?(\mathbf{Q}_p, \mathbf{A}^{\dagger}) \subset \mathrm{H}^1(\mathbf{Q}_p, \mathbf{A}^{\dagger})$ to be the orthogonal complement of $\mathrm{H}^1_?(\mathbf{Q}_p, \mathbf{V}^{\dagger})$ under the local Tate duality

$$\mathrm{H}^1(\mathbf{Q}_p, \mathbf{V}^{\dagger}) \times \mathrm{H}^1(\mathbf{Q}_p, \mathbf{A}^{\dagger}) \rightarrow \mathbf{Q}_p/\mathbb{Z}_p.$$

Similarly as above, we then define the balanced and $\boldsymbol{f}$-unbalanced Selmer groups with coefficients in $\mathbf{A}^{\dagger}$ by

$$\mathrm{Sel}^?(\mathbf{Q}, \mathbf{A}^{\dagger}) := \ker\left\{\mathrm{H}^1(G_{\mathbf{Q},S}, \mathbf{A}^{\dagger}) \rightarrow \frac{\mathrm{H}^1(\mathbf{Q}_p, \mathbf{A}^{\dagger})}{\mathrm{H}^1_?(\mathbf{Q}_p, \mathbf{A}^{\dagger})} \times \prod_{\ell \in S \smallsetminus \{p\}} \mathrm{H}^1(\mathbf{Q}_\ell, \mathbf{A}^{\dagger})\right\},$$

and let $X^?(\mathbf{Q}, \mathbf{A}^\dagger) = \mathrm{Hom}_{\mathbb{Z}_p}(\mathrm{Sel}^?(\mathbf{Q}, \mathbf{A}^\dagger), \mathbf{Q}_p/\mathbb{Z}_p)$ denote the Pontryagin dual of $\mathrm{Sel}^?(\mathbf{Q}, \mathbf{A}^\dagger)$.

3.2 *Anticyclotomic Selmer Groups for Modular Forms*

Let $f \in S_2(\Gamma_0(pN_f))$ be an ordinary p-stabilised newform and $K/\mathbf{Q}$ an imaginary quadratic field as in Sect. 2.3.

Let V_f be the p-adic Galois representation associated to f. We adopt the convention that if f corresponds to the isogeny class of an elliptic curve $E/\mathbf{Q}$, then $V_f \simeq V_pE$ (rather than its dual). By p-ordinarity, restricted to $G_{\mathbf{Q}_p}$ the representation V_f fits into a short exact sequence

$$0 \to V_f^+ \to V_f \to V_f^- \to 0$$

with $V_f^\pm$ both 1-dimensional, and with the $G_{\mathbf{Q}_p}$-action on V_f^- given by the unramified character sending arithmetic Frobenius to $\alpha_p \in \mathcal{O}^\times$, the U_p-eigenvalue of f.

Below we fix Σ to be any finite set of places of K containing ∞ and the prime dividing pN_f, and for any field extension L/K let $G_{L,\Sigma}$ the Galois group of the maximal extension of L unramified outside Σ.

Definition 3.2.1 Let L be a finite field extension of K, and let $\mathscr{F} = \{\mathscr{F}_v(V_f)\}_{v|p}$ be a collection a G_{K_v}-stable subpaces $\mathscr{F}_v(V_f) \subset V_f$ for $v \mid p$. We define the *Greenberg Selmer group* $\mathrm{Sel}_{\mathscr{F}}(L, V_f)$ by

$$\mathrm{Sel}_{\mathscr{F}}(L, V_f) := \ker\left\{\mathrm{H}^1(G_{L,\Sigma}, V_f) \to \prod_w \frac{\mathrm{H}^1(L_w, V_f)}{\mathrm{H}^1_{\mathscr{F}}(L_w, V_f)}\right\},$$

where w runs over the finite primes of L lying above a prime $v \in \Sigma$, and

$$\mathrm{H}^1_{\mathscr{F}}(L_w, V_f) := \begin{cases} \ker\{\mathrm{H}^1(L_w, V_f) \to \mathrm{H}^1(L_w^{\mathrm{ur}}, V_f)\} & \text{if } w \nmid p, \\ \mathrm{im}\{\mathrm{H}^1(L_w, \mathscr{F}_v(V_f)) \to \mathrm{H}^1(L_w, V_f)\} & \text{if } w \mid v \mid p. \end{cases}$$

We shall be particularly interested in the following choices of $\mathscr{F}$:

- The *relaxed-strict* Selmer group $\mathrm{Sel}_{\emptyset,0}(L, V_f)$ obtained by taking

$$\mathscr{F}_v(V_f) = \begin{cases} V_f & \text{if } v = \mathfrak{p}, \\ 0 & \text{if } v = \overline{\mathfrak{p}}. \end{cases}$$

- The *ordinary* Selmer group $\mathrm{Sel}(L, V_f)$ obtained by taking $\mathscr{F}_v(V_f) = V_f^+$ for all $v \mid p$.

For a G_K-stable lattice $T_f \subset V_f$, we let $\mathrm{H}^1_{\mathscr{F}}(L_w, T_f)$ be the inverse image of $\mathrm{H}^1_{\mathscr{F}}(L_w, V_f)$ under the natural map $\mathrm{H}^1(L_w, T_f) \to \mathrm{H}^1(L_w, V_f)$, and define $\mathrm{Sel}_{\mathscr{F}}(L, T_f)$ by the same recipe as above. Then, for $A_f := \mathrm{Hom}_{\mathbb{Z}_p}(T_f, \mu_{p^\infty})$, we define the *dual Selmer group* $\mathrm{Sel}_{\mathscr{F}^*}(L, A_f)$ by

$$\mathrm{Sel}_{\mathscr{F}^*}(L, A_f) := \ker\left\{\mathrm{H}^1(G_{L,\Sigma}, A_f) \to \prod_w \frac{\mathrm{H}^1(L_w, A_f)}{\mathrm{H}^1_{\mathscr{F}^*}(L_w, A_f)}\right\},$$

where $\mathrm{H}^1_{\mathscr{F}^*}(L_w, A_f)$ is the orthogonal complement of $\mathrm{H}^1_{\mathscr{F}}(L_w, T_f)$ under local Tate duality

$$\mathrm{H}^1(L_w, T_f) \times \mathrm{H}^1(L_w, A_f) \to \mathbf{Q}_p/\mathbf{Z}_p.$$

Write K_m^- for the subextension of the the anticyclotomic $\mathbb{Z}_p$-extension K_∞^- with $[K_m^- : K] = p^m$, and put

$$\mathrm{Sel}_{\mathscr{F}}(K_\infty^-, T_f) := \varprojlim_m \mathrm{Sel}_{\mathscr{F}}(K_m^-, T_f), \quad \mathrm{Sel}_{\mathscr{F}}(K_\infty^-, A_f) := \varinjlim_m \mathrm{Sel}_{\mathscr{F}}(K_m^-, A_f),$$

where the limits are with respect to the corestriction and restriction maps, respectively. Let $\Psi : G_K \to \mathscr{O}[\![\Gamma^-]\!]^\times$ be the character arising from the projection $G_K \to \Gamma^-$. Writing $T_f \otimes \Psi^{-1}$ (resp. $A_f \otimes \Psi$) for the module $T_f \otimes_{\mathscr{O}} \mathscr{O}[\![\Gamma^-]\!]$ (resp. $A_f \otimes_{\mathscr{O}} \mathscr{O}[\![\Gamma^-]\!]$) with G_K-action on the second factor given by Ψ^{-1} (resp. Ψ), we then have natural $\mathscr{O}[\![\Gamma^-]\!]$-module pseudo-isomorphisms

$$\mathrm{Sel}_{\mathscr{F}}(K_\infty^-, T_f) \sim \mathrm{Sel}_{\mathscr{F}}(K, T_f \otimes \Psi^{-1}), \quad \mathrm{Sel}_{\mathscr{F}}(K_\infty^-, A_f) \sim \mathrm{Sel}_{\mathscr{F}}(K, A_f \otimes \Psi), \tag{11}$$

where the Selmer groups on the right-hand side are defined is the same way as in Definition 3.2.1, with $\mathscr{F}_v(T_f \otimes \Psi^{-1}) := (\mathscr{F}_v(V_f) \cap T_f) \otimes \Psi^{-1}$ and $\mathscr{F}_v(A_f \otimes \Psi) := \mathrm{Hom}_{\mathbb{Z}_p}(T_f/(\mathscr{F}_v(V_f) \cap T_f), \mu_{p^\infty}) \otimes \Psi$.

3.3 *Decomposition of Triple Product Selmer Groups*

Suppose now that $\boldsymbol{f}$ is the Hida family passing through the p-stabilised newform $f \in S_2(\Gamma_0(pN_f))$, and $(\boldsymbol{g}, \boldsymbol{g}^*) = (\boldsymbol{\theta}_\psi(S_1), \boldsymbol{\theta}_{\psi^{-1}}(S_2))$ are CM Hida families as in (9).

For any primitive Hida family $\boldsymbol{\phi}$, we now let $V_{\boldsymbol{\phi}}$ be the realisation of $\rho_{\boldsymbol{\phi}}$ arising in the p-adic étale cohomology of the p-tower of modular curves as in Ohta's works [42, 43], following the conventions in [34, Sect. 7.2] (except that a "Hida family" for us is a "branch" in their sense).

In the case of $\boldsymbol{g}$ (and similarly $\boldsymbol{g}^*$), when $\chi := \psi/\psi^{\mathbf{c}}$ satisfies

$$\bar{\chi}|_{G_{K_{\mathfrak{p}}}} \neq 1, \tag{12}$$

where $G_{K_\mathfrak{p}} \subset G_K$ is a decomposition group at $\mathfrak{p}$, it follows from a slight extension of the isomorphism in [38, Corrollary 5.2.5] (see [8, Sect. 3.2.3]) that

$$V_g \simeq \mathrm{Ind}_K^{\mathbf{Q}}(\psi^{-1}\Psi_{T_1}),$$

where $T_1 = \mathbf{v}^{-1}(1+S_1) - 1$ and Ψ_{T_1} is the associated universal character (7). Suppose $Q_0 : \mathbb{I} \to \mathscr{O}$ is such that f is the specialisation of $\boldsymbol{f}$ at Q_0, and write $\mathbb{V}^\dagger_{f\boldsymbol{g}\boldsymbol{g}^*}$ for the resulting specialisation of $\mathbf{V}^\dagger$. Similarly setting $T_2 = \mathbf{v}^{-1}(1+S_2) - 1$ and noting that $(\det V_g)(\det V_{g^*}) = \Psi_{T_1}\Psi_{T_2} \circ \mathscr{V}$, we then have

$$\begin{aligned} \mathbb{V}^\dagger_{f\boldsymbol{g}\boldsymbol{g}^*} &\simeq T_f \otimes \mathrm{Ind}_K^{\mathbf{Q}}(\psi^{-1}\Psi_{T_1}) \otimes \mathrm{Ind}_K^{\mathbf{Q}}(\psi\Psi_{T_2}) \otimes (\Psi_{T_1}^{1/2}\Psi_{T_2}^{1/2} \circ \mathscr{V})^{-1} \\ &\simeq \big(T_f \otimes \mathrm{Ind}_K^{\mathbf{Q}} \Psi_{W_1}^{1-\mathbf{c}}\big) \oplus \big(T_f \otimes \mathrm{Ind}_K^{\mathbf{Q}} \chi^{-1}\Psi_{W_2}^{1-\mathbf{c}}\big), \end{aligned} \tag{13}$$

where T_f is the specialisation of $V_{\boldsymbol{f}}$ at Q_0 and

$$W_1 = \mathbf{v}^{-1}(1+S_1)^{1/2}(1+S_2)^{1/2} - 1, \quad W_2 = (1+S_1)^{1/2}(1+S_2)^{-1/2} - 1.$$

In particular, together with Shapiro's lemma this gives

$$\mathrm{H}^1(\mathbf{Q}, \mathbb{V}^\dagger_{f\boldsymbol{g}\boldsymbol{g}^*}) \simeq \mathrm{H}^1(K, T_f \otimes \Psi_{W_1}^{1-\mathbf{c}}) \oplus \mathrm{H}^1(K, T_f \otimes \chi^{-1}\Psi_{W_2}^{1-\mathbf{c}}) \tag{14}$$

Proposition 3.3.1 *Suppose ψ is a ray class character of K such that $\chi = \psi/\psi^{\mathbf{c}}$ satisfies (12). Then under the isomorphism (14), the balanced Selmer group decomposes as*

$$\mathrm{Sel}^{\mathrm{bal}}(\mathbf{Q}, \mathbb{V}^\dagger_{f\boldsymbol{g}\boldsymbol{g}^*}) \simeq \mathrm{Sel}_{\emptyset,0}(K, T_f \otimes \Psi_{W_1}^{1-\mathbf{c}}) \oplus \mathrm{Sel}(K, T_f \otimes \chi^{-1}\Psi_{W_2}^{1-\mathbf{c}}),$$

and the $\boldsymbol{f}$-unbalanced Selmer group decomposes as

$$\mathrm{Sel}^{\boldsymbol{f}}(\mathbf{Q}, \mathbb{V}^\dagger_{f\boldsymbol{g}\boldsymbol{g}^*}) \simeq \mathrm{Sel}(K, T_f \otimes \Psi_{W_1}^{1-\mathbf{c}}) \oplus \mathrm{Sel}(K, T_f \otimes \chi^{-1}\Psi_{W_2}^{1-\mathbf{c}}).$$

Proof This is shown in [14, Proposition 5.3.1]; we recall the details for the convenience of the reader. From (13), we see that

$$\mathbb{V}^\dagger_{f\boldsymbol{g}\boldsymbol{g}^*}|_{G_{\mathbf{Q}_p}} \simeq \big(T_f \otimes \Psi_{W_1}^{1-\mathbf{c}}\big) \oplus \big(T_f \otimes \Psi_{W_1}^{\mathbf{c}-1}\big) \oplus \big(T_f \otimes \chi^{-1}\Psi_{W_2}^{1-\mathbf{c}}\big) \oplus \big(T_f \otimes \chi\Psi_{W_2}^{\mathbf{c}-1}\big)$$

noting that $\chi^{\mathbf{c}} = \chi^{-1}$, and we find that the balanced local condition is given by

$$\mathscr{F}_p^{\mathrm{bal}}(\mathbb{V}^\dagger_{f\boldsymbol{g}\boldsymbol{g}^*}) \simeq \big(T_f \otimes \Psi_{W_1}^{1-\mathbf{c}}\big) \oplus \big(T_f^+ \otimes \chi^{-1}\Psi_{W_1}^{1-\mathbf{c}}\big) \oplus \big(T_f^+ \otimes \chi\Psi_{W_2}^{\mathbf{c}-1}\big), \tag{15}$$

where $T_f^+ \subset T_f$ is the specialisation of $V_{\boldsymbol{f}}^+$ at Q_0. Put $\widetilde{\mathbb{V}}^\dagger_{f\boldsymbol{g}\boldsymbol{g}^*} = (T_f \otimes \Psi_{W_1}^{1-\mathbf{c}}) \oplus (T_f \otimes \chi^{-1}\Psi_{W_2}^{1-\mathbf{c}})$, and note that

$$\mathrm{H}^1(\mathbf{Q}, \mathbb{V}^\dagger_{f\boldsymbol{g}\boldsymbol{g}^*}) \simeq \mathrm{H}^1(K, \widetilde{\mathbb{V}}^\dagger_{f\boldsymbol{g}\boldsymbol{g}^*}) \tag{16}$$

by Shapiro's lemma. For $\mathfrak{q} \in \{\mathfrak{p}, \overline{\mathfrak{p}}\}$, letting $\mathscr{F}_{\mathfrak{q}}^{\mathrm{bal}}(\widetilde{\mathbb{V}}_{fgg^*}^{\dagger})$ be the submodule of $\widetilde{\mathbb{V}}_{fgg^*}^{\dagger}$ corresponding to $\mathscr{F}_{p}^{\mathrm{bal}}(\mathbb{V}_{fgg^*}^{\dagger})$, from (15) we have

$$\begin{cases} \mathscr{F}_{\mathfrak{p}}^{\mathrm{bal}}(\widetilde{\mathbb{V}}_{fgg^*}^{\dagger}) \simeq (T_f \otimes \Psi_{W_1}^{1-\mathbf{c}}) \oplus (T_f^+ \otimes \chi^{-1}\Psi_{W_2}^{1-\mathbf{c}}), \\ \mathscr{F}_{\overline{\mathfrak{p}}}^{\mathrm{bal}}(\widetilde{\mathbb{V}}_{fgg^*}^{\dagger}) \simeq T_f^+ \otimes \chi^{-1}\Psi_{W_2}^{1-\mathbf{c}}, \end{cases} \tag{17}$$

using again that $\chi^{\mathbf{c}} = \chi^{-1}$ for the second isomorphism. As a result, under the isomorphism (16) we have

$$\mathrm{Sel}^{\mathrm{bal}}(\mathbf{Q}, \mathbb{V}_{fgg^*}^{\dagger}) \simeq$$

$$\ker\left\{\mathrm{H}^1(G_{K,\Sigma}, \widetilde{\mathbb{V}}_{fgg^*}^{\dagger}) \rightarrow \frac{\mathrm{H}^1(K_{\mathfrak{p}}, \widetilde{\mathbb{V}}_{fgg^*}^{\dagger})}{\mathrm{H}^1(K_{\mathfrak{p}}, (T_f \otimes \Psi_{W_1}^{1-\mathbf{c}}) \oplus (T_f^+ \otimes \chi^{-1}\Psi_{W_2}^{1-\mathbf{c}}))} \times \frac{\mathrm{H}^1(K_{\overline{\mathfrak{p}}}, \widetilde{\mathbb{V}}_{fgg^*}^{\dagger})}{\mathrm{H}^1(K_{\overline{\mathfrak{p}}}, T_f^+ \otimes \chi^{-1}\Psi_{W_2}^{1-\mathbf{c}})}\right\},$$

yielding the result in this case. Similarly, we find that the f-balanced local condition is given by

$$\begin{cases} \mathscr{F}_{\mathfrak{p}}^{f}(\widetilde{\mathbb{V}}_{fgg^*}^{\dagger}) \simeq (T_f^+ \otimes \Psi_{W_1}^{1-\mathbf{c}}) \oplus (T_f^+ \otimes \chi^{-1}\Psi_{W_2}^{1-\mathbf{c}}), \\ \mathscr{F}_{\overline{\mathfrak{p}}}^{f}(\widetilde{\mathbb{V}}_{fgg^*}^{\dagger}) \simeq (T_f^+ \otimes \Psi_{W_1}^{1-\mathbf{c}}) \oplus (T_f^+ \otimes \chi^{-1}\Psi_{W_2}^{1-\mathbf{c}}), \end{cases}$$

and as above we arrive at the claimed description of $\mathrm{Sel}^f(\mathbf{Q}, \mathbb{V}_{fgg^*}^{\dagger})$. □

As a consequence, we also obtain decompositions for corresponding Selmer groups with coefficients in $\mathbb{A}_{fgg^*}^{\dagger} = \mathrm{Hom}_{\mathbb{Z}_p}(\mathbb{V}_{fgg^*}^{\dagger}, \mu_{p^\infty})$, mirroring in the case of $\mathrm{Sel}^f(\mathbf{Q}, \mathbb{A}_{fgg^*}^{\dagger})$ the factorisation of p-adic L-functions in Proposition 2.4.1. Put $A_f = \mathrm{Hom}_{\mathbb{Z}_p}(T_f, \mu_{p^\infty})$.

Corollary 3.3.2 *Under the asumption in Proposition 3.3.1, we have the decompositions*

$$\mathrm{Sel}^{\mathrm{bal}}(\mathbf{Q}, \mathbb{A}_{fgg^*}^{\dagger}) \simeq \mathrm{Sel}_{0,\emptyset}(K, A_f \otimes \Psi_{W_1}^{\mathbf{c}-1}) \oplus \mathrm{Sel}(K, A_f \otimes \chi\Psi_{W_2}^{\mathbf{c}-1}),$$
$$\mathrm{Sel}^{f}(\mathbf{Q}, \mathbb{A}_{fgg^*}^{\dagger}) \simeq \mathrm{Sel}(K, A_f \otimes \Psi_{W_1}^{\mathbf{c}-1}) \oplus \mathrm{Sel}(K, A_f \otimes \chi\Psi_{W_2}^{\mathbf{c}-1}).$$

Proof This is immediate from Proposition 3.3.1 and local Tate duality. □

4 Iwasawa Main Conjectures

Keeping the setting from Sect. 3.3, and put $\Lambda_2 = \mathscr{O}[\![S_1, S_2]\!]$ and

$$\mathbb{V}_{f\underline{gg}^*}^{\dagger} = \mathbb{V}_{fgg^*}^{\dagger} \hat{\otimes}_{\Lambda_2} \Lambda_2/(S_1 - S_2),$$

where $\Lambda_2 = \mathscr{O}[\![S_1, S_2]\!]$. In this section we recall the diagonal cycle main conjecture formulated in [1] specialised to our setting, and explain its relation with the anticyclotomic Iwasawa main conjecture for f.

4.1 Diagonal Cycle Main Conjecture

Let $\kappa(\boldsymbol{f}, \boldsymbol{g}, \boldsymbol{g}^*) \in \mathrm{H}^1(\mathbf{Q}, \mathbf{V}^\dagger)$ be the *big diagonal class* constructed in [11, Sect. 8.1], where $\boldsymbol{f}$ is the primitive Hida family passing through f, and denote by

$$\kappa(f, \underline{\boldsymbol{g}\boldsymbol{g}^*}) \in \mathrm{H}^1(\mathbf{Q}, \mathbb{V}^\dagger_{f\underline{\boldsymbol{g}\boldsymbol{g}^*}}) \tag{18}$$

its specialisation. More precisely, from *loc. cit.* one directly obtains a class as above with coefficient in a representation $\mathbf{V}^\dagger(N)$ isomorphic (non-canonically) to finitely many copies of $\mathbf{V}^\dagger$, where $N = \mathrm{lcm}(N_f, N_g, N_{g^*}) = \mathrm{lcm}(N_f, D_K\ell)$; to obtain (18) we apply the projection $\mathbf{V}^\dagger(N) \to \mathbf{V}^\dagger$ corresponding to the choice of level-N test vectors for $(\boldsymbol{f}, \boldsymbol{g}, \boldsymbol{g}^*)$ furnished by [31, Theorem A].

From the decompositions in Proposition 3.3.1 we obtain

$$\begin{aligned} \mathrm{Sel}^{\mathrm{bal}}(\mathbf{Q}, \mathbb{V}^\dagger_{f\underline{\boldsymbol{g}\boldsymbol{g}^*}}) &\simeq \mathrm{Sel}_{\emptyset,0}(K, T_f \otimes \Psi_T^{1-\mathbf{c}}) \oplus \mathrm{Sel}(K, T_f \otimes \chi^{-1}\Psi_{W_2}^{1-\mathbf{c}})_{/W_2}, \\ \mathrm{Sel}^{f}(\mathbf{Q}, \mathbb{V}^\dagger_{f\underline{\boldsymbol{g}\boldsymbol{g}^*}}) &\simeq \mathrm{Sel}(K, T_f \otimes \Psi_T^{1-\mathbf{c}}) \oplus \mathrm{Sel}(K, T_f \otimes \chi^{-1}\Psi_{W_2}^{1-\mathbf{c}})_{/W_2}, \end{aligned} \tag{19}$$

where $T = T_1 = \mathbf{v}^{-1}(1 + S_1) - 1$ and the subscript $/W_2$ denotes the cokernel of multiplication by W_2. Put

$$\mathscr{F}_p^3(\mathbb{V}^\dagger_{f\boldsymbol{g}\boldsymbol{g}^*}) := T_f^+ \hat{\otimes}_{\mathscr{O}} V_f^+ \hat{\otimes}_{\mathscr{O}} V_{g^*}^+ \otimes \mathcal{X}^{-1},$$

and denote by $\mathscr{F}_p^3(\mathbb{V}^\dagger_{f\underline{\boldsymbol{g}\boldsymbol{g}^*}})$ the resulting submodule of $\mathbb{V}^\dagger_{f\underline{\boldsymbol{g}\boldsymbol{g}^*}}$. Similarly defining $\mathscr{F}_p^{\mathrm{bal}}(\mathbb{V}^\dagger_{f\underline{\boldsymbol{g}\boldsymbol{g}^*}})$ from $\mathscr{F}_p^{\mathrm{bal}}(\mathbb{V}^\dagger_{f\boldsymbol{g}\boldsymbol{g}^*}) = \mathscr{F}_p^2(\mathbb{V}^\dagger_{f\boldsymbol{g}\boldsymbol{g}^*})$, in terms of the description in (17) we find that

$$\begin{cases} \mathscr{F}_{\mathfrak{p}}^{\mathrm{bal}}(\widetilde{\mathbb{V}}^\dagger_{f\underline{\boldsymbol{g}\boldsymbol{g}^*}})/\mathscr{F}_{\mathfrak{p}}^3(\widetilde{\mathbb{V}}^\dagger_{f\underline{\boldsymbol{g}\boldsymbol{g}^*}}) \simeq \left(T_f^- \otimes \Psi_T^{1-\mathbf{c}}\right) \oplus \left(T_f^+ \otimes \chi^{-1}\Psi_{W_2}^{1-\mathbf{c}}\right)_{/W_2}, \\ \mathscr{F}_{\overline{\mathfrak{p}}}^{\mathrm{bal}}(\widetilde{\mathbb{V}}^\dagger_{f\underline{\boldsymbol{g}\boldsymbol{g}^*}})/\mathscr{F}_{\overline{\mathfrak{p}}}^3(\widetilde{\mathbb{V}}^\dagger_{f\underline{\boldsymbol{g}\boldsymbol{g}^*}}) \simeq (T_f^+ \otimes \chi^{-1}\Psi_{W_2}^{1-\mathbf{c}})_{/W_2}, \end{cases} \tag{20}$$

where $T_f^- := T_f/T_f^+$.

It follows from [11, Corrollary 8.2] that $\kappa(f, \underline{\boldsymbol{g}\boldsymbol{g}^*})$ lies in the balanced Selmer group $\mathrm{Sel}^{\mathrm{bal}}(\mathbf{Q}, \mathbb{V}^\dagger_{f\underline{\boldsymbol{g}\boldsymbol{g}^*}})$. Therefore, viewing this class in $\mathrm{H}^1(K, \widetilde{\mathbb{V}}^\dagger_{f\underline{\boldsymbol{g}\boldsymbol{g}^*}})$ via the isomorphism (16), we can consider the image of $\mathrm{loc}_{\mathfrak{p}}(\kappa(f, \underline{\boldsymbol{g}\boldsymbol{g}^*}))$ under the natural map

$$p_f^- : \mathrm{H}^1\big(K_{\mathfrak{p}}, \mathscr{F}_{\mathfrak{p}}^{\mathrm{bal}}(\widetilde{\mathbb{V}}_{f\underline{gg}^*}^{\dagger})\big) \to \mathrm{H}^1\big(K_{\mathfrak{p}}, \mathscr{F}_{\mathfrak{p}}^{\mathrm{bal}}(\widetilde{\mathbb{V}}_{f\underline{gg}^*}^{\dagger})/\mathscr{F}_{\mathfrak{p}}^{3}(\widetilde{\mathbb{V}}_{f\underline{gg}^*}^{\dagger})\big) \to \mathrm{H}^1(K_{\mathfrak{p}}, T_f^- \otimes \Psi_T^{1-\mathrm{c}})$$

arising from the projection onto the first factor in (20).

Put

$$\Lambda = \mathscr{O}[\![T]\!].$$

Let $u := 1 + p$, and for any Λ-module M and integer k, denote by M_k the specialisation of M at $T = u^{k-2} - 1$. Letting Φ be the field of fractions of $\mathscr{O}$, it is then easy to see that the Bloch–Kato logarithm and dual exponential maps define isomorphisms

$$\begin{aligned} \log_{\mathfrak{p}} &: \mathrm{H}^1(K_{\mathfrak{p}}, T_f^- \otimes \Psi_T^{1-\mathrm{c}})_k \otimes \mathbf{Q}_p \to \Phi, \quad k \geq 3, \\ \exp_{\mathfrak{p}}^* &: \mathrm{H}^1(K_{\mathfrak{p}}, T_f^- \otimes \Psi_T^{1-\mathrm{c}})_k \otimes \mathbf{Q}_p \to \Phi, \quad k = 2. \end{aligned} \tag{21}$$

The following fundamental result due to Bertolini–Seveso–Venerucci and Darmon–Rotger relates $p_f^-(\mathrm{loc}_{\mathfrak{p}}(\kappa(f, \underline{gg}^*)))$ to the restricted triple product p-adic L-function $\mathscr{L}_p^f(f, \underline{gg}^*)$ in (10).

Theorem 4.1.1 (Explicit reciprocity law) *There is an injective Λ-module homomorphism*

$$\mathrm{Log}^f : \mathrm{H}^1(K_{\mathfrak{p}}, T_f^- \otimes \Psi_T^{1-\mathrm{c}}) \to \Lambda$$

with pseudo-null cokernel satisfying for any $\mathfrak{Z} \in \mathrm{H}^1(K_{\mathfrak{p}}, T_f^- \otimes \Psi_T^{1-\mathrm{c}})$ the interpolation property

$$\mathrm{Log}^f(\mathfrak{Z})_k = c_k \times \begin{cases} \log_{\mathfrak{p}}(\mathfrak{Z}_k) & \text{if } k \geq 3, \\ \exp_{\mathfrak{p}}(\mathfrak{Z}_k) & \text{if } k = 2, \end{cases}$$

where c_k is an explicit nonzero constant, and such that

$$\mathrm{Log}^f\big(p_f^-(\mathrm{loc}_{\mathfrak{p}}(\kappa(f, \underline{gg}^*)))\big) = \mathscr{L}_p^f(f, \underline{gg}^*).$$

Proof The map Log^f is obtained by specialising the three-variable big regulator map of [11, Sect. 7.1], and the explicit reciprocity law is a consequence of the results of [11, Theorem A] and [24, Theorem 10]. The details are explained in [1, Proposition 7.3] and [14, Theorem 5.1.1]. □

The next result shows the equivalence between a "diagonal cycle main conjecture" in the spirit of [44, Conj. B] and the Iwasawa–Greenberg main conjecture for $\mathscr{L}_p^f(f, \underline{gg}^*)$.

Proposition 4.1.2 *The following statements (1)-(2) are equivalent:*

(1) $\kappa(f, \underline{gg}^*)$ *is not Λ-torsion,*

$$\mathrm{rank}_{\Lambda}\big(\mathrm{Sel}^{\mathrm{bal}}(\mathbf{Q}, \mathbb{V}_{f\underline{gg}^*}^{\dagger})\big) = \mathrm{rank}_{\Lambda}\big(X^{\mathrm{bal}}(\mathbf{Q}, \mathbb{A}_{f\underline{gg}^*}^{\dagger})\big) = 1,$$

and the following equality holds in $\Lambda \otimes \mathbf{Q}_p$*:*

$$\mathrm{char}_\Lambda\big(X^{\mathrm{bal}}(\mathbf{Q}, \mathbb{A}^\dagger_{f\underline{\boldsymbol{g}\boldsymbol{g}^*}})_{\mathrm{tors}}\big) = \mathrm{char}_\Lambda\big(\mathrm{Sel}^{\mathrm{bal}}(\mathbf{Q}, \mathbb{V}^\dagger_{f\underline{\boldsymbol{g}\boldsymbol{g}^*}})/(\kappa(f, \underline{\boldsymbol{g}\boldsymbol{g}^*}))\big)^2,$$

where the subscript tors *denotes the* Λ*-torsion submodule.*

(2) $\mathscr{L}_p^f(f, \underline{\boldsymbol{g}\boldsymbol{g}^*})$ *is nonzero, the modules* $\mathrm{Sel}^f(\mathbf{Q}, \mathbb{V}^\dagger_{f\underline{\boldsymbol{g}\boldsymbol{g}^*}})$ *and* $X^f(\mathbf{Q}, \mathbb{A}^\dagger_{f\underline{\boldsymbol{g}\boldsymbol{g}^*}})$ *are both* Λ*-torsion, and*

$$\mathrm{char}_\Lambda\big(X^f(\mathbf{Q}, \mathbb{A}^\dagger_{f\underline{\boldsymbol{g}\boldsymbol{g}^*}})\big) = \big(\mathscr{L}_p^f(f, \underline{\boldsymbol{g}\boldsymbol{g}^*})^2\big)$$

in $\Lambda \otimes \mathbf{Q}_p$*.*

Proof This follows from an argument by now standard using Theorem 4.1.1 and global duality as in [1, Theorem 7.15]. □

4.2 Anticyclotomic Main Conjecture for Modular Forms

For our later use, here we record some known results on the anticyclotomic Iwasawa main conjecture for modular forms.

Let $\mathbf{F}$ be the residue field of $\mathscr{O}$, and denote by

$$\bar{\rho}_f : G_{\mathbf{Q}} \to \mathrm{GL}_2(\mathbf{F})$$

the residual representation associated to f. Factor $N_f = N^+N^-$ as in (8). Following [36], we say that *Assumption (A)* holds if:

- $\bar{\rho}_f$ is absolutely irreducible,
- $\bar{\rho}_f$ is ramified at every prime $q \mid N^-$ with $q \equiv \pm 1 \pmod{p}$,
- either $p > 5$ or the image of $\bar{\rho}_f$ contains a conjugate of $\mathrm{SL}_2(\mathbf{F}_p)$.

Note that this is the same as Assumption (A) from [36], except that the latter also includes the non-anomalous condition $a_p(f) \not\equiv 1 \pmod{p}$; that the next result holds without this condition follows from recent advances on Ihara's lemma, [40].

Theorem 4.2.1 *Suppose that the ordinary p-stabilised newform* $f \in S_2(\Gamma_0(pN_f))$ *is p-old, and that* N^- *is the squarefree product of an odd number of primes. Under Assumption (A), the module* $\mathrm{Sel}(K, A_f \otimes \Psi_T^{\mathfrak{c}-1})$ *is* Λ*-cotorsion, and*

$$\mathrm{char}_\Lambda\big(\mathrm{Sel}(K, A_f \otimes \Psi_T^{\mathfrak{c}-1})^\vee\big) = \big(\Theta_{f/K}(T)^2\big)$$

in $\Lambda \otimes \mathbf{Q}_p$*.*

Proof In view of (11) and Remark 2.3.1, this follows from the combination of the main result of Bertolini–Darmon [3], as later refined by several authors (see [15, 36, 45]), showing the Λ-cotornionness $\mathrm{Sel}(K, A_f \otimes \Psi_T^{\mathfrak{c}-1})$ and the integral divisibility

$$\mathrm{char}_\Lambda\big(\mathrm{Sel}(K, A_f \otimes \Psi_T^{\mathbf{c}-1})^\vee\big) \supset \big(\Theta_{f/K}(T)^2\big), \tag{22}$$

and the converse divisibility that follows from the two-variable main conjecture for f/K proved in the work of Skinner–Urban and Wan [47, 50] (after inverting p in the latter case) restricted to the anticyclotomic line. □

Remark 4.2.2 By using the Euler system divisibility from [14, Theorem 5.6.1] (using diagonal cycles rather than congruence arguments and Heegner points), the ramification hypothesis on $\bar{\rho}_f$ in Theorem 4.2.1 can be replaced by the "big image" hypothesis in [*op. cit.*, Sect. 3.3.2].

5 Main Result

In this section we state and prove the main result of this note toward the nonvanishing conjectures of [22] in the setting of rank two elliptic curves.

5.1 Generalised Kato Classes

Let $E/\mathbf{Q}$ be an elliptic curve of conductor N_f, and let $p > 3$ a prime of good ordinary reduction for E. Let $f \in S_2(\Gamma_0(pN_f))$ be the ordinary p-stabilisation of the newform associated to E, and let $(\boldsymbol{g}, \boldsymbol{g}^*) = (\boldsymbol{\theta}_\psi(S_1), \boldsymbol{\theta}_{\psi^{-1}}(S_2))$ be a dual pair of primitive CM Hida families as in (9). When $\chi = \psi/\psi^{\mathbf{c}}$ satisfies (12), we have

$$\mathrm{H}^1(\mathbf{Q}, \mathbb{V}^\dagger_{f\underline{\boldsymbol{g}\boldsymbol{g}^*}}) \simeq \mathrm{H}^1(K, T_f \otimes \Psi_T^{1-\mathbf{c}}) \oplus \mathrm{H}^1(K, T_f \otimes \chi^{-1}\Psi_{W_2}^{1-\mathbf{c}})_{/W_2} \tag{23}$$

by specialising (14). The following is a special case of the construction of generalised Kato classes by Darmon–Rotger [22] in the *adjoint case*.

Definition 5.1.1 Let $\kappa_p(E)$ be the image of $\kappa(f, \underline{\boldsymbol{g}\boldsymbol{g}^*})$ under the composition

$$\mathrm{H}^1(\mathbf{Q}, \mathbb{V}^\dagger_{f\underline{\boldsymbol{g}\boldsymbol{g}^*}}) \to \mathrm{H}^1(K, T_f \otimes \Psi_T^{1-\mathbf{c}}) \to \mathrm{H}^1(K, T_f) \simeq \mathrm{H}^1(K, T_pE),$$

where the first arrow is the projection onto the first direct summand in (23), and the second is induced by the multiplication by T on $T_f \otimes \Psi_T^{1-\mathbf{c}}$.

By (19), the inclusion $\kappa(f, \underline{\boldsymbol{g}\boldsymbol{g}^*}) \in \mathrm{Sel}^{\mathrm{bal}}(\mathbf{Q}, \mathbb{V}^\dagger_{f\underline{\boldsymbol{g}\boldsymbol{g}^*}})$ from [11, Corrollary 8.2] implies that

$$\kappa_p(E) \in \mathrm{Sel}_{\emptyset,0}(K, T_pE). \tag{24}$$

In the following we shall view $\kappa_p(E)$ as a class with coefficients in V_pE.

Proposition 5.1.2 *The following implication holds*

$$L(f/K, 1) = 0 \implies \kappa_p(E) \in \mathrm{Sel}(K, V_p E).$$

Proof By (24), the inclusion $\kappa_p(E) \in \mathrm{Sel}(K, V_p E)$ holds if and only if $\mathrm{res}_{\mathfrak{p}}(\kappa_p(E))$ vanishes under the natural map $\mathrm{H}^1(K_{\mathfrak{p}}, V_f) \to \mathrm{H}^1(K_{\mathfrak{p}}, V_f^-)$. Since from the combination of Theorem 4.1.1, Proposition 2.4.1, and Theorem 2.3.2 yield the equivalence

$$\exp_{\mathfrak{p}}(\mathrm{res}_{\mathfrak{p}}(\kappa_p(E))) = 0 \iff L(f/K, 1) \cdot L(f/K \otimes \chi, 1) = 0,$$

and the Bloch–Kato dual in (21) is an isomorphism, the result follows. □

Denote by E^K the twist of E by the quadratic character corresponding to $K/\mathbf{Q}$.

Corollary 5.1.3 *Suppose* $L(E, 1) = 0$ *and* $L(E^K, 1) \neq 0$. *Then* $\kappa_p(E) \in \mathrm{Sel}(\mathbf{Q}, V_p E)$.

Proof Since Kato's results [33] show that $\mathrm{Sel}(\mathbf{Q}, V_p E^K) = 0$ when $L(E^K, 1) \neq 0$, this follows from Proposition 5.1.2 and the isomorphism $\mathrm{Sel}(K, V_p E) \simeq \mathrm{Sel}(\mathbf{Q}, V_p E) \oplus \mathrm{Sel}(\mathbf{Q}, V_p E^K)$. □

Remark 5.1.4 The proof of Proposition 5.1.2 show that when $L(f/K, 1) = 0$, the class $\kappa_p(E)$ lies in fact in the kernel of the restriction map on $\mathrm{Sel}(K, V_p E)$ at the primes above p. In particular, in the setting of Corollary 5.1.3, $\kappa_p(E)$ lands in the *strict* Selmer group

$$\mathrm{Sel}_0(\mathbf{Q}, V_p E) = \ker\left\{\mathrm{Sel}(\mathbf{Q}, V_p E) \xrightarrow{\mathrm{loc}_p} E(\mathbf{Q}_p)\hat{\otimes}\mathbf{Q}_p\right\}.$$

5.2 *Nonvanishing of $\kappa_p(E)$ in Rank Two*

Denote by

$$\rho_f : G_{\mathbf{Q}} \to \mathrm{Aut}_{\mathbf{F}_p}(E[p]) \simeq \mathrm{GL}_2(\mathbf{F}_p)$$

the mod p Galois representation associated to E, and assume that:

(h1) $\bar{\rho}_f$ is absolutely irreducible,
(h2) there exists a prime $q \| N_f$, and if $q \equiv \pm 1 \pmod{p}$ then $\bar{\rho}_f$ is ramified at q,
(h3) either $p > 5$ of the image of $\bar{\rho}_f$ contains a conjugate of $\mathrm{SL}_2(\mathbf{F}_p)$.

We shall consider generalised Kato classes as in Sect. 5.1 attached to the following.

Choice 5.2.1 Let $q \| N_f$ be a prime as in (h2) above, and let $\ell \nmid 6pN_f$ be a prime. Choose an imaginary quadratic field K and a ring class character χ of conductor dividing $\ell^\infty \mathcal{O}_K$ such that:

(1) q is inert in K,
(2) every prime factor of N_f/q splits in K,
(3) ℓ splits in K,
(4) ℓ is ordinary for E,
(5) $L(E^K, 1) \neq 0$ and $L(E/K, \chi, 1) \neq 0$,
(6) $\chi|_{G_{K_\mathfrak{p}}} \neq 1$.

Remark 5.2.2 The existence of (infinitely many) K satisfying (1)–(3) and such that $L(E^K, 1) \neq 0$ follows from [6]; for any such K, the results of [49] (see also [16, Theorem D]) ensure that $L(E/K, \chi, 1) \neq 0$ for all but finitely many χ of ℓ-power conductor.

Fix (K, χ) as in Choice 5.2.1. Write $\chi = \psi/\psi^{\mathbf{c}}$ with ψ a ray class character modulo $\ell^m \mathcal{O}_K$ for some $m > 0$, let $(\boldsymbol{g}, \boldsymbol{g}^*) = (\boldsymbol{\theta}_\psi(S_1), \boldsymbol{\theta}_{\psi^{-1}}(S_2))$ be the associated primitive CM Hida families as in (9), and let $\kappa_p(E) \in \mathrm{H}^1(K, V_pE)$ the corresponding generalised Kato class.

By Corollary 5.1.3, when $L(E, 1) = 0$ we have $\kappa_p(E) \in \mathrm{Sel}(\mathbf{Q}, V_pE)$. Moreover, by our choice of K and χ, when further $E/\mathbf{Q}$ has sign $+1$ (so in particular $\mathrm{ord}_{s=1} L(E, s) \geq 2$), the nonvanishing conjecture of Darmon–Rotger [22, Conj. 3.2] (as specialised in [*op. cit.*, Sect. 4.5.3] to the setting of rank two elliptic curves) predicts the equivalence

$$\kappa_p(E) \neq 0 \quad \overset{?}{\Longleftrightarrow} \quad \dim_{\mathbf{Q}_p} \mathrm{Sel}(\mathbf{Q}, V_pE) = 2.$$

The following is the main result of this note in the direction of this conjecture.

Theorem 5.2.3 *Suppose that $L(E, s)$ vanishes to positive even order at $s = 1$. Then*

$$\kappa_p(E) \neq 0 \quad \Longrightarrow \quad \dim_{\mathbf{Q}_p} \mathrm{Sel}(\mathbf{Q}, V_pE) = 2.$$

Conversely, if $\dim_{\mathbf{Q}_p} \mathrm{Sel}(\mathbf{Q}, V_pE) = 2$ then $\kappa_p(E) \neq 0$ if and only if the restriction map

$$\mathrm{loc}_p : \mathrm{Sel}(\mathbf{Q}, V_pE) \to E(\mathbf{Q}_p)\hat{\otimes}\mathbf{Q}_p$$

is nonzero. Moreover, in that case $\kappa_p(E)$ spans the strict Selmer group $\mathrm{Sel}_0(\mathbf{Q}, V_pE) = \ker(\mathrm{loc}_p)$.

The rest of this section is devoted to the proof of Theorem 5.2.3.

5.3 Some Galois Cohomology

Put $K_p = K \otimes_\mathbf{Q} \mathbf{Q}_p \simeq K_\mathfrak{p} \oplus K_{\overline{\mathfrak{p}}}$. Let

$$\mathrm{loc}_p = \mathrm{loc}_\mathfrak{p} \oplus \mathrm{loc}_{\overline{\mathfrak{p}}} : \mathrm{H}^1(G_{K,\Sigma}, V_pE) \to \mathrm{H}^1(K_p, V_pE) \tag{25}$$

be the restriction map, and denote by X_f (resp. $X_{\mathfrak{p}}$) the image of $\mathrm{Sel}(K, V_pE)$ (resp. $\mathrm{Sep}_{\emptyset,0}(K, V_pE)$) under loc_p.

Lemma 5.3.1 *We have* $\dim_{\mathbf{Q}_p} \mathrm{Sel}_{\emptyset,0}(K, V_pE) = \dim_{\mathbf{Q}_p} \mathrm{Sel}(K, V_pE) + \delta$, *where*

$$\delta = \begin{cases} -1 & \text{if } X_f \neq 0, \\ 0 & \text{if } X_f = X_{\mathfrak{p}} = 0, \\ 1 & \text{if } X_f = 0 \text{ and } X_{\mathfrak{p}} \neq 0. \end{cases}$$

Proof Since X_f is invariant under the action of complex conjugation, if X_f is nonzero then the restriction maps $\mathrm{Sel}(K, V_pE) \to \mathrm{H}^1(K_{\mathfrak{q}}, V_pE)$ for $\mathfrak{q} \in \{\mathfrak{p}, \overline{\mathfrak{p}}\}$ are both nonzero, and so in this case the result follows from [18, Lemma 2.2].

Now suppose $X_f = 0$, so that $\mathrm{Sel}(K, V_pE)$ is the same as the strict Selmer group $\mathrm{Sel}_{\mathrm{str}}(K, V_pE) := \ker(\mathrm{loc}_p)$, and consider the exact sequence

$$0 \to \mathrm{Sel}_{\mathrm{str}}(K, V_pE) \to \mathrm{Sel}_{\emptyset,0}(K, V_pE) \xrightarrow{\mathrm{loc}_{\mathfrak{p}}} \mathrm{H}^1(K_{\mathfrak{p}}, V_pE). \tag{26}$$

If $X_{\mathfrak{p}} = 0$, then (26) yields the result, so it remains to consider the case $X_{\mathfrak{p}} \neq 0$. If $\dim_{\mathbf{Q}_p} X_{\mathfrak{p}} = 2$, then the image of (25) contains $X_{\mathfrak{p}} \oplus X_{\overline{\mathfrak{p}}}$, where $X_{\overline{\mathfrak{p}}}$ is the image of $X_{\mathfrak{p}}$ under complex conjugation (equivalently, the image of $\mathrm{Sel}_{0,\emptyset}(K, V_pE)$), but this contradicts [46, Lemma 2.3.1]. Therefore if $X_{\mathfrak{p}} \neq 0$ then it is one-dimensional, and the result in this case follows again from (26). □

Remark 5.3.2 Standard conjectures predict that $X_f \neq 0$ unless $\mathrm{Sel}(K, V_pE) = 0$ (indeed, the vanishing of X_f when $\mathrm{Sel}(K, V_pE) \neq 0$ would imply that $Ш(E/K)[p^\infty]$ is infinite).

Recall that we set $\Lambda = \mathcal{O}[\![T]\!]$, and for any Λ-module M denote by $M_{/T} = M/TM$ the cokernel of multiplication by T. The following is a variant of Mazur's control theorem [39].

Proposition 5.3.3 *Multiplication by T induces natural maps*

$$r^* : \mathrm{Sel}_{0,\emptyset}(K, A_f) \to \mathrm{Sel}_{0,\emptyset}(K, A_f \otimes \Psi_T^{\mathfrak{c}-1})[T],$$
$$r : \mathrm{Sel}_{\emptyset,0}(K, T_f \otimes \Psi_T^{1-\mathfrak{c}})_{/T} \to \mathrm{Sel}_{\emptyset,0}(K, T_f)$$

with finite kernel and cokernel.

Proof Letting $\mathrm{Sel}_{\mathrm{rel}}(K, A_f)$ and $\mathrm{Sel}_{\mathrm{rel}}(L, A_f \otimes \Psi_T^{\mathfrak{c}-1})$ be the Selmer groups defined in the same manner as $\mathrm{Sel}_{0,\emptyset}(K, A_f)$ and $\mathrm{Sel}_{\emptyset,0}(L, A_f \otimes \Psi_T^{\mathfrak{c}-1})$, respectively, but with the (propagated in the case of A_f) relaxed local condition at the primes above p, the map r^* fits into the commutative diagram with exact rows

$$\begin{array}{ccccccc} 0 & \longrightarrow & \mathrm{Sel}_{0,\emptyset}(K, A_f) & \longrightarrow & \mathrm{Sel}_{\mathrm{rel}}(K, A_f) & \longrightarrow & \mathrm{H}^1(K_{\mathfrak{p}}, A_f) \times \frac{\mathrm{H}^1(K_{\overline{\mathfrak{p}}}, A_f)}{\mathrm{H}^1(K_{\overline{\mathfrak{p}}}, A_f)_{\mathrm{div}}} \\ & & \downarrow{\scriptstyle r^*} & & \downarrow{\scriptstyle s^*} & & \downarrow{\scriptstyle t^*} \\ 0 & \longrightarrow & \mathrm{Sel}_{0,\emptyset}(K, A_f \otimes \Psi_T^{\mathfrak{c}-1})[T] & \longrightarrow & \mathrm{Sel}_{\mathrm{rel}}(K, A_f \otimes \Psi_T^{\mathfrak{c}-1})[T] & \longrightarrow & \mathrm{H}^1(K_{\mathfrak{p}}, A_f \otimes \Psi_T^{\mathfrak{c}-1})[T] \times \{0\}, \end{array}$$

where $\mathrm{H}^1(K_{\overline{\mathfrak{p}}}, A_f)_{\mathrm{div}}$ denotes the maximal divisible submodule of $\mathrm{H}^1(K_{\overline{\mathfrak{p}}}, A_f)$. The map t^* arises from the cohomology long exact sequence associated to

$$0 \to A_f \to A_f \otimes \Psi_T^{\mathfrak{c}-1} \xrightarrow{\times T} A_f \otimes \Psi_T^{\mathfrak{c}-1} \to 0,$$

and therefore is surjective with kernel $\mathrm{H}^0(K_{\mathfrak{p}}, A_f \otimes \Psi_T^{\mathfrak{c}-1})_{/T}$. Since [35, Lemma 2.7] implies that $\#\mathrm{H}^0(K_{\mathfrak{p}}, A_f \otimes \Psi_T^{\mathfrak{c}-1}) < \infty$, by the Snake Lemma to prove the stated property for r^* it suffices to show that s^* has finite kernel and cokernel.

The latter map fits into the commutative diagram with exact rows

$$\begin{array}{ccccccc}
0 \to & \mathrm{Sel}_{\mathrm{rel}}(K, A_f) & \to & \mathrm{H}^1(G_{K,\Sigma}, A_f) & \to & \bigoplus_{w\in\Sigma, w\nmid p} \frac{\mathrm{H}^1(K_w, A_f)}{\mathrm{H}^1_f(K_w, A_f)} & \\
& \downarrow s^* & & \downarrow u^* & & \downarrow v^* & \\
0 \to & \mathrm{Sel}_{\mathrm{rel}}(K, A_f \otimes \Psi_T^{\mathfrak{c}-1})[T] & \to & \mathrm{H}^1(G_{K,\Sigma}, A_f \otimes \Psi_T^{\mathfrak{c}-1})[T] & \to & \bigoplus_{w\in\Sigma, w\nmid p} \mathrm{H}^1(K_w, A_f \otimes \Psi_T^{\mathfrak{c}-1})[T], &
\end{array}$$

where $\mathrm{H}^1_f(K_w, A_f)$ is the natural image of $\ker\{\mathrm{H}^1(K_w, V_f) \to \mathrm{H}^1(K_w^{\mathrm{ur}}, V_f)\}$ (see Definition 3.2.1). The finiteness of $\mathrm{H}^0(K_{\mathfrak{p}}, A_f \otimes \Psi_T^{\mathfrak{c}-1})$ implies that of $\mathrm{H}^0(G_{K,\Sigma}, A_f \otimes \Psi_T^{\mathfrak{c}-1})$, and so the map u^* has finite kernel. Since u^* is clearly surjective and by [27, Lemma 3.3] (see also [32, Sect. 3.3.6]) the kernel of the map v^* is finite, by the Snake Lemma it follows that s^* has the desired properties. This shows the result for r^* and the case of r is shown similarly. □

5.4 Proof of the Main Result

Proof of Theorem 5.2.3 The decomposition in Corollary 3.3.2 gives

$$X^f(\mathbf{Q}, \mathbb{A}^{\dagger}_{f\underline{gg^*}}) \simeq \mathrm{Sel}(K, A_f \otimes \Psi_T^{\mathfrak{c}-1})^{\vee} \oplus \left(\mathrm{Sel}(K, A_f \otimes \chi\Psi_T^{\mathfrak{c}-1})[T]\right)^{\vee}. \quad (27)$$

The same argument as in the proof of Proposition 5.3.3 shows that the natural restriction map

$$\mathrm{Sel}(K, A_f \otimes \chi) \to \mathrm{Sel}(K, A_f \otimes \chi\Psi_T^{\mathfrak{c}-1})[T]$$

has finite kernel and cokernel, and therefore from [3, Corrollary 4] (a consequence of the divisibility in the anticyclotomic main conjecture proved in *op. cit.* with $p = \ell$), as extended in [15, 36, 45]), it follows that

$$L(E/K, \chi, 1) \neq 0 \implies \#\mathrm{Sel}(K, A_f \otimes \chi\Psi_T^{\mathfrak{c}-1})[T] < \infty.$$

Together with Theorem 4.2.1 and Proposition 2.4.1, in light of (27) this shows that $X^f(\mathbf{Q}, \mathbb{A}^{\dagger}_{f\underline{gg^*}})$ is Λ-torsion, with

$$\mathrm{char}_\Lambda\big(X^f(\mathbf{Q}, \mathbb{A}^\dagger_{f\underline{gg^*}})\big) = \big(\mathscr{L}_p^f(f, \underline{gg^*})^2\big)$$

in $\Lambda \otimes \mathbf{Q}_p$. By Proposition 4.1.2, it follows that $\kappa(f, \underline{gg^*})$ is not Λ-torsion, that both $\mathrm{Sel}^{\mathrm{bal}}(\mathbf{Q}, \mathbb{V}^\dagger_{f\underline{gg^*}})$ and $X^{\mathrm{bal}}(\mathbf{Q}, \mathbb{A}^\dagger_{f\underline{gg^*}})$ have Λ-rank one, and that

$$\mathrm{char}_\Lambda\big(X^{\mathrm{bal}}(\mathbf{Q}, \mathbb{A}^\dagger)_{\mathrm{tors}}\big) = \mathrm{char}_\Lambda\big(\mathrm{Sel}^{\mathrm{bal}}(\mathbf{Q}, \mathbb{V}^\dagger)/(\kappa(f, \underline{gg^*}))\big)^2$$

in $\Lambda \otimes \mathbf{Q}_p$. Denote by $\kappa_1(f, \underline{gg^*}) \in \mathrm{Sel}_{\emptyset,0}(K, T_f \otimes \Psi_T^{1-\mathbf{c}})$ the projection of $\kappa(f, \underline{gg^*})$ onto the first direct summand in (19). By Proposition 3.3.1 and Corollary 3.3.2, the above shows that both $\mathrm{Sel}_{\emptyset,0}(K, T_f \otimes \Psi_T^{1-\mathbf{c}})$ and $X_{0,\emptyset}(K, A_f \otimes \Psi_T^{\mathbf{c}-1})$ have Λ-rank one, with

$$\mathrm{char}_\Lambda\big(X_{0,\emptyset}(K, A_f \otimes \Psi_T^{\mathbf{c}-1})_{\mathrm{tors}}\big) = \mathrm{char}_\Lambda\big(\mathrm{Sel}_{\emptyset,0}(K, T_f \otimes \Psi_T^{1-\mathbf{c}})/(\kappa_1(f, \underline{gg^*}))\big)^2$$

in $\Lambda \otimes \mathbf{Q}_p$. By the control theorem of Proposition 5.3.3, this implies that

$$\mathrm{corank}_{\mathbb{Z}_p}\mathrm{Sel}_{0,\emptyset}(K, A_f) = \mathrm{rank}_{\mathbb{Z}_p}\big(X_{0,\emptyset}(K, A_f \otimes \Psi_T^{\mathbf{c}-1})_{/T}\big) = 1 + 2\,\mathrm{rank}_{\mathbb{Z}_p}(\mathfrak{Z}_{/T}), \tag{28}$$

where $\mathfrak{Z} = \mathrm{Sel}_{\emptyset,0}(K, T_f \otimes \Psi_T^{1-\mathbf{c}})/(\kappa_1(f, \underline{gg^*}))$. Thus we conclude that

$$\mathrm{corank}_{\mathbb{Z}_p}\mathrm{Sel}_{0,\emptyset}(K, A_f) = 1 \iff (\kappa_1(f, \underline{gg^*}) \bmod T) \neq 0,$$

where $(\kappa_1(f, \underline{gg^*}) \bmod T)$ is the image of $\kappa_1(f, \underline{gg^*})$ in $\mathrm{Sel}_{\emptyset,0}(K, T_f \otimes \Psi_T^{1-\mathbf{c}})_{/T}$. Since the natural map

$$\mathrm{Sel}_{\emptyset,0}(K, T_f \otimes \Psi_T^{1-\mathbf{c}})_{/T} \to \mathrm{Sel}_{\emptyset,0}(K, T_f)$$

has finite kernel by Proposition 5.3.3, and it sends $(\kappa_1(f, \underline{gg^*}) \bmod T)$ to $\kappa_p(E)$ by construction, we arrive at

$$\kappa_p(E) \neq 0 \iff \mathrm{rank}_{\mathbb{Z}_p}\mathrm{Sel}_{\emptyset,0}(K, T_f) = 1, \tag{29}$$

using the action of complex conjugation to reverse the roles of $\mathfrak{p}$ and $\overline{\mathfrak{p}}$. Now, assuming $\kappa_p(E) \neq 0$, by Theorem 4.2.1 (resp. the p-parity conjecture [41]) the case $\delta = 0$ (resp. $\delta = 1$) is excluded in Lemma 5.3.1, and so $\dim_{\mathbf{Q}_p}\mathrm{Sel}(K, V_pE) = 2$; since the nonvanishing of $L(E^K, 1)$ implies that $\mathrm{Sel}(\mathbf{Q}, V_pE) = \mathrm{Sel}(K, V_pE)$ by Kato's work, the first implication in Theorem 5.2.3 follows.

Conversely, if $\dim_{\mathbf{Q}_p}\mathrm{Sel}(\mathbf{Q}, V_pE) = 2$ (and so $\dim_{\mathbf{Q}_p}\mathrm{Sel}(K, V_pE) = 2$), from (28) we see that the case $\delta = 0$ is excluded in Lemma 5.3.1 and the case $\delta = -1$ holds (i.e., $\dim_{\mathbf{Q}_p}\mathrm{Sel}_{\emptyset,0}(K, V_pE) = 1$) if and only if $\mathrm{Sel}(\mathbf{Q}, V_pE) \neq \ker(\mathrm{loc}_p)$. By (29), this concludes the proof. □

Remark 5.4.1 Even though the approach in this paper recovers a strenghtened form of [17, Theorem A] under slightly weaker hypotheses, a noteworthy advantage of the approach in *op. cit.* is that—by relating the derived p-adic heights against $\kappa_p(E)$ to the leading coefficient of $\Theta_{f/K}(T)$ (see [17, Theorem 5.3])—it provides a method to numerically verify the nonvanishing of $\kappa_p(E)$, and to relate its rationality properties predicted by [22, Conj. 3.12] to (the leading coefficient part of) the anticyclotomic p-adic Birch–Swinnerton-Dyer conjectures formulated in [2]. (See also [10] for a more general setting).

References

1. Alonso, R., Castella, F., Rivero, Ó.: The diagonal cycle Euler system for $GL_2 \times GL_2$. J. Inst. Math, Jussieu **24**(5), 1591–1693 (2025)
2. Bertolini, M., Darmon, H.: Heegner points on Mumford-Tate curves. Invent. Math. **126**(3), 413–456 (1996)
3. Bertolini, M., Darmon, H.: Iwasawa's main conjecture for elliptic curves over anticyclotomic $\mathbb{Z}_p$-extensions. Ann. of Math. (2) **162**(1), 1–64 (2005)
4. Bertolini, M., Darmon, H., Rotger, V., Seveso, M.A., Venerucci, R.: Heegner points, Stark-Heegner points, and diagonal classes. Astérisque **2022**(434). Société Mathématique de France, Paris (2022)
5. Bertolini, M.: Selmer groups and Heegner points in anticyclotomic Z_p-extensions. Compositio Math. **99**(2), 153–182 (1995)
6. Bump, D., Friedberg, S., Hoffstein, J.: Nonvanishing theorems for L-functions of modular forms and their derivatives. Invent. Math. **102**(3), 543–618 (1990)
7. Bloch, S., Kato, K.: L-functions and Tamagawa numbers of motives. In: The Grothendieck Festschrift, Vol. I. Progr. Math., vol. 86, pp. 333–400. Birkhäuser Boston, Boston (1990)
8. Büyükboduk, K., Lei, A.: Anticyclotomic p-ordinary Iwasawa theory of elliptic modular forms. Forum Math. **30**(4), 887–913 (2018)
9. Bertolini, M., Seveso, M.A., Venerucci, R.: Diagonal classes and the Bloch-Kato conjecture. Münster J. Math. **13**(2), 317–352 (2020)
10. Bertolini, M., Seveso, M.A., Venerucci, R.: On p-adic analogues of the Birch and Swinnerton-Dyer conjecture for Garrett L-functions. Ann. Inst. Fourier. Preprint (2022)
11. Bertolini, M., Seveso, M.A., Venerucci, R.: Reciprocity laws for balanced diagonal classes. Astérisque **2022**(434), 77–174 (2022)
12. Castella, F.: On the exceptional specializations of big Heegner points. J. Inst. Math. Jussieu **17**(1), 207–240 (2018)
13. Castella, F.: Generalised Kato classes on CM elliptic curves of rank 2. Preprint, arXiv:2204.09608
14. Castella, F., Do, K.T.: Diagonal cycles and anticyclotomic Iwasawa theory of modular forms. J. Eur. Math. Soc. (JEMS). Preprint, arXiv:2303.06751 (2023)
15. Chida, M., Hsieh, M.-L.: On the anticyclotomic Iwasawa main conjecture for modular forms. Compos. Math. **151**(5), 863–897 (2015)
16. Chida, M., Hsieh, M.-L.: Special values of anticyclotomic L-functions for modular forms. J. Reine Angew. Math. **741**, 87–131 (2018)
17. Castella, F., Hsieh, M.-L.: On the nonvanishing of generalised Kato classes for elliptic curves of rank 2. Forum Math. Sigma **10**, Paper No. e12 (2022)
18. Castella, F., Hsu, C.-Y., Kundu, D., Lee, Y.-S., Liu, Z.: Derived p-adic heights and the leading coefficient of the Bertolini–Darmon–Prasanna p-adic L-functions. Trans. Amer. Math. Soc. Ser. B **12**, 748–788 (2025). Preprint, arXiv:2308.10474
19. Darmon, H., Fornea, M.: Mock plectic points. J. Inst. Math. Jussieu **24**(5), 2107–2131 (2025). Preprint, arXiv:2310.16758

20. Darmon, H., Lauder, A., Rotger, V.: Stark points and p-adic iterated integrals attached to modular forms of weight one. Forum Math. Pi **3**, e8, 95 (2015)
21. Darmon, H., Rotger, V.: Diagonal cycles and Euler systems I: A p-adic Gross-Zagier formula. Ann. Sci. Éc. Norm. Supér. (4) **47**(4), 779–832 (2014)
22. Darmon, H., Rotger, V.: Elliptic curves of rank two and generalised Kato classes. Res. Math. Sci. **3**, 3:27 (2016)
23. Darmon, H., Rotger, V.: Diagonal cycles and Euler systems II: The Birch and Swinnerton-Dyer conjecture for Hasse-Weil-Artin L-functions. J. Amer. Math. Soc. **30**(3), 601–672 (2017)
24. Darmon, H., Rotger, V.: p-Adic families of diagonal cycles. Astérisque **2022**(434), 29–75 (2022)
25. Fornea, M., Gehrmann, L.: Iwasawa theory and mock plectic points. Preprint, arXiv:2311.03100
26. Gross, B.H., Kudla, S.S.: Heights and the central critical values of triple product L-functions. Compositio Math. **81**(2), 143–209 (1992)
27. Greenberg, R.: Iwasawa theory for elliptic curves. In: Arithmetic theory of elliptic curves (Cetraro). Lecture Notes in Math., vol. 1716, pp. 51–144. Springer, Berlin (1999)
28. Gross, B.H., Schoen, C.: The modified diagonal cycle on the triple product of a pointed curve. Ann. Inst. Fourier (Grenoble) **45**(3), 649–679 (1995)
29. Hida, H.: Modules of congruence of Hecke algebras and L-functions associated with cusp forms. Amer. J. Math. **110**(2), 323–382 (1988)
30. Howard, B.: Derived p-adic heights and p-adic L-functions. Amer. J. Math. **126**(6), 1315–1340 (2004)
31. Hsieh, M.-L.: Hida families and p-adic triple product L-functions. Amer. J. Math. **143**(2), 411–532 (2021)
32. Jetchev, D., Skinner, C., Wan, X.: The Birch and Swinnerton-Dyer formula for elliptic curves of analytic rank one. Camb. J. Math. **5**(3), 369–434 (2017)
33. Kato, K.: p-adic Hodge theory and values of zeta functions of modular forms. Astérisque **2004**(295), ix, 117–290 (2004)
34. Kings, G., Loeffler, D., Zerbes, S.L.: Rankin-Eisenstein classes and explicit reciprocity laws. Camb. J. Math. **5**(1), 1–122 (2017)
35. Kobayashi, S., Ota, K.: Anticyclotomic main conjecture for modular forms and integral Perrin-Riou twists. In: Development of Iwasawa theory—the centennial of K. Iwasawa's birth. Adv. Stud. Pure Math., vol. 86, pp. 537–594. Math. Soc. Japan, Tokyo (2020)
36. Kim, C.-H., Pollack, R., Weston, T.: On the freeness of anticyclotomic Selmer groups of modular forms. Int. J. Number Theory **13**(6), 1443–1455 (2017)
37. Khare, C., Wintenberger, J.-P.: Serre's modularity conjecture. I. Invent. Math. **178**(3), 485–504 (2009)
38. Lei, A., Loeffler, D., Zerbes, S.L.: Euler systems for modular forms over imaginary quadratic fields. Compos. Math. **151**(9), 1585–1625 (2015)
39. Mazur, B.: Rational points of abelian varieties with values in towers of number fields. Invent. Math. **18**, 183–266 (1972)
40. Manning, J., Shotton, J.: Ihara's lemma for Shimura curves over totally real fields via patching. Math. Ann. **379**(1–2), 187–234 (2021)
41. Nekovář, J.: On the parity of ranks of Selmer groups. II. C. R. Acad. Sci. Paris Sér. I Math. **332**(2), 99–104 (2001)
42. Ohta, M.: Ordinary p-adic étale cohomology groups attached to towers of elliptic modular curves. Compositio Math. **115**(3), 241–301 (1999)
43. Ohta, M.: Ordinary p-adic étale cohomology groups attached to towers of elliptic modular curves. II. Math. Ann. **318**(3), 557–583 (2000)
44. Perrin-Riou, B.: Fonctions L p-adiques, théorie d'Iwasawa et points de Heegner. Bull. Soc. Math. France **115**(4), 399–456 (1987)
45. Pollack, R., Weston, T.: On anticyclotomic μ-invariants of modular forms. Compos. Math. **147**(5), 1353–1381 (2011)

46. Skinner, C.: A converse to a theorem of Gross, Zagier, and Kolyvagin. Ann. of Math. (2) **191**(2), 329–354 (2020)
47. Skinner, C., Urban, E.: The Iwasawa Main Conjectures for GL_2. Invent. Math. **195**(1), 1–277 (2014)
48. Tate, J.: Number theoretic background. In: Automorphic forms, representations and L-functions. Proc. Sympos. Pure Math., vol. XXXIII, Part 2, pp. 3–26. Amer. Math. Soc., Providence (1979)
49. Vatsal, V.: Special values of anticyclotomic L-functions. Duke Math. J. **116**(2), 219–261 (2003)
50. Wan, X.: The Iwasawa main conjecture for Hilbert modular forms. Forum Math. Sigma **3**, Paper No. e18, 95 (2015)
51. Wiles, A.: On ordinary λ-adic representations associated to modular forms. Invent. Math. **94**(3), 529–573 (1988)
52. Wiles, A.: Modular elliptic curves and Fermat's last theorem. Ann. Math. (2) **141**(3), 443–551 (1995)

Flach Classes and Generalised Hecke Eigenvalues

Henri Darmon and Alice Pozzi

To Massimo Bertolini on his 60th birthday

Abstract We describe the action of Hecke operators on generalised eigenspaces attached to certain mod p cuspidal eigenforms of weight two in terms of certain extension classes of Galois representations constructed by Matthias Flach. This description can be viewed as a partial generalisation to cusp forms of a formula of Barry Mazur for the Eisenstein series of weight two and prime level.

Keywords Modular forms · Elliptic curves · Flach elements · Selmer groups · Galois representations

1 Introduction

Given an integer $N \geq 1$, let $M_2(N)$ denote the space of modular forms of weight two on $\Gamma_0(N)$ with integer Fourier coefficients, and let $\mathbb{T}(N)$ be the Hecke algebra generated over $\mathbb{Z}$ by the prime-to-N Hecke operators in the endomorphism ring of $M_2(N)$.

When N is prime, the space $M_2(N)$ contains a unique holomorphic Eisenstein series $E_{2,N}$, with q-expansion given by

$$E_{2,N}(q) \; = \; \tfrac{N-1}{12} + \sum_{n=1}^{\infty} \sigma_{1,N}(n) q^n, \qquad \sigma_{1,N}(n) = \sum_{\substack{d \mid n, \\ (d,N)=1}} d. \tag{1}$$

H. Darmon (✉)
McGill University, Montreal, Canada
e-mail: darmon@math.mcgill.ca

A. Pozzi
University of Bristol, Bristol, UK
e-mail: alice.pozzi@bristol.ac.uk

M. Longo et al. (eds.), *Elliptic Curves and Modular Forms in Arithmetic Geometry*, Springer Proceedings in Mathematics & Statistics 527,
https://doi.org/10.1007/978-3-032-13123-2_3

In his celebrated work on the Eisenstein ideal [13], Mazur determines the possible structure of the torsion subgroup of an elliptic curve over $\mathbb{Q}$ by studying congruences between the systems of Hecke eigenvalues of $E_{2,N}$ and cusp forms modulo a prime p (which we assume for simplicity to be strictly greater than 3). Mazur shows that such congruences occur precisely when p divides $(N-1)$. At such primes, let $\mathbb{T}(N)_{p,\mathrm{eis}}$ be the localisation of $\mathbb{T}(N)\otimes\mathbb{Z}_p$ at the maximal ideal generated by $(T_\ell-(\ell+1))$ for primes $\ell\nmid N$ and p. The kernel of the morphism

$$\varphi_{\mathrm{eis}}\colon \mathbb{T}(N)_{p,\mathrm{eis}} \longrightarrow \mathbb{Z}_p$$

sending T_ℓ to $(\ell+1)$ for primes $\ell\nmid N$ is the *Eisenstein ideal*, denoted by $I_{\mathrm{eis},(N)}$. An important result towards arithmetic applications is the principality of this ideal, which is deduced from the construction of a canonical isomorphism

$$\lambda_{\mathrm{eis}}\colon I_{\mathrm{eis},(N)}/I^2_{\mathrm{eis},(N)} \xrightarrow{\sim} (\mathbb{Z}/N\mathbb{Z})^\times\otimes\mathbb{Z}_p \tag{2}$$

satisfying

$$\lambda_{\mathrm{eis}}(T_\ell-(\ell+1)) = [\ell]\otimes(\ell-1),\quad \text{for all primes } \ell\nmid Np. \tag{3}$$

In other words, after fixing an $m\geq 1$ for which p^m divides $N-1$, and a mod p^m discrete logarithm

$$\log_{N,p}\colon (\mathbb{Z}/N\mathbb{Z})^\times \longrightarrow \mathbb{Z}/p^m\mathbb{Z},$$

the mod p^m reduction of φ_{eis} lifts to a surjective homomorphism

$$\tilde{\varphi}_{\mathrm{eis}}\colon \mathbb{T}(N) \longrightarrow \left(\mathbb{Z}/p^m\mathbb{Z}\right)[\varepsilon]/(\varepsilon^2)$$

satisfying

$$\tilde{\varphi}_{\mathrm{eis}}(T_\ell)=a_\ell+a'_\ell\cdot\varepsilon,\qquad \text{with}\quad \begin{cases} a_\ell=(\ell+1)\\ a'_\ell=(\ell-1)\log_{N,p}(\ell),\end{cases}\qquad \text{for all primes } \ell\neq N. \tag{4}$$

The collection $\{a'_\ell\}_\ell$ of *generalised Hecke eigenvalues* is independent of the choice of discrete logarithm up to simultaneous rescaling. This intriguing arithmetic invariant arises precisely when the action of the Hecke operators on the generalised eigenspace attached to $E_{2,N}$ mod p is non-semisimple. (Cf. Remark 12.2.) The formulation (4) suggests that the generalised eigenvalues a'_ℓ are governed by the images of global elements in $(\mathbb{Z}/N\mathbb{Z})^\times$, their dependence on the primes N and $p\mid(N-1)$ arising only through the choice of a discrete mod p^m logarithm on $(\mathbb{Z}/N\mathbb{Z})^\times$.

The study of generalised eigenvalues was taken up subsequently from the analytic perspective of Eisenstein series and L-functions by Merel [14] and then by Lecouturier [12], where it constitutes the starting point for his theory of *higher Eisenstein elements*. It was further explored from a more Galois deformation theoretic perspective in [3] and [18] Among several applications, the notion of higher Eisenstein element plays an important role in formulating a conjecture of Harris and Venkatesh [11] on derived Hecke operators acting on the coherent cohomology of modular curves attached to spaces of weight one forms, and in the proof of this conjecture for dihedral forms [5].

It is natural to seek analogous formulae for the quantities a'_ℓ when the underlying eigenform is not an Eisenstein series. That such generalised eigenvalues might encode rich arithmetic information is suggested by a number of results already in the literature. For instance, when $E_{2,N}$ is replaced by a classical eigenform of weight one, and Φ_p by $\bar{\mathbb{Q}}_p$, the generalised eigenvalues can be expressed in terms of p-adic logarithms of algebraic numbers in the field cut out by the adjoint of the associated two-dimensional Artin representation [6, 7], a circumstance that provides a key to a better understanding of explicit class field theory for real quadratic fields [8, 9].

The present work considers the setting where the weight two Eisenstein series is replaced by a *cuspidal* newform f of weight two on $\Gamma_0(M)$. It is convenient (but entirely inessential, of course) to assume that f has integer Fourier coefficients, i.e., that it corresponds to an elliptic curve E over $\mathbb{Q}$ by the construction of Eichler and Shimura.

Given a prime p, the circumstances under which f is congruent to a modular form of level M occur somewhat sporadically: one needs to assume, essentially, that p divides the degree of the optimal modular parametrisation $\phi_E : X_0(M) \longrightarrow E$. The present work has nothing interesting to say about the generalised eigenvalues that arise from such homomorphisms. Rather, a prime $p \nmid 6M$ is fixed at the outset for which the Galois action on the p-division points $E[p] \subset E$ has full image $\mathrm{Aut}(E[p])$, and which does *not* divide the degree of ϕ_E. It is also assumed that the conductor of E is minimal among those of its quadratic twists. As recalled in Sect. 9, it then follows that the generalised eigenspace attached to f in $M_2(M) \otimes \Phi_p$ is spanned by f.

The mechanism whereby generalised Hecke eigenvalues can nonetheless be conjured from this setting involves level-raising. Namely, choose a prime $N \nmid Mp$ for which

$$p \text{ divides } (N-1) \cdot (N+1-a_N(f)) \cdot (N+1+a_N(f)), \tag{5}$$

and replace $\mathbb{T}(M)$ by the larger Hecke algebra $\mathbb{T}(MN^2)$ of level MN^2, which is endowed with a natural surjective map $\mathbb{T}(MN^2) \longrightarrow \mathbb{T}(M)$. Let $\mathbb{T}(MN^2)_{p,f}$ denote the localisation of $\mathbb{T}(MN^2) \otimes \mathbb{Z}_p$ at the maximal ideals attached to $f \pmod p$. The morphism

$$\varphi_{f,(N)} \colon \mathbb{T}(MN^2)_{p,f} \longrightarrow \mathbb{Z}_p$$

sends a Hecke operator T_ℓ to the coefficient $a_\ell(f)$ for $\ell \nmid MNp$. Its kernel, denoted by $I_{f,(N)}$, can be viewed as the analogue of the Eisenstein ideal in the elliptic setting.

The desired extension of Mazur's formula to cusp forms can be better explained by reinterpreting the latter in the language of Galois cohomology (see Sect. 4). Let T_pE be the Tate module of the elliptic curve E, and let

$$T_f := \mathrm{Sym}^2(T_pE)$$

denote its symmetric square, viewed as a $G_{\mathbb{Q}}$-module. A fundamental construction of M. Flach [10] associates to each rational prime $\ell \nmid pMN$ a global class

$$c_f[\ell] \in H^1(\mathbb{Q}, T_f)$$

which is "singular only at ℓ", i.e., is crystalline at p and minimally ramified at all primes different from ℓ and p. In particular, its restriction $\mathrm{res}_N(c_f[\ell])$ to the decomposition group at N belongs to the finite part $H^1_{\mathrm{fin}}(\mathbb{Q}_N, T_f)$ of the local cohomology at N. (Cf. Sect. 2.) The class $c_f[\ell]$ is the p-adic étale regulator of a global element in a higher Chow group of $X_0(M)^2$, as described in Sect. 10 below. It plays the same role as the class $[\ell^{(\ell-1)}]$ in Mazur's identity (3), as the following theorem illustrates.

Theorem 1.1 *There is a unique isomorphism*

$$\lambda_f : I_{f,(N)}/I^2_{f,(N)} \longrightarrow H^1_{\mathrm{fin}}(\mathbb{Q}_N, T_f)$$

characterised by

$$\lambda_f(T_\ell - a_\ell) = \mathrm{res}_N(c_f[\ell]), \textit{for all primes } \ell \nmid MNp.$$

Let us assume for simplicity that $p \nmid (N \pm 1)$. As explained in Sect. 5, the assumption that N is a level-raising prime for f implies that the local cohomology group $H^1_{\mathrm{fin}}(\mathbb{Q}_N, T_f)$ is cyclic of order p^m for some $m > 0$. In Sect. 6, an identification of $H^1_{\mathrm{fin}}(\mathbb{Q}_N, T_f)$ with $\mathbb{Z}/p^m\mathbb{Z}$ is given, depending quadratically on the choice of a mod p^m discrete elliptic logarithm on $E(\Phi_{N^2})$, which is therefore denoted

$$\log^{\otimes 2}_{E,N,p} : H^1_{\mathrm{fin}}(\mathbb{Q}_N, T_f) \xrightarrow{\sim} \mathbb{Z}/p^m\mathbb{Z}.$$

Under these assumptions, there is a surjective homomorphism

$$\tilde{\varphi}_f : \mathbb{T}(MN^2) \longrightarrow \left(\mathbb{Z}/p^m\mathbb{Z}\right)[\varepsilon]/(\varepsilon^2) \tag{6}$$

lifting the mod p^m-reduction of φ_f, and for which

$$\tilde{\varphi}_f(T_\ell) = a_\ell(f) + a'_\ell(f) \cdot \varepsilon.$$

Corollary 1.2 *After possibly rescaling the collection $\{a'_\ell(f)\}_\ell$ by a common factor, the generalised eigenvalues satisfy*

$$a'_\ell(f) = \log^{\otimes 2}_{E,N,p}(\mathrm{res}_N(c_f[\ell])), \quad \textit{for all } \ell \nmid MNp.$$

Corollary 1.2 reveals that $a'_\ell(f)$ is accounted for by a global class in a higher Chow group which depends neither on N nor on p, suggesting that the generalised eigenvalues attached to f have a motivic incarnation. In the opposite direction, the relation between Flach's classes and generalised Hecke eigenvalues gives some insights into the local behaviours of the global classes $c_f[\ell]$ as the prime ℓ varies but N and p are fixed.

2 Selmer Groups

Let V be a finite dimensional $\mathbb{Q}_p$-vector space with a continuous action of $G_{\mathbb{Q}}$ (unramified outside a finite set of places of $\mathbb{Q}$). Fix a $G_{\mathbb{Q}}$-stable $\mathbb{Z}_p$-lattice T, and write

$$A = V/T = T \otimes \mathbb{Q}_p/\mathbb{Z}_p \text{ and } A_n = p^{-n}T/T \hookrightarrow A$$

for $n \in \mathbb{Z}_{>0}$. For $W \in \{V, T, A, A_n\}$, let $H^1(\mathbb{Q}, W)$ denote the continuous Galois cohomology with coefficients in W. It is equipped, for each rational prime q, with a localisation map

$$\mathrm{res}_q : H^1(\mathbb{Q}, W) \longrightarrow H^1(\mathbb{Q}_q, W)$$

obtained by restricting a one-cocycle to a decomposition group $G_q = \mathrm{Gal}(\bar{\mathbb{Q}}_q/\mathbb{Q}_q)$ at q.

Let I_q denote the inertia subgroup of G_q. The quotient G_q/I_q is topologically procyclic with a canonical generator: the *arithmetic Frobenius element* at q, denoted σ_q, which acts as $x \mapsto x^q$ on the residue field of any unramified extension of $\mathbb{Q}_q$. For W as above, the inflation-restriction exact sequence identifies the subgroup of $H^1(\mathbb{Q}_q, W)$ of unramified cohomology classes

$$H^1_{\mathrm{ur}}(\mathbb{Q}_q, W) = \ker\left(H^1(\mathbb{Q}_q, W) \to H^1(I_q, W)\right)$$

with

$$H^1(\mathbb{Q}_q^{\mathrm{ur}}/\mathbb{Q}_q, W^{I_q}) = W^{I_q}/(\sigma_q - 1)W^{I_q} \tag{7}$$

where $\mathbb{Q}_q^{\mathrm{ur}}$ is the maximal unramified extension of $\mathbb{Q}_q$ and the superscript I_q denotes inertia invariants.

The local cohomology group $H^1(\mathbb{Q}_q, W)$ contains a distinguished subgroup $H^1_{\mathrm{fin}}(\mathbb{Q}_q, W)$. When $W = V$ is a $\mathbb{Q}_p$-vector space, this is defined as

$$H^1_{\mathrm{fin}}(\mathbb{Q}_q, V) := \begin{cases} H^1_{\mathrm{ur}}(\mathbb{Q}_q, V) & \text{if } q \neq p, \\ H^1_{\mathrm{cris}}(\mathbb{Q}_p, V) & \text{if } q = p, \end{cases}$$

where $H^1_{\mathrm{cris}}(\mathbb{Q}_p, V) = \ker\left(H^1(\mathbb{Q}_p, V) \to H^1(\mathbb{Q}_p, V \otimes B_{\mathrm{cris}}\right)$ and B_{cris} is the period ring defined by Fontaine [2]. (The assumption that $p \neq 2$ obviates the need to treat the case $q = \infty$.) When $W = T$ (resp. $W = A$), the subgroup $H^1_{\mathrm{fin}}(\mathbb{Q}, W)$ is defined as the natural preimage (resp. the image) of $H^1_{\mathrm{fin}}(\mathbb{Q}, V)$ in $H^1(\mathbb{Q}, W)$. Similarly, for $n \in \mathbb{Z}_{>0}$, the subgroup $H^1_{\text{fin}}(\mathbb{Q}, A_n)$ is the preimage of $H^1_{\text{fin}}(\mathbb{Q}, A)$ in $H^1(\mathbb{Q}, A_n)$. Note that if T is unramified at $q \neq p$, the subgroups $H^1_{\text{fin}}(\mathbb{Q}_q, W)$ and $H^1_{\text{ur}}(\mathbb{Q}_q, W)$) agree for all choices of W as above (see [15, Lemma 3.5, 3.8]).

Let

$$H^1_{\text{sing}}(\mathbb{Q}_q, W) := \frac{H^1(\mathbb{Q}_q, W)}{H^1_{\text{fin}}(\mathbb{Q}_q, W)},$$

denote the *singular quotient* of the local cohomology at q and write ∂_q for the natural map obtained by composing res_q with the projection to this quotient.

When $q \neq p$ and W is unramified at q, the map ∂_q corresponds to the restriction to I_q. Because the maximal pro p-quotient of I_q is isomorphic to $\mathbb{Z}_p(1)$ as a G_q-module, the image of ∂_q is identified with

$$H^1(I_q, W)^{G_q} = H^1(\mathbb{Z}_p(1), W)^{G_q} = W(-1)^{G_q},$$

and it will be convenient to view ∂_q as a map

$$\partial_q : H^1(\mathbb{Q}, W) \longrightarrow W(-1)^{G_q}.$$

Given a global class $c \in H^1(\mathbb{Q}_q, W)$, its restriction $\text{res}_q(c)$ belongs to $H^1_{\text{fin}}(\mathbb{Q}_q, W)$ for all but finitely many q. A *set of local conditions* for $H^1(\mathbb{Q}, W)$ is a collection $\Sigma = \{\Sigma_q\}_q$ indexed by the places of $\mathbb{Q}$, and satisfying

$$\Sigma_q = H^1_{\text{fin}}(\mathbb{Q}_q, W), \quad \text{for all but finitely many } q.$$

The *Selmer group* attached to W and Σ is defined to be

$$H^1_\Sigma(\mathbb{Q}, W) = \{\kappa \in H^1(\mathbb{Q}, W) \text{ for which } \text{res}_q(\kappa) \in \Sigma_q, \text{ for all } q\}.$$

When $\Sigma_q = H^1_{\text{fin}}(\mathbb{Q}_q, W)$ for all q, then $H^1_\Sigma(\mathbb{Q}, W)$ is just called the *Selmer group* attached to W, and is denoted $H^1_\emptyset(\mathbb{Q}, W)$. More generally, the *relaxed Selmer group* at $S \in \mathbb{Z}_{>0}$, denoted $H^1_{(S)}(\mathbb{Q}, W)$, is obtained by setting

$$\Sigma_q = \begin{cases} H^1(\mathbb{Q}_q, W) & \text{if } q|S, \\ H^1_{\text{fin}}(\mathbb{Q}_q, W) & \text{if } q \nmid S, \end{cases}$$

i.e.,

$$H^1_{(S)}(\mathbb{Q}, W) = \left\{c \in H^1(\mathbb{Q}, W) \text{ such that } \partial_q(c) = 0, \text{ for all } q \nmid S\right\}. \tag{8}$$

For any set of local conditions Σ, the Selmer groups $H^1_\Sigma(\mathbb{Q}, A_n)$ is finite, and $H^1_\Sigma(\mathbb{Q}, T)$ and is a finitely generated $\mathbb{Z}_p$-module. The Pontryagin dual of $H^1_\Sigma(\mathbb{Q}, A_n)$ is also a finitely generated $\mathbb{Z}_p$-module (cf. [15, Lemma 5.7]).

3 Local and Global Duality

Given V a finite dimensional $\mathbb{Q}_p$-vector space with a continuous $G_{\mathbb{Q}}$-action and a $G_{\mathbb{Q}}$-stable $\mathbb{Z}_p$-lattice $T \subset V$, let $T^* = \mathrm{Hom}(T, \mathbb{Z}_p(1))$ be the dual of T. Denote

$$V^* = T^* \otimes \mathbb{Q}_p, \quad A^* = V^*/T^*, \text{ and } A_n^* = p^{-n}T^*/T^*$$

for $n \in \mathbb{Z}_{>0}$.

Let q be a rational prime. The cup product combined with the tautological pairings

$$\langle\ ,\ \rangle : A_n \times A_n^* \longrightarrow \mu_{p^n} \text{ and } \langle\ ,\ \rangle : T \times A^* \longrightarrow \mu_{p^\infty}$$

give rise to the perfect local Tate pairings

$$\begin{aligned} \langle\ ,\ \rangle_q &: H^1(\mathbb{Q}_q, A_n) \times H^1(\mathbb{Q}_q, A_n^*) \longrightarrow H^2(\mathbb{Q}_q, \mu_{p^n}) \xlongequal{\mathrm{inv}_q} \tfrac{1}{p^n}\mathbb{Z}_p/\mathbb{Z}_p, \quad (9)\\ \langle\ ,\ \rangle_q &: H^1(\mathbb{Q}_q, T) \times H^1(\mathbb{Q}_q, A^*) \longrightarrow H^2(\mathbb{Q}_q, \mu_{p^\infty}) \xlongequal{\mathrm{inv}_q} \mathbb{Q}_p/\mathbb{Z}_p, \end{aligned}$$

relative to which the respective finite parts are orthogonal complements of each other when $q \neq p$. Thus, for $W \in \{T, A, A_n\}$, the Tate pairing induces a perfect duality

$$[\ ,\ ]_q : H^1_{\mathrm{sing}}(\mathbb{Q}_q, W) \times H^1_{\mathrm{fin}}(\mathbb{Q}_q, \mathrm{Hom}(W, \mu_{p^\infty})) \longrightarrow \mathbb{Q}_p/\mathbb{Z}_p. \quad (10)$$

In the special case where $q \neq p$ and W is unramified at q, the local Tate pairing is given by the following explicit formula:

$$\langle c, \kappa\rangle_q = \langle \partial_q(c), \kappa(\sigma_q)\rangle, \quad \text{for all } c \in H^1(\mathbb{Q}_q, W),\ \kappa \in H^1_{\mathrm{fin}}(\mathbb{Q}_q, \mathrm{Hom}(W, \mu_{p^\infty})), \quad (11)$$

where the pairing $\langle\ ,\ \rangle$ on the right hand side is induced from the tautological $\mathbb{Q}_p/\mathbb{Z}_p$-valued pairing between $W(-1)$ and $\mathrm{Hom}(W, \mu_{p^\infty})$, after restricting to the Frobenius invariants and co-invariants respectively.

The reciprocity law of global class field asserts that

$$\sum_q \mathrm{inv}_q(b) = 0, \quad \text{for all } b \in H^2(\mathbb{Q}, \mu_{p^\infty}),$$

and implies that

$$\sum_q \langle \mathrm{res}_q(c), \mathrm{res}_q(\kappa)\rangle_q = 0, \quad \text{for all } c \in H^1(\mathbb{Q}, W),\ \kappa \in H^1(\mathbb{Q}, \mathrm{Hom}(W, \mu_{p^\infty})), \quad (12)$$

where the sum need only be taken over the non-archimedean places in light of the running assumption that $p \neq 2$.

For $W \in \{T, A, A_n\}$, let $\Sigma = \{\Sigma_q\}_q$ be a set of local conditions for $H^1(\mathbb{Q}, W)$. The orthogonal complements

$$\Sigma_q^* \subset H^1(\mathbb{Q}_q, \mathrm{Hom}(W, \mu_{p^\infty}))$$

of Σ_q relative to the local Tate pairings (9) form a collection of local conditions. The Selmer group $H^1_{\Sigma^*}(\mathbb{Q}, \mathrm{Hom}(W, \mu_{p^\infty}))$ is called the *dual Selmer group* of $H^1_\Sigma(\mathbb{Q}, W)$. For instance, the dual of the relaxed Selmer group $H^1_{(S)}(\mathbb{Q}, W)$ is the *restricted Selmer group* at S,

$$H^1_{[S]}(\mathbb{Q}, \mathrm{Hom}(W, \mu_{p^\infty})) := \{c \in H^1_\emptyset(\mathbb{Q}, W^*) \text{ such that } \mathrm{res}_q(c) = 0, \quad \text{for all } q|S\}. \tag{13}$$

For $W = A_n$, while the size of a single Selmer group often represents a subtle global invariant, the ratio of the cardinalities of a Selmer group and its dual is accounted for by a simple explicit product of local quantities:

$$\frac{\#H^1_\Sigma(\mathbb{Q}, A_n)}{\#H^1_{\Sigma^*}(\mathbb{Q}, A_n^*)} = \frac{\#H^0(\mathbb{Q}, A_n)}{\#H^0(\mathbb{Q}, A_n^*)} \prod_q \frac{\#\Sigma_q}{\#H^0(G_q, A_n)}. \tag{14}$$

The proof of this identity rests on the Poitou–Tate long exact sequence in Galois cohomology (cf. [4, Theorem 2.19]). Notice that the ostensibly infinite product on the right is really a finite one since $\#H^1_{\mathrm{fin}}(\mathbb{Q}_q, A_n) = \#H^0(G_q, A_n)$ for all $q \neq p$.

4 Cohomological Reinterpretation of Mazur's Formula

This section recasts Mazur's formula (4) in cohomological terms. This formulation is amenable to generalisation to the elliptic setting. Recall that the Hecke eigenvalues of the Eisenstein series $E_{2,N}$ are encoded by the traces of the image of σ_ℓ of the $G_\mathbb{Q}$-representation

$$\varrho_{\mathrm{eis}} := \mathbb{Z}_p \oplus \mathbb{Z}_p(1)$$

for primes $\ell \nmid Np$.

Denote $T_\mu := \mathbb{Z}_p(1)$ and $A_\mu^* := \mathbb{Q}_p/\mathbb{Z}_p$. For each rational prime $\ell \neq p$, there is a distinguished global class

$$c_\mu[\ell] \in \mathbb{Q}^\times \otimes \mathbb{Z}_p \simeq H^1(\mathbb{Q}, T_\mu)$$

which is unramified at all primes $q \nmid p\ell$, is crystalline at p, and is ramified at ℓ. This class is simply the image of ℓ under the identification provided by classical Kummer theory. The class $c_\mu[\ell]$ is a canonical choice of generator for the Selmer group $H^1_{(\ell)}(\mathbb{Q}, T_\mu) \simeq \mathbb{Z}_p$.

Let $N \nmid p\ell$ be a prime. The Kronecker-Weber Theorem provides an isomorphism

$$H^1_{(N)}(\mathbb{Q}, A_\mu^*) \simeq (\mathbb{Z}/N\mathbb{Z})^\times \otimes \mathbb{Z}_p,$$

so that the choice of a generator κ amounts to giving a discrete logarithm $\log_{N,p}$ modulo p^m for $p^m \parallel (N-1)$. Mazur's formula can thus be rewritten in terms of local Tate duality (at least, up to sign) as

$$a'_\ell = (\ell - 1) \cdot \log_{N,p}(\mathrm{res}_N(c_\mu[\ell])) = (\ell - 1) \cdot \langle \kappa, c_\mu[\ell] \rangle_N.$$

Remark 4.1 Formulae for generalised eigenvalues in the Eisenstein setting can be obtained through the study of the deformation theory of the mod p-reduction of ϱ_{eis} (or more precisely, of the corresponding pseudo-representation), as carried out by Wake and Wang-Erickson [18]. The deceptively simple expression for generalised eigenvalues in the prime-level setting follows from the fact the first-order deformations of the residual representation that are unramified away from $\{p, N\}$, crystalline at p and Steinberg at N are reducible (cf. *op.cit.* §9.1). When studying generalised eigenvalues arising from congruences between the Eisenstein $E_{2,N}$ series and cusp forms of a more general level, one should expect formulae involving Massey products (cf. *op.cit.* Part 3).

5 The Symmetric Square and Adjoint Representations

We place ourselves in the setting of the introduction, namely, assume that f is a weight two cusp form attached to an elliptic curve E over $\mathbb{Q}$ of conductor M, and let

$$\varrho_f : G_\mathbb{Q} \longrightarrow \mathrm{Aut}(T_p E)$$

be the representation arising from the Galois action on the p-adic Tate module of E. For every prime $N \nmid Mp$, the characteristic polynomial of σ_N is given by

$$x^2 - a_N(f)x + N = (x - \alpha_N)(X - \beta_N) \tag{15}$$

for some α_N and $\beta_N \in \overline{\mathbb{Z}}$ which have complex absolute value $\sqrt{N}$ and hence are different from ± 1 and $\pm N$.

It is assumed that the mod p reduction of $\overline{\varrho}_f : G_\mathbb{Q} \to \mathrm{Aut}(E[p])$ is surjective and that p does not divide the minimal degree of a modular parametrisation $X_0(M) \longrightarrow E$. Let

$$\mathrm{Sym}^2 \varrho_f : G_\mathbb{Q} \to \mathrm{Aut}(\mathrm{Sym}^2(T_p E))$$

be the symmetric square representation of ϱ_f. The action of $G_\mathbb{Q}$ by conjugation on the module $\mathrm{Ad}^0(T_p E)$ of trace zero endomorphisms of $T_p E$ gives rise to the adjoint representation

$$\mathrm{Ad}^0 \varrho_f : G_\mathbb{Q} \to \mathrm{Aut}(\mathrm{Ad}^0(T_p E)).$$

The perfect $G_{\mathbb{Q}}$-equivariant pairing

$$\langle\ ,\ \rangle_f : \mathrm{Ad}^0(T_pE) \times \mathrm{Ad}^0(T_pE) \longrightarrow \mathbb{Z}_p, \qquad \langle A, B\rangle_f = \mathrm{Trace}(AB),$$

identifies $\mathrm{Ad}^0(T_pE)$ with its $\mathbb{Z}_p$-linear dual as a $G_{\mathbb{Q}}$-module. The classical Weil pairing $\langle\ ,\ \rangle_{\mathrm{Weil}}$ on T_pE yields a pairing

$$\langle\ ,\ \rangle_f : \mathrm{Sym}^2(T_pE) \times \mathrm{Ad}^0(T_pE) \longrightarrow \mathbb{Z}_p(1), \qquad \langle P \otimes Q, \lambda\rangle_f = \langle \lambda(P), Q\rangle_{\mathrm{Weil}}. \tag{16}$$

Following the introduction, we shall denote $T_f := \mathrm{Sym}^2\, T_pE$. The above discussion implies that the dual of T_f is

$$T_f^* \simeq \mathrm{Ad}^0(T_pE) \simeq T_f(-1).$$

We shall denote

$$A_f = T_f \otimes \mathbb{Q}_p/\mathbb{Z}_p \quad \text{and} \quad A_f^* = T_f^* \otimes \mathbb{Q}_p/\mathbb{Z}_p$$

the p-divisible groups attached to T_f and T_f^* and

$$A_{f,n} = p^{-n}T_f/T_f \quad \text{and} \quad A_{f,n}^* = p^{-n}T_f^*/T_f^*.$$

their p^n-torsion parts.

Remark 5.1 The representation T_f plays a similar role to that of T_μ in Mazur's Eisenstein ideal setting. The crucial shared feature is that they are isomorphic to their duals up to twisting by $\mathbb{Z}_p(-1)$.

6 Local Cohomology Groups for the Symmetric Square Representation

This section describes the singular and finite part of the local cohomology of the representations T_f and T_f^* for all primes $N \nmid pM$.

Lemma 6.1 *Let $N \nmid pM$ be a prime. Then $H^1_{\mathrm{fin}}(\mathbb{Q}_N, T_f)$ is the torsion subgroup of $H^1(\mathbb{Q}_N, T_f)$, and there are isomorphisms*

$$H^1_{\mathrm{sing}}(\mathbb{Q}_N, T_f) \simeq \mathbb{Z}_p, \qquad H^1_{\mathrm{fin}}(\mathbb{Q}_N, A_f^*) \simeq \mathbb{Q}_p/\mathbb{Z}_p.$$

Proof Let $V_f = T_f \otimes \mathbb{Q}_p$. It follows from (15) that the eigenvalues of σ_N on V_f are α_N^2, β_N^2 and N, with

$$\alpha_N^2, \beta_N^2 \notin \{1, N^2\}.$$

Since no eigenvalue is equal to 1, the group $H^1_{\text{fin}}(\mathbb{Q}_N, V_f)$ is trivial. Thus, $H^1_{\text{fin}}(\mathbb{Q}_N, T_f)$ is the torsion subgroup of $H^1(\mathbb{Q}_N, T_f)$. Since the representation T_f is unramified at N, the morphism ∂_N identifies

$$H^1_{\text{sing}}(\mathbb{Q}_N, T_f) \xrightarrow{\sim} T_f(-1)^{G_N}$$

and the latter is isomorphic to $\mathbb{Z}_p$. The description of $H^1_{\text{fin}}(\mathbb{Q}_N, A_f^*)$ is then obtained via local Tate duality (10). □

We now turn to describing the finite part of the local cohomology $H^1_{\text{fin}}(\mathbb{Q}_N, T_f)$, which is the target of the isomorphism of Theorem 1.1. For a prime $N \nmid pM$, denote

$$r_f^+ = (N+1) - a_N(f) \qquad \text{and} \qquad r_f^- = a_N(f) + (N+1).$$

Definition 6.2 A prime $N \nmid pM$ that is called a *level-raising prime* for (f, p) if

$$p \text{ divides } r_f^+ r_f^-(N-1).$$

The terminology level-raising prime will be justified in light of Corollary 9.3. Note that, since p is assumed to be odd, it can divide at most two of the factors r_f^+, r_f^- and $(N-1)$.

Lemma 6.3 *The cohomology groups $H^1_{\text{fin}}(\mathbb{Q}_N, T_f)$ and $H^1_{\text{sing}}(\mathbb{Q}_N, A_f^*)$ are isomorphic finite abelian groups of order equal to*

$$\#\mathbb{Z}_p/(r_f^+ r_f^-(N-1)).$$

More precisely:

(i) if $p \nmid (N \pm 1)$, or if $p \mid (N-1)$ and $p \nmid r_f^+ r_f^-$, the group $H^1_{\text{fin}}(\mathbb{Q}_N, T_f)$ is cyclic.
(ii) If $p \mid (N+1)$, there is an isomorphism

$$H^1_{\text{fin}}(\mathbb{Q}_N, T_f) \simeq \mathbb{Z}_p/(r_f^+) \oplus \mathbb{Z}_p/(r_f^-).$$

Proof By Lemma 6.1, the finite local cohomology group $H^1_{\text{fin}}(\mathbb{Q}_N, T_f)$ is a finite abelian group, so it is isomorphic to its Pontryagin dual, which can be identified with $H^1_{\text{sing}}(\mathbb{Q}_N, A_f^*)$ by (10). Since T_f is unramified at N, Equation 7 yields an isomorphism

$$H^1_{\text{fin}}(\mathbb{Q}_N, T_f) \simeq T_f/(\sigma_N - 1)T_f.$$

Thus, describing the group structure of $H^1_{\text{fin}}(\mathbb{Q}_N, T_f)$ amounts to determining the Smith normal form of $(\text{Sym}^2 \varrho_f(\sigma_N) - 1)$. In particular, the order of the group is equal to p^v, where v is the p-adic valuation of $\det(\text{Sym}^2 \varrho_f(\sigma_N) - 1)$. A direct calculation shows that the characteristic polynomial of σ_N acting on $V_f = T_f \otimes \mathbb{Q}_p$ is given by

$$x^3 - (a_N^2(f) - N)x^2 + N(a_N^2(f) - N)x^2 - N^3, \tag{17}$$

so that

$$\det(\mathrm{Sym}^2 \varrho_f(\sigma_N) - 1) = r_f^+ r_f^- \cdot (N - 1).$$

Let p be a prime dividing $r_f^+ r_f^- \cdot (N-1)$. We shall consider the following cases.

(i) If either $p \nmid (N \pm 1)$ or $p \nmid r_f^+ r_f^-$, two of the eigenvalues of the matrix $(\mathrm{Sym}^2\, \varrho_f(\sigma_N) - 1)$ are invertible in $\bar{\mathbb{Z}}_p$. On the other hand, it follows from properties of the Smith normal form that each eigenvalue of $\mathrm{Sym}^2\, \varrho_f(\sigma_N) - 1$ must divide the exponent of $T_f/(\sigma_N - 1)T_f$ (in $\bar{\mathbb{Z}}_p$). As all but one eigenvalues are invertible, this forces $T_f/(\sigma_N - 1)T_f$ to be cyclic.

(ii) Suppose $p \mid (N+1)$, and $p \mid r_f^+ r_f^-$. Under these assumptions, the characteristic polynomial of $\bar{\varrho}_f(\sigma_N)$ has distinct roots equal to ± 1 mod p; we shall assume that the eigenvalues of $\varrho_f(\sigma_N)$ are $\alpha_N, \beta_N \in \mathbb{Z}_p$ with

$$\alpha_N = 1 \mod p \quad \text{and} \quad \beta_N = -1 \mod p. \tag{18}$$

Since the eigenvalues residually distinct, σ_N acts semisimply on $T_p E$ and, as a consequence on T_f as well, with eigenvalues α_N^2, β_N^2, N on the latter. Hence,

$$T_f/(\sigma_N - 1)T_f \simeq \mathbb{Z}_p/(\alpha_N^2 - 1) \oplus \mathbb{Z}_p/(\beta_N^2 - 1) \oplus \mathbb{Z}_p/(N-1)\mathbb{Z}_p,$$

with $\mathbb{Z}_p/(N-1)\mathbb{Z}_p = 0$ under the running assumptions. Denote by v_p the p-adic valuation in $\mathbb{Z}_p$. The conclusion follows observing that

$$v_p(\alpha_N^2 - 1) = v_p(r_f^+), \quad \text{and} \quad v_p(\beta_N^2 - 1) = v_p(r_f^-),$$

in light of (18).

□

7 Local Cohomology via Discrete Elliptic Logarithms

This section describes the finite part of the cohomology group $H^1_{\mathrm{fin}}(\mathbb{Q}_N, T_f)$ in terms of the finite group $E(\Phi_{N^2})$ for every level-raising prime $p \nmid (N-1)$. This description is immaterial to the proof of Theorem 1.1. It is primarily motivated by the interpretation of generalised eigenvalues in terms of suitable discrete logarithms on the group $(\mathbb{Z}/N\mathbb{Z})^\times$ in the Eisenstein ideal setting given by (4).

Let τ be the generator of the group $\mathrm{Gal}(\Phi_{N^2}/\Phi_N)$. Since p is odd, every $\mathbb{Z}_p$-module L with an action of $\mathrm{Gal}\,(\Phi_{N^2}/\Phi_N)$ decomposes as $L = L^+ \oplus L^-$, where

$L^{\pm}$ is the ± 1-eigenspace for the action of τ. In particular, for $E(\Phi_{N^2}) \otimes \mathbb{Z}_p$, we obtain a decomposition

$$E(\Phi_{N^2}) \otimes \mathbb{Z}_p = (E(\Phi_{N^2}) \otimes \mathbb{Z}_p)^+ \oplus (E(\Phi_{N^2}) \otimes \mathbb{Z}_p)^-,$$

where $\#(E(\Phi_{N^2}) \otimes \mathbb{Z}_p)^{\pm} = \#\mathbb{Z}_p/(r_f^{\pm})$. There is a $\mathrm{Gal}(\Phi_{N^2}/\Phi_N)$-equivariant isomorphism

$$\nu\colon E(\Phi_{N^2}) \otimes \mathbb{Z}_p \xrightarrow{\sim} T_pE/(\sigma_N^2 - 1)T_pE,$$

induced by the isomorphisms given by the Kummer maps

$$E(\Phi_{N^2}) \otimes \mathbb{Z}/p^n\mathbb{Z} \xrightarrow{\sim} E[p^n]/(\sigma_N^2 - 1)E[p^n], \qquad P \mapsto Q^{\sigma_N^2} - Q$$

where Q satisfies $p^n Q = P$ and n is a positive integer.

The natural map

$$\theta\colon (T_f/(\sigma_N^2 - 1)T_f)^+ \to T_f/(\sigma_N - 1)T_f \simeq H^1_{\mathrm{fin}}(\mathbb{Q}_N, T_f). \tag{19}$$

is an isomorphism, since p is an odd prime. It will be convenient to identify $H^1_{\mathrm{fin}}(\mathbb{Q}_N, T_f)$ with the source of θ.

For any $\mathbb{Z}_p[\mathrm{Gal}\,(\bar{\Phi}_N/\Phi_N)]$-module L, let $\mathrm{Sym}^2 L$ denote its symmetric square as a $\mathbb{Z}_p$-module, with the natural $\mathrm{Gal}\,(\bar{\Phi}_N/\Phi_N)$-action. The projection

$$\mathrm{Sym}^2(T_pE) \longrightarrow \mathrm{Sym}^2\left(T_pE/(\sigma_N^2 - 1)T_pE\right)$$

gives rise to a surjective homomorphism

$$\psi\colon T_f/(\sigma_N^2 - 1)T_f \longrightarrow \mathrm{Sym}^2\left(T_pE/(\sigma_N^2 - 1)T_pE\right)$$

compatible with the action of τ.

Let

$$j\colon H^1_{\mathrm{fin}}(\mathbb{Q}_N, T_f) \longrightarrow \left(\mathrm{Sym}^2(E(\Phi_{N^2}) \otimes \mathbb{Z}_p)\right)^+.$$

be the composition $j = (\nu^{-1} \otimes \nu^{-1}) \circ \psi \circ \theta^{-1}$, where $\nu^{-1} \otimes \nu^{-1}$ denotes the map induced by ν^{-1} on the respective symmetric squares.

Proposition 7.1 *If $p \nmid (N-1)$, then j is an isomorphism.*

Proof The composition j is surjective, so it suffices to compare cardinalities. Note that the target can be described as

$$\left(\mathrm{Sym}^2(E(\Phi_{N^2}) \otimes \mathbb{Z}_p)\right)^+ \simeq \mathrm{Sym}^2((E(\Phi_{N^2}) \otimes \mathbb{Z}_p)^+) \oplus \mathrm{Sym}^2((E(\Phi_{N^2}) \otimes \mathbb{Z}_p)^-).$$

Consider the exact sequence of $\mathbb{Z}_p[\mathrm{Gal}\,(\Phi_{N^2}/\Phi_N)]$-modules

$$0 \to E[p](\Phi_{N^2}) \to E(\Phi_{N^2}) \xrightarrow{p} E(\Phi_{N^2}) \to E(\Phi_{N^2}) \otimes \Phi_p \to 0$$

induced by multiplication by p. Then $(E(\Phi_{N^2}) \otimes \Phi_p)^{\pm}$ is isomorphic to the ± 1-eigenspace of $\bar{\varrho}_f(\sigma_N)$, since they are Φ_p-vector spaces of the same cardinality. If the characteristic polynomial of $\bar{\varrho}_f(\sigma_N)$ has a double root equal to ± 1, then $p \mid (N-1)$. Thus, we shall assume that $\dim_{\Phi_p}(E(\Phi_{N^2}) \otimes \Phi_p)^{\pm} \leq 1$, which implies that $E(\Phi_{N^2})^{\pm} \otimes \mathbb{Z}_p$ is cyclic and isomorphic to $\mathbb{Z}_p/(r_f^{\pm})$; the same is true for $\mathrm{Sym}^2((E(\Phi_{N^2}) \otimes \mathbb{Z}_p)^{\pm})$. The conclusion now follows from Lemma 6.3. □

Assume that $p \nmid (N-1)$ and that N is a level raising prime for (f, p). Let $n_{\pm}$ denote the p-adic valuation of $r_f^{\pm}$. If $n^{\pm} > 0$, a discrete logarithm for $E(\Phi_{N^2}) \otimes \mathbb{Z}_p$ is a surjective group homomorphism

$$\log_{E,N,p}^{\pm} : (E(\Phi_{N^2}) \otimes \mathbb{Z}_p)^{\pm} \to \mathbb{Z}/p^{n_{\pm}}\mathbb{Z}.$$

This choice is unique up to rescaling by a unit in $(\mathbb{Z}/p^{n_{\pm}}\mathbb{Z})^{\times}$, and both $\log_{E,N,p}^{+}$ and $\log_{E,N,p}^{-}$ exist if and only if, additionally, $p \mid (N+1)$. We shall drop the sign from the notation when $p \nmid (N+1)$. By abuse of notation, let

$$(\log_{E,N,p}^{\pm})^{\otimes 2} : H^1_{\mathrm{fin}}(\mathbb{Q}_N, T_f) \to \mathbb{Z}/p^{n_{\pm}}\mathbb{Z} \tag{20}$$

denote the homomorphism sending a class c to $(\log_{E,N,p}^{\pm})^{\otimes 2}(j(c))$. It follows from Proposition 7.1 that the $\mathbb{Z}_p$-module $\mathrm{Hom}(H^1_{\mathrm{fin}}(\mathbb{Q}_N, T_f), \mathbb{Q}_p/\mathbb{Z}_p)$ is generated by either the symmetric square of the unique discrete logarithm of $E(\Phi_{N^2}) \otimes \mathbb{Z}_p$ up to rescaling if $p \nmid (N+1)$ or

$$\{(\log_{E,N,p}^{+})^{\otimes 2}, (\log_{E,N,p}^{-})^{\otimes 2}\}$$

when $p \mid (N+1)$, where we view $\mathbb{Z}/p^{n_{\pm}}\mathbb{Z}$ as embedded in $\mathbb{Q}_p/\mathbb{Z}_p$ in the standard way.

8 Selmer Groups for the Symmetric Square Representation

Proposition 8.1 *The Selmer groups $H^1_{\emptyset}(\mathbb{Q}, T_f)$ and $H^1_{\emptyset}(\mathbb{Q}, A_f^*)$ are trivial.*

Proof The triviality of $H^1_{\emptyset}(\mathbb{Q}, A_f^*)$ is the main theorem of [10, Theorem 1] in light of the running assumptions on f and on p that were made in the introduction. From the long exact sequences in cohomology induced by multiplication by p^n on A_f^* it follows that for every $n \geq 1$, the Selmer group

$$H^1_{\emptyset}(\mathbb{Q}_N, A_{f,n}^*) \simeq H^1_{\emptyset}(\mathbb{Q}, A_f^*)[p^n]$$

is also trivial. Formula (14) applied to $W = A_{f,n}$ and $\Sigma_q = H^1_{\mathrm{fin}}(\mathbb{Q}_q, A_{f,n})$ for all q gives

$$\#H^1_\emptyset(\mathbb{Q}, A_{f,n}) = \#H^1_\emptyset(\mathbb{Q}, A^*_{f,n}) = 0,$$

and the proposition follows from passing to the inverse limit. □

Proposition 8.2 *Let $N \nmid pM$ be a rational prime. Restricting to the decomposition group at N and projecting onto the singular cohomology gives rise to isomorphisms*

$$\partial_N \circ \mathrm{res}_N : H^1_{(N)}(\mathbb{Q}, T_f) \xrightarrow{\sim} H^1_{\mathrm{sing}}(\mathbb{Q}_N, T_f)$$

and

$$\partial_N \circ \mathrm{res}_N : H^1_{(N)}(\mathbb{Q}, A^*_f) \xrightarrow{\sim} H^1_{\mathrm{sing}}(\mathbb{Q}_N, A^*_f).$$

In particular, $H^1_{(N)}(\mathbb{Q}, T_f) \simeq \mathbb{Z}_p$ for every $N \nmid pM$, and $H^1_{(N)}(\mathbb{Q}, A^*_f)$ is finite; it is non-trivial if and only if N is a level-raising prime for (f, p).

Proof Let $n \in \mathbb{Z}_{>0}$, and consider the cartesian diagram

$$\begin{array}{ccc} H^1_\emptyset(\mathbb{Q}, A_{f,n}) & \longrightarrow & H^1_{(N)}(\mathbb{Q}, A_{f,n}) \\ \downarrow & & \downarrow \\ H^1_{\mathrm{fin}}(\mathbb{Q}_N, A_{f,n}) & \longrightarrow & H^1(\mathbb{Q}_N, A_{f,n}) \end{array}$$

where the vertical arrows are given by restricting to the decomposition group at N. A similar cartesian diagram is obtained for $A^*_{f,n}$. The triviality of $H^1_\emptyset(\mathbb{Q}_N, A_{f,n})$ and $H^1_\emptyset(\mathbb{Q}_N, A^*_{f,n})$ shows that the natural maps

$$H^1_{(N)}(\mathbb{Q}, A_{f,n}) \to H^1_{\mathrm{sing}}(\mathbb{Q}_N, A_{f,n}) \text{ and } H^1_{(N)}(\mathbb{Q}, A^*_{f,n}) \to H^1_{\mathrm{sing}}(\mathbb{Q}_N, A^*_{f,n})$$

are injective. Proposition 8.1 implies that the restricted Selmer groups $H^1_{[N]}(\mathbb{Q}, A_{f,n})$ and $H^1_{[N]}(\mathbb{Q}, A^*_{f,n})$ are trivial *a fortiori*, and it follows from (14) that

$$\#H^1_{(N)}(\mathbb{Q}, A_{f,n}) = \#H^1_{\mathrm{sing}}(\mathbb{Q}_N, A_{f,n}), \qquad \#H^1_{(N)}(\mathbb{Q}, A^*_{f,n}) = \#H^1_{\mathrm{sing}}(\mathbb{Q}_N, A^*_{f,n}).$$

The conclusion is obtained by passing to inverse and direct limits and invoking Lemmas 6.1 and 6.3, respectively. □

Remark 8.3 Note how the hypothesis that N is a level-raising prime for (f, p) is essential for the non-triviality of $H^1_{(N)}(\mathbb{Q}, A^*_f)$, while $H^1_{(N)}(\mathbb{Q}, T_f)$ is always non-trivial for any prime $N \nmid 6Mp$, a key fact that underlies the existence of Flach classes described in Sect. 10.

9 Deformations of Galois Representations

The celebrated theorem of Wiles [17] and Taylor-Wiles [16] identifies certain localisations of Hecke algebras at the maximal ideal attached to f with suitable deformation rings. More precisely, let

$$\mathbb{T}(M)_{p,f} \quad \text{and} \quad \mathbb{T}(MN^2)_{p,f}$$

denote the localisations of the semi-local rings $\mathbb{T}(M) \otimes \mathbb{Z}_p$ and $\mathbb{T}(MN^2) \otimes \mathbb{Z}_p$ at the maximal ideals attached to f (mod p). The morphisms

$$\varphi_{f,\emptyset} \colon \mathbb{T}(M)_{p,f} \to \mathbb{Z}_p \qquad \text{and} \qquad \varphi_{f,(N)} \colon \mathbb{T}(MN^2)_{p,f} \to \mathbb{Z}_p$$

are determined by sending the Hecke operator T_ℓ to the coefficient $a_\ell(f)$ for $\ell \nmid MNp$. Let $I_{f,\emptyset}$ and $I_{f,(N)}$ denote the kernels of the morphisms $\varphi_{f,\emptyset}$ and $\varphi_{f,(N)}$ respectively.

Let $R_{\bar{\varrho}_f,\emptyset}$ be the universal deformation ring parametrising lifts of the residual representation $\bar{\varrho}_f \colon G_{\mathbb{Q}} \to \mathrm{Aut}(E[p])$ with fixed determinant, that are minimally ramified, in the sense of [4, Sect. 2.7], at primes $q \neq p$ and crystalline at p. Denote by $R_{\bar{\varrho}_f,(N)}$ the deformation ring classifying the lifts as above for which the local condition at N is omitted. Note that, since the residual representation $\bar{\varrho}_f$ is unramified at N, the highest power of N dividing the conductor of a lift of $\bar{\varrho}_f$ is at most N^2 (cf.[4, Lemma 2.7]). The universal properties of the deformation rings $R_{\bar{\varrho}_f,(N)}$ and $R_{\bar{\varrho}_f,\emptyset}$ give rise to a commutative diagram

$$\begin{array}{ccc} R_{\bar{\varrho}_f,(N)} & \xrightarrow{\gamma_{(N)}} & \mathbb{T}(MN^2)_{p,f} \\ \downarrow & & \downarrow \\ R_{\bar{\varrho}_f,\emptyset} & \xrightarrow{\gamma_{\emptyset}} & \mathbb{T}(M)_{p,f} \end{array}$$

where the vertical arrows are surjective. The Taylor-Wiles modularity lifting theorem (proved in full generality in [1]) implies that $\gamma_{(N)}$ and $\gamma_\emptyset$ are isomorphisms.

Definition 9.1 Given a cocycle κ representing a class in $H^1_{(N)}(\mathbb{Q}, A_f^*)$, and a prime $\ell \nmid NMp$, the *generalised eigenvalue* attached to κ at ℓ is

$$a'_\ell(f,\kappa) := \mathrm{Tr}(\kappa(\sigma_\ell)\varrho_f(\sigma_\ell)).$$

Note that this is independent of the choice of the representative of the cohomology class and of the Frobenius element $\sigma_\ell \in G_\ell$. When the group $H^1_{(N)}(\mathbb{Q}, A_f^*)$ is cyclic, a choice of generator κ can be fixed throughout, and we will simply denote $a'_\ell(f) := a'_\ell(f,\kappa)$.

The term generalised eigenvalue is justified by the following immediate consequence of the modularity lifting theorem.

Proposition 9.2 *For* $\bigstar \in \{\emptyset, (N)\}$*, there is an isomorphism*

$$H^1_\bigstar(\mathbb{Q}, A_f^*) \xrightarrow{\sim} \mathrm{Hom}(I_{f,\bigstar}/I_{f,\bigstar}^2, \mathbb{Q}_p/\mathbb{Z}_p)$$

sending a class $\kappa \in H^1_\bigstar(\mathbb{Q}, A_f^*)$ *to the morphism* $F_\kappa : I_{f,\bigstar}/I_{f,\bigstar}^2 \to \mathbb{Q}_p/\mathbb{Z}_p$ *characterised by*

$$F_\kappa(T_\ell - a_\ell) = a'_\ell(f, \kappa)$$

for every $\ell \nmid MNp$.

Proof Let $J_\bigstar$ be the kernel of $\phi_{f,\bigstar} \circ \gamma_\bigstar$. A standard calculation (see, for example, [4, Lemma 2.40]) identifies $H^1_\bigstar(\mathbb{Q}, T_f^*)$ with $\mathrm{Hom}(J_\bigstar/J_\bigstar^2, \mathbb{Q}_p/\mathbb{Z}_p)$ via the map induced by

$$H^1_\bigstar(\mathbb{Q}, A_{f,n}^*) \xrightarrow{\sim} \mathrm{Hom}_{(\mathbb{Z}/p^n\mathbb{Z})-\mathrm{aug}}(R_{\bar{\varrho}_f,\bigstar}, \mathbb{Z}/p^n\mathbb{Z}[\varepsilon]/(\varepsilon^2))$$

for $n \in \mathbb{Z}_{>0}$, where the target denotes homomorphisms of $\mathbb{Z}/p^n\mathbb{Z}$-augmented rings, which sends a cohomology class κ to

$$\tilde{\varrho}_f = (1 + \varepsilon \cdot \kappa) \cdot \varrho_f.$$

Upon identifying $I_{f,\bigstar} \simeq J_\bigstar$, the conclusion follows from the fact that

$$\mathrm{Trace}(\tilde{\varrho}_f(\sigma_\ell)) = a_\ell(f) + a'_\ell(f, \kappa) \cdot \varepsilon.$$

□

Combining this result with Propositions 8.1 and 8.2, we obtain the following corollary.

Corollary 9.3 *Let* $p \nmid 6\deg(\phi_E)$ *be a prime such that* $\bar{\varrho}_f$ *is surjective. Then:*

1. *The ideal* $I_{f,\emptyset}$ *is trivial;*
2. *The ideal* $I_{f,(N)}$ *is non-trivial if and only if* N *is a level-raising prime for* (f, p)*. If* $p \nmid (N \pm 1)$ *or if* $p \mid (N - 1)$ *and* $p \nmid r_f^+ r_f^-$*, the ideal* $I_{f,(N)}$ *is cyclic.*

In particular, it follows that the generalised eigenspace attached to f in $M_2(M) \otimes \Phi_p$ is spanned by f, while it is strictly larger in $M_2(MN^2) \otimes \Phi_p$ when N is a level-raising prime for (f, p).

10 Flach Classes

The goal of this section is to introduce the Flach classes $c_f[\ell] \in H^1_{(\ell)}(\mathbb{Q}, T_f)$, following [10, Sect. 2]. These classes are the key ingredient in Flach's proof of Proposition 8.1. This section and the next complements this by reinterpreting the generalised eigenvalues $a'_\ell(f,\kappa)$ appearing in Proposition 9.2 as the local Tate pairing at ℓ between the classes κ and $c_f[\ell]$.

Let X be an irreducible regular Notherian scheme which is either of finite type over a field or is smooth over a discrete valuation ring. Of importance for the constructions of [10, Sect. 2] are the cases where:

1. X is the modular surface $X_0(M)^2$ viewed as a scheme over spec $\mathbb{Q}$;
2. $X = \mathscr{X}_0(M)^2_{\mathbb{Z}_\ell}$ is the smooth proper integral model of $X_0(M)^2$ over spec $\mathbb{Z}_\ell$ for $\ell \nmid M$.

Let $\mathscr{K}_2$ be the sheaf associated to the Quillen's second K-group functor $U \mapsto K_2(U)$. The group $H^1(X,\mathscr{K}_2)$ is the first homology of the complex

$$K_2(k(X)) \xrightarrow{\partial} \oplus_{x\in X^{(1)}} k(x)^\times \xrightarrow{\text{div}} \oplus_{x\in X^{(2)}} \mathbb{Z}, \tag{21}$$

where $X^{(n)}$ is the set of codimension n subschemes of X, and $k(x)$ is the residue field of the local ring of X at x. The map ∂ is a residue map and div sends an element $u \in k(x)^\times$ to the pushforward to X of its divisor on x.

Let $\ell \nmid M$ be a prime, and let

$$\pi_1, \pi_2 : X_0(M\ell) \longrightarrow X_0(M), \qquad \pi := (\pi_1, \pi_2) : X_0(M\ell) \longrightarrow X := X_0(M)^2$$

be the maps arising from the two standard degeneracy maps sending a pair (A_1, A_2) of elliptic curves with level M structure related by a cyclic ℓ-isogeny to the points of $X_0(M)$ attached to A_1 and A_2 respectively. The image

$$T_\ell := \pi(X_0(M\ell)) \subset X$$

is birational to $X_0(M\ell)$ and is the graph of the ℓ-th Hecke correspondence on $X_0(M)$. The modular unit $\Delta(z)/\Delta(\ell z)$ has divisor supported at the cusps of $X_0(M\ell)$, and the pushforward of this divisor to X vanishes [10, p. 317]. It follows that the element

$$\varepsilon(\ell) := (\Delta(z)/\Delta(\ell z)) \in k(T_\ell)^\times$$

belongs to the kernel of the map div of (21), and thus defines an element of $H^1(X,\mathscr{K}_2)$.

The special element $\varepsilon(\ell)$ can be parlayed into the construction of global cohomology classes

$$c[\ell] \in H^1(\mathbb{Q}, H^2_{\text{et}}(X_{\bar{\mathbb{Q}}}, \mathbb{Z}_p(2))), \qquad c_f[\ell] \in H^1(\mathbb{Q}, T_f)$$

by setting

$$c[\ell] := h \cdot \kappa(\varepsilon(\ell)), \qquad c_f[\ell] := \text{Sym}_*(\phi_E \times \phi_E)_*(c[\ell]),$$

where

(1) the map

$$\kappa : H^1(X, \mathcal{K}_2) \longrightarrow H^3_{\text{et}}(X, \mathbb{Z}_p(2))$$

is an étale regulator map. Roughly speaking, it is obtained by combining the Kummer maps

$$\delta_x : k(x)^\times \longrightarrow H^1_{\text{et}}(k(x), \mathbb{Z}_p(1))$$

of Kummer theory with the pushforward maps

$$i_{x*} : H^1_{\text{et}}(x, \mathbb{Z}_p(1)) \longrightarrow H^3_{\text{et}}(X, \mathbb{Z}_p(2))$$

in étale cohomology, induced by the inclusions $i_x : x \hookrightarrow X$;

(2) the map

$$h : H^3_{\text{et}}(X, \mathbb{Z}_p(2)) \longrightarrow H^1(\mathbb{Q}, H^2_{\text{et}}(X_{\bar{\mathbb{Q}}}, \mathbb{Z}_p(2)))$$

arises from an edge map in the Hochschild-Serre spectral sequence

$$H^p(\mathbb{Q}, H^q_{\text{et}}(X_{\bar{\mathbb{Q}}}, \mathbb{Z}_p(2))) \Rightarrow H^{p+q}_{\text{et}}(X, \mathbb{Z}_p(2)),$$

in light of the triviality of $H^3(X_{\bar{\mathbb{Q}}}, \mathbb{Z}_p(2))^{G_{\mathbb{Q}}}$ which follows from weight considerations ([10, Proposition 2.2]);

(3) the map

$$(\phi_E \times \phi_E)_* : H^1(\mathbb{Q}, H^2(X_{\bar{\mathbb{Q}}}, \mathbb{Z}_p(2))) \longrightarrow H^1(\mathbb{Q}, H^2(E^2_{\bar{\mathbb{Q}}}, \mathbb{Z}_p(2)))$$

is the pushforward induced by the map $\phi_E \times \phi_E : X \longrightarrow E^2$ arising from the modular parametrisation ϕ_E;

(4) the last map

$$\text{Sym}_* : H^1(\mathbb{Q}, H^2(E^2_{\bar{\mathbb{Q}}}, \mathbb{Z}_p(2))) \longrightarrow H^1(\mathbb{Q}, T_f)$$

is induced from the natural Kunneth projection from $H^2(E^2_{\bar{\mathbb{Q}}}, \mathbb{Z}_p(2))$ to $H^1(E_{\bar{\mathbb{Q}}}, \mathbb{Z}_p(1))^{\otimes 2}$ composed with the projection to the space

$$T_f = \text{Sym}^2(H^1_{\text{et}}(E_{\bar{\mathbb{Q}}}, \mathbb{Z}_p(1)))$$

of symmetric tensors.

Remark 10.1 When $p|(N-1)$, Chris Skinner has proposed an interesting construction of a global class in the cohomology of the adjoint representation, which is somewhat dual to Flach's construction. This class is obtained from the *Shimura class*

$$\mathfrak{S} \in H^1_{\text{et}}(X_0(MN), \mathbb{Z}/p^n\mathbb{Z})$$

arising by applying the discrete logarithm $(\mathbb{Z}/N\mathbb{Z})^\times \longrightarrow \mathbb{Z}/p^n\mathbb{Z}$ to the class of the étale covering $X_1(N) \longrightarrow X_0(N)$ with Galois group $(\mathbb{Z}/N\mathbb{Z})^\times/(\pm 1)$. The image of $\mathfrak{S}$ under pushforward by the diagonal embedding $\Delta\colon X_0(MN) \hookrightarrow X_0(MN)^2$ yields a class

$$\kappa \in H^3_{\text{et}}(X_0(MN)^2, \mathbb{Z}/p^n\mathbb{Z}(1))$$

to which steps (2)-(3)-(4) above can be applied, leading ultimately to a class in $H^1(\mathbb{Q}, A^*_{f,n})$. This class appears to play a role analogous to the class in $H^1_{(N)}(\mathbb{Q}, \mathbb{Z}/p^n\mathbb{Z})$ arising from global class field theory (or the Kronecker-Weber theorem) which also depends linearly on the choice of a discrete logarithm on $(\mathbb{Z}/N\mathbb{Z})^\times$.

11 Local Behaviour of the Flach Classes

If r is a prime that does not divide M, then the curve $X_0(M)$ extends to a smooth proper model $\mathscr{X}_0(M)_{\mathbb{Z}_r}$ over $\mathrm{spec}(\mathbb{Z}_r)$. Let $\mathscr{X}_0(M)_{\Phi_r}$ denote its special fiber, and write

$$\mathscr{X}_{\mathbb{Z}_r} := \mathscr{X}_0(M)^2_{\mathbb{Z}_r}, \qquad X_{\mathbb{Q}_r} := X_0(M)^2_{\mathbb{Q}_r}, \qquad \mathscr{X}_{\Phi_r} := \mathscr{X}_0(M)^2_{\Phi_r}.$$

There is an exact localisation sequence ([10, (17)])

$$H^1(\mathscr{X}_{\mathbb{Z}_r}, \mathscr{K}_2) \longrightarrow H^1(X_{\mathbb{Q}_r}, \mathscr{K}_2) \xrightarrow{\partial_r} \mathrm{Pic}(\mathscr{X}_{\Phi_r}),$$

where ∂_r sends $u \in k(x)^\times$ to its divisor along the special fiber. It is shown (cf. [10, (19)]) that

$$\partial_r(\varepsilon(\ell)) = \begin{cases} 0 & \text{if } r \neq \ell; \\ 6 \cdot (\Gamma_\ell - \Gamma'_\ell) & \text{if } r = \ell, \end{cases} \tag{22}$$

where $\Gamma_\ell \in \mathrm{Pic}(\mathscr{X}_{\Phi_\ell})$ is the class of the graph of the Frobenius morphism $\mathscr{X}_0(M)_{\Phi_\ell} \longrightarrow \mathscr{X}_0(M)_{\Phi_\ell}$, and Γ'_ℓ is the class of its transpose.

The elliptic curve E extends to a smooth proper model $\mathscr{E}$ over $\mathbb{Z}_\ell$, with special fiber $\mathscr{E}_{\Phi_\ell}$. Let

$$\Gamma_{\ell,E} \in \mathrm{Pic}((\mathscr{E} \times \mathscr{E})_{\Phi_\ell})$$

be the class of the graph of the Frobenius endomorphism on $\mathscr{E}_{\Phi_\ell}$, and let $\Gamma'_{\ell,E}$ be the class of its transpose. They are related to Γ_ℓ and Γ'_ℓ by the formulae

$$(\phi_E \times \phi_E)_*(\Gamma_\ell) = \deg(\phi_E) \cdot \Gamma_{\ell,E}, \qquad (\phi_E \times \phi_E)_*(\Gamma'_\ell) = \deg(\phi_E) \cdot \Gamma'_{\ell,E}. \tag{23}$$

Proposition 11.1 *Let $\ell \nmid Mp$ be a rational prime.*

1. The global class $c[\ell]$ belongs to $H^1_{(\ell)}(\mathbb{Q}, H^2_{\text{et}}(X_{\bar{\mathbb{Q}}}, \mathbb{Z}_p(2)))$, and

$$\partial_\ell(c[\ell]) = 6 \cdot \mathrm{cl}(\Gamma_\ell - \Gamma'_\ell),$$

where

$$\mathrm{cl} : \mathrm{Pic}(\mathscr{X}_{\Phi_\ell}) \longrightarrow H^2_{\mathrm{et}}(\mathscr{X}_{\bar{\Phi}_\ell}, \mathbb{Z}_p(2))^{G_{\Phi_\ell}}$$

is the étale cycle class map.

2. *The global class $c_f[\ell]$ belongs to $H^1_{(\ell)}(\mathbb{Q}, T_f)$, and*

$$\partial_\ell(c_f[\ell]) = 6 \cdot \deg(\phi_E) \cdot \mathrm{Sym}_*(\mathrm{cl}(\Gamma_{\ell,E} - \Gamma'_{\ell,E})).$$

Proof These two assertions follow from chasing through the top and bottom parts of the commutative diagram

$$\begin{array}{ccccc}
H^1(\mathscr{X}_{\mathbb{Z}_r}, \mathscr{K}_2) & \longrightarrow & H^1(X_{\mathbb{Q}_r}, \mathscr{K}_2) & \xrightarrow{\partial_r} & \mathrm{Pic}(\mathscr{X}_{\Phi_r}) \\
 & & & & \downarrow{\scriptstyle \mathrm{cl}} \\
\downarrow{\scriptstyle h\cdot\kappa} & & \downarrow{\scriptstyle h\cdot\kappa} & & H^2_{\mathrm{et}}(\mathscr{X}_{\Phi_r}, \mathbb{Z}_p(1)) \\
 & & & & \| \\
H^1_{\mathrm{fin}}(\mathbb{Q}_r, H^2_{\mathrm{et}}(X, \mathbb{Z}_p(2))) & \longrightarrow & H^1(\mathbb{Q}_r, H^2_{\mathrm{et}}(X, \mathbb{Z}_p(2))) & \xrightarrow{\partial_r} & H^2_{\mathrm{et}}(X, \mathbb{Z}_p(1))^{G_{\mathbb{Q}_r}} \\
\downarrow{\scriptstyle \mathrm{Sym}(\phi_E\times\phi_E)_*} & & \downarrow{\scriptstyle \mathrm{Sym}(\phi_E\times\phi_E)_*} & & \downarrow{\scriptstyle \mathrm{Sym}(\phi_E\times\phi_E)_*} \\
H^1_{\mathrm{fin}}(\mathbb{Q}_r, T_f) & \longrightarrow & H^1(\mathbb{Q}_r, T_f) & \xrightarrow{\partial_r} & T_f(-1)^{G_{\mathbb{Q}_r}},
\end{array}$$

and invoking (22) for the first, and (23) for the second. □

This determination of $\partial_\ell(c_f[\ell])$ can be used to compute the local Tate pairing between $c_f[\ell]$ and a global class $\kappa \in H^1_{(N)}(\mathbb{Q}, A^*_f)$.

Proposition 11.2 *For all primes $\ell \nmid MNp$, and $\kappa \in H^1_{(N)}(\mathbb{Q}, A^*_f)$*

$$\langle c_f[\ell], \kappa\rangle_\ell = 12 \cdot \deg(\phi_E) \cdot a'_\ell(f, \kappa).$$

Proof By setting $c = c_f[\ell]$ in (11), we obtain

$$\langle c_f[\ell], \kappa\rangle_\ell = \langle \partial_\ell(c_f[\ell]), \kappa(\sigma_\ell)\rangle, \tag{24}$$

where the pairing on the right is induced from the natural $\mathbb{Q}_p/\mathbb{Z}_p$-valued trace pairing between $T_f(-1)$ and $T^*_f \simeq T_f(-1) \otimes \mathbb{Q}_p/\mathbb{Z}_p$.

On the other hand, Proposition 11.1 gives

$$\partial_\ell(c_f[\ell]) = 6 \cdot \deg(\phi_E) \cdot \mathrm{Sym}_*\mathrm{cl}(\Gamma_{\ell,E} - \Gamma'_{\ell,E}) = 6 \cdot \deg(\phi_E) \cdot (\varrho_f(\sigma_\ell) - \varrho_f(\sigma_\ell)'),$$

where $\varrho_f(\sigma_\ell)$ is viewed as an element of $\mathrm{End}(T_pE) \supset \mathrm{Aut}(T_pE)$, and $\varrho_f(\sigma_\ell)'$ is the adjoint of $\varrho_f(\sigma_\ell)$ relative to the Weil pairing on T_pE, i.e., its *adjugate*

$$\varrho_f(\sigma_\ell)' = a_\ell(f) - \varrho_f(\sigma_\ell).$$

It follows that

$$\partial_\ell(c_f[\ell]) = 6 \cdot \deg(\phi_E) \cdot (2 \cdot \varrho_f(\sigma_\ell) - a_\ell(f)). \tag{25}$$

Combining (24) with (25) gives

$$\begin{aligned} \langle c_f[\ell], \kappa\rangle_\ell &= 12 \cdot \deg(\phi_E) \cdot \langle \varrho_f(\sigma_\ell), \kappa(\sigma_\ell)\rangle \\ &= 12 \cdot \deg(\phi_E) \cdot \mathrm{Trace}(\varrho_f(\sigma_\ell) \cdot \kappa(\sigma_\ell)) \end{aligned}$$

and the formula follows. □

12 An Application of Global Reciprocity

Proposition 12.1 *For all primes $\ell \nmid MNp$, and $\kappa \in H^1_{(N)}(\mathbb{Q}, A_f^*)$,*

$$\langle c_f[\ell], \kappa\rangle_N = -12 \cdot \deg(\phi_E) \cdot a'_\ell(f, \kappa).$$

Proof Since $c_f[\ell]$ belongs to $H^1_{(\ell)}(\mathbb{Q}, T_f)$, and κ belongs to $H^1_{(N)}(\mathbb{Q}, A_f^*)$, it follows that

$$\langle c_f[\ell], \kappa\rangle_q = 0 \qquad \text{for all } q \neq \ell, N.$$

Therefore, (12) implies that

$$\langle c_f[\ell], \kappa\rangle_\ell + \langle c_f[\ell], \kappa\rangle_N = 0,$$

and the result follows from Proposition 11.2. □

Proof The isomorphism defined in Proposition 9.2 sends a global class κ in $H^1_{(N)}(\mathbb{Q}, A_f^*)$ to the map $F_\kappa : I_{f,(N)}/I^2_{f,(N)} \to \mathbb{Q}_p/\mathbb{Z}_p$ characterised by the property

$$F_\kappa(T_\ell - a_\ell) = \mathrm{Trace}(\kappa(\sigma_\ell)\varrho_f[\ell]) = (12 \cdot \deg(\phi_E))^{-1} \langle c_f[\ell], \kappa\rangle_\ell$$

for every prime $\ell \nmid MNp$ by Lemma 11.2. On the other hand, by Proposition 8.2, the restriction to the decomposition group at N, combined with local Tate duality gives an isomorphism

$$H^1_{(N)}(\mathbb{Q}, A_f^*) \xrightarrow{\sim} \mathrm{Hom}(H^1_{\mathrm{fin}}(\mathbb{Q}_N, T_f), \mathbb{Q}_p/\mathbb{Z}_p)$$

sending κ to the homomorphism $\kappa \mapsto \langle c, \kappa\rangle_N$ for $c \in H^1_{\mathrm{fin}}(\mathbb{Q}_N, T_f)$. The result now follows from Proposition 12.1 by passing to Pontryagin duals. □

Proof of Corollary 1.2 This follows immediately from Theorem 1.1, and the fact that if $p \nmid (N \pm 1)$, the ideal $I_{f,(N)}$ is cyclic by Corollary 9.3. □

Remark 12.2 The generalised eigenvalues attached to a cusp form f can be easily computed numerically. For simplicity, assume that $p \parallel r_f^+ r_f^-(N-1)$. Then $I_{f,(N)}/I_{f,(N)}^2 = \mathbb{Z}/p\mathbb{Z} \cdot T$, for some Hecke operator $T \in I_{f,(N)}$. This implies that the generalised eigenspace attached to the system of eigenvalues of f in the space of mod p modular forms of weight 2 and level $\Gamma_0(MN^2)$ admits a basis $f = f_1, \ldots, f_r$ for some $r > 1$ such that

$$Tf_i = f_{i-1} \qquad \text{for every } 1 \le i \le r.$$

where we set $f_0 = 0$. The generalised eigenvalues are characterised by the property

$$(T_\ell - a_\ell(f))f_i = a'_\ell(f)\ f_{i-1} \mod (f_0, \ldots, f_{i-2}), \qquad \text{for every } \ell \nmid NMp.$$

for $i > 1$, up to rescaling by a common constant in $(\mathbb{Z}/p\mathbb{Z})^\times$ (cf. [12, Sect. 2]).

Remark 12.3 For any prime $\ell \nmid Mp$, the Flach class $c_f[\ell]$ is obtained from the image of a canonical element in motivic cohomology $H^1(\mathscr{X}_{\Phi_N}, \mathscr{K}_2)$ under the composition of the maps described in Sect. 10. An independent description of the image of the resulting local class under the morphism $\log_{E,N,p}^{\otimes 2}$, suitable for machine calculations for instance, appears somewhat elusive. The connection between $c_f[\ell]$ and the generalised eigenvalues a'_ℓ, which are more readily calculated numerically, supplies non-trivial information about the behaviour of Flach's classes as the prime ℓ varies.

References

1. Breuil, C., Conrad, B., Diamond, F., Taylor, R.: On the modularity of elliptic curves over $\mathbb{Q}$: wild 3-adic exercises. J. Am. Math. Soc. **14**(4), 843–939 (2001)
2. Bloch, S., Kato, K.: L-functions and Tamagawa numbers of motives. The Grothendieck Festschrift, Collect. Artic. in Honor of the 60th Birthday of A. Grothendieck. Vol. I, Prog. Math. **86**, 333–400 (1990)
3. Calegari, F., Emerton, M.: On the ramification of Hecke algebras at Eisenstein primes. Invent. Math. **160**(1), 97–144 (2005)
4. Darmon, H., Diamond, F., Taylor, R., Theorem, F.L., Current, I., developments in mathematics, 1995. Lectures of a seminar, held in Boston, MA, USA, May 7–8, 1995, pp. 1–107 (1–154, preliminary version,: Cambridge, p. 1995. International Press, MA (1994)
5. Darmon, H., Harris, M., Rotger, V., Venkatesh, A.: The derived Hecke algebra for dihedral weight one forms. Mich. Math. J. **72**, 145–207 (2022)
6. Darmon, H., Lauder, A., Rotger, V.: Overconvergent generalised eigenforms of weight one and class fields of real quadratic fields. Adv. Math. **283**, 130–142 (2015)
7. Darmon, H., Lauder, A., Rotger, V.: First order p-adic deformations of weight one newforms. In: L-functions and automorphic forms. LAF, Heidelberg, Germany, February 22–26, 2016, pp. 39–80. Springer, Cham (2017)

8. Darmon, H., Pozzi, A., Vonk, J.: The values of the Dedekind-Rademacher cocycle at real multiplication points. J. EMS (2023)
9. Darmon, H., Vonk, J.: Heights of RM divisors and real quadratic singular moduli. In progress
10. Flach, M.: A finiteness theorem for the symmetric square of an elliptic curve. Invent. Math. **109**(2), 307–327 (1992)
11. Harris, M., Venkatesh, A.: Derived Hecke algebra for weight one forms. Exp. Math. **28**(3), 342–361 (2019)
12. Lecouturier, E.: Higher Eisenstein elements, higher Eichler formulas and rank of Hecke algebras. Invent. Math. **223**(2), 485–595 (2021)
13. Mazur, B.: Modular curves and the Eisenstein ideal. Publ. Math. Inst. Hautes Étud. Sci. **47**, 33–186 (1977)
14. Merel, L.: The Weil pairing between the Shimura subgroup and the cuspidal subgroup of $J_0(p)$. J. Reine Angew. Math. **477**, 71–115 (1996)
15. Rubin, K.: Euler systems. (Hermann Weyl lectures), volume 147 of Ann. Math. Stud. Princeton, NJ, Princeton University Press (2000)
16. Taylor, R., Wiles, A.: Ring-theoretic properties of certain Hecke algebras. Ann. Math. (2) **141**(3), 553–572 (1995)
17. Wiles, A.: Modular elliptic curves and Fermat's Last Theorem. Ann. Math. (2) **141**(3), 443–551 (1995)
18. Wake, P., Wang-Erickson, C.: The rank of Mazur's Eisenstein ideal. Duke Math. J. **169**(1), 31–115 (2020)

On Constructing Extensions of Residually Isomorphic Characters

Samit Dasgupta

Abstract This is an exposition of our joint work with Kakde, Silliman, and Wang, in which we prove a version of Ribet's Lemma for GL_2 in the residually indistinguishable case. We suppose we are given a Galois representation taking values in the total ring of fractions of a complete reduced Noetherian local ring $\mathbf{T}$, such that the characteristic polynomial of the representation is reducible modulo some ideal $I \subset \mathbf{T}$. We assume that the two characters that arise are congruent modulo the maximal ideal of $\mathbf{T}$. We construct an associated Galois cohomology class valued in a $\mathbf{T}$-module that is "large" in the sense that its Fitting ideal is contained in I. We make some simplifying assumptions that streamline the exposition—we assume the two characters are actually trivial, and we ignore the local conditions needed in arithmetic applications.

Keywords Galois representations · construction of cohomology classes · Ribet's Lemma

1 Introduction

In 1970, Ribet proved the converse of Herbrand's Theorem [13]. When I learned about this in graduate school, Dick Gross recalled to me that the methods introduced by Ribet "were like a thunderbolt" in the number theory community at the time. Over 50 years later, Ribet's method remains a central force in algebraic number theory, particularly in Iwasawa Theory. Perhaps the seminal work on the topic is the beautiful book of Joel Bellaïche and Gaetan Chenevier [2]. This book presents a very general form of Ribet's approach and also describes deep arithmetic applications. An introduction written for a more general audience is given by Mazur [10].

The author is supported by NSF grant DMS–2200787.

S. Dasgupta (✉)
Department of Mathematics, Duke University, Durham, UK
e-mail: samit.dasgupta@duke.edu

M. Longo et al. (eds.), *Elliptic Curves and Modular Forms in Arithmetic Geometry*, Springer Proceedings in Mathematics & Statistics 527,
https://doi.org/10.1007/978-3-032-13123-2_4

The goal of Ribet's method is to construct a nontrivial Galois cohomology class from the knowledge that an L-function is appropriately divisible. Typically, we will be given a special value of an L-function, which we denote by L, lying in a ring $\mathbf{T}$ (e.g. $\mathbf{T} = \mathbf{Z}_p$ or $\mathbf{T} = \mathbf{Z}_p[[T]]$), and we assume assume that L lies in some ideal $I \subset \mathbf{T}$. The value L will be associated to two (or more) representations of the Galois group of a number field F over R, say ρ_1 and ρ_2. One then wants to construct a nontrivial class in

$$H^1(G_F, \overline{\rho}_1 \otimes \overline{\rho}_2^*),$$

where ρ_2^* is the dual of ρ_2 and $\overline{\rho}_i$ denotes the reduction of ρ_i modulo I. In Ribet's original setting, he had $\mathbf{T} = \mathbf{Z}_p$, $I = (p)$, and ρ_1, ρ_2 one dimensional characters of $G_{\mathbf{Q}}$. Specifically, ρ_1 was the trivial character and ρ_2 a nontrivial character of $\mathrm{Gal}(\mathbf{Q}(\mu_p)/\mathbf{Q})$.

To produce an extension, Ribet constructed a cusp form congruent to the Eisenstein series associated to ρ_1 and ρ_2. In his case, one can show that the cusp form is an eigenform. One therefore obtains a representation

$$\rho \colon G_{\mathbf{Q}} \longrightarrow \mathrm{GL}_2(K),$$

where K is a finite extension of $\mathbf{Q}_p$, such that

$$\mathrm{tr}(\rho) \equiv \rho_1 + \rho_2, \qquad \det(\rho) \equiv \rho_1\rho_2 \pmod{I}. \tag{1}$$

The representation ρ can be conjugated to land in $\mathrm{GL}_2(\mathcal{O}_K)$, and the Brauer–Nesbitt Theorem implies that one can choose a basis so that the reduction of ρ modulo the maximal ideal $\mathfrak{m}_K \subset \mathcal{O}_K$ has the form

$$\overline{\rho} = \begin{pmatrix} \overline{\rho}_1 & * \\ 0 & \overline{\rho}_2 \end{pmatrix}.$$

Ribet then proves an important lemma which shows that the basis can be chosen so that the image of $*$ in $\mathcal{O}_K/\mathfrak{m}_K$ defines a non-trivial class in $H^1(G_F, \overline{\rho}_1 \otimes \overline{\rho}_2^{-1})$. Furthermore, Ribet proves certain local conditions satisfied by this non-trivial class. It is elementary class field theory to prove that the existence of this class implies the converse to Herbrand's Theorem. We state Ribet's Lemma formally as follows.

Theorem 1.1 (Ribet's Lemma, Version 1) *Let* $\mathbf{T}$ *be a complete DVR, and let* $\mathfrak{m}$ *denote its maximal ideal. Let* $\mathcal{G}$ *be a compact group. Suppose we are given a continuous irreducible representation*

$$\rho \colon \mathcal{G} \longrightarrow \mathrm{GL}_2(K), \qquad K = \mathrm{Frac}(\mathbf{T}),$$

such that

$$char(\rho(g)) \equiv (x - \chi_1(g))(x - \chi_2(g)) \pmod{\mathfrak{m}} \tag{2}$$

for two characters $\chi_1, \chi_2 \colon \mathcal{G} \longrightarrow \mathbf{T}^*$. *Then there exists a non-zero cohomology class*

$$\kappa \in H^1(\mathcal{G}, \mathbf{T}/\mathfrak{m}(\chi_2^{-1}\chi_1)).$$

In more general settings, such as that employed by Mazur–Wiles [11] and Wiles [15] in their study of the Iwasawa Main Conjecture, the ring $\mathbf{T}$ cannot be assumed to be a DVR. It will usually be a complete local Noetherian ring, perhaps reduced. An example of such a ring that is not a DVR is

$$\mathbf{T} = \{(a, b) \in \mathbf{Z}_p \times \mathbf{Z}_p \colon a \equiv b \pmod{p}\}.$$

This example corresponds to two eigenforms with Hecke eigenvalues in $\mathbf{Z}_p$ that are congruent to each other modulo p. The total ring of fractions of $\mathbf{T}$ is $K = \mathrm{Frac}(\mathbf{T}) = \mathbf{Q}_p \times \mathbf{Q}_p$.

In addition, the ideal $I \subset \mathbf{T}$ modulo which the characteristic polynomial of ρ factors will in general not be the maximal ideal of $\mathbf{T}$. One then wants to construct a cohomology class that generates a module that (in a sense we will make precise in a moment) is "as large as" $\mathbf{T}/I$. The first version of Ribet's Lemma that applies in this case was proven by Mazur–Wiles and Wiles. Their work was groundbreaking and had a profound impact, leading to Wiles' theory of pseudorepresentations. In the statement below, a cohomology class κ valued in a $\mathbf{T}$-module M is called *surjective* if the image of every representative cocycle generates M as a $\mathbf{T}$-module.

Theorem 1.2 (Ribet's Lemma, Version 2) *Let* $\mathbf{T}$ *be a complete reduced local Noetherian ring, and let* $\mathfrak{m}$ *denote its maximal ideal. Let* $\mathcal{G}$ *be a topological group. Suppose we are given a representation*

$$\rho \colon \mathcal{G} \longrightarrow \mathrm{GL}_2(K), \qquad K = \mathrm{Frac}(\mathbf{T}),$$

such that for any projection onto a field $K \to k$, *the projection of* ρ *to a representation* $\mathcal{G} \to \mathrm{GL}_2(k)$ *is irreducible. Suppose that for each* $g \in \mathcal{G}$, *the characteristic polynomial* $char(\rho(g))$ *lies in* $\mathbf{T}[x]$ *and* $g \mapsto char(\rho(g))$ *is continuous. Let* $I \subset \mathbf{T}$ *be a proper ideal and suppose that for each* $g \in \mathcal{G}$, *we have*

$$char(\rho(g)) \equiv (x - \chi_1(g))(x - \chi_2(g)) \pmod{I} \tag{3}$$

for two characters $\chi_1, \chi_2 \colon \mathcal{G} \longrightarrow \mathbf{T}^*$ *satisfying* $\chi_1 \not\equiv \chi_2 \pmod{\mathfrak{m}}$. *Then there exists a fractional ideal of* $\mathbf{T}$, $B \subset K$, *and a surjective cohomology class*

$$\kappa \in H^1(G, B/IB(\chi_2^{-1}\chi_1)).$$

The assumption $\chi_1 \not\equiv \chi_2 \pmod{\mathfrak{m}}$ is essential in Wiles' approach to Theorem 1.2, and it has important consequences. In the Main Conjecture of Iwasawa Theory, one has χ_1 equal to the trivial character and χ_2 equal to a totally odd character of a totally real field. Let c denote complex conjugation. When $p \neq 2$, we have

$\chi_1(c) = 1 \not\equiv -1 = \chi_2(c) \pmod{p}$. However when $p = 2$ we may have $\chi_1 \equiv \chi_2 \pmod{p}$, and Theorem 1.2 cannot be applied. This is the main reason that Wiles sets $p \neq 2$ in his proof of the Main Conjecture.

The purpose of this exposition is to describe the main theorem of our paper [6], joint with Mahesh Kakde, Jesse Silliman, and Jiuya Wang, in which we establish a version of Ribet's Lemma that holds even if $\chi_1 \equiv \chi_2 \pmod{\mathfrak{m}}$. Here we describe the proof of the following result, a simplified form of Theorem 2.1 of *loc. cit.*

Theorem 1.3 (Ribet's Lemma, Version 3) *Let* $\mathbf{T}$ *be a complete reduced local Noetherian ring, and let* $\mathfrak{m}$ *denote its maximal ideal. Let* $\mathcal{G}$ *be a topological group. Suppose we are given a representation*

$$\rho\colon \mathcal{G} \longrightarrow \mathrm{GL}_2(K), \qquad K = \mathrm{Frac}(\mathbf{T}),$$

such that for any projection onto a field $K \to k$*, the projection of* ρ *to a representation* $\mathcal{G} \to \mathrm{GL}_2(k)$ *is irreducible. Suppose that for each* $g \in \mathcal{G}$*, the characteristic polynomial* $char(\rho(g))$ *lies in* $\mathbf{T}[x]$ *and* $g \mapsto char(\rho(g))$ *is continuous. Let* $I \subset \mathbf{T}$ *be a proper ideal and suppose that for each* $g \in \mathcal{G}$*, we have*

$$char(\rho(g)) \equiv (x - \chi_1(g))(x - \chi_2(g)) \pmod{I} \tag{4}$$

for two characters $\chi_1, \chi_2\colon \mathcal{G} \longrightarrow \mathbf{T}^*$*. Finally, suppose that the* $\mathbf{T}$*-submodule of* $M_2(K)$ *generated by the elements* $\rho(g) - \chi_1(g)$ *for* $g \in \mathcal{G}$ *is finitely generated. Then there exists a finitely generated* $\mathbf{T}$*-module* M *and a surjective cohomology class*

$$\kappa \in H^1(G, M(\chi_2^{-1}\chi_1))$$

such that

$$\mathrm{Fitt}_{\mathbf{T}}(M) \subset I.$$

The ideal $\mathrm{Fitt}_{\mathbf{T}}(M)$ is the 0th Fitting ideal of the module M, which will be defined in Sect. 4.1. Intuitively, the inclusion $\mathrm{Fitt}_{\mathbf{T}}(M) \subset I$ says that M is "large."

In [5], we prove the Brumer–Stark conjecture at $p = 2$ using a suitably generalized version of Theorem 1.3. Previously, we proved the conjecture over $\mathbf{Z}[1/2]$ in [4], with the prime $p = 2$ being avoided for reasons of residual distinguishability. We hope that further strengthenings of Theorem 1.3 (for example, to groups other than GL_2) could have other arithmetic applications.

In this paper, we simplify notation by setting χ_1 and χ_2 to be the trivial character; the case of general χ_1, χ_2 requires no extra ideas, but the notation is heavier. A more significant change in this paper relative to [6] is that all the versions of Ribet's method stated above do not include local conditions on the cohomology classes constructed. In arithmetic applications, ranging from Ribet's original proof of the converse to Herbrand's theorem to our proof of the Brumer–Stark conjecture, local conditions are always necessary. To prove the local properties we need in the Brumer–Stark context, the argument presented here is generalized in [6] using the Buchsbaum–

Rim resolution of determinantal ideals; in this paper, the simple Koszul complex suffices. In *loc. cit.* we also give a mild generalization to certain non-reduced Hecke algebras **T**.

It is an honor to contribute this article to the 60th birthday volume for Massimo Bertolini. Massimo and I collaborated (with Henri Darmon) when I was a Ph.D. student working on Stark–Heegner points, and I thank him for playing a critical role in my development as a young mathematician. The present article grew from the lectures I delivered at a summer school for graduate students during the first week of Massimo's 60th Birthday conference.

2 The DVR Case

In this section we prove Theorem 1.1, Ribet's original setting. Let **T** be a complete DVR and let $\mathfrak{m}$ denote its maximal ideal. Let $\mathcal{G}$ be a compact group. We are given a continuous irreducible representation

$$\rho\colon \mathcal{G} \longrightarrow \mathrm{GL}_2(K), \qquad K = \mathrm{Frac}(\mathbf{T}),$$

such that

$$\mathrm{char}(\rho(g)) \equiv (x - \chi_1(g))(x - \chi_2(g)) \pmod{\mathfrak{m}} \tag{5}$$

for two characters $\chi_1, \chi_2\colon \mathcal{G} \longrightarrow \mathbf{T}^*$.

Lemma 2.1 *The representation ρ may be conjugated to have image contained in* $\mathrm{GL}_2(\mathbf{T})$*, and such that the reduction $\overline{\rho}$ has the shape*

$$\overline{\rho} = \begin{pmatrix} \overline{\chi}_1 & * \\ 0 & \overline{\chi}_2 \end{pmatrix}.$$

Proof The maximal compact subgroups of $\mathrm{GL}_2(K)$ are precisely the conjugates of $\mathrm{GL}_2(\mathbf{T})$. Since $\mathcal{G}$ is compact and ρ is continuous, it follows ρ that may be conjugated to have image contained in $\mathrm{GL}_2(\mathbf{T})$. Now (5) states that

$$\mathrm{char}(\overline{\rho}(g)) = (x - \overline{\chi}_1(g))(x - \overline{\chi}_2(g)) \text{ in } \mathbf{T}/\mathfrak{m}. \tag{6}$$

The Brauer–Nesbitt Theorem [3] states that if two representations over a field have the same characteristic polynomial, then their semisimplifications are isomorphic. Therefore, (6) implies that $\overline{\rho}^{\mathrm{ss}} \cong \overline{\chi}_1 \oplus \overline{\chi}_2$. Hence we have either

$$\overline{\rho} \cong \begin{pmatrix} \overline{\chi}_1 & * \\ 0 & \overline{\chi}_2 \end{pmatrix} \quad \text{or} \quad \overline{\rho} \cong \begin{pmatrix} \overline{\chi}_2 & * \\ 0 & \overline{\chi}_1 \end{pmatrix}. \tag{7}$$

It is a pleasant exercise to prove (7) directly from (6) without reference to the full strength of the Brauer–Nesbitt Theorem. Now, suppose we are in the second case.

Then we can conjugate ρ by the matrix $\begin{pmatrix} \pi & 0 \\ 0 & 1 \end{pmatrix}$, where π is a uniformizer of $\mathbf{T}$, to obtain

$$\overline{\rho} \cong \begin{pmatrix} \overline{\chi}_2 & 0 \\ * & \overline{\chi}_1 \end{pmatrix} \cong \begin{pmatrix} \overline{\chi}_1 & * \\ 0 & \overline{\chi}_2 \end{pmatrix}$$

as desired. □

Lemma 2.2 *Suppose that the representation ρ has been conjugated so that its image lands in* $\mathrm{GL}_2(\mathbf{T})$ *and*

$$\overline{\rho} = \begin{pmatrix} \overline{\chi}_1 & \overline{b} \\ 0 & \overline{\chi}_2 \end{pmatrix}.$$

Then the function $\kappa(\sigma) = \overline{\chi}_2^{-1}(\sigma)\overline{b}(\sigma)$ is a 1-cocycle in $Z^1(\mathcal{G}, \mathbf{T}/\mathfrak{m}(\chi_2^{-1}\chi_1))$.

Proof Since $\overline{\rho}$ is a matrix representation, we have

$$\overline{b}(\sigma\tau) = \overline{\chi_1}(\sigma)\overline{b}(\tau) + \overline{b}(\sigma)\overline{\chi_2}(\tau).$$

Multiplying by $\overline{\chi}_2^{-1}(\sigma\tau)$ gives the desired 1-cocycle formula for $\kappa = \overline{\chi}_2^{-1}\overline{b}$. □

What remains to prove Theorem 1.1 is to show that after conjugating ρ further, we can arrange for the cohomology class represented by κ to be non-trivial.

Proof of Theorem 1.1 In his 2008 lectures from the Clay Summer School in Hawaii, Bellaïche gives a beautiful and conceptual proof of the fact that ρ can be chosen so that the cohomology class represented by κ is non-trivial [1]*Proposition 1.4. He attributes this proof to Serre. Here, we take the more computational approach applied by Ribet.

Since we will be applying a recursive process, let $\rho_1 = \rho$ and write

$$\rho_1(\sigma) = \begin{pmatrix} a_1(\sigma) & b_1(\sigma) \\ c_1(\sigma) & d_1(\sigma \end{pmatrix} \in \mathrm{GL}_2(\mathbf{T}).$$

If π denotes a uniformizer of $K = \mathrm{Frac}(\mathbf{T})$, then we have

$$a_1(\sigma) \equiv \chi_1(\sigma) \pmod{\pi}, \tag{8}$$
$$c_1(\sigma) \equiv 0 \pmod{\pi}, \tag{9}$$
$$d_1(\sigma) \equiv \chi_2(\sigma) \pmod{\pi} \tag{10}$$

for all $\sigma \in \mathcal{G}$. Denote the cocycle constructed in Lemma 2.2 by κ_1.

Suppose that κ_1 represents a trivial cohomology class; then there exists $x_1 \in \mathbf{T}$ such that

$$\kappa_1(\sigma) \equiv (\chi_2^{-1}\chi_1(\sigma) - 1)x_1 \pmod{\pi}$$

for all $\sigma \in \mathcal{G}$, or equivalently,

$$b_1(\sigma) \equiv (\chi_1(\sigma) - \chi_2(\sigma))x_1 \pmod{\pi}. \tag{11}$$

Conjugating the representation ρ_1, we define

$$\rho_2(\sigma) = \begin{pmatrix} a_2(\sigma) & b_2(\sigma) \\ c_2(\sigma) & d_2(\sigma) \end{pmatrix} = \begin{pmatrix} 1 & x_1 \\ 0 & \pi \end{pmatrix} \rho_1(\sigma) \begin{pmatrix} 1 & x_1 \\ 0 & \pi \end{pmatrix}^{-1}.$$

Using the congruences (8)–(11), we find that $\rho_2(\sigma) \in \mathrm{GL}_2(\mathbf{T})$ and furthermore that

$$\begin{aligned} a_2(\sigma) &\equiv \chi_1(\sigma) && \pmod{\pi}, \\ c_2(\sigma) &\equiv 0 && \pmod{\pi^2}, \\ d_2(\sigma) &\equiv \chi_2(\sigma) && \pmod{\pi}. \end{aligned}$$

We are therefore once again in the setting of Lemma 2.2 and obtain a cocycle

$$\kappa_2(\sigma) = \overline{\chi_2}^{-1}(\sigma)\overline{b_2}(\sigma) \in Z^1(\mathcal{G}, \mathbf{T}/\mathfrak{m}(\chi_2^{-1}\chi_1)).$$

If κ_2 represents a nontrivial cohomology class, we are done. If not, we may repeat this process and obtain another representation ρ_3, where now $\pi^3 \mid c_3(\sigma)$. This process continues.

If at any stage, we obtain a cocycle κ_i that represents a non-trivial class, then we are done. If the process continues forever, then one checks that by defining $x = x_1 + \pi x_2 + \pi^2 x_3 + \cdots$, conjugating the original representation ρ_1 by $\begin{pmatrix} 1 & x \\ 0 & 1 \end{pmatrix}$ leaves a representation over K with a 0 in the upper right-hand corner. This contradicts the irreducibility of ρ_1. □

3 The Residually Distinguishable Case

In this section, we prove Theorem 1.2. Our complete local ring $\mathbf{T}$ is no longer assumed to be a DVR, but we grant ourselves the assumption $\chi_1 \not\equiv \chi_2 \pmod{\mathfrak{m}}$. Fix $\tau \in \mathcal{G}$ such that $\chi_1(\tau) \not\equiv \chi_2(\tau) \pmod{\mathfrak{m}}$. By Hensel's Lemma, the congruence (3) implies that $\rho(\tau)$ has two distinct eigenvalues $\lambda_1, \lambda_2 \in \mathbf{T}$ such that

$$\lambda_i \equiv \chi_i(\tau) \pmod{I} \tag{12}$$

for $i = 1, 2$. We choose a basis for ρ over $K = \mathrm{Frac}(\mathbf{T})$ such that

$$\rho(\tau) = \begin{pmatrix} \lambda_1 & 0 \\ 0 & \lambda_2 \end{pmatrix}.$$

For $\sigma \in \mathcal{G}$, write

$$\rho(\sigma) = \begin{pmatrix} a(\sigma) & b(\sigma) \\ c(\sigma) & d(\sigma) \end{pmatrix}.$$

Note that unlike the DVR case, we cannot ensure that ρ takes values in $\mathrm{GL}_2(\mathbf{T})$, only $\mathrm{GL}_2(K)$. Nevertheless, we have the following.

Lemma 3.1 *We have $a(\sigma), d(\sigma) \in \mathbf{T}$ for all $\sigma \in \mathcal{G}$, and furthermore*

$$a(\sigma) \equiv \chi_1(\sigma) \pmod{I},$$
$$d(\sigma) \equiv \chi_2(\sigma) \pmod{I}.$$

Proof The congruence (3) implies that

$$a(\sigma) + d(\sigma) \equiv \chi_1(\sigma) + \chi_2(\sigma) \pmod{I}. \tag{13}$$

Applying this with σ replaced by $\sigma\tau$, and noting that $a(\sigma\tau) = a(\sigma)\lambda_1$ and $d(\sigma\tau) = d(\sigma)\lambda_2$ by our choice of basis, we obtain

$$a(\sigma)\lambda_1 + d(\sigma)\lambda_2 \equiv \chi_1(\sigma)\chi_1(\tau) + \chi_2(\sigma)\chi_2(\tau). \pmod{I} \tag{14}$$

Solving the congruences (13) and (14) using (12), we find that $a(\sigma), d(\sigma) \in \mathbf{T}$ and furthermore $a(\sigma) \equiv \chi_1(\sigma) \pmod{I}$ and $d(\sigma) \equiv \chi_2(\sigma) \pmod{I}$ as desired. Note that this uses the fact that $\chi_1(\tau) \not\equiv \chi_2(\tau) \pmod{\mathfrak{m}}$ in a crucial way. □

We now let B denote the $\mathbf{T}$-submodule of K spanned by the $b(\sigma)$ for $\sigma \in \mathcal{G}$. Note that the irreducibility assumption on ρ implies that B is a fractional ideal of $\mathbf{T}$. Indeed, if we write K as a product of fields, then the projection of ρ on to each field factor is irreducible, hence the projection of B onto each field factor is nonzero.

In view of the congruences of Lemma 3.1, the equation

$$b(\sigma\sigma') = a(\sigma)b(\sigma') + b(\sigma)d(\sigma')$$

yields

$$\overline{b}(\sigma\sigma') \equiv \chi_1(\sigma)\overline{b}(\sigma') + \chi_2(\sigma')\overline{b}(\sigma) \quad \text{in } B/IB.$$

As in Lemma 2.2, multiplying by $\chi_2^{-1}(\sigma\sigma')$ shows that

$$\kappa(\sigma) = \chi_2^{-1}(\sigma)\overline{b}(\sigma) \in B/IB$$

defines an element of $Z^1(\mathcal{G}, B/IB(\chi_2^{-1}\chi_1))$.

It remains to prove that the cohomology class represented by κ is surjective. Therefore let

$$\kappa'(\sigma) = \kappa(\sigma) + (\chi_2^{-1}\chi_1(\sigma) - 1)x \tag{15}$$

be a cohomologous cocycle, where $x \in B/IB$. Let B' denote the $\mathbf{T}$-submodule of B/IB generated by the values of κ'. Applying (15) to τ and recalling that $\kappa(\tau) = 0$, we see that

$$\kappa'(\tau) = (\chi_2^{-1}\chi_1(\tau) - 1)x \in B'.$$

Since the item in parentheses in a unit in $\mathbf{T}$ by assumption, we find that $x \in B'$. Going back to (15), it follows that $\kappa(\sigma) \in B'$ for all $\sigma \in \mathcal{G}$. But it is clear from the definitions of B and κ that κ generates B/IB, whence $B' = B/IB$ as desired. This completes the proof of Theorem 1.2.

4 The Residually Indistinguishable Case

In the remainder of the paper we describe the proof of Theorem 1.3. Before explaining the definition of the Fitting ideal that appears in the statement of the theorem, we remark that the methods of the previous sections are not directly applicable—in the DVR case, manipulations with the uniformizer π were essential, and in the residually distinguishable case, the basis for ρ afforded by the special element τ was critical. Indeed, in the residually indistinguishable case we do not know how to show that $a(\sigma), d(\sigma) \in \mathbf{T}$ (let alone that the congruences of Lemma 3.1 hold) in *any* basis. A separate construction and proof is required.

As mentioned in the introduction, we will assume for the rest of the paper that χ_1 and χ_2 are trivial; this offers notational simplifications, but no significant changes to the argument.

4.1 Fitting Ideal

Let R be a commutative ring. Let M be a finitely presented R-module. This means that there is a short exact sequence

$$R^n \xrightarrow{f} R^m \longrightarrow M \longrightarrow 0.$$

Definition 4.1 The 0th Fitting ideal of M over R, which we denote $\mathrm{Fitt}_R(M)$, is the ideal of R generated by all $m \times m$ minors of the matrix representing the linear map f. By convention, if $n < m$, then $\mathrm{Fitt}_R(M) = 0$.

We leave the proof of the following basic facts about Fitting ideals as an exercise for the reader (alternatively, she may consult [12]).

Proposition 4.2 *We have the following.*

- *The Fitting ideal* $\mathrm{Fitt}_R(M)$ *depends only on* M *up to* R*-module isomorphism, and not on the particular presentation taken.*
- *If* $I \subset R$ *is a finitely generated ideal, then* $\mathrm{Fitt}_R(R/I) = I$.
- *The Fitting ideal of* M *is contained in the annihilator of* M: $\mathrm{Fitt}_R(M) \subset \mathrm{Ann}_R(M)$.
- *If* $R = \mathbf{Z}$ *and* M *is a finitely generated abelian group, then* $\mathrm{Fitt}_{\mathbf{Z}}(M) = 0$ *if* M *is infinite and* $\mathrm{Fitt}_{\mathbf{Z}}(M) = (\#M)$*, the ideal generated by the size of* M*, if* M *is finite.*
- *If* $M \twoheadrightarrow M'$ *is a surjection of finitely presented* R*-modules, then* $\mathrm{Fitt}_R(M') \supset \mathrm{Fitt}_R(M)$.
- *(Base Change) If* S *is an* R*-algebra and* M *is a finitely presented* R*-module, then* $\mathrm{Fitt}_S(M \otimes_R S) = \mathrm{Fitt}_R(M) \cdot S$.

Corollary 4.3 *Let* $\mathbf{T}$ *and* B *be as in Theorem 1.2. Then* $\mathrm{Fitt}_{\mathbf{T}}(B/IB) \subset I$.

Proof Since B is a fractional ideal of $\mathbf{T}$, it is a faithful $\mathbf{T}$-module, i.e. $\mathrm{Ann}_{\mathbf{T}}(B) = 0$. It follows that $\mathrm{Fitt}_{\mathbf{T}}(B) = 0$. Therefore $\mathrm{Fitt}_{\mathbf{T}/I}(B/IB) = 0$, whence $\mathrm{Fitt}_{\mathbf{T}}(B/IB) \subset I$. □

Motivated by this corollary, one of the insights of Theorem 1.3 is its statement—in the residually indistinguishable case, instead of attempting to necessarily construct a $\mathbf{T}$-module of the form B/IB for a faithful $\mathbf{T}$-module B, one should just attempt to construct some module M such that $\mathrm{Fitt}_{\mathbf{T}}(M) \subset I$.

4.2 Construction of M

To prove Theorem 1.3, we must construct, under the assumptions of the theorem, a finitely generated $\mathbf{T}$-module M such that $\mathrm{Fitt}_{\mathbf{T}}(M) \subset I$ and a surjective cohomology class $\kappa \in H^1(\mathcal{G}, M)$. Since we have specialized to $\chi_1 = \chi_2 = 1$, the $\mathcal{G}$-action on M is trivial, whence $H^1(\mathcal{G}, M)$ is the group of continuous homomorphisms $\mathcal{G} \to M$. Since M is abelian, such a homomorphism necessarily factors through the abelianization $\mathcal{G}^{\mathrm{ab}}$ and thereby induces a $\mathbf{T}$-module map

$$\mathcal{G}^{\mathrm{ab}} \otimes \mathbf{T} \longrightarrow M. \tag{16}$$

The surjectivity of κ is simply the statement that the $\mathbf{T}$-module homomorphism (16) is surjective.

These considerations lead to a very natural construction of the module M. Let Δ denote the augmentation ideal of $\mathcal{G}$ over $\mathbf{T}$, i.e. the kernel of the $\mathbf{T}$-algebra homomorphism

$$\mathbf{T}[\mathcal{G}] \longrightarrow \mathbf{T}, \quad \sum a_g[g] \mapsto \sum a_g. \tag{17}$$

It is then well-known that we have a $\mathbf{T}$-module isomorphism

$$\Delta/\Delta^2 \cong \mathcal{G}^{\mathrm{ab}} \otimes \mathbf{T}, \qquad \sum a_g[g] \mapsto \sum g \otimes a_g.$$

In order to apply the assumptions of Theorem 1.3, we must invoke the group representation

$$\rho\colon \mathcal{G} \to \mathrm{GL}_2(K).$$

Note that ρ can be extended to a $\mathbf{T}$-algebra homomorphism (also denoted ρ)

$$\rho\colon \mathbf{T}[\mathcal{G}] \longrightarrow M_2(K).$$

We then define

$$M = \rho(\Delta)/\rho(\Delta^2),$$

so there is a canonical surjection

$$\mathcal{G}^{\mathrm{ab}} \otimes \mathbf{T} \cong \Delta/\Delta^2 \overset{\rho}{\twoheadrightarrow} M.$$

As explained above, this homomorphism represents a surjective cohomology class

$$\kappa \in H^1(\mathcal{G}, M).$$

It remains to prove that $\mathrm{Fitt}_{\mathbf{T}}(M) \subset I$, and this will take up the remainder of the paper.

4.3 Explication of Fitting Ideal

By assumptions, we are given that $\rho(\Delta)$ is finitely generated as a $\mathbf{T}$-module. Let $\rho_1, \ldots, \rho_r$ denote a set of $\mathbf{T}$-module generators for $\rho(\Delta)$, where $\rho_i = \rho(g_i - 1)$ for elements $g_i \in \mathcal{G}$. The images of the ρ_i in M are $\mathbf{T}$-module generators, and there are two types of relations for these generators in M:

- We may have relations

$$\sum_{i=1}^{r} \epsilon_i \rho_i = 0 \text{ in } M_2(K) \tag{18}$$

  for constants $\epsilon_i \in \mathbf{T}$. We call these relations of ϵ-type.
- For each pair $1 \leq i, j \leq r$, we may write

$$\rho_i \rho_j = \sum_{k=1}^{r} \delta_{ijk} \rho_k \tag{19}$$

for constants $\delta_{ijk} \in \mathbf{T}$. These expressions vanish in M; we call these relations of δ-type.

The Fitting ideal of M is the ideal generated by all determinants $\det(D)$ where D is an $r \times r$ matrix whose rows have the form $(\epsilon_1, \ldots, \epsilon_r)$ for relations of ϵ-type or $(\delta_{ij1}, \delta_{ij2}, \ldots, \delta_{ijr})$ for relations of δ-type. We need to show that $\det(D) \in I$ for each such matrix D.

4.4 Traces and Determinants

In order to prove that $\det(D) \in I$, we need to generate expressions involving the ρ_i that are known to lie in I. This is given by the following.

Lemma 4.4 *For each* $A \in \rho(\Delta)$, *we have* $\mathrm{tr}(A) \in I$ *and* $\det(A) \in I$.

Proof Let $\chi : \mathbf{T}[\mathcal{G}] \longrightarrow \mathbf{T}$ denote the augmentation map defined in (17), i.e. the trivial character of $\mathcal{G}$ extended to a $\mathbf{T}$-algebra homomorphism on $\mathbf{T}[\mathcal{G}]$. The congruence (4) implies that

$$\mathrm{tr}(\rho(g)) \equiv 2\chi(g) \pmod{I}, \qquad \det(\rho(g)) \equiv \chi(g) \pmod{I}$$

for $g \in \mathcal{G}$. A simple argument shows that these congruences extend to all $g \in \mathbf{T}[\mathcal{G}]$ (see [14]*Lemma 3.1). By definition, $\chi(g) = 0$ for $g \in \Delta$. The result follows. □

4.5 An Altered Matrix

Write

$$\rho_i = \begin{pmatrix} a_i & b_i \\ c_i & d_i \end{pmatrix}.$$

Given an $r \times r$ matrix D of ϵ-type and δ-type relations as in Sect. 4.3, we define an associated matrix D' obtained by altering the rows of D as follows.

- No change for ϵ-type rows.
- For rows of δ-type, replace δ_{ijj} by $\delta_{ijj} - a_i$, replace δ_{iji} by $\delta_{iji} - d_j$, and leave the other δ_{ijk} unchanged.

Lemma 4.5 *We have* $\det(D') = 0$.

Proof Let

$$w = (b_1, b_2, \ldots, b_r)^T \qquad \text{(column vector)}.$$

We claim that $D'w = 0$. For a row of ϵ-type, the corresponding component of $D'w$ is $\sum_{i=1}^{r} \epsilon_i b_i$. This is the upper right coefficient of the matrix $\sum_{i=1}^{r} \epsilon_i \rho_i$, and hence vanishes by the definition of the ϵ_i (see (18)). Similarly for a row of δ-type, the corresponding component of $D'w$ is

$$\left(\sum_{k=1}^{r} \delta_{ijk} b_k \right) - (a_i b_j + b_i d_j) = 0. \tag{20}$$

The quantity (20) vanishes because each expression in parentheses equals the upper right coefficient of $\rho_i \rho_j$; this follows on the left by the definition (19) of the δ_{ijk}, and on the right by the definition of matrix multiplication.

The ring $K = \mathrm{Frac}(\mathbf{T})$ is a product of fields. To prove $\det(D') = 0$, it suffices to prove this on each field factor of K. Now, in each field factor of K, the projection of some b_i must be nonzero; otherwise the projection of ρ to that field factor would be lower triangular and hence reducible, contrary to assumption. Therefore the equation $D'w = 0$ implies $\det(D') = 0$ as desired. □

Our goal is to show that $\det(D) \in I$, and we have shown that $\det(D') = 0$. It therefore suffices to show that $\det(D') - \det(D) \in I$. In other words, the alterations used to go from D to D' are small enough to leave the determinant unchanged modulo I. Let us motivate our strategy to prove this with an example.

4.6 An Example

Suppose $r = 2$. We consider a matrix with only rows of δ-type, namely

$$D = \begin{pmatrix} \delta_{121} & \delta_{122} \\ \delta_{211} & \delta_{212} \end{pmatrix}, \qquad \text{whence} \qquad D' = \begin{pmatrix} \delta_{121} - d_2 & \delta_{122} - a_1 \\ \delta_{211} - a_2 & \delta_{212} - d_1 \end{pmatrix}.$$

As shorthand, write $t_i = a_i + d_i$ for the trace of ρ_i. We write $d_{12} = c_1 b_2 + d_1 d_2$ for the "d"-component of $\rho_1 \rho_2$, and $t_{12} = a_{12} + d_{12}$ for the trace of $\rho_1 \rho_2$. By multilinearity of the determinant, we have

$$\det(D') - \det(D) = -\det \begin{pmatrix} d_2 & a_1 \\ \delta_{211} & \delta_{212} \end{pmatrix} - \det \begin{pmatrix} \delta_{121} & \delta_{122} \\ a_2 & d_1 \end{pmatrix} + \det \begin{pmatrix} d_2 & a_1 \\ a_2 & d_1 \end{pmatrix}. \tag{21}$$

We evaluate the determinants on the right using the substitution $t_i = a_i + d_i$ where convenient. Then

$$\det\begin{pmatrix} d_2 & a_1 \\ \delta_{211} & \delta_{212} \end{pmatrix} = -t_1\delta_{211} + (d_1\delta_{211} + d_2\delta_{212})$$

$$= -t_1\delta_{211} + d_{21} \tag{22}$$

$$= -t_1\delta_{211} + (c_2b_1 + d_1d_2). \tag{23}$$

Note that equation (22) uses the definition of the δ_{ijk}. Similarly

$$\det\begin{pmatrix} \delta_{121} & \delta_{122} \\ a_2 & d_1 \end{pmatrix} = -t_2\delta_{122} + (d_1\delta_{121} + d_2\delta_{122})$$

$$= -t_2\delta_{122} + d_{12}$$

$$= -t_2\delta_{122} + (c_1b_2 + d_1d_2). \tag{24}$$

Combining (21), (23), and (24), we obtain

$$\begin{aligned} \det(D') - \det(D) &= t_1\delta_{211} - (c_2b_1 + d_2d_1) + \\ &+ t_2\delta_{122} - (c_1b_2 + d_1d_2) + \\ &+ (d_1d_2 - a_1a_2) \\ &= t_1\delta_{211} + t_2\delta_{122} - t_{12}. \end{aligned} \tag{25}$$

By Lemma 4.4, the expression (25) lies in I as desired.

5 Formal Variables

We do not know how to establish *explicit* formulae such as (25) to prove that

$$\det(D') - \det(D) \in I$$

in general. Instead, we will prove *abstractly* the existence of polynomial identities such as (25) that express the difference $\det(D') - \det(D)$ in terms of traces and determinants of elements of $\rho(\Delta)$.

For this, it is convenient to shift our perspective from working with the ring $\mathbf{T}$ to working with formal polynomial rings. We will define a polynomial algebra R and a specialization homomorphism $\pi : R \longrightarrow K = \mathrm{Frac}(\mathbf{T})$. We will show that the identities we need hold in $R/\ker\pi$, so they apply in $\mathbf{T}$ as well by applying π. The advantage of working in R is that we can identify the subring generated by traces and determinants of matrices mapping to $\rho(\Delta)$ under π as the subspace of invariants under a certain group action, and use cohomological considerations to show that our element of interest lies in this subspace. We now describe this in greater detail.

5.1 The Ring R

Define

$$R_0 = \mathbf{Z}[\boldsymbol{\epsilon}_i, \boldsymbol{\delta}_{ijk}],$$

the commutative polynomial ring in r^2 free variables enumerated by the entries of the matrix D. Define

$$R = R_0[\boldsymbol{a}_i, \boldsymbol{b}_i, \boldsymbol{c}_i, \boldsymbol{d}_i]_{i=1}^r,$$

a commutative polynomial ring in $r^2 + 4r$ free variables. Define a ring homomorphism $R \longrightarrow K$ in the natural way indicated by our variable names, i.e.

$$\pi(\boldsymbol{\epsilon}_i) = \epsilon_i, \quad \pi(\boldsymbol{\delta}_{ijk}) = \delta_{ijk}, \quad \pi(\boldsymbol{a}_i) = a_i, \ldots, \quad \pi(\boldsymbol{d}_i) = d_i.$$

Note that $\pi(R_0) \subset \mathbf{T}$.

Finally, let $\boldsymbol{D}, \boldsymbol{D}' \in M_r(R)$ denote the natural matrices whose images under π are equal D, D', respectively, i.e. they are defined by making each entry bold. Our goal is to prove that

$$\pi(\det(\boldsymbol{D}') - \det(\boldsymbol{D})) = \det(D') - \det(D) \in I.$$

5.2 Relation Ideal

We now define the polynomial relations that allow us to reduce $\det(\boldsymbol{D}') - \det(\boldsymbol{D})$ to an expression involving traces and determinants, as in the example of Sect. 4.6. Define

$$\boldsymbol{\rho}_i = \begin{pmatrix} \boldsymbol{a}_i & \boldsymbol{b}_i \\ \boldsymbol{c}_i & \boldsymbol{d}_i \end{pmatrix} \in M_2(R).$$

Let $J \subset R$ be the ideal generated by the following:

- The 4 coefficients of

$$\sum_{i=1}^{r} \boldsymbol{\epsilon}_i \boldsymbol{\rho}_i \tag{26}$$

  for each row of ϵ-type in D.
- The 4 coefficients of

$$\boldsymbol{\rho}_i \boldsymbol{\rho}_j - \sum_{k=1}^{r} \boldsymbol{\delta}_{ijk} \boldsymbol{\rho}_k \tag{27}$$

  for each row of δ-type in D.

It is clear that $J \subset \ker(\pi)$.

5.3 Subring of Traces and Determinants

Let $A \subset R$ denote the R_0-subalgebra generated by the traces and determinants of all matrices in the noncommutative $\mathbf{Z}$-algebra generated by the matrices $\boldsymbol{\rho}_i$. Denote by $\overline{A}$ the image of A in R/J. We will show that in order to deduce our desired result $\pi(\det(\boldsymbol{D}') - \det(\boldsymbol{D})) \in I$, it suffices to prove that the image of $\det(\boldsymbol{D}') - \det(\boldsymbol{D})$ in R/J lies in $\overline{A}$. For this, it is important that $\det(\boldsymbol{D}') - \det(\boldsymbol{D})$ lies in the following ideal of R:

$$I_R = \langle \boldsymbol{a}_i, \boldsymbol{b}_i, \boldsymbol{c}_i, \boldsymbol{d}_i \rangle.$$

Lemma 5.1 *We have* $\det(\boldsymbol{D}') - \det(\boldsymbol{D}) \in I_R$.

Proof This follows immediately from multilinearity of the determinant since every entry of $\boldsymbol{D}' - \boldsymbol{D}$ lies in I_R. □

Lemma 5.2 *We have* $\pi(A \cap I_R) \subset I$.

Proof Let $f \in \mathbf{Z}\langle X_1, \ldots, X_r \rangle$ be an element of the free polynomial algebra in r noncommuting variables. If f has no constant term, then

$$\pi(f(\boldsymbol{\rho}_1, \ldots, \boldsymbol{\rho}_r)) = f(\rho_1, \ldots, \rho_r) \in \rho(\Delta),$$

and hence the trace and determinant of this element lies in I by Lemma 4.4. The element $f(\boldsymbol{\rho}_1, \ldots, \boldsymbol{\rho}_r)$ clearly has coefficients lying in I_R. From these considerations, to prove the lemma it suffices to prove that $\pi(R_0 \cap I_R) \subset I$. In fact, it is clear that $R_0 \cap I_R = 0$, as

$$R/I_R \cong R_0,$$

with $\boldsymbol{a}_i, \boldsymbol{b}_i, \boldsymbol{c}_i, \boldsymbol{d}_i \mapsto 0$. This concludes the proof. □

We summarize the result of this section: in order to prove the desired result

$$\det(D') - \det(D) \in I, \tag{28}$$

it suffices to prove that the image of $\det(\boldsymbol{D}') - \det(\boldsymbol{D})$ in R/J lies in the subring $\overline{A}$, the image of A in R/J. Indeed, if this condition holds, then then

$$\det(\boldsymbol{D}') - \det(\boldsymbol{D}) = a + j \tag{29}$$

for some $a \in A$, $j \in J$. Since $J \subset I_R$, we have $a \in I_R$ by Lemma 5.1 and hence $\pi(a) \in I$ by Lemma 5.2. Since $\pi(j) = 0$, the desired result (28) follows from applying π to (29).

6 Matrix Invariant Theory and Rational Cohomology

The advantage of passing to the ring of formal variables R (rather than working directly with $\mathbf{T}$ and K) is that we may identify the subring $A \subset R$ as the subspace invariant under a group action, rather than needing to write down explicit formulae. To this end, we have the following important classical result, called the fundamental theorem of matrix invariant theory.

Theorem 6.1 *Endow the ring $\mathbf{Z}[\boldsymbol{a}_i, \boldsymbol{b}_i, \boldsymbol{c}_i, \boldsymbol{d}_i]_{i=1}^r$, with an action of $\mathrm{GL}_2(\mathbf{Z})$ defined by simultaneous conjugation on the matrices $\boldsymbol{\rho}_i = \begin{pmatrix} \boldsymbol{a}_i & \boldsymbol{b}_i \\ \boldsymbol{c}_i & \boldsymbol{d}_i \end{pmatrix}$. The subring of invariant elements is generated over $\mathbf{Z}$ by the traces and determinants of all matrices in the noncommutative algebra generated by the $\boldsymbol{\rho}_i$.*

With notation as in Sect. 5.1, we have the following corollary.

Corollary 6.2 *Endow the ring R with an action of $\mathrm{GL}_2(\mathbf{Z})$ defined by simultaneous conjugation on the matrices $\rho_i = \begin{pmatrix} \boldsymbol{a}_i & \boldsymbol{b}_i \\ \boldsymbol{c}_i & \boldsymbol{d}_i \end{pmatrix}$. The subring of invariant elements is equal to A.*

Proof We may write $R = \mathbf{Z}[\boldsymbol{a}_i, \boldsymbol{b}_i, \boldsymbol{c}_i, \boldsymbol{d}_i]_{i=1}^r \otimes R_0$, where $\mathrm{GL}_2(\mathbf{Z})$ acts trivially on R_0. Since R_0 is $\mathbf{Z}$-flat, the result follows immediately from Theorem 6.1. □

In our computations, it will be convenient to work with the Borel subgroup of GL_2 consisting of lower triangular matrices. We would like restriction to the Borel to induce an isomorphism on cohomology. This is not true in general for ordinary group cohomology, so for this reason, we must work with the cohomology of algebraic groups called *rational cohomology*. Here "rational" refers to actions defined by rational functions, rather than the rational numbers; throughout, we work integrally over $\mathbf{Z}$.

6.1 Rational Cohomology

For simplicity, we will describe algebraic groups and their cohomology through their "functor of points" rather than via group schemes. We therefore view the algebraic group $\mathbf{G} = \mathrm{GL}_2$ as the functor that associates to any commutative ring S the group $\mathrm{GL}_2(S)$.

Definition 6.3 A *rational* $\mathbf{G}$-*module* is a $\mathbf{Z}$-module V endowed with a functorial action, for every ring S, of the group $\mathrm{GL}_2(S)$ on $V \otimes S$, such that the action of a matrix $\begin{pmatrix} a & b \\ c & d \end{pmatrix}$ is given by rational functions in the variables a, b, c, d. We say that V is a rational $\mathbf{G}$-module over a commutative ring R_0 if V has a structure of R_0-module that commutes with the $\mathbf{G}$-action.

Example 6.4 We denote by $\mathcal{A} = \mathbf{Z}A \oplus \mathbf{Z}B \oplus \mathbf{Z}C \oplus \mathbf{Z}D$ the $\mathbf{G}$-module given by the adjoint representation, i.e. if $g = \begin{pmatrix} a & b \\ c & d \end{pmatrix}$, then

$$\begin{pmatrix} g \cdot A & g \cdot B \\ g \cdot C & g \cdot D \end{pmatrix} = \begin{pmatrix} a & b \\ c & d \end{pmatrix}^{-1} \begin{pmatrix} A & B \\ C & D \end{pmatrix} \begin{pmatrix} a & b \\ c & d \end{pmatrix}.$$

Let $\mathbf{B} \subset \mathbf{G}$ denote the algebraic subgroup of lower triangular matrices. Note that for $g = \begin{pmatrix} x & 0 \\ y & z \end{pmatrix}$ in $\mathbf{B}$, we have

$$g \cdot A = A + \frac{y}{x} B, \qquad g \cdot B = \frac{z}{x} B, \qquad g \cdot D = D - \frac{y}{x} B.$$

In particular, $\mathbf{Z}B$ is a $\mathbf{B}$-submodule of $\mathcal{A}$.

Definition 6.5 ([9] Chap. 4) If V is a $\mathbf{G}$-module, define the invariants

$$H^0(\mathbf{G}, V) = V^{\mathbf{G}} = \{v \in V : \text{for all commutative rings } A \text{ and } g \in \mathbf{G}(A),\ g \cdot v = v\}.$$

The rational cohomology groups $H^i(\mathbf{G}, -)$ are the right derived functors of $H^0(\mathbf{G}, -)$.

The same definitions hold with $\mathbf{G}$ replaced by $\mathbf{B}$. The following is our motivation for using rational cohomology.

Theorem 6.6 *Let V be a $\mathbf{G}$-module. The restriction map*

$$H^i(\mathbf{G}, V) \to H^i(\mathbf{B}, V)$$

is an isomorphism for all $i \geq 0$.

Proof See [6]*Theorem 4.4 for the general case. In this paper we only need the case that $i = 0$ and V is a finitely generated $\mathbf{Z}$-module, so we describe the proof in this case. Injectivity is clear, so we need only prove surjectivity. Let $\boldsymbol{V}$ denote the spectrum of the tensor algebra of $V^* = \mathrm{Hom}(V, \mathbf{Z})$. This is the affine scheme whose points over a ring S are $\boldsymbol{V}(S) = V \otimes S$.

Let $x \in H^0(\mathbf{B}, V)$. If S is any ring, we can define a map $\mathbf{G}(S) \longrightarrow \boldsymbol{V}(S) = V \otimes S$ by $g \mapsto g \cdot x$. Since $x \in H^0(\mathbf{B}, V)$, this map factors through $G(S)/B(S) = \mathbf{P}^1(S)$. This yields a morphism of schemes $\mathbf{P}^1 \longrightarrow \boldsymbol{V}$. Since $\boldsymbol{V}$ is affine, this morphism must be constant, i.e. $x \in H^0(\mathbf{G}, V)$. □

Corollary 6.7 *We have $H^0(\mathbf{B}, R) = A$.*

Proof First note that $H^0(\mathbf{G}, R) = A$. Indeed, the inclusion $H^0(\mathbf{G}, R) \supset A$ is clear. Meanwhile the inclusion $H^0(\mathbf{G}, R) \subset A$ follows from Corollary 6.2 and the observation that invariance under $\mathbf{G}$ is stronger than invariance under its group of points $\mathbf{G}(\mathbf{Z})$. The corollary then follows from Theorem 6.6. □

We conclude this section by stating a key input we will need from the cohomology theory of algebraic groups. See [6]*Sect. 4.2 for a proof.

Theorem 6.8 *Let $\mathcal{A}$ be the adjoint representation defined in Example 6.4. For any nonnegative integer k, the* $\mathbf{B}$*-module* $\mathcal{A}^{\otimes k} \otimes R$ *is acyclic, i.e.* $H^i(\mathbf{B}, \mathcal{A}^{\otimes k} \otimes R) = 0$ *for* $i \geq 1$.

6.2 Roadmap

Let $\boldsymbol{e} = \det(\boldsymbol{D}') - \det(\boldsymbol{D})$ and denote by $\overline{\boldsymbol{e}}$ its image in R/J. Our strategy to prove that $\overline{\boldsymbol{e}}$ lies in $\overline{A}$ is as follows. We will first prove that the ideal J is stable under the action of $\mathbf{B}$. We will then show that

$$\overline{\boldsymbol{e}} \in H^0(\mathbf{B}, R/J).$$

The long exact sequence in rational cohomology associated to

$$0 \longrightarrow J \longrightarrow R \longrightarrow R/J \longrightarrow 0$$

yields an exact sequence

$$H^0(\mathbf{B}, R) = A \longrightarrow H^0(\mathbf{B}, R/J) \xrightarrow{c_J} H^1(\mathbf{B}, J). \tag{30}$$

The equality on the left is Corollary 6.7. Let $\beta \in H^1(\mathbf{B}, J)$ denote the image of $\overline{\boldsymbol{e}}$ under the connecting homomorphism c_J. In view of (30), the desired result $\overline{\boldsymbol{e}} \in \overline{A}$ will follow if we can show that $\beta = 0$.

For this, we will define a certain $\mathbf{B}$-stable subideal $J' \subset J$ and show that in fact

$$\overline{\boldsymbol{e}} \in H^0(\mathbf{B}, R/J').$$

(This is a slight abuse of notation, as here $\overline{\boldsymbol{e}}$ denotes the reduction of $\boldsymbol{e}$ modulo J'.)

We let $\alpha = c_{J'}(\overline{\boldsymbol{e}}) \in H^1(\mathbf{B}, J')$ defined as above. Then $\beta = \iota_*(\alpha)$ where $\iota\colon J' \to J$ is the inclusion and ι_* is the induced map on rational cohomology. To conclude, we will prove that the map

$$\iota_*\colon H^1(\mathbf{B}, J') \longrightarrow H^1(\mathbf{B}, J)$$

vanishes. Therefore $\beta = 0$, and our result follows.

6.3 Invariance

Let $J' \subset J$ denote the subideal generated by the "b" coefficients of the matrices in (26)–(27). To be precise, J' is generated by:

- The elements $\sum_{i=1}^{r} \epsilon_i \boldsymbol{b}_i$ for each row of ϵ-type in D.
- The elements $\boldsymbol{a}_i \boldsymbol{b}_j + \boldsymbol{b}_i \boldsymbol{d}_j - \sum_{k=1}^{r} \delta_{ijk} \boldsymbol{b}_k$ for each row of δ-type in D.

Lemma 6.9 *The ideals J' and J are stable under the action of* **B**. *More precisely, the* **Z**-*module spanned by each set of 4 relations in (26)–(27) is isomorphic as a* **B**-*module to a copy of the adjoint representation $\mathcal{A}$ (see Example 6.4).*

Proof The relations defining J are linear combinations of products of the $\boldsymbol{\rho}_i$, with coefficients in R_0 (on which **B** acts trivially), and the definition of our action is by simultaneous conjugation on $\boldsymbol{\rho}_i$. The second sentence of the Lemma follows. The stability of J' under **B** follows since the module of upper right entries $\mathbf{Z}B \subset \mathcal{A}$ is stable under **B** (see Example 6.4). □

The goal of the rest of this subsection is to prove the following.

Lemma 6.10 *We have $\overline{\boldsymbol{e}} \in H^0(\mathbf{B}, R/J')$.*

Proof The matrix $\boldsymbol{D}$ has coefficients in R_0, on which **B** acts trivially, so we must show that

$$\overline{\det(\boldsymbol{D}')} \in H^0(\mathbf{B}, R/J').$$

The group **B** is generated by elements of the form $\sigma_{x,y} = \begin{pmatrix} x & 0 \\ 0 & y \end{pmatrix}$ and $\tau_x = \begin{pmatrix} 1 & 0 \\ x & 1 \end{pmatrix}$. The matrix $\boldsymbol{D}'$ has coefficients in $R_0[\boldsymbol{a}_i, \boldsymbol{d}_i]$ on which $\sigma_{x,y}$ acts trivially. Therefore we must only consider the action of τ_x.

The difference $\det(\tau_x(\boldsymbol{D}')) - \det(\boldsymbol{D}')$ is a linear combination (with coefficients ± 1) of determinants of all matrices M obtained by starting with $\boldsymbol{D}'$ and replacing some nonempty subset of the rows w with $\tau_x(w) - w$. We will show $\det(M) \in J'$ for each such matrix M.

Rows of ϵ-type in $\boldsymbol{D}'$ contain only the elements $\boldsymbol{\epsilon}_i \in R_0$ that are fixed by the action of **B**. So any matrix M that contains a row $\tau_x(w) - w = 0$ for a row w of ϵ-type will have determinant 0. Therefore we need only consider matrices M that contain a row $\tau_x(w) - w$ for a row w of δ-type. Conjugation by τ_x sends $\boldsymbol{a}_i \mapsto \boldsymbol{a}_i + \boldsymbol{b}_i x$ and $\boldsymbol{d}_i \mapsto \boldsymbol{d}_i - \boldsymbol{b}_i x$. Say w is associated to a pair (i, j). Then

$$\tau_x(w) - w = (0, 0, \ldots, \boldsymbol{b}_j x, 0, \ldots, -\boldsymbol{b}_i x, 0, \ldots, 0),$$

where there is $\boldsymbol{b}_j x$ in the ith slot and $-\boldsymbol{b}_i x$ in the jth slot. Note that if $i = j$ then $\tau_x(w) - w = 0$ and hence $\det(M) = 0$. So we assume $i \neq j$. We make the following alterations to M which do not change the determinant:

- We replace $\tau_x(w) - w$ by

$$(0, 0, \ldots, x, 0, \ldots, -x, 0, \ldots, 0).$$

  At the same time, in every row other than w we multiply the jth coordinate by $\boldsymbol{b}_j$ and the ith coordinate by $\boldsymbol{b}_i$.
- We then add the new jth column to the new ith column.
- For each $1 \leq k \leq r, k \neq i, j$, we add $\boldsymbol{b}_k$ times the kth column to the new ith column.

In the matrix M that results from these changes, the ith coordinate is precisely the generator of J' associated to the row. Therefore M has an entire column with elements lying in J', so its determinant lies in J' as well. □

6.4 Koszul Complex

For simplicity, denote the r elements of J' corresponding to the rows of $\boldsymbol{D}'$ by $B_1, \ldots, B_r$, i.e.

$$\boldsymbol{D}' \begin{pmatrix} \boldsymbol{b}_1 \\ \vdots \\ \boldsymbol{b}_r \end{pmatrix} = \begin{pmatrix} B_1 \\ \vdots \\ B_r \end{pmatrix}.$$

The rank 1 modules $\mathbf{Z}B_i$ are rational $\mathbf{B}$-representations isomorphic to $\mathbf{Z}B \subset \mathcal{A}$ as in Example 6.4. Let $V = \oplus_{i=1}^r \mathbf{Z}B_i$. Consider the following complex of rational $\mathbf{B}$-representations over R:

$$0 \longrightarrow \left(\bigwedge^r V\right) \otimes R \xrightarrow{f_r} \left(\bigwedge^{r-1} V\right) \otimes R \xrightarrow{f_{r-1}} \cdots \left(\bigwedge^2 V\right) \otimes R \xrightarrow{f_2} V \otimes R \xrightarrow{f_1} R. \tag{31}$$

Here all wedge products and tensor products are over $\mathbf{Z}$. The maps

$$f_i \colon \left(\bigwedge^i V\right) \otimes R \to \left(\bigwedge^{i-1} V\right) \otimes R$$

are given by

$$B_{k_1} \wedge B_{k_2} \wedge \cdots \wedge B_{k_i} \otimes r \mapsto \sum_{j=1}^{i} (-1)^j B_{k_1} \wedge B_{k_2} \wedge \cdots \wedge \hat{B}_{k_j} \wedge \cdots B_{k_i} \otimes B_{k_j} r.$$

Noting that each term $\left(\bigwedge^r V\right) \otimes R$ in (31) is isomorphic to $\bigwedge_R^r (V \otimes R)$, the sequence (31) is precisely the Koszul complex for the free R-module $V \otimes R$. It is therefore exact if we can prove that the elements $B_1, \ldots, B_r$ form a regular sequence in R.

Lemma 6.11 *The elements $B_1, \ldots, B_r$ form a regular sequence in R.*

After enacting the change of variables for rows of δ-type:

$$\delta'_{ijk} = \begin{cases} \delta_{iji} - d_j & k = i \\ \delta_{ijj} - a_i & k = j \\ \delta_{ijk} & k \neq i, j, \end{cases}$$

we see the lemma follows from the following.

Lemma 6.12 *Let S be a commutative ring and let $T = S[x_{ij}]_{i,j=1}^r$. The elements $L_i = \sum_{j=1}^r x_{ij} y_j$ form a regular sequence in $T[y_1, \ldots, y_r]$.*

We leave the proof of Lemma 6.12 as an exercise, referring the reader to [6]*Proposition 5.13.

6.5 Embedding Into an Acyclic Complex

Theorem 6.13 *There is a commutative diagram of complexes of rational **B**-modules*

$$\begin{array}{ccccccccccc} 0 \to (\bigwedge^r V) \otimes R & \xrightarrow{f_r} & \left(\bigwedge^{r-1} V\right) \otimes R & \xrightarrow{f_{r-1}} & \cdots \left(\bigwedge^2 V\right) \otimes R & \xrightarrow{f_2} & V \otimes R & \xrightarrow{f_1} & J' \\ \downarrow \qquad \downarrow{\iota_r} & & \downarrow{\iota_{r-1}} & & \downarrow{\iota_2} & & \downarrow{\iota_1} & & \downarrow{\iota_0} \\ 0 \longrightarrow W_r \otimes R & \xrightarrow{g_r} & W_{r-2} \otimes R & \xrightarrow{g_{r-1}} & \cdots W_2 \otimes R & \xrightarrow{g_2} & W_1 \otimes R & \xrightarrow{g_1} & J, \end{array} \tag{32}$$

*where the $W_i \otimes R$ are acyclic **B**-modules for $i \geq 1$.*

Proof We have already constructed the complex in the top row (see (31)), noting that the image of f_1 is the ideal generated by the B_i, namely J'. Next we define the bottom row. Fix an index $i = 1, \ldots, r$, and suppose that the ith row of D corresponds to a pair (m, n), i.e. we have $B_i = B_{mn}$ where

$$\boldsymbol{\rho}_m \boldsymbol{\rho}_n - \sum_{k=1}^r \boldsymbol{\delta}_{mnk} \boldsymbol{\rho}_k = \begin{pmatrix} A_{mn} & B_{mn} \\ C_{mn} & D_{mn} \end{pmatrix}.$$

We then let

$$\mathcal{A}_i = \mathbf{Z}A_{mn} \oplus \mathbf{Z}B_{mn} \oplus \mathbf{Z}C_{mn} \oplus \mathbf{Z}D_{mn}$$

denote the corresponding copy of the adjoint. Define

$$W_i = \bigoplus_{\{k_1,k_2,\ldots,k_i\}\subset\{1,\ldots,r\}} \mathcal{A}_{k_1} \otimes \cdots \otimes \mathcal{A}_{k_i}.$$

The vertical maps ι_i are given by, for $k_1 < k_2 < \cdots < k_i$,

$$B_{k_1} \wedge \cdots \wedge B_{k_i} \otimes r \mapsto B_{k_1} \otimes \cdots \otimes B_{k_i} \otimes r.$$

The maps g_i are given by:

$$X_1 \otimes X_2 \otimes \cdots \otimes X_i \otimes r \mapsto \sum_{j=1}^{i} (-1)^j X_1 \otimes X_2 \otimes \cdots \hat{X}_j \cdots \otimes X_i \otimes X_j r.$$

The image of the map g_1 is precisely our ideal J. The fact that the bottom row of (32) is a complex, as well as the commutativity of the diagram, is clear. The **B**-acyclicity of the $W_i \otimes R$ follows from Theorem 6.8. □

6.6 A Cascade of Cohomology Classes

Theorem 6.14 *Let $\iota\colon J' \to J$ be the inclusion and let*

$$\iota_*\colon H^1(\mathbf{B}, J') \longrightarrow H^1(\mathbf{B}, J)$$

be the induced map on first rational cohomology groups. Then $\iota_* = 0$.

Proof Let the notation be as in Theorem 6.13. Recall $\mathrm{im}(f_1) = J'$. Let

$$\alpha_1 \in H^1(\mathbf{B}, J') = H^1(\mathbf{B}, \mathrm{im}(f_1)).$$

We need to show that $\iota_{0,*}\alpha_1 = 0$. The long exact sequence in cohomology associated to

$$0 \to \ker(f_1) \to V \otimes R \to \mathrm{im}(f_1) \to 0$$

yields a class $\alpha_2 \in H^2(\mathbf{B}, \ker(f_1))$ that represents the obstruction to lifting α_1 to a class in $H^1(\mathbf{B}, V \otimes R)$. Writing $\ker(f_1) = \mathrm{im}(f_2)$, we can view $\alpha_2 \in H^2(\mathbf{B}, \mathrm{im}(f_2))$ and repeat the process above, using the coboundary in the long exact sequence associated to

$$0 \to \ker(f_2) \to \bigwedge^2 V \otimes R \to \mathrm{im}(f_2) \to 0$$

to obtain $\alpha_3 \in H^3(\mathbf{B}, \ker(f_2))$. Continuing in this way we obtain

$$\alpha_i \in H^i(\mathbf{B}, \ker(f_{i-1})) = H^i(\mathbf{B}, \mathrm{im}(f_i))$$

for $i = 1, \ldots, r+1$. Note that $\alpha_{r+1} = 0$ since $\ker(f_r) = 0$.

For each $i = 1, \ldots, r+1$, define

$$\beta_i = \iota_{i-1,*}\alpha_i \in H^i(\mathbf{B}, \mathrm{im}(g_i)).$$

In particular, we have $\beta_1 = \iota_{0,*}\alpha_1$, which is the class we are trying to show vanishes. The bottom row of (32) is a complex but we do not claim it is exact. Nevertheless the obstruction to $\beta_i \in H^i(\mathbf{B}, \mathrm{im}(g_i))$ lifting to a class in $H^i(\mathbf{B}, W_i \otimes R)$ is precisely the image of $\beta_{i+1} \in H^{i+1}(\mathbf{B}, \mathrm{im}(g_{i+1}))$ in $H^{i+1}(\mathbf{B}, \ker(g_i))$. Now $\beta_{r+1} = 0$ since $\alpha_{r+1} = 0$, and hence we conclude that β_r lifts to a class in $H^r(\mathbf{B}, W_r \otimes R)$. However, by acyclicity we have $H^r(\mathbf{B}, W_r \otimes R) = 0$ and hence $\beta_r = 0$. Therefore, β_{r-1} lifts to a class in $H^{r-1}(\mathbf{B}, W_{r-1} \otimes R)$; again this cohomology group vanishes so $\beta_{r-1} = 0$. This downward cascading continues and we obtain $\beta_i = 0$ for all i. In particular $\beta_1 = 0$ as desired. □

As explained in Sect. 6.2, Theorem 6.14 implies that $\overline{e} \in H^0(\mathbf{B}, R/J)$ lies in $\overline{A} \subset R/J$. From the discussion of Sect. 5.3, it follows that $\det(D') - \det(D) = -\det(D) \in I$. This completes the proof of Theorem 1.3.

References

1. Bellaïche, J.: Ribet's lemma, generalizations, and pseudocharacters (2009). https://people.brandeis.edu/~jbellaic/RibetHawaii3.pdf
2. Bellaïche, J., Chenevier, G.: Families of Galois representations and Selmer groups. Astérisque **324**, xii+314 (2009)
3. Brauer, R., Nesbitt, C.: On the modular characters of groups. Ann. of Math. **42**(2), 556–590 (1941)
4. Dasgupta, S., Kakde, M.: On the Brumer-Stark conjecture. Ann. of Math. **197**(1), 289–388 (2023)
5. Dasgupta, S., Kakde, M., Silliman, J., Wang, J.: The Brumer–Stark Conjecture over **Z** (2023)
6. Dasgupta, S., Kakde, M., Silliman, J., Wang, J.: The Residually Indistinguishable Case of Ribet's Method for GL_2 (2023)
7. De Concini, C., Procesi, C.: The invariant theory of matrices. University Lecture Series, vol. 69. American Mathematical Society, Providence (2017)
8. Friedlander, E.M., Parshall, B.J.: Cohomology of Lie algebras and algebraic groups. Amer. J. Math. **108**(1), 235–253 (1986)
9. Jantzen, J.C.: Representations of algebraic groups. Mathematical Surveys and Monographs, vol. 107, 2nd edn. American Mathematical Society, Providence (2003)
10. Mazur, B.: How can we construct abelian Galois extensions of basic number fields?. Bull. Amer. Math. Soc. (N.S.) **48**(2), 155–209 (2011)
11. Mazur, B., Wiles, A.: Class fields of abelian extensions of **Q**. Invent. Math. **76**(2), 179–330 (1984)
12. Nickel, A.: Notes on noncommutative Fitting invariants. Development of Iwasawa theory–the centennial of K. Iwasawa's birth. Adv. Stud. Pure Math., vol. 86, pp. 27–60. Math. Soc. Japan, Tokyo (2020)
13. Ribet, K.A.: A modular construction of unramified p-extensions of $\mathbb{Q}(\mu_p)$. Invent. Math. **34**(3), 151–162 (1976)

14. Ophir, A., Weiss, A.: On Ribet's Lemma for GL_2 modulo prime powers (2021). https://arxiv.org/abs/2111.01559
15. Wiles, A.: The Iwasawa conjecture for totally real fields. Ann. of Math. **131**(3), 493–540 (1990)

On the Proalgebraic Fundamental Group of Topological Spaces and Amalgamated Products of Affine Group Schemes

Christopher Deninger and Michael Wibmer

Dedicated to Massimo Bertolini on the occasion of his 60th birthday.

Abstract The proalgebraic fundamental group of a connected topological space X, recently introduced by the first author, is an affine group scheme whose representations classify local systems of finite-dimensional vector spaces on X. In this article, we further develop the theory of the proalgebraic fundamental group, in particular, we establish homotopy invariance and a Seifert–van Kampen theorem. To facilitate the latter, we study amalgamated free product of affine group schemes. We also compute the proalgebraic fundamental group of the arithmetically relevant Kucharcyzk–Scholze spaces and compare it to the motivic Galois group.

Keywords Proalgebraic fundamental group · Amalgamated product of group schemes · Tannakian criterion for reducedness · Kucharczyk-Scholze space

Mathematics Subject Classification 14L15 · 11G09 · 18M25 · 14C15

In [15], Kucharczyk and Scholze introduced the étale fundamental group $\pi_1^{\text{et}}(X, x)$ of a pointed connected topological space (X, x), using Grothendieck's formalism

Christopher Deninger: Funded by the Deutsche Forschungsgemeinschaft (DFG, German Research Foundation) under Germany's Excellence Strategy EXC 2044–390685587, Mathematics Münster: Dynamics–Geometry–Structure and the CRC 1442 Geometry: Deformations and Rigidity.
Michael Wibmer: Supported by the Lise Meitner grant M 2582-N32 of the Austrian Science Fund FWF.

C. Deninger (✉)
Mathematical Institute, University of Münster, Münster, Germany
e-mail: c.deninger@uni-muenster.de
URL: https://www.uni-muenster.de/arithm/deninger/index.html

M. Wibmer
School of Mathematics, University of Leeds, Leeds, UK
e-mail: m.wibmer@leeds.ac.uk
URL: https://sites.google.com/view/wibmer

M. Longo et al. (eds.), *Elliptic Curves and Modular Forms in Arithmetic Geometry*, Springer Proceedings in Mathematics & Statistics 527,
https://doi.org/10.1007/978-3-032-13123-2_5

of Galois categories. It is a profinite group which classifies the finite coverings of X. For path-connected, locally path-connected and semi-locally simply connected spaces X, the group $\pi_1^{\text{et}}(X, x)$ is the profinite completion of the ordinary fundamental group $\pi_1(X, x)$. For a field F containing the maximal cyclotomic extension $\mathbb{Q}(\mu_\infty)$ of $\mathbb{Q}$, Kucharczyk and Scholze defined connected topological spaces $\mathfrak{X}_F(\mathbb{C})$ and X_F whose étale fundamental groups can be identified with the absolute Galois group of F. These spaces are in general not (locally) path-connected and are rather wild compared to CW-complexes for example; hence the need for the introduction of $\pi_1^{\text{et}}(X, x)$. One could hope to realize not only the Galois group of a field F, but even the motivic Galois group of F, which for motives over F with coefficients in a field k is a k-group i.e. an affine group scheme over k. The maximal pro-étale quotient of the motivic Galois group is the ordinary Galois group viewed as a k-group. For these reasons, in [12] a fundamental k-group $\pi_k(X, x)$ for connected pointed topological spaces (X, x) was introduced and studied. It is defined using the Tannakian formalism [7], as the k-group whose finite dimensional representations classify the local systems $\mathcal{E}$ of finite-dimensional k-vector spaces on X with respect to the fibre functor $\mathcal{E} \mapsto \mathcal{E}_x$. It is shown in [12] that the maximal pro-étale quotient of $\pi_k(X, x)$ is $\pi_1^{\text{et}}(X, x)$ as a k-group. For well behaved topological spaces as above, $\pi_k(X, x)$ is the proalgebraic completion of $\pi_1(X, x)$. Moreover, using ideas of Nori and a theorem of Deligne [9] on isomorphisms of fibre functors, a replacement for the ordinary universal covering space was constructed in [12] if k is algebraically closed. Only the beginnings of the theory of the proalgebraic fundamental group $\pi_k(X, x)$ were established in [12]. The most glaring omissions were discussions of homotopy invariance and of a Seifert–van Kampen theorem. Moreover the proof in [12] that $\pi_k(X, x)$ is always reduced was not based on a general Tannakian criterion but on a precise study of the special situation. Finally, the proalgebraic fundamental groups of the Kucharczyk–Scholze spaces $\mathfrak{X}_F(\mathbb{C})$ and X_F were not calculated, although this is a necessary first step to find topological spaces with π_k related to motivic Galois groups. In the present paper, we clarify all these items.

For the Seifert–van Kampen theorem, we construct and study amalgamated free products of k-groups. It is known [19] that the category of commutative Hopf algebras over k is complete. In particular, it has fibre products and so amalgamated free products of k-groups exist. However, this existence result offers no insight into the structure and the algebraic properties of amalgamated free products of k-groups. Here we offer two constructions of amalgamated free products of k-groups. The first construction is based on the formalism of Tannakian categories applied to fibre products of categories. It has the advantage that it leads to a proof of the Seifert–van Kampen theorem for the proalgebraic fundamental group rather directly. The second construction is based on an explicit, albeit quite involved construction with Hopf algebras. It has the advantage that it allows us to deduce properties of the amalgamated free product from properties of the factors. For example, we show that the amalgamated free product of two geometrically reduced k-groups is geometrically reduced.

We conclude the introduction with an overview of the article. In Sect. 1 we construct the amalgamated free product of k-groups using the formalism of Tannakian categories applied to fibre products of categories. In Sect. 2 we present an explicit construction for the Hopf algebras of free amalgamated products of k-groups and use this to derive several properties of amalgamated free products. The proof in Sect. 3 of the Seifert–van Kampen theorem for the proalgebraic fundamental group follows immediately from our categorial construction of the amalgamated free product of k-groups. In Sect. 4 we prove Tannakian criteria for an affine group scheme over a perfect field k of positive characteristic to be (geometrically) reduced and even stronger to be perfect. Although these criteria are not stated explicitely in [6], all the necessary ideas can be found there. As an application we show that the proalgebraic fundamental group $\pi_k(X, x)$ of a connected pointed topological space (X, x) is always perfect. This improves the result of [12] asserting that $\pi_k(X, x)$ is reduced with an easier and more conceptual proof. In the last Sect. 5 we first prove homotopy invariance of $\pi_k(X, x)$ for all fields k. The proof requires some care since we make no assumptions on the topological space X apart from being connected. We then recall the construction of the spaces $\mathfrak{X}_F(\mathbb{C})$ and X_F for fields $F \supset \mathbb{Q}(\mu_\infty)$ in [15] and compute π_k along the same lines as π_1^{et} in [15]. Whereas π_1^{et} vanishes if F is algebraically closed this is not true for π_k. We find that $\pi_k(X_{\overline{F}}, \overline{x})$ is the identity component $\pi_k(X_F, x)^0$ which turns out to be commutative. The pro-unipotent part of $\pi_k(X_F, x)^0$ is related to the pro-unipotent part of the motivic Galois group for motives over F with coefficients in $k \supset \mathbb{Q}$. For the pro-reductive quotients there does not seem to be a relation. This shows again that as already noted in the introduction to [15] the spaces are not yet the optimal ones (in [15] it was noted that the Steinberg relations do not hold in the rational cohomology of X_F).

The authors are grateful to Andy Magid and Hans Porst for helpful comments. The second author is grateful for the invitation to the Cluster of Excellency, Mathematics Münster, where this work originated.

Conventions. Throughout this article k is a field. All schemes are assumed to be over k unless the contrary is indicated. All rings and algebras are assumed to be unital and commutative. In particular, all Hopf algebras are assumed to be commutative. Unadorned tensor products of k-vector spaces are understood to be over k. For simplicity, we use the term "k-group" for "affine group scheme over k". For a k-group G we denote its Hopf algebra of global sections by $k[G]$. A morphism $\phi\colon G \to H$ of k-groups is a quotient map if it is faithfully flat (equivalently, the dual map $\phi^*\colon k[H] \to k[G]$ is injective).

1 Fibre Products of Tannakian Categories

In this section we discuss fibre products of neutralized Tannakian categories. These fibre products can be used to establish the existence of amalgamated free products of k-groups (Corollary 1.12) and they play a crucial role in the proof of our Seifert–van Kampen theorem for the proalgebraic fundamental group (Theorem 3.2).

1.1 Fibre Products of Categories

While the product category of two categories is a well-known construction, the fibre product of two categories may be less familiar. We recall the definition from [23, Tag 003O, Example 4.31.3] or [3, Chap. VII, Sect. 3].

Let $F\colon \mathcal{A} \to \mathcal{C}$ and $G\colon \mathcal{B} \to \mathcal{C}$ be functors. The *fibre product* $\mathcal{A} \times_{\mathcal{C}} \mathcal{B}$ is the category defined as follows: Objects are triples (A, B, c), where A, B are objects of $\mathcal{A}$, $\mathcal{B}$ and $c\colon F(A) \to G(B)$ is an isomorphism. A morphism $(A, B, c) \to (A', B', c')$ in $\mathcal{A} \times_{\mathcal{C}} \mathcal{B}$ is a pair (a, b) of morphisms $a\colon A \to A'$, $b\colon B \to B'$ such that

$$\begin{array}{ccc} F(A) & \xrightarrow{F(a)} & F(A') \\ {\scriptstyle c}\big\downarrow & & \big\downarrow{\scriptstyle c'} \\ G(B) & \xrightarrow{G(b)} & G(B') \end{array}$$

commutes. The composition of two morphisms $(a, b)\colon (A, B, c) \to (A', B', c')$ and $(a', b')\colon (A', B', c') \to (A'', B'', c'')$ in $\mathcal{A} \times_{\mathcal{C}} \mathcal{B}$ is $(a'a, b'b)$. If $\mathcal{C}$ is the trivial category with one object and one morphism (the identity), then there are unique functors F and G, and $\mathcal{A} \times_{\mathcal{C}} \mathcal{B} = \mathcal{A} \times \mathcal{B}$ is the product of $\mathcal{A}$ and $\mathcal{B}$.

Example 1.1 For an (abstract) group G, let $\mathrm{Rep}(G)$ denote the category of finite dimensional k-linear representations of G. Let $\phi_1\colon H \to G_1$ and $\phi_2\colon H \to G_2$ be morphisms of groups. We then have restriction functors $\mathrm{Res}_{\phi_1}\colon \mathrm{Rep}(G_1) \to \mathrm{Rep}(H)$ and $\mathrm{Res}_{\phi_2}\colon \mathrm{Rep}(G_2) \to \mathrm{Rep}(H)$, which we can use to form the fibre product $\mathrm{Rep}(G_1) \times_{\mathrm{Rep}(H)} \mathrm{Rep}(G_2)$. The free product $G_1 *_H G_2$ of G_1 and G_2 with amalgamation over H, comes with a commutative diagram

$$\begin{array}{ccccc} & & G_1 *_H G_2 & & \\ & {\scriptstyle i_1}\nearrow & & \nwarrow{\scriptstyle i_2} & \\ G_1 & & & & G_2 \\ & {\scriptstyle \phi_1}\nwarrow & & \nearrow{\scriptstyle \phi_2} & \\ & & H & & \end{array}$$

Define a functor

$$F\colon \mathrm{Rep}(G_1 *_H G_2) \to \mathrm{Rep}(G_1) \times_{\mathrm{Rep}(H)} \mathrm{Rep}(G_2),\ V \rightsquigarrow (\mathrm{Res}_{i_1}(V), \mathrm{Res}_{i_2}(V), \mathrm{id}_V)$$

by $F(f) = (\mathrm{Res}_{i_1}(f), \mathrm{Res}_{i_2}(f))$ for a morphism $f\colon V \to W$ in $\mathrm{Rep}(G_1 *_H G_2)$. A k-linear map $V \to W$ between $G_1 *_H G_2$-representation is a $G_1 *_H G_2$-morphism if and only if it is a G_1-morphism and a G_2-morphism. Thus F is fully faithful. To see that F is essentially surjective, consider an object (V_1, V_2, c) of $\mathrm{Rep}(G_1) \times_{\mathrm{Rep}(H)}$

$\mathrm{Rep}(G_2)$. We can use the k-linear isomorphism $c\colon V_1 \to V_2$ to define a G_2-representation on V_1. Explicitly, we have $g_2(v_1) = c^{-1}(g_2(c(v_1)))$ for $v_1 \in V_1$ and $g_2 \in G_2$. Because c is an H-morphism, an $h \in H$ acts in this new representation on V_1 in the same fashion as it would act through the original G_1 representation on V_1. That is, the morphisms $G_1 \to \mathrm{GL}(V_1)$ and $G_2 \to \mathrm{GL}(V_1)$ have the same restriction to H. The universal property of $G_1 *_H G_2$ yields a morphism $G_1 *_H G_2 \to \mathrm{GL}(V_1)$, i.e., V_1 is a $G_1 *_H G_2$-representation with $\mathrm{Res}_{i_1}(V_1)$ the original G_1-representation on V_1 and $\mathrm{Res}_{i_2}(V_1) \simeq V_2$ as G_2-representation via c. So $F(V_1) = (V_1, \mathrm{Res}_{i_2}(V_1), \mathrm{id}_{V_1})$ is isomorphic to (V_1, V_2, c) via the isomorphism (id_{V_1}, c) in $\mathrm{Rep}(G_1) \times_{\mathrm{Rep}(H)} \mathrm{Rep}(G_2)$. Thus F is an equivalence of categories. Note that for the trivial group H, the category Rep H is trivial and the free product $G_1 * G_2$ corresponds to $\mathrm{Rep}(G_1) \times \mathrm{Rep}(G_2)$.

Example 1.2 For a topological space X, let $\mathrm{Sh}(X)$ denote the category of sheaves of Abelian groups on X. Let U_1, U_2 be open subset of X such that $X = U_1 \cup U_2$. We have restriction functors $\mathrm{Sh}(U_1) \to \mathrm{Sh}(\mathrm{U}_1 \cap \mathrm{U}_2)$, $\mathcal{F} \rightsquigarrow \mathcal{F}|_{U_1 \cap U_2}$ and $\mathrm{Sh}(U_2) \to \mathrm{Sh}(\mathrm{U}_1 \cap \mathrm{U}_2)$, $\mathcal{F} \rightsquigarrow \mathcal{F}|_{U_1 \cap U_2}$, which we can use to form the fibre product $\mathrm{Sh}(U_1) \times_{\mathrm{Sh}(U_1 \cap U_2)} \mathrm{Sh}(U_2)$. Basic sheaf theory [23, Tag 00AK] implies that

$$\mathrm{Sh(X)} \to \mathrm{Sh}(U_1) \times_{\mathrm{Sh}(U_1 \cap U_2)} \mathrm{Sh}(U_2),\ \mathcal{F} \rightsquigarrow (\mathcal{F}|_{U_1}, \mathcal{F}|_{U_2}, \mathrm{id}_{\mathcal{F}|_{U_1 \cap U_2}})$$

is an equivalence of categories. Note that since $\mathcal{F}(\emptyset) = 0$ for any sheaf of abelian groups, $\mathrm{Sh}(\emptyset)$ is the trivial category. So in case $U_1 \cap U_2$ is empty, the above construction yields the product category.

We return to the discussion of the general case. We have a functor $P_1 : \mathcal{A} \times_{\mathcal{C}} \mathcal{B} \to \mathcal{A}$ given by $P_1(A, B, c) = A$ and $P_1(a, b) = a$. Similarly, $P_2 \colon \mathcal{A} \times_{\mathcal{C}} \mathcal{B} \to \mathcal{B}$ is given by $P_2(A, B, c) = B$ and $P_2(a, b) = b$. The diagram

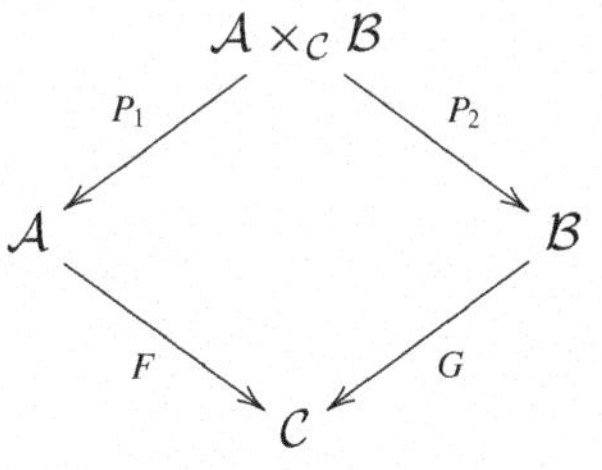

2-commutes in the sense that the functors FP_1 and GP_2 are isomorphic via an isomorphism $\psi\colon FP_1 \simeq GP_2$. In fact, for any object (A, B, c) of $\mathcal{A} \times_{\mathcal{C}} \mathcal{B}$ we have an isomorphism

$$\psi_{(A,B,c)}\colon (FP_1)(A, B, c) = F(A) \xrightarrow{c} G(B) = (GP_2)(A, B, c).$$

To state the (2-categorical) universal property satisfied by the fibre product $\mathcal{A} \times_{\mathcal{C}} \mathcal{B}$, we formally introduce 2-commutative diagrams as follows. A *2-commutative diagram* (with respect to $F\colon \mathcal{A} \to \mathcal{C}$ and $G\colon \mathcal{B} \to \mathcal{C}$) is a quadruple $(\mathcal{D}, Q_1, Q_2, \alpha)$, where $Q_1 : \mathcal{D} \to \mathcal{A}$ and $Q_2 : \mathcal{D} \to \mathcal{B}$ are functors and $\alpha : FQ_1 \simeq GQ_2$ is an isomorphism of functors. A morphism $(\mathcal{D}, Q_1, Q_2, \alpha) \to (\mathcal{D}', Q'_1, Q'_2, \alpha')$ of 2-commutative diagrams is a triple (H, β_1, β_2), with $H : \mathcal{D} \to \mathcal{D}'$ a functor and $\beta_1 : Q_1 \to Q'_1 H$ and $\beta_2 : Q_2 \to Q'_2 H$ isomorphisms of functors, such that

$$\begin{array}{ccc} FQ_1 & \xrightarrow{\mathrm{id}_F \star \beta_1} & FQ'_1H \\ {\scriptstyle \alpha}\downarrow & & \downarrow{\scriptstyle \alpha' \star \mathrm{id}_H} \\ GQ_2 & \xrightarrow{\mathrm{id}_G \star \beta_2} & GQ'_2H \end{array}$$

commutes. An isomorphism between two morphisms (H, β_1, β_2), (H', β'_1, β'_2) (with the same source and target) is an isomorphism $\theta : H \to H'$ such that

$$\begin{array}{ccc} Q_1 & \xrightarrow{\beta_1} & Q'_1H \\ & {\scriptstyle \beta'_1}\searrow & \downarrow{\scriptstyle \mathrm{id}_{Q'_1} \star \theta} \\ & & Q'_1H' \end{array} \qquad \text{and} \qquad \begin{array}{ccc} Q_2 & \xrightarrow{\beta_2} & Q'_2H \\ & {\scriptstyle \beta'_2}\searrow & \downarrow{\scriptstyle \mathrm{id}_{Q'_2} \star \theta} \\ & & Q'_2H' \end{array}$$

commute. The fibre product of categories satisfies the following (2-categorical) universal property.

Lemma 1.3 *Let $F : \mathcal{A} \to \mathcal{C}$ and $G : \mathcal{B} \to \mathcal{C}$ be functors. For every 2-commutative diagram $(\mathcal{D}, Q_1, Q_2, \alpha)$ there exists a morphism $(H, \beta_1, \beta_2) : (\mathcal{D}, Q_1, Q_2, \alpha) \to (\mathcal{A} \times_{\mathcal{C}} \mathcal{B}, P_1, P_2, \psi)$ of 2-commutative diagrams. If (H', β'_1, β'_2) is another such morphism, then there exists a unique isomorphism θ between (H, β_1, β_2) and (H', β'_1, β'_2).*

Proof This is [23, Tag 02X9]. For what follows, it will be helpful to have explicit formulas for (H, β_1, β_2) and θ. On objects let H be given by $H(D) = (Q_1(D), Q_2(D), \alpha_D)$ and for a morphism $f : D \to D'$ in $\mathcal{D}$ set $H(f) = (Q_1(f), Q_2(f)) : (Q_1(D), Q_2(D), \alpha_D) \to (Q_1(D'), Q_2(D'), \alpha_{D'})$. As $P_1H = Q_1$, $\beta_1 : Q_1 \simeq P_1H$ can be chosen to be the identity. Similarly, $\beta_2 : Q_2 \simeq P_2H$ is the identity.

For another morphism $(H', \beta'_1, \beta'_2) : (\mathcal{D}, Q_1, Q_2, \alpha) \to (\mathcal{A} \times_{\mathcal{C}} \mathcal{B}, P_1, P_2, \psi)$ of 2-commutative diagrams, the commutativity of

$$\begin{array}{ccc} FQ_1 & \xrightarrow{\mathrm{id}_F \star \beta'_1} & FP_1H' \\ {\scriptstyle \alpha}\downarrow & & \downarrow{\scriptstyle \psi \star \mathrm{id}_{H'}} \\ GQ_2 & \xrightarrow{\mathrm{id}_G \star \beta'_2} & GP_2H' \end{array}$$

shows that for any object D of $\mathcal{D}$ we have a morphism

$$\theta_D = (\beta'_{1,D}, \beta'_{2,D})\colon H(D) = (Q_1(D), Q_2(D), \alpha_D) \\ \longrightarrow H'(D) = (P_1(H'(D)), P_2(H'(D)), \alpha'_D)$$

in $\mathcal{A} \times_{\mathcal{C}} \mathcal{B}$. □

Before considering fibre products of Tannakian categories, we need to a take a brief look at fibre products of abelian categories.

Lemma 1.4 *If $F\colon \mathcal{A} \to \mathcal{C}$ and $G\colon \mathcal{B} \to \mathcal{C}$ are exact additive functors between abelian categories, then $\mathcal{A} \times_{\mathcal{C}} \mathcal{B}$ is an abelian category.*

Proof Let us begin by defining the addition on the hom-sets of $\mathcal{A} \times_{\mathcal{C}} \mathcal{B}$. Let $(a, b), (a', b')\colon (A, B, c) \to (A', B', c')$ be two morphisms in $\mathcal{A} \times_{\mathcal{C}} \mathcal{B}$. Adding the identities $G(b)c = c'F(a)$ and $G(b')c = c'F(a')$ and using the additivity of F and G yields $G(b + b')c = c'F(a + a')$. So we can define $(a, b) + (a', b') = (a + a', b + b')$. This defines the structure of an abelian group on the hom-sets such that composition is bilinear, i.e., $\mathcal{A} \times_{\mathcal{C}} \mathcal{B}$ is a pre-additive category.

A zero object of $\mathcal{A} \times_{\mathcal{C}} \mathcal{B}$ is given by $(0_{\mathcal{A}}, 0_{\mathcal{B}}, 0)$, where $0_{\mathcal{A}}$ and $0_{\mathcal{B}}$ are zero objects of $\mathcal{A}$ and $\mathcal{B}$ respectively and $0\colon F(0_{\mathcal{A}}) \to G(0_{\mathcal{B}})$ is the zero morphism. The direct sum of (A, B, c) and (A', B', c') is $(A \oplus A', B \oplus B', c \oplus c')$. So $\mathcal{A} \times_{\mathcal{C}} \mathcal{B}$ is an additive category.

Let $(a, b)\colon (A, B, c) \to (A', B', c')$ be a morphism in $\mathcal{A} \times_{\mathcal{C}} \mathcal{B}$. Since F and G preserve kernels, there exists a unique isomorphism $c_{\ker}\colon F(\ker(a)) \to G(\ker(b))$ such that

$$\begin{array}{ccccc} F(\ker(a)) & \longrightarrow & F(A) & \xrightarrow{F(a)} & F(A') \\ \Big\downarrow{\scriptstyle c_{\ker}} & & \Big\downarrow{\scriptstyle c} & & \Big\downarrow{\scriptstyle c'} \\ G(\ker(b)) & \longrightarrow & G(B) & \xrightarrow{G(b)} & G(B') \end{array}$$

commutes. The resulting morphism $(\ker(a), \ker(b), c_{\ker}) \to (A, B, c)$ in $\mathcal{A} \times_{\mathcal{C}} \mathcal{B}$ is then a kernel of (a, b). A similar construction shows that cokernels exist in $\mathcal{A} \times_{\mathcal{C}} \mathcal{B}$. Since kernels and cokernels are constructed componentwise, it follows that for any morphism (a, b) in $\mathcal{A} \times_{\mathcal{C}} \mathcal{B}$ the morphism $\mathrm{Coim}(a, b) \to \mathrm{Im}(a, b)$ is an isomorphism. Thus $\mathcal{A} \times_{\mathcal{C}} \mathcal{B}$ is abelian. □

1.2 Tannakian Categories

Our next goal is to show that the fibre product of Tannakian categories is naturally a Tannakian category. To this end, we first recall the basic facts concerning Tannakiann categories. For more background see [7, 8].

A *tensor category* is a tuple $(\mathcal{A}, \otimes, \phi, \psi)$, where $\mathcal{A}$ is a category, $\otimes\colon \mathcal{A} \times \mathcal{A} \to \mathcal{A}$, $(X, Y) \rightsquigarrow X \otimes Y$ is a functor and ϕ, ψ are isomophisms of functors expressing the associativity and commutativity of $\otimes$. More precisely, the *associativity constraint*

ϕ has components $\phi_{X,Y,Z}\colon X\otimes(Y\otimes Z)\to(X\otimes Y)\otimes Z$ and the *commutativity constraint* ψ has components $\psi_{X,Y}\colon X\otimes Y\to Y\otimes X$. The functorial isomorphisms ϕ and ψ are required to satisfy three *coherence conditions* (see [7, Sect. 1] for details). It is also required that there exists an *identity object* $(\mathbb{1},u)$. This means that $\mathcal{A}\to\mathcal{A},\ X\rightsquigarrow\mathbb{1}\otimes X$ is an equivalence of categories and $u\colon\mathbb{1}\to\mathbb{1}\otimes\mathbb{1}$ is an isomorphism.

Then there exists a unique isomorphism of functors l with component $l_X\colon X\to\mathbb{1}\otimes X$ such that certain properties are satisfied [7, Proposition 1.3 (a)]. Similarly, there is functorial isomorphism r with components $r_X\colon X\to X\otimes\mathbb{1}$.

We will usually omit ϕ and ψ from the notation and refer to $(\mathcal{A},\otimes)$, or even $\mathcal{A}$, as a tensor category. A *tensor functor* $(\mathcal{A},\otimes)\to(\mathcal{A}',\otimes')$ is a pair (F,α), where $F\colon\mathcal{A}\to\mathcal{A}'$ is a functor and α is an isomorphism of functors with components $\alpha_{X,Y}\colon F(X)\otimes F(Y)\to F(X\otimes Y)$ such that three natural properties involving respectively the associativity constraints, the commutativity constraints and the identity objects, are satisfied [7, Definition 1.8]. We will often omit the functorial isomorphism α from the notation and simply refer to F as a tensor functor.

A tensor category $(\mathcal{A},\otimes)$ is *rigid*, if for every object X of $\mathcal{A}$ the exists an object $X^\vee$ (a *dual* of X) of $\mathcal{A}$ together with morphisms $\mathrm{ev}\colon X\otimes X^\vee\to\mathbb{1}$ and $\delta\colon\mathbb{1}\to X^\vee\otimes X$ such that

$$X\xrightarrow{r}X\otimes\mathbb{1}\xrightarrow{\mathrm{id}\otimes\delta}X\otimes(X^\vee\otimes X)\xrightarrow{\phi}(X\otimes X^\vee)\otimes X\xrightarrow{\mathrm{ev}\otimes\mathrm{id}}\mathbb{1}\otimes X\xrightarrow{l^{-1}}X$$

and

$$X^\vee\xrightarrow{l}\mathbb{1}\otimes X^\vee\xrightarrow{\delta\otimes\mathrm{id}}(X^\vee\otimes X)\otimes X^\vee\xrightarrow{\phi^{-1}}X^\vee\otimes(X\otimes X^\vee)\xrightarrow{\mathrm{id}\otimes\mathrm{ev}}X^\vee\otimes\mathbb{1}\xrightarrow{r^{-1}}X^\vee$$

are the identity. The above definition of rigid is equivalent to the one used in [7] by [8, Sect. 2]. Given X, $X^\vee$ and ev such that there exists a δ verifying the above conditions, then δ is uniquely determined. Moreover, the pair $(X^\vee,\mathrm{ev})$ is unique up to a unique isomorphism. For a morphism $f\colon X\to Y$ the *transpose* ${}^tf\colon Y^\vee\to X^\vee$ of f is defined as the composition

$${}^tf\colon Y^\vee\simeq\mathbb{1}\otimes Y^\vee\xrightarrow{\delta\otimes\mathrm{id}}X^\vee\otimes X\otimes Y^\vee\xrightarrow{\mathrm{id}\otimes f\otimes\mathrm{id}}X^\vee\otimes Y\otimes Y^\vee\xrightarrow{\mathrm{id}\otimes\mathrm{ev}}X^\vee\otimes\mathbb{1}\simeq X^\vee.$$

For f invertible one sets $f^\vee=({}^tf)^{-1}\colon X^\vee\to Y^\vee$.

Lemma 1.5 *Let $F\colon\mathcal{A}\to\mathcal{C}$ and $G\colon\mathcal{B}\to\mathcal{C}$ be tensor functors between tensor categories. Then $\mathcal{A}\times_{\mathcal{C}}\mathcal{B}$ naturally acquires the structure of a tensor category such that the projections $\mathcal{A}\times_{\mathcal{C}}\mathcal{B}\to\mathcal{A}$ and $\mathcal{A}\times_{\mathcal{C}}\mathcal{B}\to\mathcal{B}$ are tensor functors. Moreover, if $\mathcal{A}$, $\mathcal{B}$ and $\mathcal{C}$ are rigid, then $\mathcal{A}\otimes_{\mathcal{C}}\mathcal{B}$ is rigid.*

Proof Cf. the discussion on page 360 of [3]. We define the tensor product $\otimes\colon(\mathcal{A}\times_{\mathcal{C}}\mathcal{B})\times(\mathcal{A}\times_{\mathcal{C}}\mathcal{B})\to\mathcal{A}\times_{\mathcal{C}}\mathcal{B}$ on objects by

$$(A, B, c) \otimes (A', B', c') = (A \otimes A', B \otimes B', c \otimes c')$$

and on morphisms by $(a, b) \otimes (a', b') = (a \otimes a', b \otimes b')$. The associativity and commutativity constraints on $\mathcal{A}$ and $\mathcal{B}$ induce associativity and commutativity constraints on $\mathcal{A} \times_{\mathcal{C}} \mathcal{B}$. The identity object of $\mathcal{A} \times_{\mathcal{C}} \mathcal{B}$ is $(\mathbb{1}_{\mathcal{A}}, \mathbb{1}_{\mathcal{B}}, c)$, with $c\colon F(\mathbb{1}_{\mathcal{A}}) \to G(\mathbb{1}_{\mathcal{B}})$ the unique isomorphism derived from the fact that $F(\mathbb{1}_{\mathcal{A}})$ and $G(\mathbb{1}_{\mathcal{B}})$ are identity objects of $\mathcal{C}$.

If (A, B, c) is an object of $\mathcal{A} \times_{\mathcal{C}} \mathcal{B}$ and $A^\vee$ and $B^\vee$ are duals of A and B respectively, then $(A^\vee, B^\vee, c^\vee)$ is a dual of (A, B, c). Implicit in this notation are the identifications $F(A^\vee) \simeq F(A)^\vee$ and $G(B^\vee) \simeq G(B)^\vee$ derived from the fact that tensor functors preserve duals. □

A rigid tensor category $(\mathcal{A}, \otimes)$ is *abelian* if the underlying category $\mathcal{A}$ is abelian. In this case $\otimes$ is automatically bi-additive [7, Proposition 1.16].

A basic example of a tensor category is the category Mod_R of R-modules over a commutative ring R, with the usual tensor product and the usual associativity and commutativity constraints. The most basic example of a rigid abelian tensor category is the category Vec_k of finite dimensional k-vector spaces.

A *neutral Tannakian category* over k is a rigid abelian tensor category $(\mathcal{A}, \otimes)$ together with an isomorphism $k \simeq \mathrm{End}(\mathbb{1})$ such that there exists an exact k-linear tensor functor $\omega\colon \mathcal{A} \to \mathrm{Vec}_k$. Any such ω is called a *(neutral) fibre functor*. We note that by [8, Corollary 2.10] a fibre functor is automatically faithful.

For tensor functors $F, G\colon (\mathcal{A}, \otimes) \to (\mathcal{B}, \otimes)$, a morphism $\alpha\colon F \to G$ of functors, is a *morphism of tensor functors* if

(i) $\alpha_{X \otimes Y} = \alpha_X \otimes \alpha_Y$ for any two objects X, Y of $\mathcal{A}$ under the identifications $F(X) \otimes F(Y) \simeq F(X \otimes Y)$ and $G(X) \otimes G(Y) \simeq G(X \otimes Y)$ and
(ii) the diagram

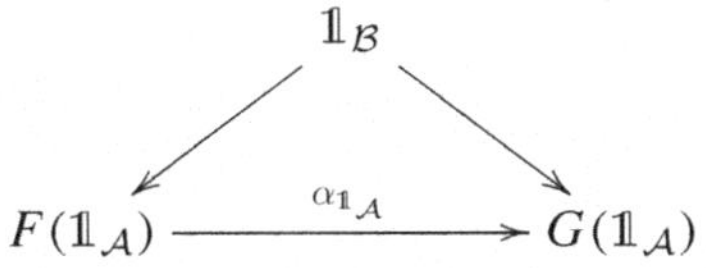

commutes.

For a neutral Tannakian category $\mathcal{A}$ with fibre functor $\omega\colon \mathcal{A} \to \mathrm{Vec}_k$, one defines a functor $\underline{\mathrm{Aut}}^{\otimes}(\omega)$ from the category of k-algebras to the category of groups by $\underline{\mathrm{Aut}}^{\otimes}(\omega)(R) = \mathrm{Aut}^{\otimes}(\omega_R)$, the group of tensor automorphisms of the tensor functor ω_R, defined as the composition of ω with the tensor functor $\mathrm{Vec}_k \to \mathrm{Mod}_R$, $V \rightsquigarrow V \otimes_k R$ for any k-algebra R.

The standard example of a neutral Tannakian category is the category $\mathrm{Rep}(G)$ of finite dimensional k-linear representations of a k-group G with the usual tensor product of representations and the forgetful functor $\omega_G\colon \mathrm{Rep}(G) \to \mathrm{Vec}_k$ as fibre functor. The main Tannaka reconstruction theorem shows that this is essentially the only example:

Theorem 1.6 *([7, Theorem 2.11]) Let* $(\mathcal{A}, \otimes)$ *be a neutral Tannakian category with fibre functor* $\omega\colon \mathcal{A} \to \mathrm{Vec}_k$*. Then* $G = \underline{\mathrm{Aut}}^{\otimes}(\omega)$ *is a* k*-group and* ω *induces an equivalence of tensor categories* $\mathcal{A} \to \mathrm{Rep}(G)$.

The k-group $G_{\mathcal{A}} = \underline{\mathrm{Aut}}^{\otimes}(\omega)$ is called the *fundamental group* of $\mathcal{A}$ or the *Tannakian dual* of $\mathcal{A}$. A *neutralized Tannakian category* is a pair $(\mathcal{A}, \omega_{\mathcal{A}})$ consisting of a neutral Tannakian category $\mathcal{A}$ and a fibre functor $\omega_{\mathcal{A}}\colon \mathcal{A} \to \mathrm{Vec}_k$. Abusing notation, a *tensor functor* $(F, \gamma)\colon (\mathcal{A}, \omega_{\mathcal{A}}) \to (\mathcal{B}, \omega_{\mathcal{B}})$ between neutralized Tannakian categories is a pair (F, γ) consisting of a k-linear tensor functor $F\colon \mathcal{A} \to \mathcal{B}$ and an isomorphism $\gamma\colon \omega_{\mathcal{B}} F \simeq \omega_{\mathcal{A}}$ of tensor functors. Note that this implies that F is exact since $\omega_{\mathcal{A}}$ and $\omega_{\mathcal{B}}$ are exact. Such a tensor functor induces a morphism $\varphi\colon \underline{\mathrm{Aut}}^{\otimes}(\omega_{\mathcal{B}}) \to \underline{\mathrm{Aut}}^{\otimes}(\omega_{\mathcal{A}})$ of k-groups as follows: For every k-algebra R, an element $g \in \underline{\mathrm{Aut}}^{\otimes}(\omega_{\mathcal{B}})(R) = \mathrm{Aut}^{\otimes}(\omega_{\mathcal{B},R})$ gives rises to a tensor automorphism $g \star \mathrm{id}_F\colon \omega_{\mathcal{B},R} F \simeq \omega_{\mathcal{B},R} F$, that yields a tensor automorphism of $\omega_{\mathcal{A},R}$ via the tensor isomorphism $\gamma_R = \mathrm{id}_{-\otimes R} \star \gamma\colon \omega_{\mathcal{B},R} F \simeq \omega_{\mathcal{A},R}$, i.e., $\varphi_R(g) = \gamma_R (g \star \mathrm{id}_F) \gamma_R^{-1}$. Note that in the context of Theorem 1.6 we have an isomorphism $(\mathcal{A}, \omega) \to (\mathrm{Rep}(G), \omega_G)$ of neutralized Tannakian categories.

The composition $(G, \delta)(F, \gamma)$ of two tensor functors $(F, \gamma)\colon (\mathcal{A}, \omega_{\mathcal{A}}) \to (\mathcal{B}, \omega_{\mathcal{B}})$ and $(G, \delta)\colon (\mathcal{B}, \omega_{\mathcal{B}}) \to (\mathcal{C}, \omega_{\mathcal{C}})$ is defined as $(G, \delta)(F, \gamma) = (GF, \omega_{\mathcal{C}} GF \xrightarrow{\delta \star \mathrm{id}_F} \omega_{\mathcal{B}} F \xrightarrow{\gamma} \omega_{\mathcal{A}})$.

Let $(F, \gamma), (F', \gamma')\colon (\mathcal{A}, \omega_{\mathcal{A}}) \to (\mathcal{B}, \omega_{\mathcal{B}})$ be tensor functors. An isomorphism between (F, γ) and (F', γ') is an isomorphism $\delta\colon F \to F'$ of tensor functors such that

$$\begin{array}{ccc} \omega_{\mathcal{B}} F & \xrightarrow{\mathrm{id}_{\omega_{\mathcal{B}}} \star \delta} & \omega_{\mathcal{B}} F' \\ & {\scriptstyle\gamma}\searrow \quad \swarrow{\scriptstyle\gamma'} & \\ & \omega_{\mathcal{A}} & \end{array}$$

commutes. Note that isomorphic (F, γ), (F', γ') induce the same morphism $\underline{\mathrm{Aut}}^{\otimes}(\omega_{\mathcal{B}}) \to \underline{\mathrm{Aut}}^{\otimes}(\omega_{\mathcal{A}})$.

Morally, the category of k-groups is anti-equivalent to the category of neutralized Tannakian categories over k. In reality, this is not quite true since a category of categories should be regarded as a 2-category. The following precise statement [18, Corollary 3.2] is sufficient for our purposes.

Lemma 1.7 *Let* $(\mathcal{A}, \omega_{\mathcal{A}})$ *and* $(\mathcal{B}, \omega_{\mathcal{B}})$ *be neutralized Tannakian categories with fundamental groups* $G_{\mathcal{A}}$ *and* $G_{\mathcal{B}}$ *respectively. Then* $\mathrm{Hom}(G_{\mathcal{B}}, G_{\mathcal{A}})$ *is in bijection with the isomorphism classes of tensor functors* $(\mathcal{A}, \omega_{\mathcal{A}}) \to (\mathcal{B}, \omega_{\mathcal{B}})$.

As one may expect, the fibre product of neutralized Tannakian categories inherits all of the structure:

Lemma 1.8 *Let* $(F, \gamma)\colon (\mathcal{A}, \omega_{\mathcal{A}}) \to (\mathcal{C}, \omega_{\mathcal{C}})$ *and* $(G, \delta)\colon (\mathcal{B}, \omega_{\mathcal{B}}) \to (\mathcal{C}, \omega_{\mathcal{C}})$ *be tensor functors between neutralized Tannakian categories. Then* $\mathcal{A} \times_{\mathcal{C}} \mathcal{B}$ *inherits the structure of a neutralized Tannakian category with fibre functor* $\omega_{\mathcal{A} \times_{\mathcal{C}} \mathcal{B}}$ *given by* $\omega_{\mathcal{A} \times_{\mathcal{C}} \mathcal{B}}(A, B, c) = \omega_{\mathcal{A}}(A)$.

Proof We already saw in Lemma 1.5 that $\mathcal{A} \times_{\mathcal{C}} \mathcal{B}$ is naturally a rigid tensor category. As F and G are exact and additive, Lemma 1.4 shows that $\mathcal{A} \times_{\mathcal{C}} \mathcal{B}$ is abelian.

The map $k \to \mathrm{End}(\mathbb{1}_{\mathcal{A}\times_{\mathcal{C}}\mathcal{B}})$ is an isomorphism because an endomorphism of $\mathbb{1}_{\mathcal{A}\times_{\mathcal{C}}\mathcal{B}} = (\mathbb{1}_{\mathcal{A}}, \mathbb{1}_{\mathcal{B}}, c)$ is a pair (a, b) with $a \in \mathrm{End}(\mathbb{1}_{\mathcal{A}}) \simeq k$ and $b \in \mathrm{End}(\mathbb{1}_{\mathcal{B}}) \simeq k$ such that a and b agree as elements of k.

Note that the projection $P_1 : \mathcal{A} \times_{\mathcal{C}} \mathcal{B} \to \mathcal{A}$ is an exact k-linear tensor functor (the implied isomorphism $P_1(-) \otimes P_1(-) \simeq P_1(- \otimes -)$ is the identity). Therefore, also the composition $\omega_{\mathcal{A}\times_{\mathcal{C}}\mathcal{B}}$ of ω with P_1 is an exact k-linear tensor functor. □

Remark 1.9 There are four possibilities to define the fibre functor $\omega_{\mathcal{A}\times_{\mathcal{C}}\mathcal{B}}$ in Lemma 1.8: On an object (A, B, c) of $\mathcal{A} \times_{\mathcal{C}} \mathcal{B}$ we can define it as $\omega_{\mathcal{A}}(A)$, $\omega_{\mathcal{B}}(B)$, $\omega_{\mathcal{C}}(F(A))$ or $\omega_{\mathcal{C}}(G(B))$. The choice is irrelevant, since all four functors are isomorphic.

1.3 Amalgamated Free Products

In this section we recall the definition of amalgamated free products and discuss their relation with fibre products of neutralized Tannakian categories.

Let $\phi_1 : H \to G_1$ and $\phi_2 : H \to G_2$ be morphisms of k-groups. In line with the common notation in group theory, we denote a pushout of ϕ_1 and ϕ_2 in the category of k-groups with $G_1 *_H G_2$ and call it a *free product of* G_1 *and* G_2 *with amalgamation over* H or simply an *amalgamated free product*. So $G_1 *_H G_2$ comes with two morphisms $i_1 : G_1 \to G_1 *_H G_2$ and $i_2 : G_2 \to G_1 *_H G_2$ such that

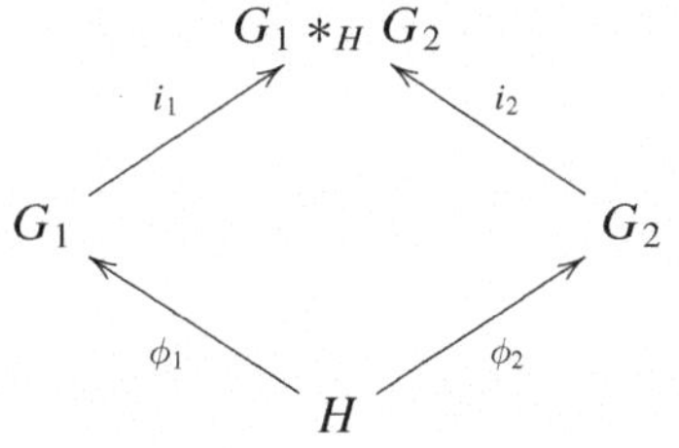

commutes and satisfies the following universal property: If $j_1 : G_1 \to G$ and $j_2 : G_2 \to G$ are morphisms of k-groups such that $j_1\phi_1 = j_2\phi_2$, then there exists a unique morphism $G_1 *_H G_2 \to G$ such that

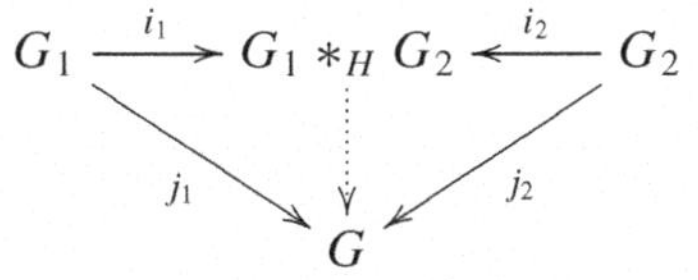

commutes. For $H = 1$ the trivial group, one obtains the *free product* $G_1 * G_2$ of G_1 and G_2.

It is known [19, Proposition 22 (2)] that the category of (commutative) Hopf algebras over a field is complete. In particular, it has fibre products. Therefore amalgamated free products exist. Beyond this existence result, the only literature on amalgamated free products of k-groups we could find is [1], which discusses amalgamated free products of finite flat group schemes over complete discrete valuation rings. We will provide two existence proofs for amalgamated free products of k-groups independent from [19]. The first one (Corollary 1.12), based on the fibre product of neutralized Tannakian categories, leads to a proof of the Seifert–van Kampen Theorem for the proalgebraic fundamental group rather directly. The second one (Proposition 2.4), based on an explicit construction with Hopf algebras, is helpful for determining properties of the amalgamated free product.

We next discuss the universal property of the fibre product constructed in Lemma 1.8. As before, let $(F, \gamma)\colon (\mathcal{A}, \omega_{\mathcal{A}}) \to (\mathcal{C}, \omega_{\mathcal{C}})$ and $(G, \delta)\colon (\mathcal{B}, \omega_{\mathcal{B}}) \to (\mathcal{C}, \omega_{\mathcal{C}})$ be tensor functors between neutralized Tannakian categories. We consider $\mathcal{A} \times_{\mathcal{C}} \mathcal{B}$ as a neutralized Tannakian category as described in Lemma 1.8. The projections $P_1\colon \mathcal{A} \times_{\mathcal{C}} \mathcal{B} \to \mathcal{A}$ and $P_2\colon \mathcal{A} \times_{\mathcal{C}} \mathcal{B} \to \mathcal{B}$ are k-linear tensor functors.

In fact, P_1 and P_2 can be regarded as tensor functors between neutralized Tannakian categories as follows. With $\chi_1 = \mathrm{id}\colon \omega_{\mathcal{A}} P_1 \to \omega_{\mathcal{A} \times_{\mathcal{C}} \mathcal{B}}$ we obtain a tensor functor $(P_1, \chi_1)\colon (\mathcal{A} \times_{\mathcal{C}} \mathcal{B}, \omega_{\mathcal{A} \times_{\mathcal{C}} \mathcal{B}}) \to (\mathcal{A}, \omega_{\mathcal{A}})$ between neutralized Tannakian categories. We define an isomorphism $\chi_2\colon \omega_{\mathcal{B}} P_2 \to \omega_{\mathcal{A} \times_{\mathcal{C}} \mathcal{B}}$ by

$$\chi_{2,(A,B,c)}\colon (\omega_{\mathcal{B}} P_2)(A, B, c) = \omega_{\mathcal{B}}(B) \xrightarrow{\delta_B^{-1}} \omega_{\mathcal{C}}(G(B)) \xrightarrow{\omega_{\mathcal{C}}(c)^{-1}} \omega_{\mathcal{C}}(F(A)) \xrightarrow{\gamma_A} \omega_{\mathcal{A}}(A) = \omega_{\mathcal{A} \times_{\mathcal{C}} \mathcal{B}}(A, B, c)$$

for any object (A, B, c) of $\mathcal{A} \times_{\mathcal{C}} \mathcal{B}$. Then also $(P_2, \chi_2)\colon (\mathcal{A} \times_{\mathcal{C}} \mathcal{B}, \omega_{\mathcal{A} \times_{\mathcal{C}} \mathcal{B}}) \to (\mathcal{B}, \omega_{\mathcal{B}})$ is a tensor functor between neutralized Tannakian categories. The isomorphism $\psi\colon FP_1 \to GP_2$, introduced on page 4, is an isomorphism of tensor functors $(F, \gamma)(P_1, \chi_1) \to (G, \delta)(P_2, \chi_2)$ since the diagram

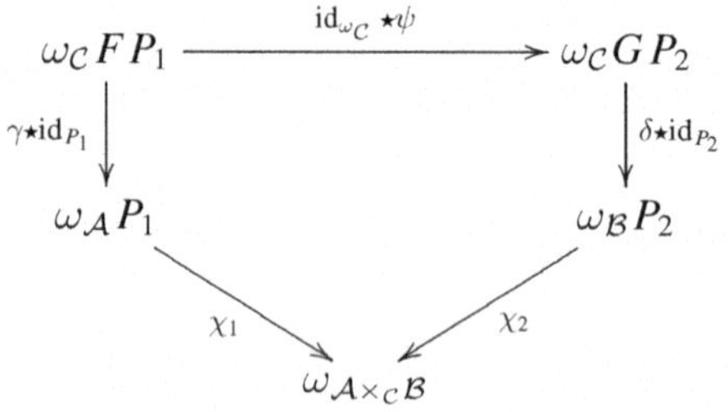

commutes by definition of χ_2.

To state the universal property of the fibre product $(\mathcal{A} \times_{\mathcal{C}} \mathcal{B}, \omega_{\mathcal{A} \times_{\mathcal{C}} \mathcal{B}})$ we need 2-commutative diagrams of neutralized Tannakian categories. Let $(F, \gamma)\colon (\mathcal{A}, \omega_{\mathcal{A}}) \to$

$(\mathcal{C}, \omega_{\mathcal{B}})$ and $(G, \delta)\colon (\mathcal{B}, \omega_{\mathcal{B}}) \to (\mathcal{C}, \omega_{\mathcal{C}})$ be tensor functors between neutralized Tannakian categories. A 2-commutative diagram (with respect to (F, γ) and (G, δ)) is a quadruple $((\mathcal{D}, \omega_{\mathcal{D}}), (Q_1, \epsilon_1), (Q_2, \epsilon_2), \alpha)$ consisting of a neutralized Tannakian category $(\mathcal{D}, \omega_{\mathcal{D}})$, tensor functors $(Q_1, \epsilon_1)\colon (\mathcal{D}, \omega_{\mathcal{D}}) \to (\mathcal{A}, \omega_{\mathcal{A}})$ and $(Q_2, \epsilon_2)\colon (\mathcal{D}, \omega_{\mathcal{D}}) \to (\mathcal{B}, \omega_{\mathcal{B}})$ and an isomorphism $\alpha\colon (F, \gamma)(Q_1, \epsilon_1) \to (G, \delta)(Q_2, \epsilon_2)$ of tensor functors.

A morphism

$$((H, \eta), \beta_1, \beta_2)\colon ((\mathcal{D}, \omega_{\mathcal{D}}), (Q_1, \epsilon_1), (Q_2, \epsilon_2), \alpha) \\ \to ((\mathcal{D}', \omega_{\mathcal{D}'}), (Q_1', \epsilon_1'), (Q_2', \epsilon_2'), \alpha')$$

of 2-commutative diagrams consists of a tensor functor $(H, \eta)\colon (\mathcal{D}, \omega_{\mathcal{D}}) \to (\mathcal{D}', \omega_{\mathcal{D}'})$ and isomorphisms $\beta_1\colon (Q_1, \epsilon_1) \to (Q_1', \epsilon_1')(H, \eta)$ and $\beta_2\colon (Q_2, \epsilon_2) \to (Q_2', \epsilon_2')(H, \eta)$ of tensor functors such that

$$\begin{array}{ccc} FQ_1 & \xrightarrow{\mathrm{id}_F \star \beta_1} & FQ_1'H \\ {\scriptstyle\alpha}\downarrow & & \downarrow{\scriptstyle\alpha' \star \mathrm{id}_H} \\ GQ_2 & \xrightarrow{\mathrm{id}_G \star \beta_2} & GQ_2'H \end{array}$$

commutes. Finally, an isomorphism between two morphisms $((H, \eta), \beta_1, \beta_2)$, $((H', \eta'), \beta_1', \beta_2')$ of 2-commutative diagrams with the same source and target, is an isomorphism $\theta\colon (H, \eta) \to (H', \eta')$ of tensor functors such that

$$\begin{array}{ccc} Q_1 & \xrightarrow{\beta_1} & Q_1'H \\ & {\scriptstyle\beta_1'}\searrow & \downarrow{\scriptstyle\mathrm{id}_{Q_1'} \star \theta} \\ & & Q_1'H' \end{array} \qquad \text{and} \qquad \begin{array}{ccc} Q_2 & \xrightarrow{\beta_2} & Q_2'H \\ & {\scriptstyle\beta_2'}\searrow & \downarrow{\scriptstyle\mathrm{id}_{Q_2'} \star \theta} \\ & & Q_2'H' \end{array}$$

commute.

Proposition 1.10 *Let $(F, \gamma)\colon (\mathcal{A}, \omega_{\mathcal{A}}) \to (\mathcal{C}, \omega_{\mathcal{B}})$ and $(G, \delta)\colon (\mathcal{B}, \omega_{\mathcal{B}}) \to (\mathcal{C}, \omega_{\mathcal{C}})$ be tensor functors between neutralized Tannakian categories. Then $((\mathcal{A} \times_{\mathcal{C}} \mathcal{B}, \omega_{\mathcal{A} \times_{\mathcal{C}} \mathcal{B}}), (P_1, \chi_1), (P_2, \chi_2), \psi)$ satisfies the following universal property. For every 2-commutative diagram $((\mathcal{D}, \omega_{\mathcal{D}}), (Q_1, \epsilon_1), (Q_2, \epsilon_2), \alpha)$ of neutralized Tannakian categories, there exists a morphism*

$$((H, \eta), \beta_1, \beta_2)\colon ((\mathcal{D}, \omega_{\mathcal{D}}), (Q_1, \epsilon_1), (Q_2, \epsilon_2), \alpha) \\ \to ((\mathcal{A} \times_{\mathcal{C}} \mathcal{B}, \omega_{\mathcal{A} \times_{\mathcal{C}} \mathcal{B}}), (P_1, \chi_1), (P_2, \chi_2), \psi) \tag{1}$$

of 2-commutative diagrams. If $((H', \eta'), \beta_1', \beta_2')$ is another such morphism, then there exists an isomorphism between $((H, \eta), \beta_1, \beta_2)$ and $((H', \eta'), \beta_1', \beta_2')$.

Proof The functor $H\colon \mathcal{D} \to \mathcal{A} \times_{\mathcal{C}} \mathcal{B}$ defined in Lemma 1.3 can be considered as a tensor functor with functorial isomorphism $H(-) \otimes H(-) \simeq H(- \otimes -)$ induced by the functorial isomorphisms $Q_i(-) \otimes Q_i(-) \simeq Q_i(- \otimes -)$ for $i = 1, 2$. Setting $\eta\colon \omega_{\mathcal{A}\times_{\mathcal{C}}\mathcal{B}} H = \omega_{\mathcal{A}} Q_1 \xrightarrow{\epsilon_1} \omega_{\mathcal{D}}$ we obtain a tensor functor $(H, \eta)\colon (\mathcal{D}, \omega_{\mathcal{D}}) \to (\mathcal{A} \times_{\mathcal{C}} \mathcal{B}, \omega_{\mathcal{A}\times_{\mathcal{C}}\mathcal{B}})$ between neutralized Tannakian categories. As $(P_1, \chi_1)(H, \eta) = (Q_1, \epsilon_1)$, the identity $\beta_1\colon (Q_1, \epsilon_1) \to (P_1, \chi_1)(H, \eta)$ is a morphism of tensor functors. By definition of χ_2, the diagram

$$\begin{array}{ccc}
\omega_{\mathcal{C}} F P_1 & \xrightarrow{\mathrm{id}_{\omega_{\mathcal{C}}} \star \psi} & \omega_{\mathcal{C}} G P_2 \\
{\scriptstyle \gamma \star \mathrm{id}_{P_1}} \downarrow & & \downarrow {\scriptstyle \delta \star \mathrm{id}_{P_2}} \\
\omega_{\mathcal{A}} P_1 & \xleftarrow{\chi_2} & \omega_{\mathcal{B}} P_2
\end{array}$$

commutes. Therefore, the upper square in

$$\begin{array}{ccc}
\omega_{\mathcal{C}} F P_1 H & \xrightarrow{\mathrm{id}_{\omega_{\mathcal{C}}} \star \psi \star \mathrm{id}_H} & \omega_{\mathcal{C}} G P_2 H \\
{\scriptstyle \gamma \star \mathrm{id}_{P_1 H}} \downarrow & & \downarrow {\scriptstyle \delta \star \mathrm{id}_{P_2 H}} \\
\omega_{\mathcal{A}} P_1 H & \xleftarrow{\chi_2 \star \mathrm{id}_H} & \omega_{\mathcal{B}} P_2 H \\
& {\scriptstyle \epsilon_1} \searrow \quad \swarrow {\scriptstyle \epsilon_2} & \\
& \omega_{\mathcal{D}} &
\end{array} \tag{2}$$

commutes. Noting that $P_1 H = Q_1$, $P_2 H = Q_2$ and $\psi \star \mathrm{id}_H = \alpha$, we see that the outer diagram in (2) commutes because $\alpha\colon (F, \gamma)(Q_1, \epsilon_1) \to (G, \delta)(Q_2, \epsilon_2)$ is an isomorphism of tensor functors. Thus also the lower triangle in (2) commutes. This shows that $(P_2, \chi_2)(H, \eta) = (Q_2, \epsilon_2)$. Thus the identity $\beta_2\colon (Q_2, \epsilon_2) \to (P_2, \chi_2)(H, \eta)$ is an isomorphism of tensor functors. We have thus constructed the morphism (1).

If $((H', \eta'), \beta_1', \beta_2')$ is another such morphism, the isomorphism $\theta\colon H \to H'$ constructed in Lemma 1.3 is a tensor isomorphism because $\beta_1'\colon Q_1 \to P_1 H'$ and $\beta_2'\colon Q_2 \to P_2 H'$ are tensor isomorphisms. Moreover, the fact that $\beta_1'\colon (Q_1, \epsilon_1) \to (P_1, \chi_1)(H', \eta')$ is an isomorphism of tensor functors, implies that

$$\begin{array}{ccc}
\omega_{\mathcal{A}\times_{\mathcal{C}}\mathcal{B}} H = \omega_{\mathcal{A}} Q_1 & \xrightarrow{\mathrm{id}_{\omega_{\mathcal{A}\times_{\mathcal{C}}\mathcal{B}}} \star \theta = \mathrm{id}_{\omega_{\mathcal{A}}} \star \beta_1'} & \omega_{\mathcal{A}} P_1 H' = \omega_{\mathcal{A}\times_{\mathcal{C}}\mathcal{B}} H' \\
& {\scriptstyle \eta = \epsilon_1} \searrow \quad \swarrow {\scriptstyle \eta'} & \\
& \omega_{\mathcal{D}} &
\end{array}$$

commutes. So $\theta\colon (H, \eta) \to (H', \eta')$ is an isomorphism of tensor functors. □

Corollary 1.11 *Let $(\mathcal{A}_1, \omega_{\mathcal{A}_1})$, $(\mathcal{A}_2, \omega_{\mathcal{A}_2})$ and $(\mathcal{B}, \omega_{\mathcal{B}})$ be neutralized Tannakian categories with fundamental groups G_1, G_2 and H. For tensor functors $(\mathcal{A}_1, \omega_{\mathcal{A}_1}) \to$*

$(\mathcal{B}, \omega_{\mathcal{B}})$ and $(\mathcal{A}_2, \omega_{\mathcal{A}_2}) \to (\mathcal{B}, \omega_{\mathcal{B}})$, the fundamental group of the neutralized Tannakian category $(\mathcal{A}_1 \times_{\mathcal{B}} \mathcal{A}_2, \omega_{\mathcal{A}_1 \times_{\mathcal{B}} \mathcal{A}_2})$ is the amalgamated free product $G_1 *_H G_2$.

Proof Let G denote the fundamental group of $(\mathcal{A}_1 \times_{\mathcal{B}} \mathcal{A}_2, \omega_{\mathcal{A}_1 \times_{\mathcal{B}} \mathcal{A}_2})$. The 2-commutative diagram

$$\begin{array}{ccccc} & & (\mathcal{A}_1 \times_{\mathcal{B}} \mathcal{A}_2, \omega_{\mathcal{A}_1 \times_{\mathcal{B}} \mathcal{A}_2}) & & \\ & \swarrow^{(P_1, \chi_1)} & & \searrow^{(P_2, \chi_2)} & \\ (\mathcal{A}_1, \omega_{\mathcal{A}_1}) & & & & (\mathcal{A}_2, \omega_{\mathcal{A}_2}) \\ & \searrow & & \swarrow & \\ & & (\mathcal{B}, \omega_{\mathcal{B}}) & & \end{array} \tag{3}$$

of neutralized Tannakian categories dualizes to a commutative diagram

$$\begin{array}{ccccc} & & G & & \\ & \nearrow^{i_1} & & \nwarrow^{i_2} & \\ G_1 & & & & G_2 \\ & \nwarrow & & \nearrow & \\ & & H & & \end{array}$$

of k-groups. Let

$$\begin{array}{ccccc} & & G' & & \\ & \nearrow^{i'_1} & & \nwarrow^{i'_2} & \\ G_1 & & & & G_2 \\ & \nwarrow & & \nearrow & \\ & & H & & \end{array} \tag{4}$$

be a commutative diagram of k-groups. As the morphisms between fundamental groups are in bijection with isomorphism classes of tensor functors between neutralized Tannakian categories (Lemma 1.7), there exists a 2-commutative diagram

$$\begin{array}{ccccc} & & (\mathrm{Rep}(G'), \omega_{G'}) & & \\ & \swarrow^{(Q_1, \epsilon_1)} & & \searrow^{(Q_2, \epsilon_2)} & \\ (\mathcal{A}_1, \omega_{\mathcal{A}_1}) & & & & (\mathcal{A}_2, \omega_{\mathcal{A}_2}) \\ & \searrow & & \swarrow & \\ & & (\mathcal{B}, \omega_{\mathcal{B}}) & & \end{array} \tag{5}$$

of neutralized Tannakian categories that dualizes to (4). By Proposition 1.10 there exists a morphism $((H, \eta), \beta_1, \beta_2)$ of 2-commutative diagrams from (5) to (3). The morphism $(H, \eta)\colon (\mathrm{Rep}(G'), \omega_{G'}) \to (\mathcal{A}_1 \times_{\mathcal{B}} \mathcal{A}_2, \omega_{\mathcal{A}_1 \times_{\mathcal{B}} \mathcal{A}_2})$ dualizes to a morphism $\phi\colon G \to G'$ of k-groups. The 2-commutativity of

$$\begin{array}{ccc}
(\mathrm{Rep}(G'), \omega_{G'}) & \xrightarrow{(H,\eta)} & (\mathcal{A}_1 \times_{\mathcal{B}} \mathcal{A}_2, \omega_{\mathcal{A}_1 \times_{\mathcal{B}} \mathcal{A}_2}) \\
{\scriptstyle (Q_j,\epsilon_j)} \searrow & & \swarrow {\scriptstyle (P_j,\chi_j)} \\
& (\mathcal{A}_j, \omega_{\mathcal{A}_j}) &
\end{array}$$

realized by β_j implies that

$$\begin{array}{ccc}
G & \xrightarrow{\phi} & G' \\
{\scriptstyle i_j} \nwarrow & & \nearrow {\scriptstyle i'_j} \\
& G_j &
\end{array} \tag{6}$$

commutes for $j = 1, 2$. Finally, the uniqueness of ϕ (making (6) commute for $j = 1, 2$) follows from the fact that $((H, \eta), \beta_1, \beta_2)$ is unique up to an isomorphism. □

Corollary 1.12 *Let $H \to G_1$ and $H \to G_2$ be morphisms of k-groups. Then the amalgamated free product $G_1 *_H G_2$ exists.*

Proof The restriction functors $\mathrm{Rep}(G_i) \to \mathrm{Rep}(H)$ $(i = 1, 2)$ can be regarded as tensor functors $(\mathrm{Rep}(G_i), \omega_{G_i}) \to (\mathrm{Rep}(H), \omega_H)$ between neutralized Tannakian categories. By Corollary 1.11 the fundamental group of $\mathrm{Rep}(G_1) \times_{\mathrm{Rep}(H)} \mathrm{Rep}(G_2)$ is $G_1 *_H G_2$. □

2 The Hopf Algebra of the Amalgamated Free Product

In this section we offer an explicit description of the Hopf algebra $k[G_1 *_H G_2]$ of the amalgamated free product of two morphisms $H \to G_1$ and $H \to G_2$ of k-groups in terms of the Hopf algebras $k[G_1]$, $k[G_2]$ and $k[H]$. In other words, we explicitly describe the fibre product in the category of Hopf algebras. (Recall that all Hopf algebras are assumed to be commutative.) We also discuss properties of amalgamated free products, such as being connected or geometrically reduced.

2.1 Fibre Products of Hopf Algebras

To motivate our explicit construction of the fibre product of Hopf algebras, let us revisit the construction of the amalgamated free product in the category of groups from a more categorical/conceptual perspective. For now, assume that $\phi_1 : H \to G_1$ and $\phi_2 \colon H \to G_2$ are morphisms of abstract groups. The amalgamated free product $G_1 *_H G_2$ (in the category of groups) can be constructed as follows: First form the disjoint union $G_1 \uplus G_2$ and then build the free monoid $M(G_1 \uplus G_2)$ on the set $G_1 \uplus G_2$. Finally, the amalgamated free product $G_1 *_H G_2$ is obtained from $M(G_1 \uplus G_2)$, by quotienting out by the relations

$$g_1 \cdot g_1' = g_1 g_1' \quad \forall\, g_1, g_1' \in G_1,$$

$$g_2 \cdot g_2' = g_2 g_2' \quad \forall\, g_2, g_2' \in G_2,$$

$$1_{G_1} = 1_{M(G_1 \uplus G_2)},$$

$$1_{G_2} = 1_{M(G_1 \uplus G_2)},$$

$$\phi_1(h) = \phi_2(h) \quad \forall\, h \in H.$$

Here the $\cdot$ is used to denote multiplication in $M(G_1 \uplus G_2)$.

This construction can be dualized. Let $k[G_1] \to k[H]$ and $k[G_2] \to k[H]$ be morphisms of Hopf algebras. The disjoint union $G_1 \uplus G_2$ corresponds to the product algebra $k[G_1] \oplus k[G_2] = k[G_1] \times k[G_2]$. The free monoid is the left adjoint of the forgetful functor from the category of monoids to the category of sets. In the Hopf algebra world, we should thus consider the right adjoint $A \rightsquigarrow C(A)$ of the forgetful functor from the category of bialgebras to the category of algebras. Finally, taking the quotient of $M(G_1 \uplus G_2)$, corresponds to choosing an appropriate subbialgebra $k[G_1 *_H G_2]$ of $C(k[G_1] \times k[G_2])$ (that turns out to be a Hopf algebra).

To carry out the above idea, we need an explicit description of the right adjoint $A \rightsquigarrow C(A)$ of the forgetful functor from the category of bialgebras to the category of algebras. To this end, let us recall the construction of the *cofree coalgebra* $C(V)$ on a k-vector space V. See [17, Sect. 6.4] or [13]. It comes with a k-linear map $\pi \colon C(V) \to V$ characterized by the following universal property: If D is a coalgebra and $\varphi \colon D \to V$ is a k-linear map, there exists a unique morphism $\tilde{\varphi} \colon D \to C(V)$ of coalgebras making

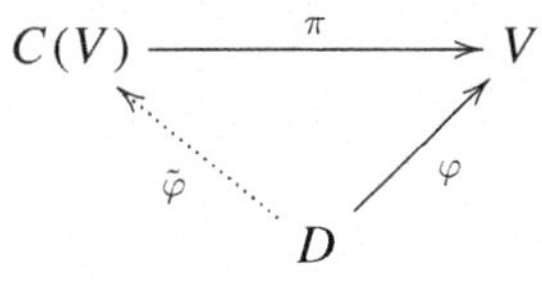

commutative. In other words, the functor $V \rightsquigarrow C(V)$ is right adjoint to the forgetful functor from coalgebras to vector spaces. We consider the tensor algebra $TV = k \oplus V \oplus V^{\otimes 2} \oplus \cdots$. However, we disregard the usual (non-commutative) algebra structure on TV. Instead, we define a coalgebra structure on TV. For every $n \in \mathbb{N}$ and $0 \leq i, j$ with $i + j = n$, we have an isomorphism $\psi_{i,j} \colon V^{\otimes n} \to V^{\otimes i} \otimes V^{\otimes j}$. Here

$V^{0\otimes} = k$ and $V^{1\otimes} = V$. For $t \in TV$ homogeneous of degree n, i.e., $t \in V^{\otimes n} \subseteq TV$, we define

$$\Delta(t) = \sum_{i=0}^{n} \psi_{i,n-i}(t) \in \bigoplus_{i=0}^{n} V^{\otimes i} \otimes V^{\otimes(n-i)} \subseteq TV \otimes TV.$$

By linearity, this assignment extends to a k-linear map $\Delta \colon TV \to TV \otimes TV$. Define $\epsilon \colon TV \to k$ as the projection onto the degree zero component. It is straight forward to check that (TV, Δ, ϵ) is a coalgebra. However, this is not yet the cofree coalgebra on V. To construct $C(V)$ we need to consider the completion $\hat{T}V = k \times V \times V^{\otimes 2} \times \cdots$ of TV. Thinking of the elements of $\hat{T}V$ as infinite sums, we have an inclusion $TV \subseteq \hat{T}V$.

One cannot extend $\Delta \colon TV \to TV \otimes TV$ directly to a map $\Delta \colon \hat{T}V \to \hat{T}V \otimes \hat{T}V$, because the natural extension of $\Delta \colon TV \to TV \otimes TV$ is a map $\Delta \colon \hat{T}V \to \hat{T}V \hat{\otimes} \hat{T}V$, where $\hat{T}V \hat{\otimes} \hat{T}V = \prod_{i,j \in \mathbb{N}} V^{\otimes i} \otimes V^{\otimes j}$. Note that $\hat{T}V \otimes \hat{T}V$ is a proper subset of $\hat{T}V \hat{\otimes} \hat{T}V$: If we think of an element of $\hat{T}V \hat{\otimes} \hat{T}V$ as an infinite matrix

$$\begin{pmatrix} f_{00} & f_{01} & f_{02} & \cdots \\ f_{10} & f_{11} & f_{12} & \cdots \\ \vdots & \vdots & & \end{pmatrix} \tag{7}$$

with $f_{ij} \in V^{\otimes i} \otimes V^{\otimes j}$, then the elements of $\hat{T}V \otimes \hat{T}V$ are finite sums of elements of the form

$$f \otimes g = \begin{pmatrix} f_0 \otimes g_0 & f_0 \otimes g_1 & f_0 \otimes g_2 & \cdots \\ f_1 \otimes g_0 & f_1 \otimes g_1 & f_1 \otimes g_2 & \cdots \\ \vdots & \vdots & & \end{pmatrix}$$

with $f = (f_0, f_1, \ldots)$, $g = (g_0, g_1, \ldots) \in \hat{T}V$. Note that with respect to this matrix representation, the map $\Delta \colon \hat{T}V \to \hat{T}V \hat{\otimes} \hat{T}V$ is given by

$$\Delta(f) = \begin{pmatrix} \psi_{00}(f_0) & \psi_{01}(f_1) & \psi_{02}(f_2) & \cdots \\ \psi_{10}(f_1) & \psi_{11}(f_2) & & \\ \psi_{21}(f_2) & & & \\ \vdots & & & \ddots \end{pmatrix}$$

for $f = (f_0, f_1, \ldots) \in \hat{T}V$. Set

$$C(V) = \{f \in \hat{T}V \mid \Delta(f) \in \hat{T}V \otimes \hat{T}V \subseteq \hat{T}V \hat{\otimes} \hat{T}V\}.$$

Following [13], we call the elements of $\hat{T}V$ that belong to $C(V)$ *representative*.

One can show [13, Theorem 3.12], that $\Delta(C(V)) \subseteq C(V) \otimes C(V) \subseteq \hat{T}V \otimes \hat{T}V$ and so $C(V)$ (together with the induced structure maps) is a coalgebra. Let $\pi\colon C(V) \to V$ be the projection to the degree one component. Then π satisfies the universal property of the cofree coalgebra [13, Theorem 3.14]. Indeed, if (D, μ, ε) is a coalgebra with a k-linear map $\varphi\colon D \to V$, the unique $\tilde{\varphi}\colon D \to C(V)$ such that $\pi\tilde{\varphi} = \varphi$ is given by $\tilde{\varphi}(d) = (\varepsilon(d), \varphi(d), \varphi^{\otimes 2}(\mu_2(d)), \varphi^{\otimes 3}(\mu_3(d)), \dots) \in C(V)$ for $d \in D$, where $\mu_i\colon D \to D^{\otimes i}$ is the $(i-1)$st iteration of μ.

If A is a k-algebra, then $C(A)$ naturally acquires the structure of a k-algebra (Cf. [17, p. 134]). Indeed, this follows from the universal property: Let $m\colon A \otimes A \to A$ be the multiplication. As $C(A) \otimes C(A)$ has the structure of a coalgebra (the product coalgebra), the map $C(A) \otimes C(A) \xrightarrow{\pi\otimes\pi} A \otimes A \xrightarrow{m} A$ lifts to a multiplication map $C(A) \otimes C(A) \to C(A)$. Similarly, as k has the trivial coalgebra structure, the unit $k \to A$ lifts to a unit $k \to C(A)$. Explicitly, the unit $k \to C(A)$ is the map $k \to C(A)$, $\lambda \mapsto (\lambda, \lambda, \lambda, \dots)$ and multiplication in $C(A)$ is componentwise, using the multiplication in $A^{\otimes i}$. One sees that $C(A)$ is a bialgebra and $\pi\colon C(A) \to A$ is a morphism of algebras. Moreover, for every bialgebra B with an algebra morphism $\varphi\colon B \to A$, the unique $\tilde{\varphi}\colon B \to C(A)$ such that $\tilde{\varphi}\pi = \varphi$ is a morphism of bialgebras. In other words, $A \rightsquigarrow C(A)$ is the sought for right adjoint of the forgetful functor from bialgebras to algebras.

Let $\alpha_1\colon k[G_1] \to k[H]$ and $\alpha_2\colon k[G_2] \to k[H]$ be morphisms of Hopf algebras. We will construct their fibre product $k[G_1] \times_{k[H]} k[G_2] = k[G_1 *_H G_2]$ as a suitable subalgebra of $C(k[G_1] \oplus k[G_2])$.

Note that for $n \geq 0$ the vector space $(k[G_1] \oplus k[G_2])^{\otimes n}$ is the direct sum of the subspaces

$$T_{i_1,\dots,i_n} = k[G_{i_1}] \otimes \cdots \otimes k[G_{i_n}]$$

with $i_1, \dots, i_n \in \{1, 2\}$. For $\ell \in \{1, \dots, n\}$ define a map

$$\psi^{\ell}_{i_1,\dots,i_n}\colon T_{i_1,\dots,i_n} \to T_{i_1,\dots,i_{\ell-1},i_\ell,i_\ell,i_{\ell+1},\dots,i_n}$$

by applying the comultiplication $\Delta_{i_\ell}\colon k[G_{i_\ell}] \to k[G_{i_\ell}] \otimes k[G_{i_\ell}]$ on the ℓth factor and the identity on all the other factors. Similarly, define

$$\delta^{\ell}_{i_1,\dots,i_n}\colon T_{i_1,\dots,i_n} \to T_{i_1,\dots,i_{\ell-1},i_{\ell+1},\dots,i_n}$$

by applying the counit $\epsilon_{i_\ell}\colon k[G_{i_\ell}] \to k$ on the ℓth factor and the identity on all the other factors. We also define

$$\theta^{\ell}_{i_1,\dots,i_n}\colon T_{i_1,\dots,i_n} \to k[G_{i_1}] \otimes \cdots \otimes k[G_{i_{\ell-1}}] \otimes k[H] \otimes k[G_{i_\ell}] \otimes \cdots \otimes k[G_{i_n}]$$

by applying $\alpha_{i_\ell}\colon k[G_{i_\ell}] \to k[H]$ on the ℓth factor and the identity on all the other factors.

An element $f \in \hat{T}(k[G_1] \oplus k[G_2])$ can uniquely be written in the form

$$f = \sum_{n=0}^{\infty} \sum_{(i_1,\ldots,i_n)\in\{1,2\}^n} f_{i_1,\ldots,i_n} \tag{8}$$

with $f_{i_1,\ldots,i_n} \in T_{i_1,\ldots,i_n}$. Note that for $n = 0$, $T_{i_1,\ldots,i_n} = k$ and in (8), the inner sum for $n = 0$ has one summand $f_\emptyset \in k$.

We say that f is *multiplicatively coherent* if

$$\psi^\ell_{i_1,\ldots,i_n}(f_{i_1,\ldots,i_n}) = f_{i_1,\ldots,i_\ell,i_\ell,\ldots,i_n} \tag{9}$$

for all $n \geq 1$, $i_1, \ldots, i_n \in \{1, 2\}$ and $\ell \in \{1, \ldots, n\}$. If

$$\delta^\ell_{i_1,\ldots,i_n}(f_{i_1,\ldots,i_n}) = f_{i_1,\ldots,i_{\ell-1},i_{\ell+1},\ldots,i_n} \tag{10}$$

for all $n \geq 1$, $i_1, \ldots, i_n \in \{1, 2\}$ and $\ell \in \{1, \ldots, n\}$, then f is *unitary coherent*. If

$$\theta^\ell_{i_1,\ldots,i_{\ell-1},1,i_{\ell+1},\ldots,i_n}(f_{i_1,\ldots,i_{\ell-1},1,i_{\ell+1},\ldots,i_n}) = \theta^\ell_{i_1,\ldots,i_{\ell-1},2,i_{\ell+1},\ldots,i_n}(f_{i_1,\ldots,i_{\ell-1},2,i_{\ell+1},\ldots,i_n}) \tag{11}$$

for all $n \geq 1$, all $\ell \in \{1, \ldots, n\}$ and $i_1, \ldots, i_{\ell-1}, i_{\ell+1}, \ldots, i_n \in \{1, 2\}$, then f is *H-coherent*. We call f *coherent* if it is multiplicatively, unitary and H-coherent.

If we think of f as a function on the words in $G_1 \uplus G_2$, multiplicative coherence means that f takes the same value on words that only differ by replacing two adjacent elements from the same group with their product. Unital coherence means that f takes the same value on words that only differ by deleting the identity element of G_1 or the identity element of G_2. The meaning of H-coherence is that if $g_1 \in G_1$ and $g_2 \in G_2$ come from the same element in H, then the value of f on a word does not change if a letter g_1 is replaced by g_2. We define

$$k[G_1 *_H G_2] = \{f \in C(k[G_1] \oplus k[G_2]) |\ f \text{ is coherent}\}.$$

As the $\psi^\ell_{i_1,\ldots,i_n}$'s, $\delta^\ell_{i_1,\ldots,i_n}$'s and $\theta^\ell_{i_1,\ldots,i_n}$'s are morphisms of k-algebras, it is clear that $k[G_1 *_H G_2]$ is a subalgebra of $C(k[G_1] \oplus k[G_2])$. On the other hand, it is not so clear that $k[G_1 *_H G_2]$ is a subcoalgebra of $C(k[G_1] \oplus k[G_2])$. To see this, we will work with a different characterization of representative elements based on duality.

We briefly go back to the general setup where V is a vector space over k. We denote with $V^* = \mathrm{Hom}_k(V, k)$ the dual vector space of functionals on V. Every element f of $\hat{T}V$ defines a functional $\mathrm{Fun}(f) \in (TV^*)^*$ on TV^*. In detail, for every $n \geq 0$ we have an evaluation map

$$V^{\otimes n} \times (V^*)^{\otimes n} \to k, \ (v, a) \mapsto a(v)$$

determined by $(v_1 \otimes \cdots \otimes v_n, a_1 \otimes \cdots \otimes a_n) \mapsto a_1(v_1) \cdots a_n(v_n)$. Then for $f = (f_0, f_1, \ldots) \in \hat{T}V$ and $a = a_0 + a_1 + \cdots + a_r \in TV^*$ one sets $\mathrm{Fun}(f)(a) = \sum_{i=0}^r a_i(f_i)$. The map $\hat{T}V \to (TV^*)^*$, $f \mapsto \mathrm{Fun}(f)$ is injective. In a similar fashion, any $f \in \hat{T}V \hat{\otimes} \hat{T}V = \prod_{i,j\in\mathbb{N}} V^{\otimes i} \otimes V^{\otimes j}$ defines a functional $\mathrm{Fun}(f)$ on

$TV^* \otimes TV^*$. The "comultiplication" Δ on $\hat{T}V$ is dual to the usual (noncommutative) multiplication in the tensor algebra TV^* in the sense that

$$\mathrm{Fun}(\Delta(f))(a \otimes b) = \mathrm{Fun}(f)(ab) \tag{12}$$

for $f \in \hat{T}V$ and $a, b \in TV^*$.

For a functional F on TV^* and $b \in TV^*$ we define a functional $R_b F$ on TV^* by $(R_b F)(a) = F(ab)$ for all $a \in TV^*$. Similarly, we define $L_a F$ by $(L_a F)(b) = F(ab)$. We abbreviate $R_b f = R_b \,\mathrm{Fun}(f)$ and $L_a f = L_a \,\mathrm{Fun}(f)$. Note that

$$Rf = \{R_b f \mid b \in TV^*\} \text{ and } Lf = \{L_a f \mid a \in TV^*\}$$

are subspaces of $(TV^*)^*$. The following lemma is essentially Theorem 3.12 in [13].

Lemma 2.1 *For $f \in \hat{T}V$ the following are equivalent:*

(i) The element f is representative.
(ii) The vector space Rf is finite dimensional.
(iii) The vector space Lf is finite dimensional.

Moreover, if these conditions are satisfied, then $\Delta(f)$ can be written in the form $\Delta(f) = \sum_{i=1}^{r} g_i \otimes h_i$ with $g_i, h_i \in \hat{T}V$ representative and $\mathrm{Fun}(g_i) \in Rf$ *and* $\mathrm{Fun}(h_i) \in Lf$ *for $i = 1, \ldots, r$.*

Proof Assume that f is representative. So $\Delta(f) = \sum_{i=1}^{r} g_i \otimes h_i$ for some $g_i, h_i \in \hat{T}V$. Using (12) we find

$$L_a f(b) = R_b f(a) = \mathrm{Fun}(f)(ab) = \sum_{i=0}^{r} \mathrm{Fun}(g_i)(a)\,\mathrm{Fun}(h_i)(b)$$

for $a, b \in TV^*$. Thus $R_b f$ lies in the subspace generated by $\mathrm{Fun}(g_1), \ldots, \mathrm{Fun}(g_r)$ and $L_a f$ lies in the subspace generated by $\mathrm{Fun}(h_1), \ldots, \mathrm{Fun}(h_r)$. Thus (i) implies (ii) and (iii). The implication (ii) $\Rightarrow$ (i) is shown in [13, Theorem 3.12] and the implication (iii) $\Rightarrow$ (i) works similar to (ii) $\Rightarrow$ (i). The final statement of the lemma is contained in the proof of [13, Theorem 3.12]. See [13, Remark 3.24]. □

We return to the discussion of the subalgebra $k[G_1 *_H G_2]$ of $C(k[G_1] \oplus k[G_2])$. For $n \geq 0$, the vector space $(k[G_1]^* \oplus k[G_2]^*)^{\otimes n}$ is the direct sum of the subspaces

$$T^*_{i_1,\ldots,i_n} = k[G_{i_1}]^* \otimes \cdots \otimes k[G_{i_n}]^*$$

with $i_1, \ldots, i_n \in \{1, 2\}$.

The comultiplication $\Delta_i \colon k[G_i] \to k[G_i] \otimes k[G_i]$ dualizes to a map $(k[G_i] \otimes k[G_i])^* \to k[G_i]^*$. Precomposing with $k[G_i]^* \otimes k[G_i]^* \to (k[G_i] \otimes k[G_i])^*$, thus yields a map $\Delta_i^* \colon k[G_i]^* \otimes k[G_i]^* \to k[G_i]^*$.

For $i_1, \ldots, i_n \in \{1, 2\}$ and $\ell \in \{1, \ldots, n\}$ we define a map

$$^*\psi^\ell_{i_1,\dots,i_n} : T^*_{i_1,\dots,i_\ell,i_\ell,\dots,i_n} \to T^*_{i_1,\dots,i_n}$$

by applying $\Delta^*_{i_\ell}$ on $k[G_{i_\ell}]^* \otimes k[G_{i_\ell}]^*$ and the identity on all the other factors. Similarly, we define a map

$$^*\delta^\ell_{i_1,\dots,i_n} : T^*_{i_1,\dots,i_{\ell-1},i_{\ell+1},\dots,i_n} \to T^*_{i_1,\dots,i_n}$$

by sending $a_1 \otimes \cdots \otimes a_{n-1}$ to $a_1 \otimes \cdots \otimes a_{\ell-1} \otimes \epsilon_{i_\ell} \otimes a_\ell, \dots, a_{n-1}$. We also define

$$^*\theta^\ell_{i_1,\dots,i_n} : k[G_{i_1}]^* \otimes \cdots \otimes k[G_{i_{\ell-1}}]^* \otimes k[H]^* \otimes k[G_{i_{\ell+1}}]^* \otimes \cdots \otimes k[G_{i_n}]^* \longrightarrow T^*_{i_1,\dots,i_n}$$

by applying $\alpha^*_{i_\ell} : k[H]^* \to k[G_{i_\ell}]^*$ on the ℓth factor and the identity on all the other factors.

A functional $F \in T(k[G_1]^* \oplus k[G_2]^*)^*$ is called *multiplicatively coherent* if

$$\begin{array}{ccc} T^*_{i_1,\dots,i_\ell,i_\ell,\dots,i_n} & \xrightarrow{\ ^*\psi^\ell_{i_1,\dots,i_n}\ } & T^*_{i_1,\dots,i_n} \\ & {\scriptstyle F}\searrow \quad \swarrow{\scriptstyle F} & \\ & k & \end{array}$$

commutes for every $^*\psi^\ell_{i_1,\dots,i_n}$. If

$$\begin{array}{ccc} T^*_{i_1,\dots,i_{\ell-1},i_{\ell+1},\dots,i_n} & \xrightarrow{\ ^*\delta^\ell_{i_1,\dots,i_n}\ } & T^*_{i_1,\dots,i_n} \\ & {\scriptstyle F}\searrow \quad \swarrow{\scriptstyle F} & \\ & k & \end{array}$$

commutes for every $^*\delta^\ell_{i_1,\dots,i_n}$, then F is *unitary coherent*. If

$$\begin{array}{ccccc} & & k[G_{i_1}]^* \otimes \cdots \otimes k[G_{i_{\ell-1}}]^* \otimes k[H]^* \otimes k[G_{i_{\ell+1}}]^* \otimes \cdots \otimes k[G_{i_n}]^* & & \\ & {\scriptstyle ^*\theta^\ell_{\mathbf{j}_1}}\swarrow & & \searrow{\scriptstyle ^*\theta^\ell_{\mathbf{j}_2}} & \\ T^*_{\mathbf{j}_1} & & & & T^*_{\mathbf{j}_2} \\ & {\scriptstyle F}\searrow & & \swarrow{\scriptstyle F} & \\ & & k & & \end{array}$$

commutes, where $\mathbf{j}_1 = (i_1, \dots, i_{\ell-1}, 1, i_{\ell+1}, \dots, i_n)$ and $\mathbf{j}_2 = (i_1, \dots, i_{\ell-1}, 2, i_{\ell+1}, \dots, i_n)$, for all $n \geq 1$, $\ell \in \{1, \dots, n\}$ and $i_1, \dots, i_{\ell-1}, i_{\ell+1}, \dots, i_n \in \{1, 2\}$, then F is H-coherent. Finally, F is *coherent* if it is multiplicatively, unitary and H-coherent.

The following lemma shows that whether or not $f \in \hat{T}(k[G_1] \oplus k[G_2])$ is coherent can be tested on $\mathrm{Fun}(f)$.

Lemma 2.2 *Let $f \in \hat{T}(k[G_1] \oplus k[G_2])$. Then f is coherent if and only if* $\mathrm{Fun}(f)$ *is coherent.*

Proof We first treat the case of multiplicative coherence. Fix $n \geq 1$, $i_1, \ldots, i_n \in \{1, 2\}^n$ and $\ell \in \{1, \ldots, n\}$. Let $a_1 \otimes \cdots \otimes a_{n+1} \in T^*_{i_1,\ldots,i_\ell,i_\ell,\ldots,i_n}$. As $\Delta^*_{i_\ell}(a_\ell \otimes a_{\ell+1})(h) = (a_\ell \otimes a_{\ell+1})(\Delta_{i_\ell}(h))$ for $h \in k[G_{i_\ell}]$, we see that

$$
\begin{aligned}
{}^*\psi^\ell_{i_1,\ldots,i_n}&(a_1 \otimes \cdots \otimes a_{n+1})(h_1 \otimes \cdots \otimes h_n)\\
&= (a_1 \otimes \cdots \otimes a_{\ell-1} \otimes \Delta^*_{i_\ell}(a_\ell \otimes a_{\ell+1}) \otimes a_{\ell+2} \otimes \cdots \otimes a_{n+1})\\
&\quad (h_1 \otimes \cdots \otimes h_n)\\
&= (a_1 \otimes \cdots \otimes a_{n+1})(h_1 \otimes \cdots \otimes \Delta_{i_\ell}(h_\ell) \otimes \cdots \otimes h_n)\\
&= (a_1 \otimes \cdots \otimes a_{n+1})(\psi^\ell_{i_1,\ldots,i_n}(h_1 \otimes \cdots \otimes h_n))
\end{aligned}
$$

for $h_1 \otimes \cdots \otimes h_n \in T_{i_1,\ldots,i_n}$. So ${}^*\psi^\ell_{i_1,\ldots,i_n}(a)(h) = a(\psi^\ell_{i_1,\ldots,i_n}(h))$ for all $a \in T^*_{i_1,\ldots,i_\ell,i_\ell,\ldots,i_n}$ and $h \in T_{i_1,\ldots,i_n}$.

Assume that f is multiplicatively coherent. Then, writing f as is in (8), we have ${}^*\psi^\ell_{i_1,\ldots,i_n}(a)(f_{i_1,\ldots,i_n}) = a(f_{i_1,\ldots,i_\ell,i_\ell,\ldots,i_n})$ for all $a \in T^*_{i_1,\ldots,i_\ell,i_\ell,\ldots,i_n}$. Thus the diagram

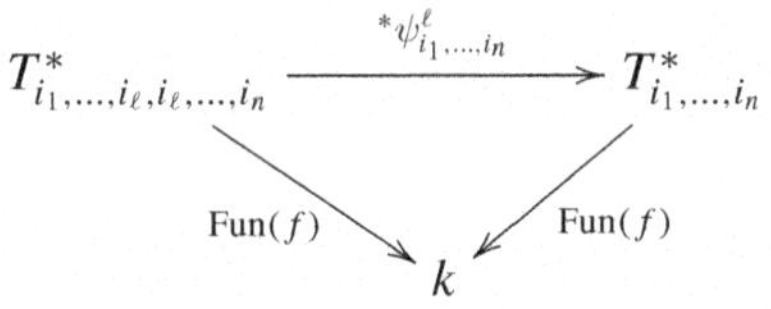

commutes and $\mathrm{Fun}(f)$ is multiplicatively coherent.

Conversely, assume that $\mathrm{Fun}(f)$ is multiplicatively coherent. Then

$$
a(\psi^\ell_{i_1,\ldots,i_n}(f_{i_1,\ldots,i_n})) = {}^*\psi^\ell_{i_1,\ldots,i_n}(a)(f_{i_1,\ldots,i_n}) = a(f_{i_1,\ldots,i_\ell,i_\ell,\ldots,i_n})
$$

for all $a \in T^*_{i_1,\ldots,i_\ell,i_\ell,\ldots,i_n}$ and so $\psi^\ell_{i_1,\ldots,i_n}(f_{i_1,\ldots,i_n}) = f_{i_1,\ldots,i_\ell,i_\ell,\ldots,i_n}$, i.e., f is multiplicatively coherent.

The case of unital coherence is similar, using that ${}^*\delta^\ell_{i_1,\ldots,i_n}(a)(h) = a(\delta^\ell_{i_1,\ldots,i_n}(h))$ for $a \in T^*_{i_1,\ldots,i_{\ell-1},i_{\ell+1},\ldots,i_n}$ and $h \in T_{i_1,\ldots,i_n}$.

For the case of H-coherence one uses that

$$
{}^*\theta^\ell_{i_1,\ldots,i_{\ell-1},i,i_{\ell+1},\ldots,i_n}(a)(h) = a(\theta^\ell_{i_1,\ldots,i_{\ell-1},i,i_{\ell+1},\ldots,i_n}(h))
$$

for

$$
a \in k[G_{i_1}]^* \otimes \cdots \otimes k[G_{i_{\ell-1}}]^* \otimes k[H]^* \otimes k[G_{i_{\ell+1}}]^* \otimes \cdots \otimes k[G_{i_n}]^*,
$$

$h \in T_{i_1,\ldots,i_{\ell-1},i,i_{\ell+1},\ldots,i_n}$ and $i = 1, 2$. □

The following lemma is the main reason, why the dual perspective is helpful for showing that $k[G_1 *_H G_2]$ is a subcoalgebra of $C(k[G_1] \oplus k[G_2])$.

Lemma 2.3 *Let F be a coherent functional on $T(k[G_1]^* \oplus k[G_2]^*)$. Then $R_b F$ is coherent for every $b \in T(k[G_1]^* \oplus k[G_2]^*)$. Similarly for $L_a F$.*

Proof We first treat the case of multiplicative coherence. As the sum of multiplicatively coherent functionals is multiplicatively coherent and $R_{b_1+b_2}F = R_{b_1}F + R_{b_2}F$, we can reduce to the case that $b = b_1 \otimes \cdots \otimes b_n$ is a pure tensor in some $T^*_{i_1,\ldots,i_n}$. Moreover, as $R_{b_1 b_2}F = R_{b_1}(R_{b_2}F)$, we are reduced to the case $b \in k[G_i]^*$ for some $i \in \{1, 2\}$. Consider the diagram

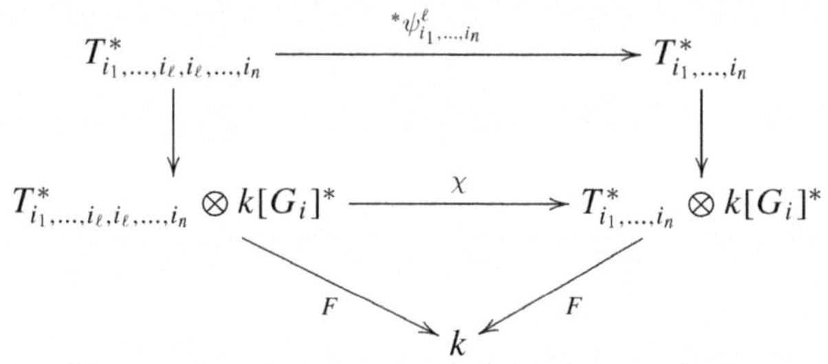

where χ is the base change of $^*\psi^\ell_{i_1,\ldots,i_n}$ and the two vertical arrows are multiplication with b (using the noncommutative multiplication of the tensor algebra). Then the top square commutes. As $\chi = {}^*\psi^\ell_{i_1,\ldots,i_n,i}$ and F is multiplicatively coherent, the lower triangle commutes. So the whole diagram commutes, showing that $R_b F$ is multiplicatively coherent.

The argument for unital coherence and H-coherence is similar. The proof for L instead of R is analogous. □

We can finally show that $k[G_1 *_H G_2]$ is a subcoalgebra of $C(k[G_1] \oplus k[G_2])$. Let $f \in k[G_1 *_H G_2]$. We have to show that

$$\Delta(f) \in k[G_1 *_H G_2] \otimes k[G_1 *_H G_2] \subseteq C(k[G_1] \oplus k[G_2]) \otimes C(k[G_1] \oplus k[G_2]).$$

From Lemma 2.1 we know that $\Delta(f)$ can be written as $\Delta(f) = \sum_{i=1}^r g_i \otimes h_i$ with $g_i, h_i \in C(k[G_1] \oplus k[G_2])$, $\mathrm{Fun}(g_1), \ldots, \mathrm{Fun}(g_r) \in Rf$ and $\mathrm{Fun}(h_1), \ldots, \mathrm{Fun}(h_i) \in Lf$. Lemma 2.3 implies that the functionals $\mathrm{Fun}(g_1), \ldots, \mathrm{Fun}(g_r), \mathrm{Fun}(h_1), \ldots, \mathrm{Fun}(h_i)$ are coherent. So Lemma 2.2 implies that $g_1, \ldots, g_r, h_1, \ldots, h_r$ are coherent, i.e., they lie in $k[G_1 *_H G_2]$ as desired.

So $k[G_1 *_H G_2]$ is a subbialgebra of $C(k[G_1] \oplus k[G_2])$. The next step is to show that $k[G_1 *_H G_2]$ is a Hopf algebra, i.e., we need to find the antipode. Define

$$S\colon \hat{T}(k[G_1] \oplus k[G_2]) \to \hat{T}(k[G_1] \oplus k[G_2])$$

by applying the antipodes $S_i\colon k[G_i] \to k[G_i]$ $(i = 1, 2)$ on each of the factors and then reversing the order of the factors. So $S\colon T_{i_1,\ldots,i_n} \to T_{i_n,\ldots,i_1}$ is given by $S(h_1 \otimes$

$\cdots \otimes h_n) = S_{i_n}(h_n) \otimes \cdots \otimes S_{i_1}(h_1)$. We would like to know that $S(C(k[G_1] \oplus k[G_2])) \subseteq C(k[G_1] \oplus k[G_2])$.

Recall that for a vector space V, we defined $\hat{T}V \hat{\otimes} \hat{T}V = \prod_{i,j \in \mathbb{N}} V^{\otimes i} \otimes V^{\otimes j}$. Define $\tau \colon \hat{T}V \hat{\otimes} \hat{T}V \to \hat{T}V \hat{\otimes} \hat{T}V$ by

$$\tau \colon V^{\otimes i} \otimes V^{\otimes j} \to V^{\otimes j} \otimes V^{i}, \; a \otimes b \mapsto b \otimes a.$$

So in the matrix notation (7), the map τ transposes the matrix and then applies the flip to every matrix entry. From this description it is clear that τ maps $\hat{T}V \otimes \hat{T}V$ to $\hat{T}V \otimes \hat{T}V$.

For $f \in \hat{T}(k[G_1] \oplus k[G_2])$, a direct computation shows that $\Delta(S(f)) = \tau((S \otimes S)(\Delta(f)))$. So if $f \in C(k[G_1] \oplus k[G_2])$, then $\Delta(S(f)) \in \hat{T}(k[G_1] \oplus k[G_2]) \otimes \hat{T}(k[G_1] \oplus k[G_2])$, i.e., $S(f) \in C(k[G_1] \oplus k[G_2])$. Thus S restricts to a map $S \colon C(k[G_1] \oplus k[G_2]) \to C(k[G_1] \oplus k[G_2])$. The commutativity of the diagrams

$$\begin{array}{ccc} T_{i_1,\dots,i_n} & \xrightarrow{\psi^{\ell}_{i_1,\dots,i_n}} & T_{i_1,\dots,i_\ell,i_\ell,\dots,i_n} \\ \downarrow S & & \downarrow S \\ T_{i_n,\dots,i_1} & \xrightarrow{\psi^{n-\ell+1}_{i_n,\dots,i_1}} & T_{i_n,\dots,i_\ell,i_\ell,\dots,i_1} \end{array}$$

$$\begin{array}{ccc} T_{i_1,\dots,i_n} & \xrightarrow{\delta^{\ell}_{i_1,\dots,i_n}} & T_{i_1,\dots,i_{\ell-1},i_{\ell+1},\dots,i_n} \\ \downarrow S & & \downarrow S \\ T_{i_n,\dots,i_1} & \xrightarrow{\delta^{n-\ell+1}_{i_1,\dots,i_n}} & T_{i_n,\dots,i_{\ell+1},i_{\ell-1},\dots,i_1} \end{array}$$

and

$$\begin{array}{ccccc} T_{i_1,\dots,1,\dots i_n} & \xrightarrow{\theta^{\ell}_{i_1,\dots,1,\dots,i_n}} & k[G_{i_1}] \otimes \cdots \otimes k[H] \otimes \cdots \otimes k[G_{i_n}] & \xleftarrow{\theta^{\ell}_{i_1,\dots,2,\dots,i_n}} & T_{i_1,\dots,2,\dots,i_n} \\ \downarrow S & & \downarrow & & \downarrow S \\ T_{i_n,\dots,1,\dots i_1} & \xrightarrow{\theta^{n-\ell+1}_{i_n,\dots,1,\dots,i_1}} & k[G_{i_n}] \otimes \cdots \otimes k[H] \otimes \cdots \otimes k[G_{i_1}] & \xleftarrow{\theta^{n-\ell+1}_{i_n,\dots,2,\dots,i_1}} & T_{i_n,\dots,2,\dots,i_1} \end{array}$$

show that indeed S restricts to a map $S \colon k[G_1 *_H G_2] \to k[G_1 *_H G_2]$. To see that S is an antipode for the bialgebra $k[G_1 *_H G_2]$, we need to show that

$$\begin{array}{ccc} k[G_1 *_H G_2] & \xrightarrow{\Delta} & k[G_1 *_H G_2] \otimes k[G_1 *_H G_2] \\ \Big\downarrow{\scriptstyle \epsilon} & & \Big\downarrow{\scriptstyle S\cdot\mathrm{id}} \\ k & \longrightarrow & k[G_1 *_H G_2] \end{array} \tag{13}$$

commutes.

Let $f \in k[G_1 *_H G_2]$. Note that the product in $\hat{T}(k[G_1] \oplus k[G_2])$ is such that for $a \in T_{i_1,\dots,i_n}$ and $b \in T_{j_1,\dots,j_m}$ we have $ab = 0$, unless $m = n$ and $i_1 = j_1, \dots, i_n = j_n$. Thus, when trying to determine the image of $\Delta(f)$ under $S \cdot \mathrm{id}$, only the components of $\Delta(f)$ in $T_{i_n,\dots,i_1} \otimes T_{i_1,\dots,i_n} \subseteq \hat{T}(k[G_1] \oplus k[G_2]) \hat{\otimes} \hat{T}(k[G_1] \oplus k[G_2])$ are relevant, as all other components map to zero under $S \cdot \mathrm{id}$. So the component $(S \cdot \mathrm{id})(\Delta(f))_{i_1,\dots,i_n}$ of $(S \cdot \mathrm{id})(\Delta(f))$ in $T_{i_1,\dots,i_n}$, is the image of the component $f_{i_n,\dots,i_1,i_1,\dots,i_n}$ of f in $T_{i_n,\dots,i_1,i_1,\dots,i_n}$ under the map

$$\xi_{i_1,\dots,i_n} \colon T_{i_n,\dots,i_1,i_1,\dots,i_n} \to T_{i_1,\dots,i_n},\ a_n \otimes \dots \otimes a_1 \otimes b_1 \otimes \dots \otimes b_n$$
$$\mapsto S_{i_1}(a_1)b_1 \otimes \dots \otimes S_{i_n}(a_n)b_n.$$

We will show, by induction on n, that $\xi_{i_1,\dots,i_n}(f_{i_n,\dots,i_1,i_1,\dots,i_n}) = \epsilon(f) \in k$ as desired. For $n = 1$, $T_{i_1,i_1} = k[G_{i_1}] \otimes k[G_{i_1}]$, $T_{i_1} = k[G_{i_1}]$ and the commutative diagram

$$\begin{array}{ccc} k[G_{i_1}] \otimes k[G_{i_1}] & \xrightarrow{\xi_{i_1,i_1} = S_{i_1} \cdot \mathrm{id}} & k[G_{i_1}] \\ \Big\uparrow{\scriptstyle \psi^1_{i_1} = \Delta_{i_1}} & & \Big\uparrow \\ k[G_{i_1}] & \xrightarrow{\epsilon_{i_1}} & k \end{array}$$

shows that $\xi_{i_1,i_1}(f_{i_1,i_1}) = \epsilon_{i_1}(f_{i_1})$. But for $n = 1$, the condition $\delta^{\ell}_{i_1,\dots,i_n}(f_{i_1,\dots,i_n}) = f_{i_1,\dots,i_{\ell-1},i_{\ell+1},\dots,i_n}$ means that $\epsilon_{i_1}(f_{i_1}) = f_{\emptyset} = \epsilon(f)$, the degree zero component of f. This resolves the $n = 1$ case. For $n > 1$, the diagram

$$\begin{array}{ccc} T_{i_n,\dots,i_1,i_1,\dots,i_n} & \searrow{\scriptstyle \xi_{i_1,\dots,i_n}} & \\ \Big\uparrow{\scriptstyle \psi^n_{i_n,\dots,i_1,\dots,i_n}} & & \\ T_{i_n,\dots,i_1,\dots,i_n} & & T_{i_1,\dots,i_n} \\ \Big\downarrow{\scriptstyle \delta^n_{i_n,\dots,i_1,\dots,i_n}} & & \Big\uparrow \\ T_{i_n,\dots,i_2,i_2,\dots,i_n} & \xrightarrow{\xi_{i_2,\dots,i_n}} & T_{i_2,\dots,i_n} \end{array}$$

commutes. (This is probably most conveniently seen by looking at what happens with the tuples of group elements when considering the dual morphisms.) So, by induction,

$$\begin{aligned}\xi_{i_1,\dots,i_n}(f_{i_n,\dots,i_1,i_1,\dots,i_n}) &= \xi_{i_1,\dots,i_n}(\psi^n_{i_n,\dots,i_1,\dots,i_n}(f_{i_n,\dots,i_1,\dots,i_n}))\\ &= \xi_{i_2,\dots,i_n}(\delta^n_{i_n,\dots,i_1,\dots,i_n}(f_{i_n,\dots,i_1,\dots,i_n})) = \xi_{i_2,\dots,i_n}(f_{i_n,\dots,i_2,i_2,\dots,i_n})\\ &= \epsilon(f).\end{aligned}$$

Thus (13) commutes and $k[G_1 *_H G_2]$ is indeed a Hopf algebra.

The map $\pi\colon C(k[G_1] \oplus k[G_2]) \to k[G_1] \oplus k[G_2]$ can be restricted to $k[G_1 *_H G_2]$ and composing with the projections on $k[G_1] \oplus k[G_2]$ yields maps $p_1 : k[G_1 *_H G_2] \to k[G_1]$ and $p_2 : k[G_1 *_H G_2] \to k[G_1]$.

It is clear that p_i $(i = 1, 2)$ is a morphism of algebras. Compatibility with the counit follows from (10) with $n = 1$, whereas compatibility with the comultiplication follows from (9) with $n = 1$. The commutativity of the diagram

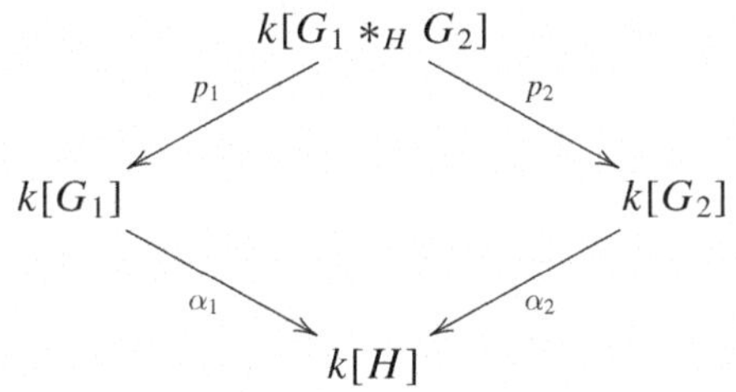

follows from (11) with $n = 1$.

Proposition 2.4 *The Hopf algebra $k[G_1 *_H G_2]$ together with the projections p_1, p_2 is the fibre product of $\alpha_1 : k[G_1] \to k[H]$ and $\alpha_2 : k[G_2] \to k[H]$ in the category of Hopf algebras.*

Proof Let $k[G]$ be a Hopf algebra with morphisms $q_1 : k[G] \to k[G_1]$ and $q_2 : k[G] \to k[G_2]$ of Hopf algebras such that $\alpha_1 q_1 = \alpha_2 q_2$. We then have a morphism $\varphi = (q_1, q_2) : k[G] \to k[G_1] \oplus k[G_2]$ of k-algebras. By the universal property of $C(k[G_1] \oplus k[G_2])$, there exists a unique morphism $\tilde{\varphi} : k[G] \to C(k[G_1] \oplus k[G_2])$ of bialgebras such that $\pi\tilde{\varphi} = \varphi$.

We claim that $\tilde{\varphi}$ maps $k[G]$ into $k[G_1 *_H G_2] \subseteq C(k[G_1] \oplus k[G_2])$. Let $f \in k[G]$. Then

$$\tilde{\varphi}(f) = (\epsilon_G(f), \varphi(f), \varphi^{\otimes 2}(\Delta^2_G(f)), \varphi^{\otimes 3}(\Delta^3_G(f)), \dots).$$

The commutative diagram

$$
\begin{array}{ccccccc}
 & & k[G]^{\otimes n} & \xrightarrow{\varphi^{\otimes n}} & (k[G_1]\oplus k[G_2])^{\otimes n} & \longrightarrow & T_{i_1,\dots,i_n} \\
 & \overset{\Delta_G^n}{\nearrow} & & & & & \\
k[G] & & & & & & \Big\downarrow {\scriptstyle \psi^{\ell}_{i_1,\dots,i_n}} \\
 & \underset{\Delta_G^{n+1}}{\searrow} & & & & & \\
 & & k[G]^{\otimes(n+1)} & \xrightarrow{\varphi^{\otimes(n+1)}} & (k[G_1]\oplus k[G_2])^{\otimes(n+1)} & \longrightarrow & T_{i_1,\dots,i_\ell,i_\ell,\dots,i_n}
\end{array}
$$

shows that $\tilde{\varphi}(f)$ is multiplicatively coherent. The commutative diagram

$$
\begin{array}{ccccccc}
 & & k[G]^{\otimes n} & \xrightarrow{\varphi^{\otimes n}} & (k[G_1]\oplus k[G_2])^{\otimes n} & \longrightarrow & T_{i_1,\dots,i_n} \\
 & \overset{\Delta_G^n}{\nearrow} & & & & & \\
k[G] & & & & & & \Big\downarrow {\scriptstyle \delta^{\ell}_{i_1,\dots,i_n}} \\
 & \underset{\Delta_G^{n-1}}{\searrow} & & & & & \\
 & & k[G]^{\otimes(n-1)} & \xrightarrow{\varphi^{\otimes(n-1)}} & (k[G_1]\oplus k[G_2])^{\otimes(n-1)} & \longrightarrow & T_{i_1,\dots,i_{\ell-1},i_{\ell+1},\dots,i_n}
\end{array}
$$

shows that $\tilde{\varphi}(f)$ is unitary coherent. Finally, the commutative diagram

$$
\begin{array}{ccccccc}
 & & k[G]^{\otimes n} & \xrightarrow{\varphi^{\otimes n}} & (k[G_1]\oplus k[G_2])^{\otimes n} & \longrightarrow & T_{i_1,\dots,1,\dots,i_n} \\
 & \overset{\Delta_G^n}{\nearrow} & & & & & \Big\downarrow {\scriptstyle \theta^{\ell}_{i_1,\dots,1,\dots,i_n}} \\
k[G] & & & & & & k[G_{i_1}]\otimes\cdots\otimes k[H]\otimes\cdots\otimes k[G_{i_n}] \\
 & \underset{\Delta_G^n}{\searrow} & & & & & \Big\uparrow {\scriptstyle \theta^{\ell}_{i_1,\dots,2,\dots,i_n}} \\
 & & k[G]^{\otimes n} & \xrightarrow{\varphi^{\otimes n}} & (k[G_1]\oplus k[G_2])^{n} & \longrightarrow & T_{i_1,\dots,2,\dots,i_n}
\end{array}
$$

shows that $\tilde{\varphi}(f)$ is H-coherent. So $\tilde{\varphi}(f)$ is coherent and we have a well-defined morphism $\tilde{\varphi}\colon k[G] \to k[G_1 *_H G_2]$ with $p_1\tilde{\varphi} = q_1$ and $p_2\tilde{\varphi} = q_2$.

If $\varphi'\colon k[G] \to k[G_1 *_H G_2]$ is another morphism with $p_1\varphi' = q_1$ and $p_2\varphi' = q_2$, then $\pi\varphi' = \varphi$ and the universal property of $C(k[G_1] \oplus k[G_2])$ yields $\varphi' = \tilde{\varphi}$. □

2.2 *Properties of Amalgamated Free Products*

In this section we discuss properties of amalgamated free products and their compatibility with other constructions like proalgebraic completion or base change. Throughout Sect. 2.2 we assume that $H \to G_1$ and $H \to G_2$ are two morphisms of k-groups.

Lemma 2.5 *The canonical morphism $G_1 * G_2 \to G_1 *_H G_2$ is a quotient map.*

Proof The canonical morphism $G_1 * G_2 \to G_1 *_H G_2$ is dual to the inclusion $k[G_1 *_H G_2] \subseteq k[G_1 * G_2]$ of subalgebras of $C(k[G_1] \oplus k[G_2])$. □

Proposition 2.6 *If G_1 and G_2 are geometrically reduced, then $G_1 *_H G_2$ is geometrically reduced.*

Proof According to Lemma 2.5 it suffices to show that $k[G_1 * G_2]$ is geometrically reduced. Let $\overline{k}$ denote the algebraic closure of k. We have to show that $k[G_1 * G_2] \otimes_k \overline{k}$ is reduced. Since $k[G_1 * G_2] \otimes_k \overline{k}$ is a subring of $\hat{T}(k[G_1] \oplus k[G_2]) \otimes_k \overline{k}$, it suffices to shows that $\hat{T}(k[G_1] \oplus k[G_2]) \otimes_k \overline{k}$ is reduced. For $i = 1, 2$ we abbreviate $\overline{k}[G_i] = k[G_i] \otimes_k \overline{k}$. As $\hat{T}(k[G_1] \oplus k[G_2]) \otimes_k \overline{k}$ injects into $\hat{T}(\overline{k}[G_1] \oplus \overline{k}[G_2])$, it suffices to see that $\hat{T}(\overline{k}[G_1] \oplus \overline{k}[G_2])$. But as a ring, $\hat{T}(\overline{k}[G_1] \oplus \overline{k}[G_2])$ is the product of the rings $(\overline{k}[G_1] \oplus \overline{k}[G_2])^{\otimes n}$. As $\overline{k}[G_1] \oplus \overline{k}[G_2] = \overline{k}[G_1] \times \overline{k}[G_2]$ is reduced, also $(\overline{k}[G_1] \oplus \overline{k}[G_2])^{\otimes n}$ is reduced [4, Chap. V, Sect. 15, No. 5, Theorem 3] and therefore $\hat{T}(\overline{k}[G_1] \oplus \overline{k}[G_2])$ is reduced. □

Lemma 2.7 *The morphisms $G_1 \to G_1 * G_2$ and $G_2 \to G_1 * G_2$ are closed embeddings.*

Proof It suffices to show that $G_1(R) \to (G_1 * G_2)(R)$ is injective for every k-algebra R. Set $\phi_1 = \mathrm{id}\colon G_1 \to G_1$ and let $\phi_2\colon G_2 \to G_1$ be the trivial morphism, i.e., the morphism that factors through the trivial group. By the universal property of $G_1 * G_2$, there exists a morphism $\phi\colon G_1 * G_2 \to G_1$ such that the composition $G_1 \to G_1 * G_2 \xrightarrow{\phi} G_1$ is the identity. In particular, $G_1(R) \to (G_1 * G_2)(R)$ is injective for every k-algebra R. □

It may seem natural to expect that $G_i \to G_1 *_H G_2$ $(i = 1, 2)$ are closed embeddings if $H \to G_i$ are closed embeddings, since the analogous statement holds for abstract groups. However, the analogous statement does not even hold for profinite groups [20, Example 9.2.10] and therefore one should not expect it for proalgebraic groups.

In a similar vein, still assuming that $H \to G_i$ are closed embeddings, one may hope that the natural map $G_1(R) *_{H(R)} G_2(R) \to (G_1 *_H G_2)(R)$ is injective for every k-algebra R, where $G_1(R) *_{H(R)} G_2(R)$ is the amalgamated free product of abstract groups. However, if $G_1(R) \to (G_1 *_H G_2)(R)$ is not injective, then also $G_1(R) *_{H(R)} G_2(R) \to (G_1 *_H G_2)(R)$ cannot be injective, because $G_1(R) \to G_1(R) *_{H(R)} G_2(R)$ is injective.

For a k-group G, let G^{et} be the maximal pro-étale quotient. It is canonically isomorphic to the k-group $\pi_0(G)$ of connected components of G. Every morphism of k-groups $H \to G^{\mathrm{et}}$ factors uniquely $H \to H^{\mathrm{et}} \to G^{\mathrm{et}}$. Hence, for k-groups H, G_1, G_2 and G, the following two commutative diagrams are equivalent

$$\begin{array}{ccc} H & \longrightarrow & G_1^{\mathrm{et}} \\ \downarrow & & \downarrow \\ G_2^{\mathrm{et}} & \longrightarrow & G \end{array} \quad \text{and} \quad \begin{array}{ccc} H^{\mathrm{et}} & \longrightarrow & G_1^{\mathrm{et}} \\ \downarrow & & \downarrow \\ G_2^{\mathrm{et}} & \longrightarrow & G. \end{array}$$

Hence we have a canonical identification

$$G_1^{\mathrm{et}} *_H G_2^{\mathrm{et}} = G_1^{\mathrm{et}} *_{H^{\mathrm{et}}} G_2^{\mathrm{et}}. \tag{14}$$

Proposition 2.8 *Given morphisms $\alpha_1 : H \to G_1$ and $\alpha_2 : H \to G_2$ of k-groups, the canonical morphism*

$$G_1 *_H G_2 \longrightarrow G_1^{\mathrm{et}} *_H G_2^{\mathrm{et}} = G_1^{\mathrm{et}} *_{H^{\mathrm{et}}} G_2^{\mathrm{et}}$$

induces an isomorphism

$$(G_1 *_H G_2)^{\mathrm{et}} \xrightarrow{\sim} (G_1^{\mathrm{et}} *_{H^{\mathrm{et}}} G_2^{\mathrm{et}})^{\mathrm{et}}. \tag{15}$$

Equivalently we have an isomorphism

$$\pi_0(G_1 *_H G_2) \xrightarrow{\sim} \pi_0(\pi_0(G_1) *_{\pi_0(H)} \pi_0(G_2)). \tag{16}$$

Proof Consider commutative diagrams of k-groups where E is pro-étale

$$\begin{array}{ccc} H & \xrightarrow{\alpha_1} & G_1 \\ {\scriptstyle \alpha_2}\downarrow & & \downarrow \\ G_2 & \longrightarrow & E. \end{array}$$

We say that the diagram is a pro-étale push-out, if for every other such diagram with pro-étale E' there is a unique morphism $E \to E'$ for which the obvious diagrams commute. A pro-étale push-out is unique up to a unique isomorphism. It is straightforward to check that both sides of (15) give rise to pro-étale push-outs E. Hence (15) is an isomorphism. □

Remark 2.9 It is not true that the amalgamated product of pro-étale k-groups is again pro-étale. Consider the representations

$$\mathbb{Z}/4 \to \mathrm{GL}_2(\mathbb{Q}),\ 1 \mapsto \left(\begin{smallmatrix} 0 & -1 \\ 1 & 0 \end{smallmatrix}\right) \quad \text{and} \quad \mathbb{Z}/6 \to \mathrm{GL}_2(\mathbb{Q}),\ 1 \mapsto \left(\begin{smallmatrix} 1 & -1 \\ 1 & 0 \end{smallmatrix}\right).$$

Their images are known to generate the infinite subgroup $\mathrm{SL}_2(\mathbb{Z})$ of $\mathrm{GL}_2(\mathbb{Q})$. We have corresponding representations of $\mathbb{Q}$-groups $(\mathbb{Z}/4)_{\mathbb{Q}} \to \mathrm{GL}_{2,\mathbb{Q}}$ and $(\mathbb{Z}/6)_{\mathbb{Q}} \to \mathrm{GL}_{2,\mathbb{Q}}$ and hence an algebraic representation of the free product

$$(\mathbb{Z}/4)_{\mathbb{Q}} * (\mathbb{Z}/6)_{\mathbb{Q}} \xrightarrow{\rho} \mathrm{GL}_{2,\mathbb{Q}}.$$

Its image on $\mathbb{Q}$-valued points contains $\mathrm{SL}_2(\mathbb{Z})$ because the latter is the image of the composition

$$\mathbb{Z}/4 * \mathbb{Z}/6 \longrightarrow ((\mathbb{Z}/4)_{\mathbb{Q}} * (\mathbb{Z}/6)_{\mathbb{Q}})(\mathbb{Q}) \xrightarrow{\rho(\mathbb{Q})} \mathrm{GL}_2(\mathbb{Q}).$$

The algebraic quotients of pro-étale group schemes are finite. Hence $(\mathbb{Z}/4)_{\mathbb{Q}} * (\mathbb{Z}/6)_{\mathbb{Q}}$ cannot be pro-étale.

Corollary 2.10 *If G_1, G_2 are connected k-groups, then $G_1 *_H G_2$ is connected as well.*

Proof By Proposition 2.8, we have

$$\pi_0(G_1 *_H G_2) = \pi_0(\pi_0(G_1) *_{\pi_0(H)} \pi_0(G_2)).$$

An affine group scheme G (over a field k) is connected if and only if $\pi_0(G) = \operatorname{Spec} k$ is the trivial group. Since the amalgamated product of two trivial k-groups is the trivial k-group by the universal property, the assertion follows. □

We next show that free products are compatible with the formation of *free proalgebraic groups*. To this end, let us recall the definition of the free proalgebraic group Γ_X on a set X [25, Sect. 2.5]. Let $\bar{k}$ denote the algebraic closure of k and let G be a k-group. A map $X \to G(\bar{k})$ *converges to* 1 if for every quotient map $G \to H$ with H algebraic (i.e., of finite type over k) the composition $X \to G(\bar{k}) \to H(\bar{k})$ maps all but finitely many elements of X to the identity element of $H(\bar{k})$. The free proalgebraic group Γ_X on X comes equipped with a map $X \to \Gamma_X(\bar{k})$ that converges to 1 and satisfies the following universal property: If $X \to G(\bar{k})$ is a map converging to 1, then there exists a unique morphism $\Gamma_X \to G$ of k-groups such that

$$\begin{array}{ccc} X & \longrightarrow & \Gamma_X(\bar{k}) \\ & \searrow \quad \swarrow & \\ & G(\bar{k}) & \end{array}$$

commutes.

Proposition 2.11 *Let X and Y be sets, then $\Gamma_X * \Gamma_Y \simeq \Gamma_{X \uplus Y}$.*

Proof We will show that $\Gamma_{X \uplus Y}$ satisfies the universal property of $\Gamma_X * \Gamma_Y$. As the compositions $X \to X \uplus Y \to \Gamma_{X \uplus Y}(\bar{k})$ and $Y \to X \uplus Y \to \Gamma_{X \uplus Y}(\bar{k})$ converge to 1, we have induced morphisms $\Gamma_X \to \Gamma_{X \uplus Y}$ and $\Gamma_Y \to \Gamma_{X \uplus Y}$. Given morphisms $\Gamma_X \to G$ and $\Gamma_Y \to G$ of k-groups, we can define a map $X \uplus Y \to G(\bar{k})$ by sending an element of X to its image under $X \to \Gamma_X(\bar{k}) \to G(\bar{k})$ and an element of Y to its image under $Y \to \Gamma_X(\bar{k}) \to G(\bar{k})$. Then $X \uplus Y \to G(\bar{k})$ converges to 1 and therefore induces a morphism $\Gamma_{X \uplus Y} \to G$. The diagrams

(17)

commute because their restrictions to X respectively Y commute. On the other hand, any two morphisms $\Gamma_{X\uplus Y} \to G$ making (17) commute, will have the same restriction to $X \uplus Y$ and therefore are equal. □

We next show that amalgamated free products are compatible with *proalgebraic completion*. For a discrete group Γ, let Γ^{alg} be the proalgebraic completion (or proalgebraic hull) of Γ over k. It is a k-group equipped with a morphism $\Gamma \to \Gamma^{\mathrm{alg}}(k)$ with the universal property that any linear representation $\Gamma \to \mathrm{GL}(V)$ on a finite dimensional k-vector space V comes from an algebraic representation of $\Gamma^{\mathrm{alg}} \to \mathrm{GL}_V$ by composition

$$\Gamma \longrightarrow \Gamma^{\mathrm{alg}}(k) \longrightarrow \mathrm{GL}_V(k) = \mathrm{GL}(V).$$

The k-group Γ^{alg} may be obtained as the Tannakian dual of the category $\mathrm{Rep}(\Gamma)$ of representations of Γ on finite dimensional k-vector spaces, with its canonical fibre functor. For more background on proalgebraic completions see e.g., [2, Sect. 2.1].

Proposition 2.12 *Given morphisms of discrete groups $\Gamma \to \Gamma_1$ and $\Gamma \to \Gamma_2$, the following canonical morphism is an isomorphism:*

$$(\Gamma_1 *_\Gamma \Gamma_2)^{\mathrm{alg}} \xrightarrow{\sim} \Gamma_1^{\mathrm{alg}} *_{\Gamma^{\mathrm{alg}}} \Gamma_2^{\mathrm{alg}}.$$

Proof We have equivalences (Sect. 1)

$$\begin{aligned}\mathrm{Rep}(\Gamma_1^{\mathrm{alg}} *_{\Gamma^{\mathrm{alg}}} \Gamma_2^{\mathrm{alg}}) &= \mathrm{Rep}(\Gamma_1^{\mathrm{alg}}) \times_{\mathrm{Rep}(\Gamma^{\mathrm{alg}})} \mathrm{Rep}(\Gamma_2^{\mathrm{alg}}) \\ &= \mathrm{Rep}(\Gamma_1) \times_{\mathrm{Rep}(\Gamma)} \mathrm{Rep}(\Gamma_2) \\ &= \mathrm{Rep}(\Gamma_1 *_\Gamma \Gamma_2) = \mathrm{Rep}((\Gamma_1 *_\Gamma \Gamma_2)^{\mathrm{alg}}).\end{aligned}$$

The assertion now follows from Tannakian duality. □

Contrary to what one might expect, the amalgamated free product may not be compatible with base change. Indeed, if $H \to G_1$ and $H \to G_2$ are morphisms of k-groups and K is a field extension of k, the commutative diagram

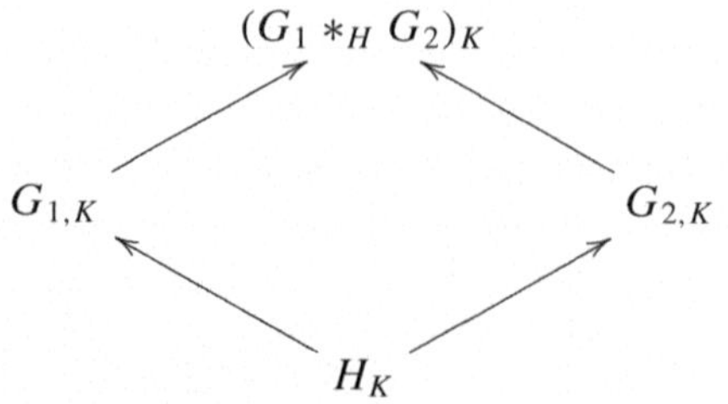

yields a morphism $G_{1,K} *_{H_K} G_{2,K} \to (G_1 *_H G_2)_K$. The dual morphism is the restriction of $\widehat{T}(k[G_1] \oplus k[G_2]) \otimes_k K \to \widehat{T}((k[G_1] \oplus k[G_2]) \otimes_k K)$. The latter map is always injective but rarely surjective [5, Chap. II, Sect. 3.7, Corollary 3]. So $(G_1 *_H G_2)_K$ is a quotient of $G_{1,K} *_{H_K} G_{2,K}$. To construct a concrete example, where the morphism $G_{1,K} *_{H_K} G_{2,K} \to (G_1 *_H G_2)_K$ fails to be an isomorphism, we need some preparation.

Assume, for now, that k is an algebraically closed field of characteristic zero. Let C_2 denote the (abstract) group with two elements. We first show that the set of isomorphism classes of irreducible finite dimensional k-linear representations of $C_2 * C_2$ has cardinality $|k|$. Note that to specify a representation of C_2 on a k-vector space V is equivalent to specifying a direct sum decomposition $V = V_1 \oplus V_{-1}$, where the non-trivial element of C_2 acts trivially on V_1 and by multiplications with -1 on V_{-1}. To specify a representation of $C_1 * C_2$ is thus equivalent to specifying two direct sum decompositions $V = V_1 \oplus V_{-1}$ and $V = V_1' \oplus V_{-1}'$. The corresponding representation of $C_2 * C_2$ is irreducible if and only if $V_i \cap V_j' = 0$ for all $i, j \in \{1, -1\}$. For $a \in k \smallsetminus \{0, 1\}$ let $V_a = k^2$ be the representation of $C_2 * C_2$ determined by $V_1 = k\begin{pmatrix}1\\0\end{pmatrix}$, $V_{-1} = k\begin{pmatrix}0\\1\end{pmatrix}$, $V_1' = k\begin{pmatrix}1\\1\end{pmatrix}$ and $V_{-1,a}' = k\begin{pmatrix}1\\a\end{pmatrix}$. The representations V_a are irreducible and pairwise non-isomorphic. Indeed, if $a, b \in k \smallsetminus \{0, 1\}$, and $f\colon V_a \to V_b$ is an isomorphism, then f must stabilize V_1 and V_{-1}, i.e., f is given by a diagonal matrix. As f also stabilizes V_1', f is in fact given by a scalar matrix. As f maps $V_{-1,a}'$ into $V_{-1,b}'$, we find that $a = b$.

Since the V_a's are irreducible and pairwise non-isomorphic as representations of $C_1 * C_2$, the V_a's are also irreducible and pairwise non-isomorphic as representations of $(C_2 * C_2)^{\mathrm{alg}} = (C_2 * C_2)^{\mathrm{alg},k}$. Every (finite dimensional) representation of a k-group G embeds into $k[G]^n$ for some $n \geq 1$ [24]. It follows that every irreducible representation of G embeds into $k[G]$. In particular, every V_a embeds into $k[(C_2 * C_2)^{\mathrm{alg}}]$. As a sum of pairwise non-isomorphic irreducible representations is always a direct sum, it follows that the dimension $\dim_k k[(C_2 * C_2)^{\mathrm{alg}}]$ of $k[(C_2 * C_2)^{\mathrm{alg}}]$ as a k-vector space is at least $|k|$.

On the other hand, the set S of all morphisms $\rho\colon C_2 * C_2 \to \mathrm{GL}_{n(\rho)}(k)$ has at most cardinality $|k|$ and $(C_2 * C_2)^{\mathrm{alg}}$ can be realized as a closed subgroup scheme of $\prod_{\rho \in S} \mathrm{GL}_{n(\rho),k}$ (see [2, Sect. 2.1]). So the dimension of $k[(C_2 * C_2)^{\mathrm{alg}}]$ is at most $|k|$. Therefore $\dim_k k[(C_2 * C_2)^{\mathrm{alg}}] = |k|$.

Let $\mu_2 = \mu_{2,k}$ denote the k-group of second roots of unity, i.e., $\mu_2(R) = \{g \in R^\times \mid g^2 = 1\}$ for every k-algebra R. Using Proposition 2.12 we find

$$\mu_2 * \mu_2 = C_2^{\mathrm{alg}} * C_2^{\mathrm{alg}} = (C_2 * C_2)^{\mathrm{alg}}.$$

So $\dim_k k[\mu_2 * \mu_2] = |k|$. Therefore, if $k \subseteq K$ is an inclusion of algebraically closed fields of characteristic zero with $|k| < |K|$, then $\mu_{2,K} * \mu_{2,K}$ and $(\mu_{2,k} * \mu_{2,k})_K$ cannot be isomorphic because the coordinate ring of $\mu_{2,K} * \mu_{2,K}$ has K-dimension $|K|$, whereas the coordinate ring of $(\mu_{2,k} * \mu_{2,k})_K$ has K-dimension $|k|$.

3 Local Systems and a Seifer–van Kampen Theorem

The main goal of this short section is to establish an analog of the Seifert–van Kampen theorem for the proalgebraic fundamental group.

Let X be a topological space. We denote with $\mathbf{Loc}(X) = \mathbf{Loc}_k(X)$ the category of local systems of finite dimensional k-vector spaces on X. An object $\mathcal{E}$ of $\mathbf{Loc}(X)$ is thus a sheaf of k-vector spaces on X, such that every point $x \in X$ has an open neighborhood U with $\mathcal{E}|_U \simeq \underline{k^r}$, where $\underline{k^r}$ is the constant sheaf on U associated to the vector space k^r (and $r \geq 0$ may depend on U).

For an open subset U of X and $\mathcal{E} \in \mathbf{Loc}(X)$, the restriction $\mathcal{E}|_U$ is in $\mathbf{Loc}(U)$. Indeed $\mathcal{E} \rightsquigarrow \mathcal{E}|_U$ is a functor $\mathbf{Loc}(X) \to \mathbf{Loc}(U)$.

Lemma 3.1 *Let X be a topological space and let U_1, U_2 be open subsets of X such that $X = U_1 \cup U_2$. Then the functor $\mathcal{E} \rightsquigarrow (\mathcal{E}|_{U_1}, \mathcal{E}|_{U_2}, \mathrm{id}_{\mathcal{E}|_{U_1 \cap U_2}})$ from $\mathbf{Loc}(X)$ to $\mathbf{Loc}(U_1) \times_{\mathbf{Loc}(U_1 \cap U_2)} \mathbf{Loc}(U_2)$ is an equivalence of categories.*

Proof As in Example 1.2 this is just the sheaf property. See [23, Tag 00AK]. □

The espace étalé of any $\mathcal{E}$ in $\mathbf{Loc}(X)$ may be viewed as a vector bundle over X with locally constant transition functions in $\mathrm{GL}_r(k)$ for some r which may vary. Equivalently, if we give k the discrete topology, these are the vector bundles with continuous $\mathrm{GL}_r(k)$-valued transition functions. The stalk $\mathcal{E}_x$ of $\mathcal{E}$ in a point $x \in X$ is isomorphic to the fibre in x of the corresponding vector bundle. We have $\mathcal{E}_x = k^r$ for some $r = r(x) \geq 0$. If X is connected, these ranks are independent of x and $\mathbf{Loc}_k(X)$ is a neutral Tannakian category. It is neutralized by the fibre functor $\omega_x : \mathbf{Loc}(X) \to \mathrm{Vec}_k$, which sends local systems and their morphisms to the respective stalks, c.f. [12, Proposition 2.1]. In [12], the *proalgebraic fundamental group* $\pi_k(X, x)$ of a connected pointed topological space (X, x) was defined to be the Tannakian dual of $(\mathbf{Loc}_k(X), \omega_x)$. Thus we have

$$\mathbf{Loc}_k(X) = \mathrm{Rep}(\pi_k(X, x)).$$

The previous constructions immediately give the following result:

Theorem 3.2 *Let U_1, U_2 be open subsets of a topological space X such that $X = U_1 \cup U_2$ and assume that X, U_1, U_2 and $U_1 \cap U_2 \neq \emptyset$ are all connected. Then, for any point $x \in U_1 \cap U_2$ the commutative diagram of k-groups*

$$\begin{array}{ccc} \pi_k(U_1 \cap U_2, x) & \longrightarrow & \pi_k(U_1, x) \\ \downarrow & & \downarrow \\ \pi_k(U_2, x) & \longrightarrow & \pi_k(X, x) \end{array} \qquad (18)$$

is a push-out diagram in the category of k-groups, i.e. we have an isomorphism

$$\pi_k(X, x) \xrightarrow{\sim} \pi_k(U_1, x) *_{\pi_k(U_1 \cap U_2, x)} \pi_k(U_2, x).$$

Proof By Lemma 3.1 the following diagram is a fibre product of categories

$$\begin{array}{ccc} \mathbf{Loc}_k(X) & \longrightarrow & \mathbf{Loc}_k(U_1) \\ \downarrow & & \downarrow \\ \mathbf{Loc}_k(U_2) & \longrightarrow & \mathbf{Loc}_k(U_1 \cap U_2). \end{array}$$

Note that all (restriction) functors commute with the respective fibre functors ω_x on the respective categories. Hence the diagram of representation categories induced by (18) is a fibre product as well

$$\begin{array}{ccc} \mathbf{Rep}(\pi_k(X, x)) & \longrightarrow & \mathbf{Rep}(\pi_k(U_1, x)) \\ \downarrow & & \downarrow \\ \mathbf{Rep}(\pi_k(U_2, x)) & \longrightarrow & \mathbf{Rep}(\pi_k(U_1 \cap U_2, x)). \end{array}$$

By Corollary 1.11 in the situation of Corollary 1.12 it follows that (18) is a push-out diagram. □

4 The Proalgebraic Fundamental Group of a Topological Space Is Perfect

Throughout Sect. 4 we assume that k is a perfect field of characteristic $p > 0$. An affine group scheme $G = \operatorname{Spec} A$ over k is *perfect* if the p-power map $(\ )^p$ is bijective on A. In this section we give Tannakian criteria with applications for when the Tannaka dual group of a neutralized Tannakian category $(\mathcal{T}, \omega)$ over k is reduced resp. perfect. The main tool is the functor Fr_+ from [6] and our criteria are implicitely contained in [6]. For convenience we give short self-contained proofs of the required properties of Fr_+. In fact, there are two versions of Fr_+ depending on the choice of a subgroup $S \subseteq \mathcal{S}_p$, the symmetric group on p letters, and either can be used in the following. The two choices are $S = \mathcal{S}_p$ and $S = \langle \sigma \rangle = \mathbb{Z}/p$ where σ is the cyclic permutation $\sigma = (1, 2, \dots, p)$. For an object X of $\mathcal{T}$ there is a natural action of S on $X^{\otimes p}$ and $\mathrm{Fr}_+(X)$ is the image in $\mathcal{T}$ of the composition:

$$H^0(S, X^{\otimes p}) \hookrightarrow X^{\otimes p} \overset{[\,]}{\twoheadrightarrow} H_0(S, X^{\otimes p}).$$

Here one sets

$$H^0(S, X^{\otimes p}) = \ker\left((\sigma - \mathrm{id})_\sigma : X^{\otimes p} \to \prod_{\sigma \in S} X^{\otimes p}\right)$$

and

$$H_0(S, X^{\otimes p}) = \operatorname{coker}\Big(\sum_{\sigma}(\sigma - \mathrm{id})_\sigma : \bigoplus_{\sigma \in S} X^{\otimes p} \to X^{\otimes p}\Big).$$

For a morphism φ in $\mathcal{T}$ there is a canonical induced morphism $\mathrm{Fr}_+(\varphi)$ and Fr_+ becomes a functor.

Let $\mathcal{T}^{(p)}$ be the category with the same objects as $\mathcal{T}$ and morphisms

$$\mathrm{Hom}_{\mathcal{T}^{(p)}}(X, Y) = \mathrm{Hom}_{\mathcal{T}}(X, Y) \otimes_{k,()^p} k.$$

These are k-vector spaces via multiplication on the second factor. We view Fr_+ as a functor from $\mathcal{T}^{(p)}$ to $\mathcal{T}$ by setting $\mathrm{Fr}_+(\varphi \otimes \lambda) = \lambda \, \mathrm{Fr}_+(\varphi)$. Note that this is well defined since for $\mu \in k$ we have $\mathrm{Fr}_+(\mu\varphi) = \mu^p \, \mathrm{Fr}_+(\varphi)$. Thus $\mathrm{Fr}_+ : \mathcal{T}^{(p)} \to \mathcal{T}$ is compatible with scalar multiplication. It is not difficult to see that Fr_+ is also additive and hence k-linear. E.g. for $S = \mathcal{S}_p$ and morphisms $\varphi, \psi : X \to Y$ we have $\mathrm{Fr}_+(\varphi + \psi) = \mathrm{Fr}_+(\varphi) + \mathrm{Fr}_+(\psi)$ because as morphisms from $H_0(\mathcal{S}_p, X^{\otimes p})$ to $H_0(\mathcal{S}_p, Y^{\otimes p})$ we have:

$$[(\varphi + \psi)^{\otimes p}] = \sum_{i=0}^{p} \binom{p}{i} [\varphi^{\otimes i} \otimes \psi^{\otimes p-i}] = [\varphi^{\otimes p}] + [\psi^{\otimes p}].$$

Define a faithful tensor functor $\omega^{(p)} : \mathcal{T}^{(p)} \to \mathrm{Vec}_k$ by setting $\omega^{(p)}(X) = \omega(X) \otimes_{k,()^p} k$ and $\omega^{(p)}(\varphi \otimes \lambda) = \omega(\varphi) \otimes \lambda$. Then $(\mathcal{T}^{(p)}, \omega^{(p)})$ becomes a neutralized Tannakian category over k. If $G = \operatorname{Spec} A$ is the Tannakian dual of $(\mathcal{T}, \omega)$, then $G^{(p)} = \operatorname{Spec} A^{(p)}$ with $A^{(p)} = A \otimes_{k,()^p} k$ may be identified with the Tannakian dual of $(\mathcal{T}^{(p)}, \omega^{(p)})$ because a k-linear $G^{(p)}$-action on $\omega^{(p)}(X)$ can be identified with a k-linear G-action on $\omega(X)$.

Let $F_{\mathrm{rel}} : G \to G^{(p)}$ be the relative Frobenius morphism. It is the k-linear morphism corresponding to the map of k-algebras $F_{\mathrm{rel}}^\sharp : A^{(p)} \to A$ sending $a \otimes \lambda$ to λa^p. The following facts can be found in [6]:

Proposition 4.1

(i) For V in Vec_k the natural map

$$t(V) : V \otimes_{k,()^p} k \xrightarrow{\sim} \mathrm{Fr}_+(V), \; v \otimes \lambda \longmapsto \lambda[v \otimes \cdots \otimes v]$$

is a k-linear isomorphism.

(ii) The maps in a) induce a natural isomorphism of functors $\omega^{(p)} = \omega \circ \mathrm{Fr}_+ : \mathcal{T}^{(p)} \to \mathrm{Vec}_k$ and the functor $\mathrm{Fr}_+ : \mathcal{T}^{(p)} \to \mathcal{T}$ can be identified with the functor $\mathrm{Rep}(F_{\mathrm{rel}}) : \mathrm{Rep}(G^{(p)}) \to \mathrm{Rep}(G)$ between the corresponding representation categories.

(iii) Fr_+ is an exact tensor functor.

Proof (i) For $S = \mathcal{S}_p$ the map $t(V)$ is additive by the binomial theorem. Additivity in the case $S = \langle\sigma\rangle = \mathbb{Z}/p$ is also not difficult to see. Compatibility with scalar multiplications is obvious so that $t(V)$ is k-linear. Since $t(k)$ is an isomorphism and the functors $\_ \otimes_{k,()^p} k$ and Fr_+ are additive on Vec_k, it follows that $t(V)$ is an isomorphism for all V.

(ii) For X in $\mathrm{Ob}\, \mathcal{T}^{(p)} = \mathrm{Ob}\, \mathcal{T}$ the isomorphisms $t(\omega(X))$ from a) give the first of the isomorphisms $\omega^{(p)} = \mathrm{Fr}_+ \circ \omega = \omega \circ \mathrm{Fr}_+$. The second one follows since ω is an exact k-linear $\otimes$-functor. The $G^{(p)}$-action $\rho^{(p)}$ on $\omega^{(p)}(X)$ corresponds to a G-action ρ on $V = \omega(X)$. By functoriality, we get an induced G-action $\mathrm{Fr}_+(\rho)$ on $\mathrm{Fr}_+(V)$ and via the isomorphism $t(V)$ a G-action ρ' on $\omega^{(p)}(X) = V \otimes_{k,()^p} k$. We have to show that $\rho' = \rho^{(p)} \circ F_{\mathrm{rel}}$. Let $\Delta : V \to V \otimes A$ be the A-comodule map corresponding to ρ and $\Delta_{\mathrm{Fr}_+(V)}$ the induced A-comodule map on $\mathrm{Fr}_+(V)$. For $v \in V$ set $\Delta(v) = \sum_i v_i \otimes a_i$. Then we have on $v \otimes 1 \in V \otimes_{k,()^p} k$

$$\begin{aligned}(\Delta_{\mathrm{Fr}_+(V)} \circ t(V))(v \otimes 1) &= \Delta_{\mathrm{Fr}_+(V)}([v \otimes \cdots \otimes v]) \\ &= \sum_{i_1,\dots,i_p} [v_{i_1} \otimes \cdots \otimes v_{ip}] \otimes a_{i_1} \cdots a_{i_p} \\ &= \sum_i [v_i \otimes \cdots \otimes v_i] \otimes a_i^p .\end{aligned}$$

The last equation holds because if the stabilizer $S_{\underline{i}}$ of $\underline{i} = (i_1, \dots, i_p)$ in $S = \langle\sigma\rangle = \mathbb{Z}/p \subseteq \mathcal{S}_p$ is not S, i.e. if not all $i_1, \dots, i_p$ are equal, then $S_{\underline{i}} = 1$ and we sum $|S| = p$-times the same element in $A \otimes \mathrm{Fr}_+(V) \subseteq A \otimes H_0(S, V^{\otimes p})$ which gives zero. For $S = \mathcal{S}_p$ this is all the more true since $H_0(\mathcal{S}_p, V^{\otimes p})$ is a quotient of $H_0(\mathbb{Z}/p, V^{\otimes p})$. On the other hand, we have

$$\begin{aligned}&((t(V) \otimes \mathrm{id}_A) \circ (\mathrm{id} \otimes F_{\mathrm{rel}}^{\sharp}) \circ \Delta^{(p)})(v \otimes 1) \\ &= ((t(V) \otimes \mathrm{id}_A) \circ (\mathrm{id} \otimes F_{\mathrm{rel}}^{\sharp}))\Big(\sum_i (v_i \otimes 1) \otimes (a_i \otimes 1)\Big) \\ &= (t(V) \otimes \mathrm{id}_A)\Big(\sum_i (v_i \otimes 1) \otimes a_i^p\Big) = \sum_i [v_i \otimes \cdots \otimes v_i] \otimes a_i^p .\end{aligned}$$

Thus we have checked that $\rho' = \rho^{(p)} \circ F_{\mathrm{rel}}$ and b) is proved.

(iii) These assertions follow from the interpretation of Fr_+ as $\mathrm{Rep}(F_{\mathrm{rel}})$ in (ii). □

Corollary 4.2 *For a neutralized Tannakian category $(\mathcal{T}, \omega)$ over a perfect field k of characteristic $p > 0$ let G be its Tannakian dual.*

(i) The group scheme G is reduced if and only if the functor $\mathrm{Fr}_+ : \mathcal{T}^{(p)} \to \mathcal{T}$ is fully faithful and every subobject of $\mathrm{Fr}_+(T)$ for some T in $\mathrm{Ob}\, \mathcal{T}^{(p)}$ is isomorphic to $\mathrm{Fr}_+(T')$ for some subobject T' of T.

(ii) The group scheme G is perfect if and only if Fr_+ is an equivalence of categories.

Proof (i) Note that G is reduced if and only if $(\ )^p$ is injective on A. This condition is equivalent to the map $F_{\mathrm{rel}}^{\sharp} : A^{(p)} \to A$ being injective, i.e. to the relative Frobenius $F_{\mathrm{rel}} : G \to G^{(p)}$ being faithfully flat, c.f. [24, Theorem 14.1]. By [7, Proposition 2.21] and Proposition 4.1, (ii) this translates into the stated properties of $\mathrm{Fr}_+ = \mathrm{Rep}(F_{\mathrm{rel}})$.

(ii) G is perfect if and only if $(\ )^p$ is an automorphism of A as an $\mathbb{F}_p$-algebra. Since $A^{(p)} \cong A$ as $\mathbb{F}_p$-algebras this is equivalent to $F_{\mathrm{rel}}^{\sharp}$ being an isomorphism of k-algebras i.e. to F_{rel} being an isomorphism of k-group schemes. By Tannakian duality and Proposition 4.1 (ii) this is equivalent to $\mathrm{Fr}_+ = \mathrm{Rep}(F_{\mathrm{rel}})$ being an equivalence of categories. □

Corollary 4.3 *Let k be a perfect field of positive characteristic p.*

(i) For a discrete group Γ the proalgebraic completion Γ^{alg} over k is a perfect group scheme over k.

(ii) Let X be a connected topological space and $x_0 \in X$ a point. Then the proalgebraic fundamental group $\pi_k(X, x_0)$ over k is a perfect group scheme over k.

Example 4.4 For $\Gamma = \mathbb{Z}$ and k algebraically closed of characteristic $p > 0$ one may find the calculation of $\mathbb{Z}^{\mathrm{alg}}$ in [21, Annex A]. There is a canonical isomorphism:

$$\mathbb{Z}^{\mathrm{alg}} = \operatorname{Spec} k[k^{\times}] \times \mathbb{Z}_p.$$

Here $\operatorname{Spec} k[k^{\times}]$ is the diagonalizable group scheme on $k^{\times}$ over k and $\mathbb{Z}_p$ denotes the unipotent, pro-étale group scheme obtained as the projective limit of the constant group schemes $\mathbb{Z}/p^n$. Since k is perfect, it is clear that $\mathbb{Z}^{\mathrm{alg}}$ is perfect as well in accordance with the corollary.

Proof (i) By definition, Γ^{alg} is the Tannakian dual of the category $\mathrm{Rep}(\Gamma)$ of representations of Γ on finite dimensional k-vector spaces V. The functor Fr_+ sends the Γ-representation V to the induced Γ-representation on the vector space $\mathrm{Fr}_+(V)$. The k-linear automorphism $t(V)$ in Proposition 4.1 (i) is Γ-equivariant. It follows that the functor $\mathrm{Fr}_+ : \mathrm{Rep}(\Gamma)^{(p)} \to \mathrm{Rep}(\Gamma)$ is isomorphic to the functor $\_ \otimes_{k,(\ )^p} k$ which is an equivalence of categories. Now we use Corollary 4.2 (ii).

(ii) By definition $\pi_k(X, x_0)$ is the Tannakian dual of the category $\mathbf{Loc}_k(X)$ of locally constant sheaves $\mathcal{E}$ of finite dimensional k-vector spaces on X together with the fibre functor $\omega(\mathcal{E}) = \mathcal{E}_{x_0}$. The functor Fr_+ sends $\mathcal{E}$ to the locally constant sheaf $\mathrm{Fr}_+(\mathcal{E})$. We define a map of sheaves

$$t(\mathcal{E}) : \mathcal{E} \otimes_{k,(\ )^p} k \longrightarrow \mathrm{Fr}_+(\mathcal{E})$$

by sending $s \otimes \lambda$ for a local section s of $\mathcal{E}$ to the image in $\mathrm{Fr}_+(\mathcal{E})$ of $\lambda[s \otimes \cdots \otimes s]$. For the stalk at a point $x \in X$ we have $t(\mathcal{E})_x = t(\mathcal{E}_x)$. Since the $t(\mathcal{E}_x)$ are isomorphisms of k-vector spaces by Proposition 4.1 (i), it follows that $t(\mathcal{E})$

is a k-linear isomorphism of sheaves. Hence the functor $\mathrm{Fr}_+ : \mathbf{Loc}_k(X)^{(p)} \to \mathbf{Loc}_k(X)$ is isomorphic to the functor $\_ \otimes_{k,()^p} k$ which is an equivalence of categories. Now we use Corollary 4.2 (ii). □

Remark 4.5 In the remark after [12, Theorem 2.5] we had previously shown that $\pi_k(X, x_0)$ was reduced.

5 The Proalgebraic Fundamental Group of the Kucharczyk–Scholze Space

Given a field F of characteristic zero containing all roots of unity, and an injective homomorphism $\mu(F) \hookrightarrow \mathbb{C}^\times$, Kucharczyk and Scholze have defined a scheme $\mathfrak{X}_F$ over $\mathbb{C}$ and a canonical isomorphism [15, Corollary 4.15]

$$\pi_1^{\mathrm{et}}(\mathfrak{X}_F(\mathbb{C}), x) = G_F. \tag{19}$$

Here $\mathfrak{X}_F(\mathbb{C})$ carries the complex topology and $x \in \mathfrak{X}_F(\mathbb{C})$ is a point. The étale fundamental group $\pi_1^{\mathrm{et}}(Z, z)$ of a pointed connected topological space (Z, z) is defined in [15, Sect. 2.2] using the formalism of Grothendieck categories. It is a profinite group which classifies the finite coverings of Z. Finally G_F is a version of the absolute Galois group of F defined via a fibre functor related to $(\mathfrak{X}_F, x)$. The group G_F is isomorphic to the usual Galois group $\mathrm{Aut}_F(\overline{F})$ where $\overline{F}$ is an algebraic closure of F, the isomorphism being unique up to an inner isomorphism. By [12, Theorem 3.6], the étale fundamental group $\pi_1^{\mathrm{et}}(Z, z)$ viewed as a group scheme over a field k is canonically isomorphic to the maximal pro-étale quotient of the proalgebraic group $\pi_k(Z, z)$ i.e.

$$\pi_k(Z, z)^{\mathrm{et}} \xrightarrow{\sim} \pi_1^{\mathrm{et}}(Z, z).$$

In particular, we have

$$\pi_k(\mathfrak{X}_F(\mathbb{C}), x)^{\mathrm{et}} \xrightarrow{\sim} \pi_1^{\mathrm{et}}(\mathfrak{X}_F(\mathbb{C}), x) = G_F.$$

Thus the group of connected components of $\pi_k(\mathfrak{X}(\mathbb{C}), x)$ may be identified with G_F. The connected component $\pi_k(\mathfrak{X}_F(\mathbb{C}), x)^0$ is abelian and we determine its structure. We also discuss relations with the motivic Galois groups of F and $\overline{F}$. For the calculation of $\pi_k(\mathfrak{X}_F(\mathbb{C}), x)$ we use the same approach as in [15] applied to π_k instead of π_1^{et}.

Given a connected topological space Z and a continuous path $\gamma : [0, 1] \to Z$ with $\gamma(0) = z_0$ and $\gamma(1) = z_1$ we obtain a $\otimes$-isomorphism $\omega_\gamma : \omega_{z_0} \to \omega_{z_1}$ between the fibre functors $\omega_{z_i} : \mathbf{Loc}_k(Z) \to \mathrm{Vec}_k$ for $i = 0, 1$ as follows: For $\mathcal{E}$ in $\mathbf{Loc}_k(Z)$ the locally constant sheaf $\gamma^{-1}\mathcal{E}$ is constant on $[0, 1]$. Evaluation therefore provides isomorphisms

$$\omega_{z_0}(\mathcal{E}) = \mathcal{E}_{z_0} \underset{\mathrm{ev}_{z_0}}{\overset{\sim}{\leftarrow}} \Gamma([0,1], \gamma^{-1}\mathcal{E}) \underset{\mathrm{ev}_{z_1}}{\overset{\sim}{\rightarrow}} \mathcal{E}_{z_1} \cong \omega_{z_1}(\mathcal{E}).$$

The natural transformation $\omega_\gamma = \mathrm{ev}_{z_1} \circ \mathrm{ev}_{z_0}^{-1}$ depends only on the homotopy class of γ. Let

$$\mathrm{Aut}^{\otimes}(\omega_{z_0}) \underset{L_{\omega_\gamma}}{\overset{\sim}{\rightarrow}} \mathrm{Iso}^{\otimes}(\omega_{z_0}, \omega_{z_1}) \underset{R_{\omega_\gamma}}{\overset{\sim}{\leftarrow}} \mathrm{Aut}^{\otimes}(\omega_{z_1})$$

be defined by left resp. right translation of the point $\omega_\gamma \in \mathrm{Iso}^{\otimes}(\omega_{z_0}, \omega_{z_1})(k)$. We obtain an isomorphism of k-groups

$$\gamma_k = R_{\omega_\gamma}^{-1} \circ L_{\omega_\gamma} : \pi_k(Z, z_0) \overset{\sim}{\rightarrow} \pi_k(Z, z_1).$$

Proposition 5.1 *Let $f, g : X \to Z$ be continuous maps between connected topological spaces. Let $H : X \times [0,1] \to Z$ be a homotopy between f and g. For a fixed point $\tilde{x} \in X$ consider the path $\gamma : [0,1] \to Z$ from $f(\tilde{x})$ to $g(\tilde{x})$ defined by $\gamma(t) = H(\tilde{x}, t)$. Then the following diagram of k-groups commutes*

$$\begin{array}{ccc} & & \pi_k(Z, f(\tilde{x})) \\ & \overset{f_*}{\nearrow} & \\ \pi_k(X, \tilde{x}) & & \downarrow \wr\, \gamma_k \\ & \underset{g_*}{\searrow} & \\ & & \pi_k(Z, g(\tilde{x})). \end{array}$$

For the proof we need a lemma on local systems. The formulation is a little bit involved because apart from connectedness the topological spaces X, Z in Proposition 5.1 are arbitrary.

Lemma 5.2 *Consider continuous maps $Y \underset{s}{\overset{\pi}{\rightleftarrows}} X$ between connected topological spaces such that $\pi \circ s = \mathrm{id}$. Assume that for all $x \in X$ the following conditions hold:*

(i) The fibre $\pi^{-1}(x)$ is a quasi-compact subspace of Y and any two different points of $\pi^{-1}(x)$ have disjoint neighborhoods in Y.
(ii) For any open subset V of Y with $\pi^{-1}(x) \subset V$ there is an open neighborhood $x \in U \subset X$ with $\pi^{-1}(U) \subset V$.
(iii) All local systems in $\mathbf{Loc}_k(\pi^{-1}(x))$ are constant.

Then for any local system $\mathcal{F}$ in $\mathbf{Loc}_k(Y)$ the natural map of sheaves $\pi_*\mathcal{F} \to s^{-1}\mathcal{F}$ is an isomorphism.

Example 5.3 Let K be a compact, path-connected, locally path-connected and simply connected topological space. Let X be a connected topological space. Then conditions (i)–(iii) in Lemma 5.2 are satisfied for $Y = X \times K \underset{s}{\overset{\pi}{\rightleftarrows}} X$, where s is any continuous section of the projection π.

Namely, (i) is clear and (iii) follows from the equivalences

$$\mathbf{Loc}_k(\pi^{-1}(x)) \cong \mathbf{Loc}_k(K) \cong \mathrm{Rep}(\pi_1(K,t)),\ \ t \in K$$

which are valid by our hypothesis, and the fact that $\pi_1(K, t)$ is trivial. As for (ii), let $V \subset Y$ be open with $V \supset \pi^{-1}(x) = \{x\} \times K$. For any point $t \in K$, there are open neighborhoods U_t of x in X and W_t of t in K such that $V \supset U_t \times W_t$. Since K is compact, there are finitely many points $t_1, \ldots, t_n \in K$ such that $W_{t_1}, \ldots, W_{t_n}$ cover K. Then $U = U_{t_1} \cap \ldots \cap U_{t_n}$ is an open neighborhood of x in X with $V \supset U \times W_{t_i}$ for all i, and hence $V \supset U \times K = \pi^{-1}(U)$.

Proof of Lemma 5.2. We recall the definition of the map $\pi_*\mathcal{F} \to s^{-1}\mathcal{F}$. Let $s^0\mathcal{F}$ be the presheaf on X with sections

$$(s^0\mathcal{F})(U) = \mathrm{colim}_{V \supset s(U)}\, \mathcal{F}(V).$$

The relation $\pi \circ s = \mathrm{id}$ implies that $\pi^{-1}(U) \supset s(U)$. The resulting natural maps

$$(\pi_*\mathcal{F})(U) = \mathcal{F}(\pi^{-1}(U)) \longrightarrow (s^0\mathcal{F})(U)$$

give a map of presheaves $\pi_*\mathcal{F} \to s^0\mathcal{F}$ and hence a map of sheaves $\pi_*\mathcal{F} \to s^{-1}\mathcal{F}$. It suffices to check that the induced maps on the stalks are isomorphisms

$$(\pi_*\mathcal{F})_x = \mathrm{colim}_{U \ni x}\, \mathcal{F}(\pi^{-1}(U)) \longrightarrow (s^{-1}\mathcal{F})_x = \mathcal{F}_{s(x)}.$$

For this, note that because of (ii) we have

$$(\pi_*\mathcal{F})_x = \mathrm{colim}_{V \supset \pi^{-1}(x)}\, \mathcal{F}(V).$$

Condition (i) allows us to apply [23, Tag 09V3] to deduce that

$$\mathrm{colim}_{V \supset \pi^{-1}(x)}\, \mathcal{F}(V) = \Gamma(\pi^{-1}(x), \mathcal{F}|_{\pi^{-1}(x)}).$$

Finally, since $\mathcal{F}|_{\pi^{-1}(x)}$ is a constant local system by condition (iii), the evaluation map in the point $s(x) \in \pi^{-1}(x)$ gives an isomorphism

$$\Gamma(\pi^{-1}(x), \mathcal{F}|_{\pi^{-1}(x)}) \underset{\mathrm{ev}_{s(x)}}{\overset{\sim}{\rightarrow}} \mathcal{F}_{s(x)} = (s^{-1}\mathcal{F})_x.$$

The composition of these three isomorphisms is the map $(\pi_*\mathcal{F})_x \to (s^{-1}\mathcal{F})_x$ above and we are done. □

Now we can turn to the *proof* of Proposition 5.1. Using the homotopy H we first construct an isomorphism of $\otimes$-functors

$$\underline{H} : f^{-1} \cong g^{-1} : \mathbf{Loc}_k(Z) \longrightarrow \mathbf{Loc}_k(X).$$

For $\mathcal{E}$ in $\mathbf{Loc}_k(Z)$ the isomorphism

$$\underline{H}(\mathcal{E}) : f^{-1}(\mathcal{E}) \xrightarrow{\sim} g^{-1}(\mathcal{E})$$

is obtained as follows. Consider the diagram

$$X \times [0,1] \underset{\xleftarrow[s_1]{}}{\overset{\xleftarrow{s_0}}{\xrightarrow{\pi}}} X$$

where π is the projection and $s_0(x) = (x, 0)$, $s_1(x) = (x, 1)$. Setting $Y = X \times [0, 1]$ and $s = s_0$ or s_1, by Example 5.3 we may apply Lemma 5.2 to the sheaf $\mathcal{F} = H^{-1}(\mathcal{E})$. We obtain canonical isomorphisms

$$f^{-1}(\mathcal{E}) = s_0^{-1}\mathcal{F} \xleftarrow{\sim} \pi_*\mathcal{F} \xrightarrow{\sim} s_1^{-1}\mathcal{F} = g^{-1}(\mathcal{E}).$$

Using these we define the isomorphism $\underline{H}(\mathcal{E})$. Its fibre in a point x is a functorial isomorphism

$$\underline{H}(\mathcal{E})_x : \mathcal{E}_{f(x)} = (f^{-1}\mathcal{E})_x \xrightarrow{\sim} (g^{-1}\mathcal{E})_x = \mathcal{E}_{g(x)},$$

so that $\underline{H}(\ )_x \in \mathrm{Iso}^{\otimes}(\omega_{f(x)}, \omega_{g(x)})$. By construction, we have $\underline{H}(\ )_{\tilde{x}} = \omega_\gamma$ as defined above, where $\gamma : [0, 1] \to Z$ is the path $\gamma(t) = H(\tilde{x}, t)$. For the commutativity of the diagram we have to show that $L_{\omega_\gamma} \circ f_* = R_{\omega_\gamma} \circ g_*$ i.e. that for any k-algebra R and any

$$\alpha \in \pi_k(X, \tilde{x})(R) = \underline{\mathrm{Aut}}^{\otimes}(\omega_{\tilde{x},R})$$

we have

$$\omega_\gamma \circ f_*(\alpha) = g_*(\alpha) \circ \omega_\gamma : \omega_{f(\tilde{x})} \longrightarrow \omega_{g(\tilde{x})}.$$

In other words, we claim that for all $\mathcal{E}$ in $\mathbf{Loc}_k(Z)$:

$$(\underline{H}(\mathcal{E})_{\tilde{x}} \otimes \mathrm{id}_R) \circ \alpha(f^{-1}(\mathcal{E})) = \alpha(g^{-1}(\mathcal{E})) \circ (\underline{H}(\mathcal{E})_{\tilde{x}} \otimes \mathrm{id}_R).$$

This holds since $\underline{H}(\mathcal{E}) : f^{-1}(\mathcal{E}) \to g^{-1}(\mathcal{E})$ is a morphism in $\mathbf{Loc}_k(X)$ and since α is a natural transformation. □

As in [15, Sect. 2.2] we obtain the following consequences:

Corollary 5.4 *Let H be a pointed homotopy of the pointed connected topological spaces $(X, \tilde{x})$ and $(Z, \tilde{z})$ and set $f = H(\ , 0)$ and $g = H(\ , 1)$. Then*

$$f_* = g_* : \pi_k(X, \tilde{x}) \longrightarrow \pi_k(Z, \tilde{z}).$$

Proof For the path γ in Proposition 5.1 we have $\gamma(t) = H(\tilde{x}, t) = \tilde{z}$ since H is a pointed homotopy. Hence the corresponding isomorphism $\omega_\gamma : \mathcal{E}_{\tilde{z}} \to \mathcal{E}_{\tilde{z}}$ is the identity for all $\mathcal{E}$ in $\mathbf{Loc}_k(Z)$ and therefore the map γ_k in Proposition 5.1 is the identity as well. □

Corollary 5.5 *Let* $(X, x) \underset{g}{\overset{f}{\rightleftarrows}} (Y, y)$ *be pointed maps of pointed topological spaces such that* $f \circ g$ *and* $g \circ f$ *are homotopic to the identities via pointed homotopies. Then* X *is connected if and only if* Y *is connected. If this is the case, the functors* $\mathbf{Loc}_k(X) \underset{f^{-1}}{\overset{g^{-1}}{\rightleftarrows}} \mathbf{Loc}_k(Y)$ *are quasi-inverse equivalences of categories and* $\pi_k(X, x) \underset{g_*}{\overset{f_*}{\rightleftarrows}} \pi_k(Y, y)$ *are isomorphisms of* k*-groups which are inverse to each other.*

Proof The assertion on connectedness is proved in the corresponding Corollary [15, Corollary 2.18]. The rest is a formal consequence of Corollary 5.4. □

Consider a field F of characteristic zero containing all roots of unity in an algebraic closure $\overline{F}$ of F and fix an injective homomorphism $\iota : \mu(F) = \mu(\overline{F}) \hookrightarrow \mathbb{C}^\times$. Let $W_{\text{rat}}(F)$ denote the ring of rational Witt vectors of F, c.f. [15, Sect. 4.1]. The Teichmüller map $[\ ] : F^\times \to W_{\text{rat}}(F)$ is multiplicative and we obtain ring homomorphisms

$$\mathbb{Z}[\mu(F)] \xrightarrow{[\ ]} W_{\text{rat}}(F) \quad \text{and} \quad \mathbb{Z}[\mu(F)] \xrightarrow{\iota} \mathbb{C},$$

defined on the group ring of $\mu(F)$. Set

$$\mathfrak{X}_F = \operatorname{Spec}(W_{\text{rat}}(F) \otimes_{\mathbb{Z}[\mu(F)]} \mathbb{C}).$$

The complex points of this scheme over Spec $\mathbb{C}$ carry the "complex topology" induced by the topology of $\mathbb{C}$, c.f. [15, Sect. 5.2]. By definition, we have

$$\mathfrak{X}_F(\mathbb{C}) = \{\alpha \in \operatorname{Hom}(W_{\text{rat}}(F), \mathbb{C}) \mid \alpha \circ [\ ] = \iota\}.$$

The topological space $\mathfrak{X}_F(\mathbb{C})$ is connected and it has the following Galois-theoretic description. Let $G_F = \operatorname{Aut}_F(\overline{F})$ be the absolute Galois group of F. The Teichmüller map $\overline{F}^\times \to W_{\text{rat}}(\overline{F})$ induces an isomorphism $\mathbb{Z}\overline{F}^\times \xrightarrow{\sim} W_{\text{rat}}(\overline{F})$ and hence an isomorphism

$$\mathbb{Z}[\overline{F}^\times]^G \xrightarrow{\sim} W_{\text{rat}}(\overline{F})^G = W_{\text{rat}}(F).$$

This implies that

$$\mathfrak{X}_{\overline{F}}(\mathbb{C}) = \{\chi : \overline{F}^\times \longrightarrow \mathbb{C}^\times \mid \chi\,|_{\mu(\overline{F})} = \iota\}.$$

By [15, Lemma 4.9], the natural map

$$\mathrm{Hom}(\mathbb{Z}[\overline{F}^\times], \mathbb{C})/G \xrightarrow{\sim} \mathrm{Hom}(\mathbb{Z}[\overline{F}^\times]^G, \mathbb{C})$$

is a bijection. This gives a bijection

$$\mathfrak{X}_F(\mathbb{C}) = \{\chi : \overline{F}^\times \to \mathbb{C}^\times \mid \chi|_{\mu(\overline{F})} = \iota\}/G.$$

Note that since $\mu(\overline{F}) = \mu(F)$, the Galois group G fixes ι. This bijection is actually a homeomorphism if the set of characters is given the topology of pointwise convergence and its quotient by G the quotient topology. Kucharczyk and Scholze prove that the categories of finite étale coverings of $\mathfrak{X}_F$ and $\mathfrak{X}_F(\mathbb{C})$ are naturally equivalent and that their isomorphic étale fundamental groups with respect to a $\mathbb{C}$-valued point x can be canonically identified with the absolute Galois group of F if the latter is defined using a certain fibre functor defined by x. We will not consider the scheme $\mathfrak{X}_F$ in the following but only the connected topological space $\mathfrak{X}_F(\mathbb{C})$ with the complex topology. Let $(\overline{F}^\times)^\vee = \mathrm{Hom}(\overline{F}^\times, \mathbb{S}^1)$ be the Pontrjagin dual of the discrete group $\overline{F}^\times$ and consider the closed subspace

$$X_{\overline{F}} = \{\chi \in (\overline{F}^\times)^\vee \mid \chi|_{\mu(F)} = \iota\}.$$

The group G_F acts by homeomorphisms and we endow the quotient $X_F = G \setminus X_{\overline{F}}$ with the quotient topology. By [15, Proposition 5.1], the G_F action on $X_{\overline{F}}$ is proper and free and the space X_F is a non-empty, connected, compact Hausdorff space. By [15, Theorem 5.7, ii] the natural G_F-equivariant projection $\mathfrak{X}_{\overline{F}}(\mathbb{C}) \to X_{\overline{F}}, \chi \mapsto \chi/|\chi|$ and the induced projection $\mathfrak{X}_F(\mathbb{C}) \to X_F$ are (strong) deformation retractions. Here we view $X_{\overline{F}}$ as a closed subspace of $\mathfrak{X}_{\overline{F}}(\mathbb{C})$ via $X_{\overline{F}} = X_{\overline{F}} \times \{0\}$ and similarly for X_F. From Corollary 5.5 we thus obtain:

Corollary 5.6 *For any point $x \in X_F \subset \mathfrak{X}_F(\mathbb{C})$ the above map $\mathfrak{X}_F(\mathbb{C}) \to X_F$ induces an isomorphism of k-groups*

$$\pi_k(\mathfrak{X}_F(\mathbb{C}), x) \xrightarrow{\sim} \pi_k(X_F, x).$$

It is sufficient therefore to determine $\pi_k(X_F, x)$. For a finite extension F'/F in $\overline{F}$ the natural map $X_{F'} \to X_F$ is a finite étale covering of compact connected Hausdorff spaces, c.f. [15, Theorem 5.2]. Let $\overline{x} \in X_{\overline{F}} \subset \mathfrak{X}_{\overline{F}}(\mathbb{C})$ be a point and let $x_{F'}$ be its image in $X_{F'}$. By [12, Theorem 3.6] we have a natural isomorphism

$$\pi_k(X_F, x)^0 = \varprojlim_{(Y,y)} \pi_k(Y, y)$$

where (Y, y) runs over the directed set of isomorphism classes of pointed finite étale coverings of (X_F, x). According to [15, Theorem 5.2], the above coverings $(X_{F'}, x_{F'}) \to (X_F, x)$ form a cofinal system, so that we obtain an isomorphism

$$\pi_k(X_F, x)^0 = \varprojlim_{F'/F} \pi_k(X_{F'}, x_{F'}). \tag{20}$$

Since we are dealing with compact Hausdorff spaces, the natural map

$$X_{\overline{F}} \longrightarrow \varprojlim_{F'/F} X_{\overline{F}}/\operatorname{Gal}(\overline{F}/F') = \varprojlim_{F'/F} X_{F'}$$

is a homeomorphism. Applying [12, Theorem 4.3] we get an isomorphism

$$\pi_k(X_{\overline{F}}, \overline{x}) = \varprojlim_{F'/F} \pi_k(X_{F'}, x_{F'}). \tag{21}$$

Combining (20) and (21) with [12, Theorem 3.6] and the [15, Theorem 5.2] identification of $\pi_1(X_F, x)$ with G_F, we obtain the following result:

Corollary 5.7 *For any point $\overline{x} \in X_{\overline{F}}$ mapping to $x \in X_F$ we have a canonical short exact sequence of affine k-groups*

$$1 \longrightarrow \pi_k(X_{\overline{F}}, \overline{x}) \longrightarrow \pi_k(X_F, x) \longrightarrow (G_F)_{/k} \longrightarrow 1.$$

There are canonical isomorphisms

$$\pi_k(X_{\overline{F}}, \overline{x}) = \pi_k(X_F, x)^0 \quad \textit{and} \quad \pi_k(X_F, x)^{\mathrm{et}} = \pi_1^{\mathrm{et}}(X_F, x)_{/k} = (G_F)_{/k}.$$

Here, for a profinite group $G = \varprojlim_i G_i$ with G_i finite, $G_{/k} = \varprojlim_i (G_{i/k})$ denotes the pro-étale k-group obtained as the limit of the finite constant k-groups $G_{i/k}$ attached to the G_i's. The global sections $\Gamma(G_{/k}, \mathcal{O})$ of G consists of the continuous maps $G \to k$ where k carries the discrete topology. Note that for the k-rational points, we have

$$\pi_k(X_F, x)^{\mathrm{et}}(k) = \pi_1^{\mathrm{et}}(X_F, x) = G_F.$$

Next we show that the connected k-group $\pi_k(X_{\overline{F}}, \overline{x})$ is abelian. The inclusion $\mu_F \hookrightarrow \overline{F}^{\times}$ induces a short exact sequence of compact abelian topological group

$$1 \longrightarrow (\overline{F}^{\times}/\mu_F)^{\vee} \longrightarrow (\overline{F}^{\times})^{\vee} \xrightarrow{\pi} \mu_F^{\vee} \longrightarrow 1.$$

Here π is the map which restricts a character $\chi : \overline{F}^{\times} \to \mathbb{S}^1$ to μ_F. All fibres of π are homeomorphic and in particular $X_{\overline{F}}$, the fibre of our fixed embedding $\iota : \mu_F \hookrightarrow \mathbb{S}^1$ is homeomorphic to $V^{\vee}$, the Pontrjagin dual of the $\mathbb{Q}$-vectorspace $V = \overline{F}^{\times}/\mu_F$. Write $V = \varinjlim_{\Gamma} \Gamma$ as a filtered union of finite free $\mathbb{Z}$-submodules. Then $V^{\vee} = \varprojlim_T T$ where $T = \Gamma^{\vee} \cong (\mathbb{S}^1)^{\operatorname{rk}\Gamma}$ is a torus. Let $y \in V^{\vee}$ be the image of $\overline{x}$ under a homeomorphism $X_{\overline{F}} \xrightarrow{\sim} V^{\vee}$ and let y_T be the image of y in T. Using [12, Theorems 4.1 and 4.3] we obtain isomorphisms of k-groups

$$\pi_k(X_{\overline{F}}, \overline{x}) \cong \pi_k(V^\vee, y) = \varprojlim_T \pi_k(T, y_T) = \varprojlim_T \pi_1(T, y_T)^{\mathrm{alg}}.$$

Since $\pi_1(T, y_T)$ is commutative, it follows that $\pi_k(X_{\overline{F}}, \overline{x})$ is commutative as well. If k is perfect, then it follows from [16, Theorem 16.13] that there is a canonical isomorphism of k groups

$$\pi_k(X_{\overline{F}}, \overline{x}) \xrightarrow{\sim} \pi_k^u(X_{\overline{F}}, \overline{x}) \times \pi_k^{\mathrm{red}}(X_{\overline{F}}, \overline{x}). \tag{22}$$

Here $\pi_k^u(X_{\overline{F}}, \overline{x})$ is the pro-unipotent radical of $\pi_k(X_{\overline{F}}, \overline{x})$ and $\pi_k^{\mathrm{red}}(X_{\overline{F}}, \overline{x})$, the maximal pro-reductive quotient of $\pi_k(X_{\overline{F}}, \overline{x})$ can be identified with the greatest k-subgroup of $\pi_k(X_{\overline{F}}, \overline{x})$ of multiplicative type. Both $\pi_k^u(X_{\overline{F}}, \overline{x})$ and $\pi_k^{\mathrm{red}}(X_{\overline{F}}, \overline{x})$ are commutative and connected. The cohomologies of $X_{\overline{F}}$ with coefficients in $\mathbb{Z}/n$ and $\mathbb{Q}$ have been calculated in [15]. We recall the argument for H^1 and coefficients in any abelian group A.

Recall the pro-torus $V^\vee = \varprojlim_T T$ above where $V = \overline{F}^\times / \mu_F$. We have

$$H^1(V^\vee, \underline{A}) = \varinjlim_T H^1(T, \underline{A}).$$

It follows that translation by an element of $V^\vee$ induces the identity on $H^1(V^\vee, \underline{A})$ since this is well known for the finite dimensional tori T. Hence the homeomorphisms $V^\vee \xrightarrow{\sim} X_{\overline{F}}, \alpha \mapsto \alpha\chi$ for $\chi \in X_{\overline{F}}$ all induce the same isomorphism, which we will view as an identification

$$H^1(X_{\overline{F}}, \underline{A}) = H^1(V^\vee, \underline{A}). \tag{23}$$

We have

$$\begin{aligned} H^1(T, \underline{A}) &= H^1(\Gamma^\vee, \underline{A}) = \mathrm{Hom}(\pi_1(\Gamma^\vee), A) \\ &= \mathrm{Hom}(\mathrm{Hom}(\Gamma, \mathbb{Z}), A) = \Gamma \otimes A \end{aligned}$$

and hence

$$H^1(X_{\overline{F}}, \underline{A}) = \varinjlim_T H^1(T, \underline{A}) = \overline{F}^\times / \mu_F \otimes A. \tag{24}$$

The category of commutative unipotent groups U over $k \supset \mathbb{Q}$ is anti-equivalent to the category of k-vector spaces by sending U to $Y(U) = \mathrm{Hom}_k(U, \mathbb{G}_a)$. Given $Y(U)$ we recover U as

$$U = \mathrm{Hom}_k(Y(U), \mathbb{G}_{a/k}),$$

where by definition for all k-schemes S we have

$$\mathrm{Hom}_k(Y(U), \mathbb{G}_{a/k})(S) := \mathrm{Hom}_k(Y(U), \Gamma(S, \mathcal{O})).$$

Here, on the right Hom_k means k-linear maps. Again using [12, Proposition 4.4], we obtain

$$Y(\pi_k^u(X_{\overline{F}}, \overline{x})) = \mathrm{Hom}_k(\pi_k(X_{\overline{F}}, \overline{x}), \mathbb{G}_{a/k}) = H^1(X_{\overline{F}}, \underline{k}), \tag{25}$$

and then

$$\pi_k^u(X_{\overline{F}}, \overline{x}) = \mathrm{Hom}_k(H^1(X_{\overline{F}}, \underline{k}), \mathbb{G}_{a/k}). \tag{26}$$

We now turn to the multiplicative part. Since we want to apply [12, Proposition 4.4] which gives a formula for $\mathrm{Hom}_k(\pi_k(X_{\overline{F}}, \overline{x}), \mathbb{G}_m)$, but not for $\mathrm{Hom}_{k^{\mathrm{sep}}}(\pi_k(X_{\overline{F}}, \overline{x}) \otimes_k k^{\mathrm{sep}}, \mathbb{G}_m)$, we restrict our discussion to the case where $k = \overline{k}$. Then the category of multiplicative groups over k is equivalent to the category of abelian groups by sending H to $X^*(H) = \mathrm{Hom}_k(H, \mathbb{G}_m)$. The group H is recovered from $X^*(H)$ as

$$H = \mathrm{Hom}(X^*(H), \mathbb{G}_{m/k}).$$

Here the right hand side is defined by its S-valued points, which are the groups

$$\mathrm{Hom}(X^*(H), \mathbb{G}_{m/k})(S) := \mathrm{Hom}(X^*(H), \Gamma(S, \mathcal{O}^\times)).$$

Using [12, Proposition 4.4], we find

$$X^*(\pi_k^{\mathrm{red}}(X_{\overline{F}}, \overline{x})) = \mathrm{Hom}_k(\pi_k(X_{\overline{F}}, \overline{x}), \mathbb{G}_{m/k}) = H^1(X_{\overline{F}}, \underline{k}^\times)$$

and therefore

$$\pi_k^{\mathrm{red}}(X_{\overline{F}}, \overline{x}) = \mathrm{Hom}(H^1(X_{\overline{F}}, \underline{k}^\times), \mathbb{G}_{m/k}). \tag{27}$$

Combining (24), (25), (26) with (27) we get the following

Corollary 5.8 *In the situation of Corollary 5.7 we have canonical isomorphisms of commutative k-groups:*

$$\pi_k(X_{\overline{F}}, \overline{x}) \xrightarrow{\sim} \pi_k^u(X_{\overline{F}}, \overline{x}) \times \pi_k^{\mathrm{red}}(X_{\overline{F}}, \overline{x}), \quad \text{if } k \text{ is perfect}$$

with

$$\pi_k^u(X_{\overline{F}}, \overline{x}) = \mathrm{Hom}_k(\overline{F}^\times / \mu_F \otimes_{\mathbb{Q}} k, \mathbb{G}_{a/k}, \quad \text{if } k \supset \mathbb{Q}$$

and

$$\pi_k^{\mathrm{red}}(X_{\overline{F}}, \overline{x}) = \mathrm{Hom}(\overline{F}^\times / \mu_F \otimes_{\mathbb{Q}} k^\times / \mu_k, \mathbb{G}_{m/k}), \quad \text{if } k = \overline{k} \supset \mathbb{Q}.$$

Remark 5.9 Since the character group $\overline{F}^\times / \mu_F \otimes_{\mathbb{Z}} k^\times = \overline{F}^\times / \mu_F \otimes_{\mathbb{Q}} k^\times / \mu_F$ of $\pi_k^{\mathrm{red}}(X_{\overline{F}}, \overline{x})$ is a $\mathbb{Q}$-vector space, it has no torsion and hence $\pi_k^{\mathrm{red}}(X_{\overline{F}}, \overline{x})$ is connected. The unipotent part $\pi_k^u(X_{\overline{F}}, \overline{x})$ being connected, it follows that $\pi_k(X_{\overline{F}}, \overline{x})$ is connected. We know this of course, since $\pi_k(X_{\overline{F}}, \overline{x}) = \pi_k(X_F, x)^0$ by Corollary 5.7.

Remark 5.10 Let $G \to G''$ be a surjective i.e. faithfully flat map of affine k-groups. Equivalently, the corresponding map of Hopf-algebras $A'' \to A$ is injective. For algebraically closed fields k the induced map $G(k) \to G''(k)$ is surjective, c.f. [11, III, Sect. 3, Corollary 7.6]. In particular, the map $G(k) \to \pi_0(G)(k)$ is surjective.

It follows from Corollary 5.7 that for $k = \overline{k}$ we have a short exact sequence of pro-discrete groups

$$1 \longrightarrow \pi_k^u(X_{\overline{F}}, \overline{x})(k) \times \pi_k^{\mathrm{red}}(X_{\overline{F}}, \overline{x})(k) \longrightarrow \pi_k(X_F, x)(k) \longrightarrow G_F \longrightarrow 1. \quad (28)$$

If additionally $\operatorname{char} k = 0$, then using Corollary 5.8, we get

$$\pi_k^u(X_{\overline{F}}, \overline{x})(k) = \operatorname{Hom}_k(\overline{F}^\times/\mu_F \otimes_{\mathbb{Q}} k, k) = \operatorname{Hom}_{\mathbb{Q}}(\overline{F}^\times/\mu_F, k)$$

and

$$\pi_k^{\mathrm{red}}(X_{\overline{F}}, \overline{x})(k) = \operatorname{Hom}(\overline{F}^\times/\mu_F \otimes_{\mathbb{Q}} k^\times/\mu_F, k^\times).$$

Going through the constructions, one sees that the G_F-action on these groups via (28) is the same as the one induced by the natural G_F-action on $\overline{F}^\times/\mu_F$. The actions are continuous for the pro-discrete topologies on $\pi_k^u(X_{\overline{F}}, \overline{x})(k)$ and $\pi_k^{\mathrm{red}}(X_{\overline{F}}, \overline{x})(k)$ but not for the discrete topologies (the G_F-orbits do not have to be finite). Thus, although these groups are $\mathbb{Q}$-vector spaces, it is not clear whether the extension (28) is trivial i.e. a semi-direct product or not. The corresponding question is open for the extension of affine groups in Corollary 5.7 as well.

The étale fundamental group of X_F is isomorphic to the Galois group of F. We now discuss the question, whether the proalgebraic fundamental group $\pi_k(X_F, x)$ is related to the motivic Galois group $G_{\mathcal{M}_F(k)}$ if $k \supset \mathbb{Q}$. We assume that our embedding $\iota : \mu_F \hookrightarrow \mathbb{C}$ is the restriction of an embedding $\iota : \overline{F} \hookrightarrow \mathbb{C}$, which then also provides a point $\overline{x}$ of $X_{\overline{F}}$. Let $\mathcal{M}_F(k)$, $\mathcal{M}_{\overline{F}}(k)$ be the Tannakian categories of mixed motives over F with coefficients in k neutralized by the Betti-realization via ι. Thus e.g. for $M = H^w(X)$, with X a variety over $\overline{F}$ the fibre functor sends M to $H_B^w(X \otimes_{\overline{F},\iota} \mathbb{C}, \underline{k})$. We can think of motives in terms of realizations as in [14]. By definition, $G_{\mathcal{M}_F(k)}$ resp. $G_{\mathcal{M}_{\overline{F}}(k)}$ are the Tannakian duals of $\mathcal{M}_F(k)$ resp. $\mathcal{M}_{\overline{F}}(k)$ with respect to this fibre functor. There is a canonical short exact sequence of k-groups

$$1 \longrightarrow G^0_{\mathcal{M}_F(k)} \longrightarrow G_{\mathcal{M}_F(k)} \longrightarrow G_{F/k} \longrightarrow 1,$$

where $G_{\mathcal{M}_{\overline{F}}(k)} = G^0_{\mathcal{M}_F(k)}$ canonically. As explained in the introduction of [15] there should be a "better" space $\tilde{X}_F$ mapping to X_F and we would expect $\pi_k(\tilde{X}_F, \tilde{x})$ to be closely related to $G_{\mathcal{M}_F(k)}$. This suggests a map $G_{\mathcal{M}_F(k)} \longrightarrow \pi_k(X_F, x)$ of k-groups and hence a map

$$G_{\mathcal{M}_{\overline{F}}(k)} = G^0_{\mathcal{M}_F(k)} \longrightarrow \pi_k(X_F, x)^0 = \pi_k(X_{\overline{F}}, \overline{x}).$$

It would map unipotent radicals to unipotent radicals and induce map between the maximal pro-reductive quotients. Since $\pi_k(X_{\overline{F}}, \overline{x})$ is commutative, we would therefore get maps

$$(G^u_{\mathcal{M}_{\overline{F}}(k)})^{\mathrm{ab}} \longrightarrow \pi^u_k(X_{\overline{F}}, \overline{x}) \tag{29}$$

and

$$(G^{\mathrm{red}}_{\mathcal{M}_{\overline{F}}(k)})^{\mathrm{ab}} \longrightarrow \pi^{\mathrm{red}}_k(X_{\overline{F}}, \overline{x}). \tag{30}$$

We will now construct a canonical non-trivial map (29) for any field $k \supset \mathbb{Q}$. To do so is equivalent to giving a map of k-vector spaces

$$Y(\pi^u_k(X_{\overline{F}}, \overline{x})) \longrightarrow Y((G^u_{\mathcal{M}_{\overline{F}}(k)})^{\mathrm{ab}}) = \mathrm{Hom}(G^u_{\mathcal{M}_{\overline{F}}(k)}, \mathbb{G}_a). \tag{31}$$

First note that by (25) and (24) we have

$$Y(\pi^u_k(X_{\overline{F}}, \overline{x})) = \overline{F}^\times / \mu_F \otimes k. \tag{32}$$

Next, any $a \in \overline{F}^\times$ defines a motive $E_a = H^1(\mathbb{G}_m, \{1, a\})(1)$ over $\overline{F}$. It can also be described as $E_a = H^1(X_a)(1)$, where X_a is the curve obtained from $\mathbb{G}_{m,\overline{F}}$ by identifying 1 and a. In Deligne's theory of 1-motives, we have $E_a = [\mathbb{Z} \xrightarrow{\phi} \mathbb{G}_m]$, where $\phi(1) = a$, c.f. [10, Sect. 10]. The motive E_a fits into a canonical extension

$$0 \longrightarrow \mathbb{Q}(0) \longrightarrow E_a \longrightarrow \mathbb{Q}(1) \longrightarrow 0 \quad \text{in } \mathcal{M}_{\overline{F}}(\mathbb{Q}).$$

Extending coefficients to k we get an injective homomorphism

$$\overline{F}^\times / \mu \otimes k \hookrightarrow \mathrm{Ext}^1_{\mathcal{M}_{\overline{F}}(k)}(k(0), k(1)), \ (a \bmod \mu) \otimes 1 \mapsto E_a \otimes k. \tag{33}$$

The motives $k(0)$, resp. $k(1)$ correspond to the trivial representation of $G_{\mathcal{M}_{\overline{F}}}(k)$ on $\mathbb{G}_a$ over k resp. to the representation denoted $\mathbb{G}_a(1)$ on $\mathbb{G}_a$ by the character $\chi : G_{\mathcal{M}_{\overline{F}}}(k) \to \mathbb{G}_{m/k}$, dual to the full embedding of $\langle k(1) \rangle^{\otimes}$ into $\mathcal{M}_{\overline{F}}(k)$. Then we have

$$\mathrm{Ext}^1_{\mathcal{M}_{\overline{F}}(k)}(k(0), k(1)) = \mathrm{Ext}^1_{G_{\mathcal{M}_{\overline{F}}}(k)}(\mathbb{G}_a, \mathbb{G}_a(1)) = H^1(G_{\mathcal{M}_{\overline{F}}(k)}, \mathbb{G}_a(1)). \tag{34}$$

In the middle, extensions are taken within the category of finite dimensional representations of $G_{\mathcal{M}_{\overline{F}}(k)}$ on k-vector spaces. For the Hochschild cohomology on the right compare [16, Chap. 15]. Consider the canonical extension of k-groups

$$1 \longrightarrow G^u_{\mathcal{M}_{\overline{F}}(k)} \longrightarrow G_{\mathcal{M}_{\overline{F}}(k)} \longrightarrow Q \longrightarrow 1.$$

Here $Q = G^{\mathrm{red}}_{\mathcal{M}_{\overline{F}}(k)}$ can be interpreted as the Tannakian dual of the category of (direct sums of) pure motives over $\overline{F}$ with coefficients in k. We now follow the argument in

[14, Appendix C4]. Since Hochschild cohomology of reductive groups over fields of characteristic zero vanishes in positive degrees, the Hochschild–Serre spectral sequence

$$H^i(Q, H^j(G^u_{\mathcal{M}_{\overline{F}}(k)}, \mathbb{G}_a(1)) \Rightarrow H^{i+j}(G_{\mathcal{M}_{\overline{F}}(k)}, \mathbb{G}_a(1))$$

degenerates into isomorphisms

$$H^j(G^u_{\mathcal{M}_{\overline{F}}(k)}, \mathbb{G}_a(1))^Q = H^j(G_{\mathcal{M}_{\overline{F}}(k)}, \mathbb{G}_a(1)).$$

In particular, using (34) we may identify:

$$\mathrm{Ext}^1_{\mathcal{M}_{\overline{F}}(k)}(k(0), k(1)) = H^1(G^u_{\mathcal{M}_{\overline{F}}(k)}, \mathbb{G}_a(1))^Q. \tag{35}$$

The character χ above factors over Q, the maximal pro-reductive quotient of $G_{\mathcal{M}_{\overline{F}}(k)}$, and hence the restriction of χ to $G^u_{\mathcal{M}_{\overline{F}}(k)}$ is trivial. Hence we have

$$H^1(G^u_{\mathcal{M}_{\overline{F}}(k)}, \mathbb{G}_a(1))^Q = H^1(G^u_{\mathcal{M}_{\overline{F}}(k)}, \mathbb{G}_a)(1)^Q = \mathrm{Hom}(G^u_{\mathcal{M}_{\overline{F}}(k)}, \mathbb{G}_a)(1)^Q. \tag{36}$$

The group on the right is the χ^{-1} eigenspace of the Q-action on $\mathrm{Hom}(G^u_{\mathcal{M}_{\overline{F}}(k)}, \mathbb{G}_a)$. In any case, we get the following inclusion by combining (35) and (36)

$$\mathrm{Ext}^1_{\mathcal{M}_{\overline{F}}(k)}(k(0), k(1)) \subset \mathrm{Hom}(G^u_{\mathcal{M}_{\overline{F}}(k)}, \mathbb{G}_a). \tag{37}$$

The promised map (31) is obtained as the following composition

$$Y(\pi^u_k(X_{\overline{F}}, \overline{x})) \overset{(32)}{=} \overline{F}^\times/\mu_F \otimes k \overset{(33)}{\hookrightarrow} \mathrm{Ext}^1_{\mathcal{M}_{\overline{F}}(k)}(k(0), k(1)) \overset{(37)}{\hookrightarrow} \mathrm{Hom}(G^u_{\mathcal{M}_{\overline{F}}(k)}, \mathbb{G}_a).$$

(The map (33) is conjectured to be an isomorphism.)

Contrary to (29), there does not seem to be a non-trivial map (30). The group $G^{\mathrm{red}}_{\mathcal{M}_{\overline{F}}}(k)$ is the motivic Galois group for pure motives over $\overline{F}$ with coefficients in k, and

$$\mathrm{Hom}_k((G^{\mathrm{red}}_{\mathcal{M}_{\overline{F}}(k)})^{\mathrm{ab}}, \mathbb{G}_m) = \mathrm{Hom}_k(G^{\mathrm{red}}_{\mathcal{M}_{\overline{F}}(k)}, \mathbb{G}_m) \tag{38}$$

is the group of isomorphism classes or rank 1 motives over $\overline{F}$ with coefficients in k. It follows that for any $k = \overline{k} \supset \mathbb{Q}$, (30) would induce a homomorphism from the $\mathbb{Q}$-vector space

$$\mathrm{Hom}(\pi^{\mathrm{red}}_k(X_{\overline{F}}, \overline{x}), \mathbb{G}_m) = \mathrm{Hom}(\pi_k(X_{\overline{F}}, \overline{x}), \mathbb{G}_m) = H^1(X_{\overline{F}}, \underline{k}^\times) = \overline{F}^\times/\mu_F \otimes k^\times$$

to the group (38), i.e. a map

$$\overline{F}^\times/\mu_F \otimes \mathbb{Q}^\times \longrightarrow \mathrm{Hom}_k((G^{\mathrm{red}}_{\mathcal{M}_{\overline{F}}(k)})^{\mathrm{ab}}, \mathbb{G}_m). \tag{39}$$

For $k = \overline{\mathbb{Q}}$, the $\overline{\mathbb{Q}}$-group

$$S = (G^{\mathrm{red}}_{\mathcal{M}_{\overline{F}}(\overline{\mathbb{Q}})})^{\mathrm{ab}}$$

is the connected Serre group of F over $\overline{\mathbb{Q}}$. Its character group $\mathrm{Hom}_{\overline{\mathbb{Q}}}(S, \mathbb{G}_m)$ is known [22, Sect. 7] and does not depend on F! It does not contain any infinitely divisible elements except zero and hence any map (39) has to be trivial. Note in this regard that as pointed out in [15] the Steinberg relations do not hold in the rational cohomology of X_F (hence the above mentioned suggestion of a better space $\tilde{X}_F$ mapping to X_F). Moreover no information about weights is built into the category $\mathbf{Loc}_k(X_F)$ as we saw in our construction of the map (29). Both deficits, and also that $\pi_k(X_F, x)^0$ is commutative show that X_F is quite far from a topological space whose proalgebraic fundamental group is more deeply related to the motivic Galois group of F.

References

1. Antei, M.: Pushout of quasi-finite and flat group schemes over a Dedekind ring. J. Algebra **371**, 314–328 (2012)
2. Bass, H., Lubotzky, A., Magid, A.R., Mozes, S.: The Proalgebraic Completion of Rigid Groups, vol. 95, Part II, pp. 19–58. Haifa (2000, 2002)
3. Bass, H.: Algebraic K-Theory. W. A. Benjamin Inc., New York, Amsterdam (1968)
4. Bourbaki, N.: Algebra II: Chapters 4–7. Elements of Mathematics. Translated from the French by Cohn, P.M., Howie, J. Springer-Verlag, Berlin (1990)
5. Bourbaki, N.: Algebra I. Chapters 1–3. Elements of Mathematics. Springer-Verlag, Berlin (1998). Translated from the French, Reprint of the 1989 English translation [MR0979982 (90d:00002)]
6. Coulembier, K.: Tannakian categories in positive characteristic. Duke Math. J. **169**(16), 3167–3219 (2020)
7. Deligne, P., Milne, J.S.: Tannakian categories. In: Hodge cycles. Motives, and Shimura varieties, volume 900 of Lecture Notes in Mathematics, pp. 101–228. Springer-Verlag, Berlin (1982)
8. Deligne, P.: Catégories tannakiennes. In: Vol. I.I., Volume 87 of Progr. Math. (eds.) The Grothendieck Festschrift, pp. 111–195. Birkhäuser Boston, Boston, MA (1990)
9. Deligne, P.: Letter to A. Vasiu, 30 Nov 2011. http://publications.ias.edu/deligne/paper/2653
10. Deligne, P.: Théorie de Hodge. III. Inst. Hautes Études Sci. Publ. Math. (44), pp. 5–77 (1974)
11. Demazure, M., Gabriel, P.: Groupes algébriques. Tome I: Géométrie algébrique, généralités, groupes commutatifs. Masson & Cie, Éditeur, Paris, (eds.): Avec un appendice ıt Corps de classes local par Michiel Hazewinkel. North-Holland Publishing Co., Amsterdam (1970)
12. Deninger, C.: A pro-algebraic fundamental group for topological spaces. Proc. Steklov Inst. Math. **320**, 62–90 (2023)
13. Hazewinkel, M.: Cofree coalgebras and multivariable recursiveness. J. Pure Appl. Algebra **183**(1–3), 61–103 (2003)
14. Jannsen, U.: Mixed motives and algebraic K-theory, volume 1400 of Lecture Notes in Mathematics, with appendices by Bloch, S., Schoen, C. Springer-Verlag, Berlin (1990)
15. Kucharczyk, R.A., Scholze, P.: Topological realisations of absolute Galois groups. In: Cohomology of Arithmetic Groups, Volume 245 of Springer Proceedings in Mathematics and Statistics, pp. 201–288. Springer, Cham (2018)
16. Milne, J.S.: Algebraic groups, volume 170 of Cambridge Studies in Advanced Mathematics. Cambridge University Press, Cambridge (2017). The theory of group schemes of finite type over a field

17. Moss, E.: Sweedler. Hopf Algebras. Mathematics Lecture Note Series. W. A. Benjamin Inc., New York (1969)
18. Ovchinnikov, A., Wibmer, M.: Tannakian categories with semigroup actions. Can. J. Math. **69**(3), 687–720 (2017)
19. Porst, H.-E.: The formal theory of Hopf algebras part II: the case of Hopf algebras. Quaest. Math. **38**(5), 683–708 (2015)
20. Ribes, L., Zalesskii, P.: Profinite groups, volume 40 of Results in Mathematics and Related Areas. 3rd Series. A Series of Modern Surveys in Mathematics, 2nd edn. Springer-Verlag, Berlin (2010)
21. Sauloy, J.: Galois theory of Fuchsian q-difference equations (2002). arXiv:math/0210221
22. Serre, J.-P.: Propriétés conjecturales des groupes de Galois motiviques et des représentations l-adiques. In: Motives. Seattle, WA (1991), vol. 55 Proceedings of Symposia in Pure Mathematics, pp. 377–400. American Mathematical Society, Providence, RI (1994)
23. The Stacks Project Authors. Stacks Project. http://stacks.math.columbia.edu (2023)
24. Waterhouse, W.C.: Introduction to Affine Group Schemes. Graduate Texts in Mathematics, vol. 66. Springer-Verlag, New York, Berlin (1979)
25. Wibmer, M.: Free proalgebraic groups. Épijournal Géom. Algébrique **4**, 36, Art. 1 (2020)

Non-Archimedean Plectic Jacobians

Michele Fornea and Lennart Gehrmann

Abstract *Plectic Stark–Heegner points* were recently introduced to explore the arithmetic of higher rank elliptic curves: the concept was inspired by Nekovář and Scholl's plectic philosophy, while the construction is based on Bertolini and Darmon's groundbreaking use of the p-adic uniformization of Shimura curves to study the Birch–Swinnerton-Dyer conjecture. In this note we present a framework based on the non-Archimedean uniformization of higher-dimensional quaternionic Shimura varieties which can be used to describe *plectic Heegner points* geometrically. To this end, we define the category of *Mumford varieties*, that is, varieties that can be uniformized by products of certain analytic subsets of $\mathbb{P}^1$, and construct the *plectic Jacobian functor* from the category of Mumford varieties to the category of topological groups extending the classical Jacobian functor on Mumford curves.

Keywords p-adic uniformization · Plectic conjectures · Jacobians

1 Introduction

Non-Archimedean uniformization of algebraic varieties has been shown to be a valuable tool in studying the arithmetic of elliptic curves. Most prominently, Bertolini and Darmon ingeniously utilized p-adic uniformizations of Shimura curves and elliptic curves to relate Heegner points to special values of p-adic L-functions or their derivatives (see [2–6]). It is important for us to recall Cerednik–Drinfeld's explicit p-adic uniformization of Shimura curves $X_{/\mathbb{Q}}$ of prime-to-p level associated to an

---

To Massimo Bertolini, on his 60th birthday

M. Fornea
Centre de Recerca Matemàtica, Barcelona, Spain
e-mail: mfornea.research@gmail.com

L. Gehrmann (✉)
Universität Bielefeld, Bielefeld, Germany
e-mail: gehrmann.math@gmail.com

M. Longo et al. (eds.), *Elliptic Curves and Modular Forms in Arithmetic Geometry*,
Springer Proceedings in Mathematics & Statistics 527,
https://doi.org/10.1007/978-3-032-13123-2_6

indefinite quaternion algebra $B/\mathbf{Q}$ ramified at p. It consists of an isomorphism of rigid analytic varieties

$$X(\mathbf{C}_p) \cong \Gamma\backslash\mathcal{H}_p$$

where $\mathcal{H}_p = \mathbb{P}^1(\mathbf{C}_p) \setminus \mathbb{P}^1(\mathbf{Q}_p)$ denotes Drinfeld's upper half plane, and $\Gamma \subset D^\times$ is an explicit p-arithmetic subgroup of the definite quaternion algebra $D/\mathbf{Q}$ obtained from $B/\mathbf{Q}$ by switching invariants at p and ∞. In other words, Cerednik and Drinfeld identified the p-adic Schottky group associated to $X_{/\mathbf{Q}_p}$, the existence of which is ensured by Mumford's general uniformization theorem. Interestingly, that result can also be parleyed into a uniformization of the Abel–Jacobi morphism $\alpha(X)$ from X to its Jacobian $\mathcal{J}_X$: there exists a lattice $\Lambda \subset \mathrm{H}^1(\Gamma, \mathbf{C}_p^\times)$ and a rigid analytic isomorphism $\mathcal{J}_X(\mathbf{C}_p) \cong \mathrm{H}^1(\Gamma, \mathbf{C}_p^\times)/\Lambda$ such that the diagram

$$\begin{array}{ccc} \Gamma\backslash\mathcal{H}_p & \longrightarrow & \mathrm{H}^1(\Gamma, \mathbf{C}_p^\times)/\Lambda \\ \downarrow\sim & & \downarrow\sim \\ X(\mathbf{C}_p) & \xrightarrow{\alpha(X)} & \mathcal{J}_X(\mathbf{C}_p) \end{array} \tag{1}$$

commutes (see for example [21], Chap. VI). Moreover, the top horizontal arrow can be described explicitly in terms of p-adic integration. Suppose now that $A_{/\mathbf{Q}}$ is an elliptic curve admitting a non-constant $\mathbf{Q}$-rational morphism $\mathcal{J}_X \to A$ from the Jacobian of $X_{/\mathbf{Q}}$. Tate's uniformization theorem then ensures the existence of a period $q \in p\mathbf{Z}_p \setminus \{0\}$ and an isomorphism of rigid analytic varieties $A(\mathbf{C}_p) \cong \mathbf{C}_p^\times/q^{\mathbf{Z}}$ fitting in the commuting diagram

$$\begin{array}{ccc} \mathrm{H}^1(\Gamma, \mathbf{C}_p^\times)/\Lambda & \longrightarrow & \mathbf{C}_p^\times/q^{\mathbf{Z}} \\ \downarrow\sim & & \downarrow\sim \\ \mathcal{J}_X(\mathbf{C}_p) & \longrightarrow & A(\mathbf{C}_p) \end{array} \tag{2}$$

where the top horizontal map arises from a surjection $\mathrm{H}^1(\Gamma, \mathbf{Z}) \to \mathbf{Z}$ induced by the modularity of $E_{/\mathbf{Q}}$. The upshot is that by choosing a p-adic embedding $\overline{\mathbf{Q}} \hookrightarrow \mathbf{C}_p$ the combination of (1) and (2) provides an explicit rigid analytic description of any modular parametrization $X(\overline{\mathbf{Q}}) \to E(\overline{\mathbf{Q}})$. As pointed out previously, this description was developed by Bertolini and Darmon to study the Birch–Swinnerton-Dyer conjecture.

During the last decade Nekovář and Scholl have been pursuing a series of *plectic conjectures* supporting the idea that higher dimensional quaternionic Shimura varieties could be used to study the Mordell group of higher rank elliptic curves (see [26–28]). The construction of *plectic Stark–Heegner points* in [16] is an attempt to provide a theoretical and computational way to test those expectations. The theory of plectic Stark-Heegner points originated from the realization that Cerednik and Drinfeld's uniformization result had been generalized to quaternionic Shimura varieties of arbitrary dimension in [10] and [30, Sect. 5]. An optimist could then believe that an explicit rigid analytic construction of *global* points on higher rank elliptic curves would be found by appropriately generalizing the top rows of diagrams (1) and (2). Eventually, plectic Stark–Heegner points were constructed by cohomological means that could deal with arbitrary quadratic extensions of number fields in a uniform fashion. That construction lacked a direct connection to geometry, but it allowed for numerical experiments [17], and to easily relate plectic Stark–Heegner points to higher derivatives of anticyclotomic p-adic L-functions ([16], Theorem A).

On the occasion of Bertolini's 60th birthday, we present a framework that can be used to describe *plectic Heegner points*—plectic Stark–Heegner points associated to quadratic CM extensions—geometrically. We hope it will open new perspectives on the results obtained in [12, 15] and [14].

The technical results of the article focus on the structure of "plectic subgroups" and their associated "Mumford varieties". To be more precise, in Sects. 2 and 3 we work over an arbitrary local field of residue characteristic p. Let $\mathbf{C}$ denote the completion of an algebraic closure of either $\mathbf{Q}_p$ or $\mathbf{F}_p(\!(T)\!)$. Section 2 introduces the notion of *plectic groups*. These are certain discrete subgroups $\Gamma \subseteq \mathrm{PGL}_2(\mathbf{C})^r$, for some integer r, whose domain of discreteness $\Omega \subseteq \mathbb{P}^1(\mathbf{C})^r$ factors as a product

$$\Omega = \Omega_1 \times \cdots \times \Omega_r$$

with each $\Omega_i \subseteq \mathbb{P}^1(\mathbf{C})$ an analytic set with compact complement. Through a reduction to the case $r = 1$, the analytic subsets Ω_i's can be identified with points of discreteness for the action of p-adic Schottky subgroups of $\mathrm{PGL}_2(\mathbf{C})$. This permits the definition of a polysimplicial complex $\mathcal{T}$—a product of trees to be more precise—on which Γ acts. Under the assumption that the quotient $\Gamma\backslash\mathcal{T}$ is a finite complex, in which case Γ is called *normal*, several finiteness properties for Γ are then deduced. For example, Γ is shown to be finitely presented and a virtual duality group. Moreover, an explicit model for its duality module is given.

Section 3 introduces the notion of *Mumford varieties* whose connected components are quotients of the form $X_\Gamma := \Gamma\backslash\Omega$ where Γ is a torsion-free normal plectic group. Mumford varieties are smooth and proper rigid analytic varieties and, in dimension one, the notion of Mumford varieties agrees with the classical notion of Mumford curves. Denote by $\mathbf{C}^\times_{\otimes,r}$ the completed tensor product of r copies of $\mathbf{C}^\times$. Inspired by the analytic uniformization of the Jacobian of Mumford curves described in (1), the *plectic Jacobian* $\mathcal{J}_\varpi(X_\Gamma)$ of the Mumford variety X_Γ is defined as a quotient of the cohomology group $\mathrm{H}^r(\Gamma, \mathbf{C}^\times_{\otimes,r})$ by some closed subgroup. Moreover, we consider $\mathcal{Z}_\varpi(X_\Gamma)$ the group of 0-cycles of degree $(0, \ldots, 0)$ of X_Γ, i.e., the group of those zero-cycles of X_Γ which have a lift to Ω of degree 0 on every factor Ω_i. Generalizing the top row of (1), we then construct a *plectic Abel–Jacobi* map

$$\alpha(X_\Gamma)\colon \mathcal{Z}_\varpi(X_\Gamma) \longrightarrow \mathcal{J}_\varpi(X_\Gamma).$$

The formation of plectic Jacobians as well as the group of zero cycles of degree $(0, \ldots, 0)$ are shown to be functorial with respect to *Hecke correspondences* of Mumford varieties. The Abel–Jacobi map is shown to be a natural transformation.

Section 4 makes sense of the top row of diagram (2) in the context of higher dimensional quaternionic Shimura varieties by applying the machinery developed in the previous sections. If F is a totally real number field and X a quaternionic Shimura variety admitting p-adic uniformization, the associated domain of discreteness Ω is a product of p-adic upper half planes indexed by a set $S = \{\mathfrak{p}_1, \ldots, \mathfrak{p}_r\}$ of p-adic places of F. Every modular elliptic $A_{/F}$ whose associated system of Hecke eigenvalues appears in the cohomology of X has multiplicative reduction for every $\mathfrak{p} \in S$ and so gives rise to a Mumford variety $A_S := \prod_{\mathfrak{p}} A_{\mathfrak{p}}$. We then obtain a natural morphism

$$\mathfrak{Z}_{\infty}(X) \longrightarrow \mathfrak{Z}_{\infty}(A_S)$$

which can be interpreted as a reformulation of [16, Sect. 4.5] in the framework developed in this article.

We hope that the constructions of this paper will be useful for studying other problems, such as:

- constructing *plectic Heegner cycles* for higher weight modular forms,
- proving *plectic comparison theorems* for quaternionic Shimura varieties with Iwahori level at p, complementing the main result of [23], and
- understanding the relation between plectic Heegner points over function fields, Heegner–Drinfeld cycles [31, 32], and the theory of higher modularity [22].

For the interested reader, we note that there is an Archimedean twin of this paper [18] where an analogue inquiry has been carried out using plectic Hodge theory.

Acknowledgements. We are grateful to the anonymous referee for their valuable comments and feedback. While working on this article the first author was supported by the MTM grant PID2020-118236GB-I00 from Universitat Autònoma de Barcelona, and the Maria de Maeztu grant from the Centre de Recerca Matemàtica. The second named author received funding from the Maria Zambrano grant for the attraction of international talent and from Deutsche Forschungsgemeinschaft (DFG, German Research Foundation) via the grant SFB-TRR 358/1 2023—491392403.

2 Plectic Subgroups

A *local étale algebra* is a finite étale algebra over a local field, or in other words, an algebra of the form

$$F_\Sigma = \prod_{\mathfrak{p} \in \Sigma} F_{\mathfrak{p}}$$

where Σ is a finite set and the fields $F_{\mathfrak{p}}$ are local fields of the same characteristic and residue characteristic. We will always assume that the residue characteristic is a fixed prime p. Denote by $\mathbf{C}$ the completion of a fixed algebraic closure of $\mathbf{Q}_p$ (resp. $\mathbf{F}_p((T))$) in case of characteristic 0 (resp. characteristic p).

Definition 2.1 An *embedding* of a local étale algebra F_Σ is a tuple $\iota = (\iota_\mathfrak{p})_{\mathfrak{p}\in\Sigma}$ of ring homomorphisms $\iota_\mathfrak{p} \colon F_\mathfrak{p} \hookrightarrow \mathbf{C}$. We call an *embedded local étale algebra* any local étale algebra with the data of an embedding.

Let (F_Σ, ι) be an embedded local étale algebra. For any $\mathfrak{p} \in \Sigma$ the valuation ring of $F_\mathfrak{p}$ is denoted by $\mathcal{O}_\mathfrak{p}$ with maximal ideal $\mathfrak{p}$. Moreover, we put

$$K_\mathfrak{p} = \mathrm{PGL}_2(\mathcal{O}_\mathfrak{p}),$$

and write

$$\mathrm{Iw}_\mathfrak{p} \subseteq \mathrm{PGL}_2(\mathcal{O}_\mathfrak{p})$$

for the Iwahori subgroup of all integral matrices which are congruent to an upper triangular matrix modulo $\mathfrak{p}$. For any subset $S \subseteq \Sigma$ we set

$$K_S = \prod_{\mathfrak{p}\in S} K_\mathfrak{p} \quad \text{and} \quad \mathrm{Iw}_S = \prod_{\mathfrak{p}\in S} \mathrm{Iw}_\mathfrak{p}.$$

Given a non-empty subset $S \subseteq \Sigma$ and a subfield $E \subseteq \mathbf{C}$ containing $\iota_\mathfrak{p}(F_\mathfrak{p})$ for all $\mathfrak{p} \in S$ we put

$$F_S = \prod_{\mathfrak{p}\in S} F_\mathfrak{p} \quad \text{and} \quad E_S = \prod_{\mathfrak{p}\in S} E.$$

We usually abbreviate $E_\mathfrak{p} := E_{\{\mathfrak{p}\}}$. The embedding ι induces a continuous action of the group $\mathrm{PGL}_2(F_S) = \prod_{\mathfrak{p}\in S} \mathrm{PGL}_2(F_\mathfrak{p})$ on the product

$$\mathbb{P}^1(E_S) = \prod_{\mathfrak{p}\in S} \mathbb{P}^1(E)$$

of projective lines. Given subsets $S' \subseteq S \subseteq \Sigma$ we often view $\mathrm{PGL}_2(F_{S'})$ as a subgroup of $\mathrm{PGL}_2(F_S)$ via the canonical embedding, and we write

$$\mathrm{pr}_{S'} \colon \mathrm{PGL}_2(F_S) \longrightarrow \mathrm{PGL}_2(F_{S'})$$

for the canonical projection.

2.1 Limit Points

For the remainder of this chapter we fix an embedded local étale algebra (F_Σ, ι). We collect basic facts about the set of limit points of the action of subgroups of $\mathrm{PGL}_2(F_S)$ on $\mathbb{P}^1(\mathbf{C}_S)$ for subsets $S \subseteq \Sigma$.

Definition 2.2 For a subgroup Γ of $\mathrm{PGL}_2(F_S)$, we define $\mathcal{L}_\Gamma^S$ to be the set of limit points of Γ in $\mathbb{P}^1(\mathbf{C}_S)$, that is, the set of elements $x \in \mathbb{P}^1(\mathbf{C}_S)$ such that there exists $y \in \mathbb{P}^1(\mathbf{C}_S)$ and distinct elements $\{\gamma_j\}_{j\geq 1} \subseteq \Gamma$ such that $\gamma_j(y)$ converges to x.

We may consider a subgroup $\Gamma \subseteq \mathrm{PGL}_2(F_S)$ also as a subgroup of $\mathrm{PGL}_2(F_\Sigma)$ via the canonical embedding. It follows directly from the definition that

$$\mathcal{L}_\Gamma^\Sigma = \mathcal{L}_\Gamma^S \times \mathbb{P}^1(\mathbf{C}_{\Sigma\setminus S}). \tag{3}$$

The compactness of the projective line implies that the set of limit points is rarely empty:

Proposition 2.1 *Let $\Gamma \subseteq \mathrm{PGL}_2(F_S)$ be a subgroup with $\mathcal{L}_\Gamma^S = \emptyset$. Then Γ is finite.*

Proof Let $E \subseteq \mathbf{C}$ be a subfield finite over F, containing the compositum of the fields $\iota_\mathfrak{p}(F_\mathfrak{p})$ for $\mathfrak{p} \in S$. By assumption, for any $x \in \mathbb{P}^1(E_S)$ the orbit $\Gamma.x$ is a discrete subset of the compact space $\mathbb{P}^1(E_S)$. Thus, it is finite. Choosing E large enough there exists a point $x \in \mathbb{P}^1(E_S)$ whose stabilizer in $\mathrm{PGL}_2(F_S)$ is finite, hence Γ is finite as well. □

Commensurable subgroups have the same set of limit points:

Lemma 2.1 *Let $\Gamma, \Gamma' \subseteq \mathrm{PGL}_2(F_S)$ be commensurable subgroups, then*

$$\mathcal{L}_{\Gamma'}^S = \mathcal{L}_\Gamma^S.$$

Proof It is enough to assume that Γ' is a finite index subgroup of Γ. Clearly $\mathcal{L}_{\Gamma'}^S \subseteq \mathcal{L}_\Gamma^S$. Let now x be an element of $\mathcal{L}_\Gamma^S$. By definition, there exists $y \in \mathbb{P}^1(\mathbf{C}_S)$ and distinct elements $\{\gamma_j\}_j \subseteq \Gamma$ such that $\gamma_j(y)$ converges to x. Since there are finitely many cosets $\Gamma'\backslash\Gamma$, up to considering a subsequence, there exists a fixed $g \in \Gamma$ such that for every index

$$\gamma_j = \delta_j \cdot g \quad \text{for some} \quad \delta_j \in \Gamma'.$$

We put $y_0 = g(y)$. Then by construction the sequence $\delta_j(y_0)$ converges to x. □

If $\Gamma \subseteq \mathrm{PGL}_2(F_S)$ is not discrete, we clearly have that $\mathcal{L}_\Gamma^S = \mathbb{P}^1(\mathbf{C}_S)$. Therefore, the set of limit points is most interesting for discrete subgroups. Let us assume for the moment that Γ is a discrete subgroup of $\mathrm{PGL}_2(F_\mathfrak{p})$ for some $\mathfrak{p} \in \Sigma$. From (1.6.4) of [21, Sect. I], we know that $\mathcal{L}_\Gamma^{\{\mathfrak{p}\}}$ is a subset of $\mathbb{P}^1(F_\mathfrak{p})$. A mild generalization of the proof of *loc. cit.* provides the following description of limits sets of discrete subgroups of $\mathrm{PGL}_2(F_S)$.

Lemma 2.2 *Let $\Gamma \subseteq \mathrm{PGL}_2(F_S)$ be a discrete subgroup, then*

$$\mathcal{L}_\Gamma^S \subseteq \bigcup_{\mathfrak{p}\in S} \left(\mathbb{P}^1(F_\mathfrak{p}) \times \mathbb{P}^1(\mathbf{C}_{S\setminus\{\mathfrak{p}\}})\right).$$

Proof Let x be an element of $\mathcal{L}_\Gamma^S$. We have to show that there exists $\mathfrak{p} \in S$ such that the $\mathfrak{p}$-th component of x is $F_\mathfrak{p}$-rational. From the definition of limit set, there exists a point $y \in \mathbb{P}^1(\mathbf{C}_S)$ and a sequence of distinct elements $\{\gamma_j\}_j \subseteq \Gamma$ such that $\gamma_j(y)$ converges to x. At the cost of replacing $\{\gamma_j\}_j$ with a subsequence, we may decompose $S = S_1 \cup S_2$ such that the sequence of elements $(\gamma_j)_\mathfrak{p}$ consists of distinct elements for all $\mathfrak{p} \in S_1$ and are constant for all $\mathfrak{p} \in S_2$. By assumption, the subset S_1 is not empty. Then, as in the proof of (1.6.4) of [21, Sect. I], we can find preimages $\widetilde{\gamma}_j$ of $\gamma_j \in \mathrm{GL}_2(F_S)$ such that

- the sequence $(\widetilde{\gamma}_j)_\mathfrak{p}$ is constant for all $\mathfrak{p} \in S_2$, and
- for every $\mathfrak{p} \in S_1$ the sequence $(\widetilde{\gamma}_j)_\mathfrak{p}$ converges in $\mathrm{M}_2(F_\mathfrak{p})$ (after a change of variables) to a matrix of the form

$$\begin{pmatrix} a_\mathfrak{p} & b_\mathfrak{p} \\ 1 & d_\mathfrak{p} \end{pmatrix}.$$

Since the subgroup Γ is discrete, the limit of the converging sequence $\{\widetilde{\gamma}_j\}_j$ is not an element of $\mathrm{GL}_2(F_S)$. In other words, there exists a $\mathfrak{p} \in S_1$ such that $a_\mathfrak{p} d_\mathfrak{p} = b_\mathfrak{p}$. In particular, the limit of the sequence $(\gamma_j(y))_\mathfrak{p}$ is either $a_\mathfrak{p}$ or ∞, which proves the claim. □

2.2 *Plectic Groups*

Consider a discrete subgroup of $\mathrm{PGL}_2(F_S)$ of the form

$$\Gamma = \prod_{\mathfrak{p} \in S} \Gamma_\mathfrak{p},$$

where $\Gamma_\mathfrak{p} \subseteq \mathrm{PGL}_2(F_\mathfrak{p})$ is a discrete subgroup for each $\mathfrak{p} \in S$. An easy calculation shows that the set of limit points of Γ is given by

$$\mathcal{L}_\Gamma^S = \bigcup_{\mathfrak{p} \in S} \left(\mathcal{L}_{\Gamma_\mathfrak{p}}^{\{\mathfrak{p}\}} \times \mathbb{P}^1(\mathbf{C}_{S \setminus \{\mathfrak{p}\}}) \right).$$

We are interested in studying the properties of all discrete subgroups of $\mathrm{PGL}_2(F_S)$ whose set of limit points has the same structure. Among those are all cocompact lattices in $\mathrm{PGL}_2\, F_S$ (see Proposition 2.2 below). A class of cocompact lattices, that are of particular interest to us, are S-arithmetic groups of definite quaternion algebras that feature in the p-adic uniformization of Shimura varieties. They will be discussed further in Sect. 4.

Definition 2.3 A subgroup $\Gamma \subseteq \mathrm{PGL}_2(F_S)$ is called *plectic* if there exist subsets $\mathcal{L}_{\Gamma,\mathfrak{p}} \subseteq \mathbb{P}^1(F_\mathfrak{p})$ for $\mathfrak{p} \in S$ such that

$$\mathcal{L}_\Gamma^S = \bigcup_{\mathfrak{p}\in S} \left(\mathcal{L}_{\Gamma,\mathfrak{p}} \times \mathbb{P}^1(\mathbf{C}_{S\setminus\{\mathfrak{p}\}})\right).$$

In that case, we define the subset of *distinguished limit points* as

$$\mathcal{L}_{\Gamma,S} := \prod_{\mathfrak{p}\in S} \mathcal{L}_{\Gamma,\mathfrak{p}}.$$

Remark 2.1 The set of distinguished limit points can be thought of as a p-adic analogue of the notion of Bergman–Shilov boundary in the theory of holomorphic functions in several complex variables (cf. [19], Chap. III–15).

Equation (3) implies that a subgroup $\Gamma \subseteq \mathrm{PGL}_2(F_S)$ is plectic if and only if it is plectic (with $\mathcal{L}_{\Gamma,\mathfrak{p}} = \emptyset$ for all $\mathfrak{p} \in \Sigma \setminus S$) when viewed as a subgroup of $\mathrm{PGL}_2(F_\Sigma)$. By Lemma 2.1 every subgroup commensurable with a plectic subgroup is plectic.

Remark 2.2 As noted earlier, every plectic subgroup is automatically discrete. However, the converse does not hold in general. For example, the subgroup $\Gamma \subseteq \mathrm{PGL}_2(\mathbf{Q}_p) \times \mathrm{PGL}_2(\mathbf{Q}_p)$ generated by the tuple

$$\left(\begin{pmatrix} p & 0 \\ 0 & 1 \end{pmatrix}, \begin{pmatrix} p & 0 \\ 0 & 1 \end{pmatrix}\right)$$

is discrete but not plectic. Indeed, its set of limit points in $\mathbb{P}^1(\mathbf{C}) \times \mathbb{P}^1(\mathbf{C})$ is

$$\mathcal{L}_\Gamma = \{(0,0), (\infty, \infty)\}.$$

Evidently, the easiest way of generating plectic subgroups is taking products of plectic subgroups:

Lemma 2.3 *Let $S = S_1 \cup S_2$ with $S_1 \cap S_2 = \emptyset$. The product $\Gamma = \Gamma_1 \times \Gamma_2$ of two plectic subgroups $\Gamma_i \subseteq \mathrm{PGL}_2(F_{S_i})$, $i = 1, 2$, is a plectic subgroup of $\mathrm{PGL}_2(F_S)$ with set of distinguished limit points given by*

$$\mathcal{L}_{\Gamma,S} = \mathcal{L}_{\Gamma_{1,S_1}} \times \mathcal{L}_{\Gamma_{2,S_2}}.$$

A more interesting collection of plectic subgroups is given by cocompact lattices in $\mathrm{PGL}_2(F_S)$:

Proposition 2.2 *Let $\Gamma \subseteq \mathrm{PGL}_2(F_S)$ be a discrete and cocompact subgroup. Then, Γ is plectic with set of distinguished limit points*

$$\mathcal{L}_{\Gamma,S} = \mathbb{P}^1(F_S).$$

Proof By Lemma 2.2 it is enough to show that

$$\bigcup_{\mathfrak{p}\in S}\left(\mathbb{P}^1(F_{\mathfrak{p}})\times\mathbb{P}^1(\mathbf{C}_{S\setminus\{\mathfrak{p}\}})\right)\subseteq\mathcal{L}_\Gamma^S.$$

So let us fix an element $\mathfrak{q}\in S$ and a point $x=(x_{\mathfrak{p}})_{\mathfrak{p}\in S}\in\mathbb{P}^1(\mathbf{C}_S)$ with $x_{\mathfrak{q}}\in\mathbb{P}^1(F_{\mathfrak{q}})$. We consider the elements $g_j\in\mathrm{GL}_2(F_{\mathfrak{q}})\subseteq\mathrm{GL}_2(F_S)$, $j\geq 1$, given by

$$g_j=\begin{pmatrix}p^j & 0\\ 1 & 1\end{pmatrix}\quad\text{if } x_{\mathfrak{q}}=0,\qquad g_j=\begin{pmatrix}1 & 0\\ x_{\mathfrak{q}}^{-1} & p^j\end{pmatrix}\quad\text{if } x_{\mathfrak{q}}\neq 0.$$

Let $y=(y_{\mathfrak{p}})_{\mathfrak{p}\in S}\in\mathbb{P}^1(\mathbf{C}_S)$ be the point given by $y_{\mathfrak{q}}=\infty$ and $y_{\mathfrak{p}}=x_{\mathfrak{p}}$ for $\mathfrak{p}\neq\mathfrak{q}$. Then, it is easy to see that

$$\lim_{j\to\infty}g_j(y)=x.$$

Since Γ is cocompact, the double coset $\Gamma\backslash\mathrm{PGL}_2(F_S)/\mathrm{Iw}_S$ is finite. Therefore, we can suppose, up to replacing $\{g_j\}_j$ with a subsequence, that there is a fixed $g\in\mathrm{GL}_2(F_S)$ such that

$$g_j=\gamma_j\cdot g\cdot r_j\qquad\text{with } \gamma_j\in\Gamma,\ r_j\in\mathrm{Iw}_S.$$

Note that the sequence $\{\gamma_j\}_j$ consists of pairwise different elements because $g_j^{-1}\cdot g_k\notin\mathrm{Iw}_S$ when $j\neq k$. As the group Iw_S is compact, so is the orbit of y under Iw_S. In particular, there exists a subsequence of $\{r_j\}_{j\geq 1}$ such that $(g\cdot r_j)(y)$ converges to some $y_\circ\in\mathbb{P}^1(\mathbf{C}_S)$. We deduce the existence of a subsequence $\{\gamma_\ell\}_\ell\subseteq\{\gamma_j\}_j$ with

$$x=\lim_{\ell\to\infty}\gamma_\ell(y_\circ),$$

that is, x is a limit point for Γ. □

Note that for arithmetic application one can produce plectic groups that are not products by considering S-arithmetic subgroups of totally definite quaternion algebras (see Sect. 4.1).

The following lemma will be used to construct a Hecke action on plectic Jacobians.

Lemma 2.4 *Let $\Gamma\subseteq\mathrm{PGL}_2(F_S)$ be a plectic subgroup and g an element of $\mathrm{PGL}_2(F_S)$. Then $g\Gamma g^{-1}$ is plectic with set of distinguished limit points*

$$\mathcal{L}_{g\Gamma g^{-1},S}=g(\mathcal{L}_{\Gamma,S}).$$

2.3 Partial Intersection with Compact Open Subgroups

In case S is a singleton $\{\mathfrak{p}\}$, the set of limits points $\mathcal{L}_{\Gamma,\mathfrak{p}}\subseteq\mathbb{P}^1(F_{\mathfrak{p}})$ of a discrete subgroup $\Gamma\subseteq\mathrm{PGL}_2(F_{\mathfrak{p}})$ is either a finite set with at most 2 elements or a perfect set, that is, a closed subset without isolated points (see (1.6.3) of [21, Chap. I]). We now prove the analogous statement for plectic subgroups for arbitrary S by reducing to the case of a singleton set.

Suppose $S_1 \subseteq S$. Given subgroups $\Gamma \subseteq \mathrm{PGL}_2(F_S)$ and $U \subseteq \mathrm{PGL}_2(F_{S\setminus S_1})$, we put

$$\widetilde{\Gamma}_U := \Gamma \cap \left(\mathrm{PGL}_2(F_{S_1}) \times U\right)$$

and

$$\Gamma_U := \mathrm{pr}_{S_1}(\widetilde{\Gamma}_U) \subseteq \mathrm{PGL}_2(F_{S_1}).$$

Lemma 2.5 *Let $\Gamma \subseteq \mathrm{PGL}_2(F_S)$ be a discrete subgroup and $U \subseteq \mathrm{PGL}_2(F_{S\setminus S_1})$ an open compact subgroup. Then $\Gamma_U \subseteq \mathrm{PGL}_2(F_{S_1})$ is a discrete subgroup, and the kernel of the projection map $\mathrm{pr}_{S_1} : \widetilde{\Gamma}_U \to \Gamma_U$ is finite.*

Proof Let $V \subseteq \mathrm{PGL}_2(F_{S_1})$ be any compact open subgroup. Since Γ is discrete, the intersection $\Gamma \cap (V \times U)$ is finite. This implies that the intersection $\Gamma_U \cap V$ is finite, which proves the first claim. The kernel of the projection from $\widetilde{\Gamma}_U$ to Γ_U is given by the intersection

$$\Gamma \cap \left(\{1_{\mathrm{PGL}_2(F_{S_1})}\} \times U\right),$$

which is both discrete and compact.Therefore, it is finite. □

Proposition 2.3 *Let $\Gamma \subseteq \mathrm{PGL}_2(F_S)$ be a plectic subgroup. For every open compact subgroup $U \subseteq \mathrm{PGL}_2(F_{S\setminus S_1})$ the subgroup $\Gamma_U \subseteq \mathrm{PGL}_2(F_{S_1})$ is plectic with*

$$\mathcal{L}_{\Gamma,\mathfrak{p}} = \mathcal{L}_{\Gamma_U,\mathfrak{p}}$$

for all $\mathfrak{p} \in S_1$.

Proof Let $U_\mathfrak{q} \subseteq \mathrm{PGL}_2(F_\mathfrak{p})$, $\mathfrak{q} \in S \setminus S_1$, be open compact subgroups with

$$U' := \prod_{\mathfrak{q}\in S\setminus S_1} U_\mathfrak{q} \subseteq U.$$

By Lemma 2.1 one may replace U by U'. Thus, by an easy inductive argument we reduce to the case that $S \setminus S_1 = \{\mathfrak{q}\}$ is a singleton set. Moreover, by invoking Lemma 2.1 again we may assume that $U = K_\mathfrak{p}$.

Now, let x be an element of $\mathcal{L}^{S_1}_{\Gamma_U}$. Choose $y \in \mathbb{P}^1(\mathbf{C}_{S_1})$ and distinct elements $\{\gamma_j\}_j \subseteq \Gamma_U$ such that $\gamma_j(y)$ converges to y. Moreover, pick preimages $\widetilde{\gamma}_j$ of $\widetilde{\gamma}_j$ in $\widetilde{\Gamma}_U$ and an element $y_\mathfrak{q} \in \mathbb{P}^1(\mathbf{C}_\mathfrak{q}) \setminus \mathbb{P}^1(F_\mathfrak{q})$. The sequence $\widetilde{\gamma}_j(y_\mathfrak{q})$ lies inside the compact set $U.x_\mathfrak{q} \subseteq \mathbb{P}^1(\mathbf{C}_\mathfrak{q}) \setminus \mathbb{P}^1(F_\mathfrak{q})$. Since $\mathbb{P}^1(\mathbf{C}_\mathfrak{q}) \setminus \mathbb{P}^1(F_\mathfrak{q})$ is a metric space, it follows that $U.x_\mathfrak{q}$ is sequentially compact. In particular, we may replace $\widetilde{\gamma}_j$ by a subsequence such that $\widetilde{\gamma}_j(y_\mathfrak{q})$ converges to an element $x_\mathfrak{q} \in \mathbb{P}^1(\mathbf{C}_\mathfrak{q}) \setminus \mathbb{P}^1(F_\mathfrak{q})$. It follows that $(x, x_\mathfrak{q})$ is an element of $\mathcal{L}^S_\Gamma$. Since Γ is plectic and $x_\mathfrak{q}$ is not an element of $\mathcal{L}_{\Gamma,\mathfrak{p}}$, it follows that $(x, x_\mathfrak{q}) \in \mathcal{L}^S_\Gamma$ for all $x_\mathfrak{q} \in \mathbb{P}^1(\mathbf{C}_\mathfrak{q})$.

On the other hand, let $x \in \mathbb{P}^1(\mathbf{C}_{S_1})$ be an element such that there exists $x_\mathfrak{q} \in \mathbb{P}^1(\mathbf{C}_\mathfrak{q}) \setminus \mathcal{L}_{\Gamma,\mathfrak{p}}$ with $(x, x_\mathfrak{q}) \in \mathcal{L}_\Gamma^S$. Then $(x, x_\mathfrak{q})$ lies in $\mathcal{L}_\Gamma^S$ for all $x_\mathfrak{q} \in \mathbb{P}^1(\mathbf{C}_\mathfrak{q})$. Consider the reduction map

$$\mathrm{red}\colon \mathbb{P}^1(\mathbf{C}_\mathfrak{q}) \setminus \mathbb{P}^1(F_\mathfrak{q}) \longrightarrow \mathcal{T}_\mathfrak{q}$$

to the Bruhat–Tits tree of $\mathrm{PGL}_2(F_\mathfrak{q})$ (see for example [20, Sect. 4.9]). The subgroup $U = K_\mathfrak{p} \subseteq \mathrm{PGL}_2(F_\mathfrak{q})$ is the stabilizer of a vertex $v \in \mathcal{T}_\mathfrak{q}$. Fix elements $x_\mathfrak{q} \in \mathrm{red}^{-1}(v)$, $y = (y_\mathfrak{p}) \in \mathbb{P}^1(\mathbf{C}_{S_1})$, $y_\mathfrak{q} \in \mathbb{P}^1(\mathbf{C}_{S_1})$, and a sequence of distinct elements $\{\gamma_j\}_j$ satisfying

$$\lim_{j\to\infty} \gamma_j(y, y_\mathfrak{q}) = (x, x_\mathfrak{q}).$$

Since $\mathrm{red}^{-1}(v)$ is an open neighbourhood of $x_\mathfrak{q}$, we may assume that $\gamma_j(y_\mathfrak{q}) \in \mathrm{red}^{-1}(v)$ for all $j \geq 0$. This implies that $\gamma_j' := \gamma_j\gamma_0^{-1} \in \Gamma_U$ for all j. By Lemma 2.5 we may choose a subsequence of the sequence $\mathrm{pr}_{S_1}(\gamma_j')$ that consists of distinct elements. It follows that x is an element of $\mathcal{L}_{\Gamma_U}^{S_1}$. □

The following characterization of limit sets of plectic subgroups can be deduced by combining Proposition 2.3 with $S_1 = \{\mathfrak{p}\}$ and equations (1.5) and (1.6.3) of [21, Chap. I].

Corollary 2.1 *Let $\Gamma \subseteq \mathrm{PGL}_2(F_S)$ be a plectic subgroup. For every $\mathfrak{p} \in S$ one of the following is true:*

(i) $\mathcal{L}_{\Gamma,\mathfrak{p}} = \emptyset$.
(ii) $|\mathcal{L}_{\Gamma,\mathfrak{p}}| = 1$ *and* $\mathrm{char}(F) = p$.
(iii) $|\mathcal{L}_{\Gamma,\mathfrak{p}}| = 2$.
(iv) $\mathcal{L}_{\Gamma,\mathfrak{p}}$ *is a perfect set.*

Case (ii) only happens in the presence of a large subgroup consisting of unipotent elements, while case (i) and (ii) imply the absence of hyperbolic elements (see Examples (1.7) of [21, Chap. I]). Remember that an element of $\mathrm{PGL}_2(F_\mathfrak{p})$ is called hyperbolic if it is represented by a matrix whose two eigenvalues have different absolute value. This motivates the following definition:

Definition 2.4 A plectic subgroup $\Gamma \subseteq \mathrm{PGL}_2(F_S)$ is called *semi-simple* if for every $\mathfrak{p} \in S$ the set $\mathcal{L}_{\Gamma,\mathfrak{p}}$ does not have cardinality 1. Given a semi-simple plectic subgroup Γ we put

$$\mathbf{S}_\Gamma := \left\{\mathfrak{p} \in S \mid \mathcal{L}_{\Gamma,\mathfrak{p}} \neq \emptyset\right\}.$$

Clearly, every subgroup commensurable to a semi-simple plectic subgroup is semi-simple. Moreover, when $S_1, S_2 \subseteq S$ are disjoint subsets and $\Gamma_i \subseteq \mathrm{PGL}_2(F_{S_i})$ are semi-simple plectic subgroups, then also the product $\Gamma_1 \times \Gamma_2$ is plectic semi-simple. Proposition 2.3 immediately implies the following:

Corollary 2.2 *Let $\Gamma \subseteq \mathrm{PGL}_2(F_S)$ be a semi-simple plectic subgroup and $U \subseteq \mathrm{PGL}_2(F_{S\setminus S_1})$ an open compact subgroup. Then Γ_U is a semi-simple plectic subgroup.*

Lemma 2.6 *Let* $\Gamma \subseteq \mathrm{PGL}_2(F_S)$ *be a torsion-free plectic subgroup. Then* Γ *is semi-simple.*

Proof This is a consequence of [21, Chap. I, (1.7)] and the fact that unipotent elements in characteristic p are torsion. □

2.4 Associated Polysimplicial Complex

Recall that an infinite halfline of a tree is called an end of that tree. The set of ends of the Bruhat–Tits tree $\mathcal{T}_\mathfrak{p}$, $\mathfrak{p} \in S$, can canonically be identified with $\mathbb{P}^1(F_\mathfrak{p})$ ([13], Lemma 1.3.6). Let $\Gamma \subseteq \mathrm{PGL}_2(F_S)$ be a semi-simple plectic subgroup. For every $\mathfrak{p} \in S$ one associates a locally finite tree $\mathcal{T}_{\mathcal{L}_{\Gamma,\mathfrak{p}}}$ to the set of limit points $\mathcal{L}_{\Gamma,\mathfrak{p}}$:

- If $\mathcal{L}_{\Gamma,\mathfrak{p}}$ is a perfect set, the construction of $\mathcal{T}_{\mathcal{L}_{\Gamma,\mathfrak{p}}}$ is given in [21, Sect. I.2], and [20, Sect. 4.9]. In a nutshell, it is the smallest subtree of the Bruhat–Tits tree $\mathcal{T}_\mathfrak{p}$ whose set of ends is $\mathcal{L}_{\Gamma,\mathfrak{p}}$. It follows directly from the construction that $\mathrm{pr}_\mathfrak{p}(\Gamma)$ acts on $\mathcal{T}_{\mathcal{L}_{\Gamma,\mathfrak{p}}}$.
- In case $|\mathcal{L}_{\Gamma,\mathfrak{p}}| = 2$, the pointwise stabilizer of $\mathcal{L}_{\Gamma,\mathfrak{p}}$ in $\mathrm{PGL}_2(F_\mathfrak{p})$ is a maximal $F_\mathfrak{p}$-split torus T and $\mathrm{pr}_\mathfrak{p}(\Gamma)$ is contained in the normalizer $N(T)$ of T. Let $K_T \subseteq T$ the maximal compact subgroup. The quotient T/K_T is an infinite cyclic group. One defines $\mathcal{T}_{\mathcal{L}_{\Gamma,\mathfrak{p}}}$ as the Cayley graph associated to a generator of T/K_T. The normalizer $N(T)$ and, thus, also $\mathrm{pr}_\mathfrak{p}(\Gamma)$ clearly act on $\mathcal{T}_{\mathcal{L}_{\Gamma,\mathfrak{p}}}$. It is possible to identify $\mathcal{T}_{\mathcal{L}_{\Gamma,\mathfrak{p}}}$ with a subtree of the Bruhat–Tits tree $\mathcal{T}_\mathfrak{p}$ of $\mathrm{PGL}_2(F_\mathfrak{p})$. More precisely, $\mathcal{T}_{\mathcal{L}_{\Gamma,\mathfrak{p}}}$ is identified with the infinite geodesic connecting the two points in $\mathcal{L}_{\Gamma,\mathfrak{p}} \subseteq \mathbb{P}^1(F_\mathfrak{p})$.
- When $\mathcal{L}_{\Gamma,\mathfrak{p}} = \emptyset$, we simply define $\mathcal{T}_{\mathcal{L}_{\Gamma,\mathfrak{p}}}$ to consist of a single vertex. Note that there is in general no canonical embedding of this vertex into the Bruhat–Tits tree.

Combining the action for all $\mathfrak{p} \in S$ we get an action of Γ on the locally finite polysimplical complex

$$\mathcal{T}_\Gamma := \prod_{\mathfrak{p} \in S} \mathcal{T}_{\mathcal{L}_{\Gamma,\mathfrak{p}}}.$$

The dimension of $\mathcal{T}_\Gamma$ is given by

$$\dim(\mathcal{T}_\Gamma) = |\mathbf{S}_\Gamma|.$$

Lemma 2.7 *Let* $\Gamma \subseteq \mathrm{PGL}_2(F_S)$ *be semi-simple plectic subgroup. The action of* Γ *on* $\mathcal{T}_\Gamma$ *is proper: the stabilizer of each simplex* σ *of* $\mathcal{T}_\Gamma$ *in* Γ *is finite.*

Proof For every $\mathfrak{p} \in \mathbf{S}_\Gamma$ the tree $\mathcal{T}_{\mathcal{L}_{\Gamma,\mathfrak{p}}}$ is a subtree of the Bruhat–Tits tree $\mathcal{T}_\mathfrak{p}$. The stabilizer in $\mathrm{PGL}_2(F_\mathfrak{p})$ of a finite subgraph of the Bruhat–Tits tree $\mathcal{T}_\mathfrak{p}$ is a compact open subgroup. Thus, for any simplex σ of $\mathcal{T}_\Gamma$ there exists exists a compact open subgroup $U \subseteq \mathrm{PGL}_2(F_{\mathbf{S}_\Gamma})$ such that

$$\mathrm{Stab}_\Gamma(\sigma) = \widetilde{\Gamma}_U.$$

By Lemma 2.3 the group Γ_U does not have any limit points and, therefore, it is finite by Proposition 2.1. Lemma 2.5 implies that $\widetilde{\Gamma}_U$ is finite, as well. □

The converse to Lemma 2.7 also holds. Indeed, by [29, Proposition 3.2], every finite order automorphism of a tree fixes a vertex or an edge. Therefore, we get:

Lemma 2.8 *Let* $\Gamma \subseteq \mathrm{PGL}_2(F_S)$ *be semi-simple plectic subgroup and* $\gamma \in \Gamma$ *an element of finite order. Then* γ *fixes a simplex of* $\mathcal{T}_\Gamma$.

In accordance with [25, Definition, p. 192], we consider the following finiteness condition:

Definition 2.5 A semi-simple plectic subgroup $\Gamma \subseteq \mathrm{PGL}_2(F_S)$ is called a *normal plectic subgroup* if the quotient $\Gamma\backslash\mathcal{T}_\Gamma$ is a finite polysimplical complex.

Every subgroup commensurable to a normal plectic subgroup is normal. Moreover, if $S_1, S_2 \subseteq S$ are disjoint subsets and $\Gamma_i \subseteq \mathrm{PGL}_2(F_{S_i})$ are normal plectic subgroup, then the product $\Gamma_1 \times \Gamma_2$ is normal.

Example 2.1 Any discrete cocompact subgroup $\Gamma \subseteq \mathrm{PGL}_2(F_S)$ is a normal plectic subgroup.

A semi-simple subgroup $\Gamma \subseteq \mathrm{PGL}_2(F_\mathfrak{p})$ is normal if and only if it is finitely generated ([21], Chap. I, Lemma 3.2.2). In higher dimension one implication still holds. Indeed, the hypothesis of [9, Theorem 4] are fulfilled by Lemma 2.7. Therefore:

Proposition 2.4 *Every normal plectic subgroup is finitely presented.*

Normal plectic subgroups also have nice cohomological finiteness properties as the next two propositions show. For cohomological notions such as groups of type (WFL) and duality groups we refer to [8, Sect. VIII].

Proposition 2.5 *Every normal plectic subgroup is of type (WFL).*

Proof By [8, Theorem 11.1] it is enough to show that every normal plectic subgroup $\Gamma \subseteq \mathrm{PGL}_2(F_S)$ is virtually torsion-free. This can be argued as in the proof of [21, Chap. I.3, Theorem 3.1]: let $\{\sigma_1, \ldots, \sigma_r\}$ be a set of simplices of $\mathcal{T}_\Gamma$ for the orbits of the Γ-action. By Lemma 2.8, every element of finite order of Γ is conjugated to an element that fixes some σ_i, $1 \leq i \leq r$. Thus, Lemma 2.7 implies that the set of conjugacy classes of finite order elements of Γ is finite. Take a set of representatives $\mathcal{R} = \{\gamma_1, \ldots, \gamma_t\}$ of these conjugacy classes. Since Γ is finitely generated, there exist subrings $R_\mathfrak{p} \subseteq F_\mathfrak{p}$, $\mathfrak{p} \in S$, which are finitely generated over their prime ring and such that $\Gamma \subseteq \prod_{\mathfrak{p}\in S} \mathrm{PGL}_2(R_\mathfrak{p})$. Choose maximal ideals $\mathfrak{m}_\mathfrak{p} \subseteq R_\mathfrak{p}$ such that the intersection of $\mathcal{R}$ with the kernel of the map

$$\Gamma \hookrightarrow \prod_{\mathfrak{p}\in S} \mathrm{PGL}_2(R_\mathfrak{p}) \twoheadrightarrow \prod_{\mathfrak{p}\in S} \mathrm{PGL}_2(R_\mathfrak{p}/\mathfrak{m}_\mathfrak{p}) \tag{4}$$

consists only of the unit element. The kernel of (4) is a torsion-free normal subgroup of Γ of finite index. □

Before stating the next proposition we need to introduce some notations.

Definition 2.6 Let $\Gamma \subseteq \mathrm{PGL}_2(F_S)$ a plectic subgroup and $\mathfrak{p}$ an element of S. The $\mathfrak{p}$-*th Steinberg module* attached to Γ is the free abelian group

$$\mathrm{St}_{\Gamma,\mathfrak{p}} = C(\mathcal{L}_{\Gamma,\mathfrak{p}}, \mathbf{Z})/\mathbf{Z}$$

of locally constant $\mathbf{Z}$-valued functions on $\mathcal{L}_{\Gamma,\mathfrak{p}}$ modulo constant functions. It naturally carries a Γ-action. The S-*Steinberg module* of Γ is defined as the tensor product

$$\mathrm{St}_{\Gamma,S} = \bigotimes_{\mathfrak{p}\in S} \mathrm{St}_{\Gamma,\mathfrak{p}}$$

of abelian groups. The S-*orientation character* of $\mathrm{PGL}(F_S)$ is the homomorphism

$$\chi_S^{\mathrm{or}}\colon \mathrm{PGL}(F_S) \longrightarrow \{\pm 1\}, \quad (g_\mathfrak{p})_{\mathfrak{p}\in S} \longmapsto (-1)^{\sum_{\mathfrak{p}\in S} \mathrm{ord}_\mathfrak{p}(\det(g_\mathfrak{p}))}.$$

Note that if $\mathcal{L}_{\Gamma,\mathfrak{p}} = \mathbb{P}^1(F_\mathfrak{p})$, the module $\mathrm{St}_{\Gamma,\mathfrak{p}}$ is the smooth $\mathbf{Z}$-valued Steinberg representation of $\mathrm{PGL}_2(F_\mathfrak{p})$; while, if $\mathcal{L}_{\Gamma,\mathfrak{p}}$ is empty, then $\mathrm{St}_\mathfrak{p}(\Gamma) = \{0\}$. This implies that $\mathrm{St}_{\Gamma,S} = 0$ unless $S = \mathbf{S}_\Gamma$. As a product of trees, the polysimplical complex $\mathcal{T}_\Gamma$ is orientable. Furthermore, the action of an element $\gamma \in \Gamma$ preserves the orientation on $\mathcal{T}_\Gamma$ if and only if $\chi_{\mathbf{S}_\Gamma}^{\mathrm{or}}(\gamma) = 1$.

Proposition 2.6 *Let Γ be a normal plectic subgroup. Then Γ is a virtual duality group of virtual cohomological dimension*

$$\mathrm{vcd}(\Gamma) = \dim(\mathcal{T}_\Gamma).$$

Its duality module is equal to $\left(\mathrm{St}_{\Gamma,\mathbf{S}_\Gamma}\right) \otimes \chi_{\mathbf{S}_\Gamma}^{\mathrm{or}}$.

Proof By Proposition 7.5 and Theorem 10.1 of [8, Chap. VIII] it is enough to prove that the integral cohomology with compact support of $\mathcal{T}_\Gamma$ vanishes outside of degree $d = \dim(\mathcal{T}_\Gamma)$ and is equal to $\mathrm{St}_{\Gamma,\mathbf{S}_\Gamma}$ in degree d. For $\mathfrak{p} \in \mathbf{S}_\Gamma$ the local tree $\mathcal{T}_{\Gamma,\mathfrak{p}}$ has no leaves and its set of endpoints is exactly $\mathcal{L}_{\Gamma,\mathfrak{p}}$. This implies that the cohomology with compact support of $\mathcal{T}_{\Gamma,\mathfrak{p}}$ is supported in degree 1 and is equal to $\mathrm{St}_{\Gamma,\mathfrak{p}}$. We now sketch the proof of this claim. Denote by $\mathcal{V}$ respectively $\mathcal{E}^{\mathrm{or}}$ the set of vertices respectively oriented edges of $\mathcal{T}_{\Gamma,\mathfrak{p}}$. Given an oriented edge $e \in \mathcal{E}^{\mathrm{or}}$ we denote by $\bar{e} \in \mathcal{E}^{\mathrm{or}}$ the edge with the opposite orientation and by $s(e) \in \mathcal{V}$ its source vertex. We have to calculate the cohomology of the complex

$$C_c(\mathcal{V}, \mathbf{Z}) \xrightarrow{\ d\ } C_c(\mathcal{E}^{\mathrm{or}}, \mathbf{Z})/\langle e + \bar{e} \mid e \in \mathcal{E}^{\mathrm{or}}\rangle, \quad v \longmapsto \sum_{s(e)=v} e,$$

that is concentrated in degree 0 and 1. Here $C_c(\mathcal{V}, \mathbf{Z})$ denotes the space of $\mathbf{Z}$-valued functions on $\mathcal{V}$ with finite support and likewise for $\mathcal{E}^{\mathrm{or}}$. Since $\mathcal{T}_{\Gamma,\mathfrak{p}}$ is infinite, one easily checks that d is injective and, therefore, that $\mathrm{H}^0_c(\mathcal{T}_{\Gamma,\mathfrak{p}}, \mathbf{Z}) = 0$. The set of all ends

$U(e)$ that start with a given oriented edge e is an open subset of $\mathcal{L}_{\Gamma,\mathfrak{p}}$. Moreover, the sets $\{U(e)\}_e$ form a basis of the topology of $\mathcal{L}_{\Gamma,\mathfrak{p}}$. It follows that the homomorphism sending an oriented edge e to the characteristic function of $U(e)$ defines a surjective homomorphism

$$C_c(\mathcal{E}^{\mathrm{or}}, \mathbf{Z}) \twoheadrightarrow C(\mathcal{L}_{\Gamma,\mathfrak{p}}, \mathbf{Z}),$$

that in turn induces an isomorphism

$$\mathrm{H}^1_c(\mathcal{T}_{\Gamma,\mathfrak{p}}, \mathbf{Z}) \xrightarrow{\sim} \mathrm{St}_{\Gamma,\mathfrak{p}}.$$

The claim on $\mathcal{T}_\Gamma$ now follows from the Künneth formula for cohomology with compact support. □

Corollary 2.3 *Let Γ be a normal plectic subgroup. Then the module of Γ-coinvariants of $\mathrm{St}_{\Gamma,S}$ is a finitely generated abelian group.*

Proof If $S \neq \mathbf{S}_\Gamma$, then $\mathrm{St}_{\Gamma,S} = 0$ and the claim is obvious. So suppose that $S = \mathbf{S}_\Gamma$. In case Γ is torsion-free, Proposition 2.6 implies that

$$\left(\mathrm{St}_{\Gamma,S}\right)_\Gamma = \mathrm{H}_0\left(\Gamma, \mathrm{St}_{\Gamma,S}\right) \cong \mathrm{H}^{|S|}\left(\Gamma, \chi_S^{\mathrm{or}}\right)$$

The right-hand side is finitely generated since Γ is of type (FL) by Proposition 2.5. In case Γ is not torsion-free, the claim follows by passing to a torsion-free finite-index subgroup and using the Hochschild–Serre spectral sequence. □

We end this section by showing that the property of being normal is preserved by intersections with compact open subgroups.

Lemma 2.9 *Let $\Gamma \subseteq \mathrm{PGL}_2(F_S)$ be a normal plectic subgroup. For every subset $\mathbf{S}_\Gamma \subseteq S_1 \subseteq S$ and every compact open subgroup $U \subseteq \mathrm{PGL}_2(F_{S\setminus S_1})$ the group Γ_U is a normal plectic subgroup.*

Proof The compact open subgroup U is commensurable with the stabilizer of every vertex in the simplicial complex $\prod_{\mathfrak{p}\in S\setminus S_1} \mathcal{T}_{\mathcal{L}_{\Gamma,\mathfrak{p}}}$. Thus, we may reduce to the case that U is equal to such a stabilizer, for which the claim is obvious. □

3 Plectic Jacobians

3.1 Mumford Varieties and Zero Cycles

Let (F_S, ι) be an embedded local étale algebra and $\Gamma \subseteq \mathrm{PGL}_2(F_S)$ a plectic subgroup. Since $\mathcal{L}_{\Gamma,\mathfrak{p}} \subseteq \mathbb{P}^1(\mathbf{C})$ is a compact subset, its complement $\Omega_{\Gamma,\mathfrak{p}} = \mathbb{P}^1(\mathbf{C}) \setminus \mathcal{L}_{\Gamma,\mathfrak{p}}$ and, thus, the product

$$\Omega_\Gamma := \prod_{\mathfrak{p}\in S} \Omega_{\Gamma,\mathfrak{p}} = \mathbb{P}^1(\mathbf{C}_S) \setminus \mathcal{L}_\Gamma,$$

carry the structure of a rigid analytic variety over $\mathbf{C}$. Under mild assumptions the quotient of Ω_Γ by the action of Γ exists in the category of rigid analytic varieties. Indeed, using methods as in Mumford's seminal paper [24] or as in [7, Lemma 4] one gets the following result:

Theorem 3.1 *Suppose that $\Gamma \subseteq \mathrm{PGL}_2(F_S)$ is a torsion-free plectic subgroup. Then there exists a pair (X_Γ, π_Γ), unique up to isomorphism, consisting of a smooth rigid analytic variety X_Γ and a surjective étale morphism*

$$\pi_\Gamma : \Omega_\Gamma \longrightarrow X_\Gamma$$

such that

- $\pi_\Gamma \circ \gamma = \pi_\Gamma$ *for all* $\gamma \in \Gamma$ *and,*
- *if* $x, y \in \Omega_\Gamma$ *satisfy* $\pi_\Gamma(x) = \pi_\Gamma(y)$, *then* $\gamma(x) = y$ *for some* $\gamma \in \Gamma$.

The rigid analytic variety X_Γ is proper if and only if Γ is normal.

Remark 3.1 One can realize a rigid analytic variety as a quotient as above in many different ways. For starters, one can enlarge F_S by choosing a finite extension of each $F_\mathfrak{p}$ together with an embedding to $\mathbf{C}$ that extends $F_\mathfrak{p}$. One can also conjugate Γ by elements of $\mathrm{PGL}_2(F_S)$ and switch factors in case $F_\mathfrak{p} \cong F_\mathfrak{q}$ for $\mathfrak{p}, \mathfrak{q} \in S$. It is not clear whether there are additional ways to change Γ (see [1] for a discussion of this question in the case of cocompact subgroups).

To keep track of the data involved in the uniformization we make the following definition.

Definition 3.1 Let (F_S, ι) be an embedded local étale algebra. A *connected Mumford variety* (over (F_S, ι)) is a triple $(\Gamma, X_\Gamma, \pi_\Gamma)$ where $\Gamma \subseteq \mathrm{PGL}_2(F_S)$ is a torsion-free normal plectic subgroup and (X_Γ, π) arises from Theorem 3.1. In particular, X_Γ is proper. A *Mumford variety* (over (F_S, ι)) is a finite tuple of connected Mumford varieties (over (F_S, ι)).

Remark 3.2 We require the plectic groups involved in the definition of Mumford varieties to be normal so that Corollary 2.3 holds. This will allow us to define plectic Jacobians in Definition 3.4.

We often suppress Γ and π_Γ from the notation and simply call X_Γ a (connected) Mumford variety. Given a rigid analytic variety X over $\mathbf{C}$ we write $\mathcal{Z}(X)$ for the group of 0-cycles on X, that is, the free abelian group on the set $X(\mathbf{C})$. Let

$$\deg \colon \mathcal{Z}(X) \longrightarrow \mathbf{Z}, \quad \sum_{x \in X(\mathbf{C})} n_x[x] \mapsto \sum_{x \in X(\mathbf{C})} n_x$$

denote the degree map, and $\mathcal{Z}_0(X) = \ker(\deg)$ the group of 0-cycles of degree zero. If X_Γ is a connected Mumford variety, the pushforward $(\pi_\Gamma)_* \colon \mathcal{Z}(\Omega_\Gamma) \to \mathcal{Z}(X_\Gamma)$ induces an isomorphism

$$\mathcal{Z}(\Omega_\Gamma)_\Gamma = \mathrm{H}_0\big(\Gamma, \mathcal{Z}(\Omega_\Gamma)\big) \xrightarrow{\sim} \mathcal{Z}(X_\Gamma). \tag{5}$$

Moreover, there is a natural Γ-equivariant isomorphism

$$\psi\colon \bigotimes_{\mathfrak{p}\in S} \mathcal{Z}(\Omega_{\Gamma,\mathfrak{p}}) \xrightarrow{\sim} \mathcal{Z}(\Omega_\Gamma)$$

sending elementary tensors $\otimes_{\mathfrak{p}\in S}[x_\mathfrak{p}]$ to S-tuples $[(x_\mathfrak{p})_{\mathfrak{p}\in S}]$. For any plectic subgroup Γ define the Γ-submodule $\mathcal{Z}_{\varpi}(\Omega_\Gamma) \subseteq \mathcal{Z}(\Omega_\Gamma)$ of 0-cycles of degree $(0,\ldots,0)$ as the image of the tensor product $\bigotimes_{\mathfrak{p}\in S} \mathcal{Z}_0(\Omega_{\Gamma,\mathfrak{p}})$ under ψ. Thus, taking Γ-coinvariants and composing with (5) yields a homomorphism

$$\Theta_\Gamma\colon \mathcal{Z}_{\varpi}(\Omega_\Gamma)_\Gamma \longrightarrow \mathcal{Z}(X_\Gamma)$$

for every Mumford variety X_Γ.

Definition 3.2 Let X_Γ be a connected Mumford variety. The group of 0-cycles of degree $(0,\ldots,0)$ of X_Γ is defined as

$$\mathcal{Z}_{\varpi}(X_\Gamma) := \mathrm{Im}(\Theta_\Gamma) \subseteq \mathcal{Z}(X_\Gamma).$$

Given a Mumford variety $X = X_{\Gamma_1} \coprod \ldots \coprod X_{\Gamma_n}$ we put

$$\mathcal{Z}_{\varpi}(X) := \bigoplus_{i=1}^{n} \mathcal{Z}_{\varpi}(X_{\Gamma_i}).$$

Remark 3.3 The group $\mathcal{Z}(X_\Gamma)$ clearly only depends on the rigid analytic variety X_Γ itself, whereas the subgroup $\mathcal{Z}_{\varpi}(X_\Gamma)$ a priori depends on chosen uniformization of X_Γ. Nevertheless, note that changing Γ by conjugation or switching factors as explained in Remark 3.1 preserves the subgroup of 0-cycles of degree $(0,\ldots,0)$. Whenever X_Γ is one-dimensional, it is clear that $\mathcal{Z}_{\varpi}(X_\Gamma)$ recovers the group of divisors of degree 0.

Let $\mathrm{Aut}(\mathbf{C}/F_\mathfrak{p})$ be the group of continuous automorphisms of $\mathbf{C}$ fixing $\iota_\mathfrak{p}(F_\mathfrak{p})$. The product

$$\mathrm{Aut}(\mathbf{C}/F_S) = \prod_{\mathfrak{p}\in S} \mathrm{Aut}(\mathbf{C}/F_\mathfrak{p}).$$

naturally acts on the product $\mathbb{P}^1(\mathbf{C}_S)$ of projective spaces and commutes with the action of $\mathrm{PGL}_2(F_S)$. Thus, for every plectic subgroup $\Gamma \subset \mathrm{PGL}_2(F_S)$ the rigid analytic variety Ω_Γ is stable under $\mathrm{Aut}(\mathbf{C}/F_S)$-action. Moreover, if Γ is torsion-free, the action descends to an action on X_Γ and thus on $\mathcal{Z}(X_\Gamma)$. For every Mumford variety X the subgroup $\mathcal{Z}_{\varpi}(X) \subseteq \mathcal{Z}(X)$ is a $\mathrm{Aut}(\mathbf{C}/F_S)$-submodule

Let $(F'_{S'}, \iota')$ be an embedded local étale algebra of the same characteristic and residue characteristic as (F_S, ι). Given Mumford varieties X and X' over (F_S, ι) and $(F'_{S'}, \iota')$ respectively, the product $X \times X'$ has the structure of a Mumford variety over $(F_S \times F'_{S'}, \iota \times \iota')$. Moreover, restricting the isomorphism

$$\begin{aligned} \mathcal{Z}(X) \otimes \mathcal{Z}(X') &\longrightarrow \mathcal{Z}(X \times X') \\ x \otimes x' &\longmapsto (x, x') \end{aligned}$$

to 0-cycles of degree $(0, \ldots, 0)$ yields an isomorphism

$$\mathcal{Z}_{\emptyset}(X) \otimes \mathcal{Z}_{\emptyset}(X') \longrightarrow \mathcal{Z}_{\emptyset}(X \times X') \tag{6}$$

of $\mathrm{Aut}(\mathbf{C}/F_S \times F'_{S'})$-modules.

3.2 Integration Theory

Let $\mathcal{L}$ be a compact, totally disconnected space and A an abelian group. We write $\mathcal{M}(\mathcal{L}, A)$ for the space of A-valued measures on $\mathcal{L}$, that is,

$$\mathcal{M}(\mathcal{L}, A) := \mathrm{Hom}_{\mathbf{Z}}(C(\mathcal{L}, \mathbf{Z}), A).$$

In order to have a well-behaved integration theory, we require A to be a finite free $\mathbf{Z}$-module. Then, if M is a second-countable prodiscrete abelian group, there exists a canonical integration pairing

$$\mathcal{M}(\mathcal{L}, A) \otimes C(\mathcal{L}, M) \longrightarrow A \otimes_{\mathbf{Z}} M \tag{7}$$

constructed as follows: first, note that for any discrete abelian group N the canonical map

$$C(\mathcal{L}, \mathbf{Z}) \otimes N \xrightarrow{\sim} C(\mathcal{L}, N)$$

is an isomorphism and, thus, there exists a canonical integration pairing

$$\mathcal{M}(\mathcal{L}, A) \otimes C(\mathcal{L}, N) \longrightarrow A \otimes N$$

that is functorial in N. For M second-countable prodiscrete, choose a basis of neighbourhoods $\{U_i\}_i$ of the identity of M consisting of open subgroups with $U_{i+1} \subseteq U_i$. Since the canonical map $C(\mathcal{L}, M) \xrightarrow{\sim} \varprojlim_i C(\mathcal{L}, M/U_i)$ is an isomorphism, and A is a finite free $\mathbf{Z}$-module, we obtain the sought after pairing by considering

$$\mathcal{M}(\mathcal{L}, A) \otimes \varprojlim_i C(\mathcal{L}, M/U_i) \longrightarrow \varprojlim_i A \otimes M/U_i = M.$$

Remark 3.4 If $\mathcal{L}$ can be written as a product

$$\mathcal{L} = \mathcal{L}_1 \times \mathcal{L}_2$$

of profinite spaces. Then the map

$$\begin{aligned} C(\mathcal{L}_1, \mathbf{Z}) \otimes_{\mathbf{Z}} C(\mathcal{L}_2, \mathbf{Z}) &\longrightarrow C(\mathcal{L}_1 \times \mathcal{L}_2, \mathbf{Z}) \\ f_1 \otimes f_2 &\longmapsto [(x_1, x_2) \mapsto f_1(x_1) \cdot f_2(x_2))] \end{aligned} \tag{8}$$

is an isomorphism of abelian groups.

Now, let us turn to the situation of a plectic subgroup $\Gamma \subseteq \mathrm{PGL}_2(F_S)$ for some embedded local étale algebra F_S. The group Γ acts on $\mathcal{L}_{\Gamma,S}$ and, thus, on $\mathcal{M}(\mathcal{L}_{\Gamma,S}, A)$. By (8) the Steinberg module $\mathrm{St}_{\Gamma,S}$ is a quotient of the function space

$$C(\mathcal{L}_{\Gamma,S}, \mathbf{Z}) \cong \bigotimes_{\mathfrak{p} \in S} C(\mathcal{L}_{\Gamma,\mathfrak{p}}, \mathbf{Z}).$$

Thus, for any abelian group A the group

$$\mathcal{M}_{\otimes}(\mathcal{L}_{\Gamma,S}, A) := \mathrm{Hom}_{\mathbf{Z}}(\mathrm{St}_{\Gamma,S}, A)$$

is a Γ-submodule of the space of A-valued measures on $\mathcal{L}_{\Gamma,S}$. More precisely, it is the space of measures vanishing on functions which are constant in at least one variable. For any second-countable prodiscrete abelian group M, the Γ-module $\mathrm{St}^{\mathrm{ct}}_{\Gamma,S}(M)$ is defined to be the quotient of $C(\mathcal{L}_{\Gamma,S}, M)$ by the submodule generated by those functions which are constant with respect to at least one variable. If A is a finite free $\mathbf{Z}$-module, the integration pairing (7) descends to a pairing

$$\langle \cdot, \cdot \rangle \colon \mathcal{M}_{\otimes}(\mathcal{L}_{\Gamma,S}, A) \times \mathrm{St}^{\mathrm{ct}}_{\Gamma,S}(M) \longrightarrow A \otimes_{\mathbf{Z}} M. \tag{9}$$

3.3 Universal Measures and Multiplicative Integrals

Let us fix a plectic subgroup $\Gamma \subseteq \mathrm{PGL}_2(F_S)$ and consider the maximal torsion-free quotient $H_{\Gamma,S}$ of the group of Γ-coinvariants $\left(\mathrm{St}_{\Gamma,S}\right)_{\Gamma}$ of $\mathrm{St}_{\Gamma,S}$. By definition, there is a canonical identification

$$\mathcal{M}_{\otimes}(\mathcal{L}_{\Gamma,S}, A)^{\Gamma} = \mathrm{Hom}_{\mathbf{Z}}(H_{\Gamma,S}, A) \tag{10}$$

for every torsion-free abelian group A.

Definition 3.3 The *universal* Γ*-invariant measure* on $\mathcal{L}_{\Gamma,S}$ is the measure

$$\mu_\Gamma^S \in \mathcal{M}_\infty(\mathcal{L}_{\Gamma,S}, H_{\Gamma,S})^\Gamma$$

corresponding under (10) to the identity on $H_{\Gamma,S}$.

The next lemma is a direct consequence of the definitions.

Lemma 3.1 *Let* $S_1, S_2 \subseteq S$ *be disjoint subsets and* $\Gamma_i \subseteq \mathrm{PGL}_2(F_{S_i})$, $i = 1, 2$, *plectic subgroups. There is a natural identification*

$$H_{\Gamma_1\times\Gamma_2, S_1\cup S_2} = H_{\Gamma_1,S_1} \otimes H_{\Gamma_2,S_2} \tag{11}$$

of abelian groups. Moreover, the universal measure of $\Gamma_1 \times \Gamma_2$ *naturally factors as*

$$\mu_{\Gamma_1\times\Gamma_2}^{S_1\cup S_2} = \mu_{\Gamma_1}^{S_1} \otimes \mu_{\Gamma_2}^{S_2}. \tag{12}$$

Given second-countable prodiscrete subgroups $M_j, i = 1, \ldots, k$, we consider the topology on $M_1 \otimes_{\mathbf{Z}} \ldots \otimes_{\mathbf{Z}} M_k$ generated by opens of the form

$$M_1 \otimes_{\mathbf{Z}} \ldots \otimes_{\mathbf{Z}} M_{\ell-1} \otimes_{\mathbf{Z}} U_\ell \otimes_{\mathbf{Z}} M_{\ell+1} \otimes \ldots \otimes_{\mathbf{Z}} M_k$$

with $1 \le \ell \le k$ and $U_\ell \subseteq M_\ell$ an open subgroup. Its completion $M_1 \widehat{\otimes} \ldots \widehat{\otimes} M_k$ is a second-countable prodiscrete group. We are mostly interested in the case

$$\mathbf{C}_{S,\otimes}^\times := \widehat{\otimes}_{\mathfrak{p}\in S} \mathbf{C}^\times.$$

As the componentwise action of $\mathrm{Aut}(\mathbf{C}/F_S)$ on $\otimes_{\mathfrak{p}\in S}\mathbf{C}^\times$ is continuous, it induces an action on the completion $\mathbf{C}_{S,\otimes}^\times$.

Lemma 3.2 *The map*

$$\Phi_S\colon \mathcal{Z}_\infty(\Omega_\Gamma) \longrightarrow \mathrm{St}_{\Gamma,S}^{\mathrm{ct}}(\mathbf{C}_{S,\otimes}^\times),$$

$$\otimes_{\mathfrak{p}\in S}([x_\mathfrak{p}] - [y_\mathfrak{p}]) \longmapsto \left[(t_\mathfrak{p})_{\mathfrak{p}\in S} \mapsto \bigotimes_{\mathfrak{p}\in S} \left(\frac{t_\mathfrak{p} - x_\mathfrak{p}}{t_\mathfrak{p} - y_\mathfrak{p}} \right) \right]$$

is a well-defined $\Gamma \times \mathrm{Aut}(\mathbf{C}/F_S)$*-equivariant homomorphism.*

Proof Note that the function

$$\left(\frac{(\cdot) - x_\mathfrak{p}}{(\cdot) - y_\mathfrak{p}} \right) : \mathcal{L}_{\Gamma,\mathfrak{p}} \longrightarrow \mathbf{C}^\times$$

is continuous for every $\mathfrak{p} \in S$. Thus Φ_S is well-defined. The equivariance with respect to the $\mathrm{Aut}(\mathbf{C}/F_S)$-action is obvious, while the equivariance with respect to the Γ-action holds because we are considering the continuous S-Steinberg module. □

Let A be a finite free $\mathbf{Z}$-module and $\mu \in \mathcal{M}_{\infty}(\mathcal{L}_{\Gamma,S}, A)$ a measure. For a 0-cycle $D \in \mathcal{Z}_{\infty}(\Omega_\Gamma)$ of degree $(0, \ldots, 0)$ the multiplicative integral of D with respect to μ is defined to be

$$\oint_D \omega_\mu := \langle \mu, \Phi_S(D) \rangle \;\in\; A \otimes_{\mathbf{Z}} \mathbf{C}_{S,\otimes}^{\times},$$

where $\langle \cdot, \cdot \rangle$ is the integration pairing (9). The multiplicative integral is $\mathrm{Aut}(\mathbf{C}/F_S)$-equivariant, where $\mathrm{Aut}(\mathbf{C}/F_S)$ acts on $H_{\Gamma,S} \otimes_{\mathbf{Z}} \mathbf{C}_{S,\otimes}^{\times}$ through the second factor.

Lemma 3.3 *Let A be a finite free $\mathbf{Z}$-module and $\mu \in \mathcal{M}_{\infty}(\mathcal{L}_{\Gamma,S}, A)$ a surjective measure. The image of the multiplicative integral*

$$\oint_{(-)} \omega_\mu \colon \mathcal{Z}_{\infty}(\Omega_\Gamma) \longrightarrow A \otimes_{\mathbf{Z}} \mathbf{C}_{S,\otimes}^{\times}$$

is dense.

Proof By hypothesis, integration over μ induces a surjective homomorphism from $\mathrm{St}_{\Gamma,S}^{\mathrm{ct}}(M)$ to M for every second-countable prodiscrete group M. Therefore, it suffices to prove that the image of the homomorphism $\Phi_{\mathfrak{p}}$ is dense for all $\mathfrak{p} \in S$. By changing coordinates on $\mathbb{P}^1(\mathbf{C})$ we may assume that $\infty \notin \mathcal{L}_{\Gamma,\mathfrak{p}}$. In this new coordinates, the image of $\Phi_{\mathfrak{p}}$ contains all polynomials with $\mathbf{C}$-coefficients, which do not vanish on $\mathcal{L}_{\Gamma,\mathfrak{p}}$. Hence, density follows from the p-adic Stone–Weierstraß theorem. □

The preceding lemma can be applied to the following situation: assume that Γ is normal, then the abelian group $H_{\Gamma,S}$ is finitely generated and free by Corollary 2.3. Thus, the universal multiplicative integral

$$\oint_D \omega_{\mu_\Gamma^S} \;\in\; H_{\Gamma,S} \otimes_{\mathbf{Z}} \mathbf{C}_{S,\otimes}^{\times}$$

exists for all $D \in \mathcal{Z}_{\infty}(\Omega_\Gamma)$ and μ_Γ^S is surjective by construction.

3.4 Plectic Jacobians and the Abel–Jacobi Map

Let (F_S, ι) be an embedded local étale algebra and X_Γ a connected Mumford variety over it. Since the universal measure μ_Γ^S is Γ-invariant, it defines a $\mathrm{Aut}(\mathbf{C}/F_S)$-equivariant homomorphism

$$\oint_{(-)} \omega_{\mu_\Gamma^S} \colon \mathcal{Z}_{\infty}(\Omega_\Gamma)_\Gamma \longrightarrow H_{\Gamma,S} \otimes_{\mathbf{Z}} \mathbf{C}_{S,\otimes}^{\times}. \tag{13}$$

In order to obtain a homomorphism from the group of 0-cycles of degree $(0, \ldots, 0)$ on X_Γ, we mod out by the relations coming from the kernel of the quotient map $\Theta_\Gamma \colon \mathcal{Z}_{\infty}(\Omega_\Gamma)_\Gamma \to \mathcal{Z}_{\infty}(X_\Gamma)$. Thus, we define

$$\Lambda_\Gamma := \times\!\!\!\!\!\!\int_{\ker(\Theta_\Gamma)} \omega_{\mu_\Gamma^S} \subseteq H_{\Gamma,S} \otimes_{\mathbf{Z}} \mathbf{C}^\times_{S,\otimes}.$$

Definition 3.4 The *plectic Jacobian* of a connected Mumford variety X_Γ is the quotient

$$\mathfrak{J}_\otimes(X_\Gamma) := \left(H_{\Gamma,S} \otimes_{\mathbf{Z}} \mathbf{C}^\times_{S,\otimes}\right) / \overline{\Lambda_\Gamma},$$

where $\overline{\Lambda_\Gamma}$ denotes the closure of Λ_Γ. The plectic Abel–Jacobi map of X_Γ is the homomorphism

$$\alpha(X_\Gamma)\colon \mathcal{Z}_\otimes(X_\Gamma) \longrightarrow \mathfrak{J}_\otimes(X_\Gamma)$$

induced by (13). Given a Mumford variety $X = X_{\Gamma_1} \coprod \ldots \coprod X_{\Gamma_n}$ we put

$$\mathfrak{J}_\otimes(X) := \mathfrak{J}_\otimes(X_{\Gamma_1}) \oplus \cdots \oplus \mathfrak{J}_\otimes(X_{\Gamma_n})$$

and

$$\alpha(X) := \alpha(X_{\Gamma_1}) \oplus \cdots \oplus \alpha(X_{\Gamma_n}).$$

The action of $\mathrm{Aut}(\mathbf{C}/F_S)$ on $H_{\Gamma,S} \otimes_{\mathbf{Z}} \mathbf{C}^\times_{S,\otimes}$ descends to an action on $\mathfrak{J}_\otimes(X_\Gamma)$. From the construction and Lemma 3.3 we deduce:

Proposition 3.1 *Let X be a Mumford variety over (F_S, ι). Its plectic Jacobian $\mathfrak{J}_\otimes(X)$ is a prodiscrete group with a continuous* $\mathrm{Aut}(\mathbf{C}/F_S)$*-action. The plectic Abel–Jacobi map $\alpha(X)$ is* $\mathrm{Aut}(\mathbf{C}/F_S)$*-equivariant and has dense image.*

Remark 3.5 When X_Γ is a one-dimensional, $\Lambda_\Gamma = \overline{\Lambda_\Gamma}$ is a discrete subgroup. Moreover, the plectic Jacobian $\mathfrak{J}_\otimes(X_\Gamma)$ of X_Γ can be identified with the set of $\mathbf{C}$-valued points of the Jacobian of X_Γ via Manin–Drinfeld's p-adic uniformization theorem (see for example [11, Sect. 2.5]), and the plectic Abel–Jacobi map coincides with the classical one. In particular, the kernel of the plectic Abel–Jacobi map is simply the subgroup of principal divisors. Is it possible to give a geometric description of the kernel of the plectic Abel–Jacobi map in higher dimensions?

Example 3.1 Let X_Γ be a connected Mumford variety over (F_S, ι) such that $\mathbf{S}_\Gamma \neq S$, then $\mathrm{St}_{\Gamma,S} = 0$ and consequently $\mathfrak{J}_\otimes(X_\infty) = 0$. This generalizes the fact that the Jacobian of the projective line is trivial.

For the remainder of this section we fix two embedded local étale algebras (F_S, ι) and $(F'_{S'}, \iota')$ of the same characteristic and residue characteristic. Given Mumford varieties X and X' over (F_S, ι) and $(F'_{S'}, \iota')$ we have already seen in (6) that the group of 0-cycles of degree $(0, \ldots, 0)$ on their product equals the tensor product of $\mathcal{Z}_\otimes(X)$ and $\mathcal{Z}_\otimes(X')$. In the following we will show that the same is true for their plectic Jacobians. First, assume that $X = X_\Gamma$ and $X' = X_{\Gamma'}$ are connected. The decomposition (11) induces an isomorphism

$$\Pi\colon \left(H_{\Gamma,S} \otimes_{\mathbf{Z}} \mathbf{C}^{\times}_{S,\otimes}\right) \widehat{\otimes} \left(H_{\Gamma',S'} \otimes_{\mathbf{Z}} \mathbf{C}^{\times}_{S',\otimes}\right) \xrightarrow{\sim} \left(H_{\Gamma\times\Gamma',S'\coprod S'}\right) \otimes_{\mathbf{Z}} \mathbf{C}^{\times}_{S\coprod S',\otimes}.$$

Theorem 3.2 *Let X_Γ and $X_{\Gamma'}$ be connected Mumford varieties over (F_S, ι) and $(F'_{S'}, \iota')$. Then Π descends to an isomorphism*

$$\mathcal{J}_{\otimes}(X_\Gamma) \mathbin{\widehat{\otimes}} \mathcal{J}_{\otimes}(X_{\Gamma'}) \xrightarrow{\sim} \mathcal{J}_{\otimes}(X_{\Gamma\times\Gamma'}) \tag{14}$$

of topological $\mathrm{Aut}(\mathbf{C}/F_S \times F'_{S'})$*-modules.*

Proof By the Künneth formula for group homology we observe that

$$\ker(\Theta_{\Gamma\times\Gamma'}) = \ker(\Theta_\Gamma) \otimes_{\mathbf{Z}} \mathcal{Z}_{\otimes}(\Omega_{\Gamma'})_{\Gamma'} \oplus \mathcal{Z}_{\otimes}(\Omega_\Gamma)_\Gamma \otimes_{\mathbf{Z}} \ker(\Theta_{\Gamma'}).$$

Hence, the product decomposition (12) of the universal measure in conjunction with Lemma 3.3 implies that

$$\overline{\Lambda_{\Gamma\times\Gamma'}} = \overline{\Lambda_\Gamma} \mathbin{\widehat{\otimes}} \left(H_{\Gamma',S'} \otimes_{\mathbf{Z}} \mathbf{C}^{\times}_{S',\otimes}\right) \oplus \left(H_{\Gamma,S} \otimes_{\mathbf{Z}} \mathbf{C}^{\times}_{S,\otimes}\right) \mathbin{\widehat{\otimes}} \overline{\Lambda_{\Gamma'}},$$

which proves the claim. □

Corollary 3.1 *Let X and X' be connected Mumford varieties over (F_S, ι) and $(F'_{S'}, \iota')$. The isomorphisms* (14) *induce an isomorphism*

$$\mathcal{J}_{\otimes}(X) \mathbin{\widehat{\otimes}} \mathcal{J}_{\otimes}(X') \xrightarrow{\sim} \mathcal{J}_{\otimes}(X \times X') \tag{15}$$

of topological $\mathrm{Aut}(\mathbf{C}/F_S \times F'_{S'})$*-modules. Moreover, the Abel–Jacobi map*

$$\alpha(X \times X') = \alpha(X) \otimes \alpha(X')$$

is compatible with the decomposition (6) *and* (15).

3.5 Hecke Correspondences

We fix an embedded local étale algebra (F_S, ι).

Definition 3.5 A morphism from a connected Mumford variety X_Γ over (F_S, ι) to another connected Mumford variety $X_{\Gamma'}$ over (F_S, ι) is a rigid analytic map of the form

$$f(g)\colon X_\Gamma \longrightarrow X_{\Gamma'}, \quad [x] \mapsto [gx],$$

where $g \in \mathrm{PGL}_2(F_S)$ is an element such that $g\Gamma g^{-1} \subseteq \Gamma'$ is a finite index subgroup.

A morphism between two Mumford varieties $X = X_{\Gamma_1} \coprod \ldots \coprod X_{\Gamma_n}$ and $X' = X_{\Gamma'_1} \coprod \ldots \coprod X_{\Gamma'_m}$ over (F_S, ι) is a rigid analytic map such that there an injection $\sigma = \sigma(f)\colon (1, \ldots, n) \hookrightarrow (1, \ldots, m)$ and a morphism of connected Mumford varieties $f_i\colon X_{\Gamma_i} \to X_{\Gamma'_{\sigma(i)}}$ for each $1 \leq i \leq n$.

The composition of two morphism of Mumford varieties is clearly a morphism of Mumford varieties. Note that a morphism f of Mumford varieties is surjective if and only if $\sigma(f)$ is surjective. Let $f\colon X \to X'$ be a morphism of Mumford varieties. In case X and X' are connected, the pushforward map

$$f_*\colon \mathcal{Z}(X) \longrightarrow \mathcal{Z}(X')$$

on 0-cycles corresponds to the canonical map between the groups of coinvariants under the identification (5). Similarly, the pullback map

$$f^*\colon \mathcal{Z}(X') \longrightarrow \mathcal{Z}(X)$$

is given by the transfer map on coinvariants. In particular, both of them restrict to $\mathrm{Aut}(\mathbf{C}/F_S)$-equivariant homomorphisms

$$f_*\colon \mathcal{Z}_{\mathbb{O}}(X) \longrightarrow \mathcal{Z}_{\mathbb{O}}(X'), \qquad f^*\colon \mathcal{Z}_{\mathbb{O}}(X') \longrightarrow \mathcal{Z}_{\mathbb{O}}(X).$$

on groups of 0-cycles of degree $(0, \ldots, 0)$. Let g be an element of $\mathrm{PGL}_2(F_S)$ and $\Gamma, \Gamma' \subseteq \mathrm{PGL}_2(F_S)$ plectic subgroup such that $g\Gamma g^{-1} \subseteq \Gamma'$ is a finite index subgroup. It follows that $\mathrm{St}_{g\Gamma g^{-1},S} = \mathrm{St}_{\Gamma',S}$. Functoriality of the operation of taking coinvariants and the existence of transfer maps for finite index subgroups gives homomorphisms

$$f_*\colon H_{\Gamma,S} \longrightarrow H_{\Gamma',S}, \qquad f^*\colon H_{\Gamma',S} \longrightarrow H_{\Gamma,S},$$

that in turn induce $\mathrm{Aut}(\mathbf{C}/F_S)$-equivariant pushforward and pullback homomorphisms between plectic Jacobians:

$$f_*\colon \mathcal{J}_{\mathbb{O}}(X) \longrightarrow \mathcal{J}_{\mathbb{O}}(X'), \qquad f^*\colon \mathcal{J}_{\mathbb{O}}(X') \longrightarrow \mathcal{J}_{\mathbb{O}}(X).$$

Going through the construction one easily verifies the following:

Lemma 3.4 *Let $f\colon X \to X'$ be a morphism of Mumford varieties over (F_S, ι). The Abel–Jacobi map commutes with pushforward and pullback, that is,*

$$\alpha(X') \circ f_* = f_* \circ \alpha(X) \quad \textit{and} \quad \alpha(X) \circ f^* = f^* \circ \alpha(X').$$

Proposition 3.2 *Let $f_1\colon X_1 \to X$ and $F_2\colon X_2 \to X$ morphisms of Mumford varieties over (F_S, ι). The fibre product of rigid analytic varieties $X_1 \times_X X_2$ is a Mumford variety over (F_S, ι). Moreover, the canonical projection maps $X_1 \times_X X_2 \to X_i$, $i = 1, 2$, are morphisms of Mumford varieties over (F_S, ι).*

Proof It is enough to consider the case where $X = X_\Gamma$, $X_1 = X_{\Gamma_1}$ and $X_2 = X_{\Gamma_2}$ are connected, that Γ_1 and Γ_2 are finite index subgroup of Γ, and that the morphisms f_1 and f_2 correspond to the inclusion maps. For $g \in \Gamma$ consider the morphism

$$X_{\Gamma_1 \cap g^{-1}\Gamma_2 g} \longrightarrow X_{\Gamma_1} \times X_{\Gamma_2}, \quad [x] \mapsto ([x], g[x]).$$

Their disjoint union induces a closed embedding

$$\coprod_{g \in \Gamma_2 \backslash \Gamma / \Gamma_1} X_{\Gamma_1 \cap g^{-1}\Gamma_2 g} \hookrightarrow X_{\Gamma_1} \times X_{\Gamma_2}$$

whose image is the fibre product of X_{Γ_1} and X_{Γ_2} over X_Γ. This proves the first claim. The second claim also follows directly from this description of the fibre product. □

Definition 3.6 Let X and X' be two Mumford varieties over (F_S, ι). A Hecke correspondence from X to X' consists of a Mumford variety $\widetilde{X}$ over (F_S, ι) and together with a surjective morphism $p\colon \widetilde{X} \to X$ and a morphism $q\colon \widetilde{X}' \to X$. Given two Hecke correspondences $(\widetilde{X}, p, q)\colon X \to X'$ and $(\widetilde{X}', p', q')\colon X' \to X''$ we define their composition via

$$(\widetilde{X}', p', q') \circ (\widetilde{X}, p, q) := (\widetilde{X} \times_{X'} \widetilde{X}', p \circ \pi_{\widetilde{X}}, q' \circ \pi_{\widetilde{X}'}),$$

where $\pi_{\widetilde{X}}\colon \widetilde{X} \times_{X'} \widetilde{X}' \to \widetilde{X}$ and $\pi_{\widetilde{X}'}\colon \widetilde{X} \times_{X'} \widetilde{X}' \to \widetilde{X}'$ denote the projection morphisms. By Proposition 3.2 the composition of Hecke correspondences is well-defined.

Write $\mathrm{Mum}_{/(F_S,\iota)}$ for the category of Mumford varieties over (F_S, ι) with Hecke correspondences as morphism. Let $\mathrm{Aut}(\mathbf{C}/F_S)$-Mod (resp. $\mathrm{Aut}(\mathbf{C}/F_S)\text{-Mod}^{\mathrm{ct}}$) be the category of $\mathrm{Aut}(\mathbf{C}/F_S)$-modules (resp. continuous $\mathrm{Aut}(\mathbf{C}/F_S)$-modules). Finally, denote by

$$\nu\colon \mathrm{Aut}(\mathbf{C}/F_S)\text{-Mod}^{\mathrm{ct}} \longrightarrow \mathrm{Aut}(\mathbf{C}/F_S)\text{-Mod}$$

the forgetful functor. The following theorem summarizes what we have proven so far:

Theorem 3.3 *Let (F_S, ι) be an embedded local étale algebra. Assigning the group of 0-cycles of degree $(0, \ldots, 0)$ (resp. the plectic Jacobian) to a Mumford variety over (F_S, ι) defines a functor*

$$\mathcal{Z}_\otimes\colon \mathrm{Mum}_{/(F_S,\iota)} \longrightarrow \mathrm{Aut}(\mathbf{C}/F_S)\text{-Mod}$$
$$(\textit{resp. } \mathfrak{J}_\otimes\colon \mathrm{Mum}_{/(F_S,\iota)} \longrightarrow \mathrm{Aut}(\mathbf{C}/F_S)\text{-Mod}^{\mathrm{ct}}).$$

Moreover, the plectic Abel–Jacobi map is natural transformation

$$\alpha\colon \mathcal{Z}_\otimes \Longrightarrow \nu \circ \mathfrak{J}_\otimes.$$

Remark 3.6 Instead of working with Mumford varieties over a fixed embedded local étale algebra one could work with the category of all Mumford varieties over $\mathbf{C}$. In this setting, Corollary 3.1 could be reformulated as saying that the Abel–Jacobi map is a natural transformation between symmetric monoidal functors.

4 Modularity

Let $\overline{\mathbf{Q}} \subseteq \mathbb{C}$ denote the algebraic closure of $\mathbf{Q}$ contained in the field of complex numbers. For a totally real field F of degree d we consider the set of its embeddings $\Sigma_\infty := \mathrm{Hom}_{\mathbf{Q}}(F, \overline{\mathbf{Q}})$ into $\overline{\mathbf{Q}}$ and the set S_p of its p-adic places. Let $\mathbf{C}_p$ denote the completion of a fixed algebraic closure of $\mathbf{Q}_p$. The choice of an embedding $\iota_p \colon \overline{\mathbf{Q}} \hookrightarrow \mathbf{C}_p$ determines a map

$$\theta \colon \Sigma_\infty \longrightarrow S_p$$

sending $\sigma \in \mathrm{Hom}_{\mathbf{Q}}(F, \overline{\mathbf{Q}})$ to the p-adic place $\mathfrak{p}_\sigma$ induced by the composition $\iota_p \circ \sigma$. Let $\Sigma \subseteq \Sigma_\infty$ be a non-empty subset such that $S := \theta(\Sigma)$ has the same cardinality as Σ. The fixed embedding ι_p induces an embedding on the local étale algebra

$$F_S := \prod_{\mathfrak{p} \in S} F_{\mathfrak{p}}.$$

Set $p_S := \prod_{\mathfrak{p} \in S} \mathfrak{p}$. Consider a square-free ideal $\mathfrak{n}^-$ coprime to p_S and write $\omega(\mathfrak{n}^-)$ for its number of prime factors. Suppose that $\omega(\mathfrak{n}^-) \equiv d \pmod 2$. Then there exists a quaternion algebra B/F ramified exactly at the places in

$$S \cup \{\mathfrak{q} \mid \mathfrak{n}^-\} \cup (\Sigma_\infty \setminus \Sigma).$$

Let $\mathcal{H}^\pm = \mathbb{P}^1(\mathbb{C}) \setminus \mathbb{P}^1(\mathbb{R})$ denote the union of the upper and lower complex half planes. For every $\sigma \in \Sigma$ we choose an isomorphism $B_\sigma := B \otimes_{F,\sigma} \mathbb{R} \cong \mathrm{M}_2(\mathbb{R})$ so that the diagonal embedding of B into $\prod_{\sigma\in\Sigma} B_\sigma$ induces an action of $B^\times$ on $\mathcal{H}^\pm_\Sigma := \prod_{\sigma\in\Sigma} \mathcal{H}^\pm$. Let E_{Gal} denote the Galois closure of $\sigma(F)$ in $\overline{\mathbf{Q}}$ for some $\sigma \in \Sigma$, and consider

$$G_\Sigma := \{\tau \in \mathrm{Gal}(E_{\mathrm{Gal}}/\mathbf{Q}) \mid \tau \circ \Sigma = \Sigma\}, \qquad E := E_{\mathrm{Gal}}^{G_\Sigma}.$$

4.1 Quaternionic Shimura Varieties

Denote by $\mathbb{A}_f$ the ring of finite adeles over $\mathbf{Q}$ and $K \subseteq B^\times(\mathbb{A}_f)$ be a compact open subgroup such that $\mathrm{nr}(K) = \widehat{\mathcal{O}}_F^\times$. The quaternionic Shimura variety

$$X_K(\mathbf{C}) = B^\times \backslash \left(\mathcal{H}_\Sigma^\pm \times B^\times(\mathbb{A}_f) \right) / K$$

has a canonical model over the reflex field E. In the following we only consider level structures of the form $K = \prod_\mathfrak{q} K_\mathfrak{q}$, where $\mathfrak{q}$ runs through the finite places of F, such that $K_\mathfrak{p}$ is the unique maximal compact subgroup of $B_\mathfrak{p}$ for all $\mathfrak{p} \in S$. We are mostly interested in level structures of type $K_0(\mathfrak{n}^+)$ where $\mathfrak{n}^+ \subseteq \mathcal{O}_F$ is an ideal coprime to $p_S \mathfrak{n}^-$ and

- $K = \prod K_\mathfrak{q}$ with $K_\mathfrak{q}$ maximal compact for $q \nmid \mathfrak{n}^+$,
- $K_\mathfrak{q}$ corresponds to the group of integral matrices that are congruent to an upper triangular matrix mod $\mathfrak{n}^+$ under an isomorphism $B_\mathfrak{q} \cong \mathrm{M}_2(K_\mathfrak{q})$ for every $\mathfrak{q} \mid \mathfrak{n}^+$.

By definition, the reflex field is a subfield of $\overline{\mathbf{Q}}$. Thus, the embedding ι_p induces a p-adic place v on E. The E_v-rigid analytic space associated to X_K admits a p-adic uniformization by [30, Sect. 5]. Concretely, let D/F be the quaternion algebra obtained from B/F by switching invariants at S and Σ, that is, D is the totally definite quaternion algebra over F that is ramified exactly at those places $\mathfrak{q}$ that divide $\mid \mathfrak{n}^-$. We choose an isomorphism $D_\mathfrak{p} \cong \mathrm{M}_2(F_\mathfrak{p})$ for every $\mathfrak{p} \in S$. The diagonal embedding of D into $\prod_{\mathfrak{p}\in S} D_\mathfrak{p}$ induces an action of $D^\times$ on $\Omega_S := \prod_{\mathfrak{p}\in S} \Omega_\mathfrak{p}$, where $\Omega_\mathfrak{p} = \mathbb{P}^1(\mathbf{C}_p) \setminus \mathbb{P}^1(F_\mathfrak{p})$ denotes Drinfeld's upper half plane associated to $F_\mathfrak{p}$. Moreover, we identify $D(\mathbb{A}_f^S)$ and $B(\mathbb{A}_f^S)$ where $\mathbb{A}_F^S$ denotes the ring of finite adeles away from S. This induces an action of $D^\times$ on $B(\mathbb{A}_f^S)$. Moreover, $D^\times$ acts on $B(F_\mathfrak{p})/K_\mathfrak{p} \cong \mathbf{Z}$ via translation by the p-adic valuation of the determinant. The p-adic uniformization theorem then states that there is an isomorphism of rigid analytic varieties

$$X_{K(\mathfrak{n})}(\mathbf{C}_p) \cong D^\times \backslash \left(\Omega_S \times B(\mathbb{A}_f)/K \right).$$

Furthermore, the group of connected components is given by the narrow class group Cl_F^+ of F. Let $\{g_i\}_i \subset B(\mathbb{A}^S)$ be elements whose reduced norms form a system of representatives for Cl_F^+. The images of the groups $\Gamma_i = g_i \cdot K \cdot g_i^{-1} \cap D^\times$ in $\mathrm{PGL}_2(F_S)$ are cocompact subgroups and thus plectic. We may write

$$X_K(\mathbf{C}_p) \cong \coprod_{i=1}^{h_F^+} X_{\Gamma_i}. \tag{16}$$

In particular, X_K carries the structure of a Mumford variety over (F_S, ι).

4.2 *Geometric Modularity*

Let $A_{/F}$ be an elliptic curve with multiplicative reduction at all $\mathfrak{p} \in S$. Let $A_\mathfrak{p}$ denote the analytification of $A_{F_\mathfrak{p}}$ seen as a rigid analytic variety over $\mathbf{C}_p$. By Tate's uniformization theorem there exists a unique element $q_\mathfrak{p} \in F_\mathfrak{p}^\times$ with positive p-adic valuation such that

$$A_{\mathfrak{p}} \cong \mathbf{C}_p^\times / q_{\mathfrak{p}}^{\mathbf{Z}}.$$

In particular, $A_{\mathfrak{p}}$ is a Mumford curve and, thus, the product

$$A_S := \prod_{\mathfrak{p} \in S} A_{\mathfrak{p}}$$

is a Mumford variety over (F_S, ι_p). More generally, given a Tate period $\widetilde{q}_{\mathfrak{p}} \in F_{\mathfrak{p}}^\times$ for every $\mathfrak{p} \in S$ we denote by $A_{\widetilde{q}_{\mathfrak{p}}}$ the corresponding elliptic curve over $\mathbf{C}_p$. By Theorem 3.2, the plectic Jacobian of the product $A_{\widetilde{q}_S} := \prod_{\mathfrak{p} \in S} A_{\widetilde{q}_p}$ is simply the completed tensor product of the groups $A_{\mathfrak{p}}(\mathbf{C}_p)$. In particular, there exists a canonical surjective homomorphism

$$\phi_{\widetilde{q}_S} : \mathbf{C}_{S,\otimes}^\times \longrightarrow \mathcal{J}_{\otimes}(A_{\widetilde{q}_S}).$$

Theorem 4.1 *Let $A_{/F}$ be a modular elliptic curve of conductor $N = p_S \cdot \mathfrak{n}^+ \cdot \mathfrak{n}^-$ for some ideal $\mathfrak{n}^+ \subseteq \mathcal{O}_F$ coprime to $p_S \cdot \mathfrak{n}^-$. There is a surjective* $\mathrm{Aut}(\mathbf{C}_p/F_S)$*-equivariant homomorphism*

$$\varphi_A : \mathcal{J}_{\otimes}(X_{K(\mathfrak{n}^+)}) \longrightarrow \mathcal{J}_{\otimes}(A_S).$$

Proof This is mostly just a reformulation of the results of [16, Sect. 4.5]. We sketch the main points. First, let $\Gamma_i \subseteq \mathrm{PGL}_2(F_S)$ be the subgroups defined in (16). The Hecke algebra acts on the direct sum $\oplus_i H_{\Gamma_i,S}$. The modularity of $A_{/F}$ implies that there is a unique $\mathbf{Z}$-free quotient H_A of $\oplus_i H_{\Gamma_i,S}$ on which the Hecke algebra acts via the character associated to A. Moreover, multiplicity one implies that H_A has rank one. The choice of a generator of H_A then yields a homomorphism

$$\bigoplus_{i=1}^{h_F^+} \left(H_{\Gamma_i,S} \otimes_{\mathbf{Z}} \mathbf{C}_{S,\otimes}^\times\right) \longrightarrow \mathbf{C}_{S,\otimes}^\times \tag{17}$$

that in turn induces the homomorphism

$$\psi_A : \bigoplus_{i=1}^{h_F^+} \mathcal{Z}_{\otimes}(\Omega_{\Gamma_i})_{\Gamma_i} \longrightarrow \mathbf{C}_{S,\otimes}^\times$$

through p-adic integration. Now, [16, Theorem 4.10] implies that there exist Tate periods $\widetilde{q}_{\mathfrak{p}} \in F_{\mathfrak{p}}$ for every $\mathfrak{p} \in S$ with the following properties:

- the subgroups $q_{\mathfrak{p}}^{\mathbf{Z}}$ and $\widetilde{q}_{\mathfrak{p}}^{\mathbf{Z}}$ are commensurable, and
- the homomorphism

$$\phi_{\widetilde{q}_A} \circ \psi_A : \bigoplus_{i=1}^{h_F^+} \mathcal{Z}_{\otimes}(\Omega_{\Gamma_i})_{\Gamma_i} \longrightarrow \mathcal{J}_{\otimes}(A_{\widetilde{q}_S})$$

factors through $\mathcal{Z}_{\oplus}(X_{K(\mathfrak{n}^+)})$ and hence through $\mathcal{J}_{\oplus}(X_{K(\mathfrak{n}^+)})$.

By choosing an isogeny $A_{\tilde{q}_\mathfrak{p}} \to A_\mathfrak{p}$ for each $\mathfrak{p} \in S$, we obtain the claimed homomorphism

$$\mathcal{J}_{\oplus}(X_{K(\mathfrak{n}^+)}) \longrightarrow \mathcal{J}_{\oplus}(A_S).$$

□

Remark 4.1 The Hecke module $\oplus_i H_{\Gamma_i,S}$ is isomorphic to the space of cusp forms on $X_{K(\mathfrak{n}^+)}$. Although the results of [16] were only stated in the case of eigenforms corresponding to elliptic curves, the methods also apply to arbitrary eigenforms. In particular, Theorem 4.1 can be reformulated as follows: the plectic Jacobian of $X_{K(\mathfrak{n}^+)}$ embeds into the plectic Jacobian of a product of Shimura curves $\{X_\mathfrak{p}\}_{\mathfrak{p}\in S}$ with each $X_\mathfrak{p}$ admitting a p-adic uniformization by a $\mathfrak{p}$-arithmetic subgroup of the definite quaternion algebra D. More precisely, the level structure used to define $X_\mathfrak{p}$ is $K_0(p_S\mathfrak{p}^{-1}n^+)$. The following question arises: let X_Γ be a connected Mumford variety over an embedded local étale algebra (F_S, ι). Do Mumford curves $\{X_\mathfrak{p}\}_{\mathfrak{p}\in S}$ exist such that

- the connected components of each $X_\mathfrak{p}$ are connected Mumford curves associated to groups of the form Γ_U for an open compact subgroup $U \subseteq \mathrm{PGL}_2(F_{S\setminus\{\mathfrak{p}\}})$,
- and there exists an $\mathrm{Aut}(\mathbf{C}/F_S)$-equivariant embedding

$$\mathcal{J}_{\oplus}(X_\Gamma) \hookrightarrow \mathcal{J}_{\oplus}(\prod_{\mathfrak{p}\in S} X_\mathfrak{p})?$$

If such a collection of Mumford curves existed, X_Γ could be seen as a braided version of the product $\prod_{\mathfrak{p}\in S} X_\mathfrak{p}$—at least, on the level of plectic Jacobians—reflecting the fact that Γ can be thought of as a braided version of the product $\prod_{\mathfrak{p}\in S} \Gamma_\mathfrak{p}$.

References

1. Alon, G.: Automorphisms of Products of Drinfeld Half Planes. preprint (2017)
2. Bertolini, M., Darmon, H.: Heegner points on Mumford-Tate curves. Invent. Math. **126**(3), 413–456 (1996)
3. Bertolini, M., Darmon, H.: Heegner points, p-adic L-functions, and the Cerednik-Drinfeld uniformization. Invent. Math. **131**(3), 453–491 (1998)
4. Bertolini, M., Darmon, H.: Hida families and rational points on elliptic curves. Invent. Math. **168**(2), 371–431 (2007)
5. Bertolini, M., Darmon, H.: The rationality of Stark-Heegner points over genus fields of real quadratic fields. Ann. Math. (2) **170**(1), 343–370 (2009)
6. Bertolini, M., Darmon, H., Prasanna, K.: Generalized Heegner cycles and p-adic Rankin L-series. Duke Math. J. **162**(6), 1033–1148 (2013). With an appendix by Brian Conrad
7. Berkovich, V.G.: The automorphism group of the Drinfeld half plane. C. R. Acad. Sci. **321**, 1127–1132 (1995)
8. Brown, K.S.: Cohomology of Groups. Graduate Texts in Mathematics. Springer-Verlag, New York (1982)

9. Brown, K.S.: Presentations for groups acting on simply-connected complexes. J. Pure Appl. Algebra **32**(1), 1–10 (1984)
10. Boutot, J.-F., Zink, T.: The p-adic uniformization of Shimura curves. preprint (1995)
11. Dasgupta, S.: Stark-Heegner points on modular Jacobians. Ann. Sci. Éc. Norm. Supér. **38**(3), 427–469 (2005)
12. Darmon, H., Fornea, M.: Mock plectic points. preprint (2023)
13. Dasgupta, S., Teitelbaum, J.: The p-adic Upper Half Plane. In: In p-adic Geometry: Lectures from the 2007 Arizona Winter School, volume 45 of University lecture series. American Mathematical Society (2008)
14. Fornea, M., Gehrmann, L.: Iwasawa theory and mock plectic points. preprint (2023)
15. Fornea, M., Gehrmann, L.: On the algebraicity of polyquadratic plectic points. Int. Math. Res. Not. (IMRN) **2023**(22), 19237–19265 (2023)
16. Fornea, M., Gehrmann, L.: Plectic Stark-Heegner points. Adv. Math. **414**, 108861 (2023)
17. Fornea, M., Guitart, X., Masdeu, M.: Plectic p-adic invariants. Adv. Math., 406: Paper No. **108484**, 26 (2022)
18. Fornea, M.: Plectic Jacobians. Q. J. Math. **75**(3), 789–808 (2024)
19. Fuks, B.A.: Special chapters in the theory of analytic functions of several complex variables. Translated from the Russian by A. Jeffrey and N. Mugibayashi, volume 14 of Transl. Math. Monogr. American Mathematical Society (1965)
20. Fresnel, J., van der Put, M.: Rigid analytic geometry and its applications. Progress in Mathematics, vol. 218. Birkhäuser Boston Inc, Boston, MA (2004)
21. Gerritzen, L., van der Put, M.: Schottky groups and Mumford curves. Lecture Notes in Mathematics, vol. 817. Springer, Berlin (1980)
22. Logan, A., Weinstein, J.: Higher modularity of elliptic curves over function fields. preprint (2022)
23. Loeffler, D., Zerbes, S.L.: Plectic structures in p-adic de Rham cohomology. J. Number Theory (2022)
24. Mumford, D.: An analytic construction of degenerating curves over complete local rings. Comp. Math. **24**(2), 129–174 (1972)
25. Mustafin, G.A.: Nonarchimedean uniformization. Math. USSR-Sb. **34**, 187–214 (1978)
26. Nekovář, J.: Hidden Symmetries in the Theory of Complex Multiplication. Algebra, Arithmetic, and Geometry: Volume II: In Honor of Y. I. Manin, pp. 399–437 (2009)
27. Nekovář, J.: Some remarks on the BSD conjecture. Rubinfest's talk at Harvard (2016)
28. Nekovář, J., Scholl, A.J.: Introduction to plectic cohomology. In: Advances in the theory of automorphic forms and their L-functions, volume 664 of Contemp. Math., pp. 321–337. Amer. Math. Soc., Providence, RI (2016)
29. Tits, J.: Sur le groupe des automorphismes d'un arbre. In: Essays on Topology and Related Topics: Memoires dédiés à Georges de Rham, pp. 188–211. Springer, Berlin Heidelberg, Berlin, Heidelberg (1970)
30. Varshavsky, Y.: p-adic uniformization of unitary Shimura varieties. II. J. Differential Geom. **49**(1), 75–113 (1998)
31. Yun, Z., Zhang, W.: Shtukas and the Taylor expansion of L-functions. Ann. Math. (2) **186**(3), 767–911 (2017)
32. Zhiwei, Y., Zhang, W.: Shtukas and the Taylor expansion of L-functions (II). Ann. Math. (2) **189**(2), 393–526 (2019)

A p-Adic Gross-Zagier Formula for the Triple p-Adic L-Function at Non-crystalline Points

Francesca Gatti and Victor Rotger

Abstract Let $E/\mathbb{Q}$ be an elliptic curve having multiplicative reduction at a prime p. Let (g, h) be a pair of eigenforms of weight 1 arising as the theta series of an imaginary quadratic field K, and assume that the triple-product L-function $L(f, g, h, s)$ is self-dual and does not vanish at the central critical point $s = 1$. The main result of this article is a formula expressing the value of the associated triple-product p-adic L-function at a point of weights $(2, 1, 1)$ lying outside the region of classical interpolation to the Kolyvagin classes associated by Bertolini and Darmon to a system of Heegner points on E.

Keywords p-adic Gross-Zagier formula · Kolyvagin classes · triple product p-adic L-functions

1 Introduction

Let $E/\mathbb{Q}$ be an elliptic curve of conductor N_f and let $f \in S_2(N_f)$ be the normalized newform attached to it by modularity. Let K be an imaginary quadratic field of discriminant $-D_K$ relatively prime to N_f.

When the sign of the functional equation of the Hasse-Weil L-series $L(E/K, s)$ of the base change of E to K is -1, the classical Gross-Zagier formula [1] expresses the first derivative $L'(E/K, 1)$ at the central critical point in terms of the Néron-Tate height of a Heegner point. Since Gross-Zagier's breakthrough, this formula has

This project has received funding from the European Research Council (ERC) under the European Union's Horizon 2020 research and innovation programme (grant agreement No 682152). The second author gratefully acknowledges Icrea for financial support through an Icrea Academia award. We are most grateful to Adrian Iovita and Xevi Guitart for their interest and extremely helpful comments and to Matteo Longo for reading our work in detail.

F. Gatti
Department de Matemàtiques, Universitat Politècnica de Catalunya, Barcelona, Spain

V. Rotger (✉)
IMTech, UPC and Centre de Recerca Matemàtiques, Barcelona, Spain
e-mail: victor.rotger@upc.edu

M. Longo et al. (eds.), *Elliptic Curves and Modular Forms in Arithmetic Geometry*, Springer Proceedings in Mathematics & Statistics 527,
https://doi.org/10.1007/978-3-032-13123-2_7

been generalized in various directions (cf. e.g. [2]) and has also found striking p-adic avatars, which typically relates the value (or derivative) of a p-adic L-function at a points outside (resp. within) the region of classical interpolation to a p-adic invariant (e.g. the Bloch-Kato logarithm or the p-adic height) associated to a motivic class (cf. e.g. [3, 4]).

The purpose of this paper is proving a formula of this guise that expresses the p-adic value of a triple-product Garrett p-adic L-function at a point lying outside the region of classical interpolation in terms of the p-adic logarithm of a Kolyvagin class. The main novelty of our formula is that the latter class is not-crystalline.

In order to state it precisely, let

$$\psi_g, \psi_h : \mathbb{A}_K^\times / K^\times \longrightarrow \mathbb{C}^\times$$

be two finite order Hecke characters of K of conductors $\mathfrak{c}_g$, $\mathfrak{c}_h$ respectively and let $g := \theta(\psi_g)$ and $h := \theta(\psi_h)$ denote the theta series associated to them. Assume that the central characters of g and h are mutually inverse. Then

$$g \in M_1(N_g, \chi), \qquad h \in M_1(N_h, \chi^{-1})$$

are weight one modular forms of level $N_g = D_K \cdot N_{K/\mathbb{Q}}(\mathfrak{c}_g)$, $N_h = D_K \cdot N_{K/\mathbb{Q}}(\mathfrak{c}_h)$ respectively. For simplicity we assume throughout that

$$\gcd(N_f, N_g N_h) = 1 \tag{1.1}$$

and we set $N := \mathrm{lcm}(N_f, N_g, N_h)$. Let p be an odd prime number such that

$$p \mid\mid N_f \text{ and } p \text{ is inert in } K. \tag{1.2}$$

Hence $E/\mathbb{Q}$ has multiplicative reduction at p and the completion K_p of K at p is the unramified quadratic extension of $\mathbb{Q}_p$. Let

$$\varphi_{\mathrm{Tate}} : \bar{\mathbb{Q}}_p^\times / q_E^{\mathbb{Z}} \xrightarrow{\cong} E(\bar{\mathbb{Q}}_p)$$

be Tate's uniformization, with $q_E \in p\mathbb{Z}_p$, and assume throughout that

$$p \nmid \mathrm{ord}_p(q_E). \tag{1.3}$$

This condition makes it simpler to describe the group of connected components of E/K along the anticyclotomic tower, as we do below in Sect. 2.1.

Set

$$a := a_p(E) = a_p(f) \in \{\pm 1\}.$$

In other words, $a = 1$ (resp. $a = -1$) according to whether E has split (non-split) multiplicative reduction at p.

Let α_g, β_g (resp. α_h, β_h) denote the roots of the p-th Hecke polynomial of g (resp. of h). Note that (1.2) implies that

$$\alpha_g = -\beta_g \text{ and } \alpha_h = -\beta_h.$$

Since the nebentype character of h is the inverse of that of g, one either has

$$(\alpha_h, \beta_h) = (1/\alpha_g, -1/\alpha_g) \text{ or } (\alpha_h, \beta_h) = (-1/\alpha_g, 1/\alpha_g).$$

Throughout this article we fix the ordering of the pair (α_h, β_h) in such a way that

$$\alpha_g \cdot \alpha_h = -a. \tag{1.4}$$

Fix throughout algebraic closures $\bar{\mathbb{Q}}$ and $\bar{\mathbb{Q}}_p$ of $\mathbb{Q}$ and $\mathbb{Q}_p$ respectively and an embedding $\bar{\mathbb{Q}} \hookrightarrow \bar{\mathbb{Q}}_p$. Let $L \subset \bar{\mathbb{Q}}$ denote the number field generated by the traces of ψ_g and ψ_h, together with the roots $\alpha_g, \beta_g, \alpha_h, \beta_h$. and let L_p denote the completion of L within $\bar{\mathbb{Q}}_p$.

Set $g_\alpha(z) := g(z) - \beta_g g(pz)$ and define h_α analogously. Let $\mathbf{f}, \mathbf{g}_\alpha, \mathbf{h}_\alpha$ be Hida families passing through f, g_α, h_α respectively. As explained in [5–7]—the latter including the current setting where $p \mid N-$, associated to any choice of Λ-adic test vectors $(\breve{\mathbf{f}}, \breve{\mathbf{g}}_\alpha, \breve{\mathbf{h}}_\alpha)$ of tame level N there is a three-variable p-adic L-function

$$\mathcal{L}_p^g(\breve{\mathbf{f}}, \breve{\mathbf{g}}_\alpha, \breve{\mathbf{h}}_\alpha).$$

This p-adic L-function interpolates the square-roots of the central values of the classical L-function $L(\breve{f}_k \otimes \breve{g}_\ell \otimes \breve{h}_m, s)$ attached to the specializations of the choice of test vectors at classical points of weights k, ℓ, m with $k, \ell, m \geq 2$ and $\ell \geq k + m$.

Note that the point $(2, 1, 1)$ corresponding to our triple of modular forms (f, g, h) lies outside the region of classical interpolation for $\mathcal{L}_p^g(\breve{\mathbf{f}}, \breve{\mathbf{g}}_\alpha, \breve{\mathbf{h}}_\alpha)$. As in [8, 9] we are interested in studying the p-adic L-value

$$I_p(f, g_\alpha, h_\alpha) := \mathcal{L}_p^g(\breve{\mathbf{f}}, \breve{\mathbf{g}}_\alpha, \breve{\mathbf{h}}_\alpha)(2, 1, 1)$$

regarded as an element of $\bar{\mathbb{Q}}_p^\times / L^\times$. As explained in [7, 8], different choices of Λ-adic test vectors yield the same p-adic value $I_p(f, g_\alpha, h_\alpha)$ up to algebraic scalars in $L^\times$. Since we are interested in this value only up to algebraic factors in $L^\times$, this entitles us to denote it simply $I_p(f, g_\alpha, h_\alpha)$, dropping from the notation this choice.

The triple-product L-function $L(f, g, h, s) = L(E, \rho, s)$ may be recast as the Hasse-Weil L-series of the twist of E by the tensor product $\rho = \rho_g \otimes \rho_h$ of the two Artin representations associated to g and h. In light of (1.1), the order of vanishing of $L(f, g, h, s)$ at $s = 1$ is always even and one thus expects the central critical value $L(E, \rho, 1)$ to be generically nonzero. We assume throughout that we fall indeed in this generic case, that is to say, we have

Assumption 1.1 $L(E, \rho, 1) \neq 0$.

The main purpose of this article is proving a formula relating the p-adic iterated integral $I_p(f, g_\alpha, h_\alpha)$ to the Kolyvagin classes associated by Bertolini and Darmon in [10].

In order to recall briefly the latter, let H be a ring class field of K of conductor relatively prime to p and for every $m \geq 1$ let F_m/H be the layer of degree p^m within the anticyclotomic $\mathbb{Z}_p$-extension of H. In [10] Bertolini and Darmon associated to a canonical collection of *Heegner points* $\{\alpha_m \in E(F_m)\}$ a global cohomology class $\mathbf{K} \in \mathrm{H}^1(H, V_f)$ that is referred to in loc. cit. as the *Kolyvagin class* associated to E/H.

Fix a prime $\mathfrak{p}$ of H above p and denote $\Phi_{m,\mathfrak{p}}$ the p-primary part of the group of connected components of the Néron model of E over the completion of F_m at the unique prime ideal above $\mathfrak{p}$. Denote $\Phi_{\infty,\mathfrak{p}} := \varprojlim_m \Phi_{m,\mathfrak{p}}$ the projective limit with respect to the natural projection maps $\Phi_{m,\mathfrak{p}} \to \Phi_{m-1,\mathfrak{p}}$.

As explained in Lemma 2.1 there is a canonical isomorphism $\varphi : \Phi_{\infty,\mathfrak{p}} \xrightarrow{\cong} \mathbb{Z}_p$ induced by Tate's uniformization. Denote by $\bar{\alpha}_m$ the image of α_m in $\Phi_{m,\mathfrak{p}}$ and set $\bar{\alpha} := (\bar{\alpha}_m)_m \in \Phi_{\infty,\mathfrak{p}}$. Define the period

$$\Pi_p := \varphi(\bar{\alpha}) \in \mathbb{Z}_p.$$

Note that $H^1(K_p, V_f)$ is naturally a $\mathrm{Gal}(K_p/\mathbb{Q}_p)$-module and we let $H^1(K_p, V_f)^{\pm}$ denote the $\mathbb{Q}_p$-subspace on which the arithmetic Frobenius element Fr_p acts with eigenvalue ± 1.

The Kolyvagin class $\mathbf{K}$ is not in general crystalline at p, that is to say, the local class $\mathrm{res}_p(\mathbf{K}) \in H^1(K_p, V_f)$ is not expected to lie in Bloch-Kato's finite subspace $H^1_f(K_p, V_f)$. Nevertheless, one can show (cf. Proposition 5.11) that $\mathrm{res}_p(\mathbf{K})^a := \mathrm{res}_p(\mathbf{K}) + a\,\mathrm{res}_p(\mathbf{K}^{\mathrm{Fr}_p})$ does lie in $H^1_f(K_p, V_f)^a$ and therefore there exists a local point

$$Q_p^a \in (E(K_p) \otimes \mathbb{Q}_p)^a \quad \text{such that} \quad \delta_p(Q_p^a) = \mathrm{res}_p(\mathbf{K})^a,$$

where $\delta_p : E(K_p) \otimes \mathbb{Q}_p \xrightarrow{\sim} H^1_f(K_p, V_f)$ stands for Kummer's isomorphism. In spite of the notation we have chosen, beware that Q_p^a is *not* expected to be the a-component of any local point $Q_p \in E(K_p) \otimes \mathbb{Q}_p$, precisely because $\mathrm{res}_p(\mathbf{K})$ does not lie in $H^1_f(K_p, V_f)$. In other words, while $\mathrm{res}_p(\mathbf{K})^a$ is crystalline, the local class $\mathrm{res}_p(\mathbf{K})^{-a}$ is not.

Let $c = \prod_{v \nmid N\infty} c_v(\breve{f}, \breve{g}_\alpha, \breve{h}_\alpha) \in L$ denote the product of local automorphic factors appearing in [8, Prop. 2.1 (a) (iii)] associated to the choice of test vectors at (2, 1, 1), and define the algebraic L-value

$$L^{\mathrm{alg}}(E \otimes \rho, 1) := \frac{L(E \otimes \rho, 1)}{\pi^4 \langle f, f \rangle^2}.$$

It follows from the work of Harris and Kudla [11] that the above ratio lies in L and is in fact non-zero by (1.1). The following is the main theorem of this note.

Theorem A. For any choice of test vectors, we have

$$I_p(f, g_\alpha, h_\alpha) = \frac{\sqrt{c} \cdot \sqrt{L^{\mathrm{alg}}(E \otimes \rho, 1)}}{\Pi_p \cdot \mathcal{L}_{g_\alpha}} \times \log_p(Q_p^a) \qquad (\mathrm{mod}\ L^\times)$$

where

- $\mathcal{L}_{g_\alpha} \in K_p$ is a period on which Fr_p acts as multiplication by -1 and only depends on g_α,
- $\log_p : E(K_p) \longrightarrow K_p$ denotes the p-adic logarithm.

Remark 1.2 Period $\mathcal{L}_{g_\alpha}$ was first introduced in [12] and is well-defined only up to scalars in $L^\times$; cf. (3.7) for more details. Let H_g be the number field cut out by the Artin representation attached to the adjoint of g and fix a completion $H_{g,p}$ of H_g at a prime above p. In [12] it is conjectured that $\mathcal{L}_{g_\alpha} = \log_p(u_{g_\alpha})$ (mod $L^\times$), where $\log_p : \mathcal{O}_{H_{g,p}}^\times \otimes L \longrightarrow H_{g,p} \otimes L$ is the usual p-adic logarithm and $u_{g_\alpha} \in \mathcal{O}_{H_g}[1/p]^\times \otimes L$ is the so-called Gross–Stark unit attached to g_α [8, Sect. 1.2]. See (3.7) for the explicit definition of $\mathcal{L}_{g_\alpha}$.

Remark 1.3 The appearance of the square-root of the central value $L^{\mathrm{alg}}(E \otimes \rho, 1)$ in the above formula might be striking, given that the point (2, 1, 1) lies *outside* the region of classical interpolation of $\mathcal{L}_p^g(\breve{\mathbf{f}}, \breve{\mathbf{g}}_\alpha, \breve{\mathbf{h}}_\alpha)$. This may be traced back to the reciprocity law governing this p-adic L-function, relating it to an (improved) family of diagonal classes, which is in turn related to $\mathcal{L}_p^f(\breve{\mathbf{f}}, \breve{\mathbf{g}}_\alpha, \breve{\mathbf{h}}_\alpha)$ along a different direction. See Proposition 4.4 for more details.

The body of the paper is devoted to the proof of this result. In Sect. 2 we relate Bloch-Kato's dual exponential map on E to the group of connected components. This is an exercise in p-adic Hodge theory which is surely well-known to experts, but we included because we did not find precise references in the literature. We are most grateful to A. Iovita for his assistance in this section. In Sect. 3 we review the theory of Hida families and Selmer groups, and culminates with the description of an explicit basis of the relaxed Selmer group associated to the triple (f, g, h). In Sect. 4 we exploit the fine work [5] of Bertolini, Seveso and Venerucci in order to prove a formula relating the iterated integral $I_p(f, g_\alpha, h_\alpha)$ in terms of the basis described in the previous section. Finally, in the last section we introduce Kolyvagin classes and prove our main result, which is found in Corollary 5.13.

Theorem A is vaguely similar to the main theorem of [8, Sect. 3], although both the statements and specially the proofs are ostensibly different. While in loc. cit. the motivic elements appearing in the statement (both the diagonal classes and Heegner points) are crystalline at p, in our article they are not and this makes the calculations substantially different. Moreover, in [8, Sect. 3] one makes crucial use of a factorization of p-adic L-series (which follows rather trivially from a comparison of critical L-values), while in our setting the analogous p-adic L-functions associated to K are not available (because in our scenario p remains inert in K). It would be of great interest to investigate whether the recent p-adic L-functions of Andreatta and Iovita [13] could be exploited in order to provide an alternative proof of our formula, although this does not appear to be a straight-forward project.

As we have already pointed out, one of the key ingredients of our proof are the results of Bertolini, Seveso and Venerucci in [5]. In loc. cit. the authors study the exceptional zero phenomena appearing in the scenario of diagonal cycles at weights $(2, 1, 1)$ when $\alpha_g\alpha_h = \pm a$. An arithmetic application of these results is obtained in [14] and [15] to the theory of Stark-Heegner points on elliptic curves, in the setting where $\alpha_g\alpha_h = a$ and an *improved* p-adic L-function plays a relevant role. The present work focus instead in the opposite setting where $\alpha_g\alpha_h = -a$ and it is rather an *improved* diagonal cohomology class (cf. [5, Sect. 8.3] and Sect. 4.2 of this note) which plays a prominent role.

2 p-Adic Hodge Theory for Elliptic Curves with Multiplicative Reduction

Let p be an odd prime. Fix an algebraic closure $\bar{\mathbb{Q}}_p$ of $\mathbb{Q}_p$ and set $G_{\mathbb{Q}_p} = \mathrm{Gal}(\bar{\mathbb{Q}}_p/\mathbb{Q}_p)$. Let $E/\mathbb{Q}_p$ be an elliptic curve of multiplicative reduction. In this section we focus on the p-adic Hodge theory of the local p-adic Galois representation associated to E. The main technical results are Proposition 2.3 and Corollary 2.7, which allow us later in Sect. 5.3 to relate the iterated integral $I_p(f, g_\alpha, h_\alpha)$ to Bertolini-Darmon's Kolyvagin classes and eventually prove Theorem A.

Let

$$\chi_E : G_{\mathbb{Q}_p} \longrightarrow \{\pm 1\}$$

be the trivial (resp. the quadratic unramified) character if E has split (resp. non-split) multiplicative reduction. For any $G_{\mathbb{Q}_p}$-module M, let $M(\chi_E)$ denote the twist of M by χ_E.

Tate's uniformisation provides a $G_{\mathbb{Q}_p}$-equivariant isomorphism

$$\varphi_{\mathrm{Tate}} : \bar{\mathbb{Q}}_p^\times(\chi_E)/q_E^{\mathbb{Z}} \xrightarrow{\cong} E(\bar{\mathbb{Q}}_p) \tag{2.1}$$

for some $q_E \in p\mathbb{Z}_p$. As in (1.3), assume throughout that $p \nmid n := \mathrm{ord}_p(q_E)$.

Let $T = T_E := \varprojlim E[p^m]$ denote the Tate module associated to E and set $V := T \otimes \mathbb{Q}_p$. Let $E_0(\bar{\mathbb{Q}}_p)$ denote the set of points in $E(\bar{\mathbb{Q}}_p)$ that stay nonsingular in the special fiber of E at p. The module

$$T^+ := \varprojlim E_0[p^m]$$

fits in an exact sequence of $\mathbb{Z}_p[G_{\mathbb{Q}_p}]$-modules

$$0 \longrightarrow T^+ \xrightarrow{\iota} T \xrightarrow{\pi} T^- \longrightarrow 0, \tag{2.2}$$

where $T^- := T/T^+$. The modules T^+ and T^- are free over $\mathbb{Z}_p$ of rank one, and Tate's uniformization (2.1) induces $G_{\mathbb{Q}_p}$-equivariant isomorphisms

$$\mathbb{Z}_p(\chi_E)(1) \xrightarrow{\cong} T^+, \quad \mathbb{Z}_p(\chi_E) = \varprojlim_m (\bar{\mathbb{Q}}_p^\times(\chi_E)/q_E^{\mathbb{Z}})[p^m]/\mu_{p^m}(\chi_E) \xrightarrow{\cong} T^-. \tag{2.3}$$

More precisely, if we fix compatible systems

$$\tilde{\varepsilon} := (\tilde{\varepsilon}^{(m)})_m \in \mathbb{Z}_p(\chi_E)(1), \qquad \tilde{q} := (q_E^{1/p^m})_m \in \varprojlim \left(\bar{\mathbb{Q}}_p^\times/q_E^{\mathbb{Z}}\right)[p^m]$$

then

$$\varepsilon := \varphi_{\mathrm{Tate}}(\tilde{\varepsilon}), \; q := \varphi_{\mathrm{Tate}}(\tilde{q})$$

form a $\mathbb{Z}_p$-basis of T. Moreover, if $\chi_{\mathrm{cyc}} : G_{\mathbb{Q}} \longrightarrow \mathbb{Z}_p^\times$ denotes the cyclotomic character, then

(1) ε is a basis of T^+ on which $G_{\mathbb{Q}_p}$ acts as the character $\chi_E \chi_{\mathrm{cyc}}$;
(2) $\bar{q} := \pi(q)$ is a basis of T^- on which $G_{\mathbb{Q}_p}$ acts via χ_E.

Let K be an imaginary quadratic field in which p is inert and let K_p denote the completion of K at p. Notice that $\chi_{E|G_{K_p}} = 1$ and thus (2.2) gives an exact sequence of $\mathbb{Q}_p[G_{K_p}]$-modules

$$0 \longrightarrow V^+ \xrightarrow{\iota} V \xrightarrow{\pi} V^- \longrightarrow 0 \tag{2.4}$$

such that $\dim_{\mathbb{Q}_p} V^+ = \dim_{\mathbb{Q}_p} V^- = 1$, G_{K_p} acts on V^+ via χ_{cyc}, and G_{K_p} acts trivially on V^-.

2.1 The Group of Connected Components

Fix a ring class field H of K of conductor c prime to p and for every integer $m \geq 1$, let $H(p^m)$ be the ring class field of K of conductor $c \cdot p^m$. The Galois group $\mathrm{Gal}(H(p^m)/H)$ is cyclic of order $e_m := (p+1)p^{m-1}$. Let F_m be the intermediate field $H \subseteq F_m \subseteq H(p^{m+1})$ such that $\mathrm{Gal}(F_m/H)$ is cyclic of order p^m. Since p is inert in K, the prime ideal $p\mathcal{O}_K$ splits completely in H. Fix once and for all a prime $\mathfrak{p}$ of H above p; it ramifies totally in $H(p^m)$ as $\mathfrak{p}\mathcal{O}_{F_m} = \mathfrak{p}_m^{p^m}$. Let us denote $F_{m,\mathfrak{p}}$ the completion of F_m at $\mathfrak{p}_m$, $\mathcal{O}_{m,\mathfrak{p}}$ its ring of integers and $\mathbb{F}_{m,\mathfrak{p}}$ its residue field.

Let $\mathcal{E}$ be the Néron model of E over $\mathbb{Z}_p$, and let $\tilde{\mathcal{E}} := \mathcal{E} \times_{\mathbb{Z}_p} \mathbb{F}_p$ denote its special fiber. For all $m \geq 1$, let $\Phi_{m,\mathfrak{p}}$ denote the p-Sylow subgroup of the group

$$\tilde{\mathcal{E}}(\mathbb{F}_{m,\mathfrak{p}})/\tilde{\mathcal{E}}_0(\mathbb{F}_{m,\mathfrak{p}}) \cong E(F_{m,\mathfrak{p}})/E_0(F_{m,\mathfrak{p}}) \tag{2.5}$$

of connected components of the base change of $\mathcal{E}$ to $\mathcal{O}_{m,\mathfrak{p}}$ and define

$$\Phi_{\infty,\mathfrak{p}} = \varprojlim_m \Phi_{m,\mathfrak{p}}$$

to be the inverse limit with respect to the natural trace maps $\Phi_{m,\mathfrak{p}} \to \Phi_{m-1,\mathfrak{p}}$.

Lemma 2.1 *There are isomorphisms of $G_{\mathbb{Q}_p}$-modules*

$$T^- \cong \mathbb{Z}_p(\chi_E) \cong \Phi_{\infty,\mathfrak{p}}.$$

Proof The first isomorphism is (2.3). Tate's uniformisation provides a description of the group (2.5) as

$$\overline{\varphi}_{\mathrm{Tate}} : F_{m,\mathfrak{p}}^\times/q_E^{\mathbb{Z}}\mathcal{O}_{m,\mathfrak{p}}^\times(\chi_E) \cong \left(\mathbb{Z}/\operatorname{ord}_{m,\mathfrak{p}}(q_E)\mathbb{Z}\right)(\chi_E) \xrightarrow{\cong} E(F_{m,\mathfrak{p}})/E_0(F_{m,\mathfrak{p}}), \tag{2.6}$$

where $\operatorname{ord}_{m,\mathfrak{p}}$ is the discrete valuation on $\mathcal{O}_{m,\mathfrak{p}}$. We have $p\mathcal{O}_{F_m} = \prod_{\mathfrak{p}_m|p} \mathfrak{p}_m^{p^m}$, where $\mathfrak{p}_m = (\pi_m)$ is the maximal ideal of $\mathcal{O}_{m,\mathfrak{p}}$. Hence $q_E = p^n\alpha = \pi_m^{np^m}\alpha'$ where $\alpha, \alpha' \in \mathcal{O}_{F_{\mathfrak{p},m}}^\times$, i.e.

$$\operatorname{ord}_{m,\mathfrak{p}}(q_E) = n \cdot p^m.$$

Under condition (1.3), then (2.6) gives the isomorphism

$$\overline{\varphi}_{\mathrm{Tate}} : \left(\mathbb{Z}/np^m\mathbb{Z}\right)(\chi_E) \cong (\mathbb{Z}/n\mathbb{Z})(\chi_E) \oplus \left(\mathbb{Z}/p^m\mathbb{Z}\right)(\chi_E) \xrightarrow{\cong} E(F_{m,\mathfrak{p}})/E_0(F_{m,\mathfrak{p}}).$$

The lemma follows after taking p-primary parts and passing to the inverse limit. □

2.2 *The G_{K_p}-Cohomology of E*

Let $\mathrm{B}_{\mathrm{dR}} = \operatorname{Frac}(\mathrm{B}_{\mathrm{dR}}^+)$ denote Fontaine's algebra of de Rham periods. For a de Rham G_{K_p}-representation W, denote

$$\mathrm{D}_{\mathrm{dR}}(W) := (W \otimes_{\mathbb{Q}_p} \mathrm{B}_{\mathrm{dR}})^{G_{K_p}}.$$

This Dieudonné module comes naturally equipped with a filtration, inherited from the filtration on B_{dR}, i.e.

$$\mathrm{Fil}^j\, \mathrm{D}_{\mathrm{dR}}(W) := (W \otimes_{\mathbb{Q}_p} \mathrm{Fil}^j\, \mathrm{B}_{\mathrm{dR}})^{G_{K_p}},$$

and in particular we have $\mathrm{Fil}^0\, \mathrm{D}_{\mathrm{dR}}(W) = (W \otimes \mathrm{B}_{\mathrm{dR}}^+)^{G_{K_p}}$. The inclusion $\alpha : W \longrightarrow W \otimes_{\mathbb{Q}_p} \mathrm{B}_{\mathrm{dR}}$ induces in cohomology the map

$$\alpha_* : \mathrm{H}^1(K_p, W) \longrightarrow \mathrm{H}^1(K_p, W \otimes \mathrm{B}_{\mathrm{dR}}).$$

Define

$$\mathrm{H}^1_g(K_p, W) := \ker(\alpha_*) \subseteq \mathrm{H}^1(K_p, W), \qquad \mathrm{H}^1_s(K_p, W) := \mathrm{H}^1(K_p, W)/\,\mathrm{H}^1_g(K_p, W).$$

Similarly, if $\mathrm{B}_{\mathrm{cris}}$ denotes the ring of crystalline periods, define

$$\mathrm{H}^1_f(K_p, W) := \ker\left(\mathrm{H}^1(K_p, W) \longrightarrow \mathrm{H}^1(K_p, W \otimes \mathrm{B}_{\mathrm{cris}})\right).$$

Let $\log_p : 1 + p\mathbb{Z}_p \longrightarrow p\mathbb{Z}_p$ be the usual p-adic logarithm, $\exp_p$ denote its inverse and set $u := \exp_p(p) \in 1 + p\mathbb{Z}_p$. Let

$$\mathrm{Art} : K_p^\times \longrightarrow G_{K_p}^{\mathrm{ab}} \tag{2.7}$$

denote the Artin map of local class field theory sending uniformizers to arithmetic Frobenii, where $G_{K_p}^{\mathrm{ab}}$ denotes the Galois group of the maximal abelian extension K_p^{ab} of K_p over K_p. Recall that (2.7) gives an isomorphism once extended to the profinite completion $\widehat{K}_p^\times$ of $K_p^\times$.

The decomposition

$$K_p^\times = p^{\mathbb{Z}} \oplus (1 + p\mathcal{O}_{K_p}) \oplus \left(\mathcal{O}_{K_p}/p\mathcal{O}_{K_p}\right)^\times,$$

corresponds via (2.7) to

$$G_{K_p}^{\mathrm{ab}} \cong \mathrm{Gal}(K_p^{\mathrm{nr}}/K_p) \times \mathrm{Gal}(K_\infty/K_p) \times \mathrm{Gal}(K'_p/K_p)$$

where K_p^{nr} is the maximal unramified extension of K_p, K_∞/K_p is a $\mathbb{Z}_p^2$-extension and K'_p/K_p is finite. More precisely, via the Artin map, $p^{\widehat{\mathbb{Z}}} \cong \mathrm{Gal}(K_p^{\mathrm{nr}}/K_p)$. Fix a prime $\mathfrak{p}$ of H above p, and denote

$$K_p(\mu_{p^\infty}) := \varinjlim K_p(\mu_{p^m}); \qquad H(p^\infty)_{\mathfrak{p}} := \varinjlim H(p^m)_{\mathfrak{p}_m},$$

where $\mathfrak{p}_m$ is the unique prime of $H(p^m)$ lying above $\mathfrak{p}$. Let

$$K_p \subseteq K_{\mathrm{cyc},p} \subseteq K_p(\mu_{p^\infty}), \qquad K_p \subseteq K_{\mathrm{ant},p} \subseteq H(p^\infty)_{\mathfrak{p}}$$

be maximal sub-extensions such that $\mathrm{Gal}(K_{\mathrm{cyc},p}/K_p) \simeq \mathrm{Gal}(K_{\mathrm{ant},\mathfrak{p}}/K_p) \simeq \mathbb{Z}_p$. Then

$$K_\infty = K_{\mathrm{cyc},p} \cdot K_{\mathrm{ant},p}.$$

Fix an element $u_\star \in 1 + p\mathcal{O}_{K_p}$ such that $\mathrm{Fr}_p\, u_\star = u_\star^{-1}$ and $\{u, u_\star\}$ is a $\mathbb{Z}_p$-basis of $1 + p\mathcal{O}_{K_p}$. Then

$$\mathrm{Gal}(K_{\infty,\mathfrak{p}}/K_p) \cong \Gamma_{\mathrm{cyc}} \times \Gamma_{\mathrm{ant}},$$

where

(1) $\Gamma_{\mathrm{cyc}} := \mathrm{Gal}(K_{\mathrm{cyc},p}/K_p)$ is generated topologically by $\sigma_{\mathrm{cyc}} := \mathrm{Art}(u)$;
(2) $\Gamma_{\mathrm{ant}} := \mathrm{Gal}(K_{\mathrm{ant},p}/K_p)$ is generated topologically by $\sigma_{\mathrm{ant}} := \mathrm{Art}(u_\star)$.

Set also $\Gamma_{\mathrm{nr}} := \mathrm{Gal}(K_p^{\mathrm{nr}}/K_p)$. Recall the cyclotomic character $\chi_{\mathrm{cyc}} : G_{\mathbb{Q}} \to \mathbb{Z}_p^\times$; by an abuse of notation, we continue to denote $\chi_{\mathrm{cyc}} : \Gamma_{\mathrm{cyc}} \longrightarrow 1 + p\mathbb{Z}_p$ its restriction to Γ_{cyc}. Let also χ_{ant} and χ_{nr} denote the characters of Γ_{ant} and Γ_{nr} such that

$$\chi_{\mathrm{ant}}(\sigma_{\mathrm{ant}}) = u, \quad \chi_{\mathrm{nr}}(\mathrm{Fr}_p) = u$$

respectively.

Note that

$$\{\xi_{\mathrm{nr}} := \log_p(\chi_{\mathrm{nr}}), \xi_{\mathrm{cyc}} := \log_p(\chi_{\mathrm{cyc}}), \xi_{\mathrm{ant}} := \log_p(\chi_{\mathrm{ant}})\}$$

is naturally a $\mathbb{Q}_p$-basis of $\mathrm{Hom}_{\mathrm{cont}}(G_{K_p}^{\mathrm{ab}}, \mathbb{Q}_p)$. After composing with the isomorphism $\mathbb{Q}_p \stackrel{\varphi_{\mathrm{Tate}}}{\longrightarrow} V^-_{f|G_{K_p}}$ provided by Tate's uniformization this further yields a $\mathbb{Q}_p$-basis of $\mathrm{H}^1(K_p, V^-)$, which in turn may be regarded as a basis of $\mathrm{H}^1(K_p, \Phi_{\infty,\mathrm{p}} \otimes \mathbb{Q}_p)$ by means of Lemma 2.1.

For a class $\xi \in \mathrm{H}^1(K_p, V^-)$, we denote $\bar{\xi}$ its image in the singular quotient $\mathrm{H}^1_s(K_p, V^-)$.

Lemma 2.2 $\{\bar{\xi}_{\mathrm{cyc}}, \bar{\xi}_{\mathrm{ant}}\}$ *is a* $\mathbb{Q}_p$*-basis for* $\mathrm{H}^1_s(K_p, V^-)$.

Proof By definition, the submodule $\mathrm{H}^1_g(K_p, \mathbb{Q}_p) = \mathrm{H}^1_f(K_p, \mathbb{Q}_p)$ of $\mathrm{H}^1(K_p, \mathbb{Q}_p) = \mathrm{Hom}_{\mathrm{cont}}(G_{K_p}, \mathbb{Q}_p)$ is given by $\mathrm{H}^1_f(K_p, \mathbb{Q}_p(\chi_E)) \cong \mathrm{Hom}_{\mathrm{cont}}(\Gamma_{\mathrm{nr}}, \mathbb{Q}_p)$ and is generated by ξ_{nr}. The lemma follows because $\mathrm{H}^1_s(K_p, \mathbb{Q}_p) = \mathrm{H}^1_s(K_p, V^-) = \mathrm{H}^1(K_p, V^-)/\mathrm{H}^1_f(K_p, V^-)$. □

Write the uniformizer of the elliptic curve $E/\mathbb{Q}_p$ as

$$q_E = p^n u^s x \in p\mathbb{Z}_p, \qquad n \geq 1, s \in \mathbb{Z}_p, x \in \mu_{p-1}, \tag{2.8}$$

so that $n = \mathrm{ord}_p(q_E)$ and $ps = \log_p(q_E)$. Let

$$\pi_* : \mathrm{H}^1(K_p, V) \longrightarrow \mathrm{H}^1(K_p, V^-)$$

be the morphism induced in G_{K_p}-cohomology by the projection $\pi : V \longrightarrow V^-$.

For a $\mathrm{Gal}(K_p/\mathbb{Q}_p)$-module M, let us write $M^\pm$ for the eigenspace on which Fr_p acts as ± 1. Set $a := a_p(E)$ as in the introduction.

Proposition 2.3 *There exist unique elements* $x_{\mathrm{cyc}}, x_{\mathrm{ant}} \in \mathrm{H}^1(K_p, V)$ *such that*

$$\pi_*(x_{\mathrm{cyc}}) = n\xi_{\mathrm{cyc}} - s\xi_{\mathrm{nr}}, \qquad \pi_*(x_{\mathrm{ant}}) = \xi_{\mathrm{ant}} \textit{ in } \mathrm{H}^1(K_p, V^-). \tag{2.9}$$

Moreover $\{\bar{x}_{\rm cyc}, \bar{x}_{\rm ant}\}$ *is a* $\mathbb{Q}_p$*-basis for* $\mathrm{H}^1_s(K_p, V)$ *and* π_* *descends to an isomorphism*

$$\bar{\pi}_* : \mathrm{H}^1_s(K_p, V) \xrightarrow{\cong} \mathrm{H}^1_s(K_p, V^-), \qquad \bar{x}_{\rm cyc} \mapsto n \cdot \bar{\xi}_{\rm cyc}, \qquad \bar{x}_{\rm ant} \mapsto \bar{\xi}_{\rm ant}.$$

The Frobenius eigenspaces $\mathrm{H}^1_s(K_p, V)^a$ *and* $\mathrm{H}^1_s(K_p, V)^{-a}$ *are generated respectively by* $\bar{x}_{\rm cyc}$ *and* $\bar{x}_{\rm ant}$.

Proof Consider the long exact sequence

$$0 \to V^- \xrightarrow{\partial^0} \mathrm{H}^1(K_p, V^+) \xrightarrow{\iota_*} \mathrm{H}^1(K_p, V) \xrightarrow{\pi_*} \mathrm{H}^1(K_p, V^-) \xrightarrow{\partial^1} \mathrm{H}^2(K_p, V^+) \to 0 \quad (2.10)$$

induced in G_{K_p}-cohomology by (2.4). The connecting homomophisms ∂^0, ∂^1 can be expressed in terms of q_E as follows. Cup product and the trace map give a pairing

$$\langle\ ,\ \rangle : \mathrm{H}^1(K_p, V^+) \times \mathrm{H}^1(K_p, V^-) \longrightarrow \mathrm{H}^2(K_p, V^+ \otimes V^-) \cong \mathrm{H}^2(K_p, \mathbb{Q}_p(1)) \cong \mathbb{Q}_p$$

that may be explicitly defined as

$$\langle \delta_p(q_E), \xi \rangle = \xi(\mathrm{Art}(q_E))$$

for all $\xi \in \mathrm{H}^1(K_p, V^-)$. If we still call ∂^1 its composition with $\mathrm{H}^2(K_p, V^+ \otimes V^-) \cong \mathbb{Q}_p$, then it coincides with the map $\langle \delta_p(q_E), \cdot \rangle$. Using the notation of (2.8), we have $\mathrm{Art}(q_E) = \mathrm{Fr}_p^n \sigma_{\rm cyc}^s$ and thus

(1) $\partial^1(\xi_{\rm cyc}) = \xi_{\rm cyc}(\mathrm{Art}(q_E)) = \xi_{\rm cyc}(\sigma_{\rm cyc}^s) = \log_p(\chi_{\rm cyc}(\sigma_{\rm cyc}^s) = s \log_p(u) = sp$;
(2) $\partial^1(\xi_{\rm ant}) = \xi_{\rm ant}(\mathrm{Art}(q_E)) = \log_p(\chi_{\rm ant}(\sigma_{\rm ant}^0)) = 0$;
(3) $\partial^1(\xi_{\rm nr}) = \xi_{\rm nr}(\mathrm{Art}(q_E)) = \xi_{\rm nr}(\mathrm{Fr}_p^n) = \log_p(\chi_{\rm nr}(\sigma_{\rm cyc})^n) = n \log_p(u) = np$.

Combining these computations with the exactness of (2.10) we deduce that the image of π_* is generated by $\{n \cdot \xi_{\rm cyc} - s \cdot \xi_{\rm nr}, \xi_{\rm ant}\}$, and this gives the existence of $x_{\rm cyc}, x_{\rm ant}$ as in the statement.

From (2.9) and Lemma 2.2 we deduce that $\bar{\pi}_*(\bar{x}_{\rm cyc}) = n \cdot \bar{\xi}_{\rm cyc}, \bar{\pi}_*(\bar{x}_{\rm ant}) = \bar{\xi}_{\rm ant}$ and that $\bar{\pi}_*$ is surjective. The map ι_* of (2.10) restricts to an isomorphism $\iota_* : \mathrm{H}^1_f(K_p, V^+) \cong \mathrm{H}^1_f(K_p, V)$, hence $\bar{\pi}_*$ is also injective.

Finally, in order to understand the action of Frobenius, note that $\mathrm{H}^1_s(K_p, V) \simeq \mathrm{H}^1_s(K_p, V^-)$ is naturally a $\mathrm{Gal}(K_p/\mathbb{Q}_p)$-module. As explained in the proof of Lemma 2.2, $\{\xi_{\rm cyc}, \xi_{\rm ant}\}$ is a basis of

$$\mathrm{H}^1_s(K_p, V^-) \cong \mathrm{Hom}(\Gamma_{\rm cyc}, \mathbb{Q}_p(\chi_E)) \oplus \mathrm{Hom}(\Gamma_{\rm ant}, \mathbb{Q}_p(\chi_E)). \quad (2.11)$$

We have

$$\xi_{\rm ant}^{\mathrm{Fr}_p}(\sigma_{\rm ant}) = \chi_E(\mathrm{Fr}_p) \cdot \xi_{\rm ant}(\mathrm{Fr}_p\, \sigma_{\rm ant}\, \mathrm{Fr}_p^{-1}) = a \cdot \xi_{\rm ant}(\sigma_{\rm ant}^{-1}) = -a \cdot \xi_{\rm ant}(\sigma_{\rm ant})$$

and via (2.11) it follows that

$$\mathrm{H}^1_s(K_p, V^-)^{-a} \cong \mathrm{Hom}(\Gamma_{\mathrm{ant}}, \mathbb{Q}_p(\chi_E))$$

is generated by ξ_{ant}. Analogously, ξ_{cyc} generates $\mathrm{H}^1_s(K_p, V^-)^a \cong \mathrm{Hom}(\Gamma_{\mathrm{cyc}}, \mathbb{Q}_p(\chi_E))$. □

2.3 Bloch–Kato Logarithm and Dual Exponential Maps

We wish to describe explicitly the Bloch–Kato logarithm and the dual exponential maps relative to the representation $V = V_f$ regarded as a G_{K_p}-module. First we need to study $\mathrm{D}_{\mathrm{dR}}(V_{K_p})$, $\mathrm{D}_{\mathrm{dR}}(V^{\pm}_{K_p})$ and their filtration, where $V_{K_p} := V_{|G_{K_p}}$. Let $\mathbb{C}_p$ be the completion of a fixed algebraic closure of $\mathbb{Q}_p$, and let

$$\mathcal{R} := \varprojlim_{x \mapsto x^p} \mathcal{O}_{\mathbb{C}_p}/p\mathcal{O}_{\mathbb{C}_p}.$$

Let $W(\mathcal{R})$ be the ring of Witt vectors of $\mathcal{R}$ and denote $[\cdot] : \mathcal{R} \longrightarrow W(\mathcal{R})$ the Teichmuller lift. There is an isomorphism

$$\varprojlim_{x \mapsto x^p} \mathcal{O}_{\mathbb{C}_p} \xrightarrow{\cong} \mathcal{R}, \qquad (x^{(n)})_n \mapsto (x_n := x^{(n)} \mod p)_n.$$

The element $\tilde{\varepsilon} := (\tilde{\varepsilon}^{(m)})_m$ introduced right before Sect. 2.1 may be regarded in $\lim_{x \mapsto x^p} \mathcal{O}_{\mathbb{C}_p} \cong \mathcal{R}$, and if $\log_{q_E}$ denotes the branch of the p-adic logarithm such that $\log_{q_E}(q_E) = 0$, then

$$t := \log_{q_E}([\tilde{\varepsilon}])$$

generates the maximal ideal $\mathrm{Fil}^1\, \mathrm{B}_{\mathrm{dR}}$ of $\mathrm{B}^+_{\mathrm{dR}} = \mathrm{Fil}^0\, \mathrm{B}_{\mathrm{dR}}$. Analogously, we can regard $\tilde{q}$ as an element of $\mathcal{R}$ and we define

$$q_E^{\#} := \log_{q_E}([\tilde{q}]/q_E) \in \mathrm{Fil}^1\, \mathrm{B}_{\mathrm{dR}}$$

Recall that the images ε, q of $\tilde{\varepsilon}, \tilde{q}$ under Tate's uniformization form a basis of V. The following lemma follows from the above definitions (see e.g. [16] for similar computations):

Lemma 2.4 *Let*

$$e_1 := q \otimes 1 - \varepsilon \otimes \frac{q_E^{\#}}{t} \in V \otimes \mathrm{Fil}^0\, \mathrm{B}_{\mathrm{dR}}, \quad e_2 := \varepsilon \otimes \frac{1}{t} \quad \in V \otimes \mathrm{Fil}^{-1}\, \mathrm{B}_{\mathrm{dR}}.$$

These elements are invariant with respect to the action of G_{K_p}, they form a K_p-basis of $\mathrm{D}_{\mathrm{dR}}(V_{K_p})$ and e_1 is a K_p-basis of $\mathrm{Fil}^0\, \mathrm{D}_{\mathrm{dR}}(V_{K_p})$. Moreover, there are isomorphisms

$$\iota_* : \mathrm{D}_{\mathrm{dR}}(V^+_{K_p})/\,\mathrm{Fil}^0\,\mathrm{D}_{\mathrm{dR}}(V^+_{K_p}) \xrightarrow{\cong} \mathrm{D}_{\mathrm{dR}}(V_{K_p})/\,\mathrm{Fil}^0\,\mathrm{D}_{\mathrm{dR}}(V_{K_p}), \qquad \varepsilon \otimes t^{-1} \mapsto e_2;$$

$$\pi_* : \mathrm{Fil}^0\,\mathrm{D}_{\mathrm{dR}}(V_{K_p}) \xrightarrow{\cong} \mathrm{Fil}^0\,\mathrm{D}_{\mathrm{dR}}(V^-_{K_p}), \qquad e_1 \mapsto \bar{q},$$

where, with a slight abuse of notation, $\bar{q} := \bar{q} \otimes 1$ *in* $(V^- \otimes \mathrm{B}^+_{\mathrm{dR}})^{G_{K_p}} = \mathrm{Fil}^0\,\mathrm{D}_{\mathrm{dR}}(V^-)$.

Kummer theory provides isomorphisms

$$\delta_p : \mathcal{O}^\times_{K_p} \otimes \mathbb{Q}_p \xrightarrow{\cong} \mathrm{H}^1_f(K_p, \mathbb{Q}_p(1)), \quad E(K_p) \otimes \mathbb{Q}_p \xrightarrow{\cong} \mathrm{H}^1_f(K_p, V),$$

which combined with (2.3) and Tate's uniformization yield canonical identifications of $\mathbb{Q}_p[G_{\mathbb{Q}_p}]$-modules

$$\mathcal{O}^\times_{K_p}(\chi_E) \otimes \mathbb{Q}_p \cong \mathrm{H}^1_f(K_p, V^+) \overset{\iota_*}{\cong} \mathrm{H}^1_f(K_p, V) = E(K_p) \otimes \mathbb{Q}_p. \tag{2.12}$$

Besides, the p-adic logarithm induces a morphism of $\mathbb{Q}_p[G_{\mathbb{Q}_p}]$-modules

$$\log_p : \mathcal{O}^\times_{K_p}(\chi_E) \otimes \mathbb{Q}_p \longrightarrow K_p(\chi_E).$$

Via the identifications (2.12), it gives rise to the map

$$\log_E : E(K_p) \longrightarrow K_p(\chi_E), \tag{2.13}$$

which is the formal group logarithm on E associated with the Néron differential $\omega_E = \omega_f$ on E (cf. Sect. 3.1 for more details on the latter). Finally, (2.12) and the isomorphism

$$\mathrm{D}_{\mathrm{dR}}(V_{K_p})/\,\mathrm{Fil}^0\,\mathrm{D}_{\mathrm{dR}}(V_{K_p}) \xrightarrow{\cong} K_p(\chi_E), \qquad e_2 \mapsto 1,$$

allow us to identify $\log_E$ with the composition of the Bloch–Kato logarithm

$$\log : \mathrm{H}^1_f(K_p, V) \longrightarrow \mathrm{D}_{\mathrm{dR}}(V_{K_p})/\,\mathrm{Fil}^0\,\mathrm{D}_{\mathrm{dR}}(V_{K_p})$$

with the pairing with ω_f.

For a p-adic de Rham representation W of G_{K_p}, let

$$\gamma : \mathrm{Fil}^0\,\mathrm{D}_{\mathrm{dR}}(W) = (W \otimes_{\mathbb{Q}_p} \mathrm{B}^+_{\mathrm{dR}})^{G_{K_p}} \longrightarrow \mathrm{H}^1(K_p, W \otimes_{\mathbb{Q}_p} \mathrm{B}^+_{\mathrm{dR}})$$

be the isomorphism defined by $x \mapsto \gamma(x)$, where $\gamma(x)$ is the cohomology class of the map

$$\sigma \mapsto x \cdot \log_p(\chi_{\mathrm{cyc}}(\sigma)).$$

Definition 2.5 The *Bloch–Kato dual exponential* of W is the map

$$\exp^* : \mathrm{H}^1(K_p, W) \xrightarrow{\alpha_*} \mathrm{H}^1(K_p, W \otimes_{\mathbb{Q}_p} \mathrm{B}_{\mathrm{dR}}^+) \xrightarrow{\gamma^{-1}} \mathrm{Fil}^0 \mathrm{D}_{\mathrm{dR}}(W).$$

We refer to [17] for more details on the above definition and its interpretation as the dual of the exponential map in p-adic Hodge theory.

The map $\mathrm{H}^1(K_p, W \otimes \mathrm{B}_{\mathrm{dR}}^+) \longrightarrow \mathrm{H}^1(K_p, W \otimes \mathrm{B}_{\mathrm{dR}})$ is injective and thus $\ker(\exp^*) = \mathrm{H}^1_g(K_p, W)$. We shall regard the dual exponential associated to W as the map

$$\exp^* : \mathrm{H}^1_s(K_p, W) = \mathrm{H}^1(K_p, W)/\mathrm{H}^1_g(K_p, W) \longrightarrow \mathrm{Fil}^0 \mathrm{D}_{\mathrm{dR}}(W).$$

Similarly as in the case of the Bloch–Kato logarithm, the dual exponential for $W = K_p$ together with the identification $\mathrm{H}^1_s(K_p, \mathbb{Q}_p)(\chi_E) = \mathrm{H}^1_s(K_p, \mathbb{Q}_p(\chi_E))$ gives rise to an isomorphism of $K_p[G_{\mathbb{Q}_p}]$-modules

$$\exp^* : \mathrm{H}^1_s(K_p, \mathbb{Q}_p(\chi_E)) \longrightarrow K_p(\chi_E).$$

It coincides with the dual exponential

$$\exp^* : \mathrm{H}^1_s(K_p, V) \longrightarrow K_p(\chi_E) \tag{2.14}$$

via the identifications

$$\mathrm{H}^1_s(K_p, \mathbb{Q}_p(\chi_E)) \cong \mathrm{H}^1_s(K_p, V^-) \cong \mathrm{H}^1_s(K_p, V)$$

explained in the previous section, where

$$\mathrm{Fil}^0 \mathrm{D}_{\mathrm{dR}}(V_{K_p}) \xrightarrow{\cong} K_p(\chi_E), \qquad e_1 \mapsto 1.$$

It follows directly from the very definitions and the $G_{\mathbb{Q}_p}$-equivariance of the dual exponential map that

$$\exp^*(\xi_{\mathrm{cyc}}) = 1 \text{ and } z := \exp^*(\xi_{\mathrm{ant}}) \in K_p^-.$$

The map (2.14) and its $\pm$-components can be explicitly described as follows.

Proposition 2.6 *The dual exponential* $\exp^* : \mathrm{H}^1_s(K_p, V) \longrightarrow K_p(\chi_E)$ *is characterized by*

$$\exp^*(\bar{x}_{\mathrm{cyc}}) = n, \;\; \exp^*(\bar{x}_{\mathrm{ant}}) = z \in K_p^-.$$

Proof By (the proof of) Proposition 2.3,

$$\mathrm{H}^1_s(K_p, \mathbb{Q}_p(\chi_E))^a \oplus \mathrm{H}^1_s(K_p, \mathbb{Q}_p(\chi_E))^{-a} \cong \mathrm{Hom}(\Gamma_{\mathrm{cyc}}, \mathbb{Q}_p(\chi_E)) \oplus \mathrm{Hom}(\Gamma_{\mathrm{ant}}, \mathbb{Q}_p(\chi_E)),$$

whose basis $\{\xi_{\mathrm{cyc}}, \xi_{\mathrm{ant}}\}$ is compatible with the above decomposition.

The claim follows by using Lemma 2.4, Proposition 2.3 and the description of (2.14) discussed before this proposition. □

Corollary 2.7 *There is a commutative diagram whose arrows are isomorphisms of* $\mathrm{Gal}(K_p/\mathbb{Q}_p)$*-modules:*

$$\begin{array}{ccc}
Hom_{\mathrm{cont}}(\Gamma_{\mathrm{cyc}}, \Phi_{\infty,\mathfrak{p}}) \oplus Hom_{\mathrm{cont}}(\Gamma_{\mathrm{ant}}, \Phi_{\infty,\mathfrak{p}}) & \xrightarrow{\ \varphi\ } & \mathrm{H}^1_s(K_p, V) \\
\Big\downarrow {\scriptstyle \frac{1}{p}\mathrm{ev}_{\mathrm{cyc}} \,\oplus\, \frac{1}{p}\mathrm{ev}_{\mathrm{ant}}} & & \Big\downarrow {\scriptstyle \exp^*} \\
\left(\Phi_{\infty,\mathfrak{p}} \otimes \mathbb{Q}_p\right) \oplus \left(\Phi_{\infty,\mathfrak{p}} \otimes \mathbb{Q}_p\right) & \xrightarrow{\ \varrho\ } & K_p(\chi_E)
\end{array}$$

where

(1) $\varphi := (\bar{\varphi}_{\mathrm{Tate}} \circ \bar{\pi}_*)^{-1}$*;*
(2) $\varrho(\bar{q}, 0) = 1$, $\varrho(0, \bar{q}) = z$, *where* $z = \exp^*(\xi_{\mathrm{ant}}) \in K_p^-$ *is as in Proposition 2.6;*
(3) $\mathrm{ev}_{\mathrm{cyc}}$ *denotes the valuation at* $\sigma_{\mathrm{cyc}} \in \Gamma_{\mathrm{cyc}}$, *and analogously for* $\mathrm{ev}_{\mathrm{ant}}$.

Moreover, φ *decomposes with respect to the action of* Fr_p *as*

$$Hom_{\mathrm{cont}}(\Gamma_{\mathrm{cyc}}, \Phi_{\infty,\mathfrak{p}}) \longrightarrow \mathrm{H}^1_s(K_p, V)^a, \qquad Hom_{\mathrm{cont}}(\Gamma_{\mathrm{ant}}, \Phi_{\infty,\mathfrak{p}}) \longrightarrow \mathrm{H}^1_s(K_p, V)^{-a}. \tag{2.15}$$

Proof By definition we have

$$\xi_{\mathrm{cyc}}(\sigma_{\mathrm{cyc}}) = \overline{\varphi}_{\mathrm{Tate}}(\log_p(\chi_{\mathrm{cyc}}(\sigma_{\mathrm{cyc}}))) = \overline{\varphi}_{\mathrm{Tate}}(\log_p(u)) = \overline{\varphi}_{\mathrm{Tate}}(p) = p \cdot \bar{q}$$

and similarly, $\xi_{\mathrm{ant}}(\sigma_{\mathrm{ant}}) = p \cdot \bar{q}$. The statement follows by combining (the proof of) Proposition 2.6, Proposition 2.3 and Lemma 2.1. □

3 Hida Families and Selmer Groups Associated to (f, g, h)

3.1 Filtrations and Differentials

As in the previous section, let p be an odd prime number. Let

$$\varphi \in S_w(M, \varepsilon)$$

be an ordinary newform of weight w, level M and character ε with coefficients in a finite extension $L/\mathbb{Q}$. Here M is a positive integer that a priori may be relatively prime to p or not. Fix an embedding $L \hookrightarrow \bar{\mathbb{Q}}_p$ and let $L_p \subset \bar{\mathbb{Q}}_p$ denote the p-adic closure of L.

Assume φ is ordinary at p with respect to the above embedding. Let V_φ denote the p-adic Galois representation attached to φ with coefficients in L_p defined e.g. as in [18, Sect. 1.2, (1.6)]. Let α_φ, β_φ be the roots of the Hecke polynomial $x^2 - a_p(\varphi)x + \varepsilon(p)p^{w-1}$ at p of V_φ. Since φ is ordinary, at least one of these roots is a p-adic unit: choose one and call it α_φ. If $w \geq 2$, this is actually the only choice. If φ has weight $w = 1$, we assume throughout that

$$\alpha_\varphi \neq \beta_\varphi.$$

As a $G_{\mathbb{Q}_p}$-representation, V_φ is equipped with a decreasing filtration of $L_p[G_{\mathbb{Q}_p}]$-modules

$$\mathrm{Fil}^2(V_\varphi) = 0 \subseteq \mathrm{Fil}^1(V_\varphi) \subseteq V_\varphi = \mathrm{Fil}^0(V_\varphi). \tag{3.1}$$

Let

$$\psi_\varphi : G_{\mathbb{Q}_p} \longrightarrow L_p^\times$$

be the unramified character such that $\psi_\varphi(\mathrm{Fr}_p) = \alpha_\varphi$. The filtration (3.1) satisfies:

(i) $V_\varphi^+ := \mathrm{Fil}^1(V_\varphi)$ and $V_\varphi^- := V_\varphi / \mathrm{Fil}^1(V_\varphi)$ have dimension one over L_p;
(ii) $G_{\mathbb{Q}_p}$ acts on V_φ^+ via $\chi_{\mathrm{cyc}}^{w-1}\varepsilon\psi_\varphi^{-1}$;
(iii) $G_{\mathbb{Q}_p}$ acts on V_φ^- via ψ_φ. In particular, it is an unramified $G_{\mathbb{Q}_p}$-module.

When $w \geq 2$, Faltings' comparison theorem yields an isomorphism

$$S_w(M)_{L_p}[\varphi] \xrightarrow{\cong} \mathrm{Fil}^1 \mathrm{D}_{\mathrm{dR}}(V_\varphi) \cong \mathrm{D}_{\mathrm{dR}}(V_\varphi^-), \tag{3.2}$$

where $S_w(M)_{L_p}[\varphi]$ denotes the φ-isotypical component of $S_w(M) \otimes L_p$. We define

$$\omega_\varphi \in \mathrm{D}_{\mathrm{dR}}(V_\varphi^-)$$

to be the image of φ via (3.2).

Denote $\varphi^* := \varphi \otimes \varepsilon^{-1} \in S_w(M, \bar{\varepsilon})$. Consider the pairing

$$\langle\, ,\, \rangle : V_\varphi \times V_{\varphi^*}(1) \longrightarrow L_p$$

given by Poincaré's duality. This in turn induces perfect pairings

$$\langle\, ,\, \rangle : \mathrm{D}_{\mathrm{dR}}(V_\varphi) \times \mathrm{D}_{\mathrm{dR}}(V_\varphi^*(1)) \longrightarrow \mathrm{D}_{\mathrm{dR}}(L_p) = L_p$$

and

$$\langle\, ,\, \rangle : \mathrm{D}_{\mathrm{dR}}(V_\varphi^+) \times S_w(M)_{L_p}[\varphi^*] \longrightarrow L_p.$$

Define

$$\eta_\varphi \in \mathrm{D}_{\mathrm{dR}}(V_\varphi^+) \subseteq \mathrm{D}_{\mathrm{dR}}(V_\varphi)$$

to be the differential characterized by

$$\langle \omega_{\varphi^*}, \eta_\varphi \rangle = 1.$$

In the setting where the weight is $w = 1$, the filtration (3.1) admits a splitting. More precisely, if we denote V_φ^α, V_φ^β the eigenspaces for the action of Fr_p on V_φ, with eigenvalues α_φ, β_φ respectively, then

$$V_\varphi = V_\varphi^\alpha \oplus V_\varphi^\beta, \quad V_\varphi^+ = V_\varphi^\beta, \qquad V_\varphi^- = V_\varphi^\alpha$$

as $G_{\mathbb{Q}_p}$-modules. In this case, Ohta's work ([19]; cf. also [12, Sect. 2]) associates to the p-stabilized eigenform φ_α differential forms

$$\omega_{\varphi_\alpha} \in \mathrm{D_{dR}}(V_\varphi^\alpha), \quad \eta_{\varphi_\alpha} \in \mathrm{D_{dR}}(V_\varphi^\beta). \tag{3.3}$$

Ohta's work also yields a pairing

$$\mathrm{D_{dR}}(V_\varphi) \times \mathrm{D_{dR}}(V_{\varphi^*}) \longrightarrow \mathrm{D_{dR}}(L_p), \tag{3.4}$$

under which

$$\langle \omega_{\varphi_\alpha^*}, \eta_{\varphi_\alpha} \rangle = 1. \tag{3.5}$$

Attached to a weight 1 modular form φ there is an Artin representation

$$\rho_\varphi : G_{\mathbb{Q}} \to \mathrm{GL}(V_\varphi^\circ)$$

where V_φ° is an L-vector space equipped with a (non-canonical) isomorphism $j_\varphi : V_\varphi^\circ \otimes_L L_p \xrightarrow{\cong} V_\varphi$. The choice of j_φ thus induces an L-structure on V_φ by $V_\varphi^L := j_\varphi(V_\varphi^\circ)$.

Let v_φ^α be an L-basis of $V_\varphi^L \cap V_\varphi^\alpha$ and v_φ^β an L-basis of $V_\varphi^L \cap V_\varphi^\beta$. Let H be the number field cut out by ρ_φ, fix once and for all an embedding $H \subseteq \bar{\mathbb{Q}}_p$ and denote H_p the p-adic completion of H inside $\bar{\mathbb{Q}}_p$. Finally, define periods

$$\Omega_{\varphi_\alpha} \in H_p^{\mathrm{Fr}_p = 1/\alpha_\varphi}, \quad \Theta_{\varphi_\alpha} \in H_p^{\mathrm{Fr}_p = 1/\beta_\varphi}$$

characterized by

$$\Omega_{\varphi_\alpha} \otimes v_\varphi^\alpha = \omega_{\varphi_\alpha} \in \mathrm{D_{dR}}(V_g^\alpha), \quad \Theta_{\varphi_\alpha} \otimes v_\varphi^\beta = \eta_{\varphi_\alpha} \in \mathrm{D_{dR}}(V_\varphi^\beta). \tag{3.6}$$

Although the above periods depend on the choice of j_φ, the ratio

$$\mathcal{L}_{g_\alpha} := \frac{\Omega_{g_\alpha}}{\Theta_{g_\alpha}} \tag{3.7}$$

is readily seen to be independent of this choice; it only depends on the choice of basis $\{v_\varphi^\alpha, v_\varphi^\beta\}$ and is thus well defined up to scalars in $L^\times$.

Let $\Lambda := \mathbb{Z}_p[[\mathbb{Z}_p^\times]]$ be the Iwasawa algebra and $\mathcal{W} := \mathrm{Spf}(\Lambda)$ be the weight space. It is the rigid analytic space whose A-points for any $\mathbb{Z}_p$-algebra A are

$$\mathcal{W}(A) = \mathrm{Hom}_{\mathbb{Z}_p-\mathrm{alg}}(\Lambda, A) = \mathrm{Hom}_{\mathrm{cts}}(\mathbb{Z}_p^\times, A^\times).$$

Recall that $\mathbb{Z}$ can be seen as a dense subset of $\mathcal{W}(\mathbb{C}_p)$ identifying an integer $k \in \mathbb{Z}$ with the character $\nu_k : (x \mapsto x^k) \in \mathrm{Hom}_{\mathrm{cts}}(\mathbb{Z}_p^\times, \mathbb{C}_p^\times)$. More generally, given a Dirichlet character ϵ of p-power conductor, we denote $\nu_{k,\epsilon}$ the point $(x \mapsto \epsilon(x)x^k) \in \mathcal{W}(\mathbb{C}_p)$. A character $\nu \in \mathcal{W}$ is called *classical* if it is of the form $\nu = \nu_{k,\epsilon}$ with $k \geq 2$ and k is called its *weight*. A classical point is said to be *crystalline* if $\epsilon = 1$.

Let $N \geq 1$ be positive integer relatively prime to p and let

$$\boldsymbol{\varphi} = \sum a_n(\boldsymbol{\varphi})q^n \in \Lambda_{\boldsymbol{\varphi}}[[q]]$$

be a Hida family of tame level N and tame character $\chi : (\mathbb{Z}/N\mathbb{Z})^\times \to \bar{\mathbb{Q}}_p^\times$, where $\Lambda_{\boldsymbol{\varphi}}$ is a finite flat extension of Λ.

Let $\kappa : \mathcal{W}_{\boldsymbol{\varphi}} := \mathrm{Spf}(\Lambda_{\boldsymbol{\varphi}}) \to \mathcal{W}$ be the *weight map* induced by the inclusion $\Lambda \subseteq \Lambda_{\boldsymbol{\varphi}}$. A point $x \in \mathcal{W}_{\boldsymbol{\varphi}}$ is called *classical* (resp. *crystalline*) if $\kappa(x)$ is, and we denote $\mathcal{W}_{\boldsymbol{\varphi}}^{\mathrm{cl}}$ (resp. $\mathcal{W}_{\boldsymbol{\varphi}}^\circ$) the set of classical (resp. crystalline) points of $\mathcal{W}_{\boldsymbol{\varphi}}$. For each $x \in \mathcal{W}_{\boldsymbol{\varphi}}^{\mathrm{cl}}$ with $\kappa(x) = \nu_{k,\epsilon}$, the specialization

$$\boldsymbol{\varphi}_x := \sum_{n\geq 1} x(a_n(\boldsymbol{\varphi}))q^n$$

of $\boldsymbol{\varphi}$ at x is the q-expansion of a classical p-ordinary eigenform of weight k and character $\chi\epsilon$. If $x \in \mathcal{W}$ is a point of weight 1, meaning that $\kappa(x) = \nu_{1,\epsilon}$ for some Dirichlet character ϵ of p-power conductor, the specialization $\boldsymbol{\varphi}_x$ may be a classical modular form or not. (Later we shall regard the p-stabilizations of g and h as weight 1 specializations of a Hida family, and these are classical by construction).

Notice that the specializations of $\boldsymbol{\varphi}$ at crystalline points have all nebentype χ, that is unramified at p. If $x \in \mathcal{W}_{\boldsymbol{\varphi}}^\circ$ is a crystalline point of weight $k > 2$, then $\boldsymbol{\varphi}_x$ is the ordinary p-stabilization of a newform $\varphi_x \in S_k(N, \chi)$. If $k = 2$, then $\boldsymbol{\varphi}_x$ may be either old or new at p. We denote $\varphi_x := \boldsymbol{\varphi}_x$ if it is new, while if $\boldsymbol{\varphi}_x$ is old at p, we denote φ_x the newform whose p-stabilization is $\boldsymbol{\varphi}_x$.

Let $V_{\boldsymbol{\varphi}}$ be the Galois representation of $G_{\mathbb{Q}}$ attached to $\boldsymbol{\varphi}$, a $\Lambda_{\boldsymbol{\varphi}}$-module of rank 2. As a $G_{\mathbb{Q}_p}$-representation there is again a filtration

$$\mathrm{Fil}^2(V_{\boldsymbol{\varphi}}) = 0 \subseteq \mathrm{Fil}^1(V_{\boldsymbol{\varphi}}) \subseteq V_{\boldsymbol{\varphi}} = \mathrm{Fil}^0(V_{\boldsymbol{\varphi}})$$

where $V_{\boldsymbol{\varphi}}^+ := \mathrm{Fil}^1(V_{\boldsymbol{\varphi}})$ has rank one, and the quotient $V_{\boldsymbol{\varphi}}^- := V_{\boldsymbol{\varphi}}/\mathrm{Fil}^1(V_{\boldsymbol{\varphi}})$ is unramified. More precisely, $G_{\mathbb{Q}_p}$ acts on $V_{\boldsymbol{\varphi}}^-$ via the unramified character

$$\psi_{\varphi} : G_{\mathbb{Q}_p} \longrightarrow \Lambda_{\varphi}^{\times}$$

such that $\psi_{\varphi}(\mathrm{Fr}_p) = a_p(\boldsymbol{\varphi})$.

If x is classical and $L_p/\mathbb{Q}_p$ is a finite extension containing the values of x, there is an isomorphism of Galois representations $V_{\varphi} \otimes_{\Lambda_{\varphi},x} L_p \cong V_{\varphi_x}$. If x is crystalline and $k \neq 2$, then V_{φ_x} is crystalline at p because the underlying newform has level relatively prime to p.

Fix now three cuspidal Hida families

$$\mathbf{f} \in \Lambda_{\mathbf{f}}[[q]], \qquad \mathbf{g} \in \Lambda_{\mathbf{g}}[[q]], \qquad \mathbf{h} \in \Lambda_{\mathbf{h}}[[q]].$$

Write $\Lambda_{\mathbf{fgh}}=\Lambda_{\mathbf{f}}\hat{\otimes}\Lambda_{\mathbf{g}}\hat{\otimes}\Lambda_{\mathbf{h}}$ and $\mathcal{O}_{\mathbf{fgh}}:=\Lambda_{\mathbf{f}}[1/p]\hat{\otimes}\Lambda_{\mathbf{g}}[1/p]\hat{\otimes}\Lambda_{\mathbf{h}}[1/p]$. Let $\mathcal{W}_{\mathbf{fgh}} = \mathrm{Spf}(\Lambda_{\mathbf{fgh}}) = \mathcal{W}_{\mathbf{f}} \times \mathcal{W}_{\mathbf{g}} \times \mathcal{W}_{\mathbf{h}}$ denote the associated three-dimensional product of weight spaces. Denote $\mathcal{W}_{\mathbf{fgh}}^{\circ}$ the set of crystalline points $(x, y, z) \in \mathcal{W}_{\mathbf{f}}^{\circ} \times \mathcal{W}_{\mathbf{g}}^{\circ} \times \mathcal{W}_{\mathbf{h}}^{\circ}$ of weights $(k, \ell, m) \in \mathbb{Z}_{\geq 2} \times \mathbb{Z}_{\geq 1} \times \mathbb{Z}_{\geq 1}$ such that $k \equiv 2 \mod (p-1)$, $\ell \equiv m \equiv 1 \mod 2(p-1)$.

Set

$$\mathbb{V} := V_{\mathbf{f}} \otimes V_{\mathbf{g}} \otimes V_{\mathbf{h}} \otimes_{\mathcal{O}_{\mathbf{fgh}}} \Xi,$$

where $\Xi : G_{\mathbb{Q}_p} \longrightarrow \mathcal{O}_{\mathbf{fgh}}^{\times}$ is the character defined in [5, Sect. 4.6.2] and is characterized by the property that, if $\underline{x} = (k, \ell, m) \in \mathcal{W}_{\mathbf{fgh}}^{\circ}$ is a triple of crystalline points and $\rho_{\underline{x}} := \nu_x \otimes \nu_y \otimes \nu_z : \mathcal{O}_{\mathbf{fgh}}^{\times} \longrightarrow L_p$ is the corresponding specialisation map, then $\rho_{\underline{x}} \circ \Xi = \chi_{\mathrm{cyc}}^{(4-k-\ell-m)/2}$. We define a filtration on $\mathbb{V}$ by

$$\mathrm{Fil}^q(\mathbb{V}) := [\oplus_{i+j+t=q} \mathrm{Fil}^i(V_{\mathbf{f}})\hat{\otimes} \mathrm{Fil}^j(V_{\mathbf{g}})\hat{\otimes} \mathrm{Fil}^t(V_{\mathbf{h}})] \otimes_{\mathcal{O}_{\mathbf{fgh}}} \Xi,$$

Since $\mathbb{V}/\mathrm{Fil}^2(\mathbb{V})$ has no $G_{\mathbb{Q}_p}$-invariants by [18, Lemma 5.3], the natural inclusion

$$\begin{aligned}\mathrm{Fil}^2(\mathbb{V}) = [V_{\mathbf{f}} \otimes V_{\mathbf{g}}^+ \otimes V_{\mathbf{h}}^+ \otimes_{\mathcal{O}_{\mathbf{fgh}}} \Xi] \oplus [V_{\mathbf{f}}^+ \otimes V_{\mathbf{g}} \otimes V_{\mathbf{h}}^+ \otimes_{\mathcal{O}_{\mathbf{fgh}}} \Xi] \\ \oplus [V_{\mathbf{f}}^+ \otimes V_{\mathbf{g}}^+ \otimes V_{\mathbf{h}} \otimes_{\mathcal{O}_{\mathbf{fgh}}} \Xi] \hookrightarrow \mathbb{V}\end{aligned}$$

induces an injective morphism in cohomology

$$\mathrm{H}^1(\mathbb{Q}_p, \mathrm{Fil}^2(\mathbb{V})) \longrightarrow \mathrm{H}^1(\mathbb{Q}_p, \mathbb{V}),$$

and henceforth we shall identify $\mathrm{H}^1(\mathbb{Q}_p, \mathrm{Fil}^2(\mathbb{V}))$ with its image in $\mathrm{H}^1(\mathbb{Q}_p, \mathbb{V})$. Defining

$$\begin{aligned}&\mathbb{V}^f := V_{\mathbf{f}}^- \otimes V_{\mathbf{g}}^+ \otimes V_{\mathbf{h}}^+ \otimes_{\mathcal{O}_{\mathbf{fgh}}} \Xi, \\ &\mathbb{V}^g := V_{\mathbf{f}}^+ \otimes V_{\mathbf{g}}^- \otimes V_{\mathbf{h}}^+ \otimes_{\mathcal{O}_{\mathbf{fgh}}} \Xi, \quad \mathbb{V}^h := V_{\mathbf{f}}^+ \otimes V_{\mathbf{g}}^+ \otimes V_{\mathbf{h}}^- \otimes_{\mathcal{O}_{\mathbf{fgh}}} \Xi, \end{aligned} \tag{3.8}$$

it follows that $\mathrm{H}^1(\mathbb{Q}_p, \mathrm{Fil}^2(\mathbb{V})/\mathrm{Fil}^3(\mathbb{V}))$ decomposes as

$$\mathrm{H}^1(\mathbb{Q}_p, \mathbb{V}^f) \oplus \mathrm{H}^1(\mathbb{Q}_p, \mathbb{V}^g) \oplus \mathrm{H}^1(\mathbb{Q}_p, \mathbb{V}^h).$$

For $\varphi \in \{f, g, h\}$, we denote

$$\pi_\varphi : \mathrm{H}^1(\mathbb{Q}_p, \mathrm{Fil}^2(\mathbb{V})) \longrightarrow \mathrm{H}^1(\mathbb{Q}_p, \mathbb{V}^\varphi)$$

the natural projection map.

3.2 Selmer Groups

Let W be a $L_p[G_\mathbb{Q}]$-module. Given a prime number ℓ, define as customary

$$\mathrm{H}^1_f(\mathbb{Q}_\ell, W) := \begin{cases} \mathrm{H}^1_{\mathrm{ur}}(\mathbb{Q}_\ell, W) := \mathrm{H}^1(\mathbb{Q}_\ell^{\mathrm{ur}}/\mathbb{Q}_\ell, W^{I_\ell}) & \ell \neq p \\ \ker\left(\mathrm{H}^1(\mathbb{Q}_p, W) \to \mathrm{H}^1(\mathbb{Q}_p, W \otimes_{\mathbb{Q}_p} \mathrm{B}_{\mathrm{cris}})\right) & \ell = p, \end{cases}$$

and

$$\mathrm{H}^1_s(\mathbb{Q}_\ell, W) := \mathrm{H}^1(\mathbb{Q}_\ell, W)/\mathrm{H}^1_f(\mathbb{Q}_\ell, W).$$

The Bloch–Kato Selmer group of W is

$$\mathrm{Sel}_p(W) := \{x \in \mathrm{H}^1(\mathbb{Q}, W) \mid \mathrm{res}_\ell(x) \in \mathrm{H}^1_f(\mathbb{Q}_\ell, W) \text{ for all } \ell\},$$

where $\mathrm{res}_\ell \colon \mathrm{H}^1(\mathbb{Q}, W) \to \mathrm{H}^1(\mathbb{Q}_\ell, W)$ denotes the restriction map in Galois cohomology.

For each prime ℓ, we denote by ∂_ℓ the composition

$$\partial_\ell : \mathrm{H}^1(\mathbb{Q}, W) \xrightarrow{\mathrm{res}_\ell} \mathrm{H}^1(\mathbb{Q}_\ell, W) \longrightarrow \mathrm{H}^1_s(\mathbb{Q}_\ell, W).$$

The relaxed Selmer group is defined as

$$\mathrm{Sel}_{(p)}(W) := \{x \in \mathrm{H}^1(\mathbb{Q}, W) \mid \mathrm{res}_\ell(x) \in \mathrm{H}^1_f(\mathbb{Q}_\ell, W) \text{ for all } \ell \neq p\} \supseteq \mathrm{Sel}_p(\mathbb{Q}, W).$$

Let now (f, g, h) be the triple of eigenforms of weights $(2, 1, 1)$ introduced at the beginning of the article. Recall Assumption 1.1 imposed on $V := V_f \otimes V_g \otimes V_h$ in the introduction, which implies

$$\mathrm{Sel}_p(V) = 0. \tag{3.9}$$

Indeed, as discussed in detail in Sect. 5.1, the representation V decomposes as a direct sum

$$V = \left[V_f \otimes V_{\psi_1}\right] \oplus \left[V_f \otimes V_{\psi_2}\right],$$

where, for $i = 1, 2$, the $G_{\mathbb{Q}}$-representation V_{ψ_i} is induced by a ring class character ψ_i of the imaginary quadratic field K. This decomposition induces factorisations

$$L(E, \rho, s) = L(E/K, \psi_1, s) \cdot L(E/K, \psi_2, s), \quad \mathrm{Sel}_p(V) = \mathrm{Sel}_p(E \otimes \psi_1) \oplus \mathrm{Sel}_p(E \otimes \psi_2).$$

Combining the factorisation of the complex L-function with Assumption 1.1 we obtain that $L(E/K, \psi_i, 1) \neq 0$ for $i = 1, 2$, which, by [10, Theorem B], implies that $\mathrm{Sel}_p(E \otimes \psi_1) = \mathrm{Sel}_p(E \otimes \psi_2) = 0$.

Lemma 3.1 *There is an isomorphism*

$$\partial_p : \mathrm{Sel}_{(p)}(V) \xrightarrow{\cong} \mathrm{H}^1_s(\mathbb{Q}_p, V).$$

Proof The proof is exactly the same as in [9, Lemma 2.5]. Note that in loc. cit. it is assumed throughout that p does not divide N_f, but this assumption does not affect the argument of the proof. □

Since g and h are theta series of a character of the imaginary quadratic field K and p is assumed to remain inert in K, the characteristic polynomials for the action of Fr_p on V_g and V_h have null trace. Hence $\alpha_g = -\beta_g$ and $\alpha_h = -\beta_h$. Moreover, since the nebentype of h is the inverse of that of g, we have

$$(\alpha_h, \beta_h) = (1/\alpha_g, -1/\alpha_g) \;\text{ or }\; (\alpha_h, \beta_h) = (-1/\alpha_g, 1/\alpha_g).$$

Using the notation of Sect. 3.1, denote $\{v_g^{\alpha}, v_g^{\beta}\}$ and $\{v_h^{\alpha}, v_h^{\beta}\}$ a pair of L-bases for V_g and V_h consisting of eigenvectors with eigenvalue $\alpha_g, \beta_g, \alpha_h, \beta_h$ respectively. Denote $V_g^{\alpha}, V_g^{\beta}$, etc. the corresponding eigenspaces. Set $V_{gh}^{\alpha,\alpha} := V_g^{\alpha} \otimes V_h^{\alpha}$ and $V^{\alpha,\alpha} := V_f \otimes V_{gh}^{\alpha,\alpha}$, and likewise for the remaining pairs of eigenvalues. There is a decomposition of $L_p[G_{\mathbb{Q}_p}]$-modules

$$V := V_{fgh} = V^{\alpha,\alpha} \oplus V^{\alpha,\beta} \oplus V^{\beta,\alpha} \oplus V^{\beta,\beta}. \tag{3.10}$$

Lemma 3.2 *Let $b, c \in \{\alpha, \beta\}$. Then*

(i) the Bloch–Kato logarithm gives an isomorphism

$$\log_{bc} : \mathrm{H}^1_f(\mathbb{Q}_p, V^{bc}) = \mathrm{H}^1_f(\mathbb{Q}_p, V_f^{+} \otimes V_{gh}^{bc}) \xrightarrow{\cong} L_p;$$

(ii) the Bloch–Kato dual exponential gives an isomorphism

$$\exp^*_{bc} : \mathrm{H}^1_s(\mathbb{Q}_p, V^{bc}) = \mathrm{H}^1_s(\mathbb{Q}_p, V_f^{-} \otimes V_{gh}^{bc}) \xrightarrow{\cong} L_p; .$$

Proof Let $W := V_{gh}^{bc}$. By [15, Lemma 5.1]

$$\mathrm{H}^1_f(\mathbb{Q}_p, V^{bc}) = \begin{cases} \mathrm{H}^1_f(\mathbb{Q}_p, V_f^+ \otimes V_{gh}^{bc}) & \text{if} b \cdot c \cdot a = +1 \\ \mathrm{H}^1(\mathbb{Q}_p, V_f^+ \otimes V_{gh}^{bc}) & \text{if} b \cdot c \cdot a = -1. \end{cases} \tag{3.11}$$

Moreover, in the case in which $b \cdot c \cdot a = -1$, we have

$$\mathrm{H}^1(\mathbb{Q}_p, V_f^+ \otimes V_{gh}^{bc}) = \mathrm{H}^1_f(\mathbb{Q}_p, V_f^+ \otimes V_{gh}^{bc})$$

(see [18, Example 1.4]). Since $V_f^+ \cong \mathbb{Q}_p(1)(\chi_E)$, $i)$ follows from the fact that

$$V_f^+ \otimes V_{gh}^{bc} = L_p(\zeta\chi_{\mathrm{cyc}})$$

where $\zeta : G_{\mathbb{Q}_p} \longrightarrow \mathbb{Q}_p^\times$ is the unramified character given by $\zeta(\mathrm{Fr}_p) = b \cdot c \cdot a$ (for more details, see e.g. [18, Example 1.4]). The statement on the singular quotients is proven similarly using (3.11) and the fact that

$$V_f^- \otimes V_{gh}^{bc} = L_p(\zeta).$$

□

Set $L_p^4 = L_p \oplus L_p \oplus L_p \oplus L_p$. We obtain from the previous lemma isomorphisms

$$\log : \mathrm{H}^1_f(\mathbb{Q}_p, V) \longrightarrow L_p^4, \qquad \exp^* : \mathrm{H}^1_s(\mathbb{Q}_p, V) \longrightarrow L_p^4.$$

where $\log = \log_{\alpha\alpha} \oplus \log_{\alpha\beta} \oplus \log_{\beta\alpha} \oplus \log_{\beta\beta}$ and $\exp^* = \exp^*_{\alpha\alpha} \oplus \exp^*_{\alpha\beta} \oplus \exp^*_{\beta\alpha} \oplus \exp^*_{\beta\beta}$.

Combining Lemma 3.1 with Lemma 3.2 we obtain the following result on the L_p-structure of the relaxed Selmer group attached to V.

Corollary 3.3 *There are isomorphisms*

$$\mathrm{Sel}_{(p)}(V) \xrightarrow{\partial_p} \mathrm{H}^1_s(\mathbb{Q}_p, V) \xrightarrow{\exp^*} L_p^4.$$

We now turn to pin down a particular basis of $\mathrm{Sel}_{(p)}(V)$ which is canonical up to multiplication by elements in $L^\times$. In order to do this, we need a more precise description of the finite and singular parts of $\mathrm{H}^1(\mathbb{Q}_p, V^{cb})$ in terms of $\mathrm{H}^1(K_p, V_f)$. Note that the latter space is equipped with a natural action of $\mathrm{Gal}(K_p/\mathbb{Q}_p)$.

In what follows, for any $\mathrm{Gal}(K_p/\mathbb{Q}_p)$-module M we denote $M^\pm$ the subspace of M on which Fr_p acts as multiplication by ± 1. Note that $\alpha_g\alpha_h, \alpha_g\beta_h, \beta_g\alpha_h, \beta_g\beta_h \in \{\pm 1\}$ and thus it makes sense to consider $M^{\alpha_g\alpha_h}$, etc.

Lemma 3.4 *Let $b, c \in \{\alpha, \beta\}$. There are canonical isomorphisms of L_p-vector spaces*

$$\mathrm{H}^1_f(\mathbb{Q}_p, V_f \otimes V_{gh}^{b,c}) \cong \mathrm{H}^1_f(K_p, V_f)^{bc} \otimes V_{gh}^{b,c}$$

and

$$\mathrm{H}^1_s(\mathbb{Q}_p, V_f \otimes V_{gh}^{b,c}) \cong \mathrm{H}^1_s(K_p, V_f)^{bc} \otimes V_{gh}^{b,c}.$$

Proof Let χ be the quadratic character of $G_{\mathbb{Q}_p}$ with $\chi(\mathrm{Fr}_p) = -1$. We describe the isomorphisms in the case in which E has split multiplicative reduction at p and $b \cdot c = -1$. In this setting,

$$\begin{aligned}
\mathrm{H}^1_f(\mathbb{Q}_p, V^{b,c}) &= \mathrm{H}^1_f(\mathbb{Q}_p, V_f^+ \otimes V_{gh}^{b,c}) && \text{by Lemma 3.2}\\
&= \mathrm{H}^1_f(\mathbb{Q}_p, \mathbb{Q}_p(1) \otimes L_p(\chi)) && \text{by Lemma 3.1 (ii)}\\
&= \mathrm{H}^1_f(K_p, \mathbb{Q}_p(1) \otimes L_p(\chi))^{G_{\mathbb{Q}_p}} && \text{by the inflation-restriction exact sequence}\\
&= (\mathrm{H}^1_f(K_p, \mathbb{Q}_p(1)) \otimes L_p(\chi))^{G_{\mathbb{Q}_p}} && \text{because } \chi_{|G_{K_p}} = 1\\
&= (\mathrm{H}^1_f(K_p, V_f) \otimes L_p(\chi))^{G_{\mathbb{Q}_p}} && \text{by 2.12.}
\end{aligned}$$

The claim follows, because $V_{gh}^{b,c} = L_p(\chi)$ as $G_{\mathbb{Q}_p}$-modules and the subspace of $G_{\mathbb{Q}_p}$-invariants of $\mathrm{H}^1_f(K_p, V_f) \otimes V_{gh}^{b,c}$ is $\mathrm{H}^1_f(K_p, V_f)^- \otimes V_{gh}^{b,c}$. Using similar computations as above, one also verifies that

$$\begin{aligned}
&\mathrm{H}^1_s(\mathbb{Q}_p, V^{b,c}) \cong (\mathrm{H}^1(K_p, V_f) \otimes V_{gh}^{b,c})^{G_{\mathbb{Q}_p}} / (\mathrm{H}^1_f(K_p, V_f) \otimes V_{gh}^{b,c})^{G_{\mathbb{Q}_p}}\\
&\cong (\mathrm{H}^1_s(K_p, V_f) \otimes V_{gh}^{b,c})^{G_{\mathbb{Q}_p}}.
\end{aligned}$$

The remaining cases are proven similarly. □

Remark 3.5 The logarithm maps of Lemma 3.2 are twisted versions of (2.13). More precisely, $\log_{\alpha\beta}$ may be accordingly recast as

$$\begin{array}{ccc}
H^1_f(\mathbb{Q}_p, V_f \otimes V_{gh}^{\alpha,\beta}) = E(K_p)^{\beta\alpha} \otimes V_{gh}^{\alpha,\beta} & \longrightarrow & K_p(\chi_E)^{\beta\alpha} \otimes K_p^{\alpha\beta} \otimes L_p \cong L_p \\
x \otimes v_g^\alpha v_h^\beta & \longmapsto & \log_E(x) \cdot \langle v_g^\alpha v_h^\beta, \eta_{g_\alpha} \omega_{h_\alpha} \rangle.
\end{array} \tag{3.12}$$

Using Proposition 2.6 we can define the following generators k^+ and k^- of $K_p(\chi_E)^+$ and $K_p(\chi_E)^-$ respectively:

$$k^+ := \begin{cases} 1 & \text{if } a = +1; \\ z & \text{if } a = -1 \end{cases} \quad \text{and } k^- := \begin{cases} z & \text{if } a = +1; \\ 1 & \text{if } a = -1. \end{cases}$$

Using (3.5) and (3.6), the latter pairing is

$$\langle v_g^\alpha v_h^\beta, \eta_{g_\alpha} \omega_{h_\alpha} \rangle = \frac{1}{\Omega_{g_\alpha} \Theta_{h_\alpha}} \langle \omega_{g_\alpha} \eta_{h_\alpha}, \eta_{g_\alpha} \omega_{h_\alpha} \rangle = \frac{1}{\Omega_{g_\alpha} \Theta_{h_\alpha}} \otimes 1 \in K_p^{\alpha\beta} \otimes L_p,$$

and the right-most isomorphism of (3.12) is given by

$$\begin{array}{ccccc} K_p(\chi_E)^{\beta\alpha}\otimes K_p^{\alpha\beta} & \xrightarrow{\cong} & K_p(\chi_E)^+ & \xrightarrow{\cong} & \mathbb{Q}_p \\ x\otimes y & \longmapsto & xy & & \\ & & k^+ & \longmapsto & 1. \end{array}$$

Analogously, the dual exponential maps of Lemma 3.2 are twisted versions of (2.14), and $\exp^*_{\alpha\beta}$ may be recast under the identification provided by Lemma 3.4 as

$$\begin{array}{rcl} \exp^*_{\alpha\beta}:\mathrm{H}^1_s(K_p,V_f)^{\beta\alpha}\otimes V_{gh}^{\alpha,\beta} & \longrightarrow & L_p \\ x\otimes v_g^\alpha v_h^\beta & \longmapsto & \exp^*(x)\cdot\langle v_g^\alpha v_h^\beta,\eta_{g_\alpha}\omega_{h_\alpha}\rangle. \end{array}$$

We will still denote by

$$\exp^*_\pm:\mathrm{H}^1_s(K_p,V_f)^\pm\longrightarrow\mathbb{Q}_p \tag{3.13}$$

the composition of (2.14) with the isomorphism $K_p(\chi_E)^\pm\cong\mathbb{Q}_p$ sending $k^\pm$ to 1.

Corollary 3.6 *Define* $X_\pm\in\mathrm{H}^1_s(K_p,V_f)^\pm$ *such that* $\exp^*_\pm(X_\pm)=1$. *The p-relaxed Selmer group* $\mathrm{Sel}_{(p)}(V)$ *admits a basis*

$$\{\xi^{\alpha\alpha},\xi^{\alpha\beta},\xi^{\beta\alpha},\xi^{\beta\beta}\}$$

characterized as

$$\begin{array}{ll} \partial_p\xi^{\alpha\alpha}=X_{\alpha\alpha}\otimes v_g^\alpha\otimes v_h^\alpha, & \partial_p\xi^{\alpha\beta}=X_{\alpha\beta}\otimes v_g^\alpha\otimes v_h^\beta, \\ \partial_p\xi^{\beta\alpha}=X_{\beta\alpha}\otimes v_g^\beta\otimes v_h^\alpha, & \partial_p\xi^{\beta\beta}=X_{\beta\beta}\otimes v_g^\beta\otimes v_h^\beta. \end{array}$$

Proof By Corollary 3.3 and using the decomposition (3.10), there are isomorphisms

$$\begin{aligned} \mathrm{Sel}_{(p)}(V)\xrightarrow{\partial_p}\mathrm{H}^1_s(\mathbb{Q}_p,V)\cong\mathrm{H}^1_s(\mathbb{Q}_p,V^{\alpha,\alpha})\oplus\mathrm{H}^1_s(\mathbb{Q}_p,V^{\alpha,\beta}) \\ \oplus\mathrm{H}^1_s(\mathbb{Q}_p,V^{\beta,\alpha})\oplus\mathrm{H}^1_s(\mathbb{Q}_p,V^{\beta,\beta}). \end{aligned} \tag{3.14}$$

By Lemma 3.4, each of the four components in (3.14) is isomorphic to

$$\begin{aligned} &\mathrm{H}^1_s(K_p,V_f)^{\beta\beta}\otimes V_{gh}^{\alpha,\alpha},\ \mathrm{H}^1_s(K_p,V_f)^{\beta,\alpha}\otimes V_{gh}^{\alpha,\beta},\ \mathrm{H}^1_s(K_p,V_f)^{\alpha,\beta} \\ &\otimes V_{gh}^{\beta,\alpha},\ \mathrm{H}^1_s(K_p,V_f)^{\alpha,\alpha}\otimes V_{gh}^{\beta\beta}, \end{aligned}$$

respectively. These are in turn isomorphic to $V_{gh}^{\alpha,\alpha}\oplus V_{gh}^{\alpha,\beta}\oplus V_{gh}^{\beta,\alpha}\oplus V_{gh}^{\beta,\beta}$ via $\exp^*_{\alpha\alpha}\oplus\exp^*_{\alpha\beta}\oplus\exp^*_{\beta\alpha}\oplus\exp^*_{\beta\beta}$. □

4 A Special Value Formula for the Triple Product p-Adic L-Function in Rank 0

The aim of this section is to describe the p-adic L-value $I_p(f, g_\alpha, h_\alpha)$ in terms of the basis $\{\xi^{\alpha\alpha}, \xi^{\alpha\beta}, \xi^{\beta\alpha}, \xi^{\beta\beta}\}$ of $\mathrm{Sel}_{(p)}(V)$ appearing in Corollary 3.6. This section lies within the framework of the *exceptional setting* of [5] and we recall here the notation and the main results from loc. cit. that we shall use.

4.1 *The Triple Product p-Adic L-Function*

If $\varphi = \sum a_n(\varphi)q^n$ is a modular form of weight w, level M and nebentype character χ_φ and $p \nmid M$, the Hecke polynomial at p of φ is

$$x^2 - a_p(\varphi)x + \chi_\varphi(p)p^{w-1} = (x - \alpha_\varphi)(x - \beta_\varphi),$$

where we label the eigenvalues so that $\mathrm{ord}_p(\alpha_\varphi) \leq \mathrm{ord}_p(\beta_\varphi)$. Recall that if φ is ordinary at p, then α_φ is a p-adic unit. If $p \mid M$ we have $\alpha_\varphi = a_p(\varphi)$.

Since g has weight $w = 1$, both α_g and β_g are p-adic units. Recall that we are assuming that $\alpha_g \neq \beta_g$, thus g has two different ordinary p-stabilisations. We denote g_α and g_β the stabilisation satisfying $U_p g_\alpha = \alpha_g g_\alpha$ and $U_p g_\beta = \beta_g g_\beta$ respectively. The same holds for h, and we denote similarly its p-stabilisations. Since E has multiplicative reduction at p, the level of the newform is divisible by p and it is p-stabilised, meaning that $f_\alpha = f$.

Let $\mathbf{f} = \sum_{n\geq 1} a_n(\mathbf{f}) \in \Lambda_{\mathbf{f}}[[q]]$, $\mathbf{g} \in \Lambda_{\mathbf{g}}[[q]]$, $\mathbf{h} \in \Lambda_{\mathbf{h}}[[q]]$ be the Hida families passing through f, g_α, h_α respectively, where $\Lambda_{\mathbf{f}}$, $\Lambda_{\mathbf{g}}$ and $\Lambda_{\mathbf{h}}$ are finite flat extensions of the Iwasawa algebra Λ. These Hida families have tame level N_f/p, N_g, N_h and tame character $1, \chi, \bar{\chi}$ respectively.

A *test vector* for $(\mathbf{f}, \mathbf{g}, \mathbf{h})$ is a triple of Hida families $(\breve{\mathbf{f}}, \breve{\mathbf{g}}, \breve{\mathbf{h}})$ such that $\breve{\mathbf{f}}$ is of the form $\sum \lambda_d \mathbf{f}(q^d) \in \Lambda_{\mathbf{f}}[[q]]$ with $\lambda_d \in \Lambda_{\mathbf{f}}$, where d runs over the divisors of N/N_f, and similarly for $\breve{\mathbf{g}}$ and $\breve{\mathbf{h}}$ with $d \mid N/pN_g$ and $d \mid N/pN_h$ respectively.

Recall from Sect. 3.1 the set $\mathcal{W}^\circ_{\mathbf{fgh}}$ of crystalline points. Denote $\mathcal{W}^f_{\mathbf{fgh}} := \{(k, \ell, m) \in \mathcal{W}^\circ_{\mathbf{fgh}} \mid k \geq \ell + m\}$ and define $\mathcal{W}^g_{\mathbf{fgh}}$ and $\mathcal{W}^h_{\mathbf{fgh}}$ analogously. Triplets of points in $\mathcal{W}^f_{\mathbf{fgh}} \cup \mathcal{W}^g_{\mathbf{fgh}} \cup \mathcal{W}^h_{\mathbf{fgh}}$ are called *unbalanced*, and points in the complement $\mathcal{W}^{\mathrm{bal}}_{\mathbf{fgh}} := \mathcal{W}^\circ_{\mathbf{fgh}} \setminus (\mathcal{W}^f_{\mathbf{fgh}} \cup \mathcal{W}^g_{\mathbf{fgh}} \cup \mathcal{W}^h_{\mathbf{fgh}})$ are called *balanced*.

Let

$$\mathcal{L}^f_p(\breve{\mathbf{f}}, \breve{\mathbf{g}}, \breve{\mathbf{h}}),\ \mathcal{L}^g_p(\breve{\mathbf{f}}, \breve{\mathbf{g}}, \breve{\mathbf{h}}),\ \mathcal{L}^h_p(\breve{\mathbf{f}}, \breve{\mathbf{g}}, \breve{\mathbf{h}}) : \mathcal{W}_{\mathbf{f}} \times \mathcal{W}_{\mathbf{g}} \times \mathcal{W}_{\mathbf{h}} \longrightarrow \mathbb{C}_p$$

be the triple product p-adic L-functions attached to $(\breve{\mathbf{f}}, \breve{\mathbf{g}}, \breve{\mathbf{h}})$ constructed in [6].

4.2 Exceptional Cases and Improved Euler Systems

As introduced in (3.1), let

$$\mathbb{V} = V_{\mathbf{f}} \otimes V_{\mathbf{g}} \otimes V_{\mathbf{h}} \otimes_{\mathcal{O}_{\mathbf{fgh}}} \Xi$$

and let

$$\boldsymbol{\kappa} := \boldsymbol{\kappa}(\mathbf{f}, \mathbf{g}, \mathbf{h}) \in \mathrm{H}^1(\mathbb{Q}, \mathbb{V})$$

be the diagonal class constructed in [5, 18]. By [5, Corollary 4.7.1] and [18, Proposition 5.8], the local class

$$\boldsymbol{\kappa}_p := \mathrm{res}_p(\boldsymbol{\kappa}) \in \mathrm{H}^1(\mathbb{Q}_p, \mathbb{V})$$

belongs to $\mathrm{H}^1(\mathbb{Q}_p, \mathrm{Fil}^2(\mathbb{V}))$. Moreover, for $\varphi \in \{f, g, h\}$, the p-adic L-function $\mathcal{L}_p^{\varphi}(\breve{\mathbf{f}}, \breve{\mathbf{g}}, \breve{\mathbf{h}})$ can be recast as the image of $\boldsymbol{\kappa}_p$ under the idoneous Perrin-Riou's Λ-adic logarithm: more precisely, [5, Theorem A], [18, Theorem 5.1] assert that

$$\mathcal{L}_p^{\varphi}(\breve{\mathbf{f}}, \breve{\mathbf{g}}, \breve{\mathbf{h}}) = \mathcal{L}_{\varphi}(\mathrm{res}_p(\boldsymbol{\kappa})), \tag{4.1}$$

where $\mathcal{L}_{\varphi} : \mathrm{H}^1(\mathbb{Q}_p, \mathrm{Fil}^2(\mathbb{V})) \longrightarrow \mathcal{O}_{\mathbf{fgh}}$ is the homomorphism described in [5, Proposition 7.3], [18, Proposition 5.7].

Let $\mathcal{S}$ denote the surface cut out by the equation $k = 2 + \ell - m$ in $\mathcal{W}_{\mathbf{f}} \times \mathcal{W}_{\mathbf{g}} \times \mathcal{W}_{\mathbf{h}}$ and let $\mathcal{F}_{\mathcal{S}}$ be the fraction field of the ring of Iwasawa functions on $\mathcal{S}$.

Define the two-variable meromorphic Iwasawa function

$$\boldsymbol{\mathcal{E}}_g := \mathcal{E}_g(\mathbf{f}, \mathbf{g}, \mathbf{h}) := 1 - \frac{a_p(\mathbf{g}_\ell)}{\chi(p) a_p(\mathbf{f}_{\ell-m+2}) a_p(\mathbf{h}_m)} \in \mathcal{F}_{\mathcal{S}}.$$

Recall from the introduction that we have $p \mid\mid N_f$, $p \nmid N_g N_h$, and $\alpha_g \alpha_h = -a$. More explicitly, we have

$$\beta_g = -\alpha_g, \qquad (\alpha_h, \beta_h) = \begin{cases} (-1/\alpha_g, 1/\alpha_g) & \text{if } a = +1; \\ (1/\alpha_g, -1/\alpha_g) & \text{if } a = -1. \end{cases} \tag{4.2}$$

Hence

$$\boldsymbol{\mathcal{E}}_g(2, 1, 1) = 1 - \frac{\alpha_g \beta_h}{a} = 0.$$

This fact forces the vanishing of the class $\boldsymbol{\kappa}(2, 1, 1)$, as explained in [5] (cf. also Prop. 4.1 below).

Let $\rho_{\mathcal{S}} : \mathcal{F}_{\mathbf{fgh}} \longrightarrow \mathcal{F}_{\mathcal{S}}$ be the map taking a function $F(k, \ell, m)$ to its restriction $F(2 + \ell - m, \ell, 1)$ to the plane $\mathcal{S}$. Denote $\mathbb{V}_{|\mathcal{S}} := \mathbb{V} \otimes_{\mathcal{F}_{\mathbf{fgh}}, \rho_{\mathcal{S}}} \mathcal{F}_{\mathcal{S}}$ and let

$$\boldsymbol{\kappa}_{|\mathcal{S}} := \rho_{\mathcal{S},\star}(\boldsymbol{\kappa}) \in \mathrm{H}^1(\mathbb{Q}, \mathbb{V}_{|\mathcal{S}})$$

denote the restriction of $\boldsymbol{\kappa}$ to $\mathcal{S}$.

Proposition 4.1 *There exists a global cohomology class*

$$\boldsymbol{\kappa}_g^* \in \mathrm{H}^1(\mathbb{Q}, \mathbb{V}_{|\mathcal{S}})$$

satisfying, for each $\ell, m \in \mathbb{Z}_{\geq 1}$*:*

$$\boldsymbol{\kappa}(2+\ell-m, \ell, m) = \boldsymbol{\mathcal{E}}_g(2+\ell-m, \ell, m) \cdot \boldsymbol{\kappa}_g^*(2+\ell-m, \ell, m). \tag{4.3}$$

Proof This is shown in [5, Sect. 9.3] □

Let $\mathcal{C} \subset \mathcal{S}$ denote the curve given on which the set of points of weights $(\ell+1, \ell, 1)$ for $\ell \in \mathbb{Z}_{\geq 1}$ is dense.

Proposition 4.2 *There exists an analytic function* $\mathcal{L}_p^{f,*}$ *on* $\mathcal{C}$ *satisfying, for each* $\ell \in \mathbb{Z}_{\geq 1}$*:*

$$\mathcal{L}_p^{f,*}(\ell) = \frac{\langle w_N(f_{\ell+1})^{(p)}, h\mathbf{g}_\ell \rangle}{\langle w_N(f_{\ell+1})^{(p)}, w_N(f_{\ell+1})^{(p)} \rangle}$$

and

$$\mathcal{L}_p^f(\breve{\mathbf{f}}, \breve{\mathbf{g}}, \breve{\mathbf{h}})(\ell+1, \ell, 1) = \left(1 - \frac{a_p(\mathbf{g}_\ell)\alpha_h}{a_p(\mathbf{f}_{\ell+1})}\right) \boldsymbol{\mathcal{E}}_g(\ell+1, \ell, 1) \mathcal{L}_p^{f,*}(\ell) \mod L^\times.$$

Moreover,

$$\mathcal{L}_p^{f,*}(1) = \frac{1}{2(1-1/p)} \exp_{\beta\beta}^*(\pi_{\beta\beta} \partial_p \boldsymbol{\kappa}_g^*(2,1,1)) = \frac{1}{2(1-1/p)} \exp_{\beta\beta}^*(\partial_p \boldsymbol{\kappa}_g^*(2,1,1)). \tag{4.4}$$

Proof The first part is proved in [5, Lemma 9.8]. The second part follows from Lemma 9.8 of loc. cit. □

Remark 4.3 By (4.2) the factor $1 - \dfrac{a_p(\mathbf{g}_\ell)\alpha_h}{a_p(\mathbf{f}_{\ell+1})}$ does not vanish in a neighborhood of $(2,1,1)$ in $\mathcal{C}$.

4.3 $I_p(f, g_\alpha, h_\alpha)$ in Terms of the Basis

In this section we finally obtain, in Theorem 4.7, the formula for $I_p(f, g_\alpha, h_\alpha)$. Recall from the introduction the Galois representation

$$V = V_f \otimes V_g \otimes V_h.$$

Note that the character Ξ introduced in Sect. 3.1 specializes at $(2, 1, 1)$ to the trivial character, and thus the specialization of the improved Λ-adic cohomology class of Prop. 4.1 yields a global cohomology class

$$\boldsymbol{\kappa}_g^*(2, 1, 1) \in \mathrm{Sel}_{(p)}(V).$$

Recall the periods

$$\Omega_{g_\alpha} \in K_p^{1/\alpha_g}, \quad \Theta_{g_\alpha} \in K_p^{1/\beta_g}, \quad \mathcal{L}_{g_\alpha} := \frac{\Omega_{g_\alpha}}{\Theta_{g_\alpha}} \in K_p^{\beta_g/\alpha_g} \tag{4.5}$$

introduced in Sect. 3.1.

Proposition 4.4 *We have*

$$\boldsymbol{\kappa}_g^*(2, 1, 1) = \Theta_{g_\alpha}\Theta_{h_\alpha} \frac{2(1 - 1/p)\sqrt{c}\sqrt{L(E \otimes \rho, 1)}}{\pi^2\langle f, f\rangle} \cdot \xi^{\beta\beta},$$

where $c \in L^\times$ is the product of the local terms appearing in [8, Proposition 2.1 (iii)].

Proof According to [5, Sect. 9.3] one has $\mathrm{res}_p(\boldsymbol{\kappa}_g^*(2, 1, 1)) \in \mathrm{H}^1(\mathbb{Q}_p, \mathrm{Fil}^2(V))$. Recall that

$$\mathrm{H}^1(\mathbb{Q}_p, \mathrm{Fil}^2(V)/\mathrm{Fil}^3(V)) = \mathrm{H}^1(\mathbb{Q}_p, V_f^- \otimes V_{gh}^{\beta\beta}) \oplus \mathrm{H}^1(\mathbb{Q}_p, V_f^+ \otimes V_{gh}^{\alpha\beta})$$
$$\oplus \mathrm{H}^1(\mathbb{Q}_p, V_f^+ \otimes V_{gh}^{\beta\alpha})$$

and hence $\partial_p\boldsymbol{\kappa}_g^*(2, 1, 1) \in \mathrm{H}^1_s(\mathbb{Q}_p, V) = \mathrm{H}^1_s(\mathbb{Q}_p, V_f^- \otimes V_{gh})$ lies in $\mathrm{H}^1_s(\mathbb{Q}_p, V^{\beta\beta})$. Then, by definition of the basis of $\mathrm{Sel}_{(p)}(V)$ described in Corollary 3.6,

$$\boldsymbol{\kappa}_g^*(2, 1, 1) = \frac{\exp^*_{\beta\beta}(\partial_p\boldsymbol{\kappa}_g^*(2, 1, 1))}{\exp^*_{\beta\beta}(\partial_p\xi^{\beta\beta})} \cdot \xi^{\beta\beta}.$$

We need to compute the numerator and denominator of the previous formula. By (4.4),

$$\begin{aligned}\exp^*_{\beta\beta}(\partial_p\boldsymbol{\kappa}_g^*(2, 1, 1)) &= 2(1 - 1/p)\mathcal{L}_p^{f*}(1)\\ &= \frac{2(1 - 1/p)\langle w_N(f), hg\rangle}{\langle w_N(f), w_N(f)\rangle}\\ &= \frac{2(1 - 1/p)\sqrt{c}\sqrt{L(E \otimes \rho, 1)}}{\pi^2\langle f, f\rangle},\end{aligned}$$

where the last equality is given by [20]. Besides,

$$\begin{aligned}\exp^*_{\beta\beta}(\partial_p\xi^{\beta\beta}) &= \langle \exp^*_{\beta\beta}\partial_p\xi^{\beta\beta}, \eta_f\omega_{g_\alpha}\omega_{h_\alpha}\rangle \\ &= \exp^*(X_{\beta\beta})\langle v_g^\beta \otimes v_h^\beta, \omega_{g_\alpha}\otimes\omega_{h_\alpha}\rangle \\ &= \frac{1}{\Theta_{g_\alpha}\Theta_{h_\alpha}}\langle \eta_{g_\alpha}\otimes\eta_{h_\alpha}, \omega_{g_\alpha}\otimes\omega_{h_\alpha}\rangle \\ &= \frac{1}{\Theta_{g_\alpha}\Theta_{h_\alpha}}.\end{aligned}$$

The proposition follows. □

Let $\Lambda_{\mathrm{cyc}} = \mathbb{Z}_p[[j+1]]$ denote the usual Iwasawa algebra regarded as the ring of bounded analytic functions on an open disc centered at $j=-1$, and let $\boldsymbol{\chi}_{\mathrm{cyc}} : G_{\mathbb{Q}} \longrightarrow \Lambda_{\mathrm{cyc}}^\times$ denote the Λ-adic cyclotomic character characterized by the property that $\nu_j(\boldsymbol{\chi}_{\mathrm{cyc}}) = \chi_{\mathrm{cyc}}^{-j}$ for all $j \in \mathbb{Z}$.

Recall the three-variable Iwasawa algebra $\Lambda_{\mathbf{fgh}}$ and set

$$\bar{\Lambda}_{\mathbf{fgh}} = \Lambda_{\mathbf{fgh}}\hat{\otimes}\Lambda_{\mathrm{cyc}}, \quad \overline{\mathcal{W}}_{\mathbf{fgh}} := \mathrm{Spf}(\bar{\Lambda}_{\mathbf{fgh}}) = \mathcal{W}_{\mathbf{fgh}} \times \mathcal{W}, \quad \overline{\mathcal{W}}^\circ_{\mathbf{fgh}} := \mathcal{W}^\circ_{\mathbf{fgh}} \times \mathcal{W}^\circ.$$

Let $\Psi : G_{\mathbb{Q}_p} \longrightarrow \Lambda_{\mathbf{fgh}}^\times$ denote the unramified character taking Fr_p to $\dfrac{a_p(\mathbf{g})}{a_p(\mathbf{f})a_p(\mathbf{h})\chi(p)}$. Define

$$\mathbb{M}^g := \Lambda_{\mathbf{fgh}}(\Psi), \qquad \overline{\mathbb{M}}^g := \mathbb{M}^g \otimes \Lambda_{\mathrm{cyc}}(\boldsymbol{\chi}_{\mathrm{cyc}})$$

and note that $\mathbb{M}^g$ is the unramified twist of the local Galois representation $\mathbb{V}^g$ introduced in (3.8).

Let $\theta : \bar{\Lambda}_{\mathbf{fgh}} \longrightarrow \Lambda_{\mathbf{fgh}}$ denote the homomorphism taking a function $F(k,\ell,m,j)$ to its restriction $F(k,\ell,m,(\ell-k-m)/2)$. As shown in the proof of [5, Prop. 7.3], there is an isomorphism

$$\theta : \overline{\mathbb{M}}^g \otimes_\theta \mathcal{O}_{\mathbf{fgh}} \cong \mathbb{V}^g. \tag{4.6}$$

Define also the Λ-adic Dieudonné modules

$$\mathbb{D}^g := (\mathbb{M}^g\hat{\otimes}\hat{\mathbb{Z}}_p^{\mathrm{ur}})^{G_{\mathbb{Q}_p}}, \quad \overline{\mathbb{D}}^g := \mathbb{D}^g \otimes \Lambda_{\mathrm{cyc}},$$

where $\hat{\mathbb{Z}}_p^{\mathrm{ur}}$ is the ring of integers of the p-adic completion $\hat{\mathbb{Q}}_p^{\mathrm{ur}}$ of the maximal unramified extension of $\mathbb{Q}_p$.

As it directly follows from the above definitions, the specialization of $\overline{\mathbb{M}}^g$ at a point $\underline{x} = (k,\ell,m,j) \in \overline{\mathcal{W}}^\circ_{\mathbf{fgh}}$ is

$$\overline{\mathbb{M}}^g_{\underline{x}} := \overline{\mathbb{M}}^g \otimes_{\underline{x}} L_p = L_p(\Psi_{(k,\ell,m)})(-j) = V^+_{\mathbf{f}_k} \otimes V^-_{\mathbf{g}_\ell} \otimes V^+_{\mathbf{h}_m}(2-k-m-j).$$

Bloch–Kato's logarithm and dual exponential maps give rise to an isomorphism

$$\log_{\underline{x}} : \mathrm{H}^1(\mathbb{Q}_p, L_p(\Psi_{(k,\ell,m)})(-j)) \longrightarrow \mathrm{D}_{\mathrm{dR}}(L_p(\Psi_{(k,\ell,m)})(-j)) \qquad \text{if } j < 0 \tag{4.7}$$

$$\exp^*_{\underline{x}} : \mathrm{H}^1(\mathbb{Q}_p, L_p(\Psi_{(k,\ell,m)})(-j)) \longrightarrow \mathrm{D}_{\mathrm{dR}}(L_p(\Psi_{(k,\ell,m)})(-j)) \qquad \text{if } j \geq 0. \tag{4.8}$$

In particular, at $\underline{x} = (2, 1, 1, -1)$ this isomorphism becomes the map

$$\log_{\alpha\beta} : \mathrm{H}^1_f(\mathbb{Q}_p, V^{\alpha\beta}) \longrightarrow L_p,$$

after taking the pairing with the differential $\eta_{g_\alpha}\omega_{h_\alpha}$ (cf. Remark 3.5).

Proposition 4.5 *There exists a unique homomorphism of* $\bar{\Lambda}_{\mathbf{fgh}}$*-modules*

$$\bar{\mathcal{L}}_g : \mathrm{H}^1(\mathbb{Q}_p, \overline{\mathbb{M}}^g) \longrightarrow \overline{\mathbb{D}}^g,$$

such that for all $\mathfrak{Z} \in \mathrm{H}^1(\mathbb{Q}_p, \overline{\mathbb{M}}^g)$ *and* $\underline{x} = (k, \ell, m, j) = (w, j) \in \overline{\mathcal{W}}^\circ_{\mathbf{fgh}}$ *with*

$$1 - \Psi_w(\mathrm{Fr}_p)p^{-j-1} \neq 0,$$

we have

$$\bar{\mathcal{L}}_g(\mathfrak{Z})(\underline{x}) = \left(1 - \frac{p^j}{\Psi_w(\mathrm{Fr}_p)}\right)(1 - \Psi_w(\mathrm{Fr}_p)p^{-j-1})^{-1} \cdot \begin{cases} \dfrac{(-1)^{j+1}}{(-j-1)!}\log_{\underline{x}}(\mathfrak{Z}(\underline{x})) & j < 0; \\ j!\,\exp^*_{\underline{x}}(\mathfrak{Z}(\underline{x})) & j \geq 0. \end{cases}$$

Proof The proposition as stated here is [5, Proposition 7.1] for **g**, and it is a consequence of [21]. □

By the proof of [5, Proposition 7.3], the relation between $\bar{\mathcal{L}}_g$ and the logarithm $\mathcal{L}_g$ that appears in (4.1) is given, for all $\mathfrak{Z} \in \mathrm{H}^1(\mathbb{Q}_p, \overline{\mathbb{M}}^g)$ by

$$\mathcal{L}_g(\theta_*(\mathfrak{Z}))(k, \ell, m) = \langle \bar{\mathcal{L}}_g(\mathfrak{Z})(k, \ell, m, (\ell - k - m)/2), \omega_{\mathbf{f}_k}\eta_{\mathbf{g}_\ell}\omega_{\mathbf{h}_m} \rangle. \tag{4.9}$$

Here, if $\underline{x} = (k, \ell, m, (\ell - k - m)/2)$, then

$$\omega_{\mathbf{f}_k}\eta_{\mathbf{g}_\ell}\omega_{\mathbf{h}_m} \in \overline{\mathbb{D}}^g \otimes_{\underline{x}} L_p \cong \mathrm{D}_{\mathrm{dR}}(V^+_{\mathbf{f}_k} \otimes V^-_{\mathbf{g}_\ell} \otimes V^+_{\mathbf{h}_m}((\ell - k - m + 4)/2))$$

are the differentials defined as in [5, Sect. 7.1.1] and are natural generalizations of the ones introduced in Sect. 3.1.

When j is a fixed integer, note that the Euler-like factors appearing above vary analytically. As it will suffice for our purposes, set $j = -1$ and recall the plane $\mathcal{S}$ parametrized by points of weights $(2 + \ell - m, \ell, m)$ in $\mathcal{W}_{\mathbf{fgh}}$. By a slight abuse of notation we also regard $\mathcal{S}$ as the plane in $\bar{\mathcal{W}}_{\mathbf{fgh}}$ parametrizing points of weights $(2 + \ell - m, \ell, m, -1)$.

Define the homomorphism

$$\bar{\mathcal{L}}_g^* : \mathrm{H}^1(\mathbb{Q}_p, \overline{\mathbb{M}}^g_{|\mathcal{S}}) \longrightarrow \overline{\mathbb{D}}^g_{|\mathcal{S}}$$

of $\mathcal{O}_{\mathcal{S}}$-modules given by

$$\bar{\mathcal{L}}_g^* = \boldsymbol{\mathcal{E}}_g \times \bar{\mathcal{L}}_{g|\mathcal{S}}. \tag{4.10}$$

In light of the above proposition, for any $\mathbf{Z} \in \mathrm{H}^1(\mathbb{Q}_p, \overline{\mathbb{M}}^g_{|\mathcal{S}})$ and $\ell \geq 1$ one then has

$$\bar{\mathcal{L}}_g^*(\mathbf{Z})(\ell) = (1 - \frac{a_p(\mathbf{g}_\ell)\beta_h}{a_p(\mathbf{f}_{\ell+1})p}) \cdot \langle \log_{(\ell+1,\ell,1,-1)}(\mathfrak{Z}(\ell+1, \ell, 1, -1)), \omega_{\mathbf{f}_{\ell+1}} \eta_{\mathbf{g}_\ell} \omega_{h_\alpha} \rangle. \tag{4.11}$$

Proposition 4.6 *We have*

$$I_p(f, g_\alpha, h_\alpha) = \left(1 - \frac{1}{p}\right) \log_{\alpha\beta}(\pi_{\alpha\beta} \boldsymbol{\kappa}_g^*(2, 1, 1)) \mod L^\times.$$

Proof To avoid notational clutter, we continue to denote $\mathcal{S}$ the plane in $\overline{\mathcal{W}}_{\mathbf{fgh}}$ containing the points $(2 + \ell - m, \ell, m, -1)$ as a dense subset. Let $\widetilde{\boldsymbol{\kappa}}^*_{g,p} \in \mathrm{H}^1(\mathbb{Q}_p, \overline{\mathbb{M}}^g_{|\mathcal{S}})$ be any lift of $\mathrm{res}_p(\boldsymbol{\kappa}_g^*)$, whose existence is assured by the fact that the map

$$\theta_* : \mathrm{H}^1(\mathbb{Q}_p, \overline{\mathbb{M}}^g) \otimes_\theta \mathcal{O}_{\mathbf{fgh}} \longrightarrow \mathrm{H}^1(\mathbb{Q}_p, \mathbb{V}^g)$$

induced by (4.6) is an isomorphism (cf. the proof of [5, Proposition 7.3]).

Combining Proposition 4.5 with (4.1), (4.3), (4.9) we deduce

$$\begin{aligned} \mathcal{L}_p^g(\breve{\mathbf{f}}, \breve{\mathbf{g}}, \breve{\mathbf{h}})_{|\mathcal{S}} &= \mathcal{L}_{g|\mathcal{S}}(\mathrm{res}_p(\boldsymbol{\kappa}_{|\mathcal{S}})) = \boldsymbol{\mathcal{E}}_{g|\mathcal{S}} \cdot \mathcal{L}_{g|\mathcal{S}}(\mathrm{res}_p(\boldsymbol{\kappa}_g^*)) \\ &= \boldsymbol{\mathcal{E}}_{g|\mathcal{S}} \cdot \langle \bar{\mathcal{L}}_{g|\mathcal{S}}(\widetilde{\boldsymbol{\kappa}}^*_{g,p}), \omega_{\mathbf{f}_{\ell+1}} \eta_{\mathbf{g}_\ell} \omega_{h_\alpha} \rangle. \end{aligned}$$

By (4.10), (4.11) and (4.7), the latter quantity is equal to

$$\langle \bar{\mathcal{L}}_g^*(\widetilde{\boldsymbol{\kappa}}^*_{g,p}), \omega_{\mathbf{f}_{\ell+1}} \eta_{\mathbf{g}_\ell} \omega_{h_\alpha} \rangle = (1 - \frac{a_p(\mathbf{g}_\ell)\beta_h}{a_p(\mathbf{f}_{\ell+1})p}) \cdot \log_{\alpha\beta}(\pi_{\alpha\beta} \boldsymbol{\kappa}_g^*(\ell+1, \ell, 1)).$$

Since $I_p(f, g_\alpha, h_\alpha) = \mathcal{L}_p^g(\breve{\mathbf{f}}, \breve{\mathbf{g}}, \breve{\mathbf{h}})(2, 1, 1)$ by definition, the proposition follows by using (4.2). □

Define the local point

$$P_{\alpha\beta} \in E(K_p)_{L_p}^{\alpha\beta}$$

as the one satisfying

$$\pi_{\alpha\beta} \, \mathrm{res}_p(\xi^{\beta\beta}) = \delta_p(P_{\alpha\beta}) \otimes v_g^\alpha \otimes v_h^\beta \in \left(\mathrm{H}^1_f(K_p, V_f) \otimes V_{gh}^{\alpha\beta}\right)^{G_{\mathbb{Q}_p}}, \tag{4.12}$$

where $\delta_p : E(K_p) \longrightarrow \mathrm{H}^1_f(K_p, V_f)$ is the Kummer map and $E(K_p)^{\alpha\beta}_{L_p} := E(K_p)^{\alpha\beta} \otimes L_p$.

Theorem 4.7 *We have*

$$I_p(f, g_\alpha, h_\alpha) = \frac{2(1-1/p)^2\sqrt{c}}{\pi^2\langle f, f\rangle} \times \sqrt{L(E\otimes\rho, 1)} \times \frac{\log_p(P_{\alpha\beta})}{\mathcal{L}_{g_\alpha}} \quad \bmod L^\times.$$

Proof By Proposition 4.4, $\boldsymbol{\kappa}^*_g(2,1,1) = \dfrac{\Theta_{g_\alpha}\Theta_{h_\alpha}2(1-1/p)\sqrt{c}\sqrt{L(E\otimes\rho,1)}}{\pi^2\langle f, f\rangle}\xi^{\beta\beta}$. It thus follows from Proposition 4.6 that

$$\begin{aligned}
I_p(f, g_\alpha, h_\alpha) &= \left(1-\frac{1}{p}\right)\log_{\alpha\beta}(\pi_{\alpha\beta}\boldsymbol{\kappa}^*_g(2,1,1)) \\
&= \frac{\Theta_{g_\alpha}\Theta_{h_\alpha}\sqrt{c}2(1-1/p)^2\sqrt{L(E\otimes\rho,1)}}{\pi^2\langle f, f\rangle}\log_{\alpha\beta}(\pi_{\alpha\beta}\xi^{\beta\beta}) \\
&= \frac{\Theta_{g_\alpha}\Theta_{h_\alpha}\sqrt{c}2(1-1/p)^2\sqrt{L(E\otimes\rho,1)}}{\pi^2\langle f, f\rangle}\log_p(P_{\alpha\beta})\langle v_g^\alpha\otimes v_h^\beta, \eta_{g_\alpha}\omega_{h_\alpha}\rangle \\
&= \frac{\Theta_{g_\alpha}\Theta_{h_\alpha}\sqrt{c}2(1-1/p)^2\sqrt{L(E\otimes\rho,1)}}{\Omega_{g_\alpha}\Theta_{h_\alpha}\pi^2\langle f, f\rangle}\log_p(P_{\alpha\beta}).
\end{aligned}$$

The theorem follows in light of (4.5). □

5 A Special Value Formula for $I_p(f, g_\alpha, h_\alpha)$ in Terms of Kolyvagin Classes

The aim of this section is to relate Theorem 4.7 to Bertolini-Darmon's Kolyvagin classes constructed in [10, Sect. 6].

5.1 Decomposition of the Representation V and Consequences

The underlying reason why the p-adic L-value $\mathcal{L}_p^g(\breve{\mathbf{f}}, \breve{\mathbf{g}}, \breve{\mathbf{h}})(2,1,1)$ may be related to Kolyvagin classes relies on the decomposition of the Galois representation $V_g \otimes V_h$ as the direct sum of a pair of 2-dimensional representations induced from characters of an imaginary quadratic field.

More precisely, recall that g and h are theta series of two finite order Hecke characters $\psi_g, \psi_h : G_K \longrightarrow L^\times$. As explained in the introduction, there is a decomposition

$$V_g \otimes V_h \cong V_{\psi_1} \oplus V_{\psi_2} \tag{5.1}$$

where $\psi_1 := \psi_g \psi_h$, $\psi_2 := \psi_g \psi_h'$ and $V_{\psi_i} := \mathrm{Ind}_{G_K}^{G_{\mathbb{Q}}}(\psi_i)$.

Since the central characters of ψ_g and ψ_h are assumed to be mutually inverse from the outset, it follows that ψ_1 and ψ_2 are ring class characters, i.e. they factor through the absolute Galois group of a number field H_i which is a ring class field of K of some conductor c_i. Recall we have assumed that p does not divide the conductor of ψ_g and ψ_h, hence that the characters ψ_1 and ψ_2 are unramified at p, and this implies that $p \nmid c_1 c_2$. The Artin representation

$$\rho = \rho_g \otimes \rho_h : G_{\mathbb{Q}} \to \mathrm{GL}(V_g \otimes_{L_p} V_h)$$

thus factors through $\mathrm{Gal}(H/K)$, where H is the ring class field of K of conductor $c := \mathrm{lcm}(c_1, c_2)$. Moreover, since the prime p is inert in K and $p \nmid c$, the principal ideal $p\mathcal{O}_K$ splits completely in H. Fix once and for all a prime $\mathfrak{p}$ of the field H dividing p and denote $H_{\mathfrak{p}}$ the completion of H at $\mathfrak{p}$. Then $H_{\mathfrak{p}} = K_p$ and, in particular, the restriction of ψ_i to the decomposition group $G_{K_p} = G_{H_{\mathfrak{p}}}$ is trivial for $i = 1, 2$.

Recall that Fr_p acts on V_g with eigenvalues $\alpha_g = \lambda$, $\beta_g = -\lambda$ and on V_h with eigenvalues

$$(\alpha_h, \beta_h) = \begin{cases} (-1/\lambda, 1/\lambda) & \text{if } a = +1; \\ (1/\lambda, -1/\lambda) & \text{if } a = -1, \end{cases}$$

so we always have $\alpha_g \alpha_h = -a$. As in the previous sections, choose an L-basis of eigenvectors v_g^α, v_g^β of V_g and likewise a basis v_h^α, v_h^β of V_h.

Note also that Fr_p acts on V_{ψ_1} and V_{ψ_2} with eigenvalues ± 1.

Lemma 5.1 *There is a $G_{\mathbb{Q}}$-equivariant isomorphism*

$$\Psi : V_g \otimes V_h \cong V_{\psi_1} \oplus V_{\psi_2}.$$

Moreover, V_{ψ_i} admits an L-basis $\{v_i^+, v_i^-\}$ of eigenvectors with relative eigenvalues $\{+1, -1\}$ satisfying:

- *if $a = +1$, then*

$$\Psi(v_g^\alpha \otimes v_h^\alpha) = v_1^- - v_2^-, \quad \Psi(v_g^\beta \otimes v_h^\beta) = v_1^- + v_2^-, \quad \Psi(v_g^\alpha \otimes v_h^\beta) = v_1^+ + v_2^+, \quad \Psi(v_g^\beta \otimes v_h^\alpha) = v_1^+ - v_2^+.$$

- *If $a = -1$, then*

$$\Psi(v_g^\alpha \otimes v_h^\alpha) = v_1^+ + v_2^+, \quad \Psi(v_g^\beta \otimes v_h^\beta) = v_1^+ - v_2^+, \quad \Psi(v_g^\alpha \otimes v_h^\beta) = v_1^- - v_2^-, \quad \Psi(v_g^\beta \otimes v_h^\alpha) = v_1^- + v_2^-.$$

Proof Let ψ be a finite order Hecke character of K, let $\varphi := \theta(\psi)$ and denote $\mathrm{Fr}_p \in G_{\mathbb{Q}} \setminus G_K$ a Frobenius element at p. As explained in [22, Sect. 2.2], we can choose a basis $\{u_\varphi, v_\varphi\}$ of V_φ such that, if

$$\rho_\varphi : G_{\mathbb{Q}} \longrightarrow \mathrm{GL}(V_\varphi)$$

is the ℓ-adic representation attached to φ, then $v_\varphi = \rho_\varphi(\mathrm{Fr}_p)u_\varphi$ and with respect to this basis,

$$\rho_\varphi(\sigma) = \begin{pmatrix} \psi(\sigma) & 0 \\ 0 & \psi'(\sigma) \end{pmatrix} \text{ for } \sigma \in G_K; \qquad \rho_\varphi(\tau) = \begin{pmatrix} 0 & \eta(\tau) \\ \eta'(\tau) & 0 \end{pmatrix} \text{ for } \tau \in G_{\mathbb{Q}} \smallsetminus G_K.$$

Here ψ' denotes the character defined by $\psi'(\sigma) := \psi(\mathrm{Fr}_p\, \sigma\, \mathrm{Fr}_p^{-1})$, and η is a function on $G_{\mathbb{Q}} \smallsetminus G_K$. Recall that the characteristic polynomial for the action of Fr_p on V_φ is $x^2 - a_p(\varphi)x + \chi_\varphi(p) = x^2 - \epsilon_\psi(p)$, where $\chi_\varphi = \chi_K \epsilon_\psi$ and ϵ_ψ is the central character of ψ. We can always choose $\{u_\varphi, u_\varphi\}$ such that $\rho_\varphi(\mathrm{Fr}_p) = \begin{pmatrix} 0 & \zeta \\ \zeta & 0 \end{pmatrix}$, with $\zeta^2 = \epsilon_\psi(p)$. Then a basis of eigenvectors for the action of Fr_p on V_φ is

$$\{v_\varphi^\zeta = u_\varphi + v_\varphi, \qquad v_\varphi^{-\zeta} = u_\varphi - v_\varphi\}$$

with eigenvalue $\zeta, -\zeta$ respectively. Using this notation we obtain the basis

$$(u_g, v_g), \ (u_h, v_h), \ (u_1, v_1), \ (u_2, v_2)$$

of $V_g, V_h, V_{\psi_1}, V_{\psi_2}$ where we denoted $u_i := u_{\theta(\psi_i)}$ and $v_i := v_{\theta(\psi_i)}$. The corresponding matrix $\rho_\varphi(\mathrm{Fr}_p)$ is

$$\begin{pmatrix} 0 & \lambda \\ \lambda & 0 \end{pmatrix}, \ \begin{pmatrix} 0 & 1/\lambda \\ 1/\lambda & 0 \end{pmatrix}, \ \begin{pmatrix} 0 & 1 \\ 1 & 0 \end{pmatrix}, \ \begin{pmatrix} 0 & 1 \\ 1 & 0 \end{pmatrix}.$$

Consider the basis

$$(u_g \otimes u_h, u_g \otimes v_h, v_g \otimes u_h, v_g \otimes v_h)$$

of $V_g \otimes V_h$, and the basis

$$(u_1, v_1, u_2, v_2)$$

of $V_{\psi_1} \oplus V_{\psi_2}$. Comparing the matrices above, we conclude that the isomorphism Ψ is given by the rule

$$u_g \otimes u_h \mapsto u_1, \quad u_g \otimes v_h \mapsto u_2, \quad v_g \otimes u_h \mapsto v_2, \quad v_g \otimes v_h \mapsto v_1.$$

Moreover, since the eigenvalues of Fr_p acting on V_g and V_h are $\lambda, -\lambda, 1/\lambda, 1/\lambda$, a basis of eigenvectors for $V_g \otimes V_h$ is

$$\begin{aligned} v_g^\lambda \otimes v_h^{1/\lambda} &= (u_g + v_g) \otimes (u_h + v_h), & v_g^\lambda \otimes v_h^{-1/\lambda} &= (u_g + v_g) \otimes (u_h - v_h), \\ v_g^{-\lambda} \otimes v_h^{1/\lambda} &= (u_g - v_g) \otimes (u_h + v_h), & v_g^{-\lambda} \otimes v_h^{-1/\lambda} &= (u_g - v_g) \otimes (u_h - v_h). \end{aligned}$$

with eigenvalues $+1, -1, -1, +1$ respectively. On the other hand, a basis of eigenvectors for $V_{\psi_1} \oplus V_{\psi_2}$ is

$$(v_1^+ := u_1 + v_1, \quad v_1^- := u_1 - v_1, \quad v_2^+ := u_2 + v_2, \quad v_2^- := u_2 - v_2)$$

In conclusion,

$$v_g^{\lambda} \otimes v_h^{1/\lambda} \mapsto v_1^+ + v_2^+, \quad v_g^{\lambda} \otimes v_h^{-1/\lambda} \mapsto v_1^- - v_2^-, \quad v_g^{-\lambda} \otimes v_h^{1/\lambda} \mapsto v_1^- + v_2^-,$$
$$v_g^{-\lambda} \otimes v_h^{-1/\lambda} \mapsto v_1^+ - v_2^+.$$

□

Recall that

$$V_{\psi_i} = \mathrm{Ind}_{G_K}^{G_{\mathbb{Q}}}(\psi_i) = \{v : G_{\mathbb{Q}} \longrightarrow L(\psi) \mid v(\sigma\tau) = \psi_i(\sigma)v(\tau)\ \forall \sigma \in G_K, \tau \in G_{\mathbb{Q}}\}$$
$$= L \oplus L \cdot \mathrm{Fr}_p.$$

The proof of the previous lemma shows that we may choose the vectors $v_i^{\pm}$ such that

$$v_i^+(1) = v_i^-(1) = 1. \tag{5.2}$$

Any other choice of basis would be of the form $(a_i v_i^+, b_i v_i^-)$ for some scalars $a_i, b_i \in L^\times$, and would yield the same formula for $I_p(f, g_\alpha, h_\alpha)$ up to a scalar in $L^\times$. This is fine because that is precisely the ambiguity we are working with in our framework, as explained in the introduction.

As in previous sections, denote

$$V := V_f \otimes V_g \otimes V_h, \quad V_1 := V_f \otimes V_{\psi_1}, \quad V_2 := V_f \otimes V_{\psi_2}$$

The map Ψ of Lemma 5.1 induces the isomorphism

$$\Psi : V \xrightarrow{\cong} V_1 \oplus V_2. \tag{5.3}$$

There is then a decomposition of the Selmer group

$$\Psi_* : \mathrm{Sel}_p(V) \xrightarrow{\cong} \mathrm{Sel}_p(V_1) \oplus \mathrm{Sel}_p(V_2), \tag{5.4}$$

and the analogous decomposition holds for the relaxed and the strict Selmer groups associated to V. By Artin's formalism, decomposition (5.3) yields the factorization

$$L(E, \rho, s) = L(E, \psi_1, s)L(E, \psi_2, s)$$

of classical L-series. As explained in (3.9) in Sect. 3.2, Assumption 1.1 on the analytic rank of $E \otimes \rho$ implies that

$$\mathrm{Sel}_p(V_1) = \mathrm{Sel}_p(V_2) = 0.$$

Denote $V_{\psi_i}^{\pm}$ the eigenspace of V_{ψ_i} on which Fr_p acts as ± 1, and let

$$V_i^{\pm} := V_f \otimes V_{\psi_i}^{\pm}.$$

Lemma 5.2 *There are isomorphisms*

$$\mathrm{H}^1_f(\mathbb{Q}_p, V_i^{\pm}) \cong \mathrm{H}^1_f(K_p, V_f)^{\pm} \otimes V_{\psi_i}^{\pm} \text{ and } \mathrm{H}^1_s(\mathbb{Q}_p, V_i^{\pm}) \cong \mathrm{H}^1_s(K_p, V_f)^{\pm} \otimes V_{\psi_i}^{\pm}. \tag{5.5}$$

The Bloch–Kato logarithm and dual exponential yield isomorphisms

$$\mathrm{H}^1_f(\mathbb{Q}_p, V_i^{\pm}) \xrightarrow{\log_{\pm}} L_p, \qquad \mathrm{H}^1_s(\mathbb{Q}_p, V_i^{\pm}) \xrightarrow{\exp^*_{\pm}} L_p. \tag{5.6}$$

Moreover, for $i \in \{1, 2\}$, there are isomorphisms

$$\mathrm{Sel}_{(p)}(V_i) \xrightarrow{\partial_p} \mathrm{H}^1_s(\mathbb{Q}_p, V_i) \xrightarrow{\exp^*} L_p^2. \tag{5.7}$$

Proof The isomorphisms are obtained as in the proofs of Lemma 3.4, Lemma 3.2 and Corollary 3.3, using (1.4). □

Similarly as in Remark 3.5, the Bloch–Kato maps (5.6) are related to the logarithm and the dual exponential of the representation V_f as follows. Since $\psi_i = \psi_i^*$, the Galois representation V_{ψ_i} attached to the theta series $\theta(\psi_i)$ is self-dual and (3.4) yields a pairing

$$\langle\ ,\ \rangle : \mathrm{D_{dR}}(V_{\psi_i}^+) \times \mathrm{D_{dR}}(V_{\psi_i}^-) \longrightarrow L_p. \tag{5.8}$$

Since V_{ψ_i} is unramified, we have

$$\mathrm{D_{dR}}(V_{\psi_i}^{\pm}) = (V_{\psi_i}^{\pm} \otimes K_p)^{G_{\mathbb{Q}_p}} = V_{\psi_i}^{\pm} \otimes K_p^{\pm}.$$

One can introduce differentials $\omega_i^{\pm} \in \mathrm{D_{dR}}(V_{\psi_i}^{\pm})$ as in (3.3), and elements $\Omega_i^{\pm} \in K_p^{\pm}$ characterised by the following relation

$$v_i^{\pm} \otimes \Omega_i^{\pm} = \omega_i^{\pm}.$$

Thus (5.8) in turn induces a pairing

$$\langle\ ,\ \rangle : V_{\psi_i}^+ \times \mathrm{D_{dR}}(V_{\psi_i}^-) \longrightarrow L_p$$

by setting

$$\langle v_i^+, \omega\rangle := \frac{1}{\Omega_i^+}\langle \omega_i^+, \omega\rangle$$

for all $\omega \in \mathrm{D_{dR}}(V_{\psi_i}^{-})$. Hence we have

$$\exp_{\pm}^{*}(x \otimes v) = \exp_{V_f}^{*}(x) \cdot \langle v, \omega_i^{\mp} \rangle \tag{5.9}$$

for any $x \in \mathrm{H}_s^1(K_p, V_f)^{\pm}$, $v \in V_{\psi_i}^{\pm}$.

Recall from Lemma 5.1 the basis $\{v_i^+, v_i^-\}$ for V_{ψ_i}, and from Corollary 3.6 the local classes $X_{\pm} \in \mathrm{H}_s(K_p, V_f)_{L_p}^{\pm}$ such that $\exp_{\pm}^{*}(X_{\pm}) = 1$.

Corollary 5.3 *For $i \in \{1, 2\}$, $\mathrm{Sel}_{(p)}(V_i)$ admits a basis $\{\xi_i^+, \xi_i^-\}$ characterized as*

$$\partial_p \xi_i^{\pm} = X_{\pm} \otimes v_i^{\pm}.$$

under the identifications given in (5.5).

Proof The statement follows after applying Lemma 5.2 and using the same argument as in the proof of Corollary 3.6. □

Proposition 5.4 *Let*

$$\Psi_* : \mathrm{Sel}_{(p)}(V) \xrightarrow{\cong} \mathrm{Sel}_{(p)}(V_1) \oplus \mathrm{Sel}_{(p)}(V_2)$$

denote the isomorphism induced by (5.4).

- *If $a = +1$ then*

$$\Psi_*\xi^{\alpha\alpha} = \xi_1^- - \xi_2^-, \quad \Psi_*\xi^{\alpha\beta} = \xi_1^+ + \xi_2^+, \quad \Psi_*\xi^{\beta\alpha} = \xi_1^+ - \xi_2^+,$$
$$\Psi_*\xi^{\beta\beta} = \xi_1^- + \xi_2^-.$$

- *If $a = -1$ then*

$$\Psi_*\xi^{\alpha\alpha} = \xi_1^+ + \xi_2^+, \quad \Psi_*\xi^{\alpha\beta} = \xi_1^- - \xi_2^-, \quad \Psi_*\xi^{\beta\alpha} = \xi_1^- + \xi_2^-,$$
$$\Psi_*\xi^{\beta\beta} = \xi_1^+ - \xi_2^+.$$

Proof Combining (5.7) and Lemma 3.1, we obtain a commutative diagram

$$\begin{array}{ccc}
\mathrm{Sel}_{(p)}(V) & \xrightarrow{\Psi_*} & \mathrm{Sel}_{(p)}(V_1) \oplus \mathrm{Sel}_{(p)}(V_2) \\
\downarrow{\scriptstyle \partial_p} & & \downarrow{\scriptstyle \partial_p \oplus \partial_p} \\
\mathrm{H}_s^1(\mathbb{Q}_p, V) & \xrightarrow{\Psi_*} & \mathrm{H}_s^1(\mathbb{Q}_p, V_1) \oplus \mathrm{H}_s^1(\mathbb{Q}_p, V_2)
\end{array}$$

where each arrow is an isomorphism. Moreover, by Lemma 3.4 and Lemma 5.2, we also have the commutative diagram

$$\begin{array}{ccc} \mathrm{H}^1_s(\mathbb{Q}_p, V) & \xrightarrow{\Psi_*} & \mathrm{H}^1_s(\mathbb{Q}_p, V_1) \oplus \mathrm{H}^1_s(\mathbb{Q}_p, V_2) \\ \downarrow & & \downarrow \\ (\mathrm{H}^1_s(K_p, V_f) \otimes V_{gh})^{G_{\mathbb{Q}_p}} & \longrightarrow & (\mathrm{H}^1_s(K_p, V_f) \otimes V_1)^{G_{\mathbb{Q}_p}} \oplus (\mathrm{H}^1_s(K_p, V_f) \otimes V_2)^{G_{\mathbb{Q}_p}} \end{array}$$

where each arrow is an isomorphism and the bottom horizontal arrow is $1 \otimes \Psi$. By definition,

$$\partial_p \xi^{\alpha\alpha} = X_{\alpha\alpha} \otimes v_g^\alpha \otimes v_h^\alpha \in (\mathrm{H}^1_s(K_p, V_f) \otimes V_{gh}^{\alpha\alpha})^{G_{\mathbb{Q}_p}} \subseteq (\mathrm{H}^1_s(K_p, V_f) \otimes V_{gh})^{G_{\mathbb{Q}_p}}.$$

By (1.4) we have

$$X_{\alpha\alpha} = \begin{cases} X_+ & \text{if } \alpha_g \cdot \alpha_h = +1; \\ X_- & \text{if } \alpha_g \cdot \alpha_h = -1, \end{cases}$$

hence we deduce

$$(1 \otimes \Psi)(\partial_p \xi^{\alpha\alpha}) = \begin{cases} X_{\alpha\alpha} \otimes (v_1^+ + v_2^+) = \partial_p \xi_1^+ + \partial_p \xi_2^+ & \text{if } \alpha_g \cdot \alpha_h = +1 \\ X_{\alpha\alpha} \otimes (v_1^- - v_2^-) = \partial_p \xi_1^- - \partial_p \xi_2^- & \text{if } \alpha_g \cdot \alpha_h = -1. \end{cases}$$

Similar computations prove the remaining cases. □

Lemma 5.5 *Let $\delta_p : E(K_p) \otimes \mathbb{Q}_p \longrightarrow \mathrm{H}^1_f(K_p, V_f)$ denote Kummer's map.*

- *If $a = +1$, there is a local point $P^+ \in E(K_p)^+ \otimes \mathbb{Q}_p$ such that*
$$\pi_1^+ \operatorname{res}_p(\xi_1^-) = \delta_p P^+ \otimes v_1^+, \quad \pi_2^+ \operatorname{res}_p(\xi_2^-), = \delta_p P^+ \otimes v_2^+ \quad \pi_{\alpha\beta} \operatorname{res}_p(\xi^{\beta\beta}) = \delta_p P^+ \otimes v_g^\alpha \otimes v_h^\beta, \tag{5.10}$$
- *If $a = -1$, there is a local point $P^- \in E(K_p)^- \otimes \mathbb{Q}_p$ such that*
$$\pi_1^- \operatorname{res}_p(\xi_1^+) = \delta_p P^- \otimes v_1^-, \quad \pi_2^- \operatorname{res}_p(\xi_2^+) = \delta_p P^- \otimes v_2^-, \quad \pi_{\alpha\beta} \operatorname{res}_p(\xi^{\beta\beta}) = \delta_p P^- \otimes v_g^\alpha \otimes v_h^\beta; \tag{5.11}$$

Proof We prove the claim when $a = -1$, as the other case works similarly. By Proposition 5.4,

$$\Psi_* \operatorname{res}_p(\xi^{\beta\beta}) = \operatorname{res}_p(\xi_1^+) - \operatorname{res}_p(\xi_2^+) \in \mathrm{H}^1(\mathbb{Q}_p, V_1) \oplus \mathrm{H}^1(\mathbb{Q}_p, V_2)$$

Write

$$\pi_{\alpha\beta} \operatorname{res}_p(\xi^{\beta\beta}) = Q_- \otimes v_g^\alpha \otimes v_h^\beta \in \mathrm{H}^1_f(K_p, V_f)^- \otimes V_{gh}^{\alpha\beta}.$$

Then, by Proposition 5.4 and (5.5)

$$\Psi_* \pi_{\alpha\beta} \operatorname{res}_p(\xi^{\beta\beta}) = Q_- \otimes (v_1^- - v_2^-) = Q_- \otimes v_1^- - Q_- \otimes v_2^- \tag{5.12}$$

in

$$(\mathrm{H}^1_f(K_p, V_f) \otimes (V_{\psi_1}^- \oplus V_{\psi_2}^-))^{G_{\mathbb{Q}_p}} \cong \mathrm{H}^1_f(K_p, V_1^-) \oplus \mathrm{H}^1_f(K_p, V_2^-).$$

On the other hand, by definition of the basis $\xi_i^{\pm}$ and using (5.5),

$$\pi_1^- \operatorname{res}_p(\xi_1^+) - \pi_2^- \operatorname{res}_p(\xi_2^+) \in \mathrm{H}^1_f(\mathbb{Q}_p, V_1^-) \oplus \mathrm{H}^1_f(\mathbb{Q}_p, V_2^-)$$
$$\cong (\mathrm{H}^1_f(K_p, V_f) \otimes (V_{\psi_1}^- \oplus V_{\psi_2}^-))^{G_{\mathbb{Q}_p}}.$$

More precisely, we can write

$$\pi_1^- \operatorname{res}_p(\xi_1^+) - \pi_2^- \operatorname{res}_p(\xi_2^+) = P_1^- \otimes v_1^- - P_2^- \otimes v_2^- \tag{5.13}$$

for points $P_i^- \in E(K_p)^-$. Hence, comparing (5.13) and (5.12) we deduce that $Q_- = P_1^- = P_2^-$. □

Theorem 5.6 *Let $P^{\pm} \in E(K_p)^{\pm}$ be the local points of Lemma 5.5. Then*

$$I_p(f, g_\alpha, h_\alpha) = \frac{\sqrt{c} \cdot 2(1-1/p)^2 \sqrt{L(E \otimes \rho, 1)}}{\pi^2 \langle f, f \rangle} \times \frac{1}{\mathcal{L}_{g_\alpha}} \times \log_p(P^a) \mod L^{\times}.$$

Proof Theorem 4.7 asserts that

$$I_p(f, g_\alpha, h_\alpha) = \frac{2(1-1/p)^2 \sqrt{c}}{\pi^2 \langle f, f \rangle} \times \sqrt{L(E \otimes \rho, 1)} \times \frac{\log_p(P_{\alpha\beta})}{\mathcal{L}_{g_\alpha}} \mod L^{\times}.$$

The assertion now follows directly from the combination of (4.12), Lemma 5.5 and the relation between the Bloch–Kato logarithms given in (3.12). □

5.2 *Kolyvagin Classes*

In this section we recall briefly the construction and main properties of the Kolyvagin classes defined in [10, Sect. 6].

Assume for simplicity that $\mathcal{O}_K^{\times} = \{\pm 1\}$. Let H be a ring class field of K of conductor prime to p, and recall the fields $H \subseteq F_{m-1} \subseteq H(p^m)$ defined in Sect. 2.1. For each m, fix a generator σ_m of $G_m := \mathrm{Gal}(F_m/H)$ such that the image of σ_m via the projection

$$G_m \longrightarrow G_m/\mathrm{Gal}(F_m/F_{m-1}) \cong G_{m-1} \tag{5.14}$$

equals σ_{m-1}. Moreover, the elements $\{\sigma_m\}_m$ can be chosen so that $\sigma_{\mathrm{ant}} = (\sigma_{m|G_{F_{m,\mathfrak{p}}}})_m$, where σ_{ant} is the generator of Γ_{ant} we fixed in Sect. 2.2. In [23], Bertolini and Darmon constructed a collection of points $\{\tilde{\alpha}_m \in E(H(p^m))\}_m$ (in the notation of [23], $\tilde{\alpha}_m = \alpha_m(1)$) such that

$$\mathrm{N}_{H(p^m)/H}\, \tilde{\alpha}_m = 0. \tag{5.15}$$

For each $m \geq 1$, let

$$\alpha_m := \mathrm{N}_{H(p^{m+1})/F_n}\, \tilde{\alpha}_{m+1} \in E(F_m).$$

To each point α_m, in [10] the authors attached a Kolyvagin class $\mathbf{K}_m \in \mathrm{H}^1(H, E[p^m])$ in the following way. The *Kolyvagin derivative operator* is the element $D_m := \sum_{i=1}^{p^m-1} i\sigma_m^i \in \mathbb{Z}[G_m]$.

Lemma 5.7 *The Kolyvagin derivative satisfies the following equality in* $\mathbb{Z}[G_m]$

$$(\sigma_m - 1)D_m = p^m - \mathrm{N}_{F_m/H}\,. \tag{5.16}$$

Proof Note that $D_n = -\sum_{i=1}^{p^n} \dfrac{\sigma_n^i - 1}{\sigma_n - 1}$. Then $(\sigma_n - 1)D_n = -\sum_{i=1}^{p^n}(\sigma_n - 1) = -\mathrm{N}_{F_n/H} + p^n$. □

Define $P_m := D_m\alpha_m \in E(F_m)$.

Lemma 5.8 *(1) The point* $P_m \in E(F_m)$ *only depends on the choice of the generator* σ_m *of* G_m *up to the multiplication by an element* $a_m \in \{1, \ldots, p^m - 1\}$. *Moreover, if we choose the elements* $\{\sigma_m\}_m$ *to be compatible in the sense of* (5.14), *then* $a_{m-1} \equiv a_m \mod p^{m-1}$, *i.e.* $(a_m)_m \in \varprojlim_m (\mathbb{Z}/p^m\mathbb{Z})^\times = \mathbb{Z}_p^\times$.
(2) The class $[P_m]$ *of* P_m *in* $E(F_m)/p^m E(F_m)$ *is fixed by* $G_m = \mathrm{Gal}(F_m/H)$.

Proof (1) Let σ'_m any generator of $G_m \cong \mathbb{Z}/p^m\mathbb{Z}$ and define $D'_m := \sum i(\sigma'_m)^i$. Then there exists an element $\overline{a_m} \in (\mathbb{Z}/p^m\mathbb{Z})^\times$ such that $\sigma'_m = \sigma_m^{\overline{a_m}}$, where $a_m \in \{1, \ldots, p^m - 1\}$. Then $D'_m = \sum i\sigma_n^{\overline{a_m}i} = a_m^{-1}D_m$.
(2) It suffices to prove that the class of $D_m\alpha_m$ is fixed by the generator σ_m of G_m, i.e. that $\sigma_n D_n\alpha_n - D_n\alpha_n \in p^n E(F_n)$. Combining (5.16) and (5.15) we obtain $(\sigma_m - 1)D_m\alpha_m = p^m\alpha_m - \mathrm{N}_{F_m/H}\,\alpha_m = p^m\alpha_m$. □

The inflation-restriction and Kummer exact sequences yield the following commutative diagram

$$\begin{array}{ccccccccc}
 & & & & \mathrm{H}^1(F_m/H, E(F_m)[p^m]) & & & & \\
 & & & & \downarrow{\scriptstyle \mathrm{Inf}} & & & & \\
0 & \longrightarrow & \big(E(H)/p^m E(H)\big) & \xrightarrow{\delta_p} & \mathrm{H}^1(H, E[p^m]) & \longrightarrow & \mathrm{H}^1(H, E)[p^m] & \longrightarrow & 0 \\
 & & \downarrow{\scriptstyle \mathrm{Res}} & & \downarrow{\scriptstyle \mathrm{Res}} & & \downarrow{\scriptstyle \mathrm{Res}} & & \\
0 & \longrightarrow & \big(E(F_m)/p^m E(F_m)\big)^{G_m} & \xrightarrow{\delta_p} & \mathrm{H}^1(F_m, E[p^m])^{G_m} & \longrightarrow & \mathrm{H}^1(F_m, E)[p^m]^{G_m} & & \\
 & & & & \downarrow & & & & \\
 & & & & \mathrm{H}^2(F_m/H, E(F_m)[p^m]). & & & &
\end{array}$$

whose rows and columns are exact. Let C be a constant annihilating $E(F_m)[p^m]$, which can be chosen to be independent of m by [10, Lemma 6.3]. By Lemma 5.8, we can consider the element $[P_m] \in (E(F_m)/p^m E(F_m))^{G_m}$, and we define

$$\mathbf{K}_m := C\mathbf{K}_m^\circ \in \mathrm{H}^1(H, E[p^m]),$$

where $\mathbf{K}_m^\circ$ is a preimage via Res of $-C\delta_p[P_m]$. For each m, denote

$$\mathcal{G}_m := \mathrm{Gal}(F_m/K) \supseteq G_m := \mathrm{Gal}(F_m/H)$$

and fix a complete set $S = S_m = \{\sigma_1^{(m)}, \ldots, \sigma_h^{(m)}\}$ of representatives for the G_m-cosets in $\mathcal{G}_m$. We require these sets to satisfy the following compatibility condition. Let $\pi_m : \mathcal{G}_m \longrightarrow \mathcal{G}_{m-1}$ denote the projection induced by the inclusion $F_{m-1} \subseteq F_m$. Then

$$\pi_m(\sigma_i^{(m)}) = \sigma_i^{(m-1)}$$

for all i and all m. In other words, under this compatibility condition we can consider a complete set $\tilde{S} = \{\tilde{\sigma}_1, \ldots, \tilde{\sigma}_m\}$ of representatives of G_H-cosets in G_K such that, under the projection $\tilde{\pi}_m : G_K \longrightarrow \mathrm{Gal}(F_m/K)$ we have

$$\tilde{\pi}_m(\tilde{\sigma}_i) = \sigma_i^{(m)}.$$

Lemma 5.9 *The classes $\mathbf{K}_m$ form a compatible system*

$$\mathbf{K} := (\mathbf{K}_m)_m \in \varprojlim_m \mathrm{H}^1(H, E[p^m]) = \mathrm{H}^1(H, T_pE) \subset \mathrm{H}^1(H, V_f).$$

Proof The norm map $\mathrm{N}_{F_{m+1}/F_m} : F_{m+1} \longrightarrow F_m$ induces a map $\mathrm{N}_{F_{m+1}/F_m} : E(F_{m+1}) \longrightarrow E(F_m)$. By [23, Sect. 2.45], we have

$$\mathrm{N}_{F_{m+1}/F_m}(\alpha_{m+1}) = \alpha_m. \tag{5.17}$$

Consider the composition

$$f_{m+1} : E(F_{m+1})/p^{m+1}E(F_{m+1}) \overset{\mathrm{N}_{F_{m+1}/F_m}}{\longrightarrow} E(F_m)/p^{m+1}E(F_m) \longrightarrow E(F_m)/p^m E(F_m)$$

where the second morphism is the projection given by the inclusion $p^{m+1}E(F_m) \subseteq p^m E(F_m)$. Hence (5.17) implies that

$$f_{m+1}([\alpha_{m+1}]) = [\alpha_m]. \tag{5.18}$$

Consider the following commutative diagram

$$\begin{array}{ccc} E(F_{m+1})/p^{m+1}E(F_{m+1}) & \xrightarrow{\delta_p} & \mathrm{H}^1(F_{m+1}, E[p^{m+1}]) \\ \downarrow{\scriptstyle f_{m+1}} & & \downarrow{\scriptstyle f_{m+1,*}} \\ E(F_m)/p^m E(F_m) & \xrightarrow{\delta_p} & \mathrm{H}^1(F_m, E[p^m]). \end{array}$$

Here $f_{m+1,*}$ is the map induced in cohomology by f_{m+1}, i.e. it is the composition

$$\mathrm{H}^1(F_{m+1}, E[p^{m+1}]) \xrightarrow{\mathrm{N}_{F_{m+1}/F_m}} \mathrm{H}^1(F_m, E[p^{m+1}]) \xrightarrow{p_*} \mathrm{H}^1(F_m, E[p^m]),$$

where the last map is the composition with the multiplication by p map $E[p^{m+1}] \xrightarrow{p} E[p^m]$. Then (5.18) implies that

$$f_{m+1,*}\delta_p[D_{m+1}\alpha_{m+1}] = \delta_p[D_m\alpha_m].$$

□

Recall the groups $\Phi_{m,\mathfrak{p}}$ and $\Phi_{\infty,\mathfrak{p}}$ defined in Sect. 2.1. By Lemma 2.2 and Proposition 2.3, there is an injection

$$\mathrm{Hom}_{\mathrm{cont}}(\Gamma_{\mathrm{ant}}, \Phi_{\infty,\mathfrak{p}} \otimes \mathbb{Q}_p) \subseteq \mathrm{H}^1_s(K_p, V_f).$$

Recall the choice σ_m of generator of G_m; if we still denote σ_m its restriction to $\mathrm{Gal}(F_{m,\mathfrak{p}}/H_\mathfrak{p}) = \mathrm{Gal}(F_{m,\mathfrak{p}}/K_p)$, then Γ_{ant} is generated by $\sigma_{\mathrm{ant}} := (\sigma_m)_m$.

Recall the map $\partial_p : \mathrm{H}^1(H, V_f) \to \mathrm{H}^1_s(K_p, V_f)$ onto the singular quotient.

Proposition 5.10 *The class* $\mathbf{K}$ *lies in* $\mathrm{Sel}_{(p)}(H, V_f)$ *and* $\partial_p\mathbf{K} \in \mathrm{Hom}_{\mathrm{cont}}(\Gamma_{\mathrm{ant}}, \Phi_{\infty,\mathfrak{p}} \otimes \mathbb{Q}_p)$. *Moreover, if* $\bar{\alpha}_m$ *denotes the image of* α_m *in* $\Phi_{m,\mathfrak{p}}$ *and* $\bar{\alpha} := (\bar{\alpha}_m)_m \in \Phi_{\infty,\mathfrak{p}}$, *then*

$$\partial_p\mathbf{K}(\sigma_{\mathrm{ant}}) = \bar{\alpha}.$$

Proof [10, Proposition 6.9, 1,2]. □

5.3 $I_p(f, g_\alpha, h_\alpha)$ and Kolyvagin Classes

Recall the ring class characters ψ_1, ψ_2 appearing in the decomposition (5.1). Since they are ring class characters unramified at p, they factor through the Galois group $\mathrm{Gal}(H/K)$, where H is a ring class field of conductor prime to p.

Using the Kolyvagin classes described in Sect. 5.2, in this section we define elements $\mathbf{K}^{\psi_i}$ in the relaxed Selmer groups $\mathrm{Sel}_{(p)}(K, V_f \otimes \psi_i)$ for $i = 1, 2$. The aim of this section is to compare the classes $\mathbf{K}^{\psi_1}, \mathbf{K}^{\psi_2}$ with the local points in $E(K_p)^\pm$ appearing in Theorem 5.6, in order to obtain a formula for the value $I_p(f, g_\alpha, h_\alpha)$ in terms of these Kolyvagin classes. Using the notation of the previous sections, consider the dual exponential

$$\exp^*_\pm : \mathrm{H}^1_s(K_p, V_f)^\pm \longrightarrow \mathbb{Q}_p$$

of (3.13). As explained in Sect. 2, Tate's uniformisation induces the isomorphism $\varphi : \Phi_{\infty,\mathfrak{p}} \xrightarrow{\cong} \mathbb{Z}_p$ (in the notation of Lemma 2.1, $\varphi := \overline{\varphi}_{\mathrm{Tate}}$ maps $\bar{q}$ to 1). Recall the point $\bar{\alpha} \in \Phi_{\infty,\mathfrak{p}}$ defined in Proposition 5.10, the sign

$$a := a_p(E) \in \{\pm 1\}$$

and the period

$$\Pi_p := \varphi(\bar{\alpha}) \in \mathbb{Z}_p.$$

Proposition 5.11 *We have*

$$\partial_p \mathbf{K} \in \mathrm{H}^1_s(K_p, V_f)^{-a} \quad \textit{and} \quad \exp^*_{-a} \partial_p \mathbf{K} = \frac{\mathrm{ord}_p(q_E)}{p} \Pi_p.$$

In particular

$$\mathrm{res}_p(\mathbf{K})^a := \mathrm{res}_p(\mathbf{K}) + a\, \mathrm{res}_p(\mathbf{K}^{\mathrm{Fr}_p})$$

is crystalline, i.e. lies in $\mathrm{H}^1_f(\mathbb{Q}_p, V_f)^a$.

Proof Recall we can regard $\mathrm{Hom}_{\mathrm{cont}}(\Gamma_{\mathrm{ant}}, \Phi_{\infty,\mathrm{p}} \otimes \mathbb{Q}_p)$ as a subspace of $\mathrm{H}^1_s(K_p, V_f)$. By (2.15) Fr_p acts on $\mathrm{Hom}_{\mathrm{cont}}(\Gamma_{\mathrm{ant}}, \Phi_{\infty,\mathrm{p}} \otimes \mathbb{Q}_p)$ as multiplication by $-a$. Hence $\mathrm{res}_p(\mathbf{K})$ belongs to $\mathrm{H}^1_s(K_p, V_f)^{-a}$ by Proposition 5.10, and the formula for $\exp^*_{-a} \partial_p \mathbf{K}$ follows from Corollary 2.7. □

For $i = 1, 2$ let

$$\mathrm{Tr}_i = \sum_{\sigma \in \tilde{S}} \psi_i(\sigma)\sigma : \mathrm{H}^1(H, V_f) \longrightarrow \mathrm{H}^1(K, V_f \otimes \psi_i)$$

denote the trace map onto the ψ_i-isotypic component and consider the isomorphism given by Shapiro's Lemma

$$\mathrm{Sh} : \mathrm{H}^1(K, V_f \otimes \psi_i) \xrightarrow{\cong} \mathrm{H}^1(\mathbb{Q}, V_f \otimes \mathrm{Ind}_{G_K}^{G_{\mathbb{Q}}}(\psi_i)) =: \mathrm{H}^1(\mathbb{Q}, V_i). \tag{5.19}$$

Define

$$\mathbf{K}^{\psi_i} := \mathrm{Sh}(\mathrm{Tr}_i(\mathbf{K})) \in \mathrm{H}^1(\mathbb{Q}, V_i).$$

It follows from Proposition 5.10 that $\mathbf{K}^{\psi_i}$ lies in $\mathrm{Sel}_{(p)}(V_i)$.

Frobenius element $\mathrm{Fr}_p \in G_{\mathbb{Q}_p}$ acts on $\mathrm{H}^1(K_p, V_f)$ as an involution and we may consider the decomposition in $\pm$-eigenspaces

$$\mathrm{H}^1(K_p, V_f \otimes L_p) = \mathrm{H}^1(K_p, V_f \otimes L_p)^+ \oplus \mathrm{H}^1(K_p, V_f \otimes L_p)^-.$$

As one readily verifies, Shapiro's isomorphism restricts to

$$\mathrm{Sh}_p : \mathrm{H}^1(K_p, V_f \otimes L_p)^{\pm} \xrightarrow{\cong} \mathrm{H}^1(\mathbb{Q}_p, V_f \otimes V_{\psi_i}^{\pm}). \tag{5.20}$$

Corollary 5.12 *Set* $h = [H : K]$. *The Kolyvagin class* $\mathbf{K}^{\psi_i}$ *satisfies*

$$\mathbf{K}^{\psi_i} = \frac{h\, \mathrm{ord}_p(q_E)\Pi_p}{p} \cdot \xi_i^{-a}.$$

Proof Since $\psi_{i|G_{K_p}} = 1$, the restriction of Tr_i to $\mathrm{H}^1(H_{\mathfrak{p}}, V_f \otimes L_p) = \mathrm{H}^1(K_p, V_f \otimes L_p)$ is multiplication by h. Hence $\mathrm{res}_p(\mathbf{K}^{\psi_i}) = h \cdot \mathrm{Sh}_p(\mathrm{res}_p(\mathbf{K}))$.

By Proposition 5.11 we have $\partial_p \mathbf{K} \in \mathrm{H}^1_s(K_p, V_f)^{-a}$, and by (5.20), it follows that $\partial_p \mathbf{K}^{\psi_i} \in \mathrm{H}^1_s(\mathbb{Q}_p, V_{\psi_i}^{-a})$. Recall the choice of basis $\xi_i^{\pm}$ made in Corollary 5.3. We may thus write

$$\mathbf{K}^{\psi_i} = \frac{\exp^*_{-a}(\partial_p \mathbf{K}^{\psi_i})}{\exp^*_{-a}(\partial_p \xi_i^{-a})} \cdot \xi_i^{-a}.$$

By definition of the basis $\{\xi_i^+, \xi_i^-\}$ and by (5.9), the denominator in the above expression is $\exp^*(X_{-a})\langle v_i^{-a}, \omega_i^{-a}\rangle = \langle v_i^{-a}, \omega_i^{-a}\rangle$. Let

$$R \otimes v_i^{-a} \in \mathrm{H}^1_s(K_p, V_f) \otimes V_{\psi_i}^{-a}$$

denote the image of $\partial_p \mathbf{K}^{\psi_i}$ via the isomorphism (5.5). Then

$$\mathbf{K}^{\psi_i} = \frac{\exp^*(R)\langle v_i^{-a}, \omega_i^{-a}\rangle}{\langle v_i^{-a}, \omega_i^{-a}\rangle} \cdot \xi_i^{-a} = \exp^*(R) \cdot \xi_i^{-a}.$$

In order to compute the dual exponential of R, we need the following explicit expression for (5.22) for Sh_p. Recall that, as a $L_p[G_{\mathbb{Q}_p}]$-module,

$$V_{\psi_i} = \mathrm{Ind}_{G_{K_p}}^{G_{\mathbb{Q}_p}}(1) = \{v : G_{\mathbb{Q}_p} \longrightarrow L_p \mid v(\sigma\tau) = v(\tau)\ \forall \sigma \in G_{K_p}, \tau \in G_{\mathbb{Q}_p}\}.$$

Consider the map

$$ev : V_{\psi_i} \longrightarrow L_p, \qquad v \mapsto v(1).$$

It is an equivariant G_{K_p}-morphism which is compatible with the inclusion $G_{K_p} \hookrightarrow G_{\mathbb{Q}_p}$, so it induces a morphism

$$\mathrm{Sh}_p^{-1} : \mathrm{H}^1(\mathbb{Q}_p, V_f \otimes V_{\psi_i}) \longrightarrow \mathrm{H}^1(K_p, V_f \otimes L_p) = \mathrm{H}^1(K_p, V_f \otimes \psi_i), \qquad (5.21)$$

which is the inverse of Shapiro's isomorphism (5.19) restricted to G_{K_p}. More explicitly, if $\xi : G_{\mathbb{Q}_p} \longrightarrow V_f \otimes V_{\psi_i}$ represents a class in $\mathrm{H}^1(\mathbb{Q}_p, V_f \otimes V_{\psi_i})$, then

$$\mathrm{Sh}_p^{-1}(\xi) := (\mathrm{id} \otimes ev) \circ \xi_{|G_{K_p}} \qquad (5.22)$$

represents its image via (5.21). Recall that $R \otimes v_i^{-a}$ is the image of $h \cdot \partial_p \mathbf{K}$ via the composition

$$\mathrm{H}^1_s(K_p, V_f \otimes L_p)^{-a} \xrightarrow{\mathrm{Sh}_p} \mathrm{H}^1_s(\mathbb{Q}_p, V_f \otimes V_{\psi_i}^{-a}) \xrightarrow{\mathrm{Res}} \mathrm{H}^1_s(K_p, V_f \otimes V_{\psi_i}^{-a})^{G_{\mathbb{Q}_p}}$$
$$\cong \mathrm{H}^1_s(K_p, V_f)^{-a} \otimes V_{\psi_i}^{-a}.$$

We conclude that the element $R \otimes v_i^{-a}$ is represented by the cocycle

$$h \cdot \mathrm{Res}(\mathrm{Sh}_p(\partial_p \mathbf{K})) : G_{K_p} \longrightarrow V_f \otimes V_{\psi_i}^{-a}$$

satisfying

$$R \cdot (v_i^{-a}(1)) = h \cdot \partial_p \mathbf{K}.$$

In other words, by (5.2), we have

$$\exp^*(R) = h \cdot \exp^*(\partial_p \mathbf{K}).$$

The statement follows by applying Proposition 5.11. □

Corollary 5.13 *We have*

$$I_p(f, g_\alpha, h_\alpha) = \frac{\sqrt{c} \cdot 2p(1-1/p)^2}{\mathrm{ord}_p(q_E)} \cdot \frac{\sqrt{L(E \otimes \rho, 1)}}{\pi^2 \langle f, f \rangle \Pi_p} \cdot \frac{1}{\mathcal{L}_{g_\alpha}} \times \log_p(Q_p^a) \mod L^\times,$$

where $Q_p^a \in E(K_p)^a$ is characterized by $\delta_p(Q_p^a) = \mathrm{res}_p(\mathbf{K})^a \in \mathrm{H}^1_f(K_p, V_f)^a$.

Proof By (5.20), if $\pi_a : \mathrm{H}^1(\mathbb{Q}_p, V_i) \longrightarrow \mathrm{H}^1(\mathbb{Q}_p, V_i^a)$ denotes the natural projection, then

$$\pi_a \, \mathrm{res}_p(\mathbf{K}^{\psi_i}) = h \cdot \mathrm{Sh}_p(\mathrm{res}_p(\mathbf{K})^a) \in \mathrm{H}^1_f(\mathbb{Q}_p, V_i^a). \tag{5.23}$$

Arguing as in the proof of Corollary 5.12, write

$$A \otimes v_i^a \in \mathrm{H}^1_f(K_p, V_f)^a \otimes V_{\psi_i}^a$$

for the image of (5.23) via the isomorphism (5.5). Then

$$A = \frac{h \cdot \mathrm{res}_p(\mathbf{K})^a}{v_i^a(1)} = h \cdot \mathrm{res}_p(\mathbf{K})^a.$$

Combining this with Corollary 5.12 and Lemma 5.5, we obtain

$$h \cdot \mathrm{res}_p(\mathbf{K})^a \otimes v_i^a = \pi_a \, \mathrm{res}_p(\mathbf{K}^{\psi_i}) = \frac{h \, \mathrm{ord}_p(q_E) \Pi_p}{p} \delta_p(P^a) \otimes v_i^a.$$

Then

$$\delta_p(P^a) = \frac{p}{\mathrm{ord}_p(q_E) \Pi_p} \cdot \mathrm{res}_p(\mathbf{K})^a,$$

and the thesis follows by applying Theorem 5.6. □

References

1. Gross, B.H., Zagier, D.: Heegner points and derivatives of L-series. Invent. Math. **84**(2), 225–320
2. Yuan, X., Zhang, Z., Zhang, W.: The Gross-Zagier Formula on Shimura Curves. Princeton University Press, Annals of Mathematics Studies (2012)
3. Bertolini, M., Darmon, M., Prasanna, M.: Generalised Heegner cycles and p-adic Rankin L-series. Duke Math. J. **162**(6), 1033–1148 (2013)
4. Disegni, D.: The p-adic Gross-Zagier formula on Shimura curves. Compos. Math. **153**, 1987–2074 (2017)
5. Bertolini, M., Seveso, M.A., Venerucci, R.: Reciprocity laws for balanced diagonal classes, Heegner points, Stark-Heegner points and diagonal classes. Astérisque **434**, 77–174 (2022)
6. Darmon, H., Rotger, V.: Diagonal cycles and Euler systems I: a p-adic gross-zagier formula. Ann. Sci. Ec. Norm. Supér. **47** (4), 779–832
7. Darmon, H., Rotger, V.: Diagonal cycles and Euler systems II: The Birch and Swinnerton-Dyer conjecture for Hasse-Weil-Artin L-functions. J. Amer. Math. Soc. **30** (3), 601–672
8. Darmon, H., Lauder, A., Rotger, V.: Stark Points and p-adic Iterated Integrals Attached to Modular forms of Weight One. Forum Math. Pi **3**, e8
9. Gatti, F., Guitart, X., Masdeu, M., Rotger, V.: Special values of triple-product p-adic L-functions and non-crystalline diagonal classes to appear. J. Théor. Nombres Bordeaux
10. Bertolini, M., Darmon, H.: A rigid analytic Gross-Zagier formula and arithmetic applications. Annals Math. **146**(1), 111–147
11. Harris, M., Kudla, S.: The central critical value of a triple product L-function. Ann. of Math. (2) **133**(3), 605–672
12. Darmon, H., Rotger, V.: Elliptic curves of rank two and generalised Kato classes. Res. Math. Sci. **3**(1)
13. Andreatta, F., Iovita, A.: Katz Type p-Adic L-Functions for Primes p Non-split in the CM Field, Preprint (2021)
14. Bertolini, M., Seveso, M.A., Venerucci, R.: Balanced diagonal classes and rational points on elliptic curves, Heegner points, Stark-Heegner points and diagonal classes. Astérisque **434**, 175–201 (2022)
15. Darmon, H., Rotger, V.: Stark-Heegner points and diagonal classes, in Heegner Points, Stark-Heegner Points and Diagonal Classes, to Appear in Astérisque
16. Bertolini, M., Seveso, M.A., Venerucci, R.: On Exceptional Zeros of Garrett-Hida p-adic L-Functions. Ann. Math, Québec (2021)
17. Kato, K.: Lectures on the approach to Iwasawa theory for Hasse-Weil L-functions via BdR. Part I, in Arithmetic Algebraic Geometry, Lecture Notes Math, p. 1553. Springer
18. Darmon, H., Rotger, V.: p-adic families of diagonal cycles, in Heegner Points, Stark Heegner Points and Diagonal Classes, to Appear in Astérisque
19. Ohta, M.: On the p-adic Eichler-Shimura isomorphism for Λ-adic cusp forms. Crelle **463**, 49–98 (1995)
20. Ichino, A.: Trilinear forms and the central values of triple product L-functions. Duke Math. J. **145**(2), 281–307
21. Loeffler, D., Zerbes, S.: Iwasawa theory and p-adic L-functions over $\mathbb{Z}_p^2$-extensions. Int. J. Number Theory **10**(8), 2045–2095
22. Darmon, H., Lauder, A., Rotger, V.: First order p-adic deformations of weight one newforms, in L-functions and automorphic forms. Contrib. Math. Comput. Sci. **10**, 39–80. Springer
23. Bertolini, M., Darmon, H.: Heegner points on Mumford-tate curves. Invent. Math. **126** (3), 413–456

On the Anticyclotomic Mazur–Tate Conjecture for Elliptic Curves with Supersingular Reduction

Chan-Ho Kim

Abstract In this short note, we study the anticyclotomic analogue of the "weak" main conjecture of Mazur–Tate on Fitting ideals of Selmer groups for elliptic curves with supersingular reduction.

Keywords Refined Iwasawa theory · Anticyclotomic Iwasawa theory

2000 Mathematics Subject Classification 11F67 · 11G40 · 11R23

1 Introduction

1.1 The Statement of the Main Result

Let E be an elliptic curve of conductor N over $\mathbb{Q}$ and $p \geq 5$ be a prime of supersingular reduction for E, so $a_p(E) = 0$.

Let K be an imaginary quadratic field with $(D_K, Np) = 1$ where p splits and and write $p = \mathfrak{p} \cdot \overline{\mathfrak{p}}$ in K. Write

$$N = N^+ \cdot N^-$$

where a prime divisor of N^+ splits in K and a prime divisor of N^- is inert in K.

This research was partially supported by a KIAS Individual Grant (SP054103) via the Center for Mathematical Challenges at Korea Institute for Advanced Study, by the National Research Foundation of Korea(NRF) grant funded by the Korea government(MSIT) (No. 2018R1C1B6007009, 2019R1A6A1A11051177), and by Global-Learning & Academic research institution for Master's·Ph.D. Students, and Postdocs (LAMP) Program of the National Research Foundation of Korea (NRF) funded by the Ministry of Education (No. RS-2024-00443714).

C.-H. Kim (✉)
Department of Mathematics and Institute of Pure and Applied Mathematics, Jeonbuk National University, Jeollabuk-do, Republic of Korea
e-mail: chanho.math@gmail.com

M. Longo et al. (eds.), *Elliptic Curves and Modular Forms in Arithmetic Geometry*, Springer Proceedings in Mathematics & Statistics 527,
https://doi.org/10.1007/978-3-032-13123-2_8

Assumption 1.1 (odd) *Assume that N^- is the square-free product of an odd number of primes.*

Let K_∞ be the anticyclotomic $\mathbb{Z}_p$-extension of K and K_n be the subextension of K in K_∞ of degree p^n for $n \geq 0$. Let $\Lambda_n = \mathbb{Z}_p[\mathrm{Gal}(K_n/K)] \simeq \mathbb{Z}_p[X]/((1+X)^{p^n}-1)$ be the finite layer Iwasawa algebra. Under Assumption 1.1, denote by

$$\theta_n(E/K) = \sum_{\sigma \in \mathrm{Gal}(K_n/K)} a_\sigma \cdot \sigma \in \Lambda_n$$

Bertolini–Darmon's theta element of E over K_n reviewed in Sect. 2. Let ι be the involution on Λ_n defined by inverting group-like elements. Then we have $\iota(\theta_n(E/K)) = \sum_{\sigma \in \mathrm{Gal}(K_n/K)} a_\sigma \cdot \sigma^{-1}$ and define

$$L_p(E/K_n) = \theta_n(E/K) \cdot \iota(\theta_n(E/K)).$$

Assumption 1.2 *We assume the following hypotheses throughout this article:*

- *(Im) The mod p representation $\overline{\rho} : G_{\mathbb{Q}} = \mathrm{Gal}(\overline{\mathbb{Q}}/\mathbb{Q}) \to \mathrm{GL}_2(\mathbb{F}_p)$ is irreducible. It is surjective if $p = 5$.*
- *(ram) K_∞/K is totally ramified at every prime lying above p.*
- *(CR) If a prime ℓ divides N^- and $\ell^2 \equiv 1 \pmod{p}$, then $\overline{\rho}$ is ramified at ℓ.*

The goal of this article is to prove the following anticyclotomic analogue of the "weak" main conjecture of Mazur and Tate [20, Conj. 3].

Theorem 1.3 *Under (odd), (Im), (ram), and (CR), we have*

$$L_p(E/K_n) \in \mathrm{Fitt}_{\Lambda_n}\left(\mathrm{Sel}(K_n, E[p^\infty])^\vee\right).$$

where $(-)^\vee$ means the Pontryagin dual.

Let $\chi : \mathrm{Gal}(K_n/K) \to \overline{\mathbb{Q}}_p^\times$ be a character and we extend it to Λ_n linearly. Let $I_\chi := \ker\left(\chi : \Lambda_n \to \overline{\mathbb{Q}}_p\right) \subseteq \Lambda_n$ be the augmentation ideal at χ and $(E(K_n) \otimes \overline{\mathbb{Q}}_p)^\chi$ be the χ-isotypic subspace of $E(K_n) \otimes \overline{\mathbb{Q}}_p$. Thanks to [20, Proposition 3], we obtain the "weak" vanishing conjecture.

Corollary 1.4 *Under (odd), (Im), (ram), and (CR), we have*

$$L_p(E/K_n) \in I_\chi^{r_\chi}/I_\chi^{r_\chi+1}$$

where $r_\chi = \dim_{\overline{\mathbb{Q}}_p}(E(K_n) \otimes \overline{\mathbb{Q}}_p)^\chi$.

In particular, $L_p(E/K_n)$ encodes an upper bound of $\mathrm{rk}_{\mathbb{Z}} E(K)$ by considering the trivial character. In the ordinary case, Corollary 1.4 can be deduced from the one-sided divisibility of the anticyclotomic main conjecture more directly. In [6], Darmon studied a similar "weak" vanishing conjecture over ring class extensions of conductor prime to p when (odd) breaks down ($N^- = 1$).

Remark 1.5 In [20], Mazur and Tate formulated "Iwasawa theory for elliptic curves over finite abelian extensions" and several related conjectures. This finite layer setting is much more delicate since the linear algebra techniques for the usual Iwasawa theory (over $\mathbb{Z}_p$-extensions) are not applicable and we cannot ignore "finite errors" in the computation. The formulation of the "main conjecture" also becomes more subtle since the notion of characteristic ideals is not available. The "weak" main conjecture is formulated in terms of Fitting ideals. See [19, Conj. 0.3] for the strong version. Here, the strong version means an equality.

1.2 The Strategy of Proof

The strategy of proof is similar to that in [14]. We begin with the one-sided divisibility of the signed main conjecture. Since signed Selmer groups have no proper Λ-submodules of finite index, we can replace the characteristic ideals by the Fitting ideals. Then we can project the divisibility to finite layers since Fitting ideals are compatible with base change. On the algebraic side, we compare the Fitting ideals of dual signed Selmer groups with the Fitting ideals of dual Selmer groups at finite layers. The difference essentially comes from the failure of the control theorem for Selmer groups in the supersingular case. On the analytic side, we compare the signed p-adic L-functions at finite layers with the square of Bertolini–Darmon's theta elements, and we observe that the same difference occurs. The conclusion follows from the comparison of these differences.

One may ask whether Theorem 1.3 can be upgraded to an equality (the strong version) if we begin with the equality of the signed main conjecture. Comparing with the cyclotomic version [14, 19], the correct formulation of the strong version would require to assume that p does not divide any Tamagawa factor and to put every $L_p(E/K_m)$ with $0 \leq m \leq n$ into the analytic side. Also, since the argument for the strong version in [14] depends heavily on the nature of Kato's zeta elements, we are not sure whether the same idea would apply even after having the correct formulation of the strong version.

1.3 Organization

In Sect. 2, we review the construction of Bertolini–Darmon's theta elements and signed p-adic L-functions. In Sect. 3, we recall the basic objects of signed Iwasawa theory and review the standard results on signed Iwasawa theory. Then we observe how the finite layer specialization of the one-sided divisibility of the signed main conjecture looks like. In Sect. 4, we compare the signed and unsigned objects in finite layers explicitly and deduce the main result.

Acknowledgement It is a great pleasure to dedicate this paper to Massimo Bertolini on the occasion of his 60th birthday for his great work and wonderful inspiration. In my graduate years, the celebrated article [3] by Bertolini and Darmon was one of the very first serious research articles I read. In particular, my thesis [12] can be explained as "a Greenberg–Vatsal type result for the Bertolini–Darmon setting". It is evident that the content of [3] became an important part of my research.

I deeply thank Matteo Longo, Marco Seveso, Stefano Vigni, and Rodolfo Venerucci for inviting me for his birthday conference and for giving me a chance to contribute this volume.

Indeed, the topic of this paper was exactly my original thesis problem suggested by Rob Pollack, but I could not find how to attack it at that time. Later, I found a way to resolve it in the middle of writing the paper with Masato Kurihara [14]. I deeply appreciate Rob Pollack and Masato Kurihara. This is the sequel to [10], which studies the ordinary case.

2 Bertolini–Darmon's Theta Elements and Anticyclotomic p-Adic L-Functions

We quickly review the construction of Gross points of conductor p^n and signed anticyclotomic p-adic L-functions. See [4, 5, 13] for details.

2.1 Gross Points

Let K be the imaginary quadratic field of discriminant $-D_K < 0$. Define

$$\vartheta := \begin{cases} \dfrac{D_K - \sqrt{-D_K}}{2} & \text{if } 2 \nmid D_K \\ \dfrac{D_K - 2\sqrt{-D_K}}{4} & \text{if } 2 \mid D_K \end{cases}$$

so that $\mathcal{O}_K = \mathbb{Z} + \mathbb{Z}\vartheta$. Let B_{N^-} be the definite quaternion algebra over $\mathbb{Q}$ of discriminant N^-. Then there exists an embedding of K into B_{N^-} [25]. More explicitly, we choose a K-basis $(1, J)$ of B_{N^-} so that $B_{N^-} = K \oplus K \cdot J$ such that $\beta := J^2 \in \mathbb{Q}^\times$ with $\beta < 0$, $J \cdot t = \overline{t} \cdot J$ for all $t \in K$, $\beta \in (\mathbb{Z}_q^\times)^2$ for all $q \mid pN^+$, and $\beta \in \mathbb{Z}_q^\times$ for all $q \mid D_K$. Fix a square root $\sqrt{\beta} \in \overline{\mathbb{Q}}$ of β. For a $\mathbb{Z}$-module A, write $\widehat{A} = A \otimes \widehat{\mathbb{Z}}$. Fix an isomorphism

$$i := \prod i_q : \widehat{B}_{N^-}^{(N^-)} \simeq \mathrm{M}_2(\mathbb{A}^{(N^-\infty)})$$

as follows:

- For each finite place $q \mid N^+p$, the isomorphism $i_q : B_{N^-,q} \simeq \mathrm{M}_2(\mathbb{Q}_q)$ is defined by

$$i_q(\vartheta) = \begin{pmatrix} \mathrm{trd}(\vartheta) & -\mathrm{nrd}(\vartheta) \\ 1 & 0 \end{pmatrix}, \qquad i_q(J) = \sqrt{\beta} \cdot \begin{pmatrix} -1 & \mathrm{trd}(\vartheta) \\ 0 & 1 \end{pmatrix}$$

  where trd and nrd are the reduced trace and the reduced norm on B, respectively.
- For each finite place $q \nmid pN^+$, the isomorphism $i_q : B_{N^-,q} \simeq \mathrm{M}_2(\mathbb{Q}_q)$ is chosen so that $i_q\left(\mathcal{O}_K \otimes \mathbb{Z}_q\right) \subseteq \mathrm{M}_2(\mathbb{Z}_q)$.

Under the fixed isomorphism i, for any rational prime q, the local Gross point $\varsigma_q \in B^\times_{N^-,q}$ is defined as follows:

- $\varsigma_q := 1$ in $B^\times_{nN^-,q}$ for $q \nmid pN^+$.
- $\varsigma_q := \frac{1}{\sqrt{D_K}} \cdot \begin{pmatrix} \vartheta & \overline{\vartheta} \\ 1 & 1 \end{pmatrix} \in \mathrm{GL}_2(K_{\mathfrak{q}}) = \mathrm{GL}_2(\mathbb{Q}_q)$ for $q \mid N^+$ with $q = \mathfrak{q}\overline{\mathfrak{q}}$ in $\mathcal{O}_K$.
- $\varsigma^{(n)}_p = \begin{pmatrix} \vartheta & -1 \\ 1 & 0 \end{pmatrix} \cdot \begin{pmatrix} p^n & 0 \\ 0 & 1 \end{pmatrix} \in \mathrm{GL}_2(K_{\mathfrak{p}}) = \mathrm{GL}_2(\mathbb{Q}_p)$ where $p = \mathfrak{p}\overline{\mathfrak{p}}$ splits in K.

By using the fixed embedding of K into B_{N^-}, we define $x_n : \widehat{K}^\times \to \widehat{B}^\times_{N^-}$ by $x_n(a) = a \cdot \varsigma^{(n)} := a \cdot \left(\varsigma^{(n)}_p \times \prod_{q \neq p} \varsigma_q\right)$. The collection $\left\{x_n(a) : a \in \widehat{K}^\times\right\}$ of points is called the **Gross points of conductor** p^n **on** $\widehat{B}^\times_{N^-}$. The fixed embedding $K \hookrightarrow B_{N^-}$ also induces an optimal embedding of $\mathcal{O}_n = \mathbb{Z} + p^n\mathcal{O}_K$ into the Eichler order $B_{N^-} \cap \varsigma^{(n)} \widehat{R}_{N^+} (\varsigma^{(n)})^{-1}$ where R_{N^+} is the Eichler order of level N^+ under the fixed isomorphism i.

2.2 Theta Elements and p-Adic L-Functions

2.2.1. Let $\phi_f : B^\times_{N^-} \backslash \widehat{B}^\times_{N^-} / \widehat{R}^\times_{N^+} \to \mathbb{C}$ be the Jacquet–Langlands transfer of f. Since $B^\times_{N^-} \backslash \widehat{B}^\times_{N^-} / \widehat{R}^\times_{N^+}$ is a finite set and f is a Hecke eigenform, we are able to and do normalize

$$\phi_f : B^\times_{N^-} \backslash \widehat{B}^\times_{N^-} / \widehat{R}^\times_{N^+} \to \mathbb{Z}_p$$

such that the image of ϕ_f does not lie in $p\mathbb{Z}_p$. This normalization is closely related to the N^--new congruence ideal [12, 16, 22]. Let

$$\widetilde{\theta}_n(E/K) = \sum_{[a] \in \mathcal{G}_n} \phi_f(x_n(a)) \cdot [a] \in \mathbb{Z}_p[\mathcal{G}_n]$$

where $\mathcal{G}_n = K^\times \backslash \widehat{K}^\times / \widehat{\mathcal{O}}^\times_n$ and $[a]$ is the image of $a \in \widehat{K}^\times$ in $\mathcal{G}_n$. Then **Bertolini–Darmon's theta element** $\theta_n(E/K)$ **of** E **over** K_n is defined by the image of $\widetilde{\theta}_n(E/K)$ in Λ_n

$$\begin{array}{ccc} \mathbb{Z}_p[\mathcal{G}_n] & \longrightarrow & \Lambda_n \\ \widetilde{\theta}_n(E/K) & \longmapsto & \theta_n(E/K). \end{array}$$

It is known that $\theta_n(E/K)$ interpolates "an half of" $L(E, \chi, 1)$ where χ runs over characters on $\mathrm{Gal}(K_n/K)$ [5, 13].

2.2.2. Let $\omega_n = \omega_n(X) = (1+X)^{p^n} - 1$ and $\Phi_n(1+X) = \omega_n(X)/\omega_{n-1}(X)$ where Φ_n is the p^n-th cyclotomic polynomial. Fix a generator γ of $\mathrm{Gal}(K_\infty/K)$ and take a generator γ_n of $\mathrm{Gal}(K_n/K)$ as the image of γ. Then we have isomorphisms $\Lambda_n \simeq \mathbb{Z}_p[X]/(\omega_n(X))$ and $\Lambda \simeq \mathbb{Z}_p[\![X]\!]$ by sending the generators to $1+X$. Via the latter isomorphism, we also regard $\omega_n \in \Lambda$. Let $\omega_0^{\pm}(X) := X$, $\widetilde{\omega}_0^{\pm}(X) := 1$, and

$$\omega_n^+ = \omega_n^+(X) := X \cdot \prod_{2 \le m \le n, m:\ \text{even}} \Phi_m(1+X), \qquad \omega_n^- = \omega_n^-(X) := X \cdot \prod_{1 \le m \le n, m:\ \text{odd}} \Phi_m(1+X),$$

$$\widetilde{\omega}_n^+ = \widetilde{\omega}_n^+(X) := \prod_{2 \le m \le n, m:\ \text{even}} \Phi_m(1+X), \qquad \widetilde{\omega}_n^- = \widetilde{\omega}_n^-(X) := \prod_{1 \le m \le n, m:\ \text{odd}} \Phi_m(1+X).$$

Then we have $\omega_n(X) = \omega_n^{\pm}(X) \cdot \widetilde{\omega}_n^{\mp}(X)$, respectively. We also regard $\omega_n^{\pm}$, $\widetilde{\omega}_n^{\pm}$ as elements in Λ_n or Λ.

Proposition 2.1 *Let ϵ be the sign of $(-1)^n$. Then:*

(1) $\omega_n^{\epsilon}\theta_n(E/K) = 0$.
(2) There exists a unique element $\theta_n^{\epsilon}(E/K)$ in $\Lambda/\omega_n^{\epsilon}\Lambda$ such that

$$\theta_n(E/K) = \widetilde{\omega}_n^{-\epsilon} \cdot \theta_n^{\epsilon}(E/K). \tag{1}$$

Proof See [7, Proposition 2.8]. □

Thanks to [7, Lemma 2.9], $\{(-1)^{n/2} \cdot \theta_n^+(E/K) : n \text{ is even}\}$ and $\{(-1)^{(n+1)/2} \cdot \theta_n^-(E/K) : n \text{ is odd}\}$ form compatible sequences with respect to $\Lambda/\omega_n^+\Lambda$ with even n and $\Lambda/\omega_n^-\Lambda$ with odd n, respectively. The signed p-adic L-functions for E and K are defined by

$$\begin{aligned} L_p^+(E/K_\infty) &:= \varprojlim_{n,\ \text{even}} \left((-1)^{n/2} \cdot \theta_n^+(E/K) \cdot (-1)^{n/2} \cdot \iota(\theta_n^+(E/K))\right) \\ &= \varprojlim_{n,\ \text{even}} \left(\theta_n^+(E/K) \cdot \iota(\theta_n^+(E/K))\right) \in \Lambda, \\ L_p^-(E/K_\infty) &:= \varprojlim_{n,\ \text{odd}} \left((-1)^{(n+1)/2} \cdot \theta_n^-(E/K) \cdot (-1)^{(n+1)/2} \cdot \iota(\theta_n^-(E/K))\right) \\ &= \varprojlim_{n,\ \text{odd}} \left(\theta_n^-(E/K) \cdot \iota(\theta_n^-(E/K))\right) \in \Lambda. \end{aligned} \tag{2}$$

3 Anticyclotomic Iwasawa Theory for Elliptic Curves with Supersingular Reduction

3.1 Basic Objects of ±-Iwasawa Theory

We quickly recall the basic objects of ±-Iwasawa theory, which is initiated by S. Kobayashi [17] and R. Pollack [21]. See also [9].

3.1.1 Local Conditions at *p*

Let E be an elliptic curve over $\mathbb{Q}$ with $a_p(E) = 0$. Since $p = \mathfrak{p} \cdot \overline{\mathfrak{p}}$ in K and the (ram) condition, we are able to write

$$K_{n,p} = K_{n,\mathfrak{p}} \oplus K_{n,\overline{\mathfrak{p}}} \simeq \mathbb{Q}_{n,p} \oplus \mathbb{Q}_{n,p} \tag{3}$$

where $K_{n,p} = K_n \otimes \mathbb{Q}_p$ and $K_{n,v}$ is the completion of K_n at $v \in \{\mathfrak{p}, \overline{\mathfrak{p}}\}$. Then we define

$$E^{+}(K_{n,p}) := \{P \in E(K_{n,p}) : \mathrm{Tr}_{n/m+1}(P) \in E(K_{m,p}) \text{ for even } m \ (0 \leq m < n)\}$$
$$E^{-}(K_{n,p}) := \{P \in E(K_{n,p}) : \mathrm{Tr}_{n/m+1}(P) \in E(K_{m,p}) \text{ for odd } m \ (0 \leq m < n)\}$$

where $\mathrm{Tr}_{n/m+1} : E(K_{n,p}) \to E(K_{m+1,p})$ is the trace map.

3.1.2 The Norm Subgroups

Let $\widehat{E}$ be the formal group associated to E and $\mathfrak{m}_n$ be the maximal ideal of $\mathbb{Q}_{n,p}$. We define

$$\widehat{E}^{+}(\mathfrak{m}_n) := \{P \in \widehat{E}(\mathfrak{m}_n) : \mathrm{Tr}_{n/m+1}(P) \in \widehat{E}(\mathfrak{m}_m) \text{ for even } m \ (0 \leq m < n)\}$$
$$\widehat{E}^{-}(\mathfrak{m}_n) := \{P \in \widehat{E}(\mathfrak{m}_n) : \mathrm{Tr}_{n/m+1}(P) \in \widehat{E}(\mathfrak{m}_m) \text{ for odd } m \ (0 \leq m < n)\}$$

where $\mathrm{Tr}_{n/m+1} : \widehat{E}(\mathfrak{m}_n) \to \widehat{E}(\mathfrak{m}_{m+1})$ is the trace map.

3.1.3 ±-Selmer Groups

Following [11], we define the **±-Selmer groups of** $E[p^\infty]$ **over** K_n by

$$\mathrm{Sel}^{\pm}(K_n, E[p^\infty]) := \ker\left(\mathrm{Sel}(K_n, E[p^\infty]) \to \frac{\mathrm{H}^1(K_{n,p}, E[p^\infty])}{E^{\pm}(K_{n,p}) \otimes \mathbb{Q}_p/\mathbb{Z}_p}\right)$$

where $\mathrm{Sel}(K_n, E[p^\infty])$ is the standard Selmer group of $E[p^\infty]$ over K_n. The $\pm$-**Selmer groups of** $E[p^\infty]$ **over** K_∞ is defined by

$$\mathrm{Sel}^\pm(K_\infty, E[p^\infty]) := \varinjlim_n \mathrm{Sel}^\pm(K_n, E[p^\infty]),$$

respectively.

3.2 Signed Iwasawa Theory

We recall the Euler system divisibility of the signed main conjectures.

Theorem 3.1 *Under (odd), (Im), (ram), and (CR), we have the following statements:*

(1) $\mathrm{Sel}^\pm(K_\infty, E[p^\infty])$ *is* Λ*-cotorsion.*
(2) $\left(L^\pm_p(E/K_\infty)\right) \subseteq \mathrm{char}_\Lambda \left(\mathrm{Sel}^\pm(K_\infty, E[p^\infty])^\vee\right)$.

Proof See [1, 7, 18, 22, 24]. All the conditions in Assumption 1.2 are needed to have this divisibility. □

The signed main conjecture claims that

$$\left(L^\pm_p(E/K_\infty)\right) = \mathrm{char}_\Lambda \left(\mathrm{Sel}^\pm(K_\infty, E[p^\infty])^\vee\right) \tag{4}$$

We recall B.D.Kim's result on the non-existence of proper Λ-submodules of finite index. See [23] for the p-inert case.

Theorem 3.2 *If* $\mathrm{Sel}^\pm(K_\infty, E[p^\infty])$ *is* Λ*-cotorsion, then* $\mathrm{Sel}^\pm(K_\infty, E[p^\infty])$ *has no proper* Λ*-submodule of finite index, respectively.*

Proof See [11, Theorem 1.1]. □

Corollary 3.3 *Under (odd), (Im), (ram), and (CR),* $\mathrm{Sel}^\pm(K_\infty, E[p^\infty])$ *has no proper* Λ*-submodule of finite index; thus,*

$$\mathrm{char}_\Lambda \mathrm{Sel}^\pm(K_\infty, E[p^\infty]) = \mathrm{Fitt}_\Lambda \mathrm{Sel}^\pm(K_\infty, E[p^\infty]).$$

Proof It follows from Theorems 3.1 and 3.2. □

We recall the signed version of the control theorem.

Theorem 3.4 *The restriction map*

$$\mathrm{Sel}^\pm(K_n, E[p^\infty])[\omega^\mp_n] \to \mathrm{Sel}^\pm(K_\infty, E[p^\infty])[\omega^\mp_n]$$

is injective with the finite cokernel whose size is bounded independently of n*. If we further assume that* $p \nmid \mathrm{Tam}(E)$*, then the restriction map is an isomorphism.*

Proof This is [9, Theorem 6.8]. The $a_p(E) = 0$ condition ensures that $E(K)[p]$ is trivial, so the injectivity of the restriction map follows as explained in [17, Lemma 9.1]. Since the size of the cokernel depends only on prime-to-p local conditions, the situation coincides with the ordinary case. By [8, Lemma 3.3], the cokernel is finite and is bounded independently of n. When $p \nmid \mathrm{Tam}(E)$, the cokernel vanishes, so the restriction map becomes an isomorphism. See also [17, Theorem 9.3] and [8, Proposition 3.8] for further details. □

3.3 The Consequence

Corollary 3.5 *Under (odd), (Im), (ram), and (CR), we have*

$$\left(\widetilde{\omega}_n^{\mp} \cdot L_p^{\pm}(E/K_\infty) \ (mod\ \omega_n)\right) \subseteq \left(\widetilde{\omega}_n^{\mp}\right) \cdot \mathrm{Fitt}_{\Lambda_n}\left(\mathrm{Sel}^{\pm}(K_n, E[p^\infty])^\vee\right)$$

in Λ_n, respectively.

Proof By applying Theorem 3.1 and Corollary 3.3, we have

$$\left(L_p^{\pm}(E/K_\infty)\right) \subseteq \mathrm{Fitt}_{\Lambda}\left(\mathrm{Sel}^{\pm}(K_\infty, E[p^\infty])^\vee\right).$$

Taking the quotient by $\omega_n^{\pm}$, we obtain

$$\left(L_p^{\pm}(E/K_\infty) \ (\mathrm{mod}\ \omega_n^{\pm})\right) \subseteq \mathrm{Fitt}_{\Lambda_n/\omega_n^{\pm}}\left(\left(\mathrm{Sel}^{\pm}(K_\infty, E[p^\infty])[\omega_n^{\pm}]\right)^\vee\right)$$

in $\Lambda_n/\omega_n^{\pm}$, respectively. By the signed control theorem (Theorem 3.4), we obtain

$$\left(L_p^{\pm}(E/K_\infty) \ (\mathrm{mod}\ \omega_n^{\pm})\right) \subseteq \mathrm{Fitt}_{\Lambda_n/\omega_n^{\pm}}\left(\left(\mathrm{Sel}^{\pm}(K_n, E[p^\infty])[\omega_n^{\pm}]\right)^\vee\right)$$

in $\Lambda_n/\omega_n^{\pm}$, respectively. Since Fitting ideals are compatible with base change, we have the equality

$$\mathrm{Fitt}_{\Lambda_n/\omega_n^{\pm}}\left(\left(\mathrm{Sel}^{\pm}(K_n, E[p^\infty])[\omega_n^{\pm}]\right)^\vee\right) = \frac{\mathrm{Fitt}_{\Lambda_n}\left(\mathrm{Sel}^{\pm}(K_n, E[p^\infty])^\vee\right) + (\omega_n^{\pm})}{(\omega_n^{\pm})}$$

in $\Lambda_n/\omega_n^{\pm}$. Thus, we have inclusions

$$\left(L_p^{\pm}(E/K_\infty) \ (\mathrm{mod}\ \omega_n)\right) + (\omega_n^{\pm}) \subseteq \mathrm{Fitt}_{\Lambda_n}\left(\mathrm{Sel}^{\pm}(K_n, E[p^\infty])^\vee\right) + (\omega_n^{\pm})$$

in Λ_n, respectively. Multiplying $\widetilde{\omega}_n^{\mp}$, the conclusion immediately follows. □

Remark 3.6 If we further assume $p \nmid \mathrm{Tam}(E)$ and the $\pm$-main conjectures (4), then the inclusion in Corollary 3.5 becomes an equality.

4 The Proof of Theorem 1.3

By using the global Poitou–Tate duality, consider the exact sequence of Λ_n-modules (cf. [15, (4.2)])

$$\left(\frac{E(K_{n,p})\otimes \mathbb{Q}_p/\mathbb{Z}_p}{E^{\pm}(K_{n,p})\otimes \mathbb{Q}_p/\mathbb{Z}_p}\right)^{\vee} \xrightarrow{\iota^{\pm}} \mathrm{Sel}(K_n, E[p^{\infty}])^{\vee} \longrightarrow \mathrm{Sel}^{\pm}(K_n, E[p^{\infty}])^{\vee} \longrightarrow 0.$$

Then we have

$$\mathrm{Fitt}_{\Lambda_n}\left(\left(\frac{E(K_{n,p})\otimes \mathbb{Q}_p/\mathbb{Z}_p}{E^{\pm}(K_{n,p})\otimes \mathbb{Q}_p/\mathbb{Z}_p}\right)^{\vee}/\ker(\iota^{\pm})\right)\cdot \mathrm{Fitt}_{\Lambda_n}\left(\mathrm{Sel}^{\pm}(K_n, E[p^{\infty}])^{\vee}\right) \subseteq \mathrm{Fitt}_{\Lambda_n}\left(\mathrm{Sel}(K_n, E[p^{\infty}])^{\vee}\right)$$

thanks to the behavior of Fitting ideals in the short exact sequence. Due to the behavior of Fitting ideals under the quotient map, we also have

$$\mathrm{Fitt}_{\Lambda_n}\left(\left(\frac{E(K_{n,p})\otimes \mathbb{Q}_p/\mathbb{Z}_p}{E^{\pm}(K_{n,p})\otimes \mathbb{Q}_p/\mathbb{Z}_p}\right)^{\vee}\right) \subseteq \mathrm{Fitt}_{\Lambda_n}\left(\left(\frac{E(K_{n,p})\otimes \mathbb{Q}_p/\mathbb{Z}_p}{E^{\pm}(K_{n,p})\otimes \mathbb{Q}_p/\mathbb{Z}_p}\right)^{\vee}/\ker(\iota^{\pm})\right).$$

Applying [15, Lemma 3.14] with (3), we observe that

$$\begin{aligned}\left(\frac{E(K_{n,p})\otimes \mathbb{Q}_p/\mathbb{Z}_p}{E^{\pm}(K_{n,p})\otimes \mathbb{Q}_p/\mathbb{Z}_p}\right)^{\vee} &\simeq \left(\frac{\widehat{E}(\mathfrak{m}_n)\otimes \mathbb{Q}_p/\mathbb{Z}_p}{\widehat{E}^{\pm}(\mathfrak{m}_n)\otimes \mathbb{Q}_p/\mathbb{Z}_p}\right)^{\vee} \oplus \left(\frac{\widehat{E}(\mathfrak{m}_n)\otimes \mathbb{Q}_p/\mathbb{Z}_p}{\widehat{E}^{\pm}(\mathfrak{m}_n)\otimes \mathbb{Q}_p/\mathbb{Z}_p}\right)^{\vee}\\ &\simeq \left(\frac{\widehat{E}(\mathfrak{m}_n)}{\widehat{E}^{\pm}(\mathfrak{m}_n)}\otimes \mathbb{Q}_p/\mathbb{Z}_p\right)^{\vee} \oplus \left(\frac{\widehat{E}(\mathfrak{m}_n)}{\widehat{E}^{\pm}(\mathfrak{m}_n)}\otimes \mathbb{Q}_p/\mathbb{Z}_p\right)^{\vee}.\end{aligned}$$

Due to [9, Proposition 4.11], we have the following exact sequence

$$\begin{array}{ccccccccc} 0 & \longrightarrow & \widehat{E}(p\mathbb{Z}_p) & \xrightarrow{f} & \widehat{E}^{+}(\mathfrak{m}_n)\oplus \widehat{E}^{-}(\mathfrak{m}_n) & \xrightarrow{g} & \widehat{E}(\mathfrak{m}_n) & \longrightarrow & 0\\ & & \downarrow{\simeq} & & \downarrow{\simeq} & & \downarrow{\simeq}\ \text{[9, Proposition 5.8]} & & \\ 0 & \longrightarrow & \widetilde{\omega}^{+}_n\widetilde{\omega}^{-}_n\Lambda_n & \longrightarrow & \widetilde{\omega}^{-}_n\Lambda_n \oplus \widetilde{\omega}^{+}_n\Lambda_n & \longrightarrow & \left(\widetilde{\omega}^{+}_n, \widetilde{\omega}^{-}_n\right)\Lambda_n & \longrightarrow & 0\end{array}$$

where f is the diagonal embedding and $g : (a, b) \mapsto a - b$. Note that $\widetilde{\omega}^{+}_n\widetilde{\omega}^{-}_n\Lambda_n \simeq \Lambda_n/X\Lambda_n \simeq \mathbb{Z}_p$. Since $\widehat{E}(\mathfrak{m}_n)/\widehat{E}^{\pm}(\mathfrak{m}_n) \simeq \left(\widetilde{\omega}^{+}_n, \widetilde{\omega}^{-}_n\right)\Lambda_n/\widetilde{\omega}^{\mp}_n\Lambda_n$, we have

$$\mathrm{Fitt}_{\Lambda_n}\left(\left(\frac{\left(\widetilde{\omega}^{+}_n, \widetilde{\omega}^{-}_n\right)\Lambda_n}{\widetilde{\omega}^{\mp}_n\Lambda_n}\otimes \mathbb{Q}_p/\mathbb{Z}_p\right)^{\vee}\right) = \mathrm{Fitt}_{\Lambda_n}\left(\left(\frac{\widehat{E}(\mathfrak{m}_n)}{\widehat{E}^{\pm}(\mathfrak{m}_n)}\otimes \mathbb{Q}_p/\mathbb{Z}_p\right)^{\vee}\right).$$

Proposition 4.1

$$\mathrm{Fitt}_{\Lambda_n}\left(\left(\frac{(\widetilde{\omega}_n^+, \widetilde{\omega}_n^-)\,\Lambda_n}{\widetilde{\omega}_n^{\mp}\Lambda_n} \otimes \mathbb{Q}_p/\mathbb{Z}_p\right)^\vee\right) = \widetilde{\omega}_n^{\mp}\Lambda_n,$$

respectively.

Proof See [14, Proposition 4.1]. □

To sum up, we have

$$\begin{aligned}
&\left(\widetilde{\omega}_n^{\mp}\right)^2 \cdot \mathrm{Fitt}_{\Lambda_n}\left(\mathrm{Sel}^{\pm}(K_n, E[p^\infty])^\vee\right) \\
&= \mathrm{Fitt}_{\Lambda_n}\left(\left(\frac{(\widetilde{\omega}_n^+, \widetilde{\omega}_n^-)\,\Lambda_n}{\widetilde{\omega}_n^{\mp}\Lambda_n} \otimes \mathbb{Q}_p/\mathbb{Z}_p\right)^\vee\right)^2 \cdot \mathrm{Fitt}_{\Lambda_n}\left(\mathrm{Sel}^{\pm}(K_n, E[p^\infty])^\vee\right) \\
&= \mathrm{Fitt}_{\Lambda_n}\left(\left(\widehat{E}(\mathfrak{m}_n)/\widehat{E}^{\pm}(\mathfrak{m}_n) \otimes \mathbb{Q}_p/\mathbb{Z}_p\right)^\vee\right)^2 \cdot \mathrm{Fitt}_{\Lambda_n}\left(\mathrm{Sel}^{\pm}(K_n, E[p^\infty])^\vee\right) \\
&\subseteq \mathrm{Fitt}_{\Lambda_n}\left(\left(\frac{E(K_{n,p}) \otimes \mathbb{Q}_p/\mathbb{Z}_p}{E^{\pm}(K_{n,p}) \otimes \mathbb{Q}_p/\mathbb{Z}_p}\right)^\vee /\ker(\iota^{\pm})\right) \cdot \mathrm{Fitt}_{\Lambda_n}\left(\mathrm{Sel}^{\pm}(K_n, E[p^\infty])^\vee\right) \\
&\subseteq \mathrm{Fitt}_{\Lambda_n}\left(\mathrm{Sel}(K_n, E[p^\infty])^\vee\right).
\end{aligned}$$

By Corollary 3.5 with the multiplication by $\widetilde{\omega}_n^{\mp}$, we have

$$\left(\left(\widetilde{\omega}_n^{\mp}\right)^2 \cdot L_p^{\pm}(E/K_\infty) \pmod{\omega_n}\right) \subseteq \mathrm{Fitt}_{\Lambda_n}\left(\mathrm{Sel}(K_n, E[p^\infty])^\vee\right).$$

By using (1), (2), and the functional equation for Bertolini–Darmon's theta elements (see [2, Proposition 2.13] and [3, Lemma 1.5]), we have

$$\begin{aligned}
\left(\left(\widetilde{\omega}_n^{\mp}\right)^2 \cdot L_p^{\pm}(E/K_\infty) \pmod{\omega_n}\right) &= \left(\left(\widetilde{\omega}_n^{\mp}\right)^2 \cdot \theta^{\pm}(E/K_n) \cdot \iota(\theta^{\pm}(E/K_n))\right) \\
&= \left(\left(\widetilde{\omega}_n^{\mp}\right)^2 \cdot \theta^{\pm}(E/K_n) \cdot \theta^{\pm}(E/K_n)\right) \\
&= (\theta(E/K_n) \cdot \theta(E/K_n)) \\
&= (\theta(E/K_n) \cdot \iota(\theta(E/K_n))) \\
&= \left(L_p(E/K_n)\right).
\end{aligned}$$

Thus, Theorem 1.3 follows.

References

1. Burungale, A.A., Büyükboduk, K., Lei, A.: Anticyclotomic Iwasawa theory of abelian arieties of GL_2-type at non-ordinary primes. Adv. Math. **439**(109465), 63 pages (2024)
2. Bertolini, M., Darmon, H.: Heegner points on Mumford-Tate curves. Invent. Math. **126**(3), 413–456 (1996)
3. Bertolini, M., Darmon, H.: Iwasawa's main conjectures for elliptic curves over anticyclotomic $\mathbb{Z}_p$-extensions. Ann. of Math. (2) **162**(1), 1–64 (2005)
4. Chida, M., Hsieh, M.-L.: On the anticyclotomic Iwasawa main conjecture for modular forms. Compos. Math. **151**(5), 863–897 (2015)
5. Chida, M., Hsieh, M.-L.: Special values of anticyclotomic L-functions for modular forms. J. Reine Angew. Math. **741**, 87–131 (2018)
6. Darmon, H.: A refined conjecture of Mazur-Tate type for Heegner points. Invent. Math. **110**, 123–146 (1992)
7. Darmon, H., Iovita, A.: The anticyclotomic main conjecture for elliptic curves at supersingular primes. J. Inst. Math. Jussieu **7**(2), 291–325 (2008)
8. Greenberg, R.: Iwasawa theory for elliptic curves, in Arithmetic theory of elliptic curves (Cetraro, 1997), vol. 1716, pp. 51–144. Springer-Verlag, Berlin (1999)
9. Iovita, A., Pollack, R.: Iwasawa theory of elliptic curves at supersingular primes over $\mathbb{Z}_p$-extensions of number fields. J. Reine Angew. Math. **598**, 71–103 (2006)
10. Kim, C.-H.: An anticyclotomic Mazur–Tate conjecture for modular forms. submitted, arXiv:1612.03743 (2016)
11. Kim, B.D.: The plus/minus Selmer groups for supersingular primes. J. Aust. Math. Soc. **95**(2), 189–200 (2013)
12. Kim, C.-H.: Anticyclotomic Iwasawa invariants and congruences of modular forms. Asian J. Math. **21**(3), 499–530 (2017)
13. Kim, C.-H.: Overconvergent quaternionic forms and anticyclotomic p-adic L-functions. Publ. Mat. **63**(2), 727–767 (2019)
14. Kim, C.-H., Kurihara, M.: On the refined conjectures on Fitting ideals of Selmer groups of elliptic curves with supersingular reduction. Int. Math. Res. Not. IMRN **2021**(14), 10559–10599 (2021)
15. Kitajima, T., Otsuki, R.: On the plus and the minus Selmer groups for elliptic curves at supersingular primes. Tokyo J. Math. **41**(1), 273–303 (2018)
16. Kim, C.-H., Ota, K.: On the quantitative variation of congruence ideals and integral periods of modular forms. Res. Math. Sci. **10**(22), 34 pages (2023)
17. Kobayashi, S.: Iwasawa theory for elliptic curves at supersingular primes. Invent. Math. **152**(1), 1–36 (2003)
18. Kim, C.-H., Pollack, R., Weston, T.: On the freeness of anticyclotomic Selmer groups of modular forms. Int. J. Number Theory **13**(6), 1443–1455 (2017)
19. Kurihara, M.: On the Tate Shafarevich groups over cyclotomic fields of an elliptic curve with supersingular reduction I. Invent. Math. **149**, 195–224 (2002)
20. Mazur, B., Tate, J.: Refined conjectures of the Birch and Swinnerton-Dyer type. Duke Math. J. **54**(2), 711–750 (1987)
21. Pollack, R.: On the p-adic L-function of a modular form at a supersingular prime. Duke Math. J. **118**(3), 523–558 (2003)
22. Pollack, R., Weston, T.: On anticyclotomic μ-invariants of modular forms. Compos. Math. **147**, 1353–1381 (2011)
23. Shii, R.: On non-trivial Λ-submodules with finite index of the plus/minus Selmer group over anticyclotomic $\mathbb{Z}_p$-extension at inert primes. preprint, arXiv:2308.16384 (2023). https://link.springer.com/article/10.1007/s40316-024-00236-5
24. Vatsal, V.: Uniform distribution of Heegner points. Invent. Math. **148**(1), 1–48 (2002)
25. Vignéras, M.-F.: Arithm'etique des alg'ebres de quaternions, vol. 800. Springer, Berlin (1980)

The Bertolini-Darmon-Prasanna p-Adic L-Function via q_{dR}-Expansions

Daniel Kriz

Abstract In this note, we explain how the notion of q_{dR}-expansions from previous work of the author can be used to recover the p-adic L-function of Bertolini-Darmon-Prasanna.

1 Introduction

A key problem permeating algebraic number theory throughout the last few decades has been relating the arithmetic of abelian varieties to special values of their associated L-functions. An incarnation of such a relation is given by the Gross-Zagier formula [13]: letting E be an elliptic curve over $\mathbb{Q}$ and K an imaginary quadratic field satisfying a certain *Heegner hypothesis* with respect to E (implying that the global root number of the base change $L(E/K, s)$ is -1), the Gross-Zagier formula is of the form

$$L'(E/K, 1) = C \cdot \langle P_K, P_K \rangle_{\mathrm{NT}} \tag{1}$$

where $C \in \mathbb{C}^\times$ is an explicit constant depending on the pair (E, K). Here $L(E/K, s)$ is known to have an analytic continuation to all of $\mathbb{C}$ by the modularity theorem [6, 34, 35], since $L(E/K, s) = L(E/\mathbb{Q}, s)L(E^{d_K}/\mathbb{Q}, s)$ where E^{d_K} is the quadratic twist of E by the squarefree radical d_K of the fundamental discriminant D_K of K. The point $P_K \in E(K)$ is a certain *Heegner point* and $\langle \cdot, \cdot \rangle_{\mathrm{NT}}$ is the Néron-Tate height. In particular, by the work of Kolyvagin [19], Formula (1) gives us the implication toward the Birch and Swinnerton-Dyer conjecture

$$L'(E/K, 1) \neq 0 \implies \mathrm{rank}_{\mathbb{Z}} E(K) = 1.$$

This project has received funding from the European Union's Horizon 2020 research and innovation programme under the Marie Skłodowska-Curie grant agreement No. 101034255.

D. Kriz (✉)
Dipartimento di Matematica "Federigo Enriques", Università degli Studi di Milano, Milano, Italy
e-mail: daniel.kriz@unimi.it

M. Longo et al. (eds.), *Elliptic Curves and Modular Forms in Arithmetic Geometry*, Springer Proceedings in Mathematics & Statistics 527,
https://doi.org/10.1007/978-3-032-13123-2_9

However, it is often difficult to establish $L'(E/K,1) \neq 0$ except through numerical means. A key question since the appearance of Gross-Zagier's formula has thus been to establish p-adic versions of the formula, in which the left-hand side of (1) would be replaced by a p-adic L-value (or the value of a derivative of a p-adic L-function). Such formulas have the appeal of providing an easier criterion for establishing non-vanishing, namely testing non-vanishing through successive mod p^n congruences. An elegant answer to this question is provided by the groundbreaking work of Bertolini-Darmon-Prasanna [2], in which they construct a p-adic L-function $\mathcal{L}_p^{\mathrm{BDP}}(E/K,s)$ for $s \in \mathbb{Z}/(p-1) \times \mathbb{Z}_p$ satisfying the *p-adic Gross-Zagier formula* of Theorem 5.13 of op. cit.[1]

$$\mathcal{L}_p^{\mathrm{BDP}}(E/K,1) = C_p \cdot \log_E(P_K), \tag{2}$$

where $C_p \in \overline{\mathbb{Q}}_p^{\times}$ is an explicit constant depending on E and $\log_E : E(\mathbb{C}_p) \to \mathbb{C}_p$ is the p-adic formal logarithm of E [4]. In particular $\log_E(P) = 0$ if and only if $P \in E(\mathbb{C}_p)$ is torsion, and so $\log_E$ can be viewed as a p-adic analogue of $\langle \cdot, \cdot \rangle_{\mathrm{NT}}$. Therefore (2) implies $\mathcal{L}_p^{\mathrm{BDP}}(E/K,1) \neq 0$ implies $P_K \in E(K)$ is non-torsion, and thus $\mathrm{rank}_{\mathbb{Z}} E(K) = 1$ by [19].

In this note, we explain how the construction of q_{dR}-expansions from [22], which were used in op. cit. in order to construct supersingular analogues of $\mathcal{L}_p^{\mathrm{BDP}}$ (i.e. where p is ramified or inert in $K/\mathbb{Q}$), can also be used to recover the construction of $\mathcal{L}_p^{\mathrm{BDP}}$ in the ordinary case (where p is split in $K/\mathbb{Q}$). The Bertolini-Darmon-Prasanna p-adic L-function [2], its generalizations [5, 7] and [23], and its supersingular analogues in [1, 20, 21] and [22] have yielded numerous geometric and arithmetic applications; for example in [3] the BDP p-adic L-function was used to carry out a deep study of the arithmetic of CM elliptic curves and in [22] they were used in order to prove Sylvester's conjecture from 1879 on primes expressible as a sum of two rational cubes [32].

2 The Bertolini-Darmon-Prasanna p-Adic L-Function

In this section, we recall the construction of the Bertolini-Darmon-Prasanna p-adic L-function via Serre-Tate coordinates, closely following [2, 7] and [37, Introduction]. Our presentation here is slightly more general than that of [2] and [7].

---

[1] More precisely, Bertolini-Darmon-Prasanna's p-adic L-function [2] is the square $(\mathcal{L}_p^{\mathrm{BDP}})^2$ of $\mathcal{L}_p^{\mathrm{BDP}}$. It is $(\mathcal{L}_p^{\mathrm{BDP}})^2$ which directly interpolates Rankin-Selberg central L-values, whereas $\mathcal{L}_p^{\mathrm{BDP}}$ interpolates toric periods (the squares of which are related to Rankin-Selberg L-values by Waldspurger's formula).

2.1 Background

Henceforth, fix algebraic closures $\overline{\mathbb{Q}}$ of $\mathbb{Q}$ and $\overline{\mathbb{Q}}_p$ of $\mathbb{Q}_p$. Fix embeddings $i_p : \overline{\mathbb{Q}} \hookrightarrow \overline{\mathbb{Q}}_p$ and $i_\infty : \overline{\mathbb{Q}} \hookrightarrow \mathbb{C}$. Given any number field L, let $\mathbb{N}_L : L^\times \backslash \mathbb{A}_L^\times \to \mathbb{C}^\times$ be the norm Hecke character, normalized so that $\mathbb{N}_\mathbb{Q}$ has infinity type $\mathbb{R}^\times \to \mathbb{R}_{>0}$, $x \mapsto |x|^{-1}$ and $\mathbb{N}_L = \mathbb{N}_\mathbb{Q} \circ \mathrm{Nm}_{L/\mathbb{Q}}$ where $\mathrm{Nm}_{L/\mathbb{Q}} : \mathbb{A}_L^\times \to \mathbb{A}_\mathbb{Q}^\times$ is the relative norm. Let $\mathbb{A}_L$ (resp. $\mathbb{A}_L^\times$) denote the adèles (resp. idèles) of L, and given a product S of places of L, let $\mathbb{A}_L^{(S)}$ (resp. $\mathbb{A}_L^{\times,(S)}$) denote the adèles (resp. idèles) with entry 1 at each place dividing S. Henceforth, given an algebraic Hecke character $\rho : L^\times \backslash \mathbb{A}_L^\times \to \mathbb{C}^\times$ and a place v of L, let $\rho_v : L_v^\times \to \mathbb{C}^\times$ denote the local character of ρ at v.

Henceforth fix $N \in \mathbb{Z}_{\geq 1}$. Let $f \in S_k(\Gamma_1(N), \epsilon_f)$ be a newform of level $\Gamma_1(N)$, where $\epsilon_f : (\mathbb{Z}/N)^\times \to \overline{\mathbb{Z}}^\times$ is the nebentype (which we view here as a possibly imprimitive Dirichlet character of conductor N). Let K be an imaginary quadratic field, and suppose that $\chi : K^\times \backslash \mathbb{A}_K^\times \to \overline{\mathbb{Z}}^\times$ is an algebraic Hecke character over K of infinity type $(k, 0)$ such that we have a factorization

$$\chi|_{\mathbb{A}_\mathbb{Q}^\times} \mathbb{N}_\mathbb{Q}^{-k} : \mathbb{Q}^\times \backslash \mathbb{A}_\mathbb{Q}^\times \to \hat{\mathbb{Z}}^\times \to \prod_{\ell | N} (\mathbb{Z}_\ell / N\mathbb{Z}_\ell)^\times \cong (\mathbb{Z}/N)^\times \to \overline{\mathbb{Z}}^\times \tag{3}$$

where the last arrow is equal to $\epsilon_f^{-1} : (\mathbb{Z}/N)^\times \to \overline{\mathbb{Z}}^\times$. Let π_f denote the (unitary) automorphic representation of GL_2 attached to f and let π_χ denote the (unitary) automorphic representation of GL_2 attached to the theta series θ_χ. For a quaternion algebra B over $\mathbb{Q}$, let $\varepsilon_\ell(B)$ denote the Hasse invariant of $B \otimes_\mathbb{Q} \mathbb{Q}_\ell$ over $\mathbb{Q}_\ell$. Given a place v of $\mathcal{O}_K$, let $\varepsilon_v(\pi_{f,K}, \chi^{-1}, 1/2)$ denote the local root number at v of the L-function $L(\pi_{f,K} \otimes \chi^{-1}, s)$, where $\pi_{f,K}$ denotes the base change to K of the $GL_2(\mathbb{A}_\mathbb{Q})$-automorphic form π_f attached to f. Given a place ℓ of $\mathbb{Z}$ let $\varepsilon_\ell(\pi_{f,K}, \chi^{-1}, 1/2) = \prod_{v|\ell} \varepsilon_v(\pi_{f,K}, \chi^{-1}, 1/2)$ where $v|\ell$ ranges over all places of $\mathcal{O}_K$ above ℓ. Henceforth, fix $B/\mathbb{Q}$ to be the unique indefinite quaternion algebra such that the pair (f, χ^{-1}) satisfies the Saito-Tunnell conditions [37, Chapter 1.3.2], [38] with respect to B and K for every finite place ℓ of $\mathbb{Q}$:

$$\varepsilon_\ell(\pi_{f,K}, \chi^{-1}, 1/2) = \chi_\ell^{-1}(-1)\varepsilon_\ell(B). \tag{4}$$

Here $\chi_\ell^{-1} = \prod_{v|\ell} \chi_v^{-1}$ is the product of local characters χ_v^{-1} for all places v of K above ℓ and $\varepsilon_\ell(B)$ denotes the Hasse invariant of $B \otimes_\mathbb{Q} \mathbb{Q}_\ell$. Note that the generalized Heegner hypothesis of [7] is satisfied if and only if $\epsilon_f = 1$. Let $\mathrm{disc}(B)$ be the discriminant of $B/\mathbb{Q}$. Note that

$$\prod_{v|\ell} \chi_v^{-1}(-1) = \chi|_{\mathbb{A}_\mathbb{Q}^\times,\ell}^{-1}(-1) \overset{(3)}{=} \epsilon_{f,\ell}(-1).$$

Moreover, the ramification set of $\pi_{f,K} \otimes \chi^{-1}$ is supported in N, and so $\epsilon_\ell(\pi_{f,K}, \chi^{-1}, 1/2) = 1$ for $\ell \nmid N$. In particular, from (4) we see that for $\ell \nmid N$ we have $\varepsilon_\ell(B) = 1$.

Henceforth fix a maximal order $\mathcal{O}_B$ of B. Fix an isomorphism $\iota_\infty : B \otimes_{\mathbb{Q}} \mathbb{R} \xrightarrow{\sim} M_2(\mathbb{R})$ (recalling that such an isomorphism exists because B is split at ∞). Then $GL_2(\mathbb{R})$ acts on $\mathcal{H}^+ := \{\mathrm{Im}(\tau) > 0\}$ and on $\mathcal{H}^\pm = \mathbb{C} \setminus \mathbb{R}$ by the left modular action

$$\begin{pmatrix} a & b \\ c & d \end{pmatrix} \cdot \tau = \frac{a\tau + b}{c\tau + d}.$$

Let S be a scheme such that $\mathrm{disc}(B) \in \mathcal{O}_S(S)^\times$. Recall that a false elliptic curve A over S (with respect to $\mathcal{O}_B$) is an abelian surface together with an embedding

$$i_A : \mathcal{O}_B \hookrightarrow \mathrm{End}_S(A)$$

where $\mathrm{End}_S(A)$ denotes the ring of abelian surface isogenies $A \to A$ commuting with $A \to S$. Given false elliptic curves A and A' and S, a *false isogeny* $A \to A'$ over S is an abelian surface isogeny commuting with $A \to S$ and $A' \to S$ and commuting with the actions of $\mathcal{O}_B$. The false endomorphism ring $\mathrm{End}_S(A)$ is the ring of false isogenies $A \to A$ over S.

Suppose $M \in \mathbb{Z}_{\geq 1}$ such that $(M, \mathrm{disc}(B)) = 1$. Fix an isomorphism $\iota_M : \mathcal{O}_B \otimes_{\mathbb{Z}} \mathbb{Z}/M \xrightarrow{\sim} M_2(\mathbb{Z}/M)$. For any finite-index congruence subgroup $\Gamma \subset (\mathcal{O}_B \otimes_{\mathbb{Z}} \hat{\mathbb{Z}})^\times$ of level M (i.e. Γ is the inverse image of a subgroup $\Gamma_0 \subset (\mathcal{O}_B \otimes_{\mathbb{Z}} \mathbb{Z}/M)^\times$ under the reduction map $\mathcal{O}_B \otimes_{\mathbb{Z}} \hat{\mathbb{Z}} \to \mathcal{O}_B \otimes_{\mathbb{Z}} \mathbb{Z}/M$), let $\mathbb{Y}(\Gamma)$ be the Shimura curve over $\mathbb{Q}$ [24, 31] such that

$$\mathbb{Y}(\Gamma)(\mathbb{C}) = B^\times \backslash (B \otimes_{\mathbb{Q}} \mathbb{A}_{\mathbb{Q}}^{(\infty)})^\times \times \mathcal{H}^\pm / \Gamma. \tag{5}$$

Here $\mathcal{H}^\pm = \mathbb{C} \setminus \mathbb{R}$. For example, consider the congruence subgroup

$$\Gamma(M) := \left\{ \gamma \in \left(\mathcal{O}_B \otimes_{\mathbb{Z}} \hat{\mathbb{Z}}\right)^\times : \iota_M(\gamma) \equiv \begin{pmatrix} 1 & 0 \\ 0 & 1 \end{pmatrix} \pmod{M\hat{\mathbb{Z}}} \right\}.$$

Let S be a scheme such that $\mathrm{disc}(B) \in \mathcal{O}_S(S)^\times$. Recall that a *false elliptic curve* A over S with respect to $\mathcal{O}_B$ is an abelian surface A over S together with an embedding $\mathcal{O}_B \hookrightarrow \mathrm{End}_S(A)$. We often write such data as "A/S".

Definition 2.1 Let $M \in \mathbb{Z}_{\geq 0}$ with $M \in \mathcal{O}_S(S)^\times$. Given a false elliptic curve A/S, a $\Gamma(M)$-level structure on a false elliptic curve A/S is an embedding of group schemes

$$P : \mathcal{O}_B \otimes_{\mathbb{Z}} \mathbb{Z}/M \xrightarrow{\sim} A[M]$$

commuting with the action of $\mathcal{O}_B \hookrightarrow \mathrm{End}_S(A)$.

Let

$$N_- := \prod_{\ell|\text{disc}(B)} \ell^{\text{ord}_\ell(N)}$$

and

$$N_+ := N/N_-,$$

so that in particular $(N_+, \text{disc}(B)) = 1$. Henceforth, let

$$\mathbb{Y} := \mathbb{Y}(\Gamma(N_+)).$$

Henceforth assume

$$N_+ \geq 3$$

so that $\mathbb{Y}$ represents the fine moduli space classifies tuples (A, P, i) where A is a false elliptic curve [with embedding of rings $i : \mathcal{O}_B \to \text{End}(A)$], $P : \mathcal{O}_B \otimes (\mathbb{Z}/N_+) \xrightarrow{\sim} A[N_+]$ is a $\Gamma(N_+)$-level structure. Thus there is a universal object $\mathcal{A} \to \mathbb{Y}$.

2.2 CM Points

We will recall some of the theory of CM points associated to ring class fields over an imaginary quadratic field K. Let c be a positive integer such that $(c, pN\text{disc}(B)) = 1$. Let $\mathcal{O}_c = \mathbb{Z} + c\mathcal{O}_K$ be the order of $\mathcal{O}_K$ of conductor c.

For $(g, \tau) \in (B \otimes_{\mathbb{Q}} \mathbb{A}_{\mathbb{Q}}^{(\infty)})^\times \times \mathcal{H}^\pm$, let $[g, \tau] \in \mathbb{Y}(\mathbb{C})$ denote the corresponding point in the double quotient (5) (with $\Gamma = \Gamma(N_+)$). For any embedding $\iota : K \hookrightarrow B$ such that $\iota(\mathcal{O}_c) = \mathcal{O}_B \cap \iota(K)$ and let $\tau \in \mathcal{H}^+$ be the unique point such that $\iota_\infty(\iota(K^\times)) \cdot \tau = \tau$. Conversely, given $\tau \in \mathcal{H}^+$ with stabilizer subgroup $\text{Stab}_{\mathcal{O}_B}(\tau)$ inside $\iota_\infty(\mathcal{O}_B)$ (acting on $\mathcal{H}^+$ via the above modular action) isomorphic to $\mathcal{O}_c$, we get an associated embedding

$$\iota_\tau : \mathcal{O}_c \cong \text{Stab}_{\mathcal{O}_B}(\tau) \subset \mathcal{O}_B$$

which induces $\iota_\tau : K \hookrightarrow B$ such that $\iota_\tau(\mathcal{O}_c) = \mathcal{O}_B \cap \iota(K)$ and $\iota_\infty(\iota_\tau(K^\times)) \cdot \tau = \tau$.

Henceforth, fix an embedding $\iota : K \hookrightarrow B$ such that $\iota(\mathcal{O}_c) = \mathcal{O}_B \cap \iota(K)$, and let $\tau_c \in \mathcal{H}^+$ be the associated point. Thus $\iota_{\tau_c} = \iota$ in the notation of the previous paragraph.

Define the false elliptic curve over $\mathbb{C}$

$$A_c := A_{\tau_c} = \mathbb{C}^{\oplus 2}/\iota_\infty(\mathcal{O}_B) \cdot \begin{pmatrix} \tau_c \\ 1 \end{pmatrix}.$$

Then the ring of false endomorphisms $\text{End}_{\mathbb{C}}(A_c)$ is isomorphic to $\mathcal{O}_c$, where the map $\mathcal{O}_c \hookrightarrow \text{End}_{\mathbb{C}}(A_c)$ is given by right multiplication on A_c by $\iota(\mathcal{O}_c)$. The fact that this map is surjective follows because $\iota(K)$ is the centralizer of B and $\iota(K) \cap \mathcal{O}_B = \mathcal{O}_c$. Thus A_c is said to have complex multiplication by $\mathcal{O}_c$.

There is a natural norm map $\mathrm{Nm} : B \to \mathbb{Q}$. Let $\Gamma = \mathcal{O}_B^{\times,\mathrm{Nm}=1}$. Then Γ acts via conjugation on B (i.e. for $\gamma \in \Gamma$, we have a map $\gamma : B \to B$ given by $x \mapsto \gamma x \gamma^{-1}$). Thus Γ acts on the set of embeddings $\iota : K \hookrightarrow B$ via postcomposition:

$$\gamma\iota : K \overset{\iota}{\hookrightarrow} B \xrightarrow{x \mapsto \gamma x \gamma^{-1}} B.$$

For any $\tau \in \mathcal{H}^+$, recalling that $\iota_\tau(K^\times) \cdot \tau = \tau$, note that $\gamma\iota_\tau$ satisfies

$$\begin{aligned}\iota_\infty(\gamma\iota_\tau(k)) \cdot (\iota_\infty(\gamma) \cdot \tau) &= \iota_\infty(\gamma)\iota_\infty(\iota_\tau(k))\iota_\infty(\gamma)^{-1} \cdot (\iota_\infty(\gamma) \cdot \tau)\\ &= \iota_\infty(\gamma)\iota_\infty(\iota_\tau(k)) \cdot \tau = \iota_\infty(\gamma) \cdot \tau,\end{aligned}$$

and hence

$$\gamma\iota_\tau = \iota_{\iota_\infty(\gamma)\cdot\tau}.$$

Let $\mathfrak{a}$ be an integral ideal of $\mathcal{O}_c$ which is prime to c. We define the left ideal of $\mathcal{O}_B$

$$\mathfrak{a}_B = \mathcal{O}_B(\iota(\mathfrak{a})) \subset \mathcal{O}_B.$$

Since B is unramified at ∞ and defined over $\mathbb{Q}$, B has class number 1 and thus $\mathfrak{a}_B$ is principal, thus generated by some $\alpha_\mathfrak{a} \in B$. Right multiplication by $\alpha_\mathfrak{a}$ induces a false isogeny

$$\pi_\mathfrak{a} : A_c \to A_c/A_c[\mathfrak{a}] = A_{\alpha_\mathfrak{a}^{-1}\tau_c} =: A_{\mathfrak{a}c} \tag{6}$$

where $A_c[\mathfrak{a}]$ is the subgroup of A_c consisting of elements annihilated by every element of $\mathfrak{a} \subset \mathcal{O}_c \cong \mathrm{End}_\mathbb{C}(A_c)$. Supposing that $\mathfrak{a}$ is prime to $M \in \mathbb{Z}_{\geq 1}$, if P is a $\Gamma(M)$-level structure on A then $\pi_\mathfrak{a}(P)$ is a $\Gamma(M)$-level structure on $A_{\alpha_\mathfrak{a}^{-1}\tau_c}$.

Consider $[1, \tau_c] \in \mathbb{Y}(\Gamma(N_+))(\mathbb{C})$. Shimura's reciprocity law [7, 31] says that in fact [37, 1.3.1]

$$[1, \tau_c] \in \mathbb{Y}(\Gamma(N_+))(K^{\mathrm{ab}}), \tag{7}$$

where K^{ab} is the maximal abelian extension of K, and the residue field of $[1, \tau_c]$ contains the ring class field $K(\mathcal{O}_c)$ attached to $\mathcal{O}_c$. Moreover, if $\mathfrak{a} \subset \mathcal{O}_c$ is a prime-to-$N\mathrm{disc}(B)$ ideal with associated Artin symbol $\sigma_\mathfrak{a} = (\mathfrak{a}, K(\mathcal{O}_c)/K) \in \mathrm{Gal}(K(\mathcal{O}_c)/K)$, we have the equality

$$[1, \tau_c]^{\sigma_\mathfrak{a}} = [1, \alpha_\mathfrak{a}^{-1}\tau_c] \in \mathbb{Y}(K^{\mathrm{ab}}).$$

Because $[1, \alpha_\mathfrak{a}^{-1}\tau_c]$ is independent of the choice of $\alpha_\mathfrak{a}$, we may write $[1, \alpha_\mathfrak{a}^{-1}\tau_c] = [1, \tau_{\mathfrak{a}c}]$, and if (A_c, i_{A_c}, P) represents the isomorphism class of triples associated to $[1, \tau_c]$, then let $\mathfrak{a} \star (A_c, P) = (\mathfrak{a} \star A_c, \mathfrak{a} \star P)$ denote any element in the isomorphism class of triples associated to $[1, \tau_{\mathfrak{a}c}]$. Note that $A_{\mathfrak{a}c} := A_c/A_c[\mathfrak{a}]$ denotes the underlying false elliptic curve underlying $[1, \tau_{\mathfrak{a}c}]$ (i.e. $A_{\mathfrak{a}c} = \mathfrak{a} \star A_c$).

2.3 *Assumptions on the Prime p*

Let $p \nmid \mathrm{disc}(B)$ be a prime number. Fix an isomorphism $\iota_p : B \otimes_{\mathbb{Q}} \mathbb{Q}_p \cong M_2(\mathbb{Q}_p)$ which induces $\iota_p : \mathcal{O}_B \otimes_{\mathbb{Z}} \mathbb{Z}_p \cong M_2(\mathbb{Z}_p)$. Let

$$e^1 := \iota_p^{-1}\left(\begin{pmatrix} 1 & 0 \\ 0 & 0 \end{pmatrix}\right) \in \mathcal{O}_B \otimes_{\mathbb{Z}} \mathbb{Z}_p. \tag{8}$$

Assumption 2.2 Henceforth assume that $\epsilon_{f,p}(-1) = \epsilon_p(\pi_{f,K}, \chi^{-1}, 1/2)$. Thus, since

$$\chi_p(-1) = \prod_{v|p} \chi_v(-1) \overset{(3)}{=} \epsilon_{f,p}^{-1}(-1),$$

by (4) we have $\epsilon_p(B) = 1$ [i.e. $p \nmid \mathrm{disc}(B)$ and so $B \otimes_{\mathbb{Q}} \mathbb{Q}_p \cong M_2(\mathbb{Q}_p)$]. Note that if $p \nmid N$ then $\epsilon_{f,p}(-1) = \epsilon_p(\pi_{f,K}, \chi^{-1}, 1/2) = 1$, and so the assumption is automatically satisfied.

Recall that $K/\mathbb{Q}$ was a fixed imaginary quadratic field. We make the following ordinariness assumption on K.

Assumption 2.3 Henceforth, assume that p splits in $K/\mathbb{Q}$.

2.4 *The Ordinary Igusa Tower*

Henceforth we will let

$$\mathbb{Y} := \mathbb{Y}(\Gamma(N_+)).$$

Let $\mathbb{Y}^+/\mathbb{Z}_{(p)}$ be the Katz-Mazur model [8, 18] over $\mathbb{Z}_{(p)}$ [a regular scheme over $\mathrm{Spec}(\mathbb{Z}_{(p)})$]. Let Y^+ be the completion of $\mathbb{Y}^+$ along its special fiber $\mathbb{Y}^+ \times_{\mathrm{Spec}(\mathbb{Z}_{(p)})} \mathrm{Spec}(\mathbb{F}_p)$; thus Y^+ is a noetherian formal scheme over $\mathrm{Spf}(\mathbb{Z}_p)$. Let $Y^+(\overline{\mathbb{F}}_p)^{\mathrm{ss}} \subset Y^+(\overline{\mathbb{F}}_p)$ be the subset of supersingular points (a finite, and thus Zariski closed, set), and let $Y^{\mathrm{ord},+} = Y^+ \setminus Y^+(\overline{\mathbb{F}}_p)^{\mathrm{ss}}$ be the resulting open formal subscheme.

Let $\hat{A}$ be the formal completion of the universal object $\mathcal{A}^+ \to Y^+$ (i.e. the universal false elliptic curve together with an embedding of $\mathcal{O}_B$ into $\mathrm{End}_Y(\mathcal{A})$ and $\Gamma(N_+)$-Drinfeld level structure) along its identity section. Define the *ordinary Igusa tower* as

$$Y^{\mathrm{Ig},+} := \varprojlim_n \mathrm{Isom}_{\text{group schemes}}(e^1\hat{\mathcal{A}}[p^n], \mu_{p^n,Y^+})$$

where e^1 is as in (8). This is a pro-finite étale covering of formal schemes $Y^{\mathrm{Ig},+} \to Y^{\mathrm{ord},+}$ over $\mathrm{Spf}(\mathbb{Z}_p)$, with canonical isomorphisms of étale Galois groups (see [5, Sect. 7])

$$\mathrm{Gal}(Y^{\mathrm{Ig},+}/Y^{\mathrm{ord},+}) \cong \mathbb{Z}_p^\times. \tag{9}$$

Let Y^{ord} and Y^{Ig} denote the adic generic fibers of $Y^{\mathrm{ord},+}$ and $Y^{\mathrm{Ig},+}$; then by the moduli interpretation we have a canonical identification $\mathrm{Gal}(Y^{\mathrm{Ig}}/Y^{\mathrm{ord}}) = \mathrm{Gal}(Y^{\mathrm{Ig},+}/Y^{\mathrm{ord},+})$

2.5 Serre-Tate Coordinates and Serre-Tate Expansions

Henceforth, when denoting a point (A, i, P) consisting of a false elliptic curve A over S with embedding $i : \mathcal{O}_B \hookrightarrow \mathrm{End}_S(A)$ and a level structure P, we will omit i (which will be obvious from context) and simply write (A, P). Henceforth, given a ring R, let $\hat{\mathbb{G}}_{m,R} = \mathrm{Spf}(R[\![T]\!])$ denote the formal torus over R. That is, $\hat{\mathbb{G}}_{m,R}$ is the completion along the identity section $T = 0$ of the torus $\mathbb{G}_{m,R} = \mathrm{Spec}(R[1+T, (1+T)^{-1}])$. This definition extends to $\hat{\mathbb{G}}_{m,S}$ for any scheme S (by gluing over an affine chart of S), and then again to any formal scheme S by functoriality of formal completion.

Theorem 2.4 *(Main Theorem of [16]) Let $(A_0, P_0) \in Y^{\mathrm{ord},+}(\overline{\mathbb{F}}_p)$, i.e. A_0 is a false elliptic curve over $\overline{\mathbb{F}}_p$ and $P_0 : \mathcal{O}_B \otimes_{\mathbb{Z}} \mathbb{Z}/N_+ \xrightarrow{\sim} A_0[N_+]$ is a $\Gamma(N_+)$-level structure on A_0 (recall $p \nmid N_+$). Let $M^{\mathrm{ord},+}(A_0, P_0)$ be the formal completion of $Y^{\mathrm{ord},+}$ along (A_0, P_0). Letting $k(A_0, P_0) \subset \overline{\mathbb{F}}_p$ denote the residue field of (A_0, P_0), we have that $M^{\mathrm{ord},+}(A_0, P_0)$ is a formal scheme defined over* $\mathrm{Spf}(W(k(A_0, P_0)))$*, where $W(R)$ denotes the Witt vectors of R. There is a canonical isomorphism of formal schemes over* $\mathrm{Spf}(WK(A_0, P_0))$.

$$\mathrm{Hom}_{\mathbb{Z}_p}(e^1 T_p A_0(\overline{\mathbb{F}}_p) \otimes_{\mathbb{Z}_p} e^1 T_p A_0(\overline{\mathbb{F}}_p), \hat{\mathbb{G}}_{m,\mathbb{Z}_p}) \cong M^{\mathrm{ord},+}(A_0, P_0), \tag{10}$$

where e^1 is as in (8) and $\mathrm{Hom}_{\mathbb{Z}_p}$ *denotes the set of homomorphisms in the category of $\mathbb{Z}_p$-modules.*

Let $M^{\mathrm{Ig},+}(A_0, P_0) = Y^{\mathrm{Ig},+} \times_{Y^{\mathrm{ord},+}} M^{\mathrm{ord},+}(A_0, P_0)$. Let $\mathcal{A}^{(+)} \to Y^{\mathrm{Ig},(+)}$ denote the universal object, where again "(+)" denotes the optional presence of a "+". Let $\mathbb{Z}_{p,Y^{\mathrm{Ig},+}}$ denote the constant sheaf associated to $\mathbb{Z}_p$ on $Y^{\mathrm{Ig},+}$. Note that there is a canonical trivialization

$$\mathbb{Z}_{p,Y^{\mathrm{Ig},+}} \xrightarrow{\sim} e^1 T_p \mathcal{A}^+|_{Y^{\mathrm{Ig},+}}$$

which induces a trivialization

$$\mathbb{Z}_{p,Y^{\mathrm{Ig},+}} \xrightarrow{\sim} e^1 T_p \mathcal{A}^+|_{Y^{\mathrm{Ig},+}} \otimes_{\mathbb{Z}_p} e^1 T_p \mathcal{A}^+|_{Y^{\mathrm{Ig},+}}.$$

Thus as a consequence of Theorem 2.4 we have a canonical isomorphism of formal schemes over $\mathrm{Spf}(W(\overline{\mathbb{F}}_p))$

$$\mathbf{e} : M^{\mathrm{Ig},+}(A_0, P_0) \xrightarrow{\sim} \bigsqcup_{\pi_0(M^{\mathrm{Ig},+}(A_0,P_0))} \hat{\mathbb{G}}_{m,W(\overline{\mathbb{F}}_p)} \tag{11}$$

where $\pi_0(M^{\mathrm{Ig},+}(A_0, P_0))$ denotes the set of geometric connected components of $M^{\mathrm{Ig},+}(A_0, P_0)$. The elements of $\pi_0(M^{\mathrm{Ig},+}(A_0, P_0))$ are in one-to-one correspondence with trivializations

$$\mathbf{e} : e^1\hat{A}_0[p^\infty] \xrightarrow{\sim} \hat{\mathbb{G}}_{m,\overline{\mathbb{F}}_p}[p^\infty].$$

Letting $T \in \mathcal{O}_{\hat{\mathbb{G}}_{m,W(k(A_0,P_0))}}(\hat{\mathbb{G}}_{m,W(k(A_0,P_0))})$ be a choice of formal parameter, we then get a coordinate

$$q_{(A_0,P_0)} := 1 + \mathbf{e}^*T \in \mathcal{O}_{M^{\mathrm{Ig},+}(A_0,P_0)}(M^{\mathrm{Ig},+}(A_0, P_0)) \tag{12}$$

via pullback by the above isomorphism. This parameter $q_{(A_0,P_0)}$ is called the *Serre-Tate parameter on* $M^{\mathrm{Ig},+}(A_0, P_0)$.

Convention 2.5 *Henceforth, given a point* $y \in Y^{\mathrm{Ig}}(\overline{\mathbb{Q}}_p, \overline{\mathbb{Z}}_p)$, *let* $\mathcal{O}_{Y^{\mathrm{Ig}}}(y)$ *denote the residue field at* y. *Given any open* $U \subset Y$ *and section* $f \in \mathcal{O}_{Y^{\mathrm{Ig}}}(U)$ *such that* $y \in U(\overline{\mathbb{Q}}_p, \overline{\mathbb{Z}}_p)$, *let* $f(y)$ *denote the image of* f *under the map* $\mathcal{O}_{Y^{\mathrm{Ig}}}(U) \to \mathcal{O}_{Y^{\mathrm{Ig}}}(y)$.

Definition 2.6 Suppose $f \in \mathcal{O}_{Y^{\mathrm{Ig}}}(Y^{\mathrm{Ig}})$. The *Serre-Tate expansion of* f is defined to be the Taylor expansion of f along $q_{(A_0,P_0)}$ from (12) at $q_{(A_0,P_0)} = 1$, i.e.

$$\begin{aligned} f(q_{(A_0,P_0)}) &:= \sum_{n=0}^{\infty} a_{n,0} \cdot (q_{(A_0,P_0)} - 1)^n \in \left(\prod_{\pi_0(M^{\mathrm{Ig}}(A_0,P_0))} W(\overline{\mathbb{F}}_p) \right) [\![q_{(A_0,P_0)} - 1]\!], \\ a_{n,0} &:= \left(\frac{1}{n!} \left(\frac{d}{dq_{(A_0,P_0)}} \right)^n f \right) \Big|_{q_{(A_0,P_0)}=1} \in \prod_{\pi_0(M^{\mathrm{Ig}}(A_0,P_0))} W(\overline{\mathbb{F}}_p). \end{aligned} \tag{13}$$

Given $y \in M^{\mathrm{Ig},+}(\overline{\mathbb{Z}}_p)$, the *Serre-Tate expansion of* f *centered at* y is

$$a_n := \left(\frac{1}{n!} q^n_{(A_0,P_0)} \left(\frac{d}{dq_{(A_0,P_0)}} \right)^n f \right) \in \mathcal{O}_{Y^{\mathrm{Ig}}}(M^{\mathrm{Ig}}(A_0, P_0)).$$

Here $q_{(A_0,P_0)}(y)$ and $a_n(y)$ are written in the notation of Convention 2.5 with respect to the sections $q_{(A_0,P_0)}$ and a_n, respectively. These constructions immediately generalize to $f \in \mathcal{O}_{Y^{\mathrm{Ig}}}(Y^{\mathrm{Ig}}) \otimes_{\mathbb{Q}_p} \overline{\mathbb{Q}}_p$.

2.6 The Katz Canonical Differential

Recall our universal objects $\pi^+ : \mathcal{A}^+ \to Y^+$ and $\pi : \mathcal{A} \to Y$, whose completions along the identity sections $Y^+ \to \mathcal{A}^+$ and $Y \to \mathcal{A}$ we denote by $\pi^+ : \hat{\mathcal{A}}^+ \to Y^+$ and $\pi : \hat{\mathcal{A}} \to Y$, respectively. Recall e^1 from (8) and let

$$\omega^+ := \pi_*^+ \Omega^1_{e^1\mathcal{A}^+[p^\infty]/Y^+} = \pi_*^+ \Omega^1_{e^1\hat{\mathcal{A}}^+/Y^+}, \quad \omega := \pi_* \Omega^1_{e^1\mathcal{A}[p^\infty]/Y} = \pi_* \Omega^1_{e^1\hat{\mathcal{A}}/Y} \tag{14}$$

be the Hodge line bundles on Y^+ and Y, respectively. By definition, on $Y^{\mathrm{Ig},+} \to Y^{\mathrm{ord},+} \to Y^+$ we have a trivialization of connected height 1 p-divisible groups

$$\mathbf{e} : e^1\hat{\mathcal{A}}|_{Y^{\mathrm{Ig},+}}[p^\infty] \xrightarrow{\sim} \hat{\mathbb{G}}_{m,Y^{\mathrm{Ig},+}}[p^\infty]$$

which by [33, Proposition 1] induces a trivialization of associated connected height 1 formal groups

$$\mathbf{e} : e^1\hat{\mathcal{A}}|_{Y^{\mathrm{Ig},+}} \xrightarrow{\sim} \hat{\mathbb{G}}_{m,Y^{\mathrm{Ig},+}}.$$

Definition 2.7 Fix a formal parameter T on $\hat{\mathbb{G}}_{m,Y^{\mathrm{Ig},+}}$. Then $dT/(1+T) \in \Omega^1_{\hat{\mathbb{G}}_{m,Y^{\mathrm{Ig},+}}/Y^{\mathrm{Ig},+}}$ is the associated normalized differential on $Y^{\mathrm{Ig},+}$. We let

$$w_{\mathrm{can}}^{\mathrm{Katz}} := \mathbf{e}^* \frac{dT}{1+T} \in \pi_*^+ \Omega^1_{\hat{\mathcal{A}}/Y^+}(Y^{\mathrm{Ig},+}) = \omega^+(Y^{\mathrm{Ig},+}) \tag{15}$$

denote the *Katz canonical differential*. This section was first introduced [16, Sect. 3.3] (where it is denoted "$\omega(\alpha)$" with $\alpha = \mathbf{e}$ in loc. cit.).

As $\mathrm{Gal}(Y^{\mathrm{Ig},+}/Y^{\mathrm{ord},+})$ acts on $\mathbf{e}$ through multiplication by a character $\kappa : \mathrm{Gal}(Y^{\mathrm{Ig},+}/Y^{\mathrm{ord},+}) \hookrightarrow \mathbb{Z}_p^\times$, i.e.

$$\sigma^*\mathbf{e} = \kappa(\sigma) \cdot \mathbf{e}$$

for all $\sigma \in \mathrm{Gal}(Y^{\mathrm{Ig},+}/Y^{\mathrm{ord},+})$, we have

$$\sigma^* w_{\mathrm{can}}^{\mathrm{Katz}} = \kappa(\sigma) \cdot w_{\mathrm{can}}^{\mathrm{Katz}} \tag{16}$$

for all $\sigma \in \mathrm{Gal}(Y^{\mathrm{Ig},+}/Y^{\mathrm{ord},+})$.

2.7 Katz p-Adic Modular Forms

Katz [15, 16] defined a notion of *p-adic modular form of weight k* as a function on Y^{Ig} [more precisely, an element of $\mathcal{O}_{Y^{\mathrm{Ig},+}}(Y^{\mathrm{Ig},+})$] satisfying certain properties under the action of $\mathrm{Gal}(Y^{\mathrm{Ig},+}/Y^{\mathrm{ord},+})$. We briefly recall this definition, referring to the above cited works for details.

Definition 2.8 A function $F \in \mathcal{O}_{Y^{\mathrm{Ig}}} Y^{\mathrm{Ig}}$ is a *Katz p-adic modular form of weight* $k \in \mathbb{Z}_p^\times$ if it satisfies

$$\sigma^* F = \kappa^{-k}(\sigma) \cdot F$$

for every $\sigma \in \mathrm{Gal}(Y^{\mathrm{Ig},+}/Y^{\mathrm{ord},+})$. Let V_k denote the $\mathbb{Z}_p$-module of weight k Katz p-adic modular forms.

Proposition 2.9 *For any* $k \in \mathbb{Z}_{\geq 0}$ *have a natural additive map*

$$\omega^{\otimes k}(Y^{\mathrm{ord}}) \to V_k$$

given by $w \mapsto \frac{w}{w_{\mathrm{can}}^{\otimes k}}$.

Proof The required equivariance property immediately follows from (16) and the fact that $w \in \omega^{\otimes k}(Y^{\mathrm{ord}})$ is invariant under the action of $\mathrm{Gal}(Y^{\mathrm{Ig},+}/Y^{\mathrm{ord},+})$. □

2.8 *The BDP p-Adic L-Function*

In this section we recall the Bertolini-Darmon-Prasanna p-adic L-function, which was first introduced in [2, Sect. 5]; we closely follow the presentation of [10, Sect. 3].

Let τ_c be as in Sect. 2.2, and let

$$y_c := [1, \tau_c] = (A_c, P_c)$$

and let F_c be the minimal field of definition of y_c; thus $y_c \in \mathbb{Y}(F_c)$, $F_c \subset K^{\mathrm{ab}}$ by (7) and $K(\mathcal{O}_c) \subset F_c$ by Shimura's reciprocity law [31, 37, Chapter 3.1.2]. Also let

$$y_{\mathfrak{a}c} := [1, \tau_{\mathfrak{a}c}] = \mathfrak{a} \star (A_c, P_c) \in \mathbb{Y}(F_c)$$

as in the notation of that section. Since A_c has complex multiplication, it attains good reduction at all primes above p over some finite extension F_c'/F_c. Let $\mathfrak{p}_{F_c'}$ be the prime ideal of $\mathcal{O}_{F_c'}$ fixed by our choice of embedding i_p from the beginning of Sect. 2.1. Then A_c and $A_{\mathfrak{a}c}$ for all prime-to-c integral ideals $\mathfrak{a}$ of $\mathcal{O}_c$ have good integral models over $\mathcal{O}_{F_c', \mathfrak{p}_{F_c'}}$. Let

$$(A_{0c}, P_{0c}) := (A_c, P_c) \pmod{\mathfrak{p}_{F_c}} \in \mathbb{Y}^+(\overline{\mathbb{F}}_p)$$

and

$$(A_{\mathfrak{a}c}, P_{\mathfrak{a}c}) := \mathfrak{a} \star (A_{0c}, P_{0c}) = \mathfrak{a} \star (A_c, P_c) = (A_{\mathfrak{a}c}, \pi_{\mathfrak{a}}(P_c)) \pmod{\mathfrak{p}_{F_c'}} \in \mathbb{Y}^+(\overline{\mathbb{F}}_p).$$

Choice 2.10 *Choose a full set of elements* $S \subset \mathbb{A}_K^{\times,(pcN\infty)} \cap (\mathcal{O}_K \otimes_{\mathbb{Z}} \hat{\mathbb{Z}})$ *such that the composition*

$$S \subset \mathbb{A}_K^{\times} \xrightarrow{\text{rec}} \mathrm{Gal}(K^{\mathrm{ab}}/K) \twoheadrightarrow \mathrm{Gal}(F_c/K)$$

is a bijection. Given $\mathfrak{a} \in S$*, we will also let* $\mathfrak{a} \in C\ell(\mathcal{O}_c)$ *denote its image under this latter map.*

Choice 2.11 *Recall* e^1 *from (8). Fix a trivialization*

$$e_c : \mathbb{Z}_p \xrightarrow{\sim} \varprojlim_n e^1 A_c[\overline{\mathfrak{p}}^n] := e^1 T_{\overline{\mathfrak{p}}} A_c.$$

Thus we get a point

$$(y_c, e_c) \in Y^{\mathrm{Ig},+}(\overline{\mathbb{Z}}_p).$$

Recalling $\pi_{\mathfrak{a}} : A_c \to A_c/A_c[\mathfrak{a}]$ *as defined in (6), for each* $\mathfrak{a} \in S$ *we get a point*

$$(y_{\mathfrak{a}c}, e_{c,\mathfrak{a}}) := (y_{\mathfrak{a}c}, \pi_{\mathfrak{a}}(e_c)) \in Y^{\mathrm{Ig},+}(\overline{\mathbb{Z}}_p) \subset Y^{\mathrm{Ig}}(\overline{\mathbb{Q}}_p).$$

Let

$$F_w := \frac{w}{w_{\mathrm{can}}^{\mathrm{Katz},\otimes k}} \in \mathcal{O}_{Y^{\mathrm{Ig}}}(Y^{\mathrm{Ig}})$$

be the weight k p-adic modular form attached to $w \in \omega^{\otimes k}(Y^{\mathrm{ord}}) \otimes_{\mathbb{Q}_p} \overline{\mathbb{Q}}_p$, the restriction to Y^{ord} of the Jacquet-Langlands transfer of f''. We consider the p-depletion $w^{\flat} = w|_{1-UV}$ [2, Sect. 3.8] and let

$$F_w^{\flat} := \frac{w^{\flat}}{w_{\mathrm{can}}^{\mathrm{Katz},\otimes k}} \in \mathcal{O}_{Y^{\mathrm{Ig}}}(Y^{\mathrm{Ig}}). \tag{17}$$

This is also a weight k p-adic modular form.

Let $\chi : K^{\times}\backslash\mathbb{A}_K^{\times} \to \mathbb{C}^{\times}$ be the algebraic Hecke character as in Sect. 2.1, and let $\check{\chi} : K^{\times}\backslash\mathbb{A}_K^{\times} \to \mathbb{C}_p^{\times}$ be its p-adic avatar.

Definition 2.12 (The Bertolini-Darmon-Prasanna (BDP) p-adic L-function [2, 5, 7, 10]) Define a $W(\overline{\mathbb{F}}_p)$-algebra isomorphism

$$\rho_{\mathfrak{a}} : W(\overline{\mathbb{F}}_p)[[q_{\mathfrak{a}\star(A_{0c},P_{0c})} - 1]] \to W(\overline{\mathbb{F}}_p)[[q-1]],$$

$$\rho_{\mathfrak{a}}\left(\sum_{n=0}^{\infty} c_n (q_{\mathfrak{a}\star(A_{0c},P_{0c})} - 1)^n\right) = \sum_{n=0}^{\infty} c_n (q-1)^n.$$

Let W be the finite extension of $W(\overline{\mathbb{F}}_p)$ generated by the values of $\check{\chi}$ and the ring of integers of the finite extension of $\mathbb{Q}_p$ over which w is defined, and $\otimes_{W(\overline{\mathbb{F}}_p)} W$-linearly extend $\rho_{\mathfrak{a}}$. Define

$$\mathcal{L}^{\mathrm{BDP}}(w) := \frac{1}{[F_c : K(\mathcal{O}_c)]} \sum_{\mathfrak{a} \in S} \check{\chi}^{-1}(\mathfrak{a}) \cdot \rho_{\mathfrak{a}} \left(F_w^{\flat}(y_{\mathfrak{a}c}, \pi_{\mathfrak{a}}(e_c))(q_{\mathfrak{a}\star(A_{0c}, P_{0c})}) \right)$$
$$\in W[\![q-1]\!][1/p]. \tag{18}$$

Remark 2.13 $\mathcal{L}^{\mathrm{BDP}}(w)$ satisfies an interpolation property relating it to the square root of classical Rankin-Selberg L-values, hence the name p-adic L-function. See [2, 5.2.3–5.2.4] and [10, Proposition 3.8]. When the classical Heegner hypothesis is satisfied $\mathcal{L}^{\mathrm{BDP}}(w)^2$ agrees with the classical Bertolini-Darmon-Prasanna p-adic L-function of [2, Sect. 5].

3 The Construction of the BDP p-Adic L-Function via q_{dR}-Expansions

The purpose of this section is to give an alternate construction of the BDP p-adic L-function via one of the constructions in [22] called q_{dR}*-expansions* in particular, their compatibility with Serre-Tate expansions and the resulting fact that q_{dR}-expansions may be viewed as analytic continuations of Serre-Tate expansions to the supersingular locus. In Sect. 4 we indicate how these generalizations to the supersingular locus are accomplished in op. cit.

3.1 Period Sheaves on the Perfected Igusa Tower

We will need to work on the adic generic fibers of the formal schemes recalled above. Recall that there is a fully faithful functor from formal schemes (that are defined over the ring of integers of a complete nonarchimedean field and admit a finitely generated ideal of definition) to adic spaces, which sends $\mathrm{Spf}(R) \mapsto \mathrm{Spa}(R, R)$ (see [30, Proposition 2.2.1]). We will henceforth use this functor in order to view formal schemes as adic spaces. In particular our noetherian formal schemes over $\mathrm{Spf}(\mathbb{Z}_p)$ defined in Sect. 2.4 will be viewed as adic spaces over $\mathrm{Spa}(\mathbb{Z}_p, \mathbb{Z}_p)$. Henceforth let

$$Y = Y^+ \times_{\mathrm{Spa}(\mathbb{Z}_p,\mathbb{Z}_p)} \mathrm{Spa}(\mathbb{Q}_p, \mathbb{Z}_p), \quad Y^{\mathrm{ord}} = Y^{\mathrm{ord},+} \times_{\mathrm{Spa}(\mathbb{Z}_p,\mathbb{Z}_p)} \mathrm{Spa}(\mathbb{Q}_p, \mathbb{Z}_p),$$
$$Y^{\mathrm{Ig}} = Y^{\mathrm{Ig},+} \times_{\mathrm{Spa}(\mathbb{Z}_p,\mathbb{Z}_p)} \mathrm{Spa}(\mathbb{Q}_p, \mathbb{Z}_p)$$

denote the adic generic fibers of those formal schemes. Let $Y_{\text{proét}}$ denote the proétale site of [27] on the adic space Y over $\mathrm{Spa}(\mathbb{Q}_p, \mathbb{Z}_p)$. As in op. cit., we will denote the statement that U is an object of the underlying category of $Y_{\text{proét}}$ by $U \in Y_{\text{proét}}$. We will also let $Y_{\text{proét}}/U$ denote proétale site localized to U.

Recall that Y is the adic space over $\mathrm{Spa}(\mathbb{Q}_p, \mathbb{Z}_p)$ attached to $Y(N_+)$. (Recall that this is the uncompactified modular curve when $B = M_2(\mathbb{Q})$.) Let $\mathcal{O}_Y$ be the structure sheaf on $Y_{\text{proét}}$ and let $\mathcal{O}_Y^+$ denote the integral subsheaf. Let $\hat{\mathcal{O}}_Y^+ = \varprojlim_n \mathcal{O}_Y^+/p^n$ denote the p-adically completed structure sheaf and let $\hat{\mathcal{O}}_Y = \hat{\mathcal{O}}_Y^+[1/p]$. Let "(+)" denote the optional presence of a "+". This is taken to mean that if "(+)" is changed to "+" anywhere then it is done so everywhere. By definition there is a natural map $\mathcal{O}_Y^{(+)} \to \hat{\mathcal{O}}_Y^{(+)}$. Let $\mathbb{B}_{\mathrm{dR},Y}^{(+)}$, $\mathcal{O}\mathbb{B}_{\mathrm{dR},Y}^{(+)}$ are the period sheaves of [27, Sect. 6] and [28]; these are sheaves on the analytic proétale site $Y_{\text{proét}}$ as defined in op. cit. (That is, the proétale site on the adic space $Y \to \mathrm{Spa}(\mathbb{Q}_p, \mathbb{Z}_p)$.) These sheaves are equipped with natural filtrations and flat connections

$$\nabla : \mathcal{O}\mathbb{B}_{\mathrm{dR},Y}^{(+)} \to \mathcal{O}\mathbb{B}_{\mathrm{dR},Y}^{(+)} \otimes_{\mathcal{O}_Y} \Omega_Y^1$$

satisfying Griffiths transversality with respect to the filtration. This means that $\nabla\left(\mathrm{Fil}^i \mathcal{O}\mathbb{B}_{\mathrm{dR},Y}^{(+)}\right) \subset \mathrm{Fil}^{i-1}\left(\mathcal{O}\mathbb{B}_{\mathrm{dR},Y}^{(+)}\right) \otimes_{\mathcal{O}_Y} \Omega_Y^1$.

Moreover, we have a canonical inclusion $\mathbb{B}_{\mathrm{dR},Y}^{(+)} \subset \mathcal{O}\mathbb{B}_{\mathrm{dR},Y}^{(+)}$ as well as the equality

$$\left(\mathcal{O}\mathbb{B}_{\mathrm{dR},Y}^{(+)}\right)^{\nabla=0} = \mathbb{B}_{\mathrm{dR},Y}^{(+)}.$$

We have a natural map $\mathcal{O}_Y \to \mathcal{O}\mathbb{B}_{\mathrm{dR},Y}^+$ whose composition with ∇ is equal to $\mathcal{O}_Y \xrightarrow{d} \Omega_Y^1 \to \Omega_Y^1 \otimes_{\mathcal{O}_Y} \mathcal{O}\mathbb{B}_{\mathrm{dR},Y}^+$ where $d : \mathcal{O}_Y \to \Omega_Y^1$ is the exterior derivative. Let

$$\theta : \mathbb{B}_{\mathrm{dR},Y}^+ \twoheadrightarrow \hat{\mathcal{O}}_Y, \quad \theta_{\mathrm{dR}} : \mathcal{O}\mathbb{B}_{\mathrm{dR},Y}^+ \twoheadrightarrow \hat{\mathcal{O}}_Y$$

denote the natural projections of [27, Sect. 6]. In particular, the composition

$$\mathbb{B}_{\mathrm{dR},\mathcal{Y}^{\mathrm{Ig},0}}^+ \subset \mathcal{O}\mathbb{B}_{\mathrm{dR},\mathcal{Y}^{\mathrm{Ig},0}}^+ \xrightarrow{\theta_{\mathrm{dR}}} \hat{\mathcal{O}}_{\mathcal{Y}^{\mathrm{Ig},0}}$$

is the map $\theta : \mathbb{B}_{\mathrm{dR},\mathcal{Y}^{\mathrm{Ig},0}}^+ \to \hat{\mathcal{O}}_{\mathcal{Y}^{\mathrm{Ig},0}}$. The filtrations on $\mathbb{B}_{\mathrm{dR},Y}^+$ and $\mathcal{O}\mathbb{B}_{\mathrm{dR},Y}^+$ are defined by

$$\mathrm{Fil}^i \mathbb{B}_{\mathrm{dR},Y}^+ = (\ker\theta)^i, \quad \mathrm{Fil}^i \mathcal{O}\mathbb{B}_{\mathrm{dR},Y}^+ = (\ker\theta_{\mathrm{dR}})^i.$$

Definition 3.1 A $\Gamma(p^\infty)$-level structure on a false elliptic curve E is a compatible system of $\Gamma(p^n)$-level structures. In other words, it is an isomorphism

$$P : \mathcal{O}_B \otimes_{\mathbb{Z}} \mathbb{Z}_p \xrightarrow{\sim} T_p E = \varprojlim_n E[p^n].$$

Note that applying the idempotent $e^1 \in M_2(\mathbb{Z}_p)$ from (8) to the previous isomorphism we obtain an isomorphism

$$e^1 P : \mathbb{Z}_p^{\oplus 2} \xrightarrow{\sim} e^1 T_p E$$

which uniquely determines P by extending $\otimes_{\mathbb{Z}} \mathcal{O}_B$-linearly. We will often denote such an isomorphism by

$$(e_1, e_2) : \mathbb{Z}_p^{\oplus 2} \xrightarrow{\sim} e^1 T_p E$$

where $e_i : \mathbb{Z}_p \hookrightarrow e^1 T_p E$ corresponds to the embedding of the ith summand of $\mathbb{Z}_p^{\oplus 2}$.

We briefly recall the Hodge-Tate period map, which is not essential for our discussion, but which will give a more geometric perspective on the object $\mathcal{Y}^{\mathrm{Ig},0}$ below. Consider the infinite-level modular curve (or inverse limit of Katz-Mazur models)

$$\mathbb{Y}^{(+)}(p^\infty) := \varprojlim_n \mathbb{Y}^{(+)}(\Gamma(N_+ p^n)),$$

which is a scheme defined over $\mathrm{Spec}(\mathbb{Z}_{(p)})$; to see that this is indeed a scheme over $\mathrm{Spec}(\mathbb{Z}_{(p)})$ note that by [18, First Main Theorem 5.1.1] the transition maps in the above inverse limit are finite flat morphisms of schemes over $\mathrm{Spec}(\mathbb{Z}_{(p)})$. Let $Y^+(p^\infty)$ denote its completion along the special fiber $\mathbb{Y}^+(p^\infty) \times_{\mathrm{Spec}(\mathbb{Z}_{(p)})} \mathrm{Spec}(\mathbb{F}_p)$ [which is a formal scheme over $\mathrm{Spf}(\mathbb{Z}_p)$ which we henceforth view as an adic space over $\mathrm{Spa}(\mathbb{Z}_p, \mathbb{Z}_p)$] and let

$$Y_\infty := Y^+(p^\infty) \times_{\mathrm{Spa}(\mathbb{Z}_p, \mathbb{Z}_p)} \mathrm{Spa}(\mathbb{Q}_p, \mathbb{Z}_p) \in Y_{\text{proét}}. \tag{19}$$

Let $\hat{Y}_\infty$ denote the strong p-adic completion (in the terminology of [30, Definition 2.4.1 and Proposition 2.4.4]) of Y_∞. We note that $\mathcal{A}|_{Y_\infty} \to Y_\infty$ has a moduli interpretation: it is the universal false elliptic curve with $\Gamma(N_+)$-level structure and $\Gamma(p^\infty)$-level structure such that these level structures are compatible; here "compatible" means that the $\Gamma(p^{\mathrm{ord}_p(N_+)})$-level structures induced by both the $\Gamma(N_+)$ and $\Gamma(p^\infty)$-level structures coincide.

Let

$$\pi_{\mathrm{HT}} : \hat{Y}_\infty \to \mathbb{P}^1 \tag{20}$$

denote the *Hodge-Tate period map* of [26] (see also [11, Theorem 2.8]). This is in particular a map of adic spaces over $\mathrm{Spa}(\mathbb{Q}_p, \mathbb{Z}_p)$. Define

$$\mathcal{Y}^{\mathrm{Ig}} := \pi_{\mathrm{HT}}^{-1}(\{\infty\}). \tag{21}$$

In this definition, we use the fact that there is a canonical homeomorphism $|\hat{Y}_\infty| \cong |Y_\infty|$ of underlying topological spaces of $\hat{Y}_\infty$ and Y_∞ (see [30, Sect. 2]), and thus $\pi_{\mathrm{HT}}^{-1}(\{\infty\}) \subset |\hat{Y}_\infty| \cong |Y_\infty|$ defines a subset of Y_∞. We let

$$\mathcal{Y}^{\mathrm{Ig},0} \subset \mathcal{Y}^{\mathrm{Ig}} \tag{22}$$

denote the adic interior, which we call the *(open) perfected Igusa tower*. This is an object of $Y_{\text{proét}}$ and admits a natural sequence of pro-finite étale maps

$$\mathcal{Y}^{\mathrm{Ig},0} \to Y^{\mathrm{Ig}} \to Y^{\mathrm{ord}}$$

in the underlying category of $Y_{\text{proét}}$. In fact each morphism in this sequence is profinite étale (in $Y_{\text{proét}}$) with canonical isomorphism ([22, Sect. 2]; the isomorphism is deduced by calculating the stabilizer of (21) using (89) below and recalling (9))

$$\mathrm{Gal}(\mathcal{Y}^{\mathrm{Ig},0}/Y^{\mathrm{Ig}}) \cong \left\{ \begin{pmatrix} a & b \\ 0 & 1 \end{pmatrix} \in GL_2(\mathbb{Z}_p) : a \equiv 1 \pmod{p^{\mathrm{ord}_p(N_+)}} \right\}.$$

Here the congruence condition on a arises from the fact that $\mathrm{Gal}(\mathcal{Y}^{\mathrm{Ig},0}/Y^{\mathrm{Ig}})$ preserves the universal $\Gamma(N_+)$-level structure, which we require to be compatible with the $\Gamma(p^\infty)$-level structure per the previous paragraph. Moreover we have

$$\mathrm{Gal}(Y^{\mathrm{Ig}}/Y^{\mathrm{ord}}) = \mathrm{Gal}(Y^{\mathrm{Ig},+}/Y^{\mathrm{ord},+}) \overset{(9)}{\cong} \mathbb{Z}_p^\times.$$

The pro-adic space $\mathcal{Y}^{\mathrm{Ig},0}$ represents the moduli problem classifying ordinary false elliptic curves E together with $\mathcal{O}_B$-endomorphism structure, $\Gamma(N_+)$-level structure $P : \mathcal{O}_B \otimes_{\mathbb{Z}} \mathbb{Z}/N_+ \xrightarrow{\sim} E[N_+]$ and $\Gamma(p^\infty)$-level structure $(e_1, e_2) : \mathbb{Z}_p^{\oplus 2} \xrightarrow{\sim} e^1 T_p E$ [e^1 as in (8)] compatible with P (in the sense defined above), such that $e_1 : \mathbb{Z}_p \xrightarrow{\sim} e^1 T_p E^0$, where $T_p E^0 \subset T_p E$ denotes the connected part of the Tate module.

3.2 The Relative Weil Pairing

Recall the idempotent e^1 from (8). Henceforth, for brevity, let

$$T_p\mathcal{A} := \varprojlim_n e^1 \mathcal{A}[p^n].$$

This is a rank 2 $\hat{\mathbb{Z}}_{p,Y}$-local system. Here $\hat{\mathbb{Z}}_{p,Y} := \varprojlim_n (\mathbb{Z}/p^n)_{Y_{\text{ét}}}$ is the inverse limit of the pulllbacks under the morphism of sites $\nu : Y_{\text{proét}} \to Y_{\text{ét}}$ of the constant sheaves $(\mathbb{Z}/p^n)_{Y_{\text{ét}}}$ attached to $\mathbb{Z}/p^n$ on the (small) étale site $Y_{\text{ét}}$. (See [27, Definition 8.1] for more details.) Let $\mu_{p^n,Y} = \nu^* \mu_{p^n,Y_{\text{ét}}}$ be the pullback along ν of the constant sheaf $\mu_{p^n,Y_{\text{ét}}}$ attached to μ_{p^n} on $Y_{\text{ét}}$. We have Weil pairings

$$e^1\mathcal{A}[p^n] \times e^1\mathcal{A}[p^n] \to \mu_{p^n,Y}(1) \tag{23}$$

which form a projective system and give rise to a pairing

$$\langle \cdot, \cdot \rangle_{\mathrm{Weil}} : T_p\mathcal{A} \times T_p\mathcal{A} \to \hat{\mathbb{Z}}_{p,Y}(1) := \varprojlim_n \mu_{p^n,Y} \tag{24}$$

which is non-degenerate and so induces a canonical identification of $\hat{\mathbb{Z}}_{p,Y}$-local systems $T_p\mathcal{A} = \mathrm{Hom}_{\mathbb{Z}_p}(T_p\mathcal{A}, \hat{\mathbb{Z}}_{p,Y}(1)) = \mathrm{Hom}_{\mathbb{Z}_p}(T_p\mathcal{A}, \hat{\mathbb{Z}}_{p,Y})(1)$. Together with the canonical identification of $\hat{\mathbb{Z}}_{p,Y}$-local systems $H^1_{\text{ét}}(\mathcal{A}) = \mathrm{Hom}_{\mathbb{Z}_p}(T_p\mathcal{A}, \hat{\mathbb{Z}}_{p,Y})$ this gives a canonical identification of $\hat{\mathbb{Z}}_{p,Y}$-local systems $T_p\mathcal{A} = H^1_{\text{ét}}(\mathcal{A})(1)$. The pairings (23) also give rise to a pairing

$$e^1\mathcal{A}[p^\infty] \times T_p\mathcal{A} \to \mu_{p^\infty,Y} \tag{25}$$

and thus an identification of $\hat{\mathbb{Z}}_{p,Y}$-modules

$$T_p\mathcal{A} \cong \mathrm{Hom}_{p\text{-div grps}}(e^1\mathcal{A}[p^\infty], \mu_{p^\infty,Y}). \tag{26}$$

Let $Y_\infty \in Y_{\text{proét}}$ be as in (19). Let

$$(e_1, e_2) : \hat{\mathbb{Z}}_{p,Y_\infty}^{\oplus 2} \xrightarrow{\sim} H^1_{\text{ét}}(\mathcal{A})(1)|_{Y_\infty} = T_p\mathcal{A}|_{Y_\infty} \tag{27}$$

be the universal $\Gamma(p^\infty)$-level structure. Let

$$t := \log([\langle e_1, e_2\rangle_{\text{Weil}}]) \in \ker(\theta)(\mathcal{Y}^{\text{Ig},0}) \tag{28}$$

denote the corresponding determinant period for the $\mathrm{Gal}(\mathcal{Y}^{\text{Ig},0}/Y^{\text{ord}})$-action (this can be viewed as a geometric analogue of Fontaine's cyclotomic period, the "Fontaine $2\pi i$" [12]). In particular, by [27, Sect. 6] we have $\mathbb{B}_{\text{dR},\mathcal{Y}^{\text{Ig},0}} = \mathbb{B}^+_{\text{dR},\mathcal{Y}^{\text{Ig},0}}[1/t]$ and $\mathcal{O}\mathbb{B}_{\text{dR},\mathcal{Y}^{\text{Ig},0}} = \mathcal{O}\mathbb{B}^+_{\text{dR},\mathcal{Y}^{\text{Ig},0}}[1/t]$. We also have a canonical trivialization

$$\hat{\mathbb{Z}}_{p,Y_\infty}(n) \cong \hat{\mathbb{Z}}_{p,Y_\infty} \cdot t^n \tag{29}$$

which we henceforth treat as an identification.

Note that (27) and (28) induce an isomorphism of $\hat{\mathbb{Z}}_{p,Y_\infty}$-local systems

$$(e_1 \cdot t^{-1}, e_2 \cdot t^{-1}) = (e_1, e_2) \cdot t^{-1} : \hat{\mathbb{Z}}_{p,Y_\infty}^{\oplus 2} \xrightarrow{\sim} H^1_{\text{ét}}(\mathcal{A})|_{Y_\infty}. \tag{30}$$

3.3 The Relative de Rham Comparison Theorem and de Rham Periods

Recall the universal object $\pi : \mathcal{A} \to Y$. The Kodaira-Spencer isomorphism ([15, Appendix A1.3.17], [36]; recall in our situation $p \nmid \mathrm{disc}(B)$) is the isomorphism of $\mathcal{O}_Y$-modules

$$\mathrm{KS} : \omega^{\otimes 2} \xrightarrow{\sim} \Omega^1_Y, \quad \mathrm{KS}(w^{\otimes 2}) = \langle w, \nabla(w)\rangle_{\text{dR}} \tag{31}$$

where

$$\langle \cdot, \cdot \rangle_{\text{dR}} : H^1_{\text{dR}}(\mathcal{A}) \times H^1_{\text{dR}}(\mathcal{A}) \to \mathcal{O}_Y \tag{32}$$

is the alternating Poincaré pairing on universal de Rham cohomology

$$H^1_{\mathrm{dR}}(\mathcal{A}) := e^1 R^1\pi_*(0 \to \mathcal{O}_{\mathcal{A}} \xrightarrow{d} \Omega^1_{\mathcal{A}\backslash Y} \to 0),$$

e^1 is as in (8), and

$$\nabla : H^1_{\mathrm{dR}}(\mathcal{A}) \to H^1_{\mathrm{dR}}(\mathcal{A}) \otimes_{\mathcal{O}_Y} \Omega^1_Y$$

is the Gauss-Manin connection.

We define a descending filtration on $H^1_{\mathrm{dR}}(\mathcal{A})$ by $\mathrm{Fil}^0 H^1_{\mathrm{dR}}(\mathcal{A}) = H^1_{\mathrm{dR}}(\mathcal{A})$,

$$\mathrm{Fil}^1 H^1_{\mathrm{dR}}(\mathcal{A}) := \omega \subset H^1_{\mathrm{dR}}(\mathcal{A}) \tag{33}$$

and $\mathrm{Fil}^2 H^1_{\mathrm{dR}}(\mathcal{A}) = 0$. (Here ω is defined in (14).) This is often known as the *Hodge filtration*. We have an exact sequence of $\mathcal{O}_Y$-modules

$$0 \to \omega \to H^1_{\mathrm{dR}}(\mathcal{A}) \to \omega^{-1} \to 0$$

so that we have an isomorphism

$$\mathrm{gr}^0 H^1_{\mathrm{dR}}(\mathcal{A}) = H^1_{\mathrm{dR}}(\mathcal{A})/\omega \cong \omega^{-1} \tag{34}$$

which we henceforth treat as an identification. In particular the alternating pairing (32) descends to a the Poincaré duality pairing

$$\langle \cdot, \cdot \rangle_{\mathrm{dR}} : \omega \times \omega^{-1} \to \mathcal{O}_Y.$$

Proposition 3.2 *Suppose $U \in Y_{pro\acute{e}t}$ and $w \in \omega(U)$ is a nowhere-vanishing section. Then*

$$\tilde{\eta}(w) := \frac{\nabla(w)}{\mathrm{KS}(w^{\otimes 2})} \in H^1_{\mathrm{dR}}(\mathcal{A})(U),$$
$$\eta(w) := \tilde{\eta}(w) \pmod{\omega} \in (H^1_{\mathrm{dR}}(\mathcal{A})/\omega)(U) \overset{(34)}{=} \omega^{-1}(U) \tag{35}$$

are both nowhere-vanishing and

$$\langle w, \tilde{\eta}(w)\rangle_{\mathrm{dR}} = \langle w, \eta(w)\rangle_{\mathrm{dR}} = 1. \tag{36}$$

Proof This follows immediately from the fact that (31) is an isomorphism. □

By [27, Theorem 8.8] we have the relative de Rham comparison theorem

$$i_{\mathrm{dR}} : H^1_{\mathrm{dR}}(\mathcal{A}) \otimes_{\mathcal{O}_Y} \mathcal{O}\mathbb{B}^+_{\mathrm{dR},Y} \to H^1_{\acute{e}t}(\mathcal{A}) \otimes_{\hat{\mathbb{Z}}_{p,Y}} \mathcal{O}\mathbb{B}^+_{\mathrm{dR},Y} \tag{37}$$

compatible with connections and filtrations. This comparison map is usually stated as an isomorphism over $\mathcal{O}\mathbb{B}_{\mathrm{dR},Y}$, but one can deduce the above version as follows: By [9, Proposition 2.2.3] we have an inclusion of $\mathbb{B}^+_{\mathrm{dR},Y}$-local systems compatible with filtrations

$$\left(H^1_{\mathrm{dR}}(\mathcal{A}) \otimes_{\mathcal{O}_Y} \mathcal{O}\mathbb{B}^+_{\mathrm{dR},Y}\right)^{\nabla=0} \subset H^1_{\text{ét}}(\mathcal{A}) \otimes_{\hat{\mathbb{Z}}_{p,Y}} \mathbb{B}^+_{\mathrm{dR},Y}.$$

Tensoring this with $\otimes_{\mathbb{B}^+_{\mathrm{dR},Y}} \mathcal{O}\mathbb{B}^+_{\mathrm{dR},Y}$ and invoking [27, Theorem 7.2] (as well as the compatibility between the filtrations of $\mathbb{B}^+_{\mathrm{dR},Y}$ and $\mathcal{O}\mathbb{B}^+_{\mathrm{dR},Y}$) we get (37). The relative de Rham comparison map (37) is known to specialize to the absolute de Rham comparison map at classical points [25]. It is moreover known to be injective. (For example, this is seen as follows: it specializes to the absolute de Rham comparison map over B^+_{dR}, which is known to be injective by [12], at the dense set of classical points, and so the kernel of i_{dR} specializes to 0 on a dense subset of points of Y and thus is 0.)

Taking the gr^0 part of (37) we obtain

$$i_{\mathrm{HT}} : \omega^{-1} \otimes_{\mathcal{O}_Y} \hat{\mathcal{O}}_Y \to H^1_{\text{ét}}(\mathcal{A}) \otimes_{\hat{\mathbb{Z}}_{p,Y}} \hat{\mathcal{O}}_Y \tag{38}$$

which is known as the *relative Hodge-Tate filtration* (see [9]).

We now recall how to define de Rham periods attached to sections of $H^1_{\mathrm{dR}}(\mathcal{A})$ following [22, Sect. 3].

Definition 3.3 (1) From (37) we have

$$i_{\mathrm{dR}} : H^1_{\mathrm{dR}}(\mathcal{A}) \otimes_{\mathcal{O}_Y} \mathcal{O}\mathbb{B}^+_{\mathrm{dR},Y_\infty} \to H^1_{\text{ét}}(\mathcal{A}) \otimes_{\hat{\mathbb{Z}}_{p,Y}} \mathcal{O}\mathbb{B}^+_{\mathrm{dR},Y_\infty} \underset{\sim}{\overset{(27)}{\longrightarrow}} \mathcal{O}\mathbb{B}^{+,\oplus 2}_{\mathrm{dR},Y_\infty}$$

Given a section $h \in H^1_{\mathrm{dR}}(\mathcal{A}) \otimes_{\mathcal{O}_Y} \mathcal{O}\mathbb{B}^+_{\mathrm{dR},Y_\infty}(V)$ for some $V \in Y_{\text{proét}}/Y_\infty$, we write

$$i_{\mathrm{dR}}(h) = (x_{\mathrm{dR}}(h), y_{\mathrm{dR}}(h)) \in \mathcal{O}\mathbb{B}^{+,\oplus 2}_{\mathrm{dR},Y_\infty}(V). \tag{39}$$

In particular if $h \in \omega \otimes_{\mathcal{O}_Y} \mathcal{O}\mathbb{B}^+_{\mathrm{dR},Y_\infty}(V)$ we have

$$(x_{\mathrm{dR}}(h), y_{\mathrm{dR}}(h)) \in \left(\mathrm{Fil}^1 \mathcal{O}\mathbb{B}^+_{\mathrm{dR},Y_\infty}\right)^{\oplus 2}(V). \tag{40}$$

(2) From (38) we have

$$i_{\mathrm{HT}} : \omega^{-1} \otimes_{\mathcal{O}_Y} \hat{\mathcal{O}}_{Y_\infty} \overset{(38)}{\longrightarrow} H^1_{\text{ét}}(\mathcal{A}) \otimes_{\hat{\mathbb{Z}}_{p,Y}} \hat{\mathcal{O}}_{Y_\infty} \underset{\sim}{\overset{(30)}{\longrightarrow}} \hat{\mathcal{O}}^{\oplus 2}_{Y_\infty}.$$

Given a section $\eta \in \omega^{-1} \otimes_{\mathcal{O}_Y} \hat{\mathcal{O}}_{Y_\infty}(V)$ for some $V \in Y_{\text{proét}}/Y_\infty$, we write

$$i_{\mathrm{HT}}(\eta) = (x_{\mathrm{HT}}(\eta), y_{\mathrm{HT}}(\eta)) \in \hat{\mathcal{O}}^{\oplus 2}_{Y_\infty}(V). \tag{41}$$

(3) In particular if $\tilde{\eta} \in H^1_{\mathrm{dR}}(\mathcal{A})(V)$ satisfies $\tilde{\eta} \equiv \eta \pmod{\omega(V)}$ under (34) then we have

$$\theta_{\mathrm{dR}}(x_{\mathrm{dR}}(\tilde{\eta})) = x_{\mathrm{HT}}(\eta), \quad \theta_{\mathrm{dR}}(y_{\mathrm{dR}}(\tilde{\eta})) = y_{\mathrm{HT}}(\eta). \tag{42}$$

Recall the universal Weil pairing (24)

$$\langle\cdot,\cdot\rangle_{\mathrm{Weil}} : T_p\mathcal{A} \times T_p\mathcal{A} \to \hat{\mathbb{Z}}_{p,Y}(1).$$

Since we have a natural identification $H^1_{\text{ét}}(\mathcal{A}) = \mathrm{Hom}_{\mathbb{Z}_p}(T_p\mathcal{A}, \mathbb{Z}_p)$, the Weil pairing induces an identification

$$H^1_{\text{ét}}(\mathcal{A}) = \mathrm{Hom}_{\mathbb{Z}_p}(T_p\mathcal{A}, \mathbb{Z}_p) = T_p\mathcal{A}(-1)$$

as well as a nondegenerate pairing

$$\langle\cdot,\cdot\rangle_{\mathrm{Weil}} : H^1_{\text{ét}}(\mathcal{A}) \times H^1_{\text{ét}}(\mathcal{A}) \to \hat{\mathbb{Z}}_{p,Y}(-1). \tag{43}$$

Pulling back to $Y_\infty \to Y$ and using (30) and (29), we have have

$$\langle e_1 \cdot t^{-1}, e_2 \cdot t^{-1}\rangle_{\mathrm{Weil}} = t^{-1}. \tag{44}$$

We will sometimes use the shorthand

$$(1,0) = e_1 \cdot t^{-1}, \quad (0,1) = e_2 \cdot t^{-1}. \tag{45}$$

Then (44) can be rewritten as

$$\langle (1,0), (0,1)\rangle_{\mathrm{Weil}} = t^{-1}.$$

Recall the Poincaré pairing $\langle\cdot,\cdot\rangle$ (32). We have the following compatibility under (37)

$$\langle i_{\mathrm{dR}}(\cdot), i_{\mathrm{dR}}(\cdot)\rangle_{\mathrm{Weil}} = \langle\cdot,\cdot\rangle. \tag{46}$$

This follows from checking on fibers at classical points (where it follows, for example, from [12]) and using the density of such points in the adic topology.

3.4 A Description of $\mathcal{O}\mathbb{B}^+_{\mathbf{dR}}$ on the Perfected Igusa Tower

Via pullback along $\mathcal{Y}^{\mathrm{Ig},0} \to Y^{\mathrm{Ig}} \to Y^{\mathrm{Ig},+}$ we view the nowhere-vanishing section $w^{\mathrm{Katz}}_{\mathrm{can}}$ from (15) as a nowhere-vanishing section $w^{\mathrm{Katz}}_{\mathrm{can}} \in \omega(\mathcal{Y}^{\mathrm{Ig},0})$. From Definition 3.3 we have

$$i_{\mathrm{dR}}(w^{\mathrm{Katz}}_{\mathrm{can}}) = (x_{\mathrm{dR}}(w^{\mathrm{Katz}}_{\mathrm{can}}), y_{\mathrm{dR}}(w^{\mathrm{Katz}}_{\mathrm{can}})) \overset{(40)}{\in} \mathrm{Fil}^1 \mathcal{O}\mathbb{B}^+_{\mathrm{dR},Y}(\mathcal{Y}^{\mathrm{Ig},0})^{\oplus 2}.$$

Theorem 3.4 *(1) We have*

$$x_{\mathrm{HT}}(\eta(w^{\mathrm{Katz}}_{\mathrm{can}})) = -1, \quad y_{\mathrm{HT}}(\eta(w^{\mathrm{Katz}}_{\mathrm{can}})) = 0. \tag{47}$$

(2) We have

$$y_{\mathrm{dR}}(w^{\mathrm{Katz}}_{\mathrm{can}}) = t. \tag{48}$$

(3) For any local section w (on $Y_{pro\acute{e}t}/Y^{\mathrm{Ig}}$) of ω, we have the "p-adic Legendre relation" (cf. [15, A1.3.4])

$$x_{\mathrm{dR}}(w) \cdot y_{\mathrm{dR}}(\tilde{\eta}(w)) - y_{\mathrm{dR}}(w) \cdot x_{\mathrm{dR}}(\tilde{\eta}(w)) = t. \tag{49}$$

(4) For any $U \in Y_{pro\acute{e}t}$ and any $h \in \mathrm{Sym}^k H^1_{\mathrm{dR}}(\mathcal{A}) \otimes_{\mathcal{O}_Y} \mathcal{O}\mathbb{B}^+_{\mathrm{dR},Y}(U)$ we let $\theta_{\mathrm{dR}}(h) \in \mathrm{Sym}^k H^1_{\mathrm{dR}}(\mathcal{A}) \otimes_{\mathcal{O}_Y} \hat{\mathcal{O}}_Y(U)$ denote the image of h under the extension of scalars $\otimes_{\mathcal{O}\mathbb{B}^+_{\mathrm{dR},Y},\theta_{\mathrm{dR}}} \hat{\mathcal{O}}_Y$. We have that

$$\theta_{\mathrm{dR}}(\nabla(x_{\mathrm{dR}}(w^{\mathrm{Katz}}_{\mathrm{can}}))), \theta_{\mathrm{dR}}(\nabla(y_{\mathrm{dR}}(w^{\mathrm{Katz}}_{\mathrm{can}}))) \in \Omega^1_Y \otimes_{\mathcal{O}_Y} \hat{\mathcal{O}}_Y(\mathcal{Y}^{\mathrm{Ig},0}).$$

Moreover $\theta_{\mathrm{dR}}(\nabla(x_{\mathrm{dR}}(w^{\mathrm{Katz}}_{\mathrm{can}})))$ is nowhere-vanishing.

Proof (1) By [16, Sects. 2, 3.4], the Weil pairing (24) induces a perfect pairing of rank 1 $\hat{\mathbb{Z}}_{p,Y}$-modules

$$\langle\cdot,\cdot\rangle'_{\acute{e}t} : T_p\hat{\mathcal{A}} \times T_p\mathcal{A}^{\acute{e}t} \to \hat{\mathbb{Z}}_{p,Y}(1) \tag{50}$$

where $T_p\hat{\mathcal{A}}$ is the connected part of $T_p\mathcal{A}$ and $T_p\mathcal{A}^{\acute{e}t}$ is the maximal étale quotient of $T_p\mathcal{A}$. It also induces a perfect pairing

$$\hat{\mathcal{A}}[p^\infty] \times T_p\mathcal{A}^{\acute{e}t} \to \mu_{p^\infty,Y}$$

where $\hat{\mathcal{A}}[p^\infty]$ is the connected part of $e^1\mathcal{A}[p^\infty]$. Thus we get an isomorphism

$$T_p\mathcal{A}^{\text{ét}} \cong \text{Hom}_{p\text{-div grps}}(\hat{\mathcal{A}}[p^\infty], \mu_{p^\infty,Y}) \tag{51}$$

which we henceforth treat as an identification. Then (3.4) in turn induces an isomorphism of $\hat{\mathcal{O}}_Y$-modules

$$\text{HT}^{\text{ét}} : T_p\mathcal{A}^{\text{ét}} \otimes_{\mathbb{Z}_p} \hat{\mathcal{O}}_Y \xrightarrow{\sim} \omega \otimes_{\mathcal{O}_Y} \hat{\mathcal{O}}_Y, \quad \alpha \mapsto \alpha^* \frac{dT}{1+T} \tag{52}$$

whose precomposition with the canonical projection

$$T_p\mathcal{A} \otimes_{\hat{\mathbb{Z}}_{p,Y}} \hat{\mathcal{O}}_Y \to T_p\mathcal{A}^{\text{ét}} \otimes_{\hat{\mathbb{Z}}_{p,Y}} \hat{\mathcal{O}}_Y, \quad \alpha \mapsto \alpha^{\text{ét}}$$

is the usual *Hodge-Tate map* [11, Sect. 2], defined using (26),

$$\text{HT} : T_p\mathcal{A} \otimes_{\hat{\mathbb{Z}}_{p,Y}} \hat{\mathcal{O}}_Y \to \omega \otimes_{\mathcal{O}_Y} \hat{\mathcal{O}}_Y, \quad \alpha \mapsto \alpha^* \frac{dT}{1+T}$$

recalling that in both of the above maps T is a chosen coordinate on $\mu_{p^\infty,Y}$ (from Definition 2.7). In particular by Definition 2.7 we have

$$w_{\text{can}}^{\text{Katz}} = e_2^* \frac{dT}{1+T} = \text{HT}(e_2) = \text{HT}^{\text{ét}}(e_2^{\text{ét}}). \tag{53}$$

Taking the $\hat{\mathcal{O}}_Y$-linear dual of (52) gives an isomorphism of $\hat{\mathcal{O}}_Y$-modules

$$\begin{aligned}\text{HT}^{\text{ét},\vee} : \omega^{-1} \otimes_{\mathcal{O}_Y} \hat{\mathcal{O}}_Y \xrightarrow{\sim} \text{Hom}_{\mathbb{Z}_p}(T_p\mathcal{A}^{\text{ét}}, \mathbb{Z}_p) \otimes_{\mathbb{Z}_p} \hat{\mathcal{O}}_Y &\overset{(50)}{=} T_p\hat{\mathcal{A}}(-1) \otimes_{\hat{\mathbb{Z}}_p} \hat{\mathcal{O}}_Y \\ &\overset{(29)}{=} T_p\hat{\mathcal{A}} \cdot t^{-1} \otimes_{\hat{\mathbb{Z}}_p} \hat{\mathcal{O}}_Y.\end{aligned}$$

By the discussion at the end of [16, Sect. 3.4] we have the following (tautological) compatibility with respect to the Weil and Poincaré pairings

$$\langle \text{HT}(\alpha), \text{HT}^{\text{ét},\vee,-1}(\beta)\rangle = \langle \text{HT}^{\text{ét}}(\alpha^{\text{ét}}), \text{HT}^{\text{ét},\vee,-1}(\beta)\rangle = \langle \alpha^{\text{ét}}, \beta\rangle'_{\text{ét}} = \langle \alpha, \beta\rangle_{\text{ét}} \tag{54}$$

for all $\alpha \in T_p\mathcal{A} \otimes_{\hat{\mathbb{Z}}_{p,Y}} \hat{\mathcal{O}}_Y$ and $\beta \in T_p\hat{\mathcal{A}} \cdot t^{-1} \otimes_{\hat{\mathbb{Z}}_{p,Y}} \hat{\mathcal{O}}_Y$. It is known (see [11, Sect. 2] for example) that

$$i_{\text{HT}}|_{\mathcal{Y}^{\text{Ig},0}} = \text{HT}^{\text{ét},\vee}|_{\mathcal{Y}^{\text{Ig},0}}. \tag{55}$$

Thus we have

$$\begin{aligned}-1 &\overset{(44)}{=} \langle e_2, e_1 \cdot t^{-1}\rangle_{\text{ét}} \overset{(54)}{=} \langle \text{HT}(e_2), \text{HT}^{\text{ét},\vee,-1}(e_1 \cdot t^{-1})\rangle \\ &\overset{(53)}{=} \langle w_{\text{can}}^{\text{Katz}}, \text{HT}^{\text{ét},\vee,-1}(e_1 \cdot t^{-1})\rangle.\end{aligned}$$

Therefore (36) implies

$$\mathrm{HT}^{\text{ét},\vee,-1}(e_1 \cdot t^{-1}) = -\eta(w_{\text{can}}^{\text{Katz}}) \tag{56}$$

which implies

$$i_{\mathrm{HT}}(\eta(w_{\text{can}}^{\text{Katz}})) \overset{(55)}{=} \mathrm{HT}^{\text{ét},\vee}(\eta(w_{\text{can}}^{\text{Katz}})) \overset{(56)}{=} -e_1 \cdot t^{-1} \overset{(45)}{=} (-1, 0). \tag{57}$$

Thus

$$(x_{\mathrm{HT}}(\eta(w_{\text{can}}^{\text{Katz}})), y_{\mathrm{HT}}(\eta(w_{\text{can}}^{\text{Katz}}))) \overset{(41)}{=} i_{\mathrm{HT}}(\eta(w_{\text{can}}^{\text{Katz}})) \overset{(57)}{=} (-1, 0)$$

which gives the assertion.
From (1) we have

$$\begin{aligned}
1 &\overset{(36)}{=} \langle w_{\text{can}}^{\text{Katz}}, \eta(w_{\text{can}}^{\text{Katz}}) \rangle \overset{(54)}{=} \langle i_{\mathrm{dR}}(w_{\text{can}}^{\text{Katz}}), \mathrm{HT}^{\text{ét},\vee,-1}(\eta(w_{\text{can}}^{\text{Katz}})) \rangle_{\text{ét}} \\
&\overset{(57)}{=} \langle i_{\mathrm{dR}}(w_{\text{can}}^{\text{Katz}}), i_{\mathrm{HT}}(\eta(w_{\text{can}}^{\text{Katz}})) \rangle_{\text{ét}} \\
&= \langle (x_{\mathrm{dR}}(w_{\text{can}}^{\text{Katz}}), y_{\mathrm{dR}}(w_{\text{can}}^{\text{Katz}})), (x_{\mathrm{HT}}(\eta(w_{\text{can}}^{\text{Katz}})), y_{\mathrm{HT}}(\eta(w_{\text{can}}^{\text{Katz}}))) \rangle_{\text{ét}} \\
&\overset{(47)}{=} \langle (x_{\mathrm{dR}}(w_{\text{can}}^{\text{Katz}}), y_{\mathrm{dR}}(w_{\text{can}}^{\text{Katz}})), (-1, 0) \rangle_{\text{ét}} \\
&= y_{\mathrm{dR}}(w_{\text{can}}^{\text{Katz}}) \cdot \langle (0, 1), (-1, 0) \rangle_{\text{ét}} \\
&\overset{(44),(45)}{=} y_{\mathrm{dR}}(w_{\text{can}}^{\text{Katz}}) \cdot t^{-1}
\end{aligned}$$

whence the assertion.

(2) We first claim that it suffices to prove this identity for $w = w_{\text{can}}^{\text{Katz}}$. As ω is a locally free $\mathcal{O}_Y$-module of rank 1 and $w_{\text{can}}^{\text{Katz}} \in \omega(Y^{\text{Ig}})$ is a nowhere-vanishing section [see (15)], any local section w of ω can be written as $w = \alpha \cdot w_{\text{can}}^{\text{Katz}}$ for some local section α of $\mathcal{O}_Y$, and so by the $\mathcal{O}\mathbb{B}_{\mathrm{dR},Y}^{+}$-linearity of (37) we have $i_{\mathrm{dR}}(w) = i_{\mathrm{dR}}(\alpha \cdot w_{\text{can}}^{\text{Katz}}) = \alpha \cdot i_{\mathrm{dR}}(w_{\text{can}}^{\text{Katz}})$ which by (39) implies

$$(x_{\mathrm{dR}}(w), y_{\mathrm{dR}}(w)) = \alpha \cdot (x_{\mathrm{dR}}(w_{\text{can}}^{\text{Katz}}), y_{\mathrm{dR}}(w_{\text{can}}^{\text{Katz}})). \tag{58}$$

On the other hand we have that

$$\tilde{\eta}(w) = \tilde{\eta}(\alpha \cdot w_{\text{can}}^{\text{Katz}}) \overset{(35)}{=} \frac{\nabla(\alpha \cdot w_{\text{can}}^{\text{Katz}})}{\mathrm{KS}(\alpha \cdot w_{\text{can}}^{\text{Katz},\otimes 2})} = \frac{\nabla(\alpha) \cdot w_{\text{can}}^{\text{Katz}} + \alpha \cdot \nabla(w_{\text{can}}^{\text{Katz}})}{\alpha^2 \cdot \mathrm{KS}(w_{\text{can}}^{\text{Katz},\otimes 2})}$$

which implies

$$(x_{\mathrm{dR}}(\tilde{\eta}(w)), y_{\mathrm{dR}}(\tilde{\eta}(w))) = \frac{\nabla(\alpha)}{\alpha^2 \cdot \mathrm{KS}(w_{\mathrm{can}}^{\mathrm{Katz},\otimes 2})}(x_{\mathrm{dR}}(w), y_{\mathrm{dR}}(w)) + \alpha^{-1} \cdot (x_{\mathrm{dR}}(\tilde{\eta}(w_{\mathrm{can}}^{\mathrm{Katz}})), y_{\mathrm{dR}}(\tilde{\eta}(w_{\mathrm{can}}^{\mathrm{Katz}}))). \tag{59}$$

This implies that

$$\begin{aligned} x_{\mathrm{dR}}(w) \cdot y_{\mathrm{dR}}(\tilde{\eta}(w)) - y_{\mathrm{dR}}(w) \cdot x_{\mathrm{dR}}(\tilde{\eta}(w)) &= x_{\mathrm{dR}}(\alpha \cdot w_{\mathrm{can}}^{\mathrm{Katz}}) \cdot y_{\mathrm{dR}}(\tilde{\eta}(\alpha \cdot w_{\mathrm{can}}^{\mathrm{Katz}})) \\ &- y_{\mathrm{dR}}(\alpha \cdot w_{\mathrm{can}}^{\mathrm{Katz}}) \cdot x_{\mathrm{dR}}(\tilde{\eta}(\alpha \cdot w_{\mathrm{can}}^{\mathrm{Katz}})) \\ &\overset{(58),(59)}{=} x_{\mathrm{dR}}(w_{\mathrm{can}}^{\mathrm{Katz}}) \cdot y_{\mathrm{dR}}(\tilde{\eta}(w_{\mathrm{can}}^{\mathrm{Katz}})) \\ &- y_{\mathrm{dR}}(w_{\mathrm{can}}^{\mathrm{Katz}}) \cdot x_{\mathrm{dR}}\tilde{\eta}(w_{\mathrm{can}}^{\mathrm{Katz}}). \end{aligned}$$

This gives the claim.

We now prove the identity for $w = w_{\mathrm{can}}^{\mathrm{Katz}}$.

(3) Using the isomorphism (30) we have

$$\begin{aligned} 1 = \langle w_{\mathrm{can}}^{\mathrm{Katz}}, \eta(w_{\mathrm{can}}^{\mathrm{Katz}})\rangle = \langle w_{\mathrm{can}}^{\mathrm{Katz}}, \tilde{\eta}(w_{\mathrm{can}}^{\mathrm{Katz}})\rangle &\overset{(46)}{=} \langle i_{\mathrm{dR}}(w_{\mathrm{can}}^{\mathrm{Katz}}), i_{\mathrm{dR}}(\tilde{\eta}(w_{\mathrm{can}}^{\mathrm{Katz}}))\rangle_{\mathrm{Weil}} \\ &\overset{(30)}{=} \langle (x_{\mathrm{dR}}(w_{\mathrm{can}}^{\mathrm{Katz}}), y_{\mathrm{dR}}(w_{\mathrm{can}}^{\mathrm{Katz}})), (x_{\mathrm{dR}}(\tilde{\eta}(w_{\mathrm{can}}^{\mathrm{Katz}})), y_{\mathrm{dR}}(\tilde{\eta}(w_{\mathrm{can}}^{\mathrm{Katz}})))\rangle_{\mathrm{Weil}} \\ &\overset{(43)}{=} (x_{\mathrm{dR}}(w_{\mathrm{can}}^{\mathrm{Katz}}) \cdot y_{\mathrm{dR}}(\tilde{\eta}(w_{\mathrm{can}}^{\mathrm{Katz}})) - y_{\mathrm{dR}}(w_{\mathrm{can}}^{\mathrm{Katz}}) \cdot x_{\mathrm{dR}}(\tilde{\eta}(w_{\mathrm{can}}^{\mathrm{Katz}}))) \cdot \langle e_1 \cdot t^{-1}, e_2 \cdot t^{-1}\rangle_{\mathrm{Weil}} \\ &\overset{(44)}{=} (x_{\mathrm{dR}}(w_{\mathrm{can}}^{\mathrm{Katz}}) \cdot y_{\mathrm{dR}}(\tilde{\eta}(w_{\mathrm{can}}^{\mathrm{Katz}})) - y_{\mathrm{dR}}(w_{\mathrm{can}}^{\mathrm{Katz}}) \cdot x_{\mathrm{dR}}(\tilde{\eta}(w_{\mathrm{can}}^{\mathrm{Katz}}))) \cdot t^{-1}. \end{aligned}$$

Thus we have

$$x_{\mathrm{dR}}(w_{\mathrm{can}}^{\mathrm{Katz}}) \cdot y_{\mathrm{dR}}(\tilde{\eta}(w_{\mathrm{can}}^{\mathrm{Katz}})) - y_{\mathrm{dR}}(w_{\mathrm{can}}^{\mathrm{Katz}}) \cdot x_{\mathrm{dR}}(\tilde{\eta}(w_{\mathrm{can}}^{\mathrm{Katz}})) = t.$$

This gives the assertion.

(4) Since $w_{\mathrm{can}}^{\mathrm{Katz}} \in \omega(Y^{\mathrm{Ig}})$ we have $\nabla(w_{\mathrm{can}}^{\mathrm{Katz}}) \in H^1_{\mathrm{dR}}(\mathcal{A}) \otimes_{\mathcal{O}_Y} \Omega^1_Y(Y^{\mathrm{Ig}})$. Thus

$$\theta_{\mathrm{dR}}(\nabla(w_{\mathrm{can}}^{\mathrm{Katz}})) \in H^1_{\mathrm{dR}}(\mathcal{A}) \otimes_{\mathcal{O}_Y} \Omega^1_Y(Y^{\mathrm{Ig}}).$$

Given $U \in Y_{\mathrm{pro\acute{e}t}}$ and $h \in \mathrm{Sym}^k H^1_{\acute{\mathrm{e}}\mathrm{t}}(\mathcal{A}) \otimes_{\hat{\mathbb{Z}}_{p,Y}} \mathcal{O}\mathbb{B}^+_{\mathrm{dR},Y}(U)$ we let $\theta_{\mathrm{dR}}(h) \in \mathrm{Sym}^k H^1_{\acute{\mathrm{e}}\mathrm{t}}(\mathcal{A}) \otimes_{\hat{\mathbb{Z}}_{p,Y}} \hat{\mathcal{O}}_Y(U)$ denote the image of h under the extension of scalars $\otimes_{\mathcal{O}\mathbb{B}^+_{\mathrm{dR},Y},\theta_{\mathrm{dR}}} \hat{\mathcal{O}}_Y$. Then since i_{dR} is $\mathcal{O}\mathbb{B}^+_{\mathrm{dR},Y}$-linear we have

$$\theta_{\mathrm{dR}}(i_{\mathrm{dR}}(\nabla(w_{\mathrm{can}}^{\mathrm{Katz}}))) = i_{\mathrm{dR}}(\nabla(w_{\mathrm{can}}^{\mathrm{Katz}})) \in H^1_{\acute{\mathrm{e}}\mathrm{t}}(\mathcal{A}) \otimes_{\hat{\mathbb{Z}}_{p,Y}} \Omega^1_Y \otimes_{\mathcal{O}_Y} \hat{\mathcal{O}}_Y(Y^{\mathrm{Ig}}). \tag{60}$$

As (37) is compatible with connections we have

$$i_{\mathrm{dR}}(\nabla(w_{\mathrm{can}}^{\mathrm{Katz}}))|_{\mathcal{Y}^{\mathrm{Ig},0}} = \nabla(i_{\mathrm{dR}}(w_{\mathrm{can}}^{\mathrm{Katz}}))|_{\mathcal{Y}^{\mathrm{Ig},0}} = (\nabla(x_{\mathrm{dR}}(w_{\mathrm{can}}^{\mathrm{Katz}})), \nabla(y_{\mathrm{dR}}(w_{\mathrm{can}}^{\mathrm{Katz}}))). \tag{61}$$

Then inserting (61) into (60) we see that

$$\theta_{\mathrm{dR}}(\nabla(x_{\mathrm{dR}}(w_{\mathrm{can}}^{\mathrm{Katz}})), \nabla(y_{\mathrm{dR}}(w_{\mathrm{can}}^{\mathrm{Katz}}))) \overset{(61)}{=} \theta_{\mathrm{dR}}(i_{\mathrm{dR}}(\nabla(w_{\mathrm{can}}^{\mathrm{Katz}})))|_{\mathcal{Y}^{\mathrm{Ig},0}}$$
$$\overset{(60)}{\in} H^1_{\text{ét}}(\mathcal{A}) \otimes_{\hat{\mathbb{Z}}_{p,Y}} \Omega^1_Y \otimes_{\mathcal{O}_Y} \hat{\mathcal{O}}_Y(\mathcal{Y}^{\mathrm{Ig},0}) \underset{\sim}{\overset{(30)}{\longrightarrow}} \left(\Omega^1_Y \otimes_{\mathcal{O}_Y} \hat{\mathcal{O}}_Y(\mathcal{Y}^{\mathrm{Ig},0})\right)^{\oplus 2}$$

which gives the first assertion. It remains to show the nowhere vanishing. Note that (37) induces a map of $\mathcal{O}\mathbb{B}^+_{\mathrm{dR},Y}$-modules

$$i_{\mathrm{dR}} : \mathrm{Sym}^k H^1_{\mathrm{dR}}(\mathcal{A}) \otimes_{\mathcal{O}_Y} \mathcal{O}\mathbb{B}^+_{\mathrm{dR},Y} \to \mathrm{Sym}^k H^1_{\text{ét}}(\mathcal{A}) \otimes_{\hat{\mathbb{Z}}_{p,Y}} \mathcal{O}\mathbb{B}^+_{\mathrm{dR},Y}$$

compatible with connections and filtrations. As i_{dR} is compatible with connections we have

$$\begin{aligned}
i_{\mathrm{dR}}(\mathrm{KS}(w_{\mathrm{can}}^{\mathrm{Katz},\otimes 2})) &\overset{(31)}{=} i_{\mathrm{dR}}(\langle w_{\mathrm{can}}^{\mathrm{Katz}}, \nabla(w_{\mathrm{can}}^{\mathrm{Katz}})\rangle) \overset{(46)}{=} \langle i_{\mathrm{dR}}(w_{\mathrm{can}}^{\mathrm{Katz}}), \nabla(i_{\mathrm{dR}}(w_{\mathrm{can}}^{\mathrm{Katz}}))\rangle_{\text{ét}} \\
&\overset{(39)}{=} \langle (x_{\mathrm{dR}}(w_{\mathrm{can}}^{\mathrm{Katz}}), y_{\mathrm{dR}}(w_{\mathrm{can}}^{\mathrm{Katz}})), (\nabla(x_{\mathrm{dR}}(w_{\mathrm{can}}^{\mathrm{Katz}})), \nabla(y_{\mathrm{dR}}(w_{\mathrm{can}}^{\mathrm{Katz}})))\rangle_{\text{ét}} \\
&= \nabla(x_{\mathrm{dR}}(w_{\mathrm{can}}^{\mathrm{Katz}})) \cdot y_{\mathrm{dR}}(w_{\mathrm{can}}^{\mathrm{Katz}}) \\
&\quad - \nabla(y_{\mathrm{dR}}(w_{\mathrm{can}}^{\mathrm{Katz}})) \cdot x_{\mathrm{dR}}(w_{\mathrm{can}}^{\mathrm{Katz}}) \cdot \langle (0,1), (1,0)\rangle_{\text{ét}} \\
&\overset{(44)}{=} -(\nabla(x_{\mathrm{dR}}(w_{\mathrm{can}}^{\mathrm{Katz}})) \cdot y_{\mathrm{dR}}(w_{\mathrm{can}}^{\mathrm{Katz}}) \\
&\quad - \nabla(y_{\mathrm{dR}}(w_{\mathrm{can}}^{\mathrm{Katz}})) \cdot x_{\mathrm{dR}}(w_{\mathrm{can}}^{\mathrm{Katz}})) \cdot t^{-1} \\
&\overset{(48)}{=} -\left(\nabla(x_{\mathrm{dR}}(w_{\mathrm{can}}^{\mathrm{Katz}})) \cdot t - \nabla(t) \cdot x_{\mathrm{dR}}(w_{\mathrm{can}}^{\mathrm{Katz}})\right) \cdot t^{-1} \\
&= -\nabla(x_{\mathrm{dR}}(w_{\mathrm{can}}^{\mathrm{Katz}})).
\end{aligned} \tag{62}$$

On the other hand, since $w_{\mathrm{can}}^{\mathrm{Katz}} \in \omega(Y^{\mathrm{Ig}})$ is nowhere-vanishing then

$$\mathrm{KS}(w_{\mathrm{can}}^{\mathrm{Katz},\otimes 2}) \overset{(31)}{\in} \Omega^1_Y(Y^{\mathrm{Ig}}) \tag{63}$$

is also nowhere-vanishing (since (31) is an isomorphism) and so

$$\theta_{\mathrm{dR}}(\mathrm{KS}(w_{\mathrm{can}}^{\mathrm{Katz},\otimes 2})) = \mathrm{KS}(w_{\mathrm{can}}^{\mathrm{Katz},\otimes 2}) \tag{64}$$

and thus

$$\begin{aligned}
\theta_{\mathrm{dR}}(\nabla(x_{\mathrm{dR}}(w_{\mathrm{can}}^{\mathrm{Katz}}))) &\overset{(62)}{=} -\theta_{\mathrm{dR}}(i_{\mathrm{dR}}(\mathrm{KS}(w_{\mathrm{can}}^{\mathrm{Katz},\otimes 2}))) = -i_{\mathrm{dR}}(\theta_{\mathrm{dR}}(\mathrm{KS}(w_{\mathrm{can}}^{\mathrm{Katz},\otimes 2}))) \\
&\overset{(64)}{=} -i_{\mathrm{dR}}(\mathrm{KS}(w_{\mathrm{can}}^{\mathrm{Katz},\otimes 2})).
\end{aligned}$$

The right hand side of this equation is nowhere-vanishing since $\mathrm{KS}(w_{\mathrm{can}}^{\mathrm{Katz},\otimes 2}) \overset{(63)}{\in} \Omega^1_Y(Y^{\mathrm{Ig}})$ is and (37) is an injection. □

We can now prove a fundamental statement on the structure of $\mathcal{O}\mathbb{B}^+_{\mathrm{dR},Y}$ on $\mathcal{Y}^{\mathrm{Ig},0}$.

Theorem 3.5 *We have an isomorphism of filtered $\mathbb{B}^+_{\mathrm{dR},\mathcal{Y}^{\mathrm{Ig},0}}$-algebras on $Y_{\text{proét}}/\mathcal{Y}^{\mathrm{Ig},0}$*

$$\mathbb{B}^+_{\mathrm{dR},\mathcal{Y}^{\mathrm{Ig},0}}[\![X]\!] \cong \mathcal{O}\mathbb{B}^+_{\mathrm{dR},\mathcal{Y}^{\mathrm{Ig},0}} \tag{65}$$

sending $X \mapsto x_{\mathrm{dR}}(w^{\mathrm{Katz}}_{\mathrm{can}}) \in \mathrm{Fil}^1\mathcal{O}\mathbb{B}^+_{\mathrm{dR},Y}(\mathcal{Y}^{\mathrm{Ig},0})$ and compatible with connections.

Proof Recall that Y is a smooth adic space over $\mathrm{Spa}(\mathbb{Q}_p, \mathbb{Z}_p)$. By [27, Proposition 6.10] we can find a $U \in Y_{\text{proét}}/\mathcal{Y}^{\mathrm{Ig},0}$ and an isomorphism of $\mathbb{B}^+_{\mathrm{dR},U}$-algebras

$$\mathcal{O}\mathbb{B}^+_{\mathrm{dR},U} \cong \mathbb{B}^+_{\mathrm{dR},U}[\![T]\!] \tag{66}$$

for a formal variable T; moreover this isomorphism is compatible with filtrations [where the filtration degree on the right hand side is given by $\mathrm{Fil}^i\mathbb{B}^+_{\mathrm{dR},U}[\![T]\!] = (\mathrm{Fil}^1\mathbb{B}^+_{\mathrm{dR},U}, T)$; in fact we have $\mathrm{Fil}^1\mathbb{B}^+_{\mathrm{dR},U} = t \cdot \mathbb{B}^+_{\mathrm{dR},U}$ by (28)] and connections and $\nabla(T)$ is a generator (i.e. nowhere-vanishing global section) of Ω^1_U. Let $\theta_T : \mathbb{B}^+_{\mathrm{dR},U}[\![T]\!] \to \mathbb{B}^+_{\mathrm{dR},U}$ denote reduction modulo T. Since

$$x_{\mathrm{dR}}(w^{\mathrm{Katz}}_{\mathrm{can}})|_U \overset{(40)}{\in} \mathrm{Fil}^1\mathcal{O}\mathbb{B}^+_{\mathrm{dR},Y}(U) \overset{(66)}{\cong} (\mathrm{Fil}^1\mathbb{B}^+_{\mathrm{dR},U}(U), T)$$

we have

$$\theta_T(x_{\mathrm{dR}}(w^{\mathrm{Katz}}_{\mathrm{can}})) \in \mathrm{Fil}^1\mathbb{B}^+_{\mathrm{dR},U}(U). \tag{67}$$

Note that the composition

$$\mathbb{B}^+_{\mathrm{dR},U}[\![T]\!] \underset{\sim}{\overset{(66)}{\longrightarrow}} \mathcal{O}\mathbb{B}^+_{\mathrm{dR},U} \overset{\theta_{\mathrm{dR}}}{\longrightarrow} \hat{\mathcal{O}}_U$$

is given by reduction modulo $(\mathrm{Fil}^1\mathbb{B}^+_{\mathrm{dR},U}, T)$. We will abuse notation and also denote the above composition by θ_{dR} for the rest of the proof. By Theorem 3.4 (4) we have that

$$\theta_{\mathrm{dR}}\left(\frac{\nabla(x_{\mathrm{dR}}(w^{\mathrm{Katz}}_{\mathrm{can}}))}{\nabla(T)}\right) \in \hat{\mathcal{O}}^\times_U(U).$$

Thus via a change of variables $T \mapsto x_{\mathrm{dR}}(w^{\mathrm{Katz}}_{\mathrm{can}}) - \theta_T(x_{\mathrm{dR}}(w^{\mathrm{Katz}}_{\mathrm{can}}))$ we have that

$$\mathcal{O}\mathbb{B}^+_{\mathrm{dR},U} \underset{\sim}{\overset{(66)}{\longrightarrow}} \mathbb{B}^+_{\mathrm{dR},U}[\![T]\!] = \mathbb{B}^+_{\mathrm{dR},U}[\![x_{\mathrm{dR}}(w^{\mathrm{Katz}}_{\mathrm{can}}) - \theta_T(x_{\mathrm{dR}}(w^{\mathrm{Katz}}_{\mathrm{can}}))]\!] = \mathbb{B}^+_{\mathrm{dR},U}[\![x_{\mathrm{dR}}(w^{\mathrm{Katz}}_{\mathrm{can}})]\!] \tag{68}$$

where the last equality follows by the further change of variables $x_{\mathrm{dR}}(w^{\mathrm{Katz}}_{\mathrm{can}}) - \theta_T(x_{\mathrm{dR}}(w^{\mathrm{Katz}}_{\mathrm{can}})) \mapsto x_{\mathrm{dR}}(w^{\mathrm{Katz}}_{\mathrm{can}})$, which is well-defined since $\theta_T(x_{\mathrm{dR}}(w^{\mathrm{Katz}}_{\mathrm{can}})) \overset{(67)}{\in} \mathrm{Fil}^1\mathbb{B}^+_{\mathrm{dR},U}(U)$ and $\mathbb{B}^+_{\mathrm{dR},U}[\![T]\!]$ is $\mathrm{Fil}^1\mathbb{B}^+_{\mathrm{dR},U}$-adically complete. Noting that the left and right hand sides of (68) do not depend on T and only on $x_{\mathrm{dR}}(w^{\mathrm{Katz}}_{\mathrm{can}})|_U$, we see that the isomorphism between these two quantities glues over $U \in Y_{\text{proét}}/\mathcal{Y}^{\mathrm{Ig},0}$ giving (65). □

Convention 3.6 *We henceforth treat (65) as an identification of sheaves on $Y_{pro\acute{e}t}/\mathcal{Y}^{\mathrm{Ig},0}$, thus identifying*

$$X = x_{\mathrm{dR}}(w_{\mathrm{can}}^{\mathrm{Katz}}).$$

In particular we have the identification

$$\mathrm{Fil}^1 \mathcal{O}\mathbb{B}^+_{\mathrm{dR},\mathcal{Y}^{\mathrm{Ig},0}} = \ker\theta_{\mathrm{dR}}|_{\mathcal{Y}^{\mathrm{Ig},0}} = (t, X). \tag{69}$$

(where here (t, X) denotes the ideal in $\mathbb{B}^+_{\mathrm{dR},\mathcal{Y}^{\mathrm{Ig},0}}[\![X]\!]$ generated by t and X).

Theorem 3.7 *There is a natural $\mathbb{B}^+_{\mathrm{dR},\mathcal{Y}^{\mathrm{Ig},0}}$-linear isomorphism of sheaves on $Y_{pro\acute{e}t}/\mathcal{Y}^{\mathrm{Ig},0}$*

$$\mathcal{O}\mathbb{B}^+_{\mathrm{dR},\mathcal{Y}^{\mathrm{Ig},0}}/(X) \cong \mathbb{B}^+_{\mathrm{dR},\mathcal{Y}^{\mathrm{Ig},0}} \tag{70}$$

such that the composition

$$\mathcal{O}\mathbb{B}^+_{\mathrm{dR},\mathcal{Y}^{\mathrm{Ig},0}} \to \mathcal{O}\mathbb{B}^+_{\mathrm{dR},\mathcal{Y}^{\mathrm{Ig},0}}/(X) \cong \mathbb{B}^+_{\mathrm{dR},\mathcal{Y}^{\mathrm{Ig},0}} \xrightarrow{\theta} \hat{\mathcal{O}}_{\mathcal{Y}^{\mathrm{Ig},0}}$$

is the canonical map

$$\theta_{\mathrm{dR}} : \mathcal{O}\mathbb{B}^+_{\mathrm{dR},\mathcal{Y}^{\mathrm{Ig},0}} \to \hat{\mathcal{O}}_{\mathcal{Y}^{\mathrm{Ig},0}}.$$

Proof This follows immediately from (65). □

Definition 3.8 Let

$$\theta_X : \mathcal{O}\mathbb{B}^+_{\mathrm{dR},\mathcal{Y}^{\mathrm{Ig},0}} \to \mathcal{O}\mathbb{B}^+_{\mathrm{dR},\mathcal{Y}^{\mathrm{Ig},0}}/(X) \overset{(70)}{\cong} \mathbb{B}^+_{\mathrm{dR},\mathcal{Y}^{\mathrm{Ig},0}} \tag{71}$$

denote reduction modulo (X). By Theorem 3.7, the composition $\theta \circ \theta_X : \mathcal{O}\mathbb{B}^+_{\mathrm{dR},\mathcal{Y}^{\mathrm{Ig},0}} \to \hat{\mathcal{O}}_{\mathcal{Y}^{\mathrm{Ig},0}}$ is the map $\theta_{\mathrm{dR}} : \mathcal{O}\mathbb{B}^+_{\mathrm{dR},\mathcal{Y}^{\mathrm{Ig},0}} \to \hat{\mathcal{O}}_{\mathcal{Y}^{\mathrm{Ig},0}}$. In particular precomposing θ_X with the natural map $\mathcal{O}_{\mathcal{Y}^{\mathrm{Ig},0}} \to \mathcal{O}\mathbb{B}^+_{\mathrm{dR},\mathcal{Y}^{\mathrm{Ig},0}}$ we get a map

$$i_X : \mathcal{O}_{\mathcal{Y}^{\mathrm{Ig},0}} \to \mathbb{B}^+_{\mathrm{dR},\mathcal{Y}^{\mathrm{Ig},0}} \tag{72}$$

whose composition with $\theta : \mathbb{B}^+_{\mathrm{dR},\mathcal{Y}^{\mathrm{Ig},0}} \to \hat{\mathcal{O}}_{\mathcal{Y}^{\mathrm{Ig},0}}$

$$\theta \circ i_X = \theta_X|_{\mathcal{O}_{\mathcal{Y}^{\mathrm{Ig},0}}} : \mathcal{O}_{\mathcal{Y}^{\mathrm{Ig},0}} \to \hat{\mathcal{O}}_{\mathcal{Y}^{\mathrm{Ig},0}} \tag{73}$$

is the natural map $\mathcal{O}_{\mathcal{Y}^{\mathrm{Ig},0}} \to \hat{\mathcal{O}}_{\mathcal{Y}^{\mathrm{Ig},0}}$.

Convention 3.9 *Given a period sheaf $\mathcal{F}$ on $Y_{pro\acute{e}t}$ (i.e. one of the sheaves of [27, Sects. 5–6]), an object $U \in Y_{pro\acute{e}t}$ and a point $y \in U(\overline{\mathbb{Q}}_p, \overline{\mathbb{Z}}_p)$, let $\mathcal{F}(y)$ be the specialization to the affinoid perfectoid* $\mathrm{Spa}(\overline{\mathbb{Q}}_p, \overline{\mathbb{Z}}_p) \xrightarrow{y} U \to Y$ *as in Sect. 5 of op. cit.*

Proposition 3.10 *We have*

(1) The complete discrete valuation ring B_{dR}^+ has a unique embedding

$$i : \overline{\mathbb{Q}}_p \hookrightarrow B_{\mathrm{dR}}^+ \tag{74}$$

whose composition with $\theta : B_{\mathrm{dR}}^+ \to \mathbb{C}_p$ is the natural inclusion $\overline{\mathbb{Q}}_p \subset \mathbb{C}_p$.

(2) Suppose $U \in Y_{\text{proét}}/\mathcal{Y}^{\mathrm{Ig},0}$, $f \in \mathcal{O}_{\mathcal{Y}^{\mathrm{Ig},0}}(U)$ and $y \in U(\overline{\mathbb{Q}}_p, \overline{\mathbb{Z}}_p)$. Then in the notation of Conventions 2.5 and 3.9 we have

$$i(f(y)) = (i_X(f))(y). \tag{75}$$

Proof (1) Recall that the residue field of the complete discrete valuation ring B_{dR}^+ is $\mathbb{C}_p$ and that $\mathbb{Q}_p \subset B_{\mathrm{dR}}^+$. Thus by Hensel's lemma, there is a unique lift of $\overline{\mathbb{Q}}_p \subset \mathbb{C}_p$ to B_{dR}^+.

(2) We have $\mathbb{B}_{\mathrm{dR},\mathcal{Y}^{\mathrm{Ig},0}}^+(y) = B_{\mathrm{dR}}^+$. We will prove (75) by showing that $\theta((i_X(f))(y)) = f(y)$, which by the uniqueness property of (74) gives the assertion. To see the claim we use the fact that $\theta \circ i_X$ is the natural map $\mathcal{O}_{\mathcal{Y}^{\mathrm{Ig},0}} \to \hat{\mathcal{O}}_{\mathcal{Y}^{\mathrm{Ig},0}}$ and so $\theta((i_X(f))(y)) = \theta(i_X(f))(y) = f(y)$. □

3.5 The Period z_{dR}

Definition 3.12 below establishes the existence of a certain coordinate z_{dR} inside $\mathrm{Fil}^1\mathcal{O}\mathbb{B}_{\mathrm{dR},\mathcal{Y}^{\mathrm{Ig},0}}^+(\mathcal{Y}^{\mathrm{Ig},0}) \cdot t^{-1}$ which measures the position of the Hodge filtration $\omega \subset H_{\mathrm{dR}}^1(\mathcal{A})$ in the following sense:

Property 3.11 *At a point $y = (E, P, e_1, e_2) \in \mathcal{Y}^{\mathrm{Ig},0}(\overline{\mathbb{Q}}_p, \overline{\mathbb{Z}}_p)$, where E is a false elliptic curve, P is a $\Gamma(N_+)$-level structure and $(e_1, e_2) : \mathbb{Z}_p^{\oplus 2} \xrightarrow{\sim} e^1T_pE$ is a $\Gamma(p^\infty)$-level structure, z_{dR} specializes to the element of $\mathbb{P}^1(B_{\mathrm{dR}})$ parametrizing the line*

$$B_{\mathrm{dR}}^+ \cong \Omega_E^1 \otimes_{\overline{\mathbb{Q}}_p} B_{\mathrm{dR}}^+ = H^{1,0}(E) \otimes_{\overline{\mathbb{Q}}_p} B_{\mathrm{dR}}^+ \subset H_{\mathrm{dR}}^1(E) \otimes_{\overline{\mathbb{Q}}_p} B_{\mathrm{dR}}^+ \subset H_{ét}^1(E) \otimes_{\mathbb{Z}_p} B_{\mathrm{dR}}^+ \overset{(30)}{\cong} (B_{\mathrm{dR}}^+)^{\oplus 2}.$$

Definition 3.12 Consider the element

$$z_{\mathrm{dR}} := -\frac{x_{\mathrm{dR}}(w_{\mathrm{can}}^{\mathrm{Katz}})}{y_{\mathrm{dR}}(w_{\mathrm{can}}^{\mathrm{Katz}})} \overset{(48)}{=} -\frac{x_{\mathrm{dR}}(w_{\mathrm{can}}^{\mathrm{Katz}})}{t} \overset{(40)}{\in} \mathrm{Fil}^1\mathcal{O}\mathbb{B}_{\mathrm{dR},Y}|^+(\mathcal{Y}^{\mathrm{Ig},0}) \cdot t^{-1} \overset{(69)}{=} (t, X) \cdot t^{-1} = \ker(\theta_{\mathrm{dR}})(\mathcal{Y}^{\mathrm{Ig},0}) \cdot t^{-1}. \tag{76}$$

This satisfies Property 3.11 by (39).

Note that

$$t \cdot z_{\mathrm{dR}} \in \ker(\theta_{\mathrm{dR}})(\mathcal{Y}^{\mathrm{Ig},0}),$$

so that since $\mathcal{O}\mathbb{B}^+_{\mathrm{dR},Y}(\mathcal{Y}^{\mathrm{Ig},0})$ is $\ker(\theta_{\mathrm{dR}})(\mathcal{Y}^{\mathrm{Ig},0})$-adically complete,

$$\exp(t \cdot z_{\mathrm{dR}}) \in 1 + \ker(\theta_{\mathrm{dR}})(\mathcal{Y}^{\mathrm{Ig},0}).$$

Definition 3.13 Recall θ_X (71). Let

$$q_{\mathrm{dR}} := \exp(t \cdot z_{\mathrm{dR}}) \in 1 + \ker(\theta_{\mathrm{dR}})(\mathcal{Y}^{\mathrm{Ig},0}). \tag{77}$$

Note that by construction and (3.8) we have

$$\theta_{\mathrm{dR}}(q_{\mathrm{dR}}) = 1 \tag{78}$$

and

$$\nabla(t \cdot z_{\mathrm{dR}}) = \nabla(\log(q_{\mathrm{dR}})) = \frac{\nabla(q_{\mathrm{dR}})}{q_{\mathrm{dR}}}. \tag{79}$$

For any $(A_0, P_0) \in Y^{\mathrm{ord},+}(\overline{\mathbb{F}}_p)$, recall the Serre-Tate coordinate

$$q_{(A_0,P_0)} \in \mathcal{O}_{M^{\mathrm{Ig},+}(A_0,P_0)}(M^{\mathrm{Ig},+}(A_0, P_0))$$

from (12). By [16, Main Theorem], we have

$$\mathrm{KS}(w_{\mathrm{can}}^{\mathrm{Katz},\otimes 2})|_{M^{\mathrm{Ig},+}(A_0,P_0)} = d \log q_{(A_0,P_0)}. \tag{80}$$

We have the following key result about z_{dR}, namely that it is the solution of a differential equation involving the Serre-Tate coordinate. This is an incarnation of the fact that the Serre-Tate coordinate, like z_{dR}, measures the position of the Hodge filtration [16, 17].

Theorem 3.14 *Let $M^{\mathrm{Ig}}(A_0, P_0)$ denote the adic generic fiber of $M^{\mathrm{Ig},+}(A_0, P_0)$. We have*

$$\begin{aligned} t \cdot \nabla(z_{\mathrm{dR}})|_{M^{\mathrm{Ig}}(A_0,P_0)} &= \nabla(t \cdot z_{\mathrm{dR}})|_{M^{\mathrm{Ig}}(A_0,P_0)} = \mathrm{KS}(w_{\mathrm{can}}^{\mathrm{Katz},\otimes 2})|_{M^{\mathrm{Ig}}(A_0,P_0)} \\ &\overset{(80)}{=} d \log q_{(A_0,P_0)}|_{M^{\mathrm{Ig}}(A_0,P_0)}. \end{aligned} \tag{81}$$

Proof We have

$$\mathrm{KS}(w_{\mathrm{can}}^{\mathrm{Katz},\otimes 2}) \overset{(62)}{=} -\nabla(x_{\mathrm{dR}}(w_{\mathrm{can}}^{\mathrm{Katz}})) \overset{(76)}{=} t \cdot \nabla(z_{\mathrm{dR}}).$$

□

3.6 q_{dR}-Expansions

In this section, we recall the notion of q_{dR}-expansions from [22, Appendix C] and show they recover Serre-Tate expansions.

Definition 3.15 (1) Given any $f \in \mathcal{O}\mathbb{B}_{\mathrm{dR},\mathcal{Y}^{\mathrm{Ig},0}}(\mathcal{Y}^{\mathrm{Ig},0})$, its q_{dR}-expansion is

$$f(q_{\mathrm{dR}}) := \sum_{n=0}^{\infty} b_n \cdot (q_{\mathrm{dR}} - 1)^n,$$
$$b_n := \frac{1}{n!}\theta_X\left(\left(\frac{\nabla}{\nabla q_{\mathrm{dR}}}\right)^n f\right) \overset{(78)}{=} \left(\frac{1}{n!}\left(\frac{\nabla}{\nabla q_{\mathrm{dR}}}\right)^n f\right)|_{q_{\mathrm{dR}}=1} \in \mathbb{B}_{\mathrm{dR},\mathcal{Y}^{\mathrm{Ig},0}}(\mathcal{Y}^{\mathrm{Ig},0}). \tag{82}$$

(2) Given a point $y \in \mathcal{Y}^{\mathrm{Ig},0}(\overline{\mathbb{Q}}_p, \overline{\mathbb{Z}}_p)$, let

$$f(y)(q_{\mathrm{dR}}) := \sum_{n=0}^{\infty} b_n(y) \cdot (q_{\mathrm{dR}} - 1)^n \in \mathbb{B}_{\mathrm{dR},\mathcal{Y}^{\mathrm{Ig},0}}(y)[\![q_{\mathrm{dR}} - 1]\!] \subset B_{\mathrm{dR}}[\![q_{\mathrm{dR}} - 1]\!]$$

where $b_n(y)$ is the specialization of b_n to y as defined in Convention 3.9.

Theorem 3.16 *Suppose* $w \in \omega^{\otimes k}(Y^{\mathrm{Ig}}) \otimes_{\mathbb{Q}_p} \overline{\mathbb{Q}}_p$ *and let* $f = w/w_{\mathrm{can}}^{\mathrm{Katz},\otimes k} \in \mathcal{O}_{Y^{\mathrm{Ig}}}(Y^{\mathrm{Ig}}) \otimes_{\mathbb{Q}_p} \overline{\mathbb{Q}}_p$. *Suppose* $y \in M^{\mathrm{Ig}}(A_0, P_0)(\overline{\mathbb{Q}}_p, \overline{\mathbb{Z}}_p)$, *and let* $\tilde{y} \in \mathcal{Y}^{\mathrm{Ig},0}(\overline{\mathbb{Q}}_p, \overline{\mathbb{Z}}_p)$ *be any point mapping to* y *under the projection* $\mathcal{Y}^{\mathrm{Ig},0} \to Y^{\mathrm{Ig}}$. *Then the power series obtained by replacing the variable "*$q_{\mathrm{dR}} - 1$*" in*

$$f(\tilde{y})(q_{\mathrm{dR}}) \in \mathbb{B}_{\mathrm{dR},\mathcal{Y}^{\mathrm{Ig},0}}(y)[\![q_{\mathrm{dR}} - 1]\!] \subset B_{\mathrm{dR}}[\![q_{\mathrm{dR}} - 1]\!]$$

with "$q_{(A_0,P_0)} - 1$*" (12) is equal to the Serre-Tate expansion of* f *centered at* y.

Proof Note the elementary formal identity of differential operators

$$\frac{1}{n!}q^n\left(\frac{d}{dq}\right)^n = \binom{\frac{d}{d\log q}}{n} \tag{83}$$

where $\binom{x}{n} \in \mathbb{Q}[x]$ is the nth binomial coefficient. Note also that since $\log q_{(A_0,P_0)}(y) \in \mathbb{C}_p$, $d\log q_{(A_0,P_0)}(y) = 0$ and

$$d\log q_{(A_0,P_0)} = d\log q_{(A_0,P_0)} - d\log q_{(A_0,P_0)}(y) = d\log(q_{(A_0,P_0)}/q_{(A_0,P_0)}(y)). \tag{84}$$

Moreover, for any section $f \in \mathcal{O}_{Y^{\mathrm{Ig}}}(M^{\mathrm{Ig}}(A_0, P_0)) \otimes_{\mathbb{Q}_p} \overline{\mathbb{Q}}_p$, considering the pullback

$$f|_{M^{\mathrm{Ig}}(A_0,P_0)\times_{Y^{\mathrm{Ig}}}\mathcal{Y}^{\mathrm{Ig},0}} \in \mathcal{O}_{\mathcal{Y}^{\mathrm{Ig},0}}(M^{\mathrm{Ig}}(A_0, P_0) \times_{Y^{\mathrm{Ig}}} \mathcal{Y}^{\mathrm{Ig},0}) \otimes_{\mathbb{Q}_p} \overline{\mathbb{Q}}_p,$$

we have

$$f|_{M^{\mathrm{Ig}}(A_0,P_0)\times_{Y^{\mathrm{Ig}}}\mathcal{Y}^{\mathrm{Ig},0}}(\tilde{y}) = f(y). \tag{85}$$

Thus have

$$\begin{aligned}
b_n(\tilde{y}) &\overset{(82)}{=} \left(\frac{1}{n!}\theta_X\left(\frac{\nabla}{\nabla q_{\mathrm{dR}}}\right)^n f\right)(\tilde{y}) \overset{(78)}{=} \theta_X\left(\frac{1}{n!}q_{\mathrm{dR}}^n\left(\frac{\nabla}{\nabla q_{\mathrm{dR}}}\right)^n f\right)(\tilde{y})\\
&\overset{(79)}{=} \theta_X\left(\binom{\frac{\nabla}{\nabla(\log q_{\mathrm{dR}})}}{n} f\right)(\tilde{y}) \overset{(83)}{=} \theta_X\left(\binom{\frac{\nabla}{\nabla(t\cdot z_{\mathrm{dR}})}}{n} f\right)(\tilde{y})\\
&\overset{(81)}{=} \theta_X\left(\binom{\frac{d}{d\log q_{(A_0,P_0)}}}{n} f\right)(\tilde{y}) \overset{(84)}{=} \theta_X\left(\binom{\frac{d}{d\log(q_{(A_0,P_0)}/q_{(A_0,P_0)}(\tilde{y}))}}{n} f\right)(\tilde{y})\\
&\overset{(83)}{=} \theta_X\left(\frac{1}{n!}q_{(A_0,P_0)}(\tilde{y})^n\left(\frac{d}{dq_{(A_0,P_0)}}\right)^n f(\tilde{y})\right)\\
&\overset{(73)}{=} i_X\left(\frac{1}{n!}q_{(A_0,P_0)}(\tilde{y})^n\left(\frac{d}{dq_{(A_0,P_0)}}\right)^n f(\tilde{y})\right)\\
&\overset{(75)}{=} i\left(\left(\frac{1}{n!}q_{(A_0,P_0)}^n\left(\frac{d}{dq_{(A_0,P_0)}}\right)^n f\right)(\tilde{y})\right) \overset{(13)}{=} i(a_n(\tilde{y})) \overset{(74)}{=} a_n(\tilde{y}) \overset{(85)}{=} a_n(y).
\end{aligned}$$

Now the assertion follows from Formulas (82) and (13). □

3.7 *Defining the BDP p-Adic L-Function via q_{dR}-Expansions*

We are now ready to recover the BDP p-adic L-function from the q_{dR}-expansion. First, we define a p-adic L-function using the q_{dR}-expansion.

Choice 3.17 *Choose a set of points*

$$\mathbf{S} := \left\{\widetilde{(y_{\mathfrak{a}c}, \pi_{\mathfrak{a}}(e_c))}\right\} \subset \mathcal{Y}^{\mathrm{Ig},0}(\overline{\mathbb{Q}}_p, \overline{\mathbb{Z}}_p)$$

such that the composition

$$\left\{\widetilde{(y_{\mathfrak{a}c}, \pi_{\mathfrak{a}}(e_c))}\right\} \subset \mathcal{Y}^{\mathrm{Ig},0}(\overline{\mathbb{Q}}_p, \overline{\mathbb{Z}}_p) \to Y^{\mathrm{Ig}}(\overline{\mathbb{Q}}_p, \overline{\mathbb{Z}}_p)$$

induces a bijection of sets

$$\left\{\widetilde{(y_{\mathfrak{a}c}, \pi_{\mathfrak{a}}(e_c))}\right\} \to \{(y_{\mathfrak{a}c}, \pi_{\mathfrak{a}}(e_c))\}$$

where the target is the set of points from Choice 2.11.

Suppose $w \in \omega^{\otimes k}(Y^{\mathrm{ord}}) \otimes_{\mathbb{Q}_p} \overline{\mathbb{Q}}_p$ is the restriction of a section of $\omega^{\otimes k}(\mathbb{X})$ which is the Jacquet-Langlands transfer from $X_1(N)$ to $\mathbb{X}$ of the modular form f from

Sect. 2.1. Let $F_w^\flat$ be as in (17) and let χ be the algebraic Hecke character as above Definition 2.12, with p-adic avatar $\check{\chi}$.

Definition 3.18 Define

$$\mathbf{L}^{\mathrm{BDP}}(w) := \frac{1}{[F_c : K(\mathcal{O}_c)]} \sum_{\tilde{y} \in \mathbf{S}} \check{\chi}^{-1}(\mathfrak{a}) \cdot F_w^\flat(\tilde{y})(q_{\mathrm{dR}}) \in B_{\mathrm{dR}}[\![q_{\mathrm{dR}} - 1]\!]. \tag{86}$$

Finally, we show that this p-adic L-function coincides with the BDP p-adic L-function.

Theorem 3.19 *The power series obtained by formally replacing the variable "$q_{\mathrm{dR}} - 1$" in*

$$\mathbf{L}^{\mathrm{BDP}}(w) \in B_{\mathrm{dR}}[\![q_{\mathrm{dR}} - 1]\!]$$

from (86) with "$q - 1$" is equal to

$$\mathcal{L}^{\mathrm{BDP}}(w) \in W[\![q - 1]\!][1/p]$$

from (18).

Proof This is a direct consequence of (18), (86) and Theorem 3.16.

4 Generalizing to the Supersingular Case

In this section we briefly indicate how the work [22] generalizes the above constructions to a large open subset of Y_∞ (19) containing the supersingular locus. These generalizations are used to construct supersingular analogues of the BDP p-adic L-function (i.e. allowing p inert or ramified in the CM field K), which we outline in the final subsection.

4.1 Loci Cut Out by the Hodge-Tate Period Map on the Infinite-Level Shimura Curve

We first explain what the correct analogue of the Igusa tower (21) is outside of the ordinary locus. Recall Y_∞ from (19) and that $\mathcal{A}|_{Y_\infty} \to Y_\infty$ is the universal false elliptic curve with $\Gamma(N_+)$-level structure and compatible $\Gamma(p^\infty)$-level structure (Definition 3.1, below which "compatible" is also defined). Let

$$(e_1, e_2) : \hat{\mathbb{Z}}_{p,Y_\infty}^{\oplus 2} \xrightarrow{\sim} T_p\mathcal{A}|_{Y_\infty}$$

denote the universal $\Gamma(p^\infty)$-level structure. We can then view points in Y_∞ as being represented by tuples $(A, P, \alpha_1, \alpha_2)$ where A is a false elliptic curve, P is a $\Gamma(N_+)$-level structure and (α_1, α_2) is a $\Gamma(p^\infty)$-level structure such that the two level structures are compatible. We will sometimes suppress the $\Gamma(N_+)$-level structure P for brevity. The moduli interpretation gives a natural identification

$$\mathrm{Gal}(Y_\infty/Y) = \Gamma_p(p^{\mathrm{ord}_p(N_+)}) \tag{87}$$

where

$$\Gamma_p(p^m) = \left\{\begin{pmatrix} a & b \\ c & d\end{pmatrix} \in GL_2(\mathbb{Z}_p) : a-1, b, c, d-1 \equiv 0 \pmod{p^m}\right\} \subset GL_2(\mathbb{Z}_p).$$

Letting $x, y \in \mathcal{O}_{\mathbb{P}^1}(1)$, we get

$$z_{\mathrm{HT}} := \pi_{\mathrm{HT}}^*\left(-\frac{x}{y}\right) \in \mathbb{P}^1(\hat{\mathcal{O}}_Y(Y_\infty)). \tag{88}$$

It satisfies the fractional linear transformation law (see [22, Sect. 2] or [11, Sect. 2.4] for the calculation; note that their $\mathfrak{z}$ coincides with our $1/z_{\mathrm{HT}}$)

$$\begin{pmatrix} a & b \\ c & d\end{pmatrix}^* z_{\mathrm{HT}} = \frac{dz_{\mathrm{HT}} + b}{cz_{\mathrm{HT}} + a}, \quad \forall \begin{pmatrix} a & b \\ c & d\end{pmatrix} \in \mathrm{Gal}(Y_\infty/Y). \tag{89}$$

We note also by (55) that $z_{\mathrm{HT}}|_U = -\frac{x_{\mathrm{HT}}(\eta)}{y_{\mathrm{HT}}(\eta)}$ for any $U \in Y_{\mathrm{pro\acute{e}t}}/Y_\infty$ and any $\eta \in \omega^{-1} \otimes_{\mathcal{O}_Y} \hat{\mathcal{O}}_Y(U)$.

Letting $\overline{Y_\infty^{\mathrm{ord}}} := \pi_{\mathrm{HT}}^{-1}(\mathbb{P}^1(\mathbb{Q}_p, \mathbb{Z}_p))$ and define the *infinite-level supersingular locus* $Y_\infty^{\mathrm{ss}} := Y_\infty \setminus \overline{Y_\infty^{\mathrm{ord}}} \subset Y_\infty$, we see that

$$Y_\infty^{\mathrm{ss}} \subset \mathcal{V}_x.$$

One can think of Y_∞^{ss} as the (adic interior of the) supersingular locus; by the Rapoport-Zink uniformization it is a finite disjoint union of infinite-level Lubin-Tate towers (see [26, Discussion after Theorem III.1.2]).

The main underlying space on which the constructions of op. cit. take place is the open (in the adic topology) subset

$$\mathcal{V}_x := \{z_{\mathrm{HT}} \neq 0\} = Y_\infty \setminus \pi_{\mathrm{HT}}^{-1}(\{0\}) \subset Y_\infty. \tag{90}$$

This is naturally an object of $Y_{\mathrm{pro\acute{e}t}}$ and thus we can pull back sheaves on $Y_{\acute{e}t}$ to it. The main reason we consider $\mathcal{V}_x$ is that it is the locus on which the *fake Hasse invariant* [11, 26]

$$\mathfrak{s} := e_2^* \frac{dT}{1+T} \in \omega \otimes_{\mathcal{O}_Y} \hat{\mathcal{O}}_Y(Y_\infty) \tag{91}$$

is non-vanishing, i.e. $\mathfrak{s} \in \omega \otimes_{\mathcal{O}_Y} \hat{\mathcal{O}}_Y(\mathcal{V}_x)$ is a nowhere-vanishing section. Here we recall that $\frac{dT}{1+T}$ is the canonical differential corresponding to the choice of coordinate T on $\mu_{p^\infty,Y}$ and $e_2^*\frac{dT}{1+T}$ is defined using (26). One should view (91) as a natural analytic continuation of Katz's canonical differential $w_{\text{can}}^{\text{Katz}}$ (15), as it is evident from its definition that it pulls back to $w_{\text{can}}^{\text{Katz}}$ on the subset $\mathcal{Y}^{\text{Ig},0} \subset \mathcal{V}_x$:

$$\mathfrak{s}|_{\mathcal{Y}^{\text{Ig},0}} = w_{\text{can}}^{\text{Katz}}|_{\mathcal{Y}^{\text{Ig},0}}. \tag{92}$$

With respect to the identification (87) we can deduce from (91) the transformation law (see [11, Sect. 2.4])

$$\begin{pmatrix} a & b \\ c & d \end{pmatrix}^* \mathfrak{s} = (b(1/z_{\text{HT}}) + d) \cdot \mathfrak{s}. \tag{93}$$

Here note that we view $1/z_{\text{HT}} \in \hat{\mathcal{O}}_Y(\mathcal{V}_x)$.

On one hand one has $Y_\infty^{\text{ss}} \subset \mathcal{V}_x$. On the other one has $\mathcal{Y}^{\text{Ig},0} \subset Y_\infty$. Thus one should think of $\mathcal{V}_x$ as a large open set containing $\mathcal{Y}^{\text{Ig},0}$ (the perfected Igusa tower (22), on which Katz p-adic modular forms live) and Y_∞^{ss} (the infinite-level supersingular locus). In the next subsection we describe how one can define generalizations of Katz p-adic modular forms outside the ordinary locus.

4.2 *Generalizations of Katz p-Adic Modular Forms*

One might use $\mathfrak{s}$ to try to define generalizations of Katz's theory of p-adic modular forms (Sect. 2.7), particularly given the appealing identity (92). However an initial problem one runs into is that the map $\mathcal{V}_x \subset Y_\infty \to Y$ is not Galois [as is seen from (89) and (90)]. Instead one may consider the discs

$$D_n := \{|z_{\text{HT}}| \geq p^n\} \subset \mathcal{V}_x \subset Y_\infty$$

for $n \in \mathbb{Z}$, which are naturally open subsets (in the adic topology) of $\mathcal{V}_x$. Letting $\lambda : Y_\infty \to Y$ denote the natural projection, by (89) we have a canonical identification

$$\text{Gal}(D_n/\lambda(D_n)) = \Gamma_{0,p}(p^{\max(\text{ord}_p(N_+),n)})$$

where

$$\Gamma_{0,p}(p^m) = \left\{\begin{pmatrix} a & b \\ c & d \end{pmatrix} \in GL_2(\mathbb{Z}_p) : c \equiv 0 \pmod{p^m}\right\} \subset GL_2(\mathbb{Z}_p)$$

is the usual congruence subgroup. For $n \in \mathbb{Z}_{\geq 0}$ and $k \in \mathbb{Z}$ one can define *the space of weight k p-adic modular forms of level p^n* as

$$M_k(\hat{\mathcal{O}}_Y(D_n)) :=$$
$$\left\{F \in \hat{\mathcal{O}}_Y(D_n) : \begin{pmatrix} a & b \\ c & d \end{pmatrix}^* F = (b(1/z_{\mathrm{HT}}) + d)^{-k} \cdot F \ \forall \begin{pmatrix} a & b \\ c & d \end{pmatrix} \in \mathrm{Gal}(D_n/\lambda(D_n))\right\}. \tag{94}$$

We note that if $\max(\mathrm{ord}_p(N_+), n) \geq 1$ then $|b \cdot (1/z_{\mathrm{HT}}) + d| = 1$ on D_n and so $(b(1/z_{\mathrm{HT}}) + d)^{-k}$ is well-defined for $k \in \mathbb{Z}/(p-1) \times \mathbb{Z}_p$ (where we embed $\mathbb{Z} \hookrightarrow \mathbb{Z}/(p-1) \times \mathbb{Z}$ by $x \mapsto (x, x)$) and thus $M_k(\hat{\mathcal{O}}_Y(D_n))$ is well-defined for $k \in \mathbb{Z}/(p-1) \times \mathbb{Z}_p$. Note that by (93) we have a map

$$\omega^{\otimes k}(\lambda(D_n)) \to M_k(\hat{\mathcal{O}}_Y(D_n)), \quad w \mapsto \frac{w}{\mathfrak{s}^{\otimes k}}.$$

Using (92) one can show that there is a natural map to the space of weight k Katz p-adic modular forms V_k (Definition 2.8)

$$M_k(\hat{\mathcal{O}}_Y(D_n)) \to V_k, \quad F \mapsto F|_{\mathcal{Y}^{\mathrm{Ig},0}} \tag{95}$$

given by restricting to $\mathcal{Y}^{\mathrm{Ig},0} \subset D_n$. One can also define weight-raising (or "p-adic Maass-Shimura") operators

$$d_k^j : M_k(\hat{\mathcal{O}}_Y(D_n)) \to M_{k+2j}(A(D_n)) \tag{96}$$

for a certain $\hat{\mathcal{O}}_{\mathcal{V}_x}$-algebra A (an $\hat{\mathcal{O}}_Y$-submodule of a completed period sheaf) over which the Hodge filtration canonically splits. The above discussion, including the construction of the p-adic Maass-Shimura operators (96), is explained in detail [22, Sect. 4]. See also [11] for a related notion of overconvergent modular forms (although no weight-raising operators are constructed in op. cit.).

Instead of (91) as a candidate to extend $w_{\mathrm{can}}^{\mathrm{Katz}}$, one can also consider

$$w_{\mathrm{can}} := \frac{\mathfrak{s}}{y_{\mathrm{dR}}(\mathfrak{s})/t} \in \omega \otimes_{\mathcal{O}_Y} \mathbb{B}^+(\mathcal{V}_x)$$

for some period $y_{\mathrm{dR}}(\mathfrak{s})/t$ attached to $\mathfrak{s}$ as in (39) (although this period will live a larger sheaf than $\mathcal{O}\mathbb{B}_{\mathrm{dR},Y}$). Here $\mathbb{B}^+$ is the completed period sheaf recalled below in Sect. 4.3. The section w_{can} is constructed in [22, Appendix C] where they are called *generalized p-adic modular forms*. By (48) and (92) it notably also satisfies

$$w_{\mathrm{can}}|_{\mathcal{Y}^{\mathrm{Ig},0}} = w_{\mathrm{can}}^{\mathrm{Katz}}|_{\mathcal{Y}^{\mathrm{Ig},0}}$$

and thus analogously to (95) there is a map from weight k generalized p-adic modular forms to weight k Katz p-adic modular forms given by restriction to $\mathcal{Y}^{\mathrm{Ig},0}$.

4.3 De Rham Periods and the p-Adic Legendre Relation in the General Case

Recall the de Rham periods from Definition 3.3. The p-adic Legendre relation (49) posited that we have the equality

$$x_{\mathrm{dR}}(w) \cdot y_{\mathrm{dR}}(\tilde{\eta}(w)) - y_{\mathrm{dR}}(w) \cdot x_{\mathrm{dR}}(\tilde{\eta}(w)) = t \tag{97}$$

for any $U \in Y_{\text{proét}}/\mathcal{Y}^{\mathrm{Ig},0}$ and any $w \in \omega(U)$. In fact nothing about the proof of this relation essentially used $\mathcal{Y}^{\mathrm{Ig},0}$ other than the nowhere-vanishing section $w_{\mathrm{can}}^{\mathrm{Katz}}|_{\mathcal{Y}^{\mathrm{Ig},0}} \in \omega(\mathcal{Y}^{\mathrm{Ig},0})$ (15). With a more refined argument (see [22, Sect. 3]) one can show (97) for any $U \in Y_{\text{proét}}/\mathcal{V}_x$ and any $w \in \omega(U)$. Using this generalized p-adic Legendre relation one can prove (as is done in Sect. 3 of op. cit.) a canonical trivialization

$$\mathcal{O}\mathbb{B}_{\mathrm{dR},U}^{+} \cong \mathbb{B}_{\mathrm{dR},U}^{+}[\![x_{\mathrm{dR}}(w)]\!] \tag{98}$$

for any $U \in Y_{\text{proét}}/\mathcal{V}_x$ and any nowhere-vanishing $w \in \omega(U)$. This gives a generalization of (65) (indeed (65) is the special case $U = \mathcal{Y}^{\mathrm{Ig},0}$) and we henceforth treat it as an identification. Reducing (98) modulo $(x_{\mathrm{dR}}(w))$ (the ideal generated by $x_{\mathrm{dR}}(w)$) one gets a canonical map $\mathcal{O}_U \to \mathbb{B}_{\mathrm{dR},U}^{+}$ which does not depend on the choice of nowhere-vanishing $w \in \omega(U)$ since

$$x_{\mathrm{dR}}(\alpha w) = \alpha x_{\mathrm{dR}}(w) \tag{99}$$

[by the $\mathcal{O}\mathbb{B}_{\mathrm{dR},Y}^{+}$-linearity of (37)]. Moreover it follows that the completion of this map with $\theta : \mathbb{B}_{\mathrm{dR},U}^{+} \to \hat{\mathcal{O}}_U$ is the canonical map $\mathcal{O}_U \to \hat{\mathcal{O}}_U$. Thus gluing over $U \in Y_{\text{proét}}/\mathcal{V}_x$ we get a canonical map

$$i_{\mathcal{V}_x} : \mathcal{O}_{\mathcal{V}_x} \to \mathbb{B}_{\mathrm{dR},\mathcal{V}_x}^{+}$$

whose composition with $\theta : \mathbb{B}_{\mathrm{dR},\mathcal{V}_x}^{+} \to \hat{\mathcal{O}}_{\mathcal{V}_x}$ is the canonical map $\mathcal{O}_{\mathcal{V}_x} \to \hat{\mathcal{O}}_{\mathcal{V}_x}$. This is a generalization of (72) which is proven in Sect. 4 of op. cit.

Note that (98) allows one to define the $(x_{\mathrm{dR}}(w)/t)$-adic completion

$$\mathbb{B}_U^{+} := \mathbb{B}_{\mathrm{dR},U}^{+}[\![x_{\mathrm{dR}}(w)/t]\!]$$

of $\mathcal{O}\mathbb{B}_{\mathrm{dR},U} = \mathbb{B}_{\mathrm{dR},U}^{+}[\![x_{\mathrm{dR}}(w)]\!][1/t]$. Again by (99) we see that the ideal generated by $x_{\mathrm{dR}}(w)/t$ is independent of the choice of nowhere-vanishing $w \in \omega(U)$. Thus the sheaves $\mathbb{B}_U^{+}$ glue over $U \in Y_{\text{proét}}/\mathcal{V}_x$ to give a sheaf $\mathbb{B}^{+}$ receiving a $\mathcal{O}\mathbb{B}_{\mathrm{dR},\mathcal{V}_x}^{+}$-algebra map $\mathcal{O}\mathbb{B}_{\mathrm{dR},\mathcal{V}_x} \to \mathbb{B}^{+}$. Let $\mathbf{X} \subset \mathbb{B}^{+}$ denote the $\mathbb{B}^{+}$-ideal locally generated by $x_{\mathrm{dR}}(w)/t$.

For any $U \in Y_{\text{proét}}/\mathcal{V}_x$ and nowhere-vanishing $w \in \omega(U)$ we define

$$z_{\text{dR}}(w) := -\frac{x_{\text{dR}}(w)}{y_{\text{dR}}(w)} \in \mathbb{B}^+(U) \tag{100}$$

using the fact that $y_{\text{dR}}(w)/t$ is a unit modulo $\mathbf{X}$ by

$$\frac{y_{\text{dR}}(w)}{t} \cdot x_{\text{dR}}(\tilde{\eta}(w)) \overset{(97)}{=} \frac{x_{\text{dR}}(w)}{t} \cdot y_{\text{dR}}(\tilde{\eta}(w)) - 1 \equiv -1 \pmod{\mathbf{X}},$$

and so $y_{\text{dR}}(w)/t \in \mathbb{B}^+(U)^\times$ because $\mathbb{B}^+(U)$ is $\mathbf{X} = (x_{\text{dR}}(w)/t)$-adically complete. One may also view $z_{\text{dR}}(w) \in \mathbb{P}^1(\mathcal{O}\mathbb{B}_{\text{dR},Y}(U))$. By construction $z_{\text{dR}}(w)$ satisfies Property 3.11 except with $\mathcal{Y}^{\text{Ig},0}$ replaced by U (i.e. $z_{\text{dR}}(w)$ parametrizes the Hodge filtration on U). This moduli interpretation allows one to conclude that $z_{\text{dR}}(w)$ is independent of the choice of nowhere-vanishing $w \in \omega(U)$ and thus we can glue (100) over $U \in Y_{\text{proét}}/\mathcal{V}_x$ to get

$$z_{\text{dR}} \in \mathbb{B}^+(\mathcal{V}_x)$$

satisfying Property 3.11 except with $\mathcal{Y}^{\text{Ig},0}$ replaced by $\mathcal{V}_x$ (i.e. z_{dR} parametrizes the Hodge filtration on $\mathcal{V}_x$). For details on the construction of $\mathbb{B}^+$ and z_{dR}, see [22, Appendix B].

4.4 q_{dR}-Expansions on Large Open Subsets of the Infinite-Level Shimura Curve

Now let $\mathbb{B}$ denote the $\mathbf{X}$-adic completion of $\mathbb{B}^+$ and let

$$\theta_{\mathbf{X}} : \mathbb{B} \to \mathbb{B}_{\text{dR},\mathcal{V}_x}$$

denote reduction modulo $\mathbf{X} \cdot \mathbb{B}$. Imitating (77) we let

$$q_{\text{dR}} := \exp(t \cdot z_{\text{dR}}) \in \mathbb{B}^+(\mathcal{V}_x).$$

By construction this specializes to (77) on $\mathcal{Y}^{\text{Ig},0} \subset \mathcal{V}_x$. In particular note $q_{\text{dR}} - 1 \in \mathbf{X}$ and it follows from a simple change of variables that

$$\mathbb{B} = \mathbb{B}_{\text{dR}}[\![q_{\text{dR}} - 1]\!]$$

with respect to the identification (98). Given any $U \in Y_{\text{proét}}/\mathcal{V}_x$ and any $F \in \mathbb{B}^+(U)$ we define its q_{dR}-*expansion* by the formula

$$F(q_{\mathrm{dR}}) := \sum_{n=0}^{\infty} a_n \cdot (q_{\mathrm{dR}} - 1)^n, \quad a_n = \frac{1}{n!}\theta_{\mathbf{x}}\left(\frac{\nabla^n F}{(\nabla q_{\mathrm{dR}})^n}\right).$$

We let

$$F(y)(q_{\mathrm{dR}}) := \sum_{n=0}^{\infty} a_n(y) \cdot (q_{\mathrm{dR}} - 1)^n. \tag{101}$$

In particular when $y \in \mathcal{V}_x(\overline{\mathbb{Q}}_p, \overline{\mathbb{Z}}_p)$ then $\mathbb{B}_{\mathrm{dR},Y}(y) = B_{\mathrm{dR}}$ and thus $F(y)(q_{\mathrm{dR}}) \in B_{\mathrm{dR}}[\![q_{\mathrm{dR}} - 1]\!]$. By Theorem 3.16 we have that for $y \in \mathcal{Y}^{\mathrm{Ig},0}(\overline{\mathbb{Q}}_p, \overline{\mathbb{Z}}_p)$, $F(y)(q_{\mathrm{dR}} - 1)$ is the Serre-Tate expansion of F centered at y. Thus one can think of $F(q_{\mathrm{dR}})$ as an analytic continuation from $\mathcal{Y}^{\mathrm{Ig},0}$ to $\mathcal{V}_x$ of the Serre-Tate expansion. Since $\mathcal{V}_x \supset Y_{\infty}^{\mathrm{ss}}$ we see in particular that we can specialize q_{dR}-expansions to points in the supersingular locus.

4.5 The Supersingular Analogue of the BDP p-Adic L-Function

We now have all the ingredients to give an explicit formula for a supersingular analogue of the BDP p-adic L-function. With $A_{\mathfrak{a}c}$ and $y_{\mathfrak{a}c}$ as in Sect. 2.8 (except with p possibly not split in the CM field K), choose $\Gamma(p^{\infty})$-level structures (α_1, α_2) on A_c compatible with the $\Gamma(N_+)$-level structure P_c. Using $\pi_{\mathfrak{a}}$ (choosing representatives $\mathfrak{a}$ prime to pNc) we have that $\pi_{\mathfrak{a}}(e_1, e_2)$ is a $\Gamma(p^{\infty})$-level structure on $A_{\mathfrak{a}c}$ compatible with $\pi_{\mathfrak{a}}(P_c)$. Then let $\mathbf{S}_{\infty} := \{(y_{\mathfrak{a}c}, \pi_{\mathfrak{a}}(e_1, e_2))_{\infty}\} \subset D_{\max(\mathrm{ord}_p(N_+),n)}(\overline{\mathbb{Q}}_p, \overline{\mathbb{Z}}_p)$ denote a set of points such that the composition

$$\mathbf{S}_{\infty} \subset Y_{\infty}(\overline{\mathbb{Q}}_p, \overline{\mathbb{Z}}_p) \xrightarrow{\lambda} Y(\overline{\mathbb{Q}}_p, \overline{\mathbb{Z}}_p)$$

is a bijection onto $\{(y_{\mathfrak{a}c})\} \subset Y(\overline{\mathbb{Q}}_p, \overline{\mathbb{Z}}_p)$. Analogously to (86) we then define

$$\mathbf{L}^{\mathrm{BDP,ss}}(w) := \frac{1}{[F_c : K(\mathcal{O}_c)]} \sum_{\tilde{y}_{\infty} \in \mathbf{S}_{\infty}} \check{\chi}^{-1}(\mathfrak{a}) \cdot F_w^{\flat}(\tilde{y}_{\infty})(q_{\mathrm{dR}}).$$

For $j \in \mathbb{Z}_{\geq 0}$ we then let

$$\mathbf{L}^{\mathrm{BDP,ss}}(w)(j) := \frac{1}{[F_c : K(\mathcal{O}_c)]} \sum_{\tilde{y} \in \mathbf{S}_{\infty}} \check{\chi}_j^{-1}(\mathfrak{a}) \cdot (d_k^j F_w^{\flat})(\tilde{y})(q_{\mathrm{dR}})|_{q_{\mathrm{dR}}=1} \tag{102}$$

where d_k^j is the p-adic Maass-Shimura operator from (96) and χ_j is an appropriate central critical twist of χ as considered in [2, Sect. 4], and [22, Sects. 6–7]. In particular χ_j has infinity type $(k + 2j, 0)$. It is shown in [22, Sects. 6–7] that the function $j \mapsto \mathbf{L}^{\mathrm{BDP,ss}}(w)(j)$ extends to a continuous function $\mathbb{Z}/(p-1) \times \mathbb{Z}_p \to \mathbb{C}_p$, which can thus be seen as a supersingular analogue of (86). The proof of this relies on the fact that (101) specializes on $\mathcal{Y}^{\mathrm{Ig},0} \subset \mathcal{V}_x$ to the Serre-Tate expansion on which one has p-integrality and congruences between the $d_k^j F_w^\flat$. One shows that for the element of $GL_2(\mathbb{Q}_p)$ (which acts on Y_∞ via the moduli interpretation)

$$g := \begin{pmatrix} 1 & 0 \\ 0 & p \end{pmatrix} : D_m \to D_{m-1}, \tag{103}$$

g^* acts as a certain lift of p-power Frobenius on d_k^j, and this together with (103); this allows one to spread out p-integrality and congruences between $d_k^j F_w^\flat$ from a small neighborhood of $\mathcal{Y}^{\mathrm{Ig},0}$ to D_n where n is determined by $\mathrm{ord}_p(N_+)$. The proof of the aforementioned Frobenius lifting statement uses the fact that $d_k^j F_w^\flat \in M_{k+2j}(A(D_n))$; the modular transformation property in (94) allows one to conclude that $d_k^j F_w^\flat$ modulo the maximal ideal $\mathfrak{m}$ of $\widehat{\mathbb{Z}_p[\mu_{p^\infty}]}$ is invariant under the action of a certain subgroup Γ of $\mathrm{Gal}(D_n/\lambda(D_n))$, which allows one to verify

$$g^* d_k^j F_w^\flat|_{D_m} \equiv (d_k^j F_w^\flat)^p|_{D_m} \pmod{\mathfrak{m}}$$

for all $m > n$ via a direct computation of both sides as a Γ-invariant functions on moduli tuples $(A, P, \alpha_1, \alpha_2)$, see [22, Sect. 4]. Moreover (102) satisfies explicit interpolation properties stemming from the fact that the splitting of the Hodge filtration used to define (96) coincides with the CM splitting at the CM points $\tilde{y}_\infty \in \mathbf{S}_\infty$, and thus $(d_k^j F_w)(\tilde{y})(q_{\mathrm{dR}})$ are equal up to explicit periods to values of complex Maass-Shimura derivatives at CM points, which are in turn related to L-values by the explicit Waldspurger formulas of [2, Sect. 4].

Acknowledgements This article originated from discussions between Massimo Bertolini and the author following the former's 60th Birthday Conference at the Università degli Studi di Milano in September 2022. The author heartily thanks M. Bertolini for his warm encouragement and for suggesting writing this note.

References

1. Andreatta, F., Iovita, A.: Katz type p-adic L-functions for primes p non-split in the CM field. Comment. Math. Helv. **99**(4), 641–716 (2024)
2. Bertolini, M., Darmon, H., Prasanna, K.: Generalized Heegner cycles and p-adic Rankin L-series. With an appendix by Brian Conrad. Duke Math. J. **162**(6), 1033–1148 (2013)
3. Bertolini, M., Darmon, H., Prasanna, K.: p-adic Rankin L-series and rational points on CM elliptic curves. Pac. J. Math. **260**(2) (2012)
4. Bourbaki, N.: Lie Groups and Lie Algebras, Chapters 103, reprint of the 1975 edition. Springer, Berlin (1989)
5. Brakočević, M.: Anticyclotomic p-adic L-function of central critical Rankin-Selberg L-value. Int. Math. Res. Not. IMRN **21**, 4967–5018 (2011)

6. Breuil, C., Conrad, B., Diamond, F., Taylor, R.: On the modularity of elliptic curves over $\mathbb{Q}$âŁ¯: wild 3-adic exercises. J. Am. Math. Soc. **14**(4), 843–939 (2001)
7. Brooks, E.: Shimura curves and special values of p-adic L-functions. Int. Math. Res. Not. **2015**(12), 4177–4241 (2015)
8. Buzzard, K.: Integral models of certain Shimura curves. Duke Math. J. **87**(3), 591–612 (1997)
9. Caraiani, A., Scholze, P.: On the generic part of the cohomology of compact unitary Shimura varieties. Ann. Math. **186**(3), 649–766 (2017)
10. Castella, F., Hsieh, M.-L.: Heegner cycles and p-adic L-functions. Math. Ann. **370**, 567–628 (2018)
11. Chojecki, P., Hansen, D., Johansson, C.: Overconvergent modular forms and perfectoid Shimura curves. Doc. Math. **22**, 191–262 (2017)
12. Colmez, P.: Intégration sur les variétés p-adiques. Astérisque (248), viii.+155pp (1998)
13. Gross, B., Zagier, D.: Heegner points and derivatives of L-series. Invent. Math. **84**(2), 225–320 (1986)
14. Katz, N.: p-adic L-functions via moduli of elliptic curves. Proc. Symp. Pure Math. **29**, 479–506 (1975)
15. Katz, N.: p-adic properties of modular schemes and modular forms. In: Modular Functions of One Variable, III (Proceedings from International Summer School, University of Antwerp, Antwerp, 1972). Lecture Notes in Mathematics, 350, pp. 69–190. Springer, Berlin, New York (1973)
16. Katz, N.: Serre-Tate local moduli. In: Algebraic Surfaces (Orsay, 1976–78), pp. 138–202, Lecture Notes in Mathematics, 868. Springer, Berlin, New York (1981)
17. Katz, N.: Travaux de Dwork, exp. no. 409, pp. 167–200. Séminaire N. Bourbaki (1971–1972)
18. Katz, N., Mazur, B.: Arithmetic moduli of elliptic curves. In: Annals of Mathematics Studies, vol. 108. Princeton University Press, Princeton, NJ (1985)
19. Kolyvagin, V.: Euler systems. In: Cartier, P., et al. (eds.) The Grothendieck Festschift, vol. II. Progress in Mathematics 87, pp. 435–483. Birkhäuser, Boston (1990)
20. Kriz, D.: A new p-adic Maass-Shimura operator and supersingular Rankin-Selberg p-adic L-functions. arXiv: 1805.03605
21. Kriz, D.: Supersingular p-adic L-functions, Maass-Shimura operators and Waldspurger formulas. In: Annals of Mathematics Studies, vol. 405 (2021)
22. Kriz, D.: Supersingular main conjectures, Sylvester's conjecture and Goldfeld's conjecture. arXiv: 2002.04767
23. Liu, Y., Zhang, S., Zhang, W.: A p-adic Waldspurger formula. Duke Math. J. **167**(4), 743–833 (2018)
24. Morita, Y.: Reduction modulo p of Shimura varieties. Hokkaido Math. J. **10** (2) (1981)
25. Nizioł, W.: On uniqueness of p-adic period morphisms, II. Compos. Math. **156**(9), 1915–1964 (2020)
26. Scholze, P.: On torsion in the cohomology of locally symmetric varieties. Ann. Math. **182**(3), 945–1066 (2015)
27. Scholze, P.: p-adic Hodge theory for rigid analytic varieties. Forum Math. Pi 1 **e1**, 77 p (2013)
28. Scholze, P.: p-adic Hodge theory for rigid analytic varieties—corrigendum. Forum Math. Pi 4 **e6**, 4 p (2016)
29. Scholze, P.: Perfectoid spaces. Publ. Math. Inst. Hautes Études Sci. **116**, 245–313
30. Scholze, P., Weinstein, J.: Moduli of p-divisible groups. Camb. J. Math. **1**(2), 145–237 (2013)
31. Shimura, G.: Introduction to the arithmetic theory of automorphic functions, volume 11 of Publications of the Mathematical Society of Japan. In: Reprint of the 1971 original, Kanô Memorial Lectures, p. 1. Princeton University Press, Princeton, NJ (1994)
32. Sylvester, J.J.: On certain ternary cubic-form equations. Amer. J. Math. **2**(4), 357–393 (1879)
33. Tate, J.: p-divisible groups. In: 1967 Proceedings of a Conference on Local Fields (Driebergen, 1966), pp. 158–183. Springer, Berlin
34. Taylor, R., Wiles, A.: Ring-theoretic properties of certain Hecke algebras. Ann. Math. **41**(3), 553–572 (1995)

35. Wiles, A.: Modular elliptic curves and Fermat's last theorem. Ann. Math. **141**(3), 443–551 (1995)
36. Yuan, X.: Explicit Kodaira-Spencer map over Shimura curves. arXiv: 2205.11334
37. Yuan, X., Zhang, S., Zhang, W.: The Gross-Zagier formula on Shimura Curves. In: Annals of Mathematics Studies (2012)
38. Yuan, X.: Erratum: the Gross-Zagier formula on Shimura Curves. https://bicmr.pku.edu.cn/~yxy/preprints/erratum-GZSC.pdf

Anticyclotomic p-Adic L-Functions for Rankin–Selberg Product

Yifeng Liu

Abstract We construct p-adic L-functions for Rankin–Selberg products of automorphic forms of hermitian type in the anticyclotomic direction for both root numbers. When the root number is $+1$, the construction relies on global Bessel periods on definite unitary groups which, due to the recent advances on the global Gan–Gross–Prasad conjecture, interpolate classical central L-values. When the root number is -1, we construct an element in the Iwasawa Selmer group using the diagonal cycle on the product of unitary Shimura varieties, and conjecture that its p-adic height interpolates derivatives of cyclotomic p-adic L-functions. We also propose the non-vanishing conjecture and the main conjecture in both cases.

Keywords p-adic L-function · Bessel period · Iwasawa main conjecture

2020 Mathematics Subject Classification. 11G18 · 11G40 · 11R34

1 Introduction

In their pioneer work [2], Bertolini and Darmon constructed what they called Heegner distribution in both definite and indefinite cases for a classical weight two modular form and an imaginary quadratic field, opening the study of p-adic L-functions in the anticyclotomic direction. In the definite case, the product of Heegner distributions leads to a p-adic measure that interpolates central values of (classical) Rankin–Selberg L-functions, using the Waldspurger formula. In the indefinite case, the height of Heegner distribution leads to a p-adic measure that interpolates central derivatives of Rankin–Selberg L-functions, using the Gross–Zagier formula and its generalization. Since then, there have been numerous works addressing related problems including, for example, [3] in the definite case and [20] in the indefinite case.

Y. Liu (✉)
Institute for Advanced Study in Mathematics, Zhejiang University, Hangzhou, China
e-mail: liuyf0719@zju.edu.cn

M. Longo et al. (eds.), *Elliptic Curves and Modular Forms in Arithmetic Geometry*,
Springer Proceedings in Mathematics & Statistics 527,
https://doi.org/10.1007/978-3-032-13123-2_10

In this note, we make the first step to generalize these ideas, still along the anticyclotomic direction, to Rankin–Selberg products of higher ranks/dimensions, that is, beyond the $\mathrm{GL}_1 \times \mathrm{GL}_2$-case. We consider a CM extension E/F, and a suitable admissible representation $\Pi = \Pi_n \boxtimes \Pi_{n+1}$ of $\mathrm{GL}_n(\mathbb{A}_E^\infty) \times \mathrm{GL}_{n+1}(\mathbb{A}_E^\infty)$ with coefficients in a p-adic field $\mathbb{L}$ (see Sect. 2 for the precise setting; in particular, after base change to $\mathbb{C}$, Π is the finite part of a cohomological isobaric automorphic representation of minimal weights and with certain conjugate self-duality). Let $\mathtt{P}$ be a set of p-adic places v of F split in E such that Π_v is ordinary (Definition 4.2). Denote by $\mathcal{E}_\mathtt{P}/E$ the maximal anticyclotomic (with respect to E/F) extension of E unramified outside (places above) $\mathtt{P}$ with torsion free Galois group; and put $\Gamma_\mathtt{P} := \mathrm{Gal}(\mathcal{E}_\mathtt{P}/E)$, which is naturally a quotient of $E^\times\backslash(\mathbb{A}_E^\infty)^\times$ and also a finite free $\mathbb{Z}_p$-module. Possibly replacing $\mathbb{L}$ by a finite extension, there is a Galois representation $\mathbb{W}_\Pi$ of E with coefficients in $\mathbb{L}$ of rank $n(n+1)$ and weight -1 associated with Π (see Example 6.2).

We may attach to Π a "root number" $\epsilon(\Pi) \in \{\pm 1\}$ and call it coherent/incoherent if $\epsilon(\Pi) = +1/-1$, parallel to the definite/indefinite cases in [2].

When Π is coherent, we show in Theorem 5.2 that there exists a unique (bounded) measure $\mathscr{L}^0_{\mathcal{E}_\mathtt{P}}(\Pi) \in \mathbb{L}[[\Gamma_\mathtt{P}]]^\circ := \mathbb{Z}_p[[\Gamma_\mathtt{P}]] \otimes_{\mathbb{Z}_p} \mathbb{L}$ such that for every finite character $\chi\colon \Gamma_\mathtt{P} \to \overline{\mathbb{L}}^\times$ that is ramified at *every* place above $\mathtt{P}$ and every embedding $\iota\colon \overline{\mathbb{L}} \to \mathbb{C}$, we have

$$\iota\mathscr{L}^0_{\mathcal{E}_\mathtt{P}}(\Pi)(\chi) = \iota\omega \cdot \frac{\Delta_{n+1} \cdot L(\frac{1}{2}, \iota(\Pi_n \otimes \widetilde{\chi}) \times \iota\Pi_{n+1})}{2^{d(\Pi_n)+d(\Pi_{n+1})} \cdot L(1, \iota\Pi_n, \mathrm{As}^{(-1)^n})L(1, \iota\Pi_{n+1}, \mathrm{As}^{(-1)^{n+1}})}.$$

Here, $\omega \in \mathbb{L}^\times$ is an elementary factor depending on Π_v and the conductor of χ_v for $v \in \mathtt{P}$; Δ_{n+1} is a positive real constant in Definition 2.1(1); $\widetilde{\chi}$ is just χ but regarded as a character of $E^\times\backslash(\mathbb{A}_E^\infty)^\times$; $d(\Pi_n)$ and $d(\Pi_{n+1})$ are certain positive integers introduced in Remark 2.2(4), same as the ones appearing in the refined Gan–Gross–Prasad conjecture [18]. The proof relies on the recent advances on the global Gan–Gross–Prasad conjecture [6, 7] and a local Birch lemma [23, 27]. We then propose a conjecture predicting when $\mathscr{L}^0_{\mathcal{E}_\mathtt{P}}(\Pi)$ is nonvanishing (Conjecture 5.6), and an Iwasawa main conjecture for the Iwasawa Selmer group $\mathscr{X}(\mathcal{E}_\mathtt{P}, \mathbb{W}_\Pi)$ (Conjecture 6.4).

When Π is incoherent, we construct a collection of elements $(\kappa(\varphi))_\varphi$ in the compact Iwasawa Selmer group $\mathscr{S}(\mathcal{E}_\mathtt{P}, \mathbb{W}_\Pi)$ (Definition 6.1) of $\mathbb{W}_\Pi$ along $\Gamma_\mathtt{P}$, parameterized by ordinary vectors $\varphi \in \pi$. Here, π is a certain "descent" of Π to a unitary product group G determined by the local Gan–Gross–Prasad conjecture (now theorem). The element $\kappa(\varphi)$ is constructed by variants of special cycles on the Shimura varieties associated with G given by the natural diagonal subgroup $H \subseteq G$, generalizing the analogous construction in [1, 2] using Heegner points. We then define a p-adic measure $\mathscr{L}^1_{\mathcal{E}_\mathtt{P}}(\Pi)$ to be the Iwasawa p-adic height pairing between $\kappa(\varphi)$ and $\kappa(\varphi^\vee)$, modified by some local terms so that the result is independent of the choices of test vectors. Finally, we propose three conjectures: the nonvanishing conjecture for $\mathscr{L}^1_{\mathcal{E}_\mathtt{P}}(\Pi)$ (Conjecture 7.11), an Iwasawa main conjecture (Conjecture 7.7) and its variant (Conjecture 7.13), and the interpolation conjecture to the derivative of cyclotomic p-adic L-function (Conjecture 7.16).

In our subsequent work [30], we show in many cases one side of the divisibility for the main conjectures for both root numbers (Conjectures 6.4 and 7.7), generalizing [3] and [20] to higher ranks/dimensions.

Notation and Conventions

- Put $\mathbb{N} := \{0, 1, 2, \dots\}$.
- Denote by e a vector of length 1.
- In this article, $\mathbb{L}$ always denotes a field embeddable into $\mathbb{C}$; and $\mathbb{L}_?$ stands for an extension of $\mathbb{L}$ that is again embeddable into $\mathbb{C}$. We denote by $\overline{\mathbb{Q}}$ the algebraic closure of $\mathbb{Q}$ in $\mathbb{C}$.

We fix a CM extension E/F contained in $\mathbb{C}$. Denote by

- $\mathrm{c} \in \mathrm{Aut}(\mathbb{C}/F)$ the complex conjugation (which generates $\mathrm{Gal}(E/F)$),
- E^{ab} the maximal abelian extension of E in $\mathbb{C}$,
- $\eta_{E/F} \colon F^\times\backslash\mathbb{A}_F^\times \to \{\pm 1\}$ the associated quadratic character,
- $E^{\times -}$ the kernel of the norm map $\mathrm{Nm}_{E/F} \colon E^\times \to F^\times$,
- $\mathrm{V}_F^{(w)}$ the set of places of F above a place w of $\mathbb{Q}$,
- $\mathrm{V}_F^{\mathrm{fin}}$ the set of non-archimedean places of F,
- $\mathrm{V}_F^{\mathrm{spl}}$, $\mathrm{V}_F^{\mathrm{inert}}$ and $\mathrm{V}_F^{\mathrm{ram}}$ the subsets of $\mathrm{V}_F^{\mathrm{fin}}$ of those that are split, inert and ramified in E, respectively,
- $\mathrm{V}_E^?$ the set of places of E above $\mathrm{V}_F^?$,
- $\mathfrak{p}_v$ the maximal ideal of O_{F_v} and q_v the cardinality of $O_{F_v}/\mathfrak{p}_v$ for every $v \in \mathrm{V}_F^{\mathrm{fin}}$.

We fix a prime number p, and take a (possibly empty) subset $\mathrm{P} \subseteq \mathrm{V}_F^{(p)} \cap \mathrm{V}_F^{\mathrm{spl}}$ (which will be specified at a certain point). For every finite extension K/F, we denote by Γ_{P}^K the maximal torsion free quotient of the profinite completion of

$$K^\times\backslash(\mathbb{A}_K^\infty)^\times \Big/ \prod_{v \in \mathrm{V}_F^{\mathrm{fin}} \setminus \mathrm{P}} (O_K \otimes_{O_F} O_{F_v})^\times ,$$

which is naturally a finite free $\mathbb{Z}_p$-module.

We have a homomorphism $\mathrm{Nm}_{E/F}^- \colon \Gamma_{\mathrm{P}}^E \to \Gamma_{\mathrm{P}}^E$ sending a to a/a^{c}, whose image is denoted by Γ_{P}. Denote by $\mathcal{E}_{\mathrm{P}} \subseteq E^{\mathrm{ab}}$ the extension of E such that $\mathrm{Gal}(\mathcal{E}_{\mathrm{P}}/E) = \Gamma_{\mathrm{P}}$ under the global class field theory. For every function χ on Γ_{P}, we put $\widetilde{\chi} := \chi \circ \mathrm{Nm}_{E/F}^-$ as a function on Γ_{P}^E.

Definition 1.1 We say that a field extension $\mathcal{E}/E$ is a P-*extension* if $\mathcal{E} \subseteq \mathcal{E}_{\mathrm{P}}$; $\mathrm{Gal}(\mathcal{E}/E)$ is torsion free; and the set of primes of E ramified in $\mathcal{E}$ consists exactly those above P.

Acknowledgements. The author would like to thank Daniel Disegni and David Loeffler for helpful comments, and also the anonymous referee for careful reading and useful commments. The research of Y. L. is supported by National Key R&D Program of China No. 2022YFA1005300.

2 Gan–Gross–Prasad Conjecture

Definition 2.1 Let N be a positive integer.

(1) Put

$$\Delta_{N,v} := \prod_{i=1}^{N} L(i, \eta^i_{E/F,v}) \in \mathbb{Q}_{>0}$$

for every $v \in \mathtt{V}_F^{\mathrm{fin}}$; and put

$$\Delta_N := \prod_{i=1}^{N} L(i, (\eta^\infty_{E/F})^i) \in \mathbb{R}_{>0}.$$

(2) For every $u \in \mathtt{V}_E^{(\infty)}$, denote by $\Pi_u^{[N]}$ the (complex irreducible) principal series representation of $\mathrm{GL}_N(E_u)$ given by the characters $(\arg^{1-N}, \arg^{3-N}, \dots, \arg^{N-3}, \arg^{N-1})$, where $\arg\colon \mathbb{C}^\times \to \mathbb{C}^\times$ is the *argument character* defined by the formula $\arg(z) := z/\sqrt{z\overline{z}}$.

(3) A *relevant* $\mathbb{L}$*-representation* of $\mathrm{GL}_N(\mathbb{A}_E^\infty)$ is a representation Π with coefficients in $\mathbb{L}$ satisfying that for every embedding $\iota\colon \mathbb{L} \to \mathbb{C}$,

$$\Pi^{(\iota)} := \left(\otimes_{u \in \mathtt{V}_E^{(\infty)}} \Pi_u^{[N]}\right) \otimes \iota\Pi \tag{1}$$

is a hermitian isobaric automorphic representation of $\mathrm{GL}_N(\mathbb{A}_E)$ [6, Definition 1.5].[1]

Remark 2.2 We have the following remarks concerning Definition 2.1.

(1) A relevant $\mathbb{L}$-representation is absolutely irreducible.
(2) A relevant $\mathbb{L}$-representation is defined over a number field contained in $\mathbb{L}$.
(3) It suffices to check the condition for one embedding ι in Definition 2.1(3).
(4) For a relevant $\mathbb{L}$-representation Π, the number of isobaric summands of (1) is independent of ι, which we denote as $d(\Pi)$, which is an integer between 1 and N.
(5) If Π is a relevant $\mathbb{L}$-representation, then for every embedding $\iota\colon \mathbb{L} \to \mathbb{C}$, $\iota\Pi_u$ is tempered for every $u \in \mathtt{V}_E^{\mathrm{fin}}$ [9, Theorem 1.2].

[1] In this case, it is equivalent to saying that $\Pi^{(\iota)}$ is an isobaric sum of mutually non-isomorphic conjugate self-dual cuspidal automorphic representations.

Fix a positive integer n throughout the article. Let Π_n and Π_{n+1} be two relevant $\mathbb{L}$-representations of $\mathrm{GL}_n(\mathbb{A}_E^\infty)$ and $\mathrm{GL}_{n+1}(\mathbb{A}_E^\infty)$, respectively. Write $\Pi := \Pi_n \boxtimes \Pi_{n+1}$ as a representation of $\mathrm{GL}_n(\mathbb{A}_E^\infty) \times \mathrm{GL}_{n+1}(\mathbb{A}_E^\infty)$. By the solution of the local Gan–Gross–Prasad conjecture [16] by [4, 5], we know that for every finite place v of F, there exists a pair $(V_{n,v}, \pi_v)$, unique up to isomorphism, in which

- $V_{n,v}$ is a hermitian space over E_v of rank n;
- π_v is an irreducible admissible representation of $G_v(F_v)$ with coefficients in $\mathbb{L}$ with Π_v as its base change, satisfying $\mathrm{Hom}_{H_v(F_v)}(\pi_v, \mathbb{L}) \neq 0$.[2]

Here, we have put $G_v := \mathrm{U}(V_{n,v}) \times \mathrm{U}(V_{n+1,v})$ with $V_{n+1,v} := V_{n,v} \oplus E_v \cdot \mathrm{e}$, and denoted by $H_v \subseteq G_v$ the graph of the natural embedding $\mathrm{U}(V_{n,v}) \hookrightarrow \mathrm{U}(V_{n+1,v})$ by realizing the source as the stabilizer of e in the target.

Definition 2.3 Put

$$\epsilon(\Pi) := \prod_{v \in \mathtt{V}_F^{\mathrm{fin}}} \eta_{E/F}\left(\det V_{n,v}\right) \in \{\pm 1\}.$$

We say that Π is a *coherent* (resp. *incoherent*) if $\epsilon(\Pi)$ equals 1 (resp. -1).

Now suppose that $\chi\colon \Gamma_{\mathrm{P}} \to \mathbb{L}_\chi^\times$ is a finite character. Then we have a new pair of relevant $\mathbb{L}_\chi$-representations $\Pi_n \otimes \widetilde{\chi}$ and Π_{n+1} of $\mathrm{GL}_n(\mathbb{A}_E^\infty)$ and $\mathrm{GL}_{n+1}(\mathbb{A}_E^\infty)$, respectively. Put

$$\Pi_\chi := (\Pi_n \otimes \widetilde{\chi}) \boxtimes \Pi_{n+1} = \Pi \otimes (\widetilde{\chi} \circ \det \boxtimes 1)$$

as a representation of $\mathrm{GL}_n(\mathbb{A}_E^\infty) \times \mathrm{GL}_{n+1}(\mathbb{A}_E^\infty)$. Since Γ_{P} is torsion free, χ_v is trivial for every finite place v of F that is nonsplit in E. It follows that for every finite place v of F, the corresponding pair for Π_χ is $(V_{n,v}, (\pi_v)_{\chi_v})$, where $(\pi_v)_{\chi_v} := \pi_v \otimes (\chi_v \circ \det \boxtimes 1)$. In particular, $\epsilon(\Pi) = \epsilon(\Pi_\chi)$.

Notation 2.4 For $N = n, n+1$, we denote by

$$\rho_{\Pi_N}\colon \mathrm{Gal}(\overline{\mathbb{Q}}/E) \to \mathrm{GL}_N(\mathbb{L})$$

the Galois representation associated with Π_N satisfying $\rho_{\Pi_N}^\vee \simeq \rho_{\Pi_N}^{\mathtt{c}}(N-1)$.[3] We propose the following condition which will be referred later:

---

[2] Indeed, one can find an $\mathbb{L}$-linear model of π_v in the space of $\mathbb{L}$-valued locally constant functions on $H_v(F_v)\backslash G_v(F_v)$, using the multiplicity one property.

[3] Strictly speaking, ρ_{Π_N} can a priori only be defined over a finite extension $\mathbb{L}'$ of $\mathbb{L}$ (although it has traces in $\mathbb{L}$). However, enlarging $\mathbb{L}$ to a finite extension will do no harm to our discussion.

(∗) For every irreducible $\mathrm{Gal}(\overline{\mathbb{Q}}/E)$-subrepresentations ρ_n and ρ_{n+1} of ρ_{Π_n} and $\rho_{\Pi_{n+1}}$, respectively, $\rho_n \otimes \rho_{n+1}$ is absolutely irreducible.

In the rest of this section, we review some facts about local matrix coefficient integrals appearing in the refined Gan–Gross–Prasad conjecture. Take a finite place v of F and suppress it in the subscripts from the notation below. In particular, E is a quadratic étale F-algebra.

The lemma below will be used later.

Lemma 2.5 *Let r, r_1, r_2 be positive integers.*

(1) Let Π_1 and Π_2 be tempered (complex) irreducible admissible representations of $\mathrm{GL}_{r_1}(E)$ and $\mathrm{GL}_{r_2}(E)$, respectively. Then for every automorphism σ of $\mathbb{C}$,

$$L(s, \sigma\Pi_1 \times \sigma\Pi_2) = \sigma L(s, \Pi_1 \times \Pi_2)$$

holds for every $s \in \mathbb{N} + 1$ (resp. $s \in \mathbb{N} + \frac{1}{2}$) when r_1 and r_2 have the same parity (resp. different parities).

(2) Let Π be a tempered (complex) irreducible admissible representations of $\mathrm{GL}_r(E)$ and $\epsilon \in \{\pm 1\}$. Then for every automorphism σ of $\mathbb{C}$,

$$L(s, \sigma\Pi, \mathrm{As}^{\epsilon}) = \sigma L(s, \Pi, \mathrm{As}^{\epsilon})$$

holds for every $s \in \mathbb{N} + 1$.

Proof Denote by e_r the vector $(0, \ldots, 0, 1)$ with $r - 1$ zeros.

For (1), without lost of generality, we may assume $r_1 \leqslant r_2$ and E a field. Regard GL_{r_1} as a subgroup of GL_{r_2} via the first r_1 coordinates. Let U_{r_1} and U_{r_2} be the standard upper-triangular maximal unipotent subgroups of GL_{r_1} and GL_{r_2} respectively. Choose two generic characters $\psi_1 \colon U_{r_1}(E) \to \mathbb{C}^\times$ and $\psi_2 \colon U_{r_2}(E) \to \mathbb{C}^\times$ satisfying $\psi_1(u)\psi_2(u) = 1$ for $u \in U_{r_1}(E)$. For $\alpha = 1, 2$, denote by $\mathcal{W}(\Pi_\alpha)_{\psi_\alpha}$ the Whittaker model of Π_α with respect to $(U_{r_\alpha}, \psi_\alpha)$. Fix a rational Haar measure on $U_{r_1}(E)\backslash \mathrm{GL}_{r_1}(E)$. Denote by $\mathcal{F}(E^{r_1})$ the space of Schwartz (resp. constant) functions on E^{r_1} when $r_1 = r_2$ (resp. $r_1 < r_2$). Let q be the residue cardinality of E.

For $W_1 \in \mathcal{W}(\Pi_1)_{\psi_1}$, $W_2 \in \mathcal{W}(\Pi_2)_{\psi_2}$, and $\Phi \in \mathcal{F}(E^{r_1})$, the integral

$$Z(s, W_1, W_2, \Phi) := \int_{U_{r_1}(E)\backslash \mathrm{GL}_{r_1}(E)} W_1(g) W_2(g) \Phi(e_{r_1} g) |\det g|_F^{s - \frac{r_2 - r_1}{2}} \, dg$$

is absolutely convergent for $\mathrm{Re}\, s > 0$ and has a meromorphic extension to an element in $\mathbb{C}(q^{-s})$. By the local Rankin–Selberg theory [22], there exists a unique element $P_{\Pi_1 \times \Pi_2}(X) \in \mathbb{C}[X]$ satisfying $P_{\Pi_1, \Pi_2}(0) = 1$ such that $\{Z(s, W_1, W_2, \Phi) \mid W_1, W_2, \Phi\}$ is a $\mathbb{C}[q^s, q^{-s}]$-submodule of $\mathbb{C}(q^{-s})$ generated by $P_{\Pi_1, \Pi_2}(q^{-s})^{-1}$, which is nothing but $L(s, \Pi_1 \times \Pi_2)$. In particular, $\{Z(s, W_1, W_2, \Phi) \mid W_1, W_2, \Phi\}$ is independent of the choices of ψ_1 and ψ_2. Now let σ be an automorphism of $\mathbb{C}$. We have $\sigma W_\alpha \in \mathcal{W}(\sigma\Pi_\alpha)_{\sigma\psi_\alpha}$ for $\alpha = 1, 2$, and moreover

$$Z(s, \sigma W_1, \sigma W_2, \sigma\Phi) = \sigma Z(s, W_1, W_2, \Phi)$$

as absolutely convergent integrals for $s \in \mathbb{N} + 1$ (resp. $s \in \mathbb{N} + \frac{1}{2}$) when r_1 and r_2 have the same parity (resp. different parities). Since elements in $\mathbb{C}(q^{-s})$ are uniquely determined by their values at an arbitrary infinite set of s, it follows easily that $L(s, \sigma\Pi_1 \times \sigma\Pi_2) = \sigma L(s, \Pi_1 \times \Pi_2)$ for the same set of s.

For (2), it is similar to (1) by considering the integral

$$Z(s, W, \Phi) := \int_{U_r(F)\backslash \mathrm{GL}_r(F)} W(g)\Phi(e_r g)\eta_{E/F}(\det g)^{\frac{1-c}{2}} |\det g|_F^s \mathrm{d}g$$

for $W \in \mathcal{W}(\Pi)_\psi$ and $\Phi \in \mathcal{S}(F^r)$.

The lemma is proved. □

Choose a rational Haar measure $\mathrm{d}h$ on $H(F)$. Let $\chi \colon E^{\times-} \to \mathbb{L}_\chi^\times$ be a finite character.

Definition 2.6 Lemma 2.5 provides us with the following definitions.

(1) We define $L(\frac{1}{2}, (\Pi_n \otimes \widetilde{\chi}) \times \Pi_{n+1}) \in \mathbb{L}_\chi$ to be the unique element such that for every $\iota \colon \mathbb{L}_\chi \to \mathbb{C}$, $\iota L(\frac{1}{2}, (\Pi_n \otimes \widetilde{\chi}) \times \Pi_{n+1}) = L(\frac{1}{2}, \iota(\Pi_n \otimes \widetilde{\chi}) \times \iota\Pi_{n+1})$.
(2) For $N = n, n+1$, we define $L(1, \Pi_N, \mathrm{As}^{(-1)^N}) \in \mathbb{L}^\times$ to be the unique element such that for every $\iota \colon \mathbb{L} \to \mathbb{C}$, $\iota L(1, \Pi_N, \mathrm{As}^{(-1)^N}) = L(1, \iota\Pi_N, \mathrm{As}^{(-1)^N})$.

Let $\iota \colon \mathbb{L}_\chi \to \mathbb{C}$ be an embedding. For $\varphi \in \pi$ and $\varphi^\vee \in \pi^\vee$, the integral

$$\alpha_\iota^\chi(\varphi, \varphi^\vee) := \int_{H(F)} \iota\left(\langle \varphi^\vee, \pi(h)\varphi\rangle_\pi \cdot \chi(\det h)\right) \mathrm{d}h$$

is absolutely convergent [18] since $\iota\pi$ is tempered, hence defines an element in the $\mathbb{L}_\chi$-vector space

$$\alpha_\iota^\chi \in \mathrm{Hom}_{H(F)\times H(F)}\left(\pi_\chi \boxtimes (\pi_\chi)^\vee, \mathbb{C}\right),$$

where we regard $\mathbb{C}$ as an $\mathbb{L}_\chi$-vector space via ι. We suppress χ in the notation when it is the trivial character.

Lemma 2.7 *Let $\chi \colon E^{\times-} \to \mathbb{L}_\chi^\times$ be a finite character. For $\varphi \in \pi$ and $\varphi^\vee \in \pi^\vee$, there exists a unique element $\alpha^\chi(\varphi, \varphi^\vee) \in \mathbb{L}_\chi$ such that for every embedding $\iota \colon \mathbb{L}_\chi \to \mathbb{C}$,*

$$\iota\alpha^\chi(\varphi, \varphi^\vee) = \alpha_\iota^\chi(\varphi, \varphi^\vee)$$

holds.

Proof By Casselman's lemma [11, Proposition 4.2.3 & Theorem 4.3.3] and the argument in [39, Sect. I.2], the absolutely convergent integral defining α_ι^χ is Galois equivariant, namely, for every automorphism σ of $\mathbb{C}$,

$$\int_{H(F)} \sigma\iota\left(\langle\varphi^\vee, \pi(h)\varphi\rangle_\pi \cdot \chi(\det h)\right) dh = \sigma \int_{H(F)} \iota\left(\langle\varphi^\vee, \pi(h)\varphi\rangle_\pi \cdot \chi(\det h)\right) dh$$

holds. The lemma then follows. □

Lemma 2.8 *Assume χ trivial if E is a field. Then the $\mathbb{L}_\chi$-vector space*

$$\mathrm{Hom}_{H(F)\times H(F)}\left(\pi_\chi \boxtimes (\pi_\chi)^\vee, \mathbb{L}_\chi\right)$$

is one-dimensional of which α^χ is a basis.

Proof Without lost of generality, we may assume $\mathbb{L}_\chi = \mathbb{C}$. Then this is proved in [5] or, in a more general setting, [37]. □

Lemma 2.9 *Suppose that E/F is unramified; χ is unramified; and V_n admits an integrally self-dual lattice L_n such that both φ and $\varphi^\vee$ are fixed by $K_n \times K_{n+1}$, where K_n and K_{n+1} are the stabilizers of L_n and $L_n \oplus O_E \cdot \mathrm{e}$ respectively. Then*

$$\alpha^\chi(\varphi, \varphi^\vee) = \langle\varphi^\vee, \varphi\rangle_\pi \cdot \mathrm{vol}(K_n, dh) \cdot \frac{\Delta_{n+1} \cdot L(\frac{1}{2}, (\Pi_n \otimes \widetilde{\chi}) \times \Pi_{n+1})}{L(1, \Pi_n, \mathrm{As}^{(-1)^n})L(1, \Pi_{n+1}, \mathrm{As}^{(-1)^{n+1}})}.$$

Proof By Definition 2.6 and Lemma 2.7, we may assume $\mathbb{L}_\chi = \mathbb{C}$. Then this is [18, Theorem 2.12]. □

Now suppose that $E \simeq F \times F$. Choose a Borel subgroup B of H such that $B_H := B \cap H$ is a Borel subgroup of H, and a generic character ψ of $U(F)$ that is trivial on $(U \cap H)(F)$, where U denotes the unipotent radical of B. For every embedding $\iota\colon \mathbb{L} \to \mathbb{C}$, denote by $\mathcal{W}(\iota\pi)_\psi$ and $\mathcal{W}(\iota\pi^\vee)_{\psi^{-1}}$ the Whittaker models of $\iota\pi$ and $\iota\pi^\vee$ with respect to (U, ψ) and (U, ψ^{-1}), respectively. Define a pairing $\vartheta\colon \mathcal{W}(\iota\pi^\vee)_{\psi^{-1}} \times \mathcal{W}(\iota\pi)_\psi \to \mathbb{C}$ by the formula

$$\vartheta(W^\vee, W) := \int_{U(F)\backslash Q(F)} W^\vee(h)W(h)dh$$

where Q is a mirabolic subgroup of G containing U.

Lemma 2.10 *There exists a positive rational constant c depending only on the rational Haar measures on $U(F)\backslash Q(F)$, $H(F)$, and $(U \cap H)(F)$, such that for every finite character $\chi\colon E^{\times -} \to \mathbb{L}_\chi^\times$, every $\iota\colon \mathbb{L}_\chi \to \mathbb{C}$, and every $(W, W^\vee) \in \mathcal{W}(\iota\pi)_\psi \times \mathcal{W}(\iota\pi^\vee)_{\psi^{-1}}$,*

$$\int_{H(F)} \vartheta(W^\vee, \pi(h)W) \cdot \iota\chi(\det h)dh$$
$$= c\left(\int_{(U\cap H)(F)\backslash H(F)} W(h) \cdot \iota\chi(\det h)dh\right)\left(\int_{(U\cap H)(F)\backslash H(F)} W^\vee(h) \cdot \iota\chi(\det h)^{-1}dh\right)$$

in which both sides are absolutely convergent.

Proof This is [41, Proposition 4.10]. □

Proposition 2.11 *Suppose that $E \simeq F \times F$. There exist elements $\varphi \in \pi$ and $\varphi^\vee \in \pi^\vee$ such that for every finite unramified character $\chi\colon E^{\times-} \to \mathbb{L}_\chi^\times$,*

$$\alpha^\chi(\varphi, \varphi^\vee) = \frac{\Delta_{n+1} \cdot L(\frac{1}{2}, (\Pi_n \otimes \widetilde{\chi}) \times \Pi_{n+1})}{L(1, \Pi_n, \mathrm{As}^{(-1)^n})L(1, \Pi_{n+1}, \mathrm{As}^{(-1)^{n+1}})}$$

holds.

Proof Take an embedding $\iota_0\colon \mathbb{L} \to \mathbb{C}$. Choose $G(F)$-equivariant isomorphisms $i_1\colon \pi \otimes_{\mathbb{L},\iota_0} \mathbb{C} \xrightarrow{\sim} \mathcal{W}(\iota_0\pi)_\psi$ and $i_2 \in \pi^\vee \otimes_{\mathbb{L},\iota_0} \mathbb{C} \xrightarrow{\sim} \mathcal{W}(\iota_0\pi^\vee)_{\psi^{-1}}$ such that $\iota_0\langle\varphi^\vee, \varphi\rangle_\pi = \vartheta(i_2\varphi^\vee, i_1\varphi)$ holds for all $\varphi \in \pi$ and $\varphi^\vee \in \pi^\vee$. By the definition of Rankin–Selberg L-factors [22], there exist elements $W \in i_1(\pi)$ and $W^\vee \in i_2(\pi^\vee)$ such that

$$\int_{(U\cap H)(F)\backslash H(F)} W(h) \cdot |\det h|_F^s dh = c_1 L(\tfrac{1}{2} + s, \iota_0\pi_n \times \iota_0\pi_{n+1}),$$
$$\int_{(U\cap H)(F)\backslash H(F)} W^\vee(h) \cdot |\det h|_F^s dh = c_2 L(\tfrac{1}{2} + s, \iota_0\pi_n^\vee \times \iota_0\pi_{n+1}^\vee),$$

hold as meromorphic functions on $\mathbb{C}$ for some constants $c_1, c_2 \in \mathbb{C}^\times$. Moreover, the integrals on the left-hand side are absolutely convergent when $\mathrm{Re}\, s \geqslant 0$. Put $\varphi = i_1^{-1}W$ and $\varphi^\vee = i_2^{-1}W^\vee$. By Lemma 2.10, we have

$$\alpha_{\iota_0}(\varphi, \varphi^\vee) = cc_1c_2 \cdot L(\tfrac{1}{2}, \iota_0\Pi_n \times \iota_0\Pi_{n+1}).$$

Since $\alpha_{\iota_0}(\varphi, \varphi^\vee) \in \iota_0\mathbb{L}$ and $L(\frac{1}{2}, \iota_0\Pi_n \times \iota_0\Pi_{n+1}) \in \iota_0\mathbb{L}^\times$, we have $cc_1c_2 \in \iota_0\mathbb{L}^\times$. By Lemma 2.5(2), after rescaling, we may assume

$$cc_1c_2 = \frac{\Delta_{n+1}}{L(1, \iota_0\Pi_n, \mathrm{As}^{(-1)^n})L(1, \iota_0\Pi_{n+1}, \mathrm{As}^{(-1)^{n+1}})}.$$

Then, again by Lemma 2.10, we know that for every finite unramified character $\chi\colon E^{\times-} \to \mathbb{L}_\chi^\times$ and every $\iota\colon \mathbb{L}_\chi \to \mathbb{C}$ extending ι_0,

$$\alpha_\iota^\chi(\varphi, \varphi^\vee) = \frac{\Delta_{n+1} \cdot L(\frac{1}{2}, \iota(\Pi_n \otimes \widetilde{\chi}) \times \iota\Pi_{n+1})}{L(1, \iota\Pi_n, \mathrm{As}^{(-1)^n})L(1, \iota\Pi_{n+1}, \mathrm{As}^{(-1)^{n+1}})}$$

holds. Finally, by Definition 2.6 and Lemma 2.7, the above identity holds for every embedding $\iota\colon \mathbb{L}_\chi \to \mathbb{C}$.

The proposition is proved. □

3 p-Adic Measure

From now on, we assume that $\mathbb{L}$ is a finite extension of $\mathbb{Q}_p$, and denote by $\mathbb{L}^\circ$ its ring of integers.

Notation 3.1 For every quotient group Γ of $\Gamma_{\mathtt{P}}$, put

$$\mathbb{L}[[\Gamma]]^\circ := \mathbb{L}^\circ[[\Gamma]] \otimes_{\mathbb{L}^\circ} \mathbb{L}$$

and denote by $-^\dagger$ the involution on $\mathbb{L}^\circ[[\Gamma]]$ or $\mathbb{L}[[\Gamma]]^\circ$ induced by the inverse map on Γ.

For every $v \in \mathtt{P}$ and every positive integer $\mathfrak{c}_v$, we put

$$\mathfrak{U}_{\mathfrak{c}_v} := \left\{(x, x^{-1}) \mid x \in 1 + \mathfrak{p}_v^{\mathfrak{c}_v}\right\} \subseteq E_v^\times.$$

For a tuple $\mathfrak{c} = (\mathfrak{c}_v)_{v\in\mathtt{P}}$ of positive integers indexed by $\mathtt{P}$, we put

$$\mathfrak{U}_{\mathfrak{c}} := \prod_{v\in\mathtt{P}} \mathfrak{U}_{\mathfrak{c}_v} \subseteq \prod_{v\in\mathtt{P}} E_v^\times.$$

Then $\mathrm{Nm}^-_{E/F}$ induces a homomorphism

$$\mathfrak{U}_{\mathfrak{c}} \to \Gamma_{\mathtt{P}}$$

whose kernel is finite, and trivial when $|\mathfrak{c}| := \sum_v \mathfrak{c}_v$ is large enough. We define a $\mathfrak{c}$-*cell* of $\Gamma_{\mathtt{P}}$ to be a $\mathfrak{U}_{\mathfrak{c}}$-orbit in $\Gamma_{\mathtt{P}}$. Then for every given $\mathfrak{c}$, all $\mathfrak{c}$-cells give a disjoint cover of $\Gamma_{\mathtt{P}}$ by open compact subsets.

In what follows, we will frequently choose a collection $\varpi = (\varpi_v)_{v\in\mathtt{P}}$ of uniformizers of F_v for $v \in \mathtt{P}$. For $\mathfrak{c}$ as above, we put

$$\varpi^{\mathfrak{c}} := (\varpi_v^{\mathfrak{c}_v})_{v\in\mathtt{P}} \in \prod_{v\in\mathtt{P}} (O_{F_v} \cap F_v^\times).$$

In this article, we will encounter measures on $\Gamma_{\mathtt{P}}$ valued in a finite-dimensional $\mathbb{L}$-vector space $\mathbb{V}$, whose definition we now review. A *distribution* on $\Gamma_{\mathtt{P}}$ valued in $\mathbb{V}$ is an assignment

$$\mu\colon \{\text{open compact subsets of } \Gamma_{\mathtt{P}}\} \to \mathbb{V}$$

that is additive. It is clear that in the above definition, we may replace the source by the set of all $\mathfrak{c}$-cells of $\Gamma_{\mathtt{P}}$ for all $\mathfrak{c}$. We say that the distribution μ is a *measure* if it is bounded, namely, there exists an $\mathbb{L}^\circ$-lattice $\mathbb{V}^\circ$ of $\mathbb{V}$ such that the range of μ is contained in $\mathbb{V}^\circ$.

For a measure μ valued in $\mathbb{V}$, we may evaluate it on every continuous character $\chi\colon \Gamma_{\mathtt{P}} \to \mathbb{L}_\chi^\times$ (for a p-adic field extension $\mathbb{L}_\chi/\mathbb{L}$) by the formula

$$\mu(\chi) := \int_{\Gamma_{\mathrm{P}}} \chi \mathrm{d}\mu \in \mathbb{V} \otimes_{\mathbb{L}} \mathbb{L}_{\chi}$$

(as the limit of Riemann sums over $\mathfrak{c}$-cells for all $\mathfrak{c}$). By evaluating at all finite characters of Γ_{P}, we obtain a map

$$\{\text{measures on } \Gamma_{\mathrm{P}} \text{ valued in } \mathbb{V}\} \to \mathbb{L}[[\Gamma_{\mathrm{P}}]]^{\circ} \otimes_{\mathbb{L}} \mathbb{V},$$

which is well-known to be a bijection. As consequences, we have

- A measure is determined by its values on the set of finite characters of Γ_{P} that are ramified at every place above P.
- When $\mathbb{V}$ is a (commutative) $\mathbb{L}$-algebra, the set of measures on Γ_{P} valued in $\mathbb{V}$ is naturally a (commutative) $\mathbb{L}$-algebra.

4 Ordinary Condition

Take an element $v \in \mathrm{V}_F^{(p)} \cap \mathrm{V}_F^{\mathrm{spl}}$ and suppress it in the subscripts from the notation below. In particular, $E \simeq F \times F$.

Definition 4.1 An isomorphism $G \simeq \mathrm{GL}_{n,F} \times \mathrm{GL}_{n+1,F}$ is called *standard* if the subgroup H is identified with $(h, \mathrm{diag}(h, 1))$ for $h \in \mathrm{GL}_{n,F}$.

We fix a standard isomorphism $G \simeq \mathrm{GL}_{n,F} \times \mathrm{GL}_{n+1,F}$ (which exists). For $N = n, n+1$, denote by B_N and U_N the upper-triangular Borel and unipotent subgroups of GL_N, respectively; for every positive integer r, denote by $I_N^{(r)}$ the inverse image of $B_N(O_F/\mathfrak{p}^r)$ under the reduction map $\mathrm{GL}_N(O_F) \to \mathrm{GL}_N(O_F/\mathfrak{p}^r)$. Put $B := B_n \times B_{n+1}$, $U := U_n \times U_{n+1}$, and $I^{(r)} := I_n^{(r)} \times I_{n+1}^{(r)}$. We also write $\pi = \pi_n \boxtimes \pi_{n+1}$ in which both π_n and π_{n+1} have coefficients in $\mathbb{L}$ with base change Π_n and Π_{n+1}, respectively.

For $N = n, n+1$ and a tuple $\mu = (\mu_1, \dots, \mu_N)$ of admissible characters $F^{\times} \to \mathbb{L}^{\times}$, we have an induced character $\mu^{\natural}$ of $(F^{\times})^N$ given by

$$\mu^{\natural}(x_1, \dots, x_N) = \prod_{i=1}^{N} \mu_i(x_i)|x_i|_F^{N-i}$$

hence a character of $B_N(F)$ by inflation, and the unnormalized principal series

$$\begin{aligned} \mathcal{I}(\mu) := \{ & f \colon \mathrm{GL}_N(F) \to \mathbb{L} \text{ locally constant} \\ & | f(bg) = \mu^{\natural}(b) f(g), \forall b \in B_N(F), g \in \mathrm{GL}_N(F)\} \end{aligned}$$

as an admissible representation of $\mathrm{GL}_N(F)$ via right translation.

Definition 4.2 For $N = n, n+1$, we say that Π_N is *ordinary* if there exists a (unique) tuple $\mu = (\mu_1, \dots, \mu_N)$ of admissible characters $F^\times \to \mathbb{L}^\times$ satisfying $|x|_F^{i-1}\mu_i(x) \in \mathbb{L}^{\circ\times}$ for $1 \leqslant i \leqslant N$ and every $x \in F^\times$ such that π_N is a subrepresentation of $\mathcal{I}(\mu)$; we say that Π_N is *semi-stably ordinary* if furthermore μ_i are all unramified.

We say that Π is (semi-stably) ordinary if both Π_n and Π_{n+1} are.

Remark 4.3 Note that if π_N satisfies the property in Definition 4.2, then so does $\pi_N^\vee$ with respect to the tuple $\check{\mu} := (|\ |_F^{1-N}\mu_N^{-1}, \dots, |\ |_F^{1-N}\mu_1^{-1})$.

For $N = n, n+1$ and an element $x \in O_F \cap F^\times$, we put $[x]_N := \mathrm{diag}(x^{N-1}, \dots, x, 1) \in \mathrm{GL}_N(F)$ and define an operator V_N^x on $\pi_N^{U_N(O_F)}$ as

$$\mathrm{V}_N^x := \sum_{u \in U_N(O_F)/(U_N(O_F) \cap [x]_N U_N(O_F)[x]_N^{-1})} \pi_N(u[x]_N).$$

Lemma 4.4 *Suppose that Π_N is ordinary for some $N \in \{n, n+1\}$. There exists a unique up to scalar nonzero element $\varphi_N \in \pi_N^{U_N(O_F)}$ satisfying that*

$$\mathrm{V}_N^x \varphi_N = \left(\prod_{m=1}^{N-1} \prod_{i=1}^{m} |x|_F^{i-1}\mu_i(x) \right) \varphi_N$$

holds for every $x \in O_F \cap F^\times$. Here, μ is the tuple of characters in Definition 4.2.

Proof Since $\mu_1 \boxtimes \cdots \boxtimes \mu_N$ is a regular character of $B_N(F)/U_N(F)$, the Jacquet module (with respect to B_N) $\mathcal{I}(\mu)_{B_N}$ of $\mathcal{I}(\mu)$ is a direct sum of N distinct characters of $B_N(F)/U_N(F)$. By [19, Corollary 5.5], it suffices to show that π_{B_N} contains the character $\mu^\natural$, which follows from the Frobenius reciprocity law. The lemma is proved. □

Definition 4.5 We call the one-dimensional $\mathbb{L}$-subspace of $\pi_N^{U_N(O_F)}$ generated by φ_N in Lemma 4.4 the *ordinary line* of π_N, and a nonzero element of it an *ordinary vector*. When Π is ordinary, we have the obvious notion of ordinary line and ordinary vectors, which are contained in $\pi^{U(O_F)} = \pi_n^{U_n(O_F)} \otimes_{\mathbb{L}} \pi_{n+1}^{U_{n+1}(O_F)}$.

Remark 4.6 The notion of being (semi-stably) ordinary and the characters in Definition 4.2 are intrinsic, namely, they do not depend on the choice of the standard isomorphism $G \simeq \mathrm{GL}_{n,F} \times \mathrm{GL}_{n+1,F}$. However, the notion of ordinary line and vectors clearly depends on such choice.

Lemma 4.7 *Suppose that Π_N is ordinary for some $N \in \{n, n+1\}$. For ordinary vectors $\varphi_N \in \pi_N^{U_N(O_F)}$ and $\varphi_N^\vee \in (\pi_N^\vee)^{U_N(O_F)}$, we have $\langle \varphi_N^\vee, \varphi_N \rangle_{\pi_N} \neq 0$.*

Proof Fix a uniformizer ϖ of F. Denote by $(\pi_N)^{\mathrm{ss}}$ the invertible part of $\pi_N^{U_N(O_F)}$ with respect to the operator V_N^{ϖ} (see [19, §5.2]). Then $\langle\ ,\ \rangle_{\pi_N}$ restricts to a perfect pairing on $(\pi_N^{\vee})^{\mathrm{ss}} \times (\pi_N)^{\mathrm{ss}}$. Moreover, by [19, Proposition 5.4 & Corollary 5.5], there exists a group $\mathfrak{S}$ consisting of permutations on $\{1, \ldots, N\}$ such that $(\pi_N)^{\mathrm{ss}}$ and $(\pi_N^{\vee})^{\mathrm{ss}}$ are spanned by $\{\varphi_{N,\sigma} \mid \sigma \in \mathfrak{S}\}$ and $\{\varphi_{N,\sigma}^{\vee} \mid \sigma \in \mathfrak{S}\}$ in which $\varphi_{N,\sigma}$ and $\varphi_{N,\sigma}^{\vee}$ are nonzero eigenvectors of V_N^{ϖ} with eigenvalues

$$\prod_{m=1}^{N-1}\prod_{i=1}^{m} |\varpi|_F^{i-1}\mu_{\sigma(i)}(\varpi), \quad \prod_{m=1}^{N-1}\prod_{i=1}^{m} |\varpi|_F^{i-1}\check{\mu}_{\sigma(i)}(\varpi),$$

respectively. Now since $\langle \mathrm{V}_N^{\varpi}\ ,\ \rangle_{\pi_N} = \langle\ , \varpi^{1-N}\mathrm{V}_N^{\varpi}\ \rangle_{\pi_N}$, it follows that $\langle \varphi_{N,\sigma_2}^{\vee}, \varphi_{N,\sigma_1}\rangle_{\pi_N} = 0$ if $\sigma_1 \neq \sigma_2$. In particular, $\langle \varphi_{N,1}^{\vee}, \varphi_{N,1}\rangle_{\pi_N} \neq 0$, which proves the lemma. □

Lemma 4.8 *Suppose that Π_N is ordinary for some $N \in \{n, n+1\}$. The following are equivalence:*

(1) Π_N is semi-stably ordinary.
(2) The ordinary line is contained in $\pi_N^{I_N^{(r)}}$ for some positive integer r.
(3) The ordinary line is contained in $\pi_N^{I_N^{(1)}}$.

Proof The implication (3) ⇒ (2) is trivial.

For (2) ⇒ (1), let φ_N be an ordinary vector fixed by $I_N^{(r)}$. Then it is also the eigenvector of the operators $\mathrm{V}_{N,m}^{\varpi}$ in [24, Lemma 3.2] for $0 \leqslant m \leqslant N$ (with respect to a uniformizer ϖ) with the eigenvalue $\prod_{i=1}^{m} |\varpi|_F^{i-1}\mu_i(\varpi)$. Now since φ_N is fixed by $I_N^{(r)}$, the eigenvalues are independent of the choice of ϖ, which implies that $\mu_1, \mu_1\mu_2, \ldots, \mu_1\cdots\mu_N$ are all unramified characters, that is, Π_N is semi-stably ordinary.

For (1) ⇒ (3), note that the above discussion also shows (1) ⇒ (2), namely, the ordinary line is contained in $\pi_N^{I_N^{(r)}}$ for some $r \geqslant 1$. Then the statement follows from [24, Proposition 3.4] by noting that elements in $\pi_N^{I_N^{(r)}}$ that are annihilated by the Hecke polynomial of π_N are already contained in $\pi_N^{I_N^{(1)}}$. □

Notation 4.9 Suppose that Π is ordinary. Let μ and ν be the two tuples of characters for π_n and π_{n+1} in Definition 4.2, respectively. For every $x \in O_F \cap F^{\times}$, put

$$\omega^x(\pi) := \left(\prod_{m=1}^{n}\prod_{i=1}^{m} |x|_F^{i-1}\mu_i(x)\right)\left(\prod_{m=1}^{n}\prod_{i=1}^{m} |x|_F^{i-1}\nu_i(x)\right) \in \mathbb{L}^{\circ\times},$$

and similarly for $\omega^x(\pi^{\vee})$ (see Remark 4.3). When Π is semi-stably ordinary, we simply put $\omega(\pi) := \omega^{\varpi}(\pi)$ and $\omega(\pi^{\vee}) := \omega^{\varpi}(\pi^{\vee})$ for some hence every uniformizer ϖ of F, and put $\omega(\Pi) := \omega(\pi)\omega(\pi^{\vee})$.

Notation 4.10 We introduce two more notations.

(1) Put

$$\xi := \begin{pmatrix} & & 1 & 1 \\ & \iddots & & \vdots \\ 1 & & & 1 \\ 0 & \cdots & 0 & 1 \end{pmatrix} \in \mathrm{GL}_{n+1}(\mathbb{Z})$$

as the element introduced on the top of [23, Page 23].

(2) For $x \in O_F \cap F^\times$, put $[x] := x[x]_n \in \mathrm{GL}_n(F)$, regarded as an element of $H(F)$ (in particular, as an element of $G(F)$, $[x] = (x[x]_n, [x]_{n+1})$).

Lemma 4.11 *For every positive integer* $\mathfrak{c}$, *put*
$K_{\mathfrak{c},H} := ((1_n, \xi) \cdot [\varpi^{\mathfrak{c}}])I^{(1)}((1_n, \xi) \cdot [\varpi^{\mathfrak{c}}])^{-1} \cap H(F)$.
Then

(1) $\det(K_{\mathfrak{c},H}) = 1 + \mathfrak{p}^{\mathfrak{c}}$;
(2) $K_{\mathfrak{c}+1,H}$ *is contained in* $K_{\mathfrak{c},H}$ *of index* $q^{\frac{n(n+1)(2n+1)}{6}}$.

Proof It is clear that $K_{\mathfrak{c},H}$ does not depend on the choice of ϖ. Let $h = (h_{ij})_{1 \leqslant i,j \leqslant n} \in H(F)$ be an element written in the matrix form. Then a straightforward computation shows that the element $((1_n, \xi) \cdot [\varpi^{\mathfrak{c}}])^{-1}h((1_n, \xi) \cdot [\varpi^{\mathfrak{c}}])$ belongs to $I^{(1)}$ if and only if the following are satisfied:

- $h_{ij} \in \mathfrak{p}^{|i-j|\mathfrak{c}}$ for $i \neq j$;
- $h_{i1} + \cdots + h_{in} - 1 \in \mathfrak{p}^{i\mathfrak{c}}$ for $1 \leqslant i \leqslant n$.

Then (1) follows directly. For (2), the inclusion is clear; while for the index, we have

$$[K_{\mathfrak{c},H} : K_{\mathfrak{c}+1,H}] = q^{\sum_{i=1}^{n-1} i(n-i)} \cdot q^{\sum_{i=1}^{n-1} i(n-i)} \cdot q^{\sum_{i=1}^{n} i} = q^{\frac{n(n+1)(2n+1)}{6}}.$$

The lemma is proved. □

Proposition 4.12 *Suppose that* Π *is semi-stably ordinary. There exists a unique element* $\gamma(\Pi) \in \mathbb{L}^\times$ *depending only on* Π *such that for every pair of ordinary vectors* $\varphi \in \pi, \varphi^\vee \in \pi^\vee$, *every positive integer* $\mathfrak{c}$, *and every finite character* $\chi \colon E^{\times -} \to \mathbb{L}_\chi^\times$ *of conductor* $\mathfrak{p}^{\mathfrak{c}}$,[4]

$$\frac{\gamma(\Pi)}{\langle \varphi^\vee, \varphi \rangle_\pi} \cdot \alpha^\chi \left(\pi((1_n, \xi) \cdot [\varpi^{\mathfrak{c}}])\varphi, \pi^\vee((1_n, \xi) \cdot [\varpi^{\mathfrak{c}}])\varphi^\vee \right)$$
$$= \left(q^{-\frac{n(n+1)(2n+1)}{6}} \right)^{\mathfrak{c}} \frac{\Delta_{n+1}}{L(1, \Pi_n, \mathrm{As}^{(-1)^n}) L(1, \Pi_n + 1, \mathrm{As}^{(-1)^{n+1}})},$$

where we adopt the Haar measure on $H(F)$ *that gives a hyperspecial maximal subgroup volume* 1.

---

[4] We adopt the convention that when $q = 2$, there is no character of conductor $\mathfrak{p}$. In particular, χ must be ramified here.

Proof Choose an embedding $\iota\colon \mathbb{L} \to \mathbb{C}$ and an additive character $\psi_F\colon F \to \mathbb{C}^\times$ of conductor O_F. Let $\psi\colon U(F) \to \mathbb{C}^\times$ be the character sending $((u_{ij})_{1\leqslant i,j\leqslant n}, (v_{ij})_{1\leqslant i,j\leqslant n+1}) \in U_n(F) \times U_{n+1}(F) = U(F)$ to $\psi_F(-u_{12} - \cdots - u_{n-1n} + v_{12} + \cdots + v_{nn+1})$, which is a generic character trivial on $(U \cap H)(F)$. Choose $G(F)$-equivariant isomorphisms $i_1\colon \pi \otimes_{\mathbb{L},\iota} \mathbb{C} \xrightarrow{\sim} \mathcal{W}(\iota\pi)_\psi$ and $i_2 \in \pi^\vee \otimes_{\mathbb{L},\iota} \mathbb{C} \xrightarrow{\sim} \mathcal{W}(\iota\pi^\vee)_{\psi^{-1}}$ such that $\iota\langle\varphi^\vee, \varphi\rangle_\pi = \vartheta(i_2\varphi^\vee, i_1\varphi)$ holds for all $\varphi \in \pi$ and $\varphi^\vee \in \pi^\vee$. By Lemma 2.10, there exists a constant $c \in \mathbb{Q}^\times$ depending only on certain rational Haar measures, such that for every finite character $\chi\colon E^{-\times} \to \mathbb{C}^\times$,

$$\alpha^\chi(\varphi, \varphi^\vee) = c\left(\int_{(U\cap H)(F)\backslash H(F)} (i_1\varphi)(h)\cdot\chi(\det h)\mathrm{d}h\right)\left(\int_{(U\cap H)(F)\backslash H(F)} (i_2\varphi^\vee)(h)\cdot\chi(\det h)^{-1}\mathrm{d}h\right)$$

for every $\varphi \in \pi$ and $\varphi^\vee \in \pi^\vee$. Apply this formula to $\pi((1_n, \xi)\cdot[\varpi^\mathfrak{c}])\varphi$ and $\pi^\vee((1_n, \xi)\cdot[\varpi^\mathfrak{c}])\varphi^\vee$ with φ and $\varphi^\vee$ ordinary vectors when χ has conductor $\mathfrak{p}^\mathfrak{c}$. By [23, Corollary 2.8], we obtain

$$\begin{aligned}&\alpha^\chi\left(\pi((1_n, \xi)\cdot[\varpi^\mathfrak{c}])\varphi, \pi^\vee((1_n, \xi)\cdot[\varpi^\mathfrak{c}])\varphi^\vee\right)\\ &= c\prod_{i=1}^{n}(1-q^{-i})^{-2}\cdot\left(q^{-\frac{n(n+1)(n+2)}{3}}\right)^\mathfrak{c}\cdot G_{\psi_F}(\chi)^{\frac{n(n+1)}{2}}G_{\psi_F^{-1}}(\chi^{-1})^{\frac{n(n+1)}{2}}\cdot(i_1\varphi)(1)\cdot(i_2\varphi^\vee)(1),\end{aligned}\tag{2}$$

where $G_{\psi_F^{\pm 1}}(\chi^{\pm 1})$ denotes the Gauss sums (the one in [24, Sect. 2]) of $\chi^{\pm 1}$ with respect to $\psi_F^{\pm 1}$. As it is well-known that $G_{\psi_F}(\chi)G_{\psi_F^{-1}}(\chi^{-1}) = q^\mathfrak{c}$, we have

$$(2) = c\prod_{i=1}^{n}(1-q^{-i})^{-2}\cdot\left(q^{-\frac{n(n+1)(2n+1)}{6}}\right)^\mathfrak{c}\cdot(i_1\varphi)(1)\cdot(i_2\varphi^\vee)(1).$$

By [27, Proposition 4.12], we have $(i_1\varphi)(1)\cdot(i_2\varphi^\vee)(1) \neq 0$. Put

$$c' := \frac{\prod_{i=1}^{n}(1-q^{-i})^2\cdot\langle\varphi^\vee, \varphi\rangle_\pi}{c\cdot(i_1\varphi)(1)\cdot(i_2\varphi^\vee)(1)},$$

which belongs to $\mathbb{C}^\times$ by Lemma 4.7 and depends only on Π and certain Haar measures. Then we have

$$\frac{c'}{\langle\varphi^\vee, \varphi\rangle_\pi}\cdot\alpha^\chi\left(\pi((1_n, \xi)\cdot[\varpi^\mathfrak{c}])\varphi, \pi^\vee((1_n, \xi)\cdot[\varpi^\mathfrak{c}])\varphi^\vee\right) = \left(q^{-\frac{n(n+1)(2n+1)}{6}}\right)^\mathfrak{c}$$

for every $\mathfrak{c} \geqslant 1$ and every finite character $\chi\colon E^{-\times} \to \mathbb{C}^{\times}$ of conductor $\mathfrak{p}^{\mathfrak{c}}$. Now it is clear that c' depends only on Π. Finally, we may choose a ramified character χ that takes values in $\iota\mathbb{L}^{\times}$, which implies that $c' \in \iota\mathbb{L}^{\times}$. The proposition follows by taking

$$\gamma(\Pi) := (\iota^{-1}c') \frac{\Delta_{n+1}}{L(1, \Pi_n, \mathrm{As}^{(-1)^n})L(1, \Pi_{n+1}, \mathrm{As}^{(-1)^{n+1}})} \in \mathbb{L}^{\times}. \qquad \square$$

5 Coherent Anticyclotomic p-Adic L-Function

In this section, we take $\mathtt{P}$ to be a subset of

$$\mathtt{P}(\Pi) := \left\{ v \in \mathtt{V}_F^{(p)} \cap \mathtt{V}_F^{\mathrm{spl}} \,\middle|\, \Pi_v \text{ is semi-stably ordinary} \right\}. \tag{3}$$

We study the case where $\Pi = \Pi_n \boxtimes \Pi_{n+1}$ is coherent and aim to construct a p-adic L-function $\mathscr{L}^0_{\mathcal{E}}(\Pi)$ for every $\mathtt{P}$-extension $\mathcal{E}/E$ (Definition 1.1). In the coherent case, there exists a totally positive definite hermitian space V_n over E of rank n, unique up to isomorphism, such that $V_{n,v}$ is the prescribed hermitian space from Sect. 2 for every $v \in \mathtt{V}_F^{\mathrm{fin}}$. Put $V_{n+1} := V_n \oplus E \cdot \mathrm{e}$. Put

$$G := \mathrm{U}(V_n) \times \mathrm{U}(V_{n+1}), \quad \pi := \otimes_{v \in \mathtt{V}_F^{\mathrm{fin}}} \pi_v$$

which is an irreducible admissible representation of $G(\mathbb{A}_F^{\infty})$ with coefficients in $\mathbb{L}$.

Put

$$\mathbb{V}_{\Pi} := \mathrm{Hom}_{G(\mathbb{A}_F^{\infty})}\left(\pi, \mathcal{S}\left(G(F)\backslash G(\mathbb{A}_F^{\infty}), \mathbb{L}\right)\right), \quad \mathbb{V}_{\Pi^{\vee}} := \mathrm{Hom}_{G(\mathbb{A}_F^{\infty})}\left(\pi^{\vee}, \mathcal{S}\left(G(F)\backslash G(\mathbb{A}_F^{\infty}), \mathbb{L}\right)\right).$$

Define $\mathscr{L}^0_{\mathcal{E}}(\Pi) = 0$ when $\mathbb{V}_{\Pi} = 0$.

Now we assume $\mathbb{V}_{\Pi} \neq 0$ hence $\mathbb{V}_{\Pi^{\vee}} \neq 0$. Then by Arthur's multiplicity formula [26, 34], we have $\dim_{\mathbb{L}} \mathbb{V}_{\Pi} = \dim_{\mathbb{L}} \mathbb{V}_{\Pi^{\vee}} = 1$. Denote by $\mathcal{V}, \mathcal{V}^{\vee} \subseteq \mathcal{S}\left(G(F)\backslash G(\mathbb{A}_F^{\infty}), \mathbb{L}\right)$ the unique irreducible $\mathbb{L}[G(\mathbb{A}_F^{\infty})]$-submodules that are isomorphic to π and $\pi^{\vee}$, respectively. Using the Petersson inner product with respect to the Tamagawa measure of $G(\mathbb{A}_F)$, we have a canonical isomorphism $\pi \otimes_{\mathbb{L}} \pi^{\vee} \simeq \mathcal{V} \otimes_{\mathbb{L}} \mathcal{V}^{\vee}$.

Denote by $H \subseteq G$ the graph of the natural embedding $\mathrm{U}(V_n) \hookrightarrow \mathrm{U}(V_{n+1})$, and fix a decomposition $\mathrm{d}h = \mathrm{d}h_{\infty} \cdot \mathrm{d}h_{\mathtt{P}} \cdot \mathrm{d}h^{\infty,\mathtt{P}}$ of the Tamagawa measure on $H(\mathbb{A}_F)$ such that the volume of $H(F_{\infty})$ under $\mathrm{d}h_{\infty}$ is 1 and the volume of every hyperspecial maximal subgroup of $H(F_{\mathtt{P}})$ under $\mathrm{d}h_{\mathtt{P}}$ is 1 (so that the measure $\mathrm{d}h^{\infty,\mathtt{P}}$ is rational).

For every $v \in \mathtt{P}$, fix a standard isomorphism $G_v \simeq \mathrm{GL}_{n,F_v} \times \mathrm{GL}_{n+1,F_v}$ (Definition 4.1). For every element $\varphi = \otimes_{v \in \mathtt{V}_F^{\mathrm{fin}}} \varphi_v \in \mathcal{V}$ with $\varphi_v \in \pi_v$ that is a (nonzero) ordinary vector for $v \in \mathtt{P}$, we define a measure $\mathscr{P}_{\varphi} \in \mathbb{L}[[\Gamma_{\mathtt{P}}]]^{\circ}$ as follows: For every subset Ω of $\Gamma_{\mathtt{P}}$, we denote by H_{Ω} its inverse image under the determinant map

$H(F)\backslash H(\mathbb{A}_F^\infty) \to \Gamma_{\mathrm{P}}$. For a $\mathfrak{c}$-cell Ω of Γ_{P} for some tuple $\mathfrak{c} = (\mathfrak{c}_v)_{v\in\mathrm{P}}$ of positive integers, we put

$$\mathscr{P}_\varphi(\Omega) := \prod_{v\in\mathrm{P}} \left(\frac{q_v^{\frac{n(n+1)(2n+1)}{6}}}{\omega(\pi_v)}\right)^{\mathfrak{c}_v} \cdot \int_{H_\Omega} \varphi\left(h^\infty \cdot (1_n, \xi) \cdot [\varpi^{\mathfrak{c}}]\right) \mathrm{d}h^\infty,$$

in which $(1_n, \xi) \in \mathrm{GL}_n(F_{\mathrm{P}}) \times \mathrm{GL}_{n+1}(F_{\mathrm{P}}) = G(F_{\mathrm{P}})$ and $[\varpi^{\mathfrak{c}}] \in \mathrm{GL}_n(F_{\mathrm{P}}) = H(F_{\mathrm{P}})$. It is clear that the above integral is a finite sum of elements in $\mathbb{L}$ and is independent of the choice of ϖ by Lemma 4.8.

Proposition 5.1 *The assignment $\Omega \mapsto \mathscr{P}_\varphi(\Omega)$ defines a measure $\mathscr{P}_\varphi$ on Γ_{P} valued in $\mathbb{L}$; in other words, $\mathscr{P}_\varphi$ is an element of $\mathbb{L}[[\Gamma_{\mathrm{P}}]]^\circ$.*

Proof The additivity of $\mathscr{P}_\varphi$ follows from the same argument in the proof of [23, Theorem 4.4] (with a translation by $[\varpi^{\mathfrak{c}}]$). For the boundedness, Lemma 4.11(2) implies that there exists a positive integer C such that for every tuple $\mathfrak{c} = (\mathfrak{c}_v)_{v\in\mathrm{P}}$ of positive integers and every $\mathfrak{c}$-cell Ω, the volume of H_Ω is bounded by

$$C \cdot \prod_{v\in\mathrm{P}} \left(q_v^{-\frac{n(n+1)(2n+1)}{6}}\right)^{\mathfrak{c}_v}.$$

Thus, the boundedness of $\mathscr{P}_\varphi$ follows from the fact that $\omega(\pi_v) \in \mathbb{L}^{\circ\times}$ for $v \in \mathrm{P}$ and that φ is bounded. □

Theorem 5.2 *There exists a unique measure $\mathscr{L}^0_{\mathcal{E}_{\mathrm{P}}}(\Pi) \in \mathbb{L}[[\Gamma_{\mathrm{P}}]]^\circ$ such that for every finite character $\chi\colon \Gamma_{\mathrm{P}} \to \mathbb{L}_\chi^\times$ of conductor $\prod \mathfrak{p}_v^{\mathfrak{c}_v}$ for a tuple $\mathfrak{c} = (\mathfrak{c}_v)_{v\in\mathrm{P}}$ of positive integers indexed by P and every embedding $\iota\colon \mathbb{L}_\chi \to \mathbb{C}$, we have*

$$\iota\mathscr{L}^0_{\mathcal{E}_{\mathrm{P}}}(\Pi)(\chi) = \prod_{v\in\mathrm{P}} \left(\frac{q_v^{\frac{n(n+1)(2n+1)}{6}}}{\iota\omega(\Pi_v)}\right)^{\mathfrak{c}_v} \cdot \frac{\Delta_{n+1} \cdot L(\frac{1}{2}, \iota(\Pi_n \otimes \widetilde{\chi}) \times \iota\Pi_{n+1})}{2^{d(\Pi_n)+d(\Pi_{n+1})} \cdot L(1, \iota\Pi_n, \mathrm{As}^{(-1)^n}) L(1, \iota\Pi_{n+1}, \mathrm{As}^{(-1)^{n+1}})}.$$

Here, $d(\Pi_N)$ for $N = n, n+1$ is introduced in Remark 2.2(4).[5]

The theorem holds even when $\mathbb{V}_\Pi = 0$ since in this case $L(\frac{1}{2}, \iota(\Pi_n \otimes \widetilde{\chi}) \times \iota\Pi_{n+1})$ always vanishes; see Remark 5.7.

Proof The uniqueness is clear. Now we show the existence. Since Γ_{P} is torsion free, $\chi_v = 1$ for every finite order character χ as in the theorem and every $v \in \mathrm{V}_F^{\mathrm{fin}} \setminus \mathrm{V}_F^{\mathrm{spl}}$.

Choose a pair $\varphi = \otimes_{v\in\mathrm{V}_F^{\mathrm{fin}}}\varphi_v \in \mathcal{V}$ and $\varphi^\vee = \otimes_{v\in\mathrm{V}_F^{\mathrm{fin}}}\varphi_v^\vee \in \mathcal{V}^\vee$ with $\varphi_v \in \pi_v$ and $\varphi_v^\vee \in \pi_v^\vee$ satisfying

[5] The evaluation of $\mathscr{L}^0_{\mathcal{E}_{\mathrm{P}}}(\Pi)$ at general finite order characters has recently been carried out in [29].

(T1) for $v \in \mathtt{P}$, both φ_v and $\varphi_v^\vee$ are (nonzero) ordinary vectors;
(T2) for $v \in \mathtt{V}_F^{\mathrm{spl}} \setminus \mathtt{P}$, the pair $(\varphi_v, \varphi_v^\vee)$ satisfies the conclusion of Proposition 2.11;
(T3) for $v \in \mathtt{V}_F^{\mathrm{fin}} \setminus \mathtt{V}_F^{\mathrm{spl}}$, $\alpha(\varphi_v, \varphi_v^\vee) \neq 0$.

By Proposition 2.11 and Lemma 2.9, such choice is possible. Put

$$\mathscr{L}^0_{\mathcal{E}_{\mathtt{P}}}(\Pi) := \prod_{v\in\mathtt{P}} \frac{\gamma(\Pi_v)}{\langle\varphi_v^\vee, \varphi_v\rangle_{\pi_v}} \cdot \prod_{v\in\mathtt{V}_F^{\mathrm{fin}}\setminus\mathtt{V}_F^{\mathrm{spl}}} \frac{\alpha(\varphi_v,\varphi_v^\vee)^{-1}\cdot\Delta_{n+1,v}\cdot L(\frac{1}{2},\Pi_{n,v}\times\Pi_{n+1,v})}{L(1,\Pi_{n,v},\mathrm{As}^{(-1)^n})L(1,\Pi_{n+1,v},\mathrm{As}^{(-1)^{n+1}})} \cdot \mathscr{P}_\varphi \cdot (\mathscr{P}_{\varphi^\vee})^\dagger, \quad (4)$$

where $\gamma(\Pi_v) \in \mathbb{L}^\times$ is the constant in Proposition 4.12 and $\dagger$ is the involution introduced in Notation 3.1. Note that by Lemmas 2.9, 4.7, and (T3), the above expression makes sense.

To show the interpolation property, we suppress ι from the notation. In particular, all representations have coefficients in $\mathbb{C}$. By (4), for χ as in the theorem, we have

$$\begin{aligned}\mathscr{L}^0_{\mathcal{E}_{\mathtt{P}}}(\Pi)(\chi) = &\prod_{v\in\mathtt{P}} \frac{\gamma(\Pi_v)}{\langle\varphi_v^\vee, \varphi_v\rangle_{\pi_v}} \left(\frac{q_v^{\frac{n(n+1)(2n+1)}{3}}}{\omega(\Pi_v)}\right)^{c_v} \cdot \prod_{v\in\mathtt{V}_F^{\mathrm{fin}}\setminus\mathtt{V}_F^{\mathrm{spl}}} \frac{\alpha(\varphi_v,\varphi_v^\vee)^{-1}\cdot\Delta_{n+1,v}\cdot L(\frac{1}{2},\Pi_{n,v}\times\Pi_{n+1,v})}{L(1,\Pi_{n,v},\mathrm{As}^{(-1)^n})L(1,\Pi_{n+1,v},\mathrm{As}^{(-1)^{n+1}})}\\ &\times \int_{H(F)\backslash H(\mathbb{A}_F^\infty)} \varphi\left(h^\infty\cdot(1_n,\xi)\cdot[\varpi^{\mathfrak{c}}]\right)\chi(\det h^\infty)\mathrm{d}h^\infty\\ &\times \int_{H(F)\backslash H(\mathbb{A}_F^\infty)} \varphi^\vee\left(h^\infty\cdot(1_n,\xi)\cdot[\varpi^{\mathfrak{c}}]\right)\chi(\det h^\infty)^{-1}\mathrm{d}h^\infty.\end{aligned}$$

Now we apply the refined Gan–Gross–Prasad conjecture (that is, the Ichino–Ikeda conjecture), which has been fully proved for G in [6, 7]. Let $\mathtt{S}$ be a subset of $\mathtt{V}_F^{\mathrm{fin}}$ containing $\mathtt{P}$ such that for $v \in \mathtt{V}_F^{\mathrm{fin}} \setminus \mathtt{S}$, we have

$$\alpha^{\chi_v}(\varphi_v,\varphi_v^\vee) = \frac{\Delta_{n+1,v}\cdot L(\frac{1}{2},(\Pi_{n,v}\otimes\widetilde{\chi_v})\times\Pi_{n+1,v})}{L(1,\Pi_{n,v},\mathrm{As}^{(-1)^n})L(1,\Pi_{n+1,v},\mathrm{As}^{(-1)^{n+1}})}.$$

By the refined formula, we have

$$\begin{aligned}&\int_{H(F)\backslash H(\mathbb{A}_F^\infty)} \varphi\left(h^\infty\cdot(1_n,\xi)\cdot[\varpi^{\mathfrak{c}}]\right)\chi(\det h^\infty)\mathrm{d}h^\infty\\ &\times \int_{H(F)\backslash H(\mathbb{A}_F^\infty)} \varphi^\vee\left(h^\infty\cdot(1_n,\xi)\cdot[\varpi^{\mathfrak{c}}]\right)\chi(\det h^\infty)^{-1}\mathrm{d}h^\infty\\ &= \frac{1}{2^{d(\Pi_n)+d(\Pi_{n+1})}} \frac{\Delta_{n+1}^{\mathtt{S}}\cdot L(\frac{1}{2},(\Pi_n\otimes\widetilde{\chi})^{\mathtt{S}}\times\Pi_{n+1}^{\mathtt{S}})}{L(1,\Pi_n^{\mathtt{S}},\mathrm{As}^{(-1)^n})L(1,\Pi_{n+1}^{\mathtt{S}},\mathrm{As}^{(-1)^{n+1}})} \prod_{v\in\mathtt{S}} \alpha^{\chi_v}(\varphi_v,\varphi_v^\vee).\end{aligned}$$

Plugging in and by (T2), we have

$$\mathscr{L}^0_{\mathcal{E}_{\mathrm{P}}}(\Pi)(\chi) = \frac{1}{2^{d(\Pi_n)+d(\Pi_{n+1})}} \frac{\Delta^{\mathrm{P}}_{n+1} \cdot L(\frac{1}{2}, (\Pi_n \otimes \widetilde{\chi})^{\mathrm{P}} \times \Pi^{\mathrm{P}}_{n+1})}{L(1, \Pi^{\mathrm{P}}_n, \mathrm{As}^{(-1)^n}) L(1, \Pi^{\mathrm{P}}_{n+1}, \mathrm{As}^{(-1)^{n+1}})} \prod_{v \in \mathrm{P}} \frac{\gamma(\Pi_v)}{\langle \varphi^\vee_v, \varphi_v \rangle_{\pi_v}} \left(\frac{q_v^{\frac{n(n+1)(2n+1)}{3}}}{\omega(\Pi_v)} \right)^{\mathfrak{c}_v} \alpha^{\chi_v}(\varphi_v, \varphi^\vee_v).$$

Finally, by (T1) and Proposition 4.12, we have

$$\mathscr{L}^0_{\mathcal{E}_{\mathrm{P}}}(\Pi)(\chi) = \frac{1}{2^{d(\Pi_n)+d(\Pi_{n+1})}} \frac{\Delta_{n+1} \cdot L(\frac{1}{2}, (\Pi_n \otimes \widetilde{\chi})^{\mathrm{P}} \times \Pi^{\mathrm{P}}_{n+1})}{L(1, \Pi_n, \mathrm{As}^{(-1)^n}) L(1, \Pi_{n+1}, \mathrm{As}^{(-1)^{n+1}})} \prod_{v \in \mathrm{P}} \left(\frac{q_v^{\frac{n(n+1)(2n+1)}{6}}}{\omega(\Pi_v)} \right)^{\mathfrak{c}_v}.$$

The interpolation formula follows as $L(\frac{1}{2}, (\Pi_{n,v} \otimes \widetilde{\chi_v}) \times \Pi_{n+1,v}) = 1$ (since χ_v is ramified) for every $v \in \mathrm{P}$. □

We have the following corollary on the functional equation satisfied by $\mathscr{L}^0_{\mathcal{E}_{\mathrm{P}}}(\Pi)$.

Corollary 5.3 *We have*

$$\mathscr{L}^0_{\mathcal{E}_{\mathrm{P}}}(\Pi)^\dagger = \mathscr{L}^0_{\mathcal{E}_{\mathrm{P}}}(\Pi^\vee).$$

Proof It suffices to show that for every finite character $\chi\colon \Gamma_{\mathrm{P}} \to \mathbb{L}^\times_\chi$ of conductor $\prod \mathfrak{p}_v^{\mathfrak{c}_v}$ for a tuple $\mathfrak{c} = (\mathfrak{c}_v)_{v \in \mathrm{P}}$ of positive integers indexed by P and every embedding $\iota\colon \mathbb{L}_\chi \to \mathbb{C}$, we have $\iota\mathscr{L}^0_{\mathcal{E}_{\mathrm{P}}}(\Pi)(\chi) = \iota\mathscr{L}^0_{\mathcal{E}_{\mathrm{P}}}(\Pi^\vee)(\chi^{-1})$. By Theorem 5.2, this follows from that $\omega(\Pi_v) = \omega(\Pi^\vee_v)$ for $v \in \mathrm{P}$, that $d(\Pi_N) = d(\Pi^\vee_N)$ for $N = n, n+1$, that $L(1, \iota\Pi_N, \mathrm{As}^{(-1)^N}) = L(1, \iota\Pi^\vee_N, \mathrm{As}^{(-1)^N})$ for $N = n, n+1$, and the classical functional equation

$$L(\tfrac{1}{2}, \iota(\Pi_n \otimes \widetilde{\chi}) \times \iota\Pi_{n+1}) = L(\tfrac{1}{2}, \iota(\Pi^\vee_n \otimes \widetilde{\chi}^{-1}) \times \iota\Pi^\vee_{n+1}).$$ □

Notation 5.4 For every P-extension $\mathcal{E}/E$ (Definition 1.1), we denote by $\mathscr{L}^0_{\mathcal{E}}(\Pi)$ the image of $\mathscr{L}^0_{\mathcal{E}_{\mathrm{P}}}(\Pi)$ under the natural homomorphism $\mathbb{L}[[\Gamma_{\mathrm{P}}]]^\circ \to \mathbb{L}[[\mathrm{Gal}(\mathcal{E}/E)]]^\circ$.

Remark 5.5 We warn the readers that for a subset P' of P, $\mathscr{L}^0_{\mathcal{E}_{\mathrm{P}'}}(\Pi)$ is generally different from the image of $\mathscr{L}^0_{\mathcal{E}_{\mathrm{P}}}(\Pi)$ under the natural homomorphism $\mathbb{L}[[\Gamma_{\mathrm{P}}]]^\circ \to \mathbb{L}[[\Gamma_{\mathrm{P}'}]]^\circ$.

We propose the following nonvanishing conjecture.

Conjecture 5.6 Suppose that $\mathbb{V}_\Pi \neq 0$ and condition (∗) in Notation 2.4 holds. Then $\mathscr{L}^0_{\mathcal{E}}(\Pi) \neq 0$ for every P-extension $\mathcal{E}/E$ that contains $\mathcal{E}_{\mathrm{P}'}$ for some nonempty subset P' of P.

The conjecture is known for $n = 1$ [12].

Remark 5.7 Take an arbitrary embedding $\iota\colon \mathbb{L} \to \mathbb{C}$. Then $\epsilon(\frac{1}{2}, \Pi_n^{(\iota)} \times \Pi_{n+1}^{(\iota)})$ equals $\epsilon(\Pi)$, which we have assumed to be 1. Now $\epsilon(\frac{1}{2}, \Pi_n^{(\iota)} \times \Pi_{n+1}^{(\iota)})$ decomposes as the product of $d(\Pi_n)d(\Pi_{n+1})$ root numbers (valued in $\{\pm 1\}$) for their isobaric factors. Then by the discussion in [16, Sect. 26], $\mathbb{V}_\Pi \neq 0$ if and only if all those root numbers are equal to 1.

The proposition below will be used in [30], which can be skipped for now. Denote by $\mathtt{V}_F^\Pi$ the finite subset of $\mathtt{V}_F^{\mathrm{fin}}$ containing $\mathtt{V}_F^{\mathrm{ram}}$ away from which Π is unramified. For every $v \in \mathtt{V}_F^{\mathrm{fin}} \setminus (\mathtt{V}_F^\Pi \cup \mathtt{P})$, choose an integrally self-dual lattice $L_{n,v}$ of $V_{n,v}$ and put $K_v := K_{n,v} \times K_{n+1,v}$, where $K_{n,v}$ and $K_{n+1,v}$ are the stabilizers of $L_{n,v}$ and $L_{n,v} \oplus O_{E_v} \cdot \mathrm{e}$, respectively.

Proposition 5.8 *Suppose further that* $\mathtt{P} \cap \mathtt{V}_F^\Pi = \emptyset$. *There exists an element* $\varphi = \otimes_{v \in \mathtt{V}_F^{\mathrm{fin}}} \varphi_v \in \mathcal{V}$ *with* $\varphi_v \in \pi_v$ *an ordinary vector for* $v \in \mathtt{P}$ *and* $\varphi_v \in \pi_v^{K_v}$ *for* $v \in \mathtt{V}_F^{\mathrm{fin}} \setminus (\mathtt{V}_F^\Pi \cup \mathtt{P})$, *such that the ideals of* $\mathbb{L}[[\Gamma_\mathtt{P}]]^\circ$ *generated by* $\mathscr{L}^0_{\mathcal{E}_\mathtt{P}}(\Pi)$ *and* $\mathscr{P}^2_\varphi$ *are the same.*

Proof We fix an MVW involution $*$ on G [33] that stabilizes H. For every $v \in \mathtt{V}_F^{\mathrm{spl}}$, we

- fix a standard isomorphism $G_v \simeq \mathrm{GL}_{n,F_v} \times \mathrm{GL}_{n+1,F_v}$ in the sense of Definition 4.1, under which $*$ coincides with the transpose-inverse;
- denote by γ_v the image of any uniformizer of F_v in $\Gamma_\mathtt{P}$ with respect to the isomorphism $F_v^\times \simeq E_v^{\times -}$ induced by the isomorphism above;
- fix an additive character $\psi_v\colon F_v \to \mathbb{C}^\times$ of conductor O_{F_v};
- denote by $\mathfrak{c}(\pi_v) \geqslant 0$ the Rankin–Selberg conductor of π_v, that is, the one satisfying that for every embedding $\iota\colon \mathbb{L} \to \mathbb{C}$, $q_v^{\mathfrak{c}(\pi_v)s} \cdot \epsilon(s, \iota\pi_{n,v} \times \iota\pi_{n+1,v}, \psi_v)$ is a constant; it equals 0 for $v \notin \mathtt{V}_F^\Pi$.

Put $\gamma := \prod_{v \in \mathtt{V}_F^{\mathrm{spl}}} \gamma_v^{\mathfrak{c}(\pi_v)} \in \Gamma_\mathtt{P}$. We claim that there exists φ as in the proposition such that

$$\mathscr{P}^2_\varphi = C \cdot \gamma \cdot \mathscr{L}^0_{\mathcal{E}_\mathtt{P}}(\Pi) \tag{5}$$

for some constant $C \in \mathbb{L}^\times$.

To check (5), we may fix an embedding $\mathbb{L} \hookrightarrow \mathbb{C}$ and show that for every finite characters $\chi\colon \Gamma_\mathtt{P} \to \mathbb{C}^\times$ that is *ramified* everywhere in $\mathtt{P}$,

$$\mathscr{P}_\varphi(\chi)^2 = C \cdot \chi(\gamma) \cdot \mathscr{L}^0_{\mathcal{E}_\mathtt{P}}(\Pi)(\chi) \tag{6}$$

holds for some constant $C \in \mathbb{C}^\times$. For every $v \in \mathtt{V}_F^{\mathrm{spl}}$, we fix Whittaker models $\pi_v \otimes_\mathbb{L} \mathbb{C} \simeq \mathcal{W}(\pi_v)_{\psi_v}$ and $\pi_v^\vee \otimes_\mathbb{L} \mathbb{C} \simeq \mathcal{W}(\pi_v^\vee)_{\psi_v^{-1}}$, and will denote by W_- the corresponding Whittaker function. Note that for every $W \in \mathcal{W}(\pi_v)_{\psi_v}$, we have $\widetilde{W} \in \mathcal{W}(\pi_v^\vee)_{\psi_v^{-1}}$, where $\widetilde{W}$ satisfies $\widetilde{W}(g) = W(w\,{}^\mathrm{t}g^{-1})$ in which w denotes the anti-diagonal Weyl element in $\mathrm{GL}_n(F_v) \times \mathrm{GL}_{n+1}(F_v)$.

For every element $\varphi = \otimes_v \varphi_v \in \mathcal{V}$, we have the function φ^* defined by the formula $\varphi^*(g) = \varphi(g^*)$. Then $\varphi^* = \otimes_v \varphi_v^*$ is again decomposable and belongs to $\mathcal{V}^\vee$. Now we may choose an element $\varphi = \otimes_v \varphi_v$ as in the proposition satisfying

(1) for every $v \in \mathrm{V}_F^{\mathrm{spl}} \setminus \mathrm{P}$, there exists a constant $c_v \in \mathbb{C}^\times$ (which equals 1 if $v \notin \mathrm{V}_F^{\Pi}$) such that

$$\int_{U_n(F_v)\backslash \mathrm{GL}_n(F_v)} W_{\varphi_v}(h) \cdot |\det h|_F^s \mathrm{d}h = c_v \cdot L(\tfrac{1}{2} + s, \pi_{n,v} \times \pi_{n+1,v})$$

holds as meromorphic functions in the variable $s \in \mathbb{C}$, where U_n denotes the upper-triangular unipotent subgroup of GL_n regarded as the diagonal subgroup of $\mathrm{GL}_n \times \mathrm{GL}_{n+1}$.

(2) for every $v \in \mathrm{V}_F^{\mathrm{fin}} \setminus \mathrm{V}_F^{\mathrm{spl}}$, $\alpha(\varphi_v, \varphi_v^*) \neq 0$.

For every $v \in \mathrm{V}_F^{\mathrm{spl}}$, there exists a constant $c'_v \in \mathbb{C}^\times$ such that $W_{\varphi_v^*} = c'_v \widetilde{W_{\varphi_v}}$.

Now we compute $\mathscr{P}_\varphi(\chi)^2$ for a finite character $\chi\colon \Gamma_\mathrm{P} \to \mathbb{C}^\times$ of conductor $\prod \mathfrak{p}_v^{\mathfrak{c}_v}$ for a tuple $\mathfrak{c} = (\mathfrak{c}_v)_{v\in\mathrm{P}}$ of positive integers indexed by P. We have

$$\begin{aligned}\mathscr{P}_\varphi(\chi)^2 &= \prod_{v\in\mathrm{P}} \left(\frac{q_v^{\frac{n(n+1)(2n+1)}{6}}}{\omega(\pi_v)}\right)^{2\mathfrak{c}_v} \cdot \left(\int_{H(F)\backslash H(\mathbb{A}_F^\infty)} \varphi_\mathfrak{c}(h^\infty)\chi(\det h^\infty)\mathrm{d}h^\infty\right)^2 \\ &= \prod_{v\in\mathrm{P}} \left(\frac{q_v^{\frac{n(n+1)(2n+1)}{6}}}{\omega(\pi_v)}\right)^{2\mathfrak{c}_v} \cdot \int_{H(F)\backslash H(\mathbb{A}_F^\infty)} \varphi_\mathfrak{c}(h^\infty)\chi(\det h^\infty)\mathrm{d}h^\infty \int_{H(F)\backslash H(\mathbb{A}_F^\infty)} (\varphi_\mathfrak{c})^*(h^\infty)\chi^{-1}(\det h^\infty)\mathrm{d}h^\infty,\end{aligned}$$

where $\varphi_\mathfrak{c} := \pi((1_n, \xi) \cdot [\varpi^\mathfrak{c}])\varphi$. Then by property (2) in the choice of φ, the refined Gan–Gross–Prasad formula, and the computation in Proposition 2.11, there exists a constant $C \in \mathbb{C}^\times$, independent of χ, such that

$$\begin{aligned}\mathscr{P}_\varphi(\chi)^2 = C \cdot &\prod_{v\in\mathrm{P}} \left(\frac{q_v^{\frac{n(n+1)(2n+1)}{6}}}{\omega(\pi_v)}\right)^{2\mathfrak{c}_v} \cdot L(\tfrac{1}{2}, (\Pi_n \otimes \widetilde{\chi}) \times \Pi_{n+1}) \\ &\times \prod_{v\in\mathrm{V}_F^{\mathrm{spl}}} \frac{\int_{U_n(F_v)\backslash \mathrm{GL}_n(F_v)} W_{\varphi_{\mathfrak{c},v}}(h)\cdot\chi_v(\det h)\mathrm{d}h}{L(\frac{1}{2}, (\pi_{n,v}\otimes\chi_v)\times\pi_{n+1,v})} \frac{\int_{U_n(F_v)\backslash \mathrm{GL}_n(F_v)} \widetilde{W_{\varphi_{\mathfrak{c},v}}}(h)\cdot\chi_v^{-1}(\det h)\mathrm{d}h}{L(\frac{1}{2}, (\pi_{n,v}^\vee\otimes\chi_v^{-1})\times\pi_{n+1,v}^\vee)}.\end{aligned}$$

Now for $v \in \mathrm{V}_F^{\mathrm{spl}} \setminus \mathrm{P}$, the functional equation of the local Rankin–Selberg integrals implies that

$$\begin{aligned}&\frac{\int_{U_n(F_v)\backslash \mathrm{GL}_n(F_v)} \widetilde{W_{\varphi_{\mathfrak{c},v}}}(h)\cdot\chi_v^{-1}(\det h)\mathrm{d}h}{L(\frac{1}{2}, (\pi_{n,v}^\vee\otimes\chi_v^{-1})\times\pi_{n+1,v}^\vee)} \\ &= \varepsilon(\tfrac{1}{2}, (\pi_{n,v}\otimes\chi_v)\times\pi_{n+1,v}, \psi_v)\cdot \frac{\int_{U_n(F_v)\backslash \mathrm{GL}_n(F_v)} W_{\varphi_{\mathfrak{c},v}}(h)\cdot\chi_v(\det h)\mathrm{d}h}{L(\frac{1}{2}, (\pi_{n,v}\otimes\chi_v)\times\pi_{n+1,v})} \\ &= c''_v \cdot \chi_v(\gamma_v^{\mathfrak{c}(\pi_v)}) \cdot \frac{\int_{U_n(F_v)\backslash \mathrm{GL}_n(F_v)} W_{\varphi_{\mathfrak{c},v}}(h)\cdot\chi_v(\det h)\mathrm{d}h}{L(\frac{1}{2}, (\pi_{n,v}\otimes\chi_v)\times\pi_{n+1,v})}\end{aligned}$$

for a constant $c''_v \in \mathbb{C}^\times$ independent of χ. Then by property (1) in the choice of φ, one can modify the constant $C \in \mathbb{C}^\times$, independent of χ, such that

$$\mathscr{P}_\varphi(\chi)^2 = C \cdot \chi(\gamma) \cdot \prod_{v \in \mathrm{P}} \left(\frac{q_v^{\frac{n(n+1)(2n+1)}{6}}}{\omega(\pi_v)} \right)^{2c_v} \cdot L(\tfrac{1}{2}, (\Pi_n \otimes \widetilde{\chi}) \times \Pi_{n+1})$$

$$\times \prod_{v \in \mathrm{P}} \frac{\int_{U_n(F_v)\backslash \mathrm{GL}_n(F_v)} W_{\varphi_{c,v}}(h) \cdot \chi_v(\det h) \mathrm{d}h}{L(\frac{1}{2}, (\pi_{n,v} \otimes \chi_v) \times \pi_{n+1,v})} \frac{\int_{U_n(F_v)\backslash \mathrm{GL}_n(F_v)} \widetilde{W_{\varphi_{c,v}}}(h) \cdot \chi_v^{-1}(\det h) \mathrm{d}h}{L(\frac{1}{2}, (\pi_{n,v}^\vee \otimes \chi_v^{-1}) \times \pi_{n+1,v}^\vee)}.$$

Now for $v \in \mathrm{P}$, since π_v is unramified, a direct computation shows that

$$\varepsilon(\tfrac{1}{2}, (\pi_{n,v} \otimes \chi_v) \times \pi_{n+1,v}, \psi_v) = \left(\frac{\omega(\pi_v)}{\omega(\pi_v^\vee)} \right)^{c_v} \cdot \left(q_v^{-\frac{n(n+1)}{2}} \right)^{c_v} \cdot G_{\psi_v^{-1}}(\chi_v^{-1})^{n(n+1)}.$$

By [23, Corollary 2.8], there exists a constant $c''_v \in \mathbb{C}^\times$ such that

$$\frac{\int_{U_n(F_v)\backslash \mathrm{GL}_n(F_v)} W_{\varphi_{c,v}}(h) \cdot \chi_v(\det h) \mathrm{d}h}{L(\frac{1}{2}, (\pi_{n,v} \otimes \chi_v) \times \pi_{n+1,v})} = c''_v \cdot \left(q_v^{-\frac{n(n+1)(n+2)}{6}} \right)^{c_v} \cdot G_{\psi_v}(\chi_v)^{\frac{n(n+1)}{2}},$$

hence

$$\frac{\int_{U_n(F_v)\backslash \mathrm{GL}_n(F_v)} \widetilde{W_{\varphi_{c,v}}}(h) \cdot \chi_v^{-1}(\det h) \mathrm{d}h}{L(\frac{1}{2}, (\pi_{n,v}^\vee \otimes \chi_v^{-1}) \times \pi_{n+1,v}^\vee)} = c''_v \cdot \left(\frac{\omega(\pi_v)}{\omega(\pi_v^\vee)} \right)^{c_v} \cdot \left(q_v^{-\frac{n(n+1)(n+2)}{6}} \right)^{c_v} \cdot G_{\psi_v^{-1}}(\chi_v^{-1})^{\frac{n(n+1)}{2}}.$$

In summary, we can again modify the constant $C \in \mathbb{C}^\times$, independent of χ, such that

$$\begin{aligned}\mathscr{P}_\varphi(\chi)^2 &= C \cdot \chi(\gamma) \cdot L(\tfrac{1}{2}, (\Pi_n \otimes \widetilde{\chi}) \times \Pi_{n+1}) \cdot \prod_{v \in \mathrm{P}} \left(\frac{q_v^{\frac{n(n+1)(n-1)}{3}}}{\omega(\pi_v)\omega(\pi_v^\vee)} \right)^{c_v} \cdot G_{\psi_v}(\chi_v)^{\frac{n(n+1)}{2}} G_{\psi_v^{-1}}(\chi_v^{-1})^{\frac{n(n+1)}{2}} \\ &= C \cdot \chi(\gamma) \cdot \prod_{v \in \mathrm{P}} \left(\frac{q_v^{\frac{n(n+1)(2n+1)}{6}}}{\omega(\Pi_v)} \right)^{c_v} \cdot L(\tfrac{1}{2}, (\Pi_n \otimes \widetilde{\chi}) \times \Pi_{n+1}).\end{aligned}$$

Thus, (6) holds and the proposition is proved. □

6 Iwasawa Selmer Groups and Height Pairing

In this section, $\mathtt{P}$ is again a subset of $\mathtt{P}(\Pi)$ (3).

Take a subfield $\mathcal{E} \subseteq \mathcal{E}_{\mathtt{P}}$ containing E. Let $\mathbb{W}$ be a geometric Galois representation of E with coefficients in $\mathbb{L}$.[6] Take a $\mathrm{Gal}(\overline{\mathbb{Q}}/E)$-stable $\mathbb{L}^\circ$-lattice $\mathbb{W}^\circ$ of $\mathbb{W}$.

For every finite extension E'/E contained in $\mathbb{C}$ and every nonarchimedean place u of E', we have a distinguished subspace $\mathrm{H}^1_f(E'_u, \mathbb{W}) \subseteq \mathrm{H}^1(E'_u, \mathbb{W})$ (known as the Bloch–Kato Selmer condition) as follows:

$$\mathrm{H}^1_f(E'_u, \mathbb{W}) := \begin{cases} \mathrm{H}^1_{\mathrm{unr}}(E'_u, \mathbb{W}), & \text{if } u \text{ is not above } p, \\ \ker\left(\mathrm{H}^1(E'_u, \mathbb{W}) \to \mathrm{H}^1(E'_u, \mathbb{W} \otimes_{\mathbb{Q}_p} \mathbb{B}_{\mathrm{cris}})\right), & \text{if } u \text{is above } p. \end{cases}$$

We then define $\mathrm{H}^1_f(E'_u, \mathbb{W}/\mathbb{W}^\circ)$ (resp. $\mathrm{H}^1_f(E'_u, \mathbb{W}^\circ)$) to be the propagation of $\mathrm{H}^1_f(E'_u, \mathbb{W})$, that is, the image (resp. preimage) of $\mathrm{H}^1_f(E'_u, \mathbb{W})$ along the natural map $\mathrm{H}^1(E'_u, \mathbb{W}) \to \mathrm{H}^1(E'_u, \mathbb{W}/\mathbb{W}^\circ)$ (resp. $\mathrm{H}^1(E'_u, \mathbb{W}^\circ) \to \mathrm{H}^1(E'_u, \mathbb{W})$). The global Bloch–Kato Selmer groups are defined as

$$\mathrm{H}^1_f(E', \mathbb{W}/\mathbb{W}^\circ) := \ker\left(\mathrm{H}^1(E', \mathbb{W}/\mathbb{W}^\circ) \to \prod_{u<\infty} \frac{\mathrm{H}^1(E'_u, \mathbb{W}/\mathbb{W}^\circ)}{\mathrm{H}^1_f(E'_u, \mathbb{W}/\mathbb{W}^\circ)}\right),$$

$$\mathrm{H}^1_f(E', \mathbb{W}^\circ) := \ker\left(\mathrm{H}^1(E', \mathbb{W}^\circ) \to \prod_{u<\infty} \frac{\mathrm{H}^1(E'_u, \mathbb{W}^\circ)}{\mathrm{H}^1_f(E'_u, \mathbb{W}^\circ)}\right).$$

Finally, we put

$$\mathscr{X}(\mathcal{E}, \mathbb{W}^\circ) := \varprojlim_{E \subseteq E' \subseteq \mathcal{E}} \mathrm{Hom}_{\mathbb{L}^\circ}\left(\mathrm{H}^1_f(E', \mathbb{W}/\mathbb{W}^\circ), \mathbb{L}/\mathbb{L}^\circ\right),$$

$$\mathscr{S}(\mathcal{E}, \mathbb{W}^\circ) := \varprojlim_{E \subseteq E' \subseteq \mathcal{E}} \mathrm{H}^1_f(E', \mathbb{W}^\circ)$$

as $\mathbb{L}^\circ[[\mathrm{Gal}(\mathcal{E}/E)]]$-modules, where the transition maps are the (dual of) restriction maps and the corestriction maps, respectively.

Definition 6.1 We define

$$\mathscr{X}(\mathcal{E}, \mathbb{W}) := \mathscr{X}(\mathcal{E}, \mathbb{W}^\circ) \otimes_{\mathbb{L}^\circ} \mathbb{L},$$

$$\mathscr{S}(\mathcal{E}, \mathbb{W}) := \mathscr{S}(\mathcal{E}, \mathbb{W}) \otimes_{\mathbb{L}^\circ} \mathbb{L},$$

which are finitely generated $\mathbb{L}[[\mathrm{Gal}(\mathcal{E}/E)]]^\circ$-modules independent of the choice of $\mathbb{W}^\circ$.

---

[6] Recall that this means the representation is continuous, unramified outside a finite set of nonarchimedean places of E, and is de Rham at every p-adic places of E.

We say that $\mathbb{W}$ is *pure* if it satisfies conditions (B,D) in [35, (2.1.2)]. It is clear that if $\mathbb{W}$ is pure, then so is $\mathbb{W}^\vee(1)$.

Example 6.2 Let $\mathbb{W}_\Pi$ be the $\mathbb{L}[\mathrm{Gal}(\overline{\mathbb{Q}}/E)]$-module underlying the representation $\rho_{\Pi_n} \otimes \rho_{\Pi_{n+1}}(n)$ (Notation 2.4).

(1) First, $\mathbb{W}_\Pi$ is pure. Indeed, by [9, Theorem 1.1 & Theorem 1.2], [10, Theorem 1.1], and [38, Lemma 1.4], we know that $\rho_{\Pi_n} \otimes \rho_{\Pi_{n+1}}(n)$ is pure of weight -1 (in the sense of [38]) at every $u \in \mathtt{V}_E^{\mathrm{fin}}$, which implies that $\mathbb{W}_\Pi$ is pure.
(2) Second, we have for every finite extension E'/E contained in $\mathcal{E}_{\mathtt{P}}$ a *conjugation isomorphism* $\mathrm{H}^1(E', \mathbb{W}_\Pi) \xrightarrow{\sim} \mathrm{H}^1(E', \mathbb{W}_\Pi^{\mathtt{c}}) \simeq \mathrm{H}^1(E', \mathbb{W}_{\Pi^\vee})$ by taking $\mathtt{c}$-conjugation of extension classes; it is $\mathbb{L}$-linear, but *inverse* $\mathrm{Gal}(E'/E)$-equivariant.
(3) The system of conjugation isomorphisms induces an isomorphism

$$-^\dagger \colon \mathscr{S}(\mathcal{E}, \mathbb{W}_{\Pi^\vee}) \xrightarrow{\sim} \mathscr{S}(\mathcal{E}, \mathbb{W}_\Pi)$$

that is $(\mathbb{L}[[\mathrm{Gal}(\mathcal{E}/E)]]^\circ, \dagger)$-linear.

Notation 6.3 Suppose that $\mathrm{Gal}(\mathcal{E}/E)$ is torsion free so that $\mathbb{L}[[\mathrm{Gal}(\mathcal{E}/E)]]^\circ$ is integral. For a finitely generated $\mathbb{L}[[\mathrm{Gal}(\mathcal{E}/E)]]^\circ$-module $\mathscr{M}$, we denote by $\mathscr{M}^{\mathrm{tor}}$ its maximal $\mathbb{L}[[\mathrm{Gal}(\mathcal{E}/E)]]^\circ$-torsion submodule, and by

$$\mathrm{char}(\mathscr{M}^{\mathrm{tor}}) \subseteq \mathbb{L}[[\mathrm{Gal}(\mathcal{E}/E)]]^\circ$$

the characteristic ideal of $\mathscr{M}^{\mathrm{tor}}$ (in particular, $\mathrm{char}\,(\mathscr{M}) = \mathrm{char}(\mathscr{M}^{\mathrm{tor}})$ when $\mathscr{M}$ is already torsion).

Conjecture 6.4 (Iwasawa's main conjecture in the coherent case) Suppose that Π is coherent and $\mathcal{E}/E$ is a $\mathtt{P}$-extension (Definition 1.1).

(1) The following are equivalent:
    (a) $\mathscr{L}^0_{\mathcal{E}}(\Pi)$ is nonzero;
    (b) the $\mathbb{L}[[\mathrm{Gal}(\mathcal{E}/E)]]^\circ$-module $\mathscr{X}(\mathcal{E}, \mathbb{W}_\Pi)$ is of rank zero;
    (c) the $\mathbb{L}[[\mathrm{Gal}(\mathcal{E}/E)]]^\circ$-module $\mathscr{S}(\mathcal{E}, \mathbb{W}_\Pi)$ is of rank zero.
(2) When the equivalent conditions in (1) hold, $\mathrm{char}\,(\mathscr{X}(\mathcal{E}, \mathbb{W}_\Pi))$ is generated by $\mathscr{L}^0_{\mathcal{E}}(\Pi)$.

Remark 6.5 When $\mathtt{P} = \emptyset$ so that $\mathcal{E} = E$, part (1) of the above conjecture recovers the rank zero case of the Beilinson–Bloch–Kato conjecture; and part (2) is trivial.

At the end of this section, we use Nekovář's p-adic height pairing to produce an $\mathbb{L}[[\mathrm{Gal}(\mathcal{E}/E)]]^\circ$-sesquilinear (that is, linear in the first variable and adjoint linear in the second variable) pairing

$$\mathbf{h}_{\mathcal{E}}^{\mathbb{W}} : \mathscr{S}(\mathcal{E}, \mathbb{W}) \times \mathscr{S}(\mathcal{E}, \mathbb{W}^\vee(1)) \to \Gamma_{\mathrm{P}}^{E} \otimes_{\mathbb{Z}_p} \mathbb{L}[[\mathrm{Gal}(\mathcal{E}/E)]]^\circ \tag{7}$$

when $\mathbb{W}$ is pure.

The construction requires a choice of splittings of Hodge filtrations of $\mathbb{W}$ at every place of E above P (see [35, Theorem 2.2] for the notion of such splitting). Since $\mathbb{W}$ is geometric, we may choose a finite extension $E_\sharp/E$ contained in $\overline{\mathbb{Q}}$ such that $\mathbb{W}$ is semistable at every p-adic place of $E_\sharp$.[7] For every finite extension E'/E contained in $\mathcal{E}$, put $E'_\sharp := E' \otimes_E E_\sharp$ and denote by $\mathrm{P}'_\sharp$ the inverse image of P with respect to the finite étale extension $E'_\sharp/F$.[8] Clearly, $\mathbb{W}$ as an $\mathbb{L}[\mathrm{Gal}(\overline{\mathbb{Q}}/E'_\sharp)]$-module remains pure, and is semistable at every p-adic places of $E'_\sharp$; and the chosen splittings of Hodge filtrations induce the ones at every place in $\mathrm{P}'_\sharp$. We then have the p-adic height pairing

$$\mathrm{h}_{E'_\sharp}^{\mathbb{W}} : \mathrm{H}^1_f(E'_\sharp, \mathbb{W}) \times \mathrm{H}^1_f(E'_\sharp, \mathbb{W}^\vee(1)) \to \Gamma_{\mathrm{P}}^{E'_\sharp} \otimes_{\mathbb{Z}_p} \mathbb{L}$$

constructed in [35, Sect. 7]. Take $\mathrm{Gal}(\overline{\mathbb{Q}}/E)$-stable $\mathbb{L}^\circ$-lattices $\mathbb{W}_1$ and $\mathbb{W}_2$ of $\mathbb{W}$ and $\mathbb{W}^\vee(1)$, respectively. Then by the construction of the p-adic height pairing, there exists an integer $\delta = \delta(\mathbb{W}_1, \mathbb{W}_2)$, depending only on $\mathbb{W}_1, \mathbb{W}_2$, such that $\mathrm{h}_{E'_\sharp}^{\mathbb{W}}$ restricts to a pairing

$$\mathrm{h}_{E'_\sharp} : \mathrm{H}^1_f(E'_\sharp, \mathbb{W}_1) \times \mathrm{H}^1_f(E'_\sharp, \mathbb{W}_2) \to \Gamma_{\mathrm{P}}^{E'_\sharp} \otimes_{\mathbb{Z}_p} p^\delta \mathbb{L}^\circ$$

for every E'. Define

$$\mathrm{h}_{E'} : \mathrm{H}^1_f(E', \mathbb{W}_1) \times \mathrm{H}^1_f(E', \mathbb{W}_2) \to \Gamma_{\mathrm{P}}^{E} \otimes_{\mathbb{Z}_p} p^\delta \mathbb{L}^\circ$$

to be the composition of the restriction map from $\mathrm{H}^1_f(E', -)$ to $\mathrm{H}^1_f(E'_\sharp, -)$, the pairing $\mathrm{h}_{E'_\sharp}$, the norm map $\mathrm{Nm}_{E'_\sharp/E}$, and the multiplication by $[E_\sharp : E]^{-1}$ (after possibly changing δ). In particular, $\mathrm{h}_{E'}$ does not depend on $E_\sharp$. We define

$$\mathbf{h}_{E'} : \mathrm{H}^1_f(E', \mathbb{W}_1) \times \mathrm{H}^1_f(E', \mathbb{W}_2) \to \Gamma_{\mathrm{P}}^{E} \otimes_{\mathbb{Z}_p} p^\delta \mathbb{L}^\circ[\mathrm{Gal}(E'/E)]$$

to be the $\mathbb{L}^\circ[\mathrm{Gal}(E'/E)]$-sesquilinearization of $\mathrm{h}_{E'}$, namely,

---

[7] When $\mathbb{W} = \mathbb{W}_\Pi$ in Example 6.2, $\mathbb{W}$ is already semistable at places above P since Π_v is semistable for every $v \in \mathrm{P}$.

[8] We regard $E'_\sharp$ as a finite product of finite extensions $E^i_\sharp/E_\sharp$ contained in $\overline{\mathbb{Q}}$. Accordingly, in what follows, we understand $\mathrm{Gal}(\overline{\mathbb{Q}}/E'_\sharp)$ as $\prod_i \mathrm{Gal}(\overline{\mathbb{Q}}/E^i_\sharp)$, $\mathrm{H}^1_f(E'_\sharp, -)$ as $\bigoplus_i \mathrm{H}^1_f(E^i_\sharp, -)$, $\Gamma_{\mathrm{P}}^{E'_\sharp}$ as $\bigoplus_i \Gamma_{\mathrm{P}}^{E^i_\sharp}$, etc.

$$\mathbf{h}_{E'}(x,y)=\sum_{\varsigma\in\mathrm{Gal}(E'/E)}\mathrm{h}_{E'}(\varsigma x,y)[\varsigma^{-1}]$$

for every $(x,y)\in\mathrm{H}^1_f(E',\mathbb{W}_1)\times\mathrm{H}^1_f(E',\mathbb{W}_2)$. By the lemma below, we obtain the desired pairing (7) after inverting p.

Lemma 6.6 *The collection of pairings* $(\mathbf{h}_{E'})_{E'}$ *is compatible under the corestriction maps on the source and the natural projection maps on the target, hence defines an* $\mathbb{L}^\circ[[\mathrm{Gal}(\mathcal{E}/E)]]$*-sesquilinear pairing*

$$\mathbf{h}_{\mathcal{E}}:=\varprojlim_{E\subseteq E'\subseteq\mathcal{E}}\mathbf{h}_{E'}\colon\mathscr{S}(\mathcal{E},\mathbb{W}_1)\times\mathscr{S}(\mathcal{E},\mathbb{W}_2)\to\Gamma^E_{\mathrm{P}}\otimes_{\mathbb{Z}_p}p^\delta\mathbb{L}^\circ[[\mathrm{Gal}(\mathcal{E}/E)]].$$

Proof It suffices to show that for $E\subseteq E'\subseteq E''\subseteq\mathcal{E}$,

$$\sum_{\varsigma\in\mathrm{Gal}(E''/E')}\mathrm{h}_{E''}(\varsigma x,y)=\mathrm{h}_{E'}\left(\mathrm{Cores}^{E''}_{E'}x,\mathrm{Cores}^{E''}_{E'}y\right)\tag{8}$$

holds for every $(x,y)\in\mathrm{H}^1_f(E'',\mathbb{W}_1)\times\mathrm{H}^1_f(E'',\mathbb{W}_2)$, where $\mathrm{Res}^{E'}_{E''}$ and $\mathrm{Cores}^{E''}_{E'}$ denote the corresponding restriction and corestriction maps, respectively. Put $x':=\sum_{\varsigma\in\mathrm{Gal}(E''/E')}\varsigma x$, which is simply the restriction of $\mathrm{Cores}^{E''}_{E'}x\in\mathrm{H}^1_f(E',\mathbb{W}_1)$ to $\mathrm{H}^1_f(E'',\mathbb{W}_1)$. By definition,

$$\sum_{\varsigma\in\mathrm{Gal}(E''/E')}\mathrm{h}_{E''}(\varsigma x,y)=\mathrm{h}_{E''}(x',y)=\mathrm{Nm}_{E''_\sharp/E}\mathrm{h}_{E''_\sharp}(x',y)=\mathrm{Nm}_{E'_\sharp/E}\left(\mathrm{Nm}_{E''_\sharp/E'_\sharp}\mathrm{h}_{E''_\sharp}(x',y)\right).$$

Thus, (8) follows as

$$\mathrm{Nm}_{E''_\sharp/E'_\sharp}\mathrm{h}_{E''_\sharp}(x',y)=\mathrm{Nm}_{E''_\sharp/E'_\sharp}\mathrm{h}_{E''_\sharp}\left(\mathrm{Res}^{E'}_{E''}\mathrm{Cores}^{E''}_{E'}x,y\right)=\mathrm{h}_{E'_\sharp}\left(\mathrm{Cores}^{E''}_{E'}x,\mathrm{Cores}^{E''}_{E'}y\right)$$

by the projection formula with respect to a field extension satisfied by local height pairings at every finite place.

The lemma follows. □

7 Incoherent Anticyclotomic p-Adic L-Function

In this section, P is again a subset of $\mathrm{P}(\Pi)$ (3).

We study the case where $\Pi=\Pi_n\boxtimes\Pi_{n+1}$ is incoherent. Then there exists a hermitian space V_n over E that has signature $(n-1,1)$ at the real place of F induced by the default embedding $E\subseteq\mathbb{C}$ and $(n,0)$ at other ones, unique up to isomorphism, such that $V_{n,v}$ is the prescribed hermitian space from Sect. 2 for every $v\in\mathrm{V}^{\mathrm{fin}}_F$. Put $V_{n+1}:=V_n\oplus E\cdot\mathrm{e}$. Put

$$G:=\mathrm{U}(V_n)\times\mathrm{U}(V_{n+1}),\quad\pi:=\otimes_{v\in\mathrm{V}^{\mathrm{fin}}_F}\pi_v$$

which is an irreducible admissible representation of $G(\mathbb{A}_F^\infty)$ with coefficients in $\mathbb{L}$.

We have a system of Shimura varieties $\{X_K\}_K$ associated with $\mathrm{Res}_{F/\mathbb{Q}} G$ indexed by neat open compact subgroups K of $G(\mathbb{A}_F^\infty)$, which are quasi-projective smooth schemes over E of dimension $2n-1$ (see for example [31, Sect. 3.2] for more details). Put

$$\mathrm{H}^i_?(\overline{X}, \mathbb{L}(j)) := \varinjlim_K \mathrm{H}^i_?(\overline{X}_K, \mathbb{L}(j))$$

for $i, j \in \mathbb{Z}$ and $? \in \{\ , c\}$, where $\overline{X}_K := X_K \otimes_E \overline{\mathbb{Q}}$.

Lemma 7.1 *Let K be a neat open compact subgroup of $G(\mathbb{A}_F^\infty)$.*

(1) The natural map $H^i_c(\overline{X}_K, \mathbb{L})[\pi^K] \to H^i(\overline{X}_K, \mathbb{L})[\pi^K]$ is an isomorphism for every $i \in \mathbb{Z}$.
(2) The $\mathbb{L}[K\backslash G(\mathbb{A}_F^\infty)/K]$-module π^K appears in $H^{2n-1}_c(\overline{X}_K, \mathbb{L})$ semisimply.
(3) If $K = K_\Box K^\Box$ in which $K^\Box$ is hyperspecial maximal, then $H^i_c(\overline{X}_K, \mathbb{L})[(\pi^\Box)^{K^\Box}]$ vanishes unless $i = 2n-1$.

Proof For this lemma, we may fix an embedding $\mathbb{L} \to \mathbb{C}$ and regard that π is defined over $\mathbb{C}$.

By possibly shrinking K, we may assume that K is of the form $K_n \times K_{n+1}$ in which K_N is a neat open compact subgroup of $G_N(\mathbb{A}_F^\infty)$ for $N = n, n+1$. By the Künneth formula, it suffices to show the following statements for $N = n, n+1$:

(1') The natural map $\mathrm{H}^i_c(\overline{X}_{K_N}, \mathbb{C})[\pi_N^{K_N}] \to \mathrm{H}^i(\overline{X}_{K_N}, \mathbb{C})[\pi_N^{K_N}]$ is an isomorphism for every $i \in \mathbb{Z}$.
(2') The $\mathbb{C}[K_N\backslash G_N(\mathbb{A}_F^\infty)/K_N]$-module $\pi_N^{K_N}$ appears in $\mathrm{H}^{N-1}_c(\overline{X}_{K_N}, \mathbb{C})$ semisimply.
(3') If $K_N = K_{N\Box} K_N^\Box$ in which $K_N^\Box$ is hyperspecial maximal, then $\mathrm{H}^i_c(\overline{X}_{K_N}, \mathbb{C})[(\pi_N^\Box)^{K_N^\Box}]$ vanishes unless $i = N-1$.

Here X_{K_N} denotes the corresponding Shimura variety for G_N and $\overline{X}_{K_N} := X_{K_N} \otimes_E \overline{\mathbb{Q}}$. Denote by $X^*_{K_N}$ the unique (smooth) toroidal compactification of X_{K_N} (which is just X_{K_N} when G_N is anisotropic) and put $Z_{K_N} := X^*_{K_N} \setminus X_{K_N}$. We claim that $\mathrm{H}^i(Z_{K_N}, \mathbb{C})[(\pi_N^\Box)^{K_N^\Box}]$ vanishes for every $i \in \mathbb{Z}$. Assuming the claim, (1') is immediate; (2') follows from the Matsushima formula; and (3') follows from the Matsushima formula and the fact that every automorphic representation of $G_N(\mathbb{A}_F)$ that is nearly equivalent to π_N is tempered everywhere by the endoscopic classification [26].

For the claim itself, it has essentially been proved in [31, Lemma 6.1.11]. Indeed, let π'_N be an irreducible admissible representation of $G_N(\mathbb{A}_F^\infty)$ that appears in $\mathrm{H}^i(Z_{K_N}, \mathbb{C})$. Then π'_N is the finite part of a representation of $G_N(\mathbb{A}_F)$ that is an irreducible subquotient of the parabolic induction of a cuspidal automorphic representation of $L_N(\mathbb{A}_F)$, where L_N is the unique proper Levi subgroup of G_N up to conjugation. Since π_N is the finite part of a tempered cuspidal automorphic representation of $G_N(\mathbb{A}_F)$, π'_N is not nearly equivalent to π_N hence cannot appear in $\mathrm{H}^i(Z_{K_N}, \mathbb{C})[(\pi_N^\Box)^{K_N^\Box}]$. In other words, $\mathrm{H}^i(Z_{K_N}, \mathbb{C})[(\pi_N^\Box)^{K_N^\Box}]$ vanishes.

The lemma is proved. □

Put

$$\mathbb{V}_\Pi := \mathrm{Hom}_{G(\mathbb{A}_F^\infty)}\left(\pi^\vee, \mathrm{H}_c^{2n-1}(\overline{X}, \mathbb{L}(n))\right), \quad \mathbb{V}_{\Pi^\vee} := \mathrm{Hom}_{G(\mathbb{A}_F^\infty)}\left(\pi, \mathrm{H}_c^{2n-1}(\overline{X}, \mathbb{L}(n))\right).$$

By the Poincaré duality and Lemma 7.1(1,2), the $\mathbb{L}[\mathrm{Gal}(\overline{\mathbb{Q}}/E)]$-modules $\mathbb{V}_{\Pi^\vee}$ and $(\mathbb{V}_\Pi)^\vee(1)$ are isomorphic. We fix a $\mathrm{Gal}(\overline{\mathbb{Q}}/E)$-equivariant perfect pairing

$$\langle\!\langle\, , \,\rangle\!\rangle_\Pi \colon \mathbb{V}_{\Pi^\vee} \times \mathbb{V}_\Pi \to \mathbb{L}(1).$$

By definition, we have maps

$$\pi \otimes_{\mathbb{L}} \mathbb{V}_{\Pi^\vee} \to \mathrm{H}_c^{2n-1}(\overline{X}, \mathbb{L}(n)), \quad \pi^\vee \otimes_{\mathbb{L}} \mathbb{V}_\Pi \to \mathrm{H}_c^{2n-1}(\overline{X}, \mathbb{L}(n))$$

of $\mathbb{L}[G(\mathbb{A}_F^\infty)]$-modules. Using the Poincaré duality pairing on $\overline{X}_K$ and $\langle\!\langle\, , \,\rangle\!\rangle_\Pi$, we have induced maps

$$-_\star \colon \pi^? \to \mathrm{Hom}_{\mathbb{L}}\left(\varprojlim_K \mathrm{H}^{2n-1}(\overline{X}_K, \mathbb{L}(n)), \mathbb{V}_{\Pi^?}\right)$$

for $? \in \{\ , \vee\}$, so that the image of $(\pi^?)^K$ is contained in $\mathrm{Hom}_{\mathbb{L}}\left(\mathrm{H}^{2n-1}(\overline{X}_K, \mathbb{L}(n)), \mathbb{V}_{\Pi^?}\right)$.

Denote by $H \subseteq G$ the graph of the natural embedding $\mathrm{U}(V_n) \hookrightarrow \mathrm{U}(V_{n+1})$, and fix a rational Haar measure on $H(\mathbb{A}_F^{\infty,\mathtt{P}})$. For every $v \in \mathtt{P}$, fix a standard isomorphism $G_v \simeq \mathrm{GL}_{n,F_v} \times \mathrm{GL}_{n+1,F_v}$ (Definition 4.1) and recall the open compact subgroups $I_v^{(r)}$ of $G(F_v)$ for $r > 0$ introduced after Definition 4.1. Put $I_{\mathtt{P}}^{(r)} := \prod_{v\in\mathtt{P}} I_v^{(r)}$.

For every element $\varphi = \otimes_{v\in\mathtt{V}_F^{\mathrm{fin}}} \varphi_v \in \pi$ with $\varphi_v \in \pi_v$ that is a (nonzero) ordinary vector for $v \in \mathtt{P}$, we choose

- a neat open compact subgroup $K^{\mathtt{P}}$ of $G(\mathbb{A}_F^{\infty,\mathtt{P}})$ that fixes $\varphi^{\mathtt{P}}$,
- a Hecke operator $\mathtt{t} \in \mathbb{L}[K^{\mathtt{P}} \backslash G(\mathbb{A}_F^{\infty,\mathtt{P}})/K^{\mathtt{P}}]$ that annihilates $\mathrm{H}^{2n}(\overline{X}_{K^{\mathtt{P}} I_{\mathtt{P}}^{(1)}}, \mathbb{L}^\circ(n))$ and such that its action on $\pi^{K^{\mathtt{P}}}$ is given by the multiplication by a constant $\omega_{\mathtt{t}} \in \mathbb{L}^\times$, which is possible by Lemma 7.1(3),
- a $\mathrm{Gal}(\overline{\mathbb{Q}}/E)$-stable $\mathbb{L}^\circ$-lattice $\mathbb{V}_\Pi^\circ$ in $\mathbb{V}_\Pi$ that contains the image of $\omega_{\mathtt{t}}^{-1}\mathrm{H}^{2n-1}(\overline{X}_{K^{\mathtt{P}} I_{\mathtt{P}}^{(1)}}, \mathbb{L}^\circ(n))$ under $\varphi_\star$.

Note that by Lemma 4.8, φ is fixed by $I_{\mathtt{P}}^{(1)}$.

For every tuple $\mathfrak{c} = (\mathfrak{c}_v)_{v\in\mathtt{P}}$ of positive integers indexed by $\mathtt{P}$, we denote by $E_{\mathfrak{c}}$ the subfield of $\mathcal{E}_{\mathtt{P}}$ such that $\mathrm{Gal}(E_{\mathfrak{c}}/E) = \Gamma_{\mathtt{P}}/\mathfrak{U}_{\mathfrak{c}}$, so that the union of all $E_{\mathfrak{c}}$ is $\mathcal{E}_{\mathtt{P}}$. We define a certain element

$$\kappa(\varphi)' \in \varprojlim_{\mathfrak{c}} \mathrm{H}_g^1(E_{\mathfrak{c}}, \mathbb{V}_\Pi^\circ)$$

as follows:[9] Put $K_H^{\mathrm{P}} := K^{\mathrm{P}} \cap H(\mathbb{A}_F^{\infty,\mathrm{P}})$. Take a tuple $\mathfrak{c} = (\mathfrak{c}_v)_{v \in \mathrm{P}}$ of positive integers indexed by P. Put

$$\begin{aligned} \varphi_{\mathfrak{c}} &:= \pi((1_n, \xi) \cdot [\varpi^{\mathfrak{c}}])\varphi, \\ K_{\mathfrak{c}} &:= ((1_n, \xi) \cdot [\varpi^{\mathfrak{c}}]) I_{\mathrm{P}}^{(1)} ((1_n, \xi) \cdot [\varpi^{\mathfrak{c}}])^{-1}, \\ K_{\mathfrak{c},H} &:= K_{\mathfrak{c}} \cap H(F_{\mathrm{P}}). \end{aligned}$$

Then we have the map

$$(\varphi_{\mathfrak{c}})_\star \colon \mathrm{H}^{2n-1}(\overline{X}_{K^{\mathrm{P}}K_{\mathfrak{c}}}, \mathbb{L}(n)) \to \mathbb{V}_\Pi.$$

By definition, we have a special homomorphism

$$Y_{K_H^{\mathrm{P}} K_{\mathfrak{c},H}} \to X_{K^{\mathrm{P}} K_{\mathfrak{c}}}$$

which is finite and unramified, where $Y_\bullet$ denotes the system of Shimura varieties associated with H. By Lemma 4.11(1), the structure morphism $Y_{K_H^{\mathrm{P}} K_{\mathfrak{c},H}} \to \operatorname{Spec} E$ factors through $\operatorname{Spec} E_{\mathfrak{c}}$, which gives rise to a decomposition

$$Y_{K_H^{\mathrm{P}} K_{\mathfrak{c},H}} \otimes_E E_{\mathfrak{c}} = \coprod_{\varsigma \in \mathrm{Gal}(E_{\mathfrak{c}}/E)} {}^{\varsigma} Z_{K_H^{\mathrm{P}} K_{\mathfrak{c},H}}$$

into disjoint union of open and closed subschemes indexed by $\mathrm{Gal}(E_{\mathfrak{c}}/E)$ that is compatible with the action of $\mathrm{Gal}(E_{\mathfrak{c}}/E)$, normalized in the way that ${}^1 Z_{K_H^{\mathrm{P}} K_{\mathfrak{c},H}}$ is the fiber over the diagonal $\operatorname{Spec} E_{\mathfrak{c}} \to \operatorname{Spec}(E_{\mathfrak{c}} \otimes_E E_{\mathfrak{c}})$. We have the induced cycle class

$$[{}^1 Z_{K_H^{\mathrm{P}} K_{\mathfrak{c},H}}] \in \mathrm{CH}^n(X_{K^{\mathrm{P}} K_{\mathfrak{c}}} \otimes_E E_{\mathfrak{c}}).$$

By the choice of $\mathtt{t}$, $\mathtt{t}[{}^1 Z_{K_H^{\mathrm{P}} K_{\mathfrak{c},H}}]$ induces an element

$$\alpha(\mathtt{t}[{}^1 Z_{K_H^{\mathrm{P}} K_{\mathfrak{c},H}}]) \in \mathrm{H}^1_g(E_{\mathfrak{c}}, \mathrm{H}^{2n-1}(\overline{X}_{K^{\mathrm{P}}K_{\mathfrak{c}}}, \mathbb{L}^\circ(n)))$$

via the Abel–Jacobi map (see the proof of [17, Proposition 9.29] for the reason of being in H^1_g in a similar case).

Now put

$$\kappa(\varphi)'_{\mathfrak{c}} := \prod_{v \in \mathrm{P}} \left(\frac{1}{\omega(\pi_v)} \right)^{\mathfrak{c}_v} \cdot \omega_{\mathtt{t}}^{-1} \cdot (\varphi_{\mathfrak{c}})_\star \alpha(\mathtt{t}[{}^1 Z_{K_H^{\mathrm{P}} K_{\mathfrak{c},H}}]) \in \mathrm{H}^1_g(E_{\mathfrak{c}}, \mathbb{V}_\Pi^\circ),$$

which depends on the choice of K^{P} but not on $\mathtt{t}$.

---

[9] Here, H^1_g is defined similarly as H^1_f but replacing $\mathbb{B}_{\mathrm{cris}}$ by $\mathbb{B}_{\mathrm{dR}}$ in the Selmer condition at p-adic places.

Proposition 7.2 *The collection* $(\kappa(\varphi)'_{\mathfrak{c}})_{\mathfrak{c}}$ *is compatible under corestriction maps hence gives an element* $\kappa(\varphi)' \in \varprojlim_{\mathfrak{c}} H^1_g(E_{\mathfrak{c}}, \mathbb{V}^{\circ}_{\Pi})$.

Proof It suffices to show the compatibility under the corestriction map $\mathrm{Cores}^{E_{\mathfrak{c}'}}_{E_{\mathfrak{c}}}$ for $\mathfrak{c}'$ satisfying $\mathfrak{c}'_v = \mathfrak{c}_v + 1$ for exactly one element $v \in \mathtt{P}$ and $\mathfrak{c}'_v = \mathfrak{c}_v$ for others. For $? \in \{\mathfrak{c}, \mathfrak{c}'\}$, denote by $\mathrm{H}^{2n}(X_{K^{\mathtt{P}}K_{\mathfrak{c}'}} \otimes_E E_?, \mathbb{L}^{\circ}(n))^0$ the kernel of the restriction map

$$\mathrm{H}^{2n}(X_{K^{\mathtt{P}}K_{\mathfrak{c}'}} \otimes_E E_?, \mathbb{L}^{\circ}(n)) \to \mathrm{H}^{2n}(\overline{X}_{K^{\mathtt{P}}K_{\mathfrak{c}'}}, \mathbb{L}^{\circ}(n)).$$

Then we have the commutative diagram

$$\begin{array}{ccc} \mathrm{H}^{2n}(X_{K^{\mathtt{P}}K_{\mathfrak{c}'}} \otimes_E E_{\mathfrak{c}'}, \mathbb{L}^{\circ}(n))^0 & \xrightarrow{\alpha} & \mathrm{H}^1(E_{\mathfrak{c}'}, \mathrm{H}^{2n-1}(\overline{X}_{K^{\mathtt{P}}K_{\mathfrak{c}'}}, \mathbb{L}^{\circ}(n))) \\ \downarrow{\scriptstyle \mathrm{Tr}^{E_{\mathfrak{c}'}}_{E_{\mathfrak{c}}}} & & \downarrow{\scriptstyle \mathrm{Cores}^{E_{\mathfrak{c}'}}_{E_{\mathfrak{c}}}} \\ \mathrm{H}^{2n}(X_{K^{\mathtt{P}}K_{\mathfrak{c}'}} \otimes_E E_{\mathfrak{c}}, \mathbb{L}^{\circ}(n))^0 & \xrightarrow{\alpha} & \mathrm{H}^1(E_{\mathfrak{c}}, \mathrm{H}^{2n-1}(\overline{X}_{K^{\mathtt{P}}K_{\mathfrak{c}'}}, \mathbb{L}^{\circ}(n))) \end{array}$$

in which $\mathrm{Tr}^{E_{\mathfrak{c}'}}_{E_{\mathfrak{c}}}$ denotes the trace map along the extension $E_{\mathfrak{c}'}/E_{\mathfrak{c}}$. Thus, it suffices to show that

$$(\varphi_{\mathfrak{c}'})_{\star}\alpha(\mathtt{t}[\coprod_{\varsigma \in \mathrm{Gal}(E_{\mathfrak{c}'}/E_{\mathfrak{c}})} {}^{\varsigma}Z_{K^{\mathtt{P}}_H K_{\mathfrak{c}',H}}]) = \omega(\pi_v) \cdot (\varphi_{\mathfrak{c}})_{\star}\alpha(\mathtt{t}[{}^1 Z_{K^{\mathtt{P}}_H K_{\mathfrak{c},H}}]), \tag{9}$$

in which $\coprod_{\varsigma \in \mathrm{Gal}(E_{\mathfrak{c}'}/E_{\mathfrak{c}})} {}^{\varsigma}Z_{K^{\mathtt{P}}_H K_{\mathfrak{c}',H}}$ is regarded as an open and closed subscheme of $Y_{K^{\mathtt{P}}_H K_{\mathfrak{c}',H}} \otimes_E E_{\mathfrak{c}}$. By Definition 4.5, we have

$$\omega(\pi_v)\varphi_{\mathfrak{c}} = \sum_{u \in U(O_{F_v})/(U(O_{F_v}) \cap [\varpi_v]U(O_{F_v})[\varpi_v]^{-1})} \pi_v((1_n, \xi) \cdot [\varpi^{\mathfrak{c}}] \cdot u \cdot [\varpi_v])\varphi. \tag{10}$$

By [24, Lemma 6.5], we have for every $u \in U(O_{F_v})/(U(O_{F_v}) \cap [\varpi_v]U(O_{F_v})[\varpi_v]^{-1})$,

$$(1_n, \xi) \cdot [\varpi^{\mathfrak{c}}] \cdot u \cdot [\varpi_v] \in K_{\mathfrak{c}',H} \cdot (1_n, \xi) \cdot [\varpi^{\mathfrak{c}'}] \cdot I^{(1)}_v.$$

By Lemma 4.11(2), we have the following commutative diagram

$$\begin{array}{ccccc} Y_{K^{\mathtt{P}}_H K_{\mathfrak{c},H}} & \xleftarrow{\kappa} & Y_{K^{\mathtt{P}}_H K_{\mathfrak{c}',H}} & \longrightarrow & Y_{K^{\mathtt{P}}_H K_{\mathfrak{c}',H}} \\ \downarrow & & \downarrow & & \downarrow \\ X_{K^{\mathtt{P}}K_{\mathfrak{c}}} & \longleftarrow & X_{K^{\mathtt{P}}(K_{\mathfrak{c}} \cap K_{\mathfrak{c}'})} & \longrightarrow & X_{K^{\mathtt{P}}K_{\mathfrak{c}}} \end{array}$$

in which κ is finite of degree $q_v^{\frac{n(n+1)(2n+1)}{6}}$, which is same as the cardinality of

$$U(O_{F_v})/(U(O_{F_v}) \cap [\varpi_v]U(O_{F_v})[\varpi_v]^{-1}).$$

Since $I_v^{(1)}$ fixes φ and translations by elements in $K_{\mathfrak{c}',H}$ fix $\kappa^{-1}\,{}^1Z_{K_H^{\mathrm{P}}K_{\mathfrak{c},H}}$, (10) implies that

$$\omega(\pi_v)\cdot(\varphi_{\mathfrak{c}})_\star\alpha(\mathtt{t}[{}^1Z_{K_H^{\mathrm{P}}K_{\mathfrak{c},H}}])=(\varphi_{\mathfrak{c}'})_\star\alpha(\mathtt{t}[\kappa^{-1}\,{}^1Z_{K_H^{\mathrm{P}}K_{\mathfrak{c},H}}]).$$

Finally, since

$$\kappa^{-1}\,{}^1Z_{K_H^{\mathrm{P}}K_{\mathfrak{c},H}}=\coprod_{\varsigma\in\mathrm{Gal}(E_{\mathfrak{c}'}/E_{\mathfrak{c}})}{}^{\varsigma}Z_{K_H^{\mathrm{P}}K_{\mathfrak{c}',H}},$$

(9) follow. The proposition is proved. □

Remark 7.3 The cohomology classes $(\kappa(\varphi)'_{\mathfrak{c}})_{\mathfrak{c}}$ and their norm compatibility relation have already been essentially worked out in [32, Sect. 5.1.1], at least when $F=\mathbb{Q}$.

In terms of the above proposition, we put

$$\kappa(\varphi):=\mathrm{vol}(K_H^{\mathrm{P}})\cdot\kappa(\varphi)'\in\left(\varprojlim_{\mathfrak{c}}\mathrm{H}^1_g(E_{\mathfrak{c}},\mathbb{V}_\Pi^\circ)\right)\otimes_{\mathbb{L}^\circ}\mathbb{L},$$

which depends only on φ.

Hypothesis 7.4 Every irreducible $\mathbb{L}[\mathrm{Gal}(\overline{\mathbb{Q}}/E)]$-subquotient of $\mathbb{V}_\Pi$ is isomorphic to a (canonical) direct summand of $\mathbb{W}_\Pi$ (Example 6.2).

Remark 7.5 The above hypothesis is known when $n\leqslant 2$, or when $n>2$ and $F\neq\mathbb{Q}$ by an ongoing work of Kisin–Shin–Zhu. In fact, there is a more precise version of the above hypothesis specifying such direct summand via Arthur's multiplicity formula; see [28, Hypothesis 6.6].

Under Hypothesis 7.4, the natural map

$$\mathscr{S}(\mathcal{E}_{\mathrm{P}},\mathbb{V}_\Pi)=\left(\varprojlim_{\mathfrak{c}}\mathrm{H}^1_f(E_{\mathfrak{c}},\mathbb{V}_\Pi^\circ)\right)\otimes_{\mathbb{L}^\circ}\mathbb{L}\to\left(\varprojlim_{\mathfrak{c}}\mathrm{H}^1_g(E_{\mathfrak{c}},\mathbb{V}_\Pi^\circ)\right)\otimes_{\mathbb{L}^\circ}\mathbb{L}$$

is an isomorphism. Indeed, it follows from the same statement for $\mathbb{W}_\Pi$ as $\mathbb{W}_\Pi$ is pure. Then for every $\mathtt{P}$-extension $\mathcal{E}/E$ (Definition 1.1), we denote by $\kappa_{\mathcal{E}}(\varphi)$ the image of $\kappa(\varphi)$ under the natural map $\mathscr{S}(\mathcal{E}_{\mathrm{P}},\mathbb{V}_\Pi)\to\mathscr{S}(\mathcal{E},\mathbb{V}_\Pi)$.

Notation 7.6 Assume Hypothesis 7.4 and further that $\mathbb{V}_\Pi$ is a semisimple $\mathbb{L}[\mathrm{Gal}(\overline{\mathbb{Q}}/E)]$-module so that it is a direct submodule of $\mathbb{W}_\Pi$. Let $\mathcal{E}/E$ be a $\mathtt{P}$-extension.

(1) Denote by $\mathscr{K}(\mathcal{E},\mathbb{W}_\Pi)$ the $\mathbb{L}[[\mathrm{Gal}(\mathcal{E}/E)]]^\circ$-submodule of $\mathscr{S}(\mathcal{E},\mathbb{W}_\Pi)$ generated by $\kappa_{\mathcal{E}}(\varphi^\vee)^\dagger$ (Example 6.2(3)) for all $\varphi^\vee=\otimes_{v\in\mathtt{V}_F^{\mathrm{fin}}}\varphi_v^\vee\in\pi^\vee$ with $\varphi_v^\vee\in\pi_v^\vee$ that is an ordinary vector for $v\in\mathtt{P}$ (with respect to all standard isomorphisms (Definition 4.1) at places in $\mathtt{P}$).

(2) Put $\mathscr{T}(\mathcal{E}, \mathbb{W}_\Pi) := \mathscr{S}(\mathcal{E}, \mathbb{W}_\Pi)/\mathscr{K}(\mathcal{E}, \mathbb{W}_\Pi)$.

The conjecture below can be regarded as a higher-dimensional analogue of Perrin-Riou's Heegner point main conjecture [36].

Conjecture 7.7 (Iwasawa's main conjecture in the incoherent case) Suppose that Π is incoherent; that Hypothesis 7.4 holds; and that $\mathbb{V}_\Pi$ is a semisimple $\mathbb{L}[\mathrm{Gal}(\overline{\mathbb{Q}}/E)]$-module. Let $\mathcal{E}/E$ be a $\mathtt{P}$-extension.

(1) The following are equivalent:

  (a) $\mathscr{K}(\mathcal{E}, \mathbb{W}_\Pi)$ (Notation 6.3) is nonzero;
  (b) the $\mathbb{L}[[\mathrm{Gal}(\mathcal{E}/E)]]^\circ$-module $\mathscr{X}(\mathcal{E}, \mathbb{W}_\Pi)$ is of rank one;
  (c) the $\mathbb{L}[[\mathrm{Gal}(\mathcal{E}/E)]]^\circ$-module $\mathscr{S}(\mathcal{E}, \mathbb{W}_\Pi)$ is of rank one.

(2) When the equivalent conditions in (1) hold, we have

$$\mathrm{char}\left(\mathscr{X}(\mathcal{E}, \mathbb{W}_\Pi)^{\mathrm{tor}}\right) = \mathrm{char}\left(\mathscr{T}(\mathcal{E}, \mathbb{W}_\Pi)\right)^2.$$

Remark 7.8 The original Perrin-Riou's conjecture (including the one for Shimura curves over $\mathbb{Q}$, under some conditions) was proved in [8, 40] (with one side of the divisibiilty already proved in [15, 20]).

Let $\mathcal{E}/E$ be a $\mathtt{P}$-extension. In what follows, we define an analog of the p-adic L-function in the incoherent case, which we denote by $\mathscr{L}^1_{\mathcal{E}}(\Pi)$. Define $\mathscr{L}^1_{\mathcal{E}}(\Pi) = 0$ when $\mathbb{V}_\Pi = 0$.

Now we assume $\mathbb{V}_\Pi \neq 0$. Assume Hypothesis 7.4. We have a pairing

$$\mathbf{h}^{\mathbb{V}_\Pi}_{\mathcal{E}} : \mathscr{S}(\mathcal{E}, \mathbb{V}_\Pi) \times \mathscr{S}(\mathcal{E}, \mathbb{V}_\Pi) \to \Gamma^E_{\mathtt{P}} \otimes_{\mathbb{Z}_p} \mathbb{L}[[\mathrm{Gal}(\mathcal{E}/E)]]^\circ$$

from the discussion in the previous section, after we identify $\mathbb{V}_{\Pi^\vee}$ with $\mathbb{V}^\vee_\Pi(1)$ using $\langle\!\langle\,,\,\rangle\!\rangle_\Pi$. Note that by Hypothesis 7.4 and the fact that Π_v is semi-stably ordinary at every $v \in \mathtt{P}$, $\mathbb{V}_\Pi$ satisfies the Panchishkin condition hence admits a canonical splitting of the Hodge filtration at every $v \in \mathtt{P}$, which is the one we use to define the above pairing.

Choose a pair $\varphi = \otimes_{v \in \mathtt{V}^{\mathrm{fin}}_F} \varphi_v \in \mathcal{V}$ and $\varphi^\vee = \otimes_{v \in \mathtt{V}^{\mathrm{fin}}_F} \varphi^\vee_v \in \mathcal{V}^\vee$ with $\varphi_v \in \pi_v$ and $\varphi^\vee_v \in \pi^\vee_v$ satisfying

(T1) for $v \in \mathtt{P}$, both φ_v and $\varphi^\vee_v$ are (nonzero) ordinary vectors;
(T2) for $v \in \mathtt{V}^{\mathrm{spl}}_F \setminus \mathtt{P}$, the pair $(\varphi_v, \varphi^\vee_v)$ satisfies the conclusion of Proposition 2.11;
(T3) for $v \in \mathtt{V}^{\mathrm{fin}}_F \setminus \mathtt{V}^{\mathrm{spl}}_F$, $\alpha(\varphi_v, \varphi^\vee_v) \neq 0$.

By Proposition 2.11 and Lemma 2.9, such choice is possible. Put

$$\mathscr{L}^1_{\mathcal{E}}(\Pi) := \prod_{v \in \mathtt{P}} \frac{\gamma(\Pi_v)}{\langle \varphi^\vee_v, \varphi_v \rangle_{\pi_v}} \cdot \prod_{v \in \mathtt{V}^{\mathrm{fin}}_F \setminus \mathtt{V}^{\mathrm{spl}}_F} \frac{\alpha(\varphi_v, \varphi^\vee_v)^{-1} \cdot \Delta_{n+1,v} \cdot L(\frac{1}{2}, \Pi_{n,v} \times \Pi_{n+1,v})}{L(1, \Pi_{n,v}, \mathrm{As}^{(-1)^n}) L(1, \Pi_{n+1,v}, \mathrm{As}^{(-1)^{n+1}})}$$
$$\times \mathrm{Nm}_{E/F}\, \mathbf{h}^{\mathbb{V}_\Pi}_{\mathcal{E}} \left(\kappa_{\mathcal{E}}(\varphi), \kappa_{\mathcal{E}}(\varphi^\vee)\right) \in \Gamma^F_{\mathtt{P}} \otimes_{\mathbb{Z}_p} \mathbb{L}[[\mathrm{Gal}(\mathcal{E}/E)]]^\circ,$$

where $\gamma(\Pi_v) \in \mathbb{L}^\times$ is the constant in Proposition 4.12. Note that by Lemma 2.9, Lemma 4.7, and (T3), the above expression makes sense.

Lemma 7.9 *The element $\mathscr{L}^1_{\mathcal{E}}(\Pi)$ does not depend on the choice of $(\varphi, \varphi^\vee)$.*

Proof It suffices to show that for every finite character $\chi\colon \mathrm{Gal}(\mathcal{E}/E) \to \mathbb{L}_\chi^\times$, the element

$$\mathbf{h}_{\mathcal{E}}^{\mathbb{V}_\Pi}\left(\kappa_{\mathcal{E}}(\varphi), \kappa_{\mathcal{E}}(\varphi^\vee)\right)(\chi) \in \Gamma_{\mathtt{P}}^F \otimes_{\mathbb{Z}_p} \mathbb{L}_\chi$$

is independent of $(\varphi, \varphi^\vee)$. Without lost of generality, we may assume $\langle \varphi_v^\vee, \varphi_v\rangle_{\pi_v} = 1$ for every $v \in \mathtt{P}$.

The assignment

$$(\varphi^{\mathtt{P}}, \varphi^{\vee\mathtt{P}}) \in \pi^{\mathtt{P}} \times \pi^{\vee\mathtt{P}} \mapsto \mathbf{h}_{\mathcal{E}}^{\mathbb{V}_\Pi}\left(\kappa_{\mathcal{E}}(\varphi), \kappa_{\mathcal{E}}(\varphi^\vee)\right)(\chi)$$

defines an element in

$$\mathrm{Hom}_{H(\mathbb{A}_F^{\infty,\mathtt{P}})\times H(\mathbb{A}_F^{\infty,\mathtt{P}})}\left((\pi_\chi)^{\mathtt{P}} \boxtimes (\pi_\chi)^{\vee\mathtt{P}}, \Gamma_{\mathtt{P}}^F \otimes_{\mathbb{Z}_p} \mathbb{L}_\chi\right).$$

Thus, by Lemma 2.8, there exists a constant $c \in \Gamma_{\mathtt{P}}^F \otimes_{\mathbb{Z}_p} \mathbb{L}_\chi$, independent of the pair $(\varphi^{\mathtt{P}}, \varphi^{\vee\mathtt{P}})$, such that

$$\mathbf{h}_{\mathcal{E}}^{\mathbb{V}_\Pi}\left(\kappa_{\mathcal{E}}(\varphi), \kappa_{\mathcal{E}}(\varphi^\vee)\right)(\chi) = c \prod_{v\in\mathtt{V}_F^\infty\setminus\mathtt{P}} \left(\frac{\Delta_{n+1,v} L(\frac{1}{2}, \Pi_{n,v} \times \Pi_{n+1,v})}{L(1, \Pi_{n,v}, \mathrm{As}^{(-1)^n}) L(1, \Pi_{n+1,v}, \mathrm{As}^{(-1)^{n+1}})}\right)^{-1} \alpha(\varphi_v^{\mathtt{P}}, \varphi_v^{\vee\mathtt{P}}).$$

The lemma then follows. □

Remark 7.10 It is easy to see that $\mathscr{L}^1_{\mathcal{E}}(\Pi)$ neither depends on the choice of the standard isomorphisms $G_v \simeq \mathrm{GL}_{n,F_v} \times \mathrm{GL}_{n+1,F_v}$ for $v \in \mathtt{P}$. However, $\mathscr{L}^1_{\mathcal{E}}(\Pi)$ does depend on the choices of $\langle\!\langle\, ,\, \rangle\!\rangle_\Pi$ and a rational Haar measure on $H(\mathbb{A}_F^{\infty,\mathtt{P}})$.[10]

The conjecture and the remark below are analogs of Conjecture 5.6 and Remark 5.7, respectively.

Conjecture 7.11 Assume Hypothesis 7.4. Suppose that $\mathbb{V}_\Pi \neq 0$ and condition (∗) in Notation 2.4 holds. Then $\mathscr{L}^1_{\mathcal{E}}(\Pi) \neq 0$ for every $\mathtt{P}$-extension $\mathcal{E}/E$ that contains $\mathcal{E}_{\mathtt{P}'}$ for some nonempty subset $\mathtt{P}'$ of $\mathtt{P}$.

When $n = 1$, the conjecture can be deduced from [12].

Remark 7.12 Take an arbitrary embedding $\iota\colon \mathbb{L} \to \mathbb{C}$. Then $\epsilon(\frac{1}{2}, \Pi_n^{(\iota)} \times \Pi_{n+1}^{(\iota)})$ equals $\epsilon(\Pi)$, which we have assumed to be -1. Now $\epsilon(\frac{1}{2}, \Pi_n^{(\iota)} \times \Pi_{n+1}^{(\iota)})$ decomposes as the product of $d(\Pi_n)d(\Pi_{n+1})$ root numbers (valued in $\{\pm 1\}$) for their isobaric factors. Then by a similar discussion in [16, Sect. 26] using [28, Lemma 3.15], $\mathbb{V}_\Pi \neq 0$ if and only if those root numbers contain -1 exactly once.

[10] Choosing $\langle\!\langle\, ,\, \rangle\!\rangle_\Pi$ is equivalent to choosing an $\mathbb{L}$-valued Haar measure on $G(\mathbb{A}_F^\infty)$.

We propose the following conjecture, which is a variant of Conjecture 7.7 without using the conditionally defined module $\mathscr{K}(\mathcal{E}, \mathbb{W}_\Pi)$.

Conjecture 7.13 Suppose that Π is incoherent and that Hypothesis 7.4 holds. Let $\mathcal{E}/E$ be a $\mathtt{P}$-extension.

(1) The following are equivalent:
   (a) $\mathscr{L}^1_{\mathcal{E}}(\Pi)$ is nonzero;
   (b) the $\mathbb{L}[[\mathrm{Gal}(\mathcal{E}/E)]]^\circ$-module $\mathscr{X}(\mathcal{E}, \mathbb{W}_\Pi)$ is of rank one;
   (c) the $\mathbb{L}[[\mathrm{Gal}(\mathcal{E}/E)]]^\circ$-module $\mathscr{S}(\mathcal{E}, \mathbb{W}_\Pi)$ is of rank one.

(2) When the equivalent conditions in (1) hold, char $\left(\mathscr{X}(\mathcal{E}, \mathbb{W}_\Pi)^{\mathrm{tor}}\right)$ is generated by the image of $\phi\mathscr{L}^1_{\mathcal{E}}(\Pi)$ in $\mathbb{L}[[\mathrm{Gal}(\mathcal{E}/E)]]^\circ$ for all $\mathbb{Z}_p$-linear maps $\phi\colon \Gamma^F_{\mathtt{P}} \to \mathbb{Z}_p$.[11]

At last, we propose a conjecture relating $\mathscr{L}^1_{\mathcal{E}_{\mathtt{P}}}(\Pi)$ to the derivative of the hypothetical (full) $\mathtt{P}$-adic L-function of Π along the anticyclotomic direction.

Hypothesis 7.14 In this hypothesis, we temporarily allow Π to be either coherent or incoherent. There exists a unique measure $\mathscr{L}^E_{\mathtt{P}}(\Pi)$ on $\Gamma^E_{\mathtt{P}}$ valued in $\mathbb{L}$, such that for every finite character $\Xi\colon \Gamma^E_{\mathtt{P}} \to \mathbb{L}^\times_\Xi$ of conductor $\prod_{u|\mathtt{P}} \mathfrak{p}_u^{\mathfrak{c}_u}$ for a tuple $\mathfrak{c} = (\mathfrak{c}_u)_{u|\mathtt{P}}$ of positive integers indexed by places of E above $\mathtt{P}$ and every embedding $\iota\colon \mathbb{L}_\Xi \to \mathbb{C}$, we have

$$\iota\mathscr{L}^E_{\mathtt{P}}(\Pi)(\Xi) = G(\iota\Xi)^{\frac{n(n+1)}{2}} \prod_{u|\mathtt{P}} \left(\frac{q_u^{\frac{(n-1)n(n+1)}{6}}}{\iota\omega(\Pi_u)}\right)^{\mathfrak{c}_u} \frac{\Delta_{n+1}\cdot L(\frac{1}{2}, \iota(\Pi_n\otimes\widetilde{\chi})\times\iota\Pi_{n+1})}{2^{d(\Pi_n)+d(\Pi_{n+1})}\cdot L(1, \iota\Pi_n, \mathrm{As}^{(-1)^n})L(1, \iota\Pi_{n+1}, \mathrm{As}^{(-1)^{n+1}})},$$

where $G(\iota\Xi)$ is the global Gauss sum defined on [25, Page 460] (which depends only on $\iota\Xi$).

Remark 7.15 When $d(\Pi_n) = d(\Pi_{n+1}) = 1$, we know

- for every fixed embedding $\iota_0\colon \mathbb{L} \to \mathbb{C}$, the existence of the measure satisfying the interpolation property in the above hypothesis up to a constant in $\mathbb{C}^\times$ for ι extending ι_0 (this is due to a series of works (mainly) [23–25, 27]);
- the existence of the measure after restriction to $\Gamma^F_{\mathtt{P}}$ satisfying the interpolation property in the above hypothesis, under the condition that for every $v \in \mathtt{V}^{(p)}_F \setminus \mathtt{P}(\Pi)$, v is nonsplit in E and Π_v is unramified [14, Theorem B].

Now when Π is incoherent, $\mathscr{L}^E_{\mathtt{P}}(\Pi)$ vanishes along the homomorphism $\mathrm{Nm}^-_{E/F}\colon \Gamma^E_{\mathtt{P}} \to \Gamma_{\mathtt{P}}$. Since the conormal bundle of the rigid analytic spectrum of $\mathbb{L}[[\Gamma_{\mathtt{P}}]]^\circ$ in that of $\mathbb{L}[[\Gamma^E_{\mathtt{P}}]]^\circ$ is canonically the constant bundle $\Gamma^F_{\mathtt{P}} \otimes_{\mathbb{Z}_p} \mathbb{L}$, we obtain an element

$$\mathrm{d}\mathscr{L}^E_{\mathtt{P}}(\Pi) \in \Gamma^F_{\mathtt{P}} \otimes_{\mathbb{Z}_p} \mathbb{L}[[\Gamma_{\mathtt{P}}]]^\circ.$$

[11] If one assumes Leopoldt's conjecture, then the $\mathbb{Z}_p$-rank of $\Gamma^F_{\mathtt{P}}$ is one; hence it suffices to consider one such ϕ that is surjective.

Conjecture 7.16 Assume Hypotheses 7.4 and 7.14. Then

$$\mathscr{L}^1_{\mathcal{E}_{\mathrm{P}}}(\Pi) = C \cdot \mathrm{d}\mathscr{L}^E_{\mathrm{P}}(\Pi),$$

where C is an explicit constant in $\mathbb{L}^\times$ depending on the choices of $\langle\!\langle\ ,\ \rangle\!\rangle_\Pi$ and a rational Haar measure on $H(\mathbb{A}_F^{\infty,\mathrm{P}})$.

Disegni proved an analog [13, Theorem C.4] of the above conjecture when $n = 1$ in terms of quaternionic Shimura curves; while the special case for modular curves under the classical Heegner condition was already known by Howard [21]. For general n, the above conjecture is known [14, Theorem D] after evaluating at everywhere unramified characters of Γ_{P} under several conditions (for instance, $d(\Pi_n) = d(\Pi_{n+1}) = 1$, E/F everywhere unramified, $\mathrm{P} = \mathrm{P}(\Pi) = \mathrm{V}_F^{(p)}$, Π_v is unramified for every $v \in \mathrm{V}_F^{(p)}$, etc).

References

1. Massimo Bertolini, *Selmer groups and Heegner points in anticyclotomic $\mathbf{Z}_p$-extensions*, Compositio Math. **99** (1995), no. 2, 153–182. MR1351834
2. M. Bertolini and H. Darmon, *Heegner points on Mumford-Tate curves*, Invent. Math. **126** (1996), no. 3, 413–456. MR1419003, DOI https://doi.org/10.1007/s002220050105
3. M. Bertolini and H. Darmon, *Iwasawa's main conjecture for elliptic curves over anticyclotomic $\mathbb{Z}_p$-extensions*, Ann. of Math. (2) **162** (2005), no. 1, 1–64. MR2178960, DOI https://doi.org/10.4007/annals.2005.162.1
4. R. Beuzart-Plessis, *Endoscopie et conjecture locale raffinée de Gan-Gross-Prasad pour les groupes unitaires*, Compos. Math. **151** (2015), no. 7, 1309–1371. MR3371496, DOI https://doi.org/10.1112/S0010437X14007891
5. Raphaël Beuzart-Plessis, *La conjecture locale de Gross-Prasad pour les représentations tempérées des groupes unitaires*, Mém. Soc. Math. Fr. (N.S.) (2016), no. 149, vii+191. MR3676153, DOI https://doi.org/10.24033/msmf.457
6. Raphaël Beuzart-Plessis, Yifeng Liu, Wei Zhang, and Xinwen Zhu, *Isolation of cuspidal spectrum, with application to the Gan-Gross-Prasad conjecture*, Ann. of Math. (2) **194** (2021), no. 2, 519–584. MR4298750, DOI https://doi.org/10.4007/annals.2021.194.2.5
7. Raphaël Beuzart-Plessis, Pierre-Henri Chaudouard, and Michał Zydor, *The global Gan-Gross-Prasad conjecture for unitary groups: the endoscopic case*, Publ. Math. Inst. Hautes Études Sci. **135** (2022), 183–336. MR4426741, DOI https://doi.org/10.1007/s10240-021-00129-1
8. Ashay Burungale, Francesc Castella, and Chan-Ho Kim, *A proof of Perrin-Riou's Heegner point main conjecture*, Algebra Number Theory **15** (2021), no. 7, 1627–1653. MR4333660, DOI https://doi.org/10.2140/ant.2021.15.1627
9. Ana Caraiani, *Local-global compatibility and the action of monodromy on nearby cycles*, Duke Math. J. **161** (2012), no. 12, 2311–2413. MR2972460, DOI https://doi.org/10.1215/00127094-1723706
10. Ana Caraiani, *Monodromy and local-global compatibility for $l = p$*, Algebra Number Theory **8** (2014), no. 7, 1597–1646. MR3272276, DOI https://doi.org/10.2140/ant.2014.8.1597
11. W. Casselman, *Introduction to the theory of admissible representations of p-adic groups*, available at https://personal.math.ubc.ca/~cass/research/pdf/p-adic-book.pdf
12. Christophe Cornut and Vinayak Vatsal, *Nontriviality of Rankin-Selberg L-functions and CM points*, L-functions and Galois representations, London Math. Soc. Lecture Note Ser., vol.

320, Cambridge Univ. Press, 2007, pp. 121–186. MR2392354, DOI https://doi.org/10.1017/CBO9780511721267.005

13. Daniel Disegni, *The p-adic Gross-Zagier formula on Shimura curves*, Compos. Math. **153** (2017), no. 10, 1987–2074. MR3692745, DOI https://doi.org/10.1112/S0010437X17007308
14. Daniel Disegni and Wei Zhang, *Gan–Gross–Prasad cycles and derivatives of p-adic L-functions*, arXiv:2410.08401
15. Olivier Fouquet, *Dihedral Iwasawa theory of nearly ordinary quaternionic automorphic forms*, Compos. Math. **149** (2013), no. 3, 356–416. MR3040744, DOI https://doi.org/10.1112/S0010437X12000619
16. Gan, W.T., Gross, B.H., Prasad, D.: Symplectic local root numbers, central critical L values, and restriction problems in the representation theory of classical groups. Astérisque **no. 346**, 1-109. MR3202556 (2012)
17. Andrew Graham and Syed Waqar Ali Shah, *Anticyclotomic Euler systems for unitary groups*, Proc. Lond. Math. Soc. (3) **127** (2023), no. 6, 1577–1680. MR4673434
18. R. Neal Harris, *The refined Gross-Prasad conjecture for unitary groups*, Int. Math. Res. Not. IMRN (2014), no. 2, 303–389. MR3159075, DOI https://doi.org/10.1093/imrn/rns219
19. Haruzo Hida, *Automorphic induction and Leopoldt type conjectures for* $\mathrm{GL}(n)$, Asian J. Math. **2** (1998), no. 4, 667–710. MR1734126, DOI https://doi.org/10.4310/AJM.1998.v2.n4.a5
20. Benjamin Howard, *Iwasawa theory of Heegner points on abelian varieties of* GL_2 *type*, Duke Math. J. **124** (2004), no. 1, 1–45. MR2072210, DOI https://doi.org/10.1215/S0012-7094-04-12411-X
21. Benjamin Howard, *The Iwasawa theoretic Gross-Zagier theorem*, Compos. Math. **141** (2005), no. 4, 811–846. MR2148200, DOI https://doi.org/10.1112/S0010437X0500134X
22. H. Jacquet, I. I. Piatetskii-Shapiro, and J. A. Shalika, *Rankin-Selberg convolutions*, Amer. J. Math. **105** (1983), no. 2, 367–464. MR701565, DOI https://doi.org/10.2307/2374264
23. Fabian Januszewski, *Modular symbols for reductive groups and p-adic Rankin-Selberg convolutions over number fields*, J. Reine Angew. Math. **653** (2011), 1–45. MR2794624, DOI https://doi.org/10.1515/CRELLE.2011.018
24. Fabian Januszewski, *On p-adic L-functions for* $\mathrm{GL}(n) \times \mathrm{GL}(n-1)$ *over totally real fields*, Int. Math. Res. Not. IMRN (2015), no. 17, 7884–7949. MR3404005, DOI https://doi.org/10.1093/imrn/rnu181
25. Fabian Januszewski, *p-adic L-functions for Rankin-Selberg convolutions over number fields*, Ann. Math. Qué. **40** (2016), no. 2, 453–489. MR3529190, DOI https://doi.org/10.1007/s40316-016-0061-y
26. Tasho Kaletha, Alberto Minguez, Sug Woo Shin, and Paul-James White, *Endoscopic Classification of Representations: Inner Forms of Unitary Groups*, arXiv:1409.3731
27. D. Kazhdan, B. Mazur, and C.-G. Schmidt, *Relative modular symbols and Rankin-Selberg convolutions*, J. Reine Angew. Math. **519** (2000), 97–141. MR1739728, DOI https://doi.org/10.1515/crll.2000.019
28. Chao Li and Yifeng Liu, *Chow groups and L-derivatives of automorphic motives for unitary groups*, Ann. of Math. (2) **194** (2021), no. 3, 817–901. DOI https://doi.org/10.4007/annals.2021.194.3.6
29. Liu, D., Sun, B.: Relative completed cohomologies and modular symbols. arXiv:1709.05762
30. Liu, Y., Tian, Y., Xiao, L.: Iwasawa's main conjecture for Rankin–Selberg motives in the anticyclotomic case. arXiv:2406.00624
31. Yifeng Liu, Yichao Tian, Liang Xiao, Wei Zhang, and Xinwen Zhu, *On the Beilinson-Bloch-Kato conjecture for Rankin-Selberg motives*, Invent. Math. **228** (2022), no. 1, 107–375. MR4392458, DOI https://doi.org/10.1007/s00222-021-01088-4
32. David Loeffler, *Spherical varieties and norm relations in Iwasawa theory*, J. Théor. Nombres Bordeaux **33** (2021), no. 3, 1021–1043. MR4402388, DOI https://doi.org/10.1007/s10884-020-09844-5
33. Mœglin, C., Vignéras, M.-F., Waldspurger, J.-L.: Correspondances de Howe sur un corps p-adique. Lecture Notes in Mathematics, vol. 1291. Springer-Verlag, Berlin (1987)

34. Chung Pang Mok, *Endoscopic classification of representations of quasi-split unitary groups*, Mem. Amer. Math. Soc. **235** (2015), no. 1108, vi+248. MR3338302, DOI https://doi.org/10.1090/memo/1108
35. Jan Nekovář, *On p-adic height pairings*, Séminaire de Théorie des Nombres, Paris, 1990–91, Progr. Math., vol. 108, Birkhäuser Boston, 1993, pp. 127–202. MR1263527, DOI https://doi.org/10.1007/s10107-005-0696-y
36. Bernadette Perrin-Riou, *Fonctions L p-adiques, théorie d'Iwasawa et points de Heegner*, Bull. Soc. Math. France **115** (1987), no. 4, 399–456. MR928018
37. Yiannis Sakellaridis and Akshay Venkatesh, Periods and harmonic analysis on spherical varieties. Astérisque **no. 396**, viii+360. MR3764130 (2017)
38. Richard Taylor and Teruyoshi Yoshida, *Compatibility of local and global Langlands correspondences*, J. Amer. Math. Soc. **20** (2007), no. 2, 467–493. MR2276777, DOI https://doi.org/10.1090/S0894-0347-06-00542-X
39. J.-L. Waldspurger, *La formule de Plancherel pour les groupes p-adiques (d'après Harish-Chandra)*, J. Inst. Math. Jussieu **2** (2003), no. 2, 235–333. MR1989693, DOI https://doi.org/10.1017/S1474748003000082
40. Xin Wan, *Heegner point Kolyvagin system and Iwasawa main conjecture*, Acta Math. Sin. (Engl. Ser.) **37** (2021), no. 1, 104–120. MR4204538, DOI https://doi.org/10.1007/s10114-021-8355-7
41. Wei Zhang, *Automorphic period and the central value of Rankin-Selberg L-function*, J. Amer. Math. Soc. **27** (2014), no. 2, 541–612. MR3164988, DOI https://doi.org/10.1090/S0894-0347-2014-00784-0

Spherical Varieties and p-Adic Families of Cohomology Classes

David Loeffler, Robert Rockwood, and Sarah Livia Zerbes

Abstract We prove a "twist-compatibility" result for p-adic families of cohomology classes associated to symmetric spaces. This shows that a single family of classes (lying in a finitely-generated Iwasawa module) interpolates classical cohomology classes of many different weights, including twists by Grössencharacters of possibly non-trivial infinity-type. This subsumes and generalises a number of prior results relating to Euler systems and p-adic L-functions, and we conclude with some novel applications to Euler systems for GSp_4, $GSp_4 \times GL_2$, and $GSp_4 \times GL_2 \times GL_2$.

Keywords Spherical varieties · Iwasawa theory · Euler systems · Hida families

This paper is dedicated to Massimo Bertolini, to mark the occasion of his 60th birthday. Massimo's beautiful work has been an inspiration to us throughout our careers and we wish him all the best for the future.

We gratefully acknowledge support from the following research grants: EPSRC Standard Grant EP/S020977/1 and ERC Consolidator Grant 101001051 "ShimBSD" (Loeffler); Royal Society URF Enhancement Award RGF/EA/180212 (Rockwood).

D. Loeffler (✉)
Faculty of Mathematics and Computer Science, UniDistance Suisse, Brig, Switzerland
e-mail: david.loeffler@unidistance.ch

R. Rockwood
Department of Mathematics, King's College London, Strand, London, United Kingdom
e-mail: robert.rockwood@kcl.ac.uk

S. L. Zerbes
Department of Mathematics, ETH Zürich, Zürich, Switzerland
e-mail: sarah.zerbes@math.ethz.ch

M. Longo et al. (eds.), *Elliptic Curves and Modular Forms in Arithmetic Geometry*, Springer Proceedings in Mathematics & Statistics 527,
https://doi.org/10.1007/978-3-032-13123-2_11

1 Introduction

This article is a sequel to the article [13] of the first author. In the previous article, we constructed norm-compatible families of cohomology classes for Shimura varieties and other symmetric spaces, associated to pairs of reductive groups satisfying an "open orbit" condition. This very general construction subsumes a number of well-known results in Iwasawa theory, including the norm-compatibility of modular symbols and Heegner points for GL_2, as well as more recent constructions such as the Euler system for GSp_4 constructed in [15].

In this paper, we consider varying the algebraic weight defining the coefficient system for our cohomology. Our main result, Theorem 5.4, shows that the norm-compatible families constructed in [13] interpolate classical cohomology classes of many different weights (including twists by Grössencharacters of possibly non-trivial infinity-type). Several specific instances of this result are already known—see Remark 5.7 below—and form an important ingredient in the proofs of "explicit reciprocity laws" for Euler systems and p-adic L-functions. The general theorem proved here subsumes all of these earlier results, and we conclude with several new applications of this theory to Euler systems for GSp_4, $\mathrm{GSp}_4 \times \mathrm{GL}_2$, and $\mathrm{GSp}_4 \times \mathrm{GL}_2 \times \mathrm{GL}_2$.

2 Control Theorems for Ordinary Cohomology

The aim of this section is to prove a general control result (Theorem 2.16) for "Q-ordinary" cohomology groups associated to a general parabolic subgroup Q. Our formulation is heavily influenced by the work of Urban [28] in the setting of Coleman theory; however, unlike Urban we obtain integral results, without inverting p in the coefficients, as a consequence of imposing the stronger condition of Q-ordinarity rather than just Q-finite-slope. These results are adapted from the second author's 2022 University of Warwick PhD thesis.

2.1 Setup

Let p be a prime. We consider the following situation:

- $\mathcal{G}$ is a connected reductive group over $\mathbf{Q}$.
- We assume there exists an unramified $\mathbf{Z}_p$-model G of $\mathcal{G}$, i.e. a reductive group scheme G over $\mathbf{Z}_p$ together with an isomorphism $\mathcal{G} \times_{\mathbf{Q}} \mathbf{Q}_p \cong G \times_{\mathbf{Z}_p} \mathbf{Q}_p$, and we fix a choice of such a model.
- Q is a parabolic subgroup of G, and $\bar{Q}$ a choice of opposite parabolic; thus $L = Q \cap \bar{Q}$ is a Levi subgroup of Q, and the "big cell" $\overline{N} \times L \times N$ is an open subscheme of G over $\mathbf{Z}_p$, where N is the unipotent radical of Q (and similarly $\overline{N}$).

Note that Q and $\bar{Q}$ are local at p; we do not assume that they come from parabolic subgroups of $\mathcal{G}$ defined over $\mathbf{Q}$.

- S denotes the torus L/L^{der}, and $X^\bullet(S)$ its character lattice. We let $X^\bullet_+(S)$ be the cone of dominant weights (relative to Q).
- A is the maximal $\mathbf{Q}_p$-split subtorus in the centre of L.

We choose a torus $S^0 \subset S$, and let L^0 and Q^0 be its preimages in L and Q respectively (so $Q^0 = L^0 \ltimes N$). We write $\mathfrak{S}$ for the profinite abelian group $L(\mathbf{Z}_p)/L^0(\mathbf{Z}_p) \subseteq (S/S^0)(\mathbf{Z}_p)$. The base ring for our constructions will be the Iwasawa algebra (completed group ring) $\mathcal{O}[[\mathfrak{S}]]$, for $\mathcal{O}$ a finite extension of $\mathbf{Z}_p$. Note that $\mathfrak{S}$ is p-torsion-free (since S/S^0 splits over an unramified extension).

We shall also make the following global assumption (Milne's "Axiom (SV5)"):

Assumption 2.1 The torus $\mathcal{Z} = Z(\mathcal{G})^\circ$ (the identity compononent of the centre of $\mathcal{G}$) is isogenous to the product of a $\mathbf{Q}$-split torus and an $\mathbf{R}$-isotropic torus. Equivalently, $\mathcal{Z}(\mathbf{Q})$ is discrete in $\mathcal{Z}(\mathbf{A}_\mathrm{f})$.

Remark 2.2 The assumption (SV5) can be weakened, but at the cost of more complicated statements: very roughly, if $\mathcal{Z}$ contains a torus which is $\mathbf{R}$-split but not $\mathbf{Q}$-split, we need to assume that this torus is contained in S^0. An important example where (SV5) does *not* hold is the group $\mathcal{G} = \mathrm{Res}_{F/\mathbf{Q}} \mathrm{GL}_2$ for F a totally real field; this case will be treated in detail in forthcoming work of Arshay Sheth (see [26]). ◇

2.2 Algebraic Representations

Let us fix a maximal torus $T \subset L$, and a finite extension $K/\mathbf{Q}_p$, with ring of integers $\mathcal{O}$, such that T splits over K. We can and do assume that K is unramified (since G has a smooth integral model).

For $\lambda \in X^\bullet_+(S)$, there is a unique isomorphism class of irreducible representations $(\rho_\lambda, V_\lambda)$ of $G_{/K}$ of highest weight λ (with respect to any choice of Borel contained in Q). A representative of this isomorphism class can be constructed using the Borel–Weil theorem, as the space of all polynomials

$$\{f \in K[G] : f(\overline{n}\ell g) = \lambda(\ell) f(g) \quad \forall \overline{n} \in \overline{N}, \ell \in L, g \in G\},$$

with G acting by right-translation. In particular, the highest weight space of V_λ has a unique basis vector f_λ^{hw} whose restriction to the big Bruhat cell $\overline{N}LN$ is given by $\overline{n}\ell n \mapsto \lambda(\ell)$. This is also the highest *relative* weight space of V_λ, i.e. the highest-weight eigenspace for the action of A.

2.3 Integral Lattices

Let $\lambda \in X^\bullet_+(S)$.

Definition 2.3 An *admissible lattice* in V_λ is an $\mathcal{O}$-lattice $\mathcal{L} \subset V_\lambda$ which is invariant under $G_{/\mathcal{O}}$ (i.e. $G(R)$ preserves $\mathcal{L} \otimes R$ for every $\mathcal{O}$-algebra R), and whose intersection with the highest-weight subspace is $\mathcal{O} \cdot f_\lambda^{\mathrm{hw}}$.

We recall the following standard facts, which can be found in [11] for example:

(a) There are only finitely many admissible lattices in V_λ.
(b) Every admissible lattice is the direct sum of its intersections with the weight spaces of V_λ^G.
(c) There are unique maximal and minimal admissible lattices $\mathcal{L}^{\max}$ and $\mathcal{L}^{\min}$ such that every admissible $\mathcal{L}$ satisfies $\mathcal{L}^{\max} \supseteq \mathcal{L} \supseteq \mathcal{L}^{\min}$.
(d) If we realise V_λ via the Borel–Weil construction as above, then the maximal admissible lattice $\mathcal{L}^{\max}$ is given by the intersection of V_λ with the coordinate ring of $G_{/\mathcal{O}}$.
(e) The dual of an admissible lattice in V_λ is an admissible lattice in $V_\lambda^\vee$. In particular, the dual of the maximal lattice is the minimal lattice and vice versa.

We shall henceforth always write $V_{\lambda,\mathcal{O}}$ for the *maximal* admissible lattice, realised via the Borel–Weil construction as above, and $V_{\lambda,\mathcal{O}}^{\min}$ for the minimal lattice.

Remark 2.4 Note that if $G = \mathrm{GL}_2$ and $V_\lambda = \mathrm{Sym}^n(K^2)$ is the n-th symmetric power of the standard representation, then $\mathrm{Sym}^n(\mathcal{O}^2)$ is the maximal lattice, and the module $\mathrm{TSym}^n(\mathcal{O}^2)$ of symmetric tensors (which features prominently in [9]) is the minimal lattice. ◇

Definition 2.5 Let Σ^+ denote the monoid of cocharacters $\eta \in X_\bullet(A)$ which are dominant with respect to Q, so that $\langle \eta, \phi \rangle \geqslant 0$ for every relative root ϕ. We let $\Sigma^{++} \subset \Sigma^+$ be the ideal of elements that are strictly dominant, so $\langle \eta, \phi \rangle > 0$ for all ϕ.

Lemma 2.6 *If $\eta \in \Sigma^+$, then the endomorphism $p^{\langle \eta, \lambda \rangle} \rho_\lambda \left(\eta(p)^{-1}\right)$ of V_λ acts on every weight space as a non-negative power of p, and hence restricts to an endomorphism of $\mathcal{L}$ for any admissible lattice $\mathcal{L}$. If $\eta \in \Sigma^{++}$ then this endomorphism acts as a strictly positive power of p on every weight space except the highest one.*

Proof This is an easy explicit check. □

2.4 Cohomology of Symmetric Spaces

Let us fix a prime-to-p level group K^p, which we shall suppose to be *neat*.[1] We define $H^*(U, \mathcal{O})$, for $U \subset G(\mathbf{Q}_p)$ open compact, to mean *one* of the following:

[1] See [24, §0.1] for a definition of this condition (which implies that the intersection of K^p with any conjugate of $G(\mathbf{Q})$ is torsion-free) and a proof that any open compact subgroup contains a neat subgroup of finite index.

- Betti cohomology of the locally symmetric space $Y_G(K^pU)$, viewed as a real manifold, with coefficients in $\mathcal{O}$.
- Supposing $\mathcal{G}$ to be equipped with a choice of Shimura datum: the étale cohomology (with $\mathcal{O}$ coefficients) of the canonical model of the Shimura variety of level K^pU, base-extended from the reflex field E to $\overline{\mathbf{Q}}$ (so the resulting module has a $\mathrm{Gal}(\overline{\mathbf{Q}}/E)$-action).
- Again in the Shimura-variety setting, and supposing that the Shimura datum is of abelian type, and $\mathcal{Z}$ splits over a CM field: the étale cohomology of the canonical integral model of the Shimura variety over $\mathcal{O}_E[1/S]$, where E is the reflex field, and S is a sufficiently large finite set of primes. In this case, we may also allow arbitrary cyclotomic twists of the coefficients.

One can also consider cohomology with compact support (in any of the above settings). More generally, since we may regard an algebraic representation of G as a coefficient sheaf (for any of the above cohomology theories), we can make sense of $H^*(U, V_{\lambda,\mathcal{O}})$ for any λ, as long as we assume $U \subseteq G(\mathbf{Z}_p)$.

Remark 2.7 For the existence of canonical integral models of abelian-type Shimura varieties, see [20, Theorem 2.2.1]; it suffices to take S to contain p and all places where K^p is not hyperspecial. (In particular, S can be chosen independently of U.)

Note that in all three cases the groups $H^*(U, V_{\lambda,\mathcal{O}})$ are finitely-generated over $\mathcal{O}$. For Betti cohomology, and for étale cohomology over $\overline{\mathbf{Q}}$, this is well-known. The case of étale cohomology of $\mathcal{O}_E[1/S]$-models is slightly more delicate. We can express the étale cohomology of the integral model as the limit of a Hochschild–Serre spectral sequence whose E_2 page is $H^i(E^S/E, H^j_{\text{et}}(Y_G(K^pU)_{\overline{\mathbf{Q}}}, \dots))$, where E^S is the maximal algebraic extension of E unramified outside S. As the functors $H^i(E^S/E, -)$ map finitely-generated $\mathcal{O}$-modules to finitely-generated $\mathcal{O}$-modules, and vanish for $i \gg 0$, the result follows.

Remark 2.8 This would not work if we worked with the canonical model over E, rather than over $\mathcal{O}_E[1/S]$, since the Galois cohomology functors $H^i(E, -)$ without ramification restrictions do not send finitely-generated modules to finitely-generated modules.

2.5 *Hecke Actions*

For general λ, the functors $H^*(-, V_{\lambda,\mathcal{O}})$ are *Cartesian cohomology functors* for $G(\mathbf{Z}_p)$, in the sense of [13, Definition 2.1.3]; that is, given open compacts $U, V \subseteq G(\mathbf{Z}_p)$ and $g \in G(\mathbf{Z}_p)$ such that $g^{-1}Ug \subseteq V$, we have pullback maps $[g]^* : H^*(V, V_{\lambda,\mathcal{O}}) \to H^*(U, V_{\lambda,\mathcal{O}})$, and pushforward maps $[g]_*$ in the opposite direction, satisfying an appropriate compatibility relation involving double cosets.

However, if the coefficients are non-trivial, they are typically not cohomology functors for the whole of $G(\mathbf{Q}_p)$ since this group does not act on the lattice $V_{\lambda,\mathcal{O}}$. We shall work around this by renormalising our Hecke operators using Lemma 2.6.

Definition 2.9 Suppose $U \subseteq G(\mathbf{Z}_p)$. Then, for any $\lambda \in X^\bullet_+(S)$ and $\eta \in \Sigma^+$, we can define a normalised Hecke operator acting on $H^*(U, V_{\lambda,\mathcal{O}})$ by

$$\mathcal{T}'_\eta = p^h[U\eta(p^{-1})U], \qquad h = \langle \lambda, \eta \rangle.$$

Remark 2.10 Recall η is a cocharacter $\mathbf{G}_m \to A$, so $\eta(p^{-1}) \in A(\mathbf{Q}_p) \subseteq G(\mathbf{Q}_p)$. In the special cases of this theory we have previously worked out, the scalar factor p^h appearing above is hidden from view, since we had always chosen η and λ such that $\langle \eta, \lambda \rangle = 0$.

Definition 2.11 We say an open compact subgroup $U \subset G(\mathbf{Z}_p)$ has an *Iwahori decomposition* (with respect to Q) if we have $U = \bar{N}_U L_U N_U$, where $N_U = U \cap N(\mathbf{Z}_p)$ etc, and N_U and $\bar{N}_U$ are the products of their intersections with the relative root spaces for A.

The prototypical examples are the parahoric subgroups $J = \{g \in G(\mathbf{Z}_p) : g \bmod p \in Q(\mathbf{F}_p)\}$ and its opposite $\bar{J}$.

Proposition 2.12 *If U has an Iwahori decomposition, then the operators $\mathcal{T}'_\eta$ for varying η define an action of Σ^+ on $H^*(U, V_{\lambda,\mathcal{O}})$; and if we choose any $\eta_0 \in \Sigma^{++}$, the* anti-ordinary idempotent

$$e'_Q = \lim_{k\to\infty} (\mathcal{T}'_{\eta_0})^k$$

is independent of the choice of η_0, and cuts out the maximal direct summand on which the operators $\mathcal{T}'_\eta$ are all invertible.

Remark 2.13 We will show in Remark 2.23 below that $e'_Q\, H^*(U, V_{\lambda,\mathcal{O}})$ in fact depends only on $U \cap Q(\mathbf{Z}_p)$.

We use the term "anti-ordinary" since our Hecke operators are defined using the image of p^{-1} under a dominant cocharacter, rather than p. In the familiar case when $G = \mathrm{GL}_2$ and Q is the upper-triangular Borel, this means we do not work with the Hecke operator U_p which acts as $\sum a_n q^n \mapsto \sum a_{np} q^n$ on q-expansions, but rather its transpose U'_p.

Proposition 2.14 *For any U with an Iwahori decomposition, the natural map*

$$e'_Q\, H^i(U, V^{\min}_{\lambda,\mathcal{O}}) \to e'_Q\, H^i(U, V_{\lambda,\mathcal{O}}),$$

induced by the inclusion $V^{\min}_{\lambda,\mathcal{O}} \hookrightarrow V_{\lambda,\mathcal{O}}$, is an isomorphism.

Proof Both $V^{\min}_{\lambda,\mathcal{O}}$ and $V_{\lambda,\mathcal{O}}$ are direct sums of their weight spaces for A. The inclusion map is an isomorphism on the highest relative weight space (by definition). On each lower weight space, $p^{\langle \lambda, \eta_0 \rangle} \rho_\lambda(\eta_0(p)^{-1})$ acts as a positive power of p, so some power of this map will annihilate the quotient of the weight spaces in the two lattices. Thus, if $r \gg 0$ (depending on λ), the operator $(\mathcal{T}'_{\eta_0})^r$ annihilates the cohomology of the quotient. □

2.6 Iwasawa Cohomology and Control Theorems

Definition 2.15 For $i \geqslant 0$, and $\lambda \in X^\bullet_+(S)$, we set

$$H^i_{\mathrm{Iw}}(Q^0, V_{\lambda,\mathcal{O}}) = \varprojlim_U H^i(U, V_{\lambda,\mathcal{O}}),$$

where U varies over open compact subgroups of $G(\mathbf{Z}_p)$ containing $Q^0(\mathbf{Z}_p)$, and the inverse limit is taken relative to the pushforward maps.

One checks that the levels U having an Iwahori decomposition are cofinal in the above inverse system, and the operators $\{\mathcal{T}'_\eta : \eta \in \Sigma^+\}$ are compatible with the pushforward maps and hence act on the inverse limit.

In this section and the next, we prove the following result, generalising results due to Ohta [21, 22] for Betti or étale cohomology of modular curves to arbitrary reductive groups:

Theorem 2.16 *(i) The modules $e'_Q H^i_{\mathrm{Iw}}(Q^0, \mathcal{O})$ are finitely-generated over $\mathcal{O}[[\mathfrak{S}]]$, and vanish for $i \gg 0$.*

(ii) Let U be any open compact subgroup of $G(\mathbf{Z}_p)$ such that $U \supseteq Q^0(\mathbf{Z}_p)$ and U admits an Iwahori decomposition. For any $\lambda \in X^\bullet_+(S/S_0)$, there is a convergent spectral sequence (supported in the second quadrant $i \leqslant 0$, $j \geqslant 0$):

$$E_2^{ij} = \mathrm{Tor}_{-i}^{\mathcal{O}[[\mathfrak{S}_U]]}\left(e'_Q H^j_{\mathrm{Iw}}(Q^0, \mathcal{O}), \mathcal{O}[-\lambda]\right) \Rightarrow e'_Q H^{i+j}(U, V_{\lambda,\mathcal{O}}),$$

where $\mathfrak{S}_U$ is the image of $U \cap Q(\mathbf{Z}_p)$ in $\mathfrak{S}$, and $\mathcal{O}[-\lambda]$ denotes $\mathcal{O}$ with $\mathfrak{S}_U$ acting via the algebraic character $-\lambda$. This spectral sequence is compatible with the action of Hecke operators away from p, and with the operators $\{\mathcal{T}'_\eta : \eta \in \Sigma^+\}$.

We shall give the proof below assuming that $H^*(-)$ denotes Betti cohomology. The remaining cases follow readily from this, as follows. One may define all the spaces and maps for a finite truncation of the coefficients, using general functoriality properties of étale cohomology. Then the remaining assertions—that the inverse limits are finitely-generated, and that the moment map in (ii) is an isomorphism on the ordinary part—can be checked using Artin's comparison theorem to relate Betti cohomology and étale cohomology over $\overline{\mathbf{Q}}$, and the Hochschild–Serre spectral sequence in order to descend to the cohomology of the integral model over $\mathcal{O}_E[1/S]$.

Remark 2.17 In the above formulation, we have assumed that λ factors through the quotient S/S_0, and in particular is trivial on L^{der}. More generally, we can fix any $\nu \in X^\bullet_+(T)$ (not necessarily trivial on L^{der}) and study the cohomology groups $e'_Q \cdot \varprojlim_U H^i(U, V_{\nu,\mathcal{O}})$; one can show that these are finite-type $\mathcal{O}[[\mathfrak{S}]]$-modules which interpolate the cohomology groups $e'_Q \cdot H^i(U, V_{\lambda+\nu,\mathcal{O}})$ as λ varies over characters of S/S_0 (so $\lambda + \nu$ varies over a coset of $X^\bullet(S/S_0)$ in $X^\bullet(T)$). We shall stick to the case $\nu = 0$ here for ease of notation, but the proof in the general case is very similar. ◇

2.7 Proof of the Control Theorem

For brevity we set $R := \mathcal{O}[[\mathfrak{S}]]$. Let us choose an element $\eta \in \Sigma^{++}$ and set $\tau = \eta(p)$.

Groups Let $Q^0(\mathbf{Z}_p) \subset U \subset G(\mathbf{Z}_p)$ be an open compact subgroup admitting an Iwahori decomposition

$$U = \bar{N}_U \times L_U \times N_0,$$

and for $r \geqslant 0$ set

$$U^{(r)} = \tau^{-r} U \tau^r \cap U.$$

Note that we have $U^{(r)} = \bar{N}_U^{(r)} \times L_U \times N_0$ for all r, where $\bar{N}_U^{(r)} = \tau^{-r} \bar{N}_U \tau^r$; in particular $\bigcap_{r \geqslant 0} U^{(r)} = U \cap Q(\mathbf{Z}_p)$.

Betti cohomology complexes For an open compact subgroup $K = K^p U \subset G(\mathbf{A}_\mathrm{f})$ the symmetric space $Y(K)$ is given by a finite disjoint union

$$Y(K) = \bigsqcup_i \mathcal{X}/\Gamma_i.$$

where $\Gamma_i \subset G(\mathbf{Q})^+$ are arithmetic subgroups, and $\mathcal{X} = G(\mathbf{R})^+/\left(K_\infty Z(\mathbf{R}) \cap G^+(\mathbf{R})\right)$ with K_∞ a maximal compact subgroup of $G(\mathbf{R})$. If K is neat then $\mathcal{X}/\Gamma_i$ has the structure of a smooth manifold. This space satisfies $\pi_1(\mathcal{X}/\Gamma_i) = \Gamma_i$ and thus

$$H^j(K, V_\lambda) = \bigoplus_i H^j(\Gamma_i, V_\lambda).$$

Let M be a $\mathbf{Z}[K]$-module with K acting through its projection to U. By the theory of Borel–Serre ([2]; see also [28, Sect. 4.2.1]), the cohomology groups $H^i(\Gamma_i, M)$ vanish for $i \gg 0$, and are computed by a complex $\mathcal{C}^\bullet(\Gamma_i, M)$ of $\mathbf{Z}[\Gamma_i]$-modules, depending functorially on M, each term of which is isomorphic to a finite direct sum of copies of M. We write

$$\mathcal{C}^\bullet(U, M) := \bigoplus_i \mathcal{C}^\bullet(\Gamma_i, M).$$

This complex computes the Betti cohomology of $Y(K)$. If M has an action of τ^{-1} compatible with the K-action then we can define an action of $\mathcal{T}'_\eta$ on $\mathcal{C}^\bullet(K, M)$ lifting the action on cohomology. There are many such liftings but these differ by a chain homotopy and thus induce the same morphism in the derived category.

Modules of continuous functions For $\lambda \in X^{\bullet}_{+}(S/S_0)$ we define U-modules

$$C_{\lambda,U} = \{f : U \to \mathcal{O} : f \text{ continuous}, \ f(n\ell g) = \lambda^{-1}(\ell)f(g) \ \forall n \in N_0, \ell \in L_U, g \in U\}$$
$$C_{\mathrm{univ}} = \{f : U \to \mathcal{O} : f \text{ continuous}, \ f(n\ell g) = f(g) \ \forall n\ell \in Q^0(\mathbf{Z}_p), g \in U\}$$

with U acting by right translation.

Note that, since we have $U = N_0 \times L_U \times \bar{N}_U$, restriction to $\bar{N}_U$ is an isomorphism between $C_{\lambda,U}$ and the space of continuous functions on $\bar{N}_U$, for any λ (but the action of U on this space depends on λ). Similarly, functions in C_{univ} are uniquely determined by their restriction to $L_U \times \bar{N}_U$, and this restriction must transform trivially under left-translation by $L_U \cap L^0$, giving an isomorphism between C_{univ} and continuous functions on $\mathfrak{S}_U \times \bar{N}_U$.

We endow these spaces with an action of τ given by

$$(\tau \cdot f)(n\ell\bar{n}) = f(n\ell\tau^{-1}\bar{n}\tau),$$

and with the uniform-convergence topology. There are natural inclusions

$$V^{\max}_{\lambda^\vee,\mathcal{O}} \to C_{\lambda,U} \to C_{\mathrm{univ}}$$

intertwining the U and τ actions, where $\lambda^\vee = -w_0\lambda$ and w_0 is the long Weyl element.

Modules of measures We define modules of bounded distributions

$$D_{\lambda,U} = \mathrm{Hom}_{\mathrm{cont}}\left(C_{\lambda,U}, \mathcal{O}\right)$$
$$D_{\mathrm{univ}} = \mathrm{Hom}_{\mathrm{cont}}\left(C_{\mathrm{univ}}, \mathcal{O}\right),$$

which inherit actions of U and τ^{-1} by duality. By dualising the above sequence, these spaces admit the following maps

$$D_{\mathrm{univ}} \to D_{\lambda,U} \to V^{\min}_{\lambda,\mathcal{O}},$$

which are equivariant for the actions of U and τ^{-1}. Writing $\mathfrak{S}_U$ for the image of U in $\mathfrak{S}$, these groups are naturally modules over $R = \mathcal{O}[[\mathfrak{S}_U]]$ and this structure is given explicitly by the isomorphisms:

$$D_{\lambda,U} \cong (\mathcal{O}[-\lambda] \otimes_{\mathcal{O}} R)\hat{\otimes}_{\mathcal{O}}\mathcal{O}[[\bar{N}_U]],$$
$$D_{\mathrm{univ}} \cong R\hat{\otimes}_{\mathcal{O}}\mathcal{O}[[\bar{N}_U]],$$

where $\mathcal{O}[-\lambda]$ denotes $\mathcal{O}$ regarded as an $\mathfrak{S}$ (and accordingly $\mathcal{O}[[\mathfrak{S}]]$) module via the inverse of λ. In particular

$$D_{\text{univ}} \otimes_R \mathcal{O}[-\lambda] \cong D_{\lambda,U}$$

as R-modules. (In this case the completed tensor product and the abstract tensor product agree, since both modules are complete and $\mathcal{O}[-\lambda]$ is finitely-generated.)

Lemma 2.18 *The R-module D_{univ} is flat.*

Proof We have shown that D_{univ} is isomorphic as an R-module to a power series ring in finitely many variables over R, which is in particular flat □

Proposition 2.19 *There is an idempotent e'_Q acting on $\mathcal{C}^\bullet(U, D_{\text{univ}})$ such that*

- *For all i and all $f \in \mathcal{C}^i(U, D_{\text{univ}})$ we have $e'_Q f = \lim_{n\to\infty} \left(\mathcal{T}'_\eta\right)^{n!} f$.*
- *There is a decomposition*

$$\mathcal{C}^\bullet(U, D_{\text{univ}}) = e'_Q \mathcal{C}^\bullet(U, D_{\text{univ}}) \oplus \left(1 - e'_Q\right) \mathcal{C}^\bullet(U, D_{\text{univ}})$$

such that $\mathcal{T}'_\eta$ acts invertibly on the first component and is topologically nilpotent on the second.

Moreover, the terms of the complex $e'_Q \mathcal{C}(U, D_{\text{univ}})$ are flat R-modules.

Proof Note that D_{univ} is a profinite R-module, isomorphic to a power series ring in finitely many variables over R; and we can write it as an inverse limit of finite quotients, $D_{\text{univ}} = \varprojlim_r D_{\text{univ}}/W_r$, where each W_r is stable under the action of U and of τ^{-1}. So $\mathcal{C}^\bullet(U, D_{\text{univ}})$ is a complex of profinite R-modules, and the W_r induce subcomplexes $W_r^\bullet$ whose components are bases of of open neighbourhoods for the components of $\mathcal{C}^\bullet(U, D_{\text{univ}})$. The action of $\mathcal{T}'_\eta$ on $\mathcal{C}^\bullet(U, D_{\text{univ}})$ preserves $W_r^\bullet$ and thus descends to an action on $\mathcal{C}^\bullet(U, D_{\text{univ}})/W_r^\bullet$ for every r. This complex consists of modules of finite cardinality. We can thus use the ideas developed in [23, Sect. 2.1] to deduce the existence of e'_Q.

Since each component of $\mathcal{C}^\bullet(U, D_{\text{univ}})$ is a direct sum of copies of D_{univ} (by the Borel–Serre theory mentioned above), this is a complex of flat R-modules by the previous lemma. Since $e'_Q \mathcal{C}^\bullet(U, D_{\text{univ}})$ is a direct summand of $\mathcal{C}^\bullet(U, D_{\text{univ}})$ we conclude that it is also a complex of flat R-modules. □

Remark 2.20 Note that the idempotent e'_Q is independent of the choice of η, since the multiples of any given η are cofinal in Σ^{++}. ◇

The key input into the proof of the control theorem is the following lemma.

Lemma 2.21 *Let M be a $\mathbf{Z}[U]$-module with a compatible action of τ^{-1}. The following diagram commutes on cohomology*

$$\begin{array}{ccc} \mathcal{C}^\bullet(U^{(s)}, M) & \xrightarrow{\text{cores}} & \mathcal{C}^\bullet(U, M) \\ \downarrow{\scriptstyle (\mathcal{T}'_\eta)^s} & \swarrow{\scriptstyle [\tau^s]_*} & \downarrow{\scriptstyle (\mathcal{T}'_\eta)^s} \\ \mathcal{C}^\bullet(U^{(s)}, M) & \xrightarrow[\text{cores}]{} & \mathcal{C}^\bullet(U, M) \end{array}$$

Proof This is a standard argument in Hida theory, compare e.g. [6, Proposition 4.1] or [27, Lemma 3.1]. □

Corollary 2.22 *When $C^\bullet(U, M)$ in addition admits an ordinary idempotent e in the sense of Proposition 2.19, the corestriction maps induce isomorphisms*

$$eH^i(U^{(s)}, M) \cong eH^i(U, M),$$

for any $s \geqslant 0$.

Proof On the image of the ordinary idempotent, the operator $(\mathcal{T}'_\eta)^s$ is invertible; so it follows from the previous proposition that the corestriction map is an isomorphism on this summand. □

Remark 2.23 In particular, if U, U' are any open compacts in $G(\mathbf{Z}_p)$ which both have Iwahori decompositions and satisfy $U \cap Q(\mathbf{Z}_p) = U' \cap Q(\mathbf{Z}_p)$, then $U^{(s)}$ is contained in $U \cap U'$ for all $s \gg 0$, and hence the pushforward map $e'_Q H^*(U \cap U', M) \to e'_Q H^*(U, M)$ is an isomorphism. The same holds with U and U' interchanged; so we deduce that $e'_Q H^*(U, M)$ depends only on $U \cap Q(\mathbf{Z}_p)$, as claimed above. ◇

Proposition 2.24 *There is an isomorphism*

$$e'_Q H^i(U, D_{\lambda,U}) \cong e'_Q H^i(U, V_{\lambda,\mathcal{O}}).$$

Proof We sketch the proof. From the previous corollary, for all $s \geqslant 0$ we have

$$e'_Q H^i(U, V_{\lambda,\mathcal{O}}) = e'_Q H^i(U^{(s)}, V_{\lambda,\mathcal{O}}) = e'_Q H^i(U, \mathcal{O}[U] \otimes_{\mathcal{O}[U^{(s)}]} V_{\lambda,\mathcal{O}}).$$

(Here we are using the Cartesian property to see that the fibres of $Y_G(U^{(s)})$ over $Y_G(U)$ are naturally identified with $U/U^{(s)}$; this step would break down in the absence of condition (SV5).)

From the Iwahori decomposition, we have $U/U^{(s)} \cong \bar{N}_U/\bar{N}_U^{(s)}$; and as $\bigcap_s \bar{N}_U^{(s)} = \{1\}$, the inverse limit of the profinite U-modules $\mathcal{O}[U] \otimes_{\mathcal{O}[U^{(s)}]} V_{\lambda,\mathcal{O}}$ is identified with $\mathcal{O}[[U]] \otimes_{\mathcal{O}[[L_U N_0]]} V_{\lambda,\mathcal{O}}$. Moreover, the action of e'_Q kills all the S-eigenspaces in $V_{\lambda,\mathcal{O}}$ except the highest-weight space, which is one-dimensional with L_U acting via λ. So we obtain the cohomology of $\mathcal{O}[[U]] \otimes_{\mathcal{O}[[L_U N_0]]} \mathcal{O}[\lambda] = D_{\lambda,U}$. □

Corollary 2.25 *We have an isomorphism*

$$e'_Q H^i(U, D_{\mathrm{univ}}) \cong e'_Q H^i_{\mathrm{Iw}}(Q^0, \mathcal{O}).$$

Proof This follows in a similar fashion to the above, identifying D_{univ} with the module of measures on the quotient $U/Q^0(\mathbf{Z}_p)$. □

Define $M^\bullet$ to be the image of $e'_Q C^\bullet(U, D_{\mathrm{univ}})$ in the subcategory $\mathcal{D}^{\mathrm{flat}}(R)$ of the derived category of R-modules generated by flat objects. Note that this is independent of the choice of U, since the maps induced by shrinking U are quasi-isomorphisms from the last proposition.

Proposition 2.26 *Suppose $e'_Q C^\bullet(U, D_{\mathrm{univ}})$ is concentrated in degrees $[a, b]$. Then $M^\bullet$ is represented by a perfect complex concentrated in degrees $[a, b]$.*

Proof Let $U_1 = \{u \in U : u \bmod p \in Q^0(\mathbf{F}_p)\}$. Then the cohomology groups

$$H^i(M^\bullet \otimes_R R/J(R)) = e'_Q H^i(U_1, \mathcal{O}/p)$$

are finite, and thus we can use the same method as [23, Proposition 2.2.1] to obtain the result. (The lemma as stated is for complete Noetherian *local* rings, and our ring R is only semi-local; but R is isomorphic to the direct product of its localisations at the finitely many maximal ideals of R, so we may obtain the result for complexes of R-modules by applying the proposition to each localisation and taking the direct product.) □

Theorem 2.27 *For all $r \geqslant 1$ and $\lambda \in X^\bullet_+(S)$ there is a quasi-isomorphism*

$$M^\bullet \otimes^{\mathbb{L}}_{\mathcal{O}[[\mathfrak{S}_U]]} \mathcal{O}[-\lambda] \cong e'_Q C^\bullet(U, V_\lambda)$$

Proof Since $M^\bullet$ is represented by the flat complex $e'_Q C^\bullet(U, D_{\mathrm{univ}})$, we can compute the derived tensor product as

$$e'_Q C^\bullet(U, D_{\mathrm{univ}}) \otimes_{\mathcal{O}[[\mathfrak{S}_U]]} \mathcal{O}[-\lambda] = e'_Q C^\bullet(U, D_{\mathrm{univ}} \otimes_{\mathcal{O}[[\mathfrak{S}_U]]} \mathcal{O}[-\lambda])$$

(where we do not need to take derived tensor product, as the terms of the complex are flat). By the previous lemmas this complex is quasi-isomorphic to $e'_Q C^\bullet(U, V_\lambda)$. □

Corollary 2.28 *There is a spectral sequence*

$$E_2^{i,j} : \mathrm{Tor}_{-i}^{\mathcal{O}[[\mathfrak{S}_U]]}(e'_Q H^j_{\mathrm{Iw}}(Q^0, \mathcal{O}), \mathcal{O}[-\lambda]) \Longrightarrow e'_Q H^{i+j}(U, V_\lambda) \qquad (1)$$

supported in the second quadrant.

Proof This follows from the previous theorem using the Künneth spectral sequence for the Tor functor (see e.g. [29, Theorem 5.6.4]). □

Definition 2.29 We denote by

$$\mathrm{mom}^\lambda_U : e'_Q H^i_{\mathrm{Iw}}(Q^0, \mathcal{O}) \otimes_{\mathcal{O}[[\mathfrak{S}_U]]} \mathcal{O}[-\lambda] \longrightarrow e'_Q H^i(U, V_{\lambda,\mathcal{O}})$$

the edge map of the spectral sequence (1).

We will need below an explicit description of this map as an inverse limit. The natural edge map

$$e'_Q H^i(U, D_{\mathrm{univ}}) \otimes_{\mathcal{O}[[\mathfrak{S}_U]]} \mathcal{O}[-\lambda] \to e'_Q H^i(U, V_{\lambda,\mathcal{O}})$$

comes from the map

$$H^i(U, D_{\mathrm{univ}}) \to H^i(U, D_{\mathrm{univ}} \otimes V_{\lambda,\mathcal{O}})$$

induced by the map on coefficients given by $\mu \mapsto \mu \otimes f_\lambda^{\mathrm{hw}}$, composed with projection to level U. Since modulo p^s we have $f_\lambda^{\mathrm{hw}} \in H^0(U^{(s)}, V_{\lambda,\mathcal{O}}/(p^s))$ it's easy to see that the above map is given by the composite of

$$\begin{aligned} \mathrm{mom}^\lambda : H^i_{\mathrm{Iw}}(Q^0, \mathcal{O}) &\to H^i_{\mathrm{Iw}}(Q^0, V_{\lambda,\mathcal{O}}) \\ x &\mapsto \varprojlim_s \left(x \cup f_\lambda^{\mathrm{hw}} \bmod p^s\right). \end{aligned}$$

and projection to level U.

3 Branching Laws for Algebraic Representations

We recall the setup considered in [13], where we consider pushforward of "Eisenstein-type" cohomology classes from a group H to a larger group G. We shall fix an embedding $\mathcal{H} \hookrightarrow \mathcal{G}$ of reductive groups over $\mathbf{Q}$, extending to an embedding $H \hookrightarrow G$ of reductive group schemes over $\mathbf{Z}_p$.

3.1 Flag Varieties

We fix choices of subgroups Q, S etc as in Sect. 2.1, both for G and for H; and we distinguish between them using subscripts, so Q_G and Q_H are parabolics in G and H respectively. Note that we do not assume any direct compatibility between these (e.g. we do not suppose that $Q_H = Q_G \cap H$ etc). We are interested in the action of H, and its subgroup Q_H^0, on the flag variety $\mathcal{F} = G/\overline{Q}_G$. We assume that there exists $u \in \mathcal{F}(\mathbf{Z}_p)$ such that

(A) the Q_H^0-orbit of u is Zariski-open in $\mathcal{F}$, and
(B) the image of $\overline{Q}_G \cap u^{-1} Q_H^0 u$ under the projection $\overline{Q}_G \twoheadrightarrow L_G \twoheadrightarrow S_G$ is contained in the subtorus S_G^0.

Remark 3.1 Note that the validity of (B) is independent of the choice of representative u in (A), since $\overline{Q}_G \cap u^{-1} Q_H^0 u$ is well-defined up to conjugation (and S_G is commutative). ◇

3.2 Highest-Weight Representations

Proposition 3.2 *For any* $\lambda \in X^\bullet_+(S_G/S_G^0)$*, we have* $\dim(V_\lambda^G)^{Q_H^0} \leqslant 1$*. If this dimension is 1, then* $(V_\lambda^G)^{Q_H^0}$ *is spanned by the unique polynomial* $\mathrm{br}_\lambda \in K[G]$ *(a "branching polynomial") such that* $\mathrm{br}_\lambda\left(\overline{q}u^{-1}q'\right) = \lambda(\overline{q})$ *for* $\overline{q} \in \overline{Q}_G$ *and* $q' \in Q_H^0$.

Proof Suppose $f \in (V_\lambda^G)^{Q_H^0}$. Then the restriction of f to the subvariety $\overline{Q}_G u^{-1} Q_H^0$ of G is uniquely determined by $f(u^{-1})$. Since $\overline{Q}_G u^{-1} Q_H^0$ is open (and G is connected), this implies that f itself is uniquely determined, as an element of $K[G]$, by $f(u^{-1})$. In particular, if f is non-zero, then we can scale it so that $f(u^{-1}) = 1$. □

Remark 3.3 Conversely, if S_G^0 is exactly the image of $\overline{Q}_G \cap u^{-1} Q_H^0 u$ in S_G, then any weight $\lambda \in X^\bullet_+(S_G)$ such that $(V_\lambda^G)^{Q_H^0} \neq 0$ must lie in $X^\bullet_+(S_G/S_G^0)$. However, it is convenient not to require S_G^0 to have this minimality property. ◇

Definition 3.4 We shall say that a weight $\lambda \in X^\bullet(S_G/S_G^0)$ is Q_H^0 *-admissible* if $(V_\lambda^G)^{Q_H^0} \neq 0$.

If this holds, then Q_H/Q_H^0 must act on $(V_\lambda^G)^{Q_H^0}$ via a weight $\mu \in X^\bullet(S_H/S_H^0)$, and this weight must be dominant (since it is the highest weight of the H-representation spanned by $(V_\lambda^G)^{Q_H^0}$, which is finite-dimensional). We denote this weight by $\mu(\lambda)$, or just μ if λ is clear from context.

Remark 3.5 The natural map

$$\frac{Q_H \cap u\overline{Q}_G u^{-1}}{Q_H^0 \cap u\overline{Q}_G u^{-1}} \to Q_H/Q_H^0$$

is injective, and for dimension reasons, it must in fact be an isomorphism. The left-hand side maps naturally to S_G/S_G^0; so we can characterise $\mu(\lambda)$ as the pullback of λ to this quotient. ◇

Proposition 3.6 *If* λ *is* Q_H^0*-admissible, then there exists a unique homomorphism*

$$V_\mu^H \to (V_\lambda^G|)|_H,$$

where $\mu \in X^\bullet(L_H/L_H^0)$ *is the character by which* Q_H *acts on* $\left(V_\lambda^G\right)^{Q_H^0}$*. This is characterised by mapping the highest-weight vector* f_μ^{hw} *of* V_μ^H *to* br_λ.

Proof This follows from Frobenius reciprocity: V_μ^H is isomorphic to the co-induced module $\mathrm{Coind}_{Q_H}^H(\mu)$ (which is left adjoint to restriction of representations, while the usual induction is right adjoint). □

Proposition 3.7 *If* λ *is a* Q_H^0*-admissible weight, then* $\mathrm{br}_\lambda \in V_{\lambda,\mathcal{O}}^G$.

Proof Since $G_{/\mathcal{O}}$ is smooth over $\mathcal{O}$, it is a normal scheme, so it suffices to check that br_λ is regular away from a subvariety of codimension $\geqslant 2$ (Hartogs' lemma). By definition, it is regular on the generic fibre, and also on an open dense subvariety of $G_{/k}$, where $k = \mathcal{O}/p\mathcal{O}$; so we are done. □

Corollary 3.8 *The branching map of Proposition 3.6 restricts to a map of $\mathcal{O}$-modules*

$$V_{\mu,\mathcal{O}}^{H,\min} \to V_{\lambda,\mathcal{O}}^{G},$$

where $V_{\mu,\mathcal{O}}^{H,\min}$ is the minimal admissible lattice in V_μ^H.

Remark 3.9 Compare [15, Proposition 4.3.5]. (We do not know whether the minimal lattice maps to the minimal lattice, nor whether the maximal one maps to the maximal one.) ◇

3.3 Projection to the Highest-Weight Subspace

If λ is an Q_H^0-admissible weight, we thus have two canonical vectors in V_λ^G: namely, the highest-weight vector f_λ^{hw}, and the Q_H^0-invariant vector br_λ. We also have a canonical linear functional ψ_λ on V_λ, given by evaluation at id_G, which factors through projection from V_λ to its highest-weight space (i.e. it is a lowest-weight vector of weight $-\lambda$ for the contragredient representation $(V_\lambda^G)^\vee$).

Proposition 3.10 *We have $\psi_\lambda(f_\lambda^{\mathrm{hw}}) = \psi_\lambda(u^{-1}\cdot \mathrm{br}_\lambda)$.*

Proof Both sides are, by definition, equal to 1. □

Corollary 3.11 *Let $\eta \in \Sigma^{++}$. If λ is Q_H^0-admissible, then for any $r \geqslant 1$ we have*

$$p^{hr}\rho_\lambda(\eta(p)^{-r})\cdot(f_\lambda^{\mathrm{hw}} - u^{-1}\cdot \mathrm{br}_\lambda) \in p^r V_{\lambda,\mathcal{O}},$$

where $h = \langle \lambda, \eta\rangle$ as above.

Proof Since $V_{\lambda,\mathcal{O}}$ is stable under the action of G, it is certainly stable under the action of $\mathbf{G}_{m,/\mathbf{Z}_p}$ defined by η, and hence is equal to the direct sum of its eigenspaces for the powers of η. So to prove that $p^h\rho_\lambda(\eta(p)^{-1})$ maps $V_{\lambda,\mathcal{O}}$ to itself, it suffices to show that the eigenvalues of $\rho_\lambda(\eta(p)^{-1})$ on V_λ are in $p^{-h}\mathbf{Z}_p$. Clearly, these eigenvalues are given by $p^{-\langle\eta,\mu\rangle}$ where μ varies over the weights V_λ. However, since λ is the highest weight of V_λ, and η is dominant, all such weights satisfy $\langle\eta,\mu\rangle \leqslant \langle\eta,\lambda\rangle = h$.

To prove the second statement, it suffices to take $r = 1$. By Proposition 3.10, $f_\lambda^{\mathrm{hw}} - u^{-1}\cdot\mathrm{br}_\lambda$ has zero projection to the highest relative weight space. So it is a sum of vectors lying in relative weight spaces of strictly smaller weights, and hence $p^h\rho_\lambda(\eta(p)^{-1})$ acts on these with eigenvalues in $p\mathbf{Z}_p$. □

4 Norm-Compatible Families

4.1 Subgroups

Let us choose $\eta \in \Sigma^{++}$ (which will remain fixed throughout the construction), and set $\tau = \eta(p)$. For $r \geqslant 0$, let us write

$$N_r = \tau^r N_G(\mathbf{Z}_p)\tau^{-r}, \qquad \overline{N}_r = \tau^{-r}\overline{N}_G(\mathbf{Z}_p)\tau^r.$$

Note that N_r and $\overline{N}_r$ are contained (usually strictly) in the kernel of reduction modulo p^r. We set $L_r = \{\ell \in L_G(\mathbf{Z}_p) : \ell \bmod p^r \in L_G^0\}$. Then the "norm-compatibility machine" of [13] uses the open compact subgroups U_r and V_r of $G(\mathbf{Z}_p)$ (see Sect. 4.4 of *op.cit.*) defined for $r \geqslant 1$ by

$$U_r = \overline{N}_0 L_r N_r, \qquad V_r = \overline{N}_r L_r N_0.$$

Note that $(U_r)_{r\geqslant 1}$ is a decreasing sequence of open neighbourhoods of $\overline{Q}_G^0(\mathbf{Z}_p)$, and $(V_r)_{r\geqslant 1}$ similarly of $Q_G^0(\mathbf{Z}_p)$.

4.2 Hecke Actions

For $r \geqslant 1$, the groups U_r and V_r have Iwahori decompositions, so we have an integrally normalised Hecke operator $\mathcal{T}'_\eta$ on cohomology at these levels, defined in weight λ using the normalisation factor p^h where $h = \langle \eta, \lambda \rangle$. Similarly, we obtain an integrally normalised pushforward map

$$[\tau^r]_* : H^i(U_r, V_{\lambda,\mathcal{O}}) \to H^i(V_r, V_{\lambda,\mathcal{O}}),$$

whose effect on the coefficient sheaves is given by $p^{hr}\rho_\lambda(\tau^{-r})$. Note that this map commutes with the actions of $\mathcal{T}'_\eta$.

4.3 Pushforward from H

Definition 4.1 Let c be the integer

$$\dim_{\mathbf{R}}(Y_{\mathcal{G}}) - \dim_{\mathbf{R}}(Y_{\mathcal{H}}) - \mathrm{rk}_{\mathbf{R}}\left(\frac{Z_H}{Z_G \cap H}\right) \geqslant 0,$$

where $\mathrm{rk}_{\mathbf{R}}$ denotes the split rank over $\mathbf{R}$. (Note that this is simply $2\,(\dim_{\mathbf{C}}(Y_{\mathcal{G}}) - \dim_{\mathbf{C}} Y_{\mathcal{H}})$, if $\mathcal{H}$ and $\mathcal{G}$ admit compatible Shimura data.)

For any open compacts $U_G \subseteq G(\mathbf{Z}_p)$ and $U_H \subset H(\mathbf{Z}_p)$ with $U_H \subset H \cap U_G$, and any H-admissible weight λ, the maps of Corollary 3.8 give us morphisms

$$\iota_\star^{[\lambda]} : H^i\left(U_H, V_{\mu,\mathcal{O}}^{H,\min}\right) \to H^{i+c}\left(U_G, V_{\lambda,\mathcal{O}}^G\right).$$

(As usual, we work with some fixed tame levels $K_{\mathcal{G}}^p$ and $K_{\mathcal{H}}^p$ away from p; we shall suppose $K_{\mathcal{H}}^p = K_{\mathcal{G}}^p \cap \mathcal{H}(\mathbf{A}_f^p)$.)

Remark 4.2 In the Shimura-variety setting, $c = 2d$ is always even, and the push-forward map for étale cohomology takes the form

$$H_{\text{et}}^i\left(U_H, V_{\mu,\mathcal{O}}^{H,\min}(j)|\right) \to H_{\text{et}}^{i+2d}|\left(U, V_{\lambda,\mathcal{O}}^G(j+d)\right)$$

for any i, j, where (j) denotes a cyclotomic twist. ◇

Now let λ be a Q_H^0-admissible weight, and μ the corresponding weight for H. We suppose we are given a compatible family of classes (for some given $i \geqslant 0$),

$$z_H^{[\mu]} \in H_{\text{Iw}}^i\left(Q_H^0, V_{\mu,\mathcal{O}}^{H,\min}\right).$$

Remark 4.3 We remind the reader that the special case when $Q_H^0 = H$, $\mu = 0$, and $z_H^{[\mu]}$ is the identity class in $H^0(H(\mathbf{Z}_p), \mathcal{O})$ is permitted (and not at all trivial). ◇

Definition 4.4 We let $z_{G,r}^{[\lambda]} \in H^{i+c}\left(U_r, V_{\lambda,\mathcal{O}}^G\right)$ be the following class: it is the image of $z_H^{[\mu]}$ under the composite map

$$\begin{aligned} H_{\text{Iw}}^i\left(Q_H^0, V_{\mu,\mathcal{O}}^{H,\min}\right) &\xrightarrow{\text{pr}^\star} H_{\text{Iw}}^i\left(Q_H^0 \cap uU_ru^{-1}, V_{\mu,\mathcal{O}}^{H,\min}\right) \\ &\xrightarrow{\iota_\star^{[\lambda]}} H^{i+c}\left(uU_ru^{-1}, V_{\lambda,\mathcal{O}}^G\right) \\ &\xrightarrow{[u]_*} H^{i+c}\left(U_r, V_{\lambda,\mathcal{O}}^G\right). \end{aligned}$$

We set

$$\xi_{G,r}^{[\lambda]} = [\tau^r]_*(z_{G,r}^{[\lambda]}) \in H^{i+c}\left(V_r, V_{\lambda,\mathcal{O}}^G\right).$$

(Here $\text{pr}^\star$ is the natural pullback map, and $[\tau]_*$ the integrally normalised pushforward as above.)

Theorem 4.5 *If $\text{pr}_{V_{r+1},V_r,\star}$ denotes pushforward along the map $Y_{\mathcal{G}}(V_{r+1}) \to Y_{\mathcal{G}}(V_r)$, then we have*

$$\text{pr}_{V_{r+1},V_r,\star}\left(\xi_{G,r+1}^{[\lambda]}\right) = \mathcal{T}_\eta' \cdot \xi_{G,r}^{[\lambda]},$$

for all $r \geqslant 1$.

Proof This is one of the main technical result of [13], Proposition 4.5.2 of *op.cit.*. (Note that the notations V_n, τ have the same meaning in *op.cit.* as here, while the the $\xi_{G,r}$ of *op.cit.* is our $\xi_{G,r}^{[\lambda]}$, and the Hecke operator $\mathcal{T}$ is our $\mathcal{T}'_\eta$.) □

Definition 4.6 We define

$$\xi_{G,\infty}^{[\lambda]} := \left((\mathcal{T}'_\eta)^{-r} e'_{Q_G} \xi_{G,r}^{[\lambda]} \right)_{r \geqslant 1} \in e'_{Q_G} H_{\mathrm{Iw}}^{(i+c)} \left(V_\infty, V_{\lambda,\mathcal{O}}^G \right),$$

and we let

$$\iota_\infty^{[\lambda]} : H_{\mathrm{Iw}}^i \left(Q_H^0, V_{\mu,\mathcal{O}}^{H,\min} \right) \longrightarrow e'_{Q_G} H_{\mathrm{Iw}}^{(i+c)} \left(Q_G^0, V_{\lambda,\mathcal{O}}^G \right)$$

to be the map sending $z_H^{[\mu]}$ to $\xi_{G,\infty}^{[\lambda]}$.

Note that λ (and hence μ) are fixed in the above construction. The goal of this paper is to compare the classes $\xi_{G,\infty}^{[\lambda]}$ for different values of λ.

5 Moment Maps

5.1 Definitions

Let $r \geqslant 1$, and let $\lambda \in X_+^\bullet(S_G/S_G^0)$. (For the moment we do not assume λ is H-admissible.)

Since f_λ^{hw} is a highest-weight vector, and its weight is trivial on L_G^0, the mod p^r reduction of this vector is invariant under the group $Q_G^0(\mathbf{Z}/p^r)$, which is the image of V_r in $G(\mathbf{Z}/p^r)$. Thus the reduction of f_λ^{hw} defines a class

$$f_{\lambda,r}^{\mathrm{hw}} \in H^0(V_r, V_{\lambda,r}),$$

and these have the property

$$f_{\lambda,r+1}^{\mathrm{hw}} \bmod p^r = \mathrm{pr}_{V_{r+1},V_r}^\star \left(f_{\lambda,r}^{\mathrm{hw}} \right) \in H^0(V_{r+1}, V_{\lambda,r}). \tag{2}$$

Definition 5.1 The *moment map* of weight λ and level r is the map

$$\mathrm{mom}_G^\lambda : H_{\mathrm{Iw}}^c \left(Q_G^0, \mathcal{O} \right) \longrightarrow H_{\mathrm{Iw}}^c \left(Q_G^0, V_{\lambda,\mathcal{O}} \right)$$

defined as follows: if $\underline{x} = (x_s)_{s \geqslant 1} \in \varprojlim_s H^c(V_s, \mathcal{O})$, then $\mathrm{mom}_G^\lambda(\underline{x})$ is given by

$$\left((x_s \bmod p^s) \cup f_{\lambda,s}^{\mathrm{hw}} \right)_{s \geqslant 1} \in \varprojlim_s H^c \left(V_s, V_{\lambda,s} \right) \cong H_{\mathrm{Iw}}^c \left(Q_G^0, V_{\lambda,\mathcal{O}} \right).$$

which is well-defined by (2).

By construction this map commutes with the action of the Hecke operator $\mathcal{T}'_\eta$, and hence with the anti-ordinary projector e'_{Q_G}. (It is *not* compatible with the action of the abelian group $S_G(\mathbf{Z}_p)$.) Moreover, we can make the same definitions with the lattice $V_{\lambda,\mathcal{O}}$ replaced with $V_{\lambda,\mathcal{O}}^{\min}$ (and the moment maps for the two lattices are compatible).

5.2 *Twist-Compatibility*

We now bring the group H back into the picture. Suppose $\lambda \in X_+^\bullet(S_G)$ is Q_H^0-admissible, and $\mu = \mu(\lambda)$ as above. We consider the following diagram of maps:

$$\begin{array}{ccc} H^i_{\mathrm{Iw}}\left(Q^0_H, \mathcal{O}\right) & \xrightarrow{\iota^{[0]}_\infty} & e'_{Q_G} H^{(i+c)}_{\mathrm{Iw}}\left(Q^0_G, \mathcal{O}\right) \\ \big\downarrow \mathrm{mom}^\mu_H & & \big\downarrow \mathrm{mom}^\lambda_G \\ H^i_{\mathrm{Iw}}\left(Q^0_H, V^{H,\min}_{\mu,\mathcal{O}}\right) & \xrightarrow{\iota^{[\lambda]}_\infty} & e'_{Q_G} H^{(i+c)}_{\mathrm{Iw}}\left(Q^0_G, V^G_{\lambda,\mathcal{O}}\right). \end{array}$$

The following is the key technical result of the present paper:

Theorem 5.2 *The above diagram is commutative.*

Remark 5.3 As in Remark 2.17 above, this can be generalised so that instead of comparing trivial coefficients with coefficients in $V^G_{\lambda,\mathcal{O}}$, we compare coefficients $V_{\nu,\mathcal{O}}$ and $V_{\lambda+\nu,\mathcal{O}}$, where ν is not necessarily trivial on L^{der}. We leave the details to the interested reader. ◇

Proof From the definition of the moment maps, it suffices to prove the following claim: suppose we define $z_H^{[\mu]} = \mathrm{mom}^\mu_H(z_H^{[0]})$ and apply the constructions above in weight 0 and in weight λ. Then for each $r \geqslant 1$ we have the following equality of elements of $H^c(V_r, V_{\lambda,r})$:

$$\xi^{[\lambda]}_{G,r} \bmod p^r = (\xi^{[0]}_{G,r} \bmod p^r) \cup f^{\mathrm{hw}}_{\lambda,r}.$$

Recall that $V_r = \tau^{-r} U_r \tau^r$. We thus have an isomorphism $\tau^r : Y_G(U_r) \to Y_G(V_r)$, and our (integrally normalised) pushforward map $H^c(Y_G(U_r), V_{\lambda,r}) \to H^c(Y_G(V_r), V_{\lambda,r})$ can be written more explicitly as the composite

$$H^c(Y_G(U_r), V_{\lambda,r}) \longrightarrow H^c(Y_G(U_r), (\tau^r)^* V_{\lambda,r}) \xrightarrow[\cong]{(\tau^r)_\star} H^c(Y_G(V_r), V_{\lambda,r}),$$

where the first map is given by the morphism of sheaves $V_{\lambda,r} \to (\tau^r)^* V_{\lambda,r}$ corresponding to the endomorphism $A = p^{hr} \rho_\lambda(\tau^{-r})$ of $V_{\lambda,\mathcal{O}}$.

Since $V_{\lambda,\mathcal{O}}$ is an admissible lattice, it is the direct sum of its intersections with the weight spaces of V_λ for the action of S. On the highest-weight space, A acts as the identity, and on all other weight spaces, it acts as multiplication by p^{nr} for some $n \geqslant 1$. Hence, modulo p^r, the map A factors through projection to the highest weight space.

Since A acts trivially on the highest-weight space, we have $A(f^{\mathrm{hw}}_{\lambda,r}) = (\tau^r)^\star(f^{\mathrm{hw}}_{\lambda,r})$. So it suffices to prove that

$$z^{[\lambda]}_{G,r} \bmod p^r = \left((z^{[0]}_{G,r} \bmod p^r) \cup f^{\mathrm{hw}}_{\lambda,r}\right) \pmod{\ker A},$$

as elements of $H^{i+c}(U_r, V_{\lambda,r})$. Via adjunction between pushforward and pullback, we are reduced to showing that the classes $u^{-1} f^H_{\lambda,r}$ and $f^{\mathrm{hw}}_{\lambda,r}$, viewed as vectors in $V_{\lambda,r}$ invariant under $u^{-1} H u \cap U_r$, agree modulo $\ker(A)$; and this is precisely the statement of Corollary 3.11. □

Applying the maps in the diagram to the class $z^{[0]}_H$, and defining $z^{[\mu]}_H$ using the moment maps as above, we have proved the following:

Theorem 5.4 *Let* $\xi_{G,\infty} := \xi^{[0]}_{G,\infty} \in e'_{Q_G} H^{(i+c)}_{\mathrm{Iw}}\left(Q^0_G, \mathcal{O}\right)$. *Then, for all* Q^0_H-*admissible* λ *and all* $r \geqslant 1$, *we have*

$$\mathrm{mom}^\lambda_{G,V_r}\left(\xi_{G,\infty}\right) = (\mathcal{T}'_\eta)^{-r} e'_{Q_G} \cdot \xi^\lambda_{G,r}.$$

(We shall give a formula in the next section for the moments of $\xi_{G,\infty}$ at parahoric level, which may be more convenient in applications than the level groups V_r used above.)

Remark 5.5 Thus the single Iwasawa-cohomology class $\xi_{G,\infty}$ "knows" the classes $\xi^{[\lambda]}_{G,r}$ (or at least their projections to the Q_G-anti-ordinary part) for all r and all λ. As $\xi_{G,\infty}$ lies in the module $e'_{Q_G} H^{(i+c)}_{\mathrm{Iw}}\left(Q^0_G, \mathcal{O}\right)$, which is finitely-generated over $\mathcal{O}[[\mathfrak{S}]]$ by Theorem 2.16, this forces very strong compatibilities between the $\xi^{[\lambda]}_{G,r}$ for varying λ, which is the crucial input needed to prove explicit reciprocity laws. ◇

For the next corollary, we shall suppose that $H^*(-)$ denotes étale cohomology of the canonical model, and moreover that G and H are of PEL type, so we have a formalism of *motivic cohomology with coefficients* (see e.g. Sect. 3.1 of [10]). In this setting, we shall say that a class in $H^*_{\mathrm{et}}(Y_G(U), V_{\lambda,\mathcal{O}})$, is *motivic* if, after extending scalars to the fraction field K of $\mathcal{O}$, it lies in the image of the natural map from motivic cohomology to étale cohomology (and similarly for Y_H in place of Y_G).

Corollary 5.6 *Suppose that for all* $r \geqslant 1$ *and all* Q^0_H-*admissible* λ, *the class* $z^{[\mu(\lambda)]}_{H,r}$ *is motivic. Then* $\mathrm{mom}^\lambda_{G,V_r}\left(\xi_{G,\infty}\right)$ *is motivic for all such* r *and* λ.

Note that the motivicity assumption on the classes $z^{[\mu(\lambda)]}_{H,r}$ is automatic in the "algebraic cycle" setting, i.e. when $Q^0_H = H$ (which implies $\mu(\lambda) = 0$ for all λ) and $z^{[0]}_{H,r}$ is the identity class in H^0, since the identity class in étale cohomology is vacuously motivic (it is the image of the identity class in motivic cohomology). A more subtle example where the motivicity condition also holds is when $z^{[\mu(\lambda)]}_{H,r}$ is the GL_2 Eisenstein class, by the main result of [8].

Remark 5.7 Special cases of this computation appear in many of our previous works, such as:

- Theorem 6.3.4 of [9] is the case $\mathcal{G} = \mathrm{GL}_2 \times \mathrm{GL}_2$, $\mathcal{H} = \mathrm{GL}_2$, with Q_H and Q_G taken to be Borel subgroups.
- The case $\mathcal{G} = \mathrm{Res}_{F/\mathbf{Q}} \mathrm{GL}_2$ and $\mathcal{H} = \mathrm{GL}_2$, with F a quadratic field, is considered in [10, Theorem 8.2.3] (for F real quadratic, using étale cohomology) and [16, Proposition 5.6] (for F imaginary quadratic, using Betti cohomology). Again, we take Q_H and Q_G here to be Borel subgroups.
- Theorem 9.6.4 of [15] is an analogous result for $\mathcal{G} = \mathrm{GSp}_4$ and $\mathcal{H} = \mathrm{GL}_2 \times_{\mathrm{GL}_1} \mathrm{GL}_2$, taking Q_G to be the Siegel parabolic subgroup. In this case we did not assume L_G/L_G^0 to be commutative.
- Theorem 11.2.1 of [14] is the case $G = \mathrm{GU}(2, 1)$, $H = \mathrm{GL}_2 \times_{\mathrm{GL}_1} \mathrm{Res}_{E/\mathbf{Q}} \mathrm{GL}_1$, for E an imaginary quadratic field.

In some of these works there are various correction terms appearing (such as the factor $(-2)^{-q}$ which appears in several places in [15]) which are, in effect, a consequence of using the "wrong" normalisations for the branching maps. In the present work, we have avoided these correction terms, by normalising the branching maps for algebraic representations using the same element u used to define the pushforward maps $\iota_\infty^{[0]}$.

We briefly mention some related works by other authors in the literature. A result roughly equivalent to the above theorem in the case of Gross–Kudla–Schoen diagonal cycles (associated to $G = (\mathrm{GL}_2)^3$, $H = \mathrm{GL}_2$ embedded diagonally) has recently been proved in [4] and independently in [1]. The methods of these works (in particular the former) appear to be closely related to the present work. The case of $G = \mathrm{GL}_2$ and H a non-split torus (the "Heegner point" setting) has also been explored, initially by Castella [3] using very different methods based on p-adic L-functions, and in a more general setting by Disegni [5] using methods rather closer to those of this paper. ◇

6 Interpolating Properties

6.1 Parahoric-Level Classes

It will be convenient to work with a slightly different family of subgroups. Let $J_G = \{g \in G(\mathbf{Z}_p) : g \bmod p \in Q_G\}$ and $\bar{J}_G = \{g \in G(\mathbf{Z}_p) : g \bmod p \in \bar{Q}_G\}$. On the other hand, let $J_H^0 = \{h : h \bmod p \in Q_H^0\}$. Then the quotient $J_H^0 \backslash G(\mathbf{Z}_p)/\bar{J}_G$ is identified with the $Q_H^0(\mathbf{F}_p)$-orbits on $\mathcal{F}(\mathbf{F}_p)$, and the double coset of u, for u as in §3.1 above, represents the unique open orbit. Thus we have a pushforward map

$$u_\star \circ \iota_\star : H^i(J_H^0, V_{\mu,\mathcal{O}}^{H,\min}) \to H^{i+c}(\bar{J}_G, V_{\lambda,\mathcal{O}}^G),$$

and this map is independent of the choice of u in its orbit (except via the normalisation of the branching map).

Proposition 6.1 *The map $\mathcal{B} : H^*(J_G, V_{\lambda,\mathcal{O}}) \to H^*(\bar{J}_G, V_{\lambda,\mathcal{O}})$ given by pullback to level $\bar{J}_G \cap J_G$ composed with the trace map, commutes with the action of the operators $\mathcal{T}'_\eta$, and induces an isomorphism on the ordinary part for these operators.*

Proof Let $\mathcal{A}$ be the map $H^*(\bar{J}_G, V_{\lambda,\mathcal{O}}) \to H^*(J_G, V_{\lambda,\mathcal{O}})$, given by the double coset $[J_G \eta(p)^{-1} \bar{J}_G]$ (normalised as in Definition 2.9). Then one checks that the composite of $\mathcal{A}$ and $\mathcal{B}$, either way around, is $\mathcal{T}'_\eta$; and the result follows. □

For λ an H-admissible weight, and μ the corresponding weight of H, we have a class

$$\xi_{J_G}^{[\lambda]} = \mathcal{B}^{-1}\left(e'_Q \cdot (u_\star \circ \iota_\star)\,(z_{J_H^0}^{[\mu]})\right) \in H^{i+c}(J_G, V_{\lambda,\mathcal{O}}),$$

where $z_{J_H^0}^{[\mu]}$ is the image of $z_H^{[\mu]}$ at level J_H^0, pulled back to $J_H^0 \cap u\bar{J}_G u^{-1}$.

Proposition 6.2 *For each Q_H^0-admissible λ, we have*

$$\xi_{J_G}^{[\lambda]} = \mathrm{mom}_{G,J_G}^{[\lambda]}\left(\xi_{G,\infty}\right).$$

Proof This follows by an argument using Cartesian diagrams which is very similar to the proof of the norm-compatibility relations; we leave the details to the reader. □

There is an analogous map $\mathcal{B}^* : H^*(\bar{J}_G, V_{\lambda,\mathcal{O}}^\vee) \to H^*(J_G, V_{\lambda,\mathcal{O}}^\vee)$ defined by interchanging the roles of Q_G and $\bar{Q}_G$, and one checks that $\mathcal{B}^*$ is the transpose of $\mathcal{B}$ with respect to Poincaré duality. This commutes with the "dominant" Hecke operator $\mathcal{T}_\eta$ defined by the double coset of $\eta(p)$ (the transpose of $\mathcal{T}'_\eta$). So, if we are given some $v \in H_c^*(J_G, V_{\lambda,\mathcal{O}}^\vee)$ which is an ordinary eigenvector for $\mathcal{T}_\eta$, in the cohomological degree complementary to $i + c$, we can define $\bar{v} = (\mathcal{B}^*)^{-1}(v)$, which is a $\mathcal{T}$-eigenvector at level $\bar{J}_G$; and the Poincaré duality pairing between $\xi_{J_G}^{[\lambda]}$ and v is given by

$$\left\langle \xi_{J_G}^{[\lambda]}, v \right\rangle_{J_G} = \left\langle (u_\star \circ \iota_\star)\,(z_{J_H}^{[\mu]}), \bar{v} \right\rangle_{\bar{J}_G} = \left\langle z_{J_H^0}^{[\mu]}, \iota^* u^*(\bar{v}) \right\rangle_{J_H^0 \cap u\bar{J}_G u^{-1}}.$$

Remark 6.3 It is natural to expect that the latter expression—an "open-orbit cohomological period" at parahoric level—should be related to cohomological periods at prime-to-p level, via an appropriate Euler factor. For examples of statements of this kind for specific groups G and H, see e.g. [9, Theorem 5.7.6] for $\mathrm{GL}_2 \subset \mathrm{GL}_2 \times \mathrm{GL}_2$, and the results for GSp_4 proved in [12] and recalled in the next section. It would be interesting to attempt to find a general recipe for this Euler factor; at least in the special case when $Q_H^0 = H$ and Q_G is a Borel, this may be possible using the results of Sakellaridis, in particular [25, Theorem 1.3.1]. ◇

7 Applications to GSp_4

We now consider applications of the above theory to Euler systems for GSp_4.

7.1 A Variant of the Setting of [15]

Let $0 \leqslant q \leqslant a$, $0 \leqslant r \leqslant b$ be integers; and let $G = \mathrm{GSp}_4$. We choose a set of primes $S \ni p$, and auxiliary data $K_S, \underline{\phi}_S, W, c_1, c_2$, as in Sect. 8.4.6 of [15]; these will remain fixed throughout the following discussion. For simplicity, we also assume that the function $\underline{\phi}_S$ has the form $\phi_{1,S} \otimes \phi_{2,S}$ where each $\phi_{i,S}$ vanishes at $(0,0)$ and transforms under $\mathbf{Z}_S^\times$ by (the adelic character attached to) χ_i^{-1}, where χ_i is a Dirichlet character unramified outside S. This allows us to define classes

$$ {}_{c_1,c_2}z_{\mathrm{et}}^{[a,b,q,r]} \in H^4_{\mathrm{et}}\left(G(\mathbf{Z}_p), \mathscr{D}^{a,b}_{\mathbf{Z}_p}(3-q)\right), $$

which are given by the same construction as *loc.cit.* (with $M = 1$) but taking the test data at p to be the spherical test data, so $\underline{\phi}_p = \mathrm{ch}(\mathbf{Z}_p^4)$ and $\xi_p = \mathrm{ch}(G(\mathbf{Z}_p))$.

We now consider the group

$$ e'_{B_G} H^4_{\mathrm{et,Iw}}\left(N_G(\mathbf{Z}_p), \mathbf{Z}_p(3)\right), $$

where B_G is the upper-triangular Borel and N_G its unipotent radical. This is a finitely-generated module over the Iwasawa algebra of $T_G(\mathbf{Z}_p)$, and we can identify this group with $(\mathbf{Z}_p^\times)^3$ in such a way that the character (a, b, q) corresponds to the highest weight of $\mathscr{D}^{a,b}(-q)$. Thus the fibre of the Iwasawa cohomology at (a, b, q) maps canonically (via the edge map of Eq. 1) to the ordinary part of the cohomology of $\mathscr{D}^{a,b}(3-q)$ at Iwahori level.

Given an automorphic representation Π of G as in §10 of *op.cit.*, of weight (a, b), good ordinary at p, we can identify the Π^*-eigenspace in the ordinary cohomology at Iwahori level with the Π^*-eigenspace at prime-to-p level, via a suitable Hecke operator (compare Note 17.1.6 of [17] for Siegel-parahoric level). We label the Hecke parameters of Π as $(\alpha, \beta, \gamma, \delta)$ in increasing order of valuation, so $v_p(\alpha) = 0$ and $v_p(\beta) = a + 1$.

Theorem 7.1 *There exists a class*

$$ {}_{c_1,c_2}z_{et} \in H^4_{et,\mathrm{Iw}}\left(N_G(\mathbf{Z}_p), \mathbf{Z}_p(3)\right) \hat{\otimes} \mathbf{Z}_p[[\mathbf{Z}_p^\times]] $$

such that for all a, b, q, r as above, and any cuspidal automorphic representation Π of weight (a, b) ordinary at p, the image of ${}_{c_1,c_2}z_{et}$ under evaluation at (a, b, q, r) and projection into $H^1(\mathbf{Q}, W^*_{\Pi_f}(-q))$ is $(-2)^{-q}\mathcal{E} \cdot \mathrm{pr}_\Pi\left({}_{c_1,c_2}z_{et}^{[a,b,q,r]}\right)$, where $\mathcal{E}$ is the Euler factor

$$\left(1 - \frac{p^q}{\alpha}\right)\left(1 - \frac{\beta}{p^{1+q}}\right)\left(1 - \frac{\gamma}{p^{1+q}}\right)\left(1 - \frac{\delta}{p^{1+q}}\right)$$
$$\times \left(1 - \frac{p^{a+1+r}\chi_2(p)}{\alpha}\right)\left(1 - \frac{p^{a+1+r}\chi_2(p)}{\beta}\right)\left(1 - \frac{\gamma}{p^{a+2+r}\chi_2(p)}\right)\left(1 - \frac{\delta}{p^{a+2+r}\chi_2(p)}\right).$$

Remark 7.2 This refines [17, Theorem 17.2.2], in which (a, q) were varying and (b, r) were fixed (but only Siegel-ordinarity was assumed, rather than full Borel-ordinarity as here). In the above result, the parameter space is four-dimensional and all four parameters (a, b, q, r) are varying independently. ◇

Proof We consider the groups $\widetilde{G} = \mathrm{GSp}_4 \times \mathrm{GL}_1$, and $\widetilde{H} = (\mathrm{GL}_2 \times_{\mathrm{GL}_1} \mathrm{GL}_2) \times \mathrm{GL}_1$. There is an obvious embedding $\widetilde{H} \hookrightarrow \widetilde{G}$, extending the embedding $H \hookrightarrow G$ used in [15]. We can also view $\widetilde{H}$ as a double cover of $\mathrm{GL}_2 \times \mathrm{GL}_2$ via $(h_1, h_2, z) \mapsto (h_1, zh_2)$. All of these maps are compatible with the standard Shimura data, if we take the Shimura cocharacter to project trivially into the GL_1 factors of $\widetilde{G}$ and $\widetilde{H}$. The cohomology group $H^4_{\mathrm{et}}\left(N_G(\mathbf{Z}_p), \mathbf{Z}_p(3)\right) \hat{\otimes} \mathbf{Z}_p[[\mathbf{Z}_p^\times]]$ then has a natural interpretation as étale cohomology for $\widetilde{G}$ with level $N_{\widetilde{G}}(\mathbf{Z}_p)$.

We apply the machinery of [13] and the present paper to the groups $\widetilde{G}$ and $\widetilde{H}$, taking the input class $z^{[0]}_{\widetilde{H}}$ to be the pullback of the Eisenstein–Iwasawa class for $\mathrm{GL}_2 \times \mathrm{GL}_2$; the associated mirabolic subgroup $Q^0_{\widetilde{H}}$ is then given by the preimage of the usual mirabolic of $\mathrm{GL}_2 \times \mathrm{GL}_2$, namely

$$\left\{\left(\begin{pmatrix} x & \star \\ 0 & 1\end{pmatrix}, \begin{pmatrix} xy & \star \\ 0 & y^{-1}\end{pmatrix}, y\right) : x, y \in \mathbf{G}_m\right\}.$$

This group has an open orbit on the Borel flag variety of $\widetilde{G}$, with trivial stabiliser. We can now obtain a family of classes in the Iwahori level tower for $\widetilde{G}$. Comparing these with their analogues at prime-to-p level is done via an Euler-factor computation, for which we refer to [12] (the second of the two formulae given in §6.3 of *op.cit.*). □

7.2 *The Setting of [7]*

We now let $\mathcal{G} = \mathrm{GSp}_4 \times_{\mathrm{GL}_1} \mathrm{GL}_2$. As in *op.cit.* we can construct classes

$${}_e z^{[a,b,c,q]}_{\mathrm{et}} \in H^5_{\mathrm{et}}\left(G(\mathbf{Z}_p), \mathscr{W}^{a,b,c,*}_{\mathbf{Z}_p}(3-q)\right)$$

where $\mathscr{W}^{a,b,c}$ is a certain representation of G for $a, b, c \geqslant 0$, and $\max(a, c) \leqslant q \leqslant \min(a+b, a+c)$. As in the previous section, we are fixing some arbitrary test data away from p, including the smoothing parameter $e > 1$; and we are taking the test data at p to be the spherical data. We have also relabelled the parameters slightly, so q is $a + r$ in the notation of *op.cit.*.

Theorem 7.3 *There exists a class ${}_e z_{\mathrm{Iw}} \in e'_{B_G} H^5_{et,\mathrm{Iw}}\left(N_G(\mathbf{Z}_p), \mathbf{Z}_p(3)\right)$ whose specialisation at an automorphic representation $\Pi \times \Sigma$ of weight (a, b, c) twisted by $\det^q$,* *with* Π *and* Σ *good ordinary at p, is given by* $\mathcal{E} \cdot \mathrm{pr}_{\Pi\times\Sigma}\left(e z_{et}^{[a,b,c,q]}\right)$, *where* $\mathcal{E}$ *is the Euler factor*

$$\prod\left\{\left(1-\tfrac{\lambda}{p^{2+q}}\right) : \lambda = \beta\mathfrak{b}, \gamma\mathfrak{a}, \gamma\mathfrak{b}, \delta\mathfrak{a}, \delta\mathfrak{b}\right\} \cdot \prod\left\{\left(1-\tfrac{p^{1+q}}{\lambda}\right) : \lambda = \alpha\mathfrak{a}, \alpha\mathfrak{b}, \beta\mathfrak{a}\right\}$$

Here $\mathfrak{a}$ and $\mathfrak{b}$ are the Hecke parameters of Σ (with $v_p(\mathfrak{a}) = 0$, $v_p(\mathfrak{b}) = c+1$).

Proof We apply the machinery of the previous sections with $\mathcal{H} = \mathrm{GL}_2 \times_{\mathrm{GL}_1} \mathrm{GL}_2$ embedded in the usual way, and an Eisenstein class on the first factor of $\mathcal{H}$ alone. As before, we obtain a family of classes at Iwahori level; and these are related to their analogues at prime-to-p level via an Euler-factor computation, which is again carried out in [12] (Sect. 7.1). □

This result is an important ingredient in the proof of new cases of the Bloch–Kato conjecture for $\mathrm{GSp}_4 \times \mathrm{GL}_2$ in [19].

7.3 The Setting of [18]

Finally, we consider the triple product of cusp forms, as in [18]. This requires some care owing to the necessity of choosing a self-dual twist (compare Sect. 3.3 of op.cit.).

Let $\mathcal{G} = \mathrm{GSp}_4 \times \mathrm{GL}_2 \times \mathrm{GL}_2 \times \mathrm{GL}_1$, and let $\mathcal{H} = \mathrm{GL}_2 \times_{\mathrm{GL}_1} \mathrm{GL}_2$ embedded via

$$(h_1, h_2) \mapsto (h_1 \boxplus h_2, h_1, h_2, \det h_1^{-1}).$$

In this setting, we let $Q_H^0 = H$, and Q_G the upper-triangular Borel of G. Then one computes that $H \cap u\bar{Q}_G u^{-1}$ is a copy of $\mathbf{G}_m$, embedded as $z \mapsto (z, z, z, z^{-2})$. We take S_G^0 to be the subgroup

$$\left\{\left(\begin{pmatrix} \star & & & \\ & \star & & \\ & & z & \\ & & & z \end{pmatrix}, \begin{pmatrix} \star & \\ & z \end{pmatrix}, \begin{pmatrix} \star & \\ & z \end{pmatrix}, z^{-2}\right) : z \in \mathbf{G}_m\right\}.$$

With these choices, the group $\mathfrak{S}$ is once more isomorphic to $(\mathbf{Z}_p^\times)^4$; and a p-adic modular form for $\mathcal{G}$, of level $Q_G^0(\mathbf{Z}_p)$, is identified with a triple of p-adic modular forms for the groups GSp_4, GL_2 and GL_2 such that the product of their weight-characters is a square, together with a choice of square root of this character. Applying the machinery of the previous sections, combined with yet another zeta-integral computation from [12] (§8 of *op.cit.*), gives exactly Theorem 8.2.5 of [18], showing that diagonal-cycle cohomology classes for $\mathrm{GSp}_4 \times \mathrm{GL}_2 \times \mathrm{GL}_2$ vary in p-adic families.

Acknowledgements We would like to thank both Arshay Sheth and the anonymous referee for their valuable feedback on the original draft of this paper.

References

1. Bertolini, M., Seveso, M.A., Venerucci, R.: Reciprocity laws for balanced diagonal classes. Astérisque **434**, 77–174 (2022). http://www.ams.org/mathscinet-getitem?mr=4489473
2. Borel, A., Serre, J.-P.: Corners and arithmetic groups. Comment. Math. Helv. **48**, 436–491 (1973). https://doi.org/10.1007/BF02566134
3. Castella, F.: Heegner cycles and higher weight specializations of big Heegner points. Math. Ann. **356**(4), 1247–1282 (2013). https://doi.org/10.1007/s00208-012-0871-4
4. Darmon, H., Rotger, V.: p-adic families of diagonal cycles. Astérisque **434**, 29–75 (2022). http://www.ams.org/mathscinet-getitem?mr=4489472
5. Disegni, D.: The universal p-adic Gross-Zagier formula. Invent. Math. **230**(2), 509–649 (2022). https://doi.org/10.1007/s00222-022-01133-w
6. Hida, H.: Control theorems of p-nearly ordinary cohomology groups for $\mathrm{SL}(n)$. Bull. Soc. Math. France **123**(3), 425–475 (1995). http://www.ams.org/mathscinet-getitem?mr=1373742
7. Hsu, C.-Y., Jin, Z., Sakamoto, R.: An Euler system for $\mathrm{GSp}(4) \times \mathrm{GL}(2)$. preprint (2020). arXiv:2011.12894
8. Kings, G.: Eisenstein classes, elliptic Soulé elements and the ℓ-adic elliptic polylogarithm. In: Coates, J., Raghuram, A., Saikia, A., Sujatha, R. (eds.) The Bloch–Kato conjecture for the Riemann zeta function. London Math. Soc. Lecture Notes, vol. 418. Cambridge Univ. Press, Cambridge (2015). https://doi.org/10.1017/CBO9781316163757.013
9. Kings, G., Loeffler, D., Zerbes, S.L.: Rankin-Eisenstein classes and explicit reciprocity laws. Cambridge J. Math. **5**(1), 1–122 (2017). https://doi.org/10.4310/CJM.2017.v5.n1.a1
10. Lei, A., Loeffler, D., Zerbes, S.L.: Euler systems for Hilbert modular surfaces. Forum Math. Sigma **6**, e23 (2018). https://doi.org/10.1017/fms.2018.23
11. Lin, Z.Z.: Representations of Chevalley groups arising from admissible lattices. Proc. Amer. Math. Soc. **114**(3), 651–660 (1992). https://doi.org/10.2307/2159384
12. Loeffler, D.: On some zeta-integrals for unramified representations of $\mathrm{GSp}(4)$. preprint (2021). arXiv:2106.16076
13. Loeffler, D.: Spherical varieties and norm relations in Iwasawa theory. J. Théor. Nombres Bordeaux **33**(3), 1021–1043 (2021). https://doi.org/10.1007/s10884-020-09844-5
14. Loeffler, D., Skinner, C., Zerbes, S.L.: An Euler system for $\mathrm{GU}(2, 1)$. Math. Ann. **382**(3–4), 1091–1141 (2021). https://doi.org/10.1007/s00208-021-02224-4
15. Loeffler, D., Skinner, C., Zerbes, S.L.: Euler systems for GSp(4). J. Eur. Math. Soc. **24**(2), 669–733 (2022). https://doi.org/10.1007/s00208-021-02224-4
16. Loeffler, D., Williams, C.: p-adic Asai L-functions of Bianchi modular forms. Algebra Num. Th. **14**(7), 1669–1710 (2020). https://doi.org/10.2140/ant.2020.14.1669
17. Loeffler, D., Zerbes, S.L.: On the Bloch–Kato conjecture for GSp(4). preprint (2020). arXiv:2003.05960
18. Loeffler, D., Zerbes, S.L.: p-adic L-functions and diagonal cycles for $\mathrm{GSp}(4) \times \mathrm{GL}(2) \times \mathrm{GL}(2)$. preprint (2020). arXiv:2011.15064
19. Loeffler, D., Zerbes, S.L.: On the Bloch–Kato conjecture for $\mathrm{GSp}(4) \times \mathrm{GL}(2)$. preprint (2021). arXiv:2106.14511
20. Lovering, T.: Integral canonical models for automorphic vector bundles of abelian type. Algebra Num. Th. **11**(8), 1837–1890 (2017). https://doi.org/10.2140/ant.2017.11.1837
21. Ohta, M.: Ordinary p-adic étale cohomology groups attached to towers of elliptic modular curves. Compos. Math. **115**(3), 241–301 (1999). https://doi.org/10.1023/A:1000556212097
22. Ohta, M.: Ordinary p-adic étale cohomology groups attached to towers of elliptic modular curves. II. Math. Ann. **318**(3), 557–583 (2000). https://doi.org/10.1007/s002080000119
23. Pilloni, V.: Higher coherent cohomology and p-adic modular forms of singular weights. Duke Math. J. **169**(9), 1647–1807 (2020). https://doi.org/10.1215/00127094-2019-0075
24. Pink, R.: Arithmetical compactification of mixed Shimura varieties. Bonner Mathematische Schriften, vol. 209. Universität Bonn, Bonn (1990). http://www.ams.org/mathscinet-getitem?mr=1128753

25. Sakellaridis, Y.: Spherical functions on spherical varieties. Amer. J. Math. **135**(5), 1291–1381 (2013). https://doi.org/10.1353/ajm.2013.0046
26. Sheth, A.: Control theorems for Hilbert modular varieties. Canad. Math. Bulletin. **68**, 530–549 (2025). https://doi.org/10.4153/S0008439524000791
27. Tilouine, J., Urban, E.: Several-variable p-adic families of Siegel–Hilbert cusp eigensystems and their Galois representations. Ann. Sci. École Norm. Sup. (4) **32**(4), 499–574 (1999). https://doi.org/10.1016/S0012-9593(99)80021-4
28. Urban, E.: Sur les représentations p-adiques associées aux représentations cuspidales de $GSp_4/\mathbf{Q}$. In: Formes automorphes. II. Le cas du groupe GSp(4). Astérisque, vol. 302, pp. 151–176. Soc. Math. France (2005). http://www.ams.org/mathscinet-getitem?mr=2234861
29. Weibel, C.: An introduction to homological algebra. Cambridge Studies in Advanced Mathematics, vol. 38. Cambridge Univ. Press, Cambridge (1994). https://doi.org/10.1017/CBO9781139644136

Reciprocity Laws for Generalized Heegner Classes

Marco Adamo Seveso

To Massimo, with affection and admiration.

Abstract We exemplify in the toy case of generalized Heegner cycles how the methods of [10] yield p-adic interpolation of generalized Heegner classes and deduce a reciprocity law relating the resulting big class to the p-adic L-function constructed in [5]. We also discuss exceptional zero phenomena of these classes and how they can be improved along suitable lines.

Keywords Heegner cycles · L-functions · Reciprocity laws · Arithmetic of modular forms

1 Description and Statement of Results

1.1 *Generalized Heegner Classes*

Let $Y_1(N)$ be the moduli space of elliptic curves with $\Gamma_1(N)$-level structure and write $E_1(N)/Y_1(N)$ for the universal elliptic curve (cf. [10, Sect. 2] for conventions). Let K be an imaginary quadratic field, write H for its Hilbert ring class field and fix an elliptic curve E/H with complex multiplication by the ring of integer $\mathcal{O}_K$ of K.[1] We always assume the *Heegner hypothesis* that there is an ideal $\mathfrak{N}$ of $\mathcal{O}_K$ of norm N such that $\frac{\mathcal{O}_K}{\mathfrak{N}} = \frac{\mathbb{Z}}{N\mathbb{Z}}$. Over a finite extension F/H the Galois group G_H acts trivially on $E[\mathfrak{N}]$ and the choice of a generator t of $E[\mathfrak{N}]_{/F} = \frac{\mathcal{O}_K}{\mathfrak{N}}$ determines an F-point

[1] We always normalize the action of K on a CM elliptic curve E/L so that $\alpha \in End_{\mathbb{Z}}(E/L)$ induces the multiplication by α on the global sections $H_{\mathrm{dR}}^{1,0}(E/L)$ of $\Omega^1_{E/L}$. This determines $K = \mathbb{Q}\otimes_{\mathbb{Z}} End_{\mathbb{Z}}(E/L)$.

M. A. Seveso (✉)
Milano, Italy
e-mail: seveso.marco@gmail.com

M. Longo et al. (eds.), *Elliptic Curves and Modular Forms in Arithmetic Geometry*, Springer Proceedings in Mathematics & Statistics 527,
https://doi.org/10.1007/978-3-032-13123-2_12

$\eta_{E,t} : \mathfrak{Spec}(F) \to Y_1(N)_F$ such that $E_1(N)_\eta = E$. We fix an algebraic closure $\overline{\mathbb{Q}}$ and always view number fields L as subfields of $\overline{\mathbb{Q}}$: then $\eta_{E,t}$ determines a geometric point $\overline{\eta}_{E,t}$ and the same notation $\overline{\eta}$ will be applied to every point $\eta : \mathfrak{Spec}(L) \to Y_1(N)_L$. If

$$\mathcal{T}_1(N) := R^1(E_1(N) \to Y_1(N))_* (\mathbb{Z}_p(1))$$

denotes the relative p-adic Tate module, then $\mathcal{T}_1(N)_{\overline{\eta}_{E,t}} = H^1_{\acute{e}t}(E_{\overline{F}}, \mathbb{Z}_p(1))$ is identified with the Tate module $T := H^1_{\acute{e}t}(E_{\overline{F}}, \mathbb{Z}_p(1))$ of E and the étale fundamental group $G_{Y_1(N)_F}$ of $Y_1(N)_F$ acts on T. Let us fix a finite field extension $L/\mathbb{Q}_p$, let $r \in \mathbb{N}$ be an integer and, setting $X := T - pT$, let us write $S_r = S_r(X)$ for the space of functions $f : X \to \mathcal{O}_L$ that are restriction of polynomial functions on T such that $f(tv) = t^r f(v)$ for every $v \in T$ and $t \in \mathbb{Z}_p^\times$: via its natural action on T, the group $G_{Y_1(N)}$ acts linearly on S_r by means of the formula $(gf)(v) := f(g^{-1}v)$. We write $\mathcal{S}_r = \mathcal{S}_r(X)$ for the étale sheaf on $Y_1(N)_{\acute{e}t}$ attached to $S_r = \mathrm{Sym}^r H^1_{\acute{e}t}(E_{\overline{F}}, \mathbb{Z}_p)$ (cf. [10, Sect. 3]). We will adopt the following convention: when taking étale cohomology group, it will be always understood that we invert p in the end.

Setting $\mathcal{O}_{K,p} := \mathbb{Z}_p \otimes_{\mathbb{Z}} \mathcal{O}_K$, the action of $\mathcal{O}_{K,p}$ on T needs not be diagonalizable. We may find such a basis of eigenvectors $\{v_E, \overline{v}_E\}$ in $T_K := \mathcal{O}_K \otimes_{\mathbb{Z}} T$, that we require to be selfdual with respect to the alternating Weil pairing $\langle -, - \rangle$, i.e. $\langle v_E, \overline{v}_E \rangle = 1$. Write $T_K^\vee := \mathcal{O}_K \otimes_{\mathbb{Z}} H^1_{\acute{e}t}(E_{\overline{F}}, \mathbb{Z}_p)$ and let $\{w_E, \overline{w}_E\} \subset T_K^\vee$ be the dual basis for the alternating Poincaré pairing: $P_{E_{\overline{F}}}(v_E \otimes w_E) = P_{E_{\overline{F}}}(\overline{v}_E \otimes \overline{w}_E) = 1$ and $P_{E_{\overline{F}}}(v_E \otimes \overline{w}_E) = P_{E_{\overline{F}}}(\overline{v}_E \otimes w_E) = 0$.[2] Let us order $\{v_E, \overline{v}_E\}$ in such a way that

$$\alpha v_E = \alpha \otimes 1 \cdot v_E \text{ and } \alpha \overline{v}_E = \overline{\alpha} \otimes 1 \cdot \overline{v}_E. \tag{1}$$

The action of G_F on v_E (resp. $\overline{v}_E$) determines two characters

$$\psi_{E,p}, \overline{\psi}_{E,p} : G_F \longrightarrow \mathcal{O}_{K,p}^\times \text{ via } g v_E = \psi_{E,p}(g) v_E \text{ and } g\overline{v}_E = \overline{\psi}_{E,p}(g) \overline{v}_E.$$

(Then $\psi_{E,p}\overline{\psi}_{E,p} = \chi_{cyc}$, the p-adic cyclotomic character, $g w_E = \psi_{E,p}^{-1}(g) w_E$ and $g\overline{w}_E = \overline{\psi}_{E,p}^{-1}(g)\overline{w}_E$).[3] If $(a, b) \in \mathbb{Z} \times \mathbb{Z}$ we define the $\mathcal{O}_L[G_F]$-module $[a, b] := \mathcal{O}_L$ with the action given by the character $\psi_{E,p}^a \overline{\psi}_{E,p}^b$. Then, if $(a, b) \in \mathbb{N} \times \mathbb{N}$, we see that

$$[a, b] \simeq \mathcal{O}_L v_E^a \overline{v}_E^b \text{ and } [-a, -b] \simeq \mathcal{O}_L w_E^a \overline{w}_E^b \tag{2}$$

[2] Although $T_K = T_K^\vee$ once we forget the G_F-action, it will be convenient to keep different notations for the dual basis because they will play different roles: $\{v_E, \overline{v}_E\}$ will be used to define an invariant element giving rise to generalized Heegner classes, while $\{w_E, \overline{w}_E\}$ will be relevant in order to relate them to the generalized Heegner cycles. Of course, because the Weil pairing is the cyclotomic twist of the Poincaré pairing, we may arrange things in such a way that $w_E = \overline{v}_E$ and $\overline{w}_E = v_E$ (forgetting Galois actions).

[3] When p is split, there is no need to work with T_K in order to diagonalize the $\mathcal{O}_K$-action and $\psi_{E,p}$ and $\overline{\psi}_{E,p}$ are valued in $\mathbb{Z}_p^\times$.

via the map sending 1 to the image $v_E^a \overline{v}_E^b \in \mathrm{Sym}^{a+b} T_K$ of $v_E^{\otimes a} \otimes \overline{v}_E^{\otimes b} \in T_K^{\otimes(a+b)}$ and, respectively, the image $w_E^a \overline{w}_E^b \in \mathrm{Sym}^{a+b} T_K^\vee$ of $w_E^{\otimes a} \otimes \overline{w}_E^{\otimes b} \in T_K^{\vee\otimes(a+b)}$. Let us remark that structural morphism of $Y_1(N)_F/F$ induces a group homomorphism $G_{Y_1(N)_F} \to G_F$. If now $(a,b) \in \mathbb{Z} \times \mathbb{Z}$ and M is an $\mathcal{O}_L[G_F]$ -module (resp. a $G_{Y_1(N)_F}$-module), we define $M[a,b] := M \otimes_L [a,b]$ (resp. $M[a,b] := M \otimes_L [a,b]_{|G_{Y_1(N)_F}}$). Considering the étale sheaf on $Spec(F)_{ét}$ (resp. $Y_1(N)_{F,ét}$) attached to $[a,b]$ (resp. $[a,b]_{|G_{Y_1(N)_F}}$), the same notation $M[a,b]$ makes sense for any sheaf of $\mathcal{O}_L$-modules M on $\mathfrak{Spec}(F)_{ét}$ (resp. $Y_1(N)_{F,ét}$). More generally, this discussion applies to any number field $F' \supset F$. We write

$$S_{r;a,b} := S_r[a,b] \text{ and } \mathcal{S}_{r;a,b} := \mathcal{S}_r[a,b]. \tag{3}$$

Let us now assume that $\varphi : E \to E_\varphi$ is an isogeny defined over some field extension F_φ/F and that $\ker(\varphi) \cap E[\mathfrak{N}] = \{0\}$, implying that $(E_\varphi, \varphi \circ t)$ determines a point $\eta_\varphi := \eta_{E_\varphi, \varphi\circ t} : \mathfrak{Spec}(F_\varphi) \to Y_1(N)_{F_\varphi}$. We define

$$S^\varphi_{r;a,b} := \mathrm{Sym}^r H^1_{ét}\left(E_{\varphi,\overline{F}_\varphi}, \mathbb{Z}_p\right)[a,b] \text{ and } \mathcal{S}^\varphi_{r;a,b} := \left(\eta_\varphi^* \mathcal{S}_r\right)[a,b]$$

(Then, by the proper base change theorem, $\mathcal{S}^\varphi_{r;a,b}$ is the étale sheaf on $\mathfrak{Spec}(F_\varphi)_{ét}$ attached to $S^\varphi_{r;a,b}$). Suppose that $k \in \mathbb{N}$ is such that $r := k-2 \geq 0$. We can consider the sequence of maps[4] (cf. [10, Sect. 3]):

$$\begin{aligned} \mathrm{AJ}_{ét} : H^0_{ét}\left(\mathfrak{Spec}(F_\varphi), \mathcal{S}^\varphi_{k-2;1-l_1,1-l_2}(k-2)\right) &\xrightarrow{\eta_{\varphi *}} H^2_{ét}\left(Y_1(N)_{F_\varphi}, \mathcal{S}_{k-2;1-l_1,1-l_2}(k-1)\right) \\ &\xrightarrow{\mathrm{HS}} H^1\left(F_\varphi, H^1_{ét}\left(Y_1(N)_{\overline{F}_\varphi}, \mathcal{S}_{k-2;1-l_1,1-l_2}(k-1)\right)\right). \end{aligned} \tag{4}$$

Assuming that $L \supset K$,[5] for every $l_1, l_2 \geq 1$ such that $l_1 + l_2 = k \geq 2$, we define

$$z_{l_1,l_2} \in S_{k-2;1-l_1,1-l_2}(k-2)^{G_F} \subset S_{k-2;1-l_1,1-l_2}(k-2)^{G_{F_\varphi}} = H^0_{ét}\left(\mathfrak{Spec}(F_\varphi), \mathcal{S}_{k-2;1-l_1,1-l_2}(k-2)\right)$$

(depending on the choice of v_E) via the formula

$$z_{l_1,l_2}(xv_E + y\overline{v}_E) := x^{l_2-1} y^{l_1-1} \text{ for } xv_E + y\overline{v}_E \in X \subset T_K. \tag{5}$$

Because $\varphi : E \to E_\varphi$ is defined over F_φ, there is a push-forward map

$$\varphi_* : H^0_{ét}\left(\mathfrak{Spec}(F_\varphi), \mathcal{S}_{k-2;1-l_1,1-l_2}(k-2)\right) \longrightarrow H^0_{ét}\left(\mathfrak{Spec}(F_\varphi), \mathcal{S}^\varphi_{k-2;1-l_1,1-l_2}(k-2)\right). \tag{6}$$

As in [10, Proposition 3.2], one may prove that

---

[4] We will systematically make the following abuse of notation. If $f : X \to Y$ and $\mathcal{S}$ is a sheaf on $Y_{ét}$, then $H^i_{ét}(X, \mathcal{S}) := H^i_{ét}(X, f^{-1}\mathcal{S})$.

[5] This is not needed when p is split.

$$\mathfrak{z}_{l_1,l_2,\varphi} := \mathrm{AJ}_{\acute{e}t}\left(\varphi_* z_{l_1,l_2}\right) \in H^1\left(F_\varphi, H^1_{\acute{e}t}\left(Y_1(N)_{\overline{F}_\varphi}, \mathcal{S}_{k-2;1-l_1,1-l_2}(k-1)\right)\right)$$

is a Selmer class: this is also a consequence of Lemma 1.1 below, which relates them to the generalized Heegner cycles first introduced in [5]. For this reason, we call these elements *generalized Heegner classes.*

In order to clarify this relationship, let us consider the identification $T \overset{\sim}{\rightarrow} Hom_{\mathbb{Z}_p}(T, \mathbb{Z}_p)(1)$ induced by the Weil pairing, which then gives $\mathrm{Sym}^r(T) \overset{\sim}{\rightarrow} \mathrm{Sym}^r(Hom_{\mathbb{Z}_p}(T, \mathbb{Z}_p))(r)$. The rule that sends the image $\lambda := l_1 \cdot \ldots \cdot l_r \in \mathrm{Sym}^r(Hom_{\mathbb{Z}_p}(T, \mathbb{Z}_p))$ of $l_1 \otimes \ldots \otimes l_r$ to the function $f_\lambda(v) := l_1(v) \cdot \ldots \cdot l_r(v)$ for $v \in X$ identifies $\mathrm{Sym}^r(Hom_{\mathbb{Z}_p}(T, \mathbb{Z}_p))(r) \overset{\sim}{\rightarrow} S_r(r)$. This gives an identification

$$\mathrm{Sym}^r(\mathcal{T})(1) \overset{\sim}{\longrightarrow} \mathcal{S}_{k-2}(k-1) \tag{7}$$

of the associated (systems of étale) sheaves. Consider the étale Abel-Jacobi map

$$\mathrm{AJ}_{\acute{e}t} : CH\left(E_1^{k-2}(N) \times E^{k-2}\right) \longrightarrow H^1\left(F_\varphi, H^1_{\acute{e}t}\left(Y_1(N)_{\overline{F}_\varphi}, \mathcal{S}_{k-2}(k-1) \otimes S_{k-2}\right)\right) \tag{8}$$

obtained as the composition of the usual $H^1\left(F_\varphi, H^{2k-3}_{\acute{e}t}\left(E_1^{k-2}(N)_{\overline{F}_\varphi} \times_{\overline{F}_\varphi} E^{k-2}_{\overline{F}_\varphi}, \mathbb{Q}_p(k-1)\right)\right)$-valued Abel-Jacobi map (cf. [5, Definition 3.1]) followed by the map to $H^1\left(F_\varphi, H^1_{\acute{e}t}\left(Y_1(N)_{\overline{F}_\varphi}, \mathrm{Sym}^r(\mathcal{T})(1) \otimes S_{k-2}\right)\right)$ obtained from the canonical projection from $H^{2k-3}_{\acute{e}t}\left(E_1^{k-2}(N)_{\overline{F}_\varphi} \times_{\overline{F}_\varphi} E^{k-2}_{\overline{F}_\varphi}, \mathbb{Q}_p(k-1)\right)$ to $H^1_{\acute{e}t}\left(Y_1(N)_{\overline{F}_\varphi}, \mathrm{Sym}^r(\mathcal{T})(1) \otimes S_{k-2}\right)$ constructed in [5, (2.2.1)] and, finally, the identification (7). We remark that as a G_F-module there is a decomposition $S_{k-2} = \oplus_{a+b=k} \mathcal{O}_L w_E^{a-1} \overline{w}_E^{b-1}$ and $\mathcal{O}_L w_E^{a-1} \overline{w}_E^{b-1} \simeq [1-a, 1-b]$ via (2). Then $z_{l_1,l_2} \otimes w_E^{(l_1-1)} \overline{w}_E^{(l_2-1)}$ is a *canonical* element which does not depend on the choice of $\{v_E, \overline{v}_E\}$ and $\{w_E, \overline{w}_E\}$ subject to the above duality conditions. We write $\mathrm{AJ}^{[l_1,l_2]}_{\acute{e}t}$ to denote the composition of $\mathrm{AJ}_{\acute{e}t}$ with the map induced by the projection $S_{k-2} \rightarrow [1-l_1, 1-l_2]$ determined by isomorphism $\mathcal{O}_L w_E^{a-1} \overline{w}_E^{b-1} \simeq [1-a, 1-b]$ (thus depending on the choice of v_E).

Lemma 1.1 *Let $\Delta_\varphi \in CH\left(E_1^{k-2}(N) \times E^{k-2}\right)$ be the generalized Heegner cycle attached to $\varphi : E \rightarrow E_\varphi$ as defined in [5, Sect. 2.3]. Then*

$$\mathrm{AJ}^{[l_1,l_2]}_{\acute{e}t}\left(\Delta_\varphi\right) = \frac{1}{(l_1-1)!\,(l_2-1)!}\mathfrak{z}_{l_1,l_2,\varphi}.$$

Proof See Sect. 3 below.

1.2 Families of Generalized Heegner Classes

We will work with connected wide open disks $\mathcal{U} \subset \mathcal{W}$ and $\mathcal{V} \subset \mathcal{W}_K$ in the rigid analytic weight spaces $\mathcal{W} := Hom_{cts}\left(\mathbb{Z}_p^\times, \mathbb{G}_m\right)$ and, respectively, $\mathcal{W}_K := Hom_{cts}\left(\mathcal{O}_{K,p}^\times, \mathbb{G}_m\right)$. Let us write $\kappa_\mathcal{U} : \mathbb{Z}_p^\times \to \mathcal{O}(\mathcal{U})^\times$ and $\kappa_\mathcal{V} : \mathcal{O}_{K,p}^\times \to \mathcal{O}(\mathcal{V})^\times$ for the corresponding universal characters. For every continuous group homomorphism $\phi : \mathcal{O}(\mathcal{U})^\times \to L^\times$, setting $k := \phi \circ \kappa_\mathcal{U}$ yields an L-point $k : \mathbb{Z}_p^\times \to L^\times$ in $\mathcal{U}(L)$ and all L-valued points $\mathcal{U}(L)$ in $\mathcal{U}$ are obtained in this way: we then write $\phi : \kappa_\mathcal{U} \rightsquigarrow k$ and say that $\kappa_\mathcal{U}$ specializes to k via ϕ. The same notation applies to characters of $\mathcal{O}_{K,p}^\times$. If $k \in \mathbb{Z}$ and ε_p is an L-valued finite order character of $\mathbb{Z}_p^\times$ (resp. $\mathcal{O}_{K,p}^\times$), we write $k_{\varepsilon_p} \in \mathcal{W}(L)$ (resp. $\mathcal{W}_K(L)$) for the character $t^{k_{\varepsilon_p}} := t^{k-2}\varepsilon_p(t)$ and set $k = k_1$.

Suppose now that $\mathbf{f}$ is a p-adic Coleman family defined in the wide open disk $\mathcal{U} \subset \mathcal{W}$ of tame level $\Gamma_1(N)$ (hence N is prime to p), with tame nebentype $\varepsilon : \left(\frac{\mathbb{Z}}{N\mathbb{Z}}\right)^\times \to L^\times$ and which is an eigenvector for all the Hecke operators T_l with $l \nmid Np$ and the U_p-operator. When $k_{\varepsilon_p} \in \mathcal{U}(L)$, we have $\mathbf{f}_{k_{\varepsilon_p}} \in S_k\left(\Gamma_0(pN), \varepsilon\varepsilon_p\right)$. Then $\mathbf{f}_{k_{\varepsilon_p}}$ is not in general new at p; if it is not p-new, we let $f_{k_{\varepsilon_p}}$ be the p-new form of which $\mathbf{f}_{k_{\varepsilon_p}}$ is a p-stabilization. Let $\mathbf{f}^\#$ be the normalized Coleman newform associated to $\mathbf{f}$. We write $V\left(\mathbf{f}_{k_{\varepsilon_p}}\right)$ for the dual of the Deligne representation $V^*\left(\mathbf{f}_{k_{\varepsilon_p}}\right)$ attached to $\mathbf{f}_{k_{\varepsilon_p}}$ and apply the same notation to $f_{k_{\varepsilon_p}}$ (cf. [10, Sect. 2.4]): they are of rank two when $\mathbf{f} = \mathbf{f}^\#$ and several copies of it in general. To be more precise, the choice $\alpha_{p,k_{\varepsilon_p}}$ of one of the two roots $\alpha_{p,k_{\varepsilon_p}}$ and $\beta_{p,k_{\varepsilon_p}}$ of the Hecke polynomial at p determines a p -stabilization operator $\Pi^{\alpha*} : V\left(f_{k_{\varepsilon_p}}\right) \overset{\sim}{\to} V\left(\mathbf{f}_{k_{\varepsilon_p}}\right)$ (cf. [10, (169)]).

Let us write $\mathbb{A}_K^{p,\infty,\times} \subset \mathbb{A}_K^\times$ (resp. $\mathbb{A}_K^{p,\times}$) for the subgroup of ideles whose p and ∞ components are trivial (resp. p component is trivial) and let $U \subset \mathbb{A}_K^{p,\infty,\times}$ be an open and compact subgroup. By a p-adic family of Hecke characters of K (called overconvergent automorphic eigenform for $\mathbf{Res}_{K/\mathbb{Q}}\left(\mathbf{GL}_{1,K}\right)$ in [9]) of weight $\kappa_\mathcal{V}$ and tame level U, we mean a continuous group homomorphism $\chi : K^\times\backslash\mathbb{A}_K^\times \to \mathcal{O}(\mathcal{V})^\times$ such that $\chi_{|\mathcal{O}_{K,p}^\times} = -\kappa_\mathcal{V}$ and which contains $U \cdot \left(K_\infty^\times\right)^\circ$ in its kernel. If $\phi : \kappa_\mathcal{V} \rightsquigarrow \nu$, then $\chi_\nu := \phi \circ \chi$ is called the specialization of χ at $\phi : \kappa_\mathcal{V} \rightsquigarrow \nu$, or simply at ν. Let us fix $\sigma_p : K \subset L$, which gives rise to the group homomorphism $\sigma_p : \mathcal{O}_{K,p}^\times \to L^\times$ and let $\overline{\sigma}_p : \mathcal{O}_{K,p}^\times \to L^\times$ be the the other group homomorphism. If $\nu \in \mathcal{V}(L)$, then we say that ν is classical of infinity type (a, b) if there is a finite index subgroup $U_p \subset \mathcal{O}_{K,p}^\times$ such that $t^\nu = \sigma_p^{-a}(t)\overline{\sigma}_p^{-b}(t)$ for every $t \in U_p$. If χ is as above and $\kappa_\mathcal{V} \rightsquigarrow \nu$ with ν classical, then $\chi_\nu = \chi_p$ for a unique algebraic Hecke character χ of infinity type (a, b) (as defined in [4, (2–2)] or [24, Chap. 0, Sect. 5]), where the notation is the following. We have that $\chi : \mathbb{A}_K^\times \to L^\times$ is a character valued in a number field $F_\chi \subset L$ which is trivial on $\left(U \cap \mathcal{O}_{\mathbb{A}_K^{p,\infty}}^\times\right) U_p \cdot \left(K_\infty^\times\right)^\circ$, with U and U_p as above, $\mathcal{O}_{\mathbb{A}_K^{p,\infty}}^\times := \prod_{v \notin \{p,\infty\}} \mathcal{O}_{K_v}^\times$, and $\chi(t) = t^a\overline{t}^b$ if $t \in K^\times$ (recall we view every number field as a subfield of $\overline{\mathbb{Q}}$, hence $\sigma_\infty : K \subset \overline{\mathbb{Q}}$ has been fixed). The algebraic character $t \mapsto t^a \overline{t}^b$ gives rise to a homomorphism $\phi_{a,b} : \mathbb{A}_K^\times \to \mathbb{A}_{F_\chi}^\times$ (cf. [24, Chap. 0, Sect. 2]) and then $\chi_p(x) := \chi(x) / \phi_{a,b}(x)_p$, called the p-adic avatar of χ. Let us write $rec : \mathbb{A}_K^\times \to G_K^{ab}$ for the reciprocity law obtained by class field theory mapping a uniformizer at $\mathfrak{l}$ to the geometric Frobenius at $\mathfrak{l}$ (inverse of the Artin map). Then χ_p (resp. χ) factors through the kernel $\overline{K^\times \cdot (K_\infty^\times)^\circ}$ of the reciprocity law and, hence, it can be regarded as a continuous group homomorphism $\chi_p : G_K^{ab} \to L^\times$ (resp. $\chi : G_K^{ab} \to \mathcal{O}(\mathcal{V})^\times$).

It will be convenient to explicitly describe the p-adic families of Hecke characters as follows. According to [17, Théorème 4.1 and (4.5) Exemples 2)] (see also [25, Theorem 7.44]), possibly enlarging F by a finite extension, writing ψ_E and $\overline{\psi}_E$ for the algebraic Hecke character with p-adic avatar $\psi_{E,p}$ and, respectively, $\overline{\psi}_{E,p}$, we may write $\psi_E = \psi \circ N_{F/K}$ and $\overline{\psi}_E = \overline{\psi} \circ N_{F/K}$ for algebraic Hecke characters

$$\psi, \overline{\psi} : \mathbb{A}_K^\times \longrightarrow K^\times.$$

This means that, if we regard the associated p-adic avatar ψ_p and $\overline{\psi}_p$ as characters of G_K^{ab} valued in $\mathcal{O}_{K,p}^\times$, we have $\psi_{p|G_F} = \psi_{E,p}$ and $\overline{\psi}_{p|G_F} = \overline{\psi}_{E,p}$. We may also assume that ψ (resp. $\overline{\psi}$) is unramified at p: indeed, if ψ' (resp. $\overline{\psi}'$) has infinity type $(-1, 0)$ (resp. $(0, -1)$), the equality $\psi' = \varsigma\psi$ (resp. $\overline{\psi}' = \overline{\varsigma}\overline{\psi}$) defines a finite order character ς (resp. $\overline{\varsigma}$) for which we have, after replacing F with its finite extension cut out by the character ς (resp. $\overline{\varsigma}$), $\psi'_{p|G_F} = \psi_{E,p}$ (resp. $\overline{\psi}'_{p|G_F} = \overline{\psi}_{E,p}$); now remark that there exists ψ' (resp. $\overline{\psi}'$) as above unramified at p (see [24, Chap. 0, end of Sect. 3]). If T is a compact p-adic Lie group and $\kappa : T \to \mathcal{O}^\times$ is a continuous character valued in the topological inverse limit of affinoid L-Banach algebras, we say that κ is accessible if there is an affinoid L-algebra $\mathcal{O}'$, a continuous character $\kappa' : T \to \mathcal{O}'^\times$ and a continuous morphism of L-algebras $\phi : \mathcal{O}'^\times \to \mathcal{O}^\times$ such that $\kappa = \phi \circ \kappa'$. We will assume that there are connected affinoid disks $\mathcal{U}'$ and $\mathcal{V}'$ such that $\mathcal{U} \subset \mathcal{U}'$ and $\mathcal{V} \subset \mathcal{V}'$: then $\kappa_{\mathcal{U}}$ and $\kappa_{\mathcal{V}}$ are accessible.

It will be convenient to assume from now on that $p \neq 2$ and that, if p ramified, then we further have $p \neq 3$. It follows that $\mathcal{O}_{K,p}^\times = \Gamma \times \Delta$, where Δ is the torsion subgroup and $\Gamma := (1 + p\mathbb{Z}_p)^{\times 2} \subset \mathcal{O}_{K,p}^\times$ if p is split in K (resp. $\Gamma := 1 + \mathfrak{p}\mathcal{O}_{K,p} \subset \mathcal{O}_{K,p}^\times$, where $\mathfrak{p}$ is the prime of $\mathcal{O}_{K,p}$ above p if p is inert or ramified in K) is 2-divisible and $\simeq \mathbb{Z}_p^2$.

Lemma 1.2 *There exist continuous group homomorphisms* $\mathbf{l}_1, \mathbf{l}_2 : \mathcal{O}_{K,p}^\times \longrightarrow \mathcal{O}(\mathcal{V})^\times$ *and finite order characters* $\varkappa_0$ *and* $\varkappa_p$ *of conductor prime to p and, respectively, a power of p such that*

$$\chi = \varkappa \psi_p^{1-\mathbf{l}_1} \overline{\psi}_p^{1-\mathbf{l}_2}, \text{ where } \varkappa = \varkappa_0 \varkappa_p.$$

Furthermore, if $(l_1, l_2) \in \mathbb{Z}^2$ *and* $\varepsilon_{p,1}$ *and* $\varepsilon_{p,2}$ *are finite order character of* $\mathcal{O}_{K,p}^\times$ *such that* $\mathbf{l}_1 \rightsquigarrow l_{1,\varepsilon_{p,1}}$ *and* $\mathbf{l}_2 \rightsquigarrow l_{2,\varepsilon_{p,2}}$, *then* $\kappa_{\mathcal{V}} \rightsquigarrow v$ *of infinity type* $(l_1 - 1, l_2 - 1)$ *and* $\chi_v =$

$\varkappa\tau_{\varepsilon_{p,1},\varepsilon_{p,2}}\psi_p^{1-l_1}\overline{\psi}_p^{1-l_2}$ *with* $\tau_{\varepsilon_{p,1},\varepsilon_{p,2}} := (\varepsilon_{p,1}^{-1} \circ \psi_p)(\varepsilon_{p,2}^{-1} \circ \overline{\psi}_p)$. *All specializations at classical weights* $v \in \mathcal{V}(L)$ *are obtained in this way.*

Proof Define $\Gamma_\pm := \{\gamma \in \Gamma : \overline{\gamma} = \gamma^{\pm 1}\}$, so that $\Gamma = \Gamma_+ \times \Gamma_-$ via the projections $\chi_\pm : \Gamma \to \Gamma_\pm$ defined by the formula $\chi_\pm(\gamma) = \gamma^{\frac{1\pm c}{2}}$ where $c(\gamma) := \overline{\gamma}$ is the conjugation. As a consequence, if $\kappa : \Gamma \to R^\times$ is a continuous character, we may write $\kappa = \chi_+^{\mathbf{s}_+} \cdot \chi_-^{\mathbf{s}_-}$ for continuous characters $\mathbf{s}_\pm : \Gamma_\pm \to R^\times$. After extending $\mathbf{s}_\pm$ to a character of $\Gamma = \Gamma_+ \times \Gamma_-$ and noticing that $\chi_+ = (\psi_p\overline{\psi}_p)^{1/2}$ and $\chi_- = (\psi_p/\overline{\psi}_p)^{1/2}$ up to restricting to some open subgroup $U_p \subset \Gamma$ (they have the same infinity type),[6] we find $\kappa = \psi_p^{\frac{\mathbf{s}_+ + \mathbf{s}_-}{2}} \overline{\psi}_p^{\frac{\mathbf{s}_+ - \mathbf{s}_-}{2}}$ on U_p.

We apply the above discussion to $\kappa = \chi_{|\Gamma}$. Then, writing $1 - \mathbf{l}_1$ and $1 - \mathbf{l}_2$ for an extension of $\frac{\mathbf{s}_+ + \mathbf{s}_-}{2}$ and, respectively, $\frac{\mathbf{s}_+ + \mathbf{s}_-}{2}$ to $\mathcal{O}_{K,p}^\times = \Gamma \times \Delta$, we see that the equality

$$\chi = \varkappa\psi_p^{1-\mathbf{l}_1}\overline{\psi}_p^{1-\mathbf{l}_2}$$

defines $\varkappa$ of trivial weight on some open subgroup $U_p \subset \mathcal{O}_{K,p}^\times$, hence a finite order character (because it factors through $\frac{\mathbb{A}_K^\times}{K^\times \cdot U U_p (K_\infty^\times)^\circ}$, which is finite, for suitable open and compact subgroups $U \subset \mathbb{A}_K^{p,\infty,\times}$ and $U_p \subset \mathcal{O}_{K,p}^\times$). The claim now easily follows (note that ψ_p and $\overline{\psi}_p$ have weight $(-1, 0)$ and, respectively, $(0, -1)$, see [24, Chap. 0, Sect. 4.1.1]).

Let us write $i_{\mathcal{U}} : \mathcal{O}(\mathcal{U}) \to \mathcal{O}(\mathcal{U})\widehat{\otimes}_L\mathcal{O}(\mathcal{V})$[7] and $i_{\mathcal{V}} : \mathcal{O}(\mathcal{U}) \to \mathcal{O}(\mathcal{U})\widehat{\otimes}_L\mathcal{O}(\mathcal{V})$ for the canonical morphisms. Then we may consider the character $\mathbf{l}_1 + \mathbf{l}_2 : \mathcal{O}_{K,p}^\times \to \mathcal{O}(\mathcal{V})$ and we have

$$i_{\mathcal{U}} \circ \kappa_{\mathcal{U}}, i_{\mathcal{V}} \circ (\mathbf{l}_1 + \mathbf{l}_2)_{|\mathbb{Z}_p^\times} : \mathbb{Z}_p^\times \longrightarrow \mathcal{O}(\mathcal{U})\widehat{\otimes}_L\mathcal{O}(\mathcal{V}).$$

We let $\pi_{cc} : \mathcal{O}(\mathcal{U})\widehat{\otimes}_L\mathcal{O}(\mathcal{V}) \to \mathcal{O}_{cc}$ be the universal quotient defined by the condition that

$$\pi_{cc} \circ i_{\mathcal{U}} \circ \kappa_{\mathcal{U}} = \pi_{cc} \circ i_{\mathcal{V}} \circ (\mathbf{l}_1 + \mathbf{l}_2)_{|\mathbb{Z}_p^\times} : \mathbb{Z}_p^\times \longrightarrow \mathcal{O}_{cc}. \tag{9}$$

To be explicit, because $\mathcal{U}$ and $\mathcal{V}$ are geometrically connected and, hence, $\mathcal{U} \times \mathcal{V}$ is connected as well, this universal quotient can be described as follows. First, it is necessary for $\mathcal{O}_{cc}$ not being zero, i.e. $\mathfrak{Sp}(\mathcal{O}_{cc}) \neq \phi$, that $i_{\mathcal{U}} \circ \kappa_{\mathcal{U}|\mathbb{F}_p^\times} = i_{\mathcal{V}} \circ (\mathbf{l}_1 + \mathbf{l}_2)_{|\mathbb{F}_p^\times}$, equivalent to the condition that the images of the connected wide open disk $\mathcal{U} \times \mathcal{V}$ in $\mathcal{W}$ via the morphisms which correspond to either $i_{\mathcal{U}} \circ \kappa_{\mathcal{U}}$ or $i_{\mathcal{V}} \circ (\mathbf{l}_1 + \mathbf{l}_2)_{|\mathbb{Z}_p^\times}$ factor through the same connected component of $\mathcal{W}$. Second, we need that $i_{\mathcal{U}} \circ \kappa_{\mathcal{U}}$ and $i_{\mathcal{V}} \circ (\mathbf{l}_1 + \mathbf{l}_2)_{|\mathbb{Z}_p^\times}$ restrict to the same morphism on $1 + p\mathbb{Z}_p$. This multiplicative group is topologically generated by $1 + p$. Hence, either $\mathcal{O}_{cc} = 0$ or $i_{\mathcal{U}} \circ \kappa_{\mathcal{U}|\mathbb{F}_p^\times} =$

---

6 Indeed, we may take $U_p = \Gamma$ if p does not divide the conductor of E.

7 Here the completion is in the category of locally convex L-vector spaces. Indeed, $\mathcal{O}(\mathcal{U})$ and $\mathcal{O}(\mathcal{V})$ are Fréchet algebras and, because they are metrizable, $\mathcal{O}(\mathcal{U})\widehat{\otimes}_L\mathcal{O}(\mathcal{V})$ acquires a natural Fréchet algebra structure which identifies it with $\mathcal{O}(\mathcal{U} \times \mathcal{V})$.

$i_{\mathcal{V}} \circ (\mathbf{l}_1 + \mathbf{l}_2)_{|\mathbb{F}_p^\times}$ and then π_{cc} is the quotient of $\mathcal{O}(\mathcal{U}) \widehat{\otimes}_L \mathcal{O}(\mathcal{V})$ by the ideal generated by $\varpi_{cc} := (1+p)^{\kappa_{\mathcal{U}}} \widehat{\otimes} 1 - 1 \widehat{\otimes} (1+p)^{\mathbf{l}_1+\mathbf{l}_2}$. We will write $\mathbf{l}_i^{cc} := \pi_{cc} \circ i_{\mathcal{V}} \circ \mathbf{l}_i$ and $\kappa_{\mathcal{U}}^{cc} := \pi_{cc} \circ i_{\mathcal{U}} \circ \mathbf{k}_{\mathcal{U}}$, so that we have

$$\kappa_{\mathcal{U}}^{cc} = \left(\mathbf{l}_1^{cc} + \mathbf{l}_2^{cc}\right)_{|\mathbb{Z}_p^\times} : \mathbb{Z}_p^\times \longrightarrow \mathcal{O}_{cc}. \tag{10}$$

We remark that the natural map $\pi_{cc} \circ i_{\mathcal{V}} : \mathcal{O}(\mathcal{V}) \to \mathcal{O}_{cc}$ is an isomorphism of topological L-algebras by the defining property (9). Furthermore, from its construction one easily sees that $\kappa_{\mathcal{U}}^{cc}$ is again accessible.

We write $\Lambda_{\mathcal{U}}, \Lambda_{\mathcal{V}}, \Lambda_{\mathcal{U}\times\mathcal{V}}$ and, respectively, Λ_{cc} for the subrings of power bounded elements in $\mathcal{O}(\mathcal{U}), \mathcal{O}(\mathcal{V}), \mathcal{O}(\mathcal{U}\times\mathcal{V}) = \mathcal{O}(\mathcal{U}) \widehat{\otimes}_L \mathcal{O}(\mathcal{V})$ and, respectively, $\mathcal{O}_{cc}$: they are local rings isomorphic to $\mathcal{O}_L[[T]]$, $\mathcal{O}_L[[T_1, T_2]]$, $\mathcal{O}_L[[T, T_1, T_2]]$ and, respectively, $\mathcal{O}_L[[T_1, T_2]]$ with maximal ideals $\mathfrak{m}_{\mathcal{U}}, \mathfrak{m}_{\mathcal{V}}, \mathfrak{m}_{\mathcal{U}\times\mathcal{V}}$ and, respectively, $\mathfrak{m}_{cc}$. If Λ is any one of these ring of bounded elements, we write $\Lambda_L := L \otimes_{\mathcal{O}_L} \Lambda$. Define $I_{cc} := \Lambda_{cc}\varpi_{cc}$ and $I_{cc,L} := \Lambda_{cc,L}\varpi_{cc}$. We remark that the profiniteness $\mathbb{Z}_p^\times$ and $\mathcal{O}_{K,p}^\times$ implies that κ takes value in the subring of power bounded elements for $\kappa = \kappa_{\mathcal{U}}, \kappa_{\mathcal{V}}, \kappa_{\mathcal{U}\times\mathcal{V}}$ and, respectively, $\kappa_{\mathcal{U}}^{cc}$: from now on we will view these characters as Λ_L-values for Λ_L any one of these subrings if needed with no further mention. The same observation and notation applies to $\mathcal{O}(\mathcal{V})$-valued or L-valued characters of G_K, which is again profinite group: if χ or χ_p are as above, we write $V(\chi)$ or $V(\chi_p)$ for the representation of G_K defined by χ or χ_p with $\Lambda_{\mathcal{V},L}$ or L coefficients.

We recall that we may attach to $\mathbf{f}$ a $\Lambda_{\mathcal{U},L}$-adic representation $V(\mathbf{f})$ of $G_{\mathbb{Q}}$ which interpolates the representations $V\left(\mathbf{f}_{k_{\varepsilon_p}}\right)$ (see [10, Sect. 4.3]). Let us consider the G_K-representation $V(\mathbf{f}) \otimes_L V(\chi)$, which has $\Lambda_{\mathcal{U}\times\mathcal{V},L} = \Lambda_{\mathcal{U},L} \widehat{\otimes}_L \Lambda_{\mathcal{V},L}$ coefficients.[8] Then we define the G_K-representation $V(\mathbf{f}, \chi) := \frac{V(\mathbf{f}) \otimes_L V(\chi)}{I_{cc,L} V(\mathbf{f}) \otimes_L V(\chi)}$, which has $\Lambda_{cc,L}$-coefficients (say $\mathcal{O}_{cc} \neq 0$ to avoid trivialities from now on). Similarly, if $\xi : G_K \to L^\times$ is any character, then we may consider $V\left(\mathbf{f}_{k_{\varepsilon_p}}, \xi\right) := V\left(\mathbf{f}_{k_{\varepsilon_p}}\right) \otimes_L V(\xi)$ (resp. $V\left(f_{k_{\varepsilon_p}}, \xi\right) := V\left(f_{k_{\varepsilon_p}}\right) \otimes_L V(\xi)$).

Suppose that $\mathbf{l}_1 \rightsquigarrow l_{1,\varepsilon_{p,1}}$ and $\mathbf{l}_2 \rightsquigarrow l_{2,\varepsilon_{p,2}}$, then $\kappa_{\mathcal{V}}^{cc} \rightsquigarrow \nu$ of infinity type $(l_1 - 1, l_2 - 1)$ and $\kappa_{\mathcal{U}}^{cc} \rightsquigarrow k_{\varepsilon_p}$ such that $k = l_1 + l_2$ and $\varepsilon_p = \left(\varkappa_p \tau_{\varepsilon_{p,1},\varepsilon_{p,2}}\right)_{|\mathbb{Z}_p^\times}$. We have specialization maps

$$\rho_\nu : V(\mathbf{f}, \chi) \longrightarrow V\left(\mathbf{f}_{k_{\varepsilon_p}}, \chi_\nu\right).$$

[8] Consider the inclusions $\Lambda_{\mathcal{U}\times\mathcal{V}} \subset \Lambda_{\mathcal{U}\times\mathcal{V},L} \subset \mathcal{O}(\mathcal{U}\times\mathcal{V})$ and $\Lambda_{\mathcal{U}} \otimes_{\mathcal{O}_L} \Lambda_{\mathcal{V}} \subset \Lambda_{\mathcal{U},L} \otimes_L \Lambda_{\mathcal{V},L} \subset \mathcal{O}(\mathcal{U}) \widehat{\otimes}_L \mathcal{O}(\mathcal{V})$. Via the identification $\mathcal{O}(\mathcal{U}\times\mathcal{V}) = \mathcal{O}(\mathcal{U}) \widehat{\otimes}_L \mathcal{O}(\mathcal{V})$, we have that $\Lambda_{\mathcal{U}\times\mathcal{V}}$ (resp. $\Lambda_{\mathcal{U}\times\mathcal{V},L}$) is identified with the closure of $\Lambda_{\mathcal{U}} \otimes_{\mathcal{O}_L} \Lambda_{\mathcal{V}}$ (resp. $\Lambda_{\mathcal{U},L} \otimes_L \Lambda_{\mathcal{V},L}$) in $\mathcal{O}(\mathcal{U}\times\mathcal{V})$ and this is also that $\mathfrak{m}_{\mathcal{U}\times\mathcal{V}}$-adic completion $\Lambda_{\mathcal{U}} \widehat{\otimes}_{\mathcal{O}_L} \Lambda_{\mathcal{V}}$ of $\Lambda_{\mathcal{U}} \otimes_{\mathcal{O}_L} \Lambda_{\mathcal{V}}$ (resp. $\Lambda_{\mathcal{U},L} \widehat{\otimes}_L \Lambda_{\mathcal{V},L} := \Lambda_{\mathcal{U}} \widehat{\otimes}_{\mathcal{O}_L} \Lambda_{\mathcal{V}}[1/p]$). Also, the $\mathfrak{m}_{\mathcal{U}\times\mathcal{V}}$-adic and the $\mathfrak{m} := \mathfrak{m}_{\mathcal{U}} \otimes_{\mathcal{O}_L} \Lambda_{\mathcal{V}} + \Lambda_{\mathcal{U}} \otimes_{\mathcal{O}_L} \mathfrak{m}_{\mathcal{V}}$-adic topologies on $\Lambda_{\mathcal{U}} \otimes \Lambda_{\mathcal{V}}$ are equivalent: $\mathfrak{m}^n \subset \mathfrak{m}_{\mathcal{U}\times\mathcal{V}}^n$ and $\mathfrak{m}_{\mathcal{U}\times\mathcal{V}}^{2n} \subset \mathfrak{m}^n$.

Indeed, the Fréchet topology on $\mathcal{O}(\mathcal{U}\times\mathcal{V})$ restricts on $\Lambda_{\mathcal{U}\times\mathcal{V},L}$ to a Banach space topology which induces the $\mathfrak{m}_{\mathcal{U}\times\mathcal{V}}$-adic topology on $\Lambda_{\mathcal{U}\times\mathcal{V}}$ and similarly for $\mathcal{O}(\mathcal{U})$ and $\mathcal{O}(\mathcal{V})$ (see [10, discussion before Proposition 4.2]).

We write again ρ_ν for the induced map

$$\rho_\nu : H^1(K, V(\mathbf{f}, \chi)) \longrightarrow H^1\left(K, V\left(\mathbf{f}_{k_{\varepsilon p}}, \chi_\nu\right)\right).$$

As in Sect. 1.1, we assume that the quadratic imaginary field K satisfies the Heegner hypothesis and, furthermore, that $(\mathbf{f}, \chi)$ is (Kummer) selfdual, meaning that (beside (10) one has $\varkappa_{0|\mathbb{A}_{\mathbb{Q}}^{p,\times}} = \varepsilon$. Then

$$\Pi^{\alpha *} : V\left(\mathbf{f}_{k_{\varepsilon p}}, \chi_\nu\right) \overset{\sim}{\leftarrow} V\left(f_{k_{\varepsilon p}}, \chi_\nu\right) := V\left(f_{k_{\varepsilon p}}\right) \otimes_L V(\chi_\nu) \simeq V^*\left(f_{k_{\varepsilon p}}\right) \otimes_L V\left(\chi_\nu^{-1}(1)\right)$$

and $\chi_\nu := \chi_\nu(-1)$ has infinity type (l_1, l_2).[9] Let us write $\Sigma_{geo} \subset \mathcal{V}(L)$ for the set of those classical ν such that $l_1, l_2 \geq 1, l_1 + l_2 =: k \geq 2$ and such that χ_ν is unramified at p; hence $\chi_\nu = \varkappa_0 \psi_p^{1-l_1} \overline{\psi}_p^{1-l_2}$. Then, as explained before Lemma 1.2, $\chi_\nu = \chi_p$ for a unique algebraic Hecke character χ and $\chi_\infty(x) := \chi(x) / \phi_{l_1,l_2}(x)_\infty$ is a Hecke character unramified at p: we will write $\chi_\nu(\mathfrak{p}) := \chi_\infty(\mathfrak{p})$ and similarly for $\chi_\nu(\overline{\mathfrak{p}})$, $\chi_\nu(\mathfrak{p})$ and $\chi_\nu(\overline{\mathfrak{p}})$.

Let us fix a rational integer c prime to p and Nd_K and let $\varphi : E \to E_\varphi$ is an isogeny such that $\ker(\varphi) \cap E[\mathfrak{N}] = 0$ and $End\left(E_{\varphi,\overline{\mathbb{Q}}}\right) = \mathcal{O}_c$, where $\mathcal{O}_c := \mathbb{Z} + c\mathcal{O}_K$. We enlarge F by a finite extension so that it contains the field cut out by the character $\varkappa_0$. As $V(f_k)$ appears as a quotient of $H^1_{\acute{e}t}\left(Y_1(N)_{\overline{F}_\varphi}, \mathcal{S}_{k-2}(k-1)\right)$, we have the G_{F_φ}-equivariant morphism

$$H^1_{\acute{e}t}\left(Y_1(N)_{\overline{F}_\varphi}, \mathcal{S}_{k-2;1-l_1,1-l_2}(k-1)\right) = H^1_{\acute{e}t}\left(Y_1(N)_{\overline{F}_\varphi}, \mathcal{S}_{k-2}(k-1)\right)[1-l_1, 1-l_2] \longrightarrow V(f_k)[1-l_1, 1-l_2].$$

On the other hand,

$$H^1(F_\varphi, V(f_k)[1-l_1, 1-l_2]) = H^1\left(K, V(f_k) \otimes \mathrm{Ind}_K^{F_\varphi}([1-l_1, 1-l_2])\right)$$

and $G_{F_\varphi/K}$ acts on $\mathrm{Ind}_K^{F_\varphi}([1-l_1, 1-l_2])$: the $\varkappa_0$-isotypic component is $V(\chi_\nu)$ (cf. [24, Chap. 1, Sect. 4.1.4] and [6, Sect. 2.1]), so that there is a projection π_{χ_ν} : $\mathrm{Ind}_K^{F_\varphi}([1-l_1, 1-l_2]) \to V(\chi_\nu)$. Summarizing, there is a projection

$$\pi_{f_k,\chi_\nu} : H^1\left(F_\varphi, H^1_{\acute{e}t}\left(Y_1(N)_{\overline{F}_\varphi}, \mathcal{S}_{k-2;1-l_1,1-l_2}(k-1)\right)\right) \longrightarrow H^1(K, V(f_k, \chi_\nu)).$$

Then, for every $\nu \in \Sigma_{geo}$, we can define $\mathfrak{z}_{f_k,\chi_\nu} := \pi_{f_k,\chi_\nu}\left(\mathfrak{z}_{l_1,l_2,\varphi}\right)$.

Similarly as in [10, Sect. 7.2], the local Bloch-Kato condition at $\mathfrak{p}$ (resp. $\overline{\mathfrak{p}}$) in the region Σ_{geo} interpolates and gives rise to a sub-

---

[9] In order to clarify the discussion of Sect. 1.3 below, let us remark that, with this notation, we have

$$V\left(f_{k_{\varepsilon p}}, \chi_\nu\right) = V\left(f_{k_{\varepsilon p}}\right) \otimes_L V(\chi_\nu(1)) \simeq V^*\left(f_{k_{\varepsilon p}}\right) \otimes_L V\left(\chi_\nu^{-1}\right).$$

module $H^1_{geo}\left(K_{\mathfrak{p}}, V(\mathbf{f}, \chi)\right) \subset H^1\left(K_{\mathfrak{p}}, V(\mathbf{f}, \chi)\right)$ (resp. $H^1_{geo}\left(K_{\overline{\mathfrak{p}}}, V(\mathbf{f}, \chi)\right) \subset H^1\left(K_{\overline{\mathfrak{p}}}, V(\mathbf{f}, \chi)\right)$). Let $H^1_{geo}(K, V(\mathbf{f}, \chi)) \subset H^1(K, V(\mathbf{f}, \chi))$ be defined by the requirement that $\mathrm{res}_{\mathfrak{p}}(\mathfrak{Z}) \in H^1_{geo}\left(K_{\mathfrak{p}}, V(\mathbf{f}, \chi)\right)$ and $\mathrm{res}_{\overline{\mathfrak{p}}}(\mathfrak{Z}) \in H^1_{geo}\left(K_{\overline{\mathfrak{p}}}, V(\mathbf{f}, \chi)\right)$. We can now state our first main result, proved in Sect. 2 below, which computes the specialization of a class $\mathfrak{Z}_{\mathbf{f},\chi}$ interpolating the generalized Heegner cycles at crystalline points $\nu \in \Sigma_{geo}$ (hence χ_ν is unramified at p, $\varepsilon_p = 1$ in view of the "weight-character" selfduality assumption (10) and f_k is old at p). We suppose for simplicity that p does not ramify in K.

Theorem 1.3 *There is a unique class* $\mathfrak{Z}_{\mathbf{f},\chi} \in H^1_{geo}(K, V(\mathbf{f}, \chi))$ *such that, setting* $\mathfrak{Z}_{\mathbf{f}_k,\chi_\nu} := \rho_\nu\left(\mathfrak{Z}_{\mathbf{f},\chi}\right)$ *at the crystalline point* $\nu \in \Sigma_{geo}$*, one has*

$$\Pi^{\alpha}_{*}\left(\mathfrak{Z}_{\mathbf{f}_k,\chi_\nu}\right) = \begin{cases} \left(1 - \beta_{p,k}\chi_\nu^{-1}(\mathfrak{p})\right)\left(1 - \beta_{p,k}\chi_\nu^{-1}(\overline{\mathfrak{p}})\right)\mathfrak{z}_{f_k,\chi_\nu}, & \text{if } p \text{ is split,} \\ \left(1 - \beta^2_{p,k}\chi_\nu^{-2}(p\mathcal{O}_K)\,\varepsilon(p)p^k\right)\mathfrak{z}_{f_k,\chi_\nu}, & \text{if } p \text{ is inert.} \end{cases} \tag{11}$$

We remark that $\mathfrak{Z}_{\mathbf{f},\chi}$ may very well be zero: a natural condition to be imposed in order to avoid this occurrence is that $(\mathbf{f}, \chi)$ should be of finite type $(c, \mathfrak{N}, \varepsilon)$, as defined in Sect. 1.3 below. This definition guarantees, via a reciprocity law, the non triviality of $\mathfrak{Z}_{\mathbf{f},\chi}$ in case p is split (a running assumption of Sect. 1.3). It also makes sense to be imposed when p is non-split and one expects that a reciprocity law can be proved in this setting as well (cf. Remark 1.6 below).

Let us now look at some source of exceptional zero phenomena in the region Σ_{geo}, noticing that the slope of $\mathbf{f}$ is constant because $\mathcal{U}$ is connected and contained in an affinoid. We first remark that the (Kummer) selfduality condition implies that $\chi_\nu(p\mathcal{O}_K) = \varepsilon(p)\,p^k$ and then the Hecke polynomial at p becomes

$$\left(X - \alpha_{p,k}\right)\left(X - \beta_{p,k}\right) = X^2 - a_pX + \varepsilon(p)\,p^{k-1} = X^2 - a_pX + \chi_\nu(p\mathcal{O}_K)\,p^{-1},$$

from which we see that $\alpha_{p,k}\beta_{p,k} = \chi_\nu(p\mathcal{O}_K)\,p^{-1}$. Using this relation and recalling that $\chi_\nu = \chi_\nu \mathbf{N}_{K,p}$ as idelic characters, the norm character $\mathbf{N}_{K,\infty} = |-|^{-1}_{\mathbb{A}_K} : \mathbb{A}^\times_K \to \mathbb{Q}^\times_+$ corresponding to $\mathbf{N}_{K,p} = \chi^{-1}_{cyc}$, we see that

$$1 - \alpha^{-1}_{p,k}\chi_\nu(\overline{\mathfrak{p}}) = 1 - \alpha^{-1}_{p,k}\chi_\nu(\mathfrak{p})\,p^{-1} = 1 - \beta_{p,k}\chi_\nu^{-1}(\mathfrak{p}) \tag{12}$$

and, similarly, $1 - \alpha^{-1}_{p,k}\chi_\nu(\mathfrak{p}) = 1 - \beta_{p,k}\chi_\nu^{-1}\left(\overline{\mathfrak{p}}\right)$ and $1 - \alpha^{-2}_{p,k}\varepsilon(p)p^{k-2} = 1 - \beta^2_{p,k}\chi_\nu^{-2}(p\mathcal{O}_K)\,\varepsilon(p)p^k$. (The equation (11) will be more convenient in Sect. 1.3 below). Next, we remark that there is nothing to be said in the p inert case. Indeed, we see that we may only have an isolated exceptional zero. Furthermore, if we look for an improvement of the class, we would like the corresponding Euler factor to be analytic, hence we need to impose that k should be constant but then we have only finitely many classes $\mathfrak{z}_{f_k,\chi_\nu}$ to work with. See also Remark 2.3 below for a related observation.

Assume until the end of Sect. 1.2 that p *is split*, so that $K_p^\times = \mathbb{Q}_p^\times \times \mathbb{Q}_p^\times$. We remark that, in this case, we see from Lemma 1.2 that the p-adic valuation of $\chi_\nu(\mathfrak{p})$ is constant if, writing (l_1, l_2) for the infinity type of χ_ν, we take l_1 constant (because $\pi_\mathfrak{p} = (p, 1)$ is a uniformizer at $\mathfrak{p}$ and $\overline{\psi}_p$ factors through the second copy of $\mathbb{Q}_p^\times$, so that $\overline{\psi}_p^{1-l_2}(x, 1) = 1$). More precisely, let us fix any $l_1^o \in \mathbb{Z}_{\geq 1}$ and let $\pi_{\mathbf{I}_1^{cc}=l_1^o} : \mathcal{O}_{cc} \to \mathcal{O}_{\mathbf{I}_1^{cc}=l_1^o}$ be the universal quotient defined by the condition that $\mathbf{I}_{1|\mathbb{Z}_p^\times}^{cc} = l_1^o$. It can be constructed in a similar way as π_{cc} was defined: for it to be non zero we need that $l_{1|\mathbb{F}_p^\times}^o = \mathbf{I}_{1|\mathbb{F}_p^\times}^{cc}$ and then $\pi_{\mathbf{I}_1^{cc}=l_1^o}$ is the quotient of $\mathcal{O}_{cc}$ by the ideal generated by $\varpi_{\mathbf{I}_1^{cc}=l_1^o} := (1+p)^{\mathbf{I}_1^{cc}} - (1+p)^{l_1^o}$. Then $\pi_{\mathbf{I}_1^{cc}=l_1^o} \circ \pi_{cc} \circ i_\mathcal{V} : \mathcal{O}(\mathcal{V}) \simeq \mathcal{O}_{cc} \to \mathcal{O}_{\mathbf{I}_1^{cc}=l_1^o}$ corresponds to a line $\mathcal{H}_{\mathbf{I}_1^{cc}=l_1^o} \subset \mathcal{V}$: because ψ_p and $\overline{\psi}_p$ factor through $\mathbb{Z}_p^\times \subset \mathcal{O}_{K,p}^\times$ (in view of the fact that p is split), we see from Lemma 1.2 that $\mathcal{H}_{\mathbf{I}_1^{cc}=l_1^o}$ is the Zariski closure of the set of points in $\mathcal{V}(L)$ that are classical of weight $\left(l_1^o - 1, l_2 - 1\right)$. Setting $\chi_{\mathbf{I}_1^{cc}=l_1^o} := \pi_{\mathbf{I}_1^{cc}=l_1^o} \circ \pi_{cc} \circ i_\mathcal{V} \circ \chi$, we have $\chi_{\mathbf{I}_1^{cc}=l_1^o}(\pi_\mathfrak{p}) = \chi_0(\pi_\mathfrak{p})\, \psi_p^{1-l_1^o}(\pi_\mathfrak{p})$ and the function $\mathcal{E}_{\mathbf{I}_1^{cc}=l_1^o} := 1 - \alpha_p^{-1} \frac{\chi_0(\pi_\mathfrak{p})\psi_p^{1-l_1^o}(\pi_\mathfrak{p})\phi_{l_1^o-1,0}(\pi_\mathfrak{p})_p}{\phi_{l_1^o-1,0}(\pi_\mathfrak{p})_\infty} \in \mathcal{O}_{\mathbf{I}_1^{cc}=l_1^o}$ is such that

$$\mathcal{E}_{\mathbf{I}_1^{cc}=l_1^o}(\nu) = 1 - \alpha_{p,k}^{-1}\chi_\nu(\mathfrak{p}) \text{ for every crystalline point } \nu \in \Sigma_{geo} \cap \mathcal{H}_{\mathbf{I}_1^{cc}=l_1^o}(L).$$

The notations $\Lambda_{\mathbf{I}_1^{cc}=l_1^o}$, $\Lambda_{\mathbf{I}_1^{cc}=l_1^o,L}$, $I_{\mathbf{I}_1^{cc}=l_1^o}$ and $I_{\mathbf{I}_1^{cc}=l_1^o,L}$ have analogous meaning as the notations Λ_{cc}, $\Lambda_{cc,L}$, I_{cc} and $I_{cc,L}$ considered above. Then we define the G_K-representation $V\left(\mathbf{f}, \chi_{\mathbf{I}_1^{cc}=l_1^o}\right) := \frac{V(\mathbf{f},\chi)}{I_{\mathbf{I}_1^{cc}=l_1^o,L} V(\mathbf{f},\chi)}$, which has $\Lambda_{\mathbf{I}_1^{cc}=l_1^o,L}$-coefficients (say $\mathcal{O}_{\mathbf{I}_1^{cc}=l_1^o} \neq 0$ to avoid trivialities from now on). By construction there is $\pi_{\mathbf{I}_1^{cc}=l_1^o} : V(\mathbf{f}, \chi) \to V\left(\mathbf{f}, \chi_{\mathbf{I}_1^{cc}=l_1^o}\right)$ and compatible specialization maps for every $\nu \in \mathcal{H}_{\mathbf{I}_1^{cc}=l_1^o}(L)$. We write $\mathfrak{Z}_{|\mathcal{H}_{\mathbf{I}_1^{cc}=l_1^o}}$ for the image of $\mathfrak{Z} \in H_{geo}^1(K, V(\mathbf{f}, \chi))$ by means of the map induced in cohomology by $\pi_{\mathbf{I}_1^{cc}=l_1^o}$. We will prove the following result in Sect. 2 below.

Theorem 1.4 *There is a unique class* $\mathfrak{Z}_{\mathbf{f},\chi_{\mathbf{I}_1^{cc}=l_1^o}} \in H_{geo}^1\left(K, V\left(\mathbf{f}, \chi_{\mathbf{I}_1^{cc}=l_1^o}\right)\right)$ *such that*

$$\mathfrak{Z}_{\mathbf{f},\chi|\mathcal{H}_{\mathbf{I}_1^{cc}=l_1^o}} = \mathcal{E}_{\mathbf{I}_1^{cc}=l_1^o} \cdot \mathfrak{Z}_{\mathbf{f},\chi_{\mathbf{I}_1^{cc}=l_1^o}}$$

One may consider, more generally, a condition of the form $\mathbf{I}_1^{cc} = l_{1,\varepsilon_{p,1}^o}^o$ for a finite order character $\varepsilon_{p,1}^o$ of $\mathbb{Z}_p^\times$: everything goes unchanged but we have choosen to simplify a bit the notation. Let us remark that Theorem 1.4 is only of interest when $\mathcal{E}_{\mathbf{I}_1^{cc}=l_1^o}$ vanishes at some classical $\nu \in \mathcal{H}_{\mathbf{I}_1^{cc}=l_1^o}(L)$ (*possibly outside* the region of interpolation Σ_{geo} of the class $\mathfrak{Z}_{\mathbf{f},\chi}$). Finally, we note that exchanging the roles of σ_p and $\overline{\sigma}_p$ one get the existence of an improved class $\mathfrak{Z}_{\mathbf{f},\chi_{\mathbf{I}_2^{cc}=l_2^o}}$ along the line $\mathcal{H}_{\mathbf{I}_2^{cc}=l_2^o}$.

1.3 The Reciprocity Law

As in [5, Sect. 5.3] we assume in Sect. 1.3 that the quadratic imaginary field K has odd discriminant d_K and satisfies the Heegner hypothesis stated above, that c is an odd rational integer prime to Nd_K, that the rational prime p is split in $K/\mathbb{Q}$ and prime to Nc. Hence, the order $\mathcal{O}_c := \mathbb{Z} + c\mathcal{O}_K$ admits a cyclic ideal $\mathfrak{N}_c$ of norm N, for which we have $\frac{\mathcal{O}_K}{\mathfrak{N}} = \frac{\mathcal{O}_c}{\mathfrak{N}_c} = \frac{\mathbb{Z}}{N\mathbb{Z}}$, and the character $\varepsilon^{-1} : \left(\frac{\mathbb{Z}}{N\mathbb{Z}}\right)^{\times} \to L^{\times}$ gives rise to a character $\psi_{\varepsilon} : \widehat{\mathcal{O}}_c^{\times} \to L^{\times}$ by precomposing it with $\widehat{\mathcal{O}}_c^{\times} \to \left(\frac{\mathcal{O}_c}{\mathfrak{N}_c}\right)^{\times} = \left(\frac{\mathbb{Z}}{N\mathbb{Z}}\right)^{\times}$. As in [5, pag. 1091-1094], let $\Sigma_{cc}^{(1)}(\mathfrak{N})$ (resp. $\Sigma_{cc}^{(2)}(\mathfrak{N})$) denote the set of algebraic Hecke characters $\chi : \mathbb{A}_K^{\times} \to L^{\times}$ of infinity type (l_1, l_2) satisfying $1 \leq l_1, l_2 \leq k-1$ (resp. $k \leq l_1$ and $l_2 \leq 0$), such that $V(f_k) \otimes V(\chi(1)) \simeq V^*(f_k) \otimes V(\chi^{-1})$ is (Kummer) selfdual, meaning that $l_1 + l_2 = k$ and that $\chi_{|\mathbb{A}_{\mathbb{Q}}^{\times}} |-|_{\mathbb{A}_{\mathbb{Q},f}}^{k} sgn((-)_{\infty})^k = \chi_{\infty|\mathbb{A}_{\mathbb{Q}}^{\times}} \mathbf{N}_{\infty}^{-k} = \varepsilon$ if $\chi_{\infty}(x) := \chi(x) / \phi_{l_1,l_2}(x)_{\infty}$ and $\mathbf{N}_{\infty} = |-|_{\mathbb{A}_{\mathbb{Q}}}^{-1} : \mathbb{A}_{\mathbb{Q}}^{\times} \to \mathbb{Q}_{+}^{\times}$ is the norm character, and that are of finite type $(c, \mathfrak{N}, \varepsilon)$, meaning that f_k is new, that the conductor $\mathfrak{f}_{\chi}$ of χ is divided by c and that $\chi_{|\widehat{\mathcal{O}}_c^{\times}} = \chi_{\infty|\widehat{\mathcal{O}}_c^{\times}} = \psi_{\varepsilon}$.

Following [5, before Definition 1.10], we let $\mathrm{Isog}_c^N(E)$ be the set (of isomorphism classes over $\overline{\mathbb{Q}}$) of pairs (E_{φ}, φ) such that $\varphi : E \to E_{\varphi}$ is an isogeny such that $\ker(\varphi) \cap E[\mathfrak{N}] = 0$ and $End(E_{\varphi,\overline{\mathbb{Q}}}) = \mathcal{O}_c$. The semigroup $P(\mathcal{O}_c)$ of $\mathcal{O}_c$-submodules of K that are projective of rank 1, contained in $\mathcal{O}_c$ and relatively prime to $\mathfrak{N}_c$ acts on $\mathrm{Isog}_c^N(A)$ (see [5, (1.4.7)]). This action gives rise to a left action $(\mathfrak{a}, x) \mapsto \mathfrak{a} * x$ of $P(\mathcal{O}_c)$ on triples $x = (E_{\varphi}, \varphi, t_{\varphi}, \omega_{\varphi})$ where $(E_{\varphi}, \varphi) \in \mathrm{Isog}_c^N(A)$, $t_{\varphi} \in E_{\varphi}[\mathfrak{N}]$ and ω_{φ} is an invariant differential of E_{φ} (see [5, (1.4.8)]).

We now fix a triple (E, t, ω), where (E, t) is as above and ω is an invariant differential of E, and also we fix $(E_c, \varphi_c) \in \mathrm{Isog}_c^N(E)$ and let (E_c, t_c, ω_c) be the triple determined by $\varphi_c : E \to E_c$ (see [5, (5.1.1)]). Then we define, for every $\chi \in \Sigma_{cc}^{(2)}(\mathfrak{N})$, the quantity

$$L_{\mathrm{alg}}^{1/2}(f_k, \chi^{-1}) := \sum_{[\mathfrak{a}] \in Pic(\mathcal{O}_c)} \chi_{-l_2}^{-1}(\mathfrak{a}) \Theta^{-l_2} f(\mathfrak{a} * (E_c, t_c, \omega_c)),$$

where $\Theta \in \{\Theta_{\mathrm{Hodge}}, \Theta_{\mathrm{Frob}}\}$ as defined in [5, pag. 1048 and pag. 1050] (see [5, Proposition 1.12 (3)] for the independence of the choice of Θ_{Hodge} or Θ_{Frob}) and $\chi_{-l_2} := \chi |-|_{\mathbb{A}_{K,f}}^{l_2} \phi_{l_1-l_2,0}(-)_{\infty} = \chi_{\infty} \mathbf{N}_{K,\infty}^{-l_2}$, viewed as a character of $Pic(\mathcal{O}_c)$. It is proved in [5, Theorem 5.5 and (5.1.17)] that $L_{\mathrm{alg}}^{1/2}(f, \chi^{-1})$ is algebraic and explicitly related to (the square root of) the complex special value $L(f_k, \chi^{-1}, 0)$ (which is central for the functional equation).

Let us now fix an isomorphism $\iota : \widehat{E}_{c,\mathbb{Q}_p^{\mathrm{ur}}} = \widehat{\mathbb{G}}_{m,\mathbb{Q}_p^{\mathrm{ur}}}$, which determines a differential $\omega_{\mathrm{can}} := \iota^*(dT/T)$, and define $\Omega_p \in \mathbb{Q}_p^{\mathrm{ur}}$ by means of the rule $\omega_c = \Omega_p \omega_{\mathrm{can}}$ (see [5, (5.2.2)]). Then, writing $f^{[p]}$ for the p-depletion of f (see [5, (3.8.4)]), one has

$$\begin{aligned}L_p^{1/2}\left(f_k,\chi^{-1}\right) &:= \sum\nolimits_{[\mathfrak{a}]\in Pic(\mathcal{O}_c)} \chi_{-l_2}^{-1}(\mathfrak{a})\,\Theta^{-l_2} f^{[p]}\left(\mathfrak{a} * (E_c,t_c,\omega_{\mathrm{can}})\right)\\ &= \frac{\left(1-\alpha_{p,k}\chi_\infty^{-1}(\overline{\mathfrak{p}})\right)\left(1-\beta_{p,k}\chi_\infty^{-1}(\overline{\mathfrak{p}})\right)}{\Omega_p^{l_2-l_1}} L_{\mathrm{alg}}^{1/2}\left(f_k,\chi^{-1},0\right) \quad (13)\end{aligned}$$

if $\chi \in \Sigma_{cc}^{(2)}(\mathfrak{N})$ is of infinity type (l_1,l_2) and this extends to a continuous function $\chi \mapsto L_p^{1/2}\left(f_k,\chi^{-1}\right)$ on a suitable p-adic closure $\overline{\Sigma_{cc}^{(2)}(\mathfrak{N})}$ of $\Sigma_{cc}^{(2)}(\mathfrak{N})$ (see [5, (5.2.4) and Theorem 5.9 and Proposition 5.10]).

Let us assume that, beside the selfduality assumption that was done in Sect. 1.2, $(\mathbf{f},\chi)$ is of finite type $(c,\mathfrak{N},\varepsilon)$, meaning that $\mathbf{f}=\mathbf{f}^{\#}$ is new, that the conductor $\mathfrak{f}_{\varkappa_0}$ of $\varkappa_0$ is divided by c and that $\varkappa_{0|\hat{\mathcal{O}}_c^\times}=\psi_\varepsilon$. Then, for every classical $\nu\in\mathcal{V}(L)$ such that χ_ν is unramified, the character $\chi_\nu := \chi_\nu(-1)$ belongs to $\Sigma_{cc}^{(1)}(\mathfrak{N})$ (resp. $\Sigma_{cc}^{(2)}(\mathfrak{N})$) according to whenever $1\le l_1,l_2\le k-1$ (resp. $k\le l_1$ and $l_2\le 0$) with $k:=l_1+l_2$: we let $\Sigma_{cc}^{(1)}\subset\Sigma_{geo}$ (resp. $\Sigma_{cc}^{(2)}$) be the set of these ν's. Using the q-expansion one sees from (13) that there is a unique two variable p-adic L-function $\mathcal{L}_p(\mathbf{f},\boldsymbol{\chi})\in\Lambda_{\mathcal{V},L}$ such that, for every $\nu\in\Sigma_{cc}^{(2)}$,

$$\mathcal{L}_p(\mathbf{f},\boldsymbol{\chi})(\nu) = L_p^{1/2}\left(f_k,\chi_\nu^{-1}\right).$$

We write ω_{f_k} for the differential attached to f_k and let $\omega_c^{l_2-1}\eta_c^{l_1-1}$ be the natural image of $\omega_c^{\otimes(l_2-1)}\otimes\eta_c^{\otimes(l_1-1)}\in H_{\mathrm{dR}}^1\left(E_{\overline{F}}\right)^{\otimes(k-2)}$ in its symmetric power, where η_c is the dual of ω_c under Poincaré duality. The differential $\omega_{f_k}\otimes\omega_c^{l_2-1}\eta_c^{l_1-1}$ in the de Rham cohomology of $Y_1(N)$ with coefficients in $\mathcal{S}_{k-2}(k-1)\otimes\mathcal{S}_{k-2}$ corresponds, via that isomorphism induced by (2) in de Rham cohomology, to a differential denoted by the same symbol (depending on the choice of v_E via (2)). Then, according to [5, Theorem 5.13] and Lemma 1.1 one has,[10] for every $\nu\in\Sigma_{cc}^{(1)}$,

$$\mathcal{L}_p(\mathbf{f},\boldsymbol{\chi})(\nu) = \frac{\left(1-\alpha_{p,k}\chi_\nu^{-1}(\overline{\mathfrak{p}})\right)\left(1-\beta_{p,k}\chi_\nu^{-1}(\overline{\mathfrak{p}})\right)}{\Omega_p^{l_2-l_1}}\cdot\frac{c^{1-l_2}}{(l_2-1)!}\langle\log\left(\mathrm{res}_{\mathfrak{p}}\left(\mathfrak{z}_{f_k,\chi_\nu}\right)\right),\omega_{f_k}\otimes\omega_c^{l_2-1}\eta_c^{l_1-1}\rangle_{\mathrm{dR}}, \quad (15)$$

where log is the de Bloch-Kato logarithm and $\langle -,-\rangle_{\mathrm{dR}}$ is the Poincaré de Rham pairing (recall our convention to write $\chi_\nu(\overline{\mathfrak{p}}):=\chi_\infty(\overline{\mathfrak{p}})$ if $\chi_\nu=\chi_p$ is the p-adic avatar of the algebraic Hecke character χ).

Arguing as in [10, Proposition 7.3], one deduces from [21] the existence of a unique morphism of $\Lambda_{cc,L}$-modules

---

[10] Let us remark that the element $\omega_A^{l_2-1}\eta_A^{l_1-1}$ of loc.cit. is related to our $\omega_c^{l_2-1}\eta_c^{l_1-1}$ by $\omega_A^{l_2-1}\eta_A^{l_1-1} = (l_1-1)!\,(l_2-1)!\omega_c^{l_2-1}\eta_c^{l_1-1}$ (cf. [5, (1.4.6)]). Also, the sum $\sum_{[\mathfrak{a}]\in Pic(\mathcal{O}_c)}\chi_\nu^{-1}(\mathfrak{a})\mathbf{N}_{K/\mathbb{Q}}(\mathfrak{a})$ appearing in [5, Theorem 5.13] amounts to applying π_{f_k,χ_ν}. Hence, using Lemma 1.1 we find

$$\langle\mathrm{AJ}\left(\Delta_\varphi\right),\omega_{f_k}\otimes\omega_A^{l_2-1}\eta_A^{l_1-1}\rangle_{\mathrm{dR}} = \langle\mathrm{AJ}\left(\mathrm{res}_{\mathfrak{p}}\left(\mathfrak{z}_{f_k,\chi_\nu}\right)\right),\omega_{f_k}\otimes\omega_c^{l_2-1}\eta_c^{l_1-1}\rangle_{\mathrm{dR}}. \quad (14)$$

Finally, let us remark that, if v_E is multiplied by $\lambda\in\mathcal{O}_K$, then z_{l_1,l_2} (resp. the second isomorphism of (2)) is multiplied by $\lambda^{l_1-l_2}$ (resp. $\lambda^{l_2-l_1}$); we deduce that $\mathfrak{z}_{f_k,\chi_\nu}$ (resp. $\omega_{f_k}\otimes\omega_c^{l_2-1}\eta_c^{l_1-1}$) is multiplied by the same scalar and (14) canonical.

$$\mathcal{L}og_{\mathbf{f},\chi,\mathfrak{p}} : H^1_{geo}\left(K_\mathfrak{p}, V(\mathbf{f},\chi)\right) \longrightarrow \Lambda_{cc,L}$$

having the following interpolation property. If $\mathfrak{z} \in H^1(K, V(\mathbf{f},\chi))$ is such that $\mathrm{res}_\mathfrak{p}(\mathfrak{z}) \in H^1_{geo}\left(K_\mathfrak{p}, V(\mathbf{f},\chi)\right)$ and $\mathfrak{z}_\nu$ denotes its specialization at $\nu \in \Sigma^{(1)}_{cc}$, then writing $k = l_1 + l_2$ as usual,

$$\mathcal{L}og_{\mathbf{f},\chi,\mathfrak{p}}\left(\mathrm{res}_\mathfrak{p}(\mathfrak{z})\right)(\nu) = \frac{\left(1 - \alpha_{p,k}\chi^{-1}(\bar{\mathfrak{p}})\right)}{\left(1 - \beta_{p,k}\chi^{-1}(\mathfrak{p})\right)} \cdot \frac{(-1)^{l_2-1}}{(l_2-1)!}\left\langle \log\left(\mathrm{res}_\mathfrak{p}(\mathfrak{z}_\nu)\right), \omega_{\mathbf{f}_k} \otimes \omega^{l_2-1}_{\mathrm{can}}\eta^{l_1-1}_{\mathrm{can}}\right\rangle, \tag{16}$$

where $\omega_{\mathbf{f}_k}$ is the canonical differential attached to $\mathbf{f}_k$, η_{can} is the de Rham class dual of ω_{can} and $\omega^{l_2-1}_{\mathrm{can}}\eta^{l_1-1}_{\mathrm{can}}$ is the image of $\omega^{\otimes(l_2-1)}_{\mathrm{can}} \otimes \eta^{\otimes(l_1-1)}_{\mathrm{can}} \in H^1_{\mathrm{dR}}\left(E_{\overline{F}}\right)^{\otimes(k-2)}$. Because $\omega_{\mathbf{f}_k} = \Pi^{\alpha *}\left(\omega_{f_k}\right)$ and $\omega^{l_2-1}_{\mathrm{can}}\eta^{l_1-1}_{\mathrm{can}} = \Omega_p^{l_1-l_2}\omega_c^{l_2-1}\eta_c^{l_1-1}$, we see that

$$\begin{aligned}\left\langle \log\left(\mathrm{res}_\mathfrak{p}(\mathfrak{z}_\nu)\right), \omega_{\mathbf{f}_k} \otimes \omega^{l_2-1}_{\mathrm{can}}\eta^{l_1-1}_{\mathrm{can}}\right\rangle &= \left\langle \log\left(\mathrm{res}_\mathfrak{p}(\mathfrak{z}_\nu)\right), \Pi^{\alpha *}\omega_{f_k} \otimes \omega^{l_2-1}_{\mathrm{can}}\eta^{l_1-1}_{\mathrm{can}}\right\rangle \\ &= \Omega_p^{l_1-l_2}\left\langle \log\left(\mathrm{res}_\mathfrak{p}\left(\Pi^\alpha_* \mathfrak{z}_\nu\right)\right), \omega_{f_k} \otimes \omega_c^{l_2-1}\eta_c^{l_1}\right\rangle \end{aligned} \tag{17}$$

Inserting (11) and (17) in (16) with $\mathfrak{z} = \mathfrak{z}_{\mathbf{f},\chi}$ and comparing the resulting equation with (15) gives the following result.

Corollary 1.5 *We have*

$$\mathcal{L}og_{\mathbf{f},\chi,\mathfrak{p}}\left(\mathrm{res}_\mathfrak{p}\left(\mathfrak{z}_{\mathbf{f},\chi}\right)\right) = (-c)^{\mathrm{l}_2-1}\mathcal{L}_p(\mathbf{f},\chi) \text{ in } \Lambda_{cc,L}.$$

Remark 1.6 There is no problem to generalize Theorem 1.3 to the case of Shimura curves, thus relaxing the Heegner hypothesis that was done since the beginning (cf. [7]): if e is the idempotent defined in [7, Sect. 2.7], just replace the Tate module T with eT in the above discussion and the proof of Theorem 1.3 below (cf. Remark 2.4 below for more technical details). However, an extension of Corollary 1.5 to the p inert case is more subtle even in the modular curve case. One may expect to replace the right hand side with the p-adic L-functions constructed in [1], but there is a technical problem in the lack of a theory of big dual exponentials.

It is also possible to generalize Theorem 1.3 to Coleman families of p-power nebentype (cf. Remark 2.4 below for more technical details).

Remark 1.7 We note that Theorem 1.3 and Corollary 1.5 have been considered in several degrees of generality by [12, 13, 18, 20]. Our approach, which relies on the methods of [10], offers a different perspective. We also add the extra richness of Theorem 1.4.

2 Proof of the Results of Sect. 1.2

We will need to work with the modular curve $Y_1(N, p^r)$ of level $\Gamma_1(N, p^r) := \Gamma_1(N) \cap \Gamma_0(p^r)$. We let $\mathcal{T}_1(N, p^r)$ be the relative p-adic Tate module of the universal elliptic curve $E_1(N, p^r)/Y_1(N, p^r)$, we choose a lift (E, t, C) of (E, t), which gives $\eta_{E,t,C} : \mathfrak{Spec}(F) \to Y_1(N, p^r)_F$, and consider the geometric fibre $\mathcal{T}_1(N, p^r)_{\overline{\eta}_{E,t,C}} = H^1_{\acute{e}t}(E_{\overline{F}}, \mathbb{Z}_p(1))$, which is identified with the Tate module $T := H^1_{\acute{e}t}(E_{\overline{F}}, \mathbb{Z}_p(1))$ of E on which the étale fundamental group $G_{Y_1(N,p^r)_F}$ acts. Let us recall that there is a canonical $G_{Y_1(N,p)_F}$ -equivariant identification $\frac{T}{pT} = E[p](\overline{F})$, from which we deduce the existence of a $G_{Y_1(N,p)_F}$-equivariant surjection $\pi : T \to E[p](\overline{F})$. Define $E[p]^\times := E[p] - \{0\}$, $C^\times := C[p](\overline{F}) - \{0\}$, $\overline{C}^\times := E[p]^\times - C^\times$: then, setting $Y := \pi^{-1}\left(\overline{C}^\times\right)$ and $X := \pi^{-1}(C^\times)$, we have the $G_{Y_1(N,p^r)_F}$-equivariant disjoint decomposition

$$X = \pi^{-1}\left(E[p]^\times\right) = Y \sqcup Z. \tag{18}$$

Let us remark that, writing $\Gamma_0 \subset Aut_{\mathbb{Z}_p}(T)$ for the subgroup of those automorphisms preserving $C[p](\overline{F}) \subset \frac{T}{pT}$, we have the canonical sequence of group homomorphisms (induced by the forgetful morphisms $Y_1(N, p^r) \to Y_1(N, p^{r-1})$ and the natural action of $G_{Y_1(N,p)_F}$ on T):

$$\ldots \to G_{Y_1(N,p^r)_F} \to \ldots \to G_{Y_1(N,p)_F} \to \Gamma_0 \subset Aut_{\mathbb{Z}_p}(T).$$

Remark 2.1 We may fix a basis $\{v_1, v_2\} \subset T$ in such a way that $C = \mathbb{F}_p v_2$: then $Y = \mathbb{Z}_p^\times v_1 \times \mathbb{Z}_p v_2$ and $Z = p\mathbb{Z}_p v_1 \times \mathbb{Z}_p^\times v_2$.

Suppose that $\kappa : \mathbb{Z}_p^\times \to \mathcal{O}^\times$ is a continuous character valued in the topological inverse limit of affinoid L-Banach algebras which is accessible: because $\mathbb{Z}_p^\times$ is profinite, κ takes value in the subring $\Lambda_{\mathcal{O}} \subset \mathcal{O}$ of power bounded elements. For every integer $m \geq 1$, let $A_\kappa = A_{\kappa,m}$ (resp. $S_r(Y)$) be the space of functions $f : Y \to \Lambda_{\mathcal{O}}$ (resp. $f : Y \to \mathcal{O}_L$) that are locally analytic and with the m-convergence condition [10, (64)] imposed[11] (resp. that are restriction of polynomial functions on T) and such that $f(tv) = t^\kappa f(v)$ (resp. $f(tv) = t^r f(v)$) for every $v \in T$ and $t \in \mathbb{Z}_p^\times$: via its natural action on Y, the group $G_{Y_1(N,p^r)}$ acts $\Lambda_{\mathcal{O}}$-linearly on $A_{\kappa_{\mathcal{U}}}$ (resp. $\mathcal{O}_L$-linearly on $S_r(Y)$) by means of the formula $(gf)(v) := f\left(g^{-1}v\right)$. The restriction map $S_r = S_r(X) \to S_r(Y)$ is a $G_{Y_1(N,p^r)}$-equivariant isomorphism. Let us assume that $\Lambda_{\mathcal{O}}$ is a local ring with maximal ideal $\mathfrak{m}_{\mathcal{O}} \subset \Lambda_{\mathcal{O}}$. Key to our construction is the fact that $\frac{A_\kappa}{\mathfrak{m}^n_{\mathcal{O}} A_\kappa}$ admits an exhaustive filtration by finite Γ_0-modules, hence a fortiori by finite $G_{Y_1(N,p^r)}$-modules and, for this reason, we may attach a system of étale sheaves $\mathcal{A}_\kappa$ on $Y_1(N, p^r)_{\acute{e}t}$ to $\left(\frac{A_\kappa}{\mathfrak{m}^n_{\mathcal{O}} A_\kappa}\right)_n$: this is proved in [10, Sect. 4.2] for $\kappa = \kappa_{\mathcal{U}}$, but the definitions of loc.cit. work equally well in this more setting (the accessibility is placed in order to ensure that κ is locally analytic and, hence that [10, Remark 4.1] is in force for m large enough). This construction will be applied in the following cases:

---

[11] The role of this m-convergence condition is purely technical: we will say few more words later when needed.

$$\kappa \in \left\{\kappa_{\mathcal{U}}, \kappa_{\mathcal{U}}^{\otimes\mathcal{V}} : \mathbb{Z}_p^\times \xrightarrow{\kappa_{\mathcal{U}}} \mathcal{O}(\mathcal{U}) \xrightarrow{i_{\mathcal{U}}} \mathcal{O}(\mathcal{U}\times\mathcal{V}), \kappa_{\mathcal{U}}^{cc} : \mathbb{Z}_p^\times \longrightarrow \mathcal{O}_{cc}\right\}.$$

If (A_n) and (B_n) are two inverse systems of objects in some category, we write $(f_{N_n,n}) : (A_n) > (B_n)$ if, for every n, there exists N_n such that there is a morphism $f_{N_n,n} : A_{N_n} \to B_n$, we have $N_{n+1} \geq N_n$ and $(f_{N_n,n})$ is a morphism of inverse systems. We say that (A_n) and (B_n) are equivalent, and write $(A_n) \sim (B_n)$, if $(f_{N_n,n}) : (A_n) > (B_n)$, $(g_{N_n,n}) : (B_n) > (A_n)$ and the compositions

$$A_{N_{N_n}} \xrightarrow{f_{N_{N_n},N_n}} B_{N_n} \xrightarrow{g_{N_n,n}} A_n \text{ and } B_{N_{N_n}} \xrightarrow{g_{N_{N_n},N_n}} A_{N_n} \xrightarrow{f_{N_n,n}} B_n$$

are the morphisms that appear in the inverse system (A_n) and, respectively, (B_n). Suppose that $\mathcal{A}$ (resp. $\mathcal{B}$) is an abelian category and let $\mathcal{A}^{\mathbb{N}}$ (resp. $\mathcal{B}^{\mathbb{N}}$) be the category of inverse systems of objects in $\mathcal{A}$ (resp. $\mathcal{B}$) indexed by the natural numbers. If $F : \mathcal{A} \to \mathcal{B}$ is a left exact functor, then it induces $F : \mathcal{A}^{\mathbb{N}} \to \mathcal{B}^{\mathbb{N}}$ and, when $\mathcal{B}$ has numerable inverse limits, we may consider $\lim_{\leftarrow n} \circ F : \mathcal{A}^{\mathbb{N}} \to \mathcal{B}$. We write $R^i F$ (resp. $R_c^i F$) for the right derived functors of $F : \mathcal{A}^{\mathbb{N}} \to \mathcal{B}^{\mathbb{N}}$ (resp. $\lim_{\leftarrow n} \circ F$), assuming that $\mathcal{A}$ has enough injectives (see [19, (1.1) Proposition]). Then [19, (1.6) Proposition] is in force[12]: we have that $(A_n) \sim (B_n)$ implies $R^i F((A_n)) \sim R^i F((B_n))$, from which it follows that $\lim_{\leftarrow n} R^i F((A_n)) = \lim_{\leftarrow n} R^i F((B_n))$ and $\left(\lim_{\leftarrow n}\right)^{(1)} R^i F((A_n)) = \left(\lim_{\leftarrow n}\right)^{(1)} R^i F((B_n))$[13] and, finally, $R_c^i F((A_n)) = R_c^i F((B_n))$. This fact will be used without explicit mention.

Let us now make the following two remarks. First, we have $A_{\kappa_{\mathcal{U}}^{\otimes\mathcal{V}}-2} = A_{\kappa_{\mathcal{U}}-2} \widehat{\otimes}_{\mathcal{O}_L} \Lambda_{\mathcal{V}}$ as a $\Lambda_{\mathcal{U}\times\mathcal{V}}[\Gamma_0]$-modules and $\frac{A_{\kappa_{\mathcal{U}}^{\otimes\mathcal{V}}-2}}{\mathfrak{m}_{\mathcal{U}\times\mathcal{V}}^n A_{\kappa_{\mathcal{U}}^{\otimes\mathcal{V}}-2}} \sim \frac{A_{\kappa_{\mathcal{U}}-2}}{\mathfrak{m}_{\mathcal{U}}^n A_{\kappa_{\mathcal{U}}-2}} \otimes_{\mathcal{O}_L} \frac{\Lambda_{\mathcal{V}}}{\mathfrak{m}_{\mathcal{V}}^n}$ as filtered modules over $\frac{\Lambda_{\mathcal{U}\times\mathcal{V}}}{\mathfrak{m}_{\mathcal{U}\times\mathcal{V}}^n}[\Gamma_0] \sim \frac{\Lambda_{\mathcal{U}}}{\mathfrak{m}_{\mathcal{U}}^n} \otimes_{\mathcal{O}_L} \frac{\Lambda_{\mathcal{V}}}{\mathfrak{m}_{\mathcal{V}}^n}[\Gamma_0]$,[14] as it easily follows from [10, Remark 4.1] and the explicit description of the filtration given in [10, Sect. 4.2]. Second, because $\mathfrak{m}_{cc} = \pi_{cc}(\mathfrak{m}_{\mathcal{U}\times\mathcal{V}})$ we have $A_{\kappa_{\mathcal{U}}^{cc}-2} = \frac{A_{\kappa_{\mathcal{U}}^{\otimes\mathcal{V}}-2}}{I_{cc} A_{\kappa_{\mathcal{U}}^{\otimes\mathcal{V}}-2}}$ as $\Lambda_{\mathcal{U}\times\mathcal{V}}[\Gamma_0]$-modules and $\frac{A_{\kappa_{\mathcal{U}}^{cc}-2}}{\mathfrak{m}_{cc}^n A_{\kappa_{\mathcal{U}}^{cc}-2}} = \frac{A_{\kappa_{\mathcal{U}}^{\otimes\mathcal{V}}-2}/\mathfrak{m}_{\mathcal{U}\times\mathcal{V}}^n A_{\kappa_{\mathcal{U}}^{\otimes\mathcal{V}}-2}}{I_{cc} A_{\kappa_{\mathcal{U}}-2} + \mathfrak{m}_{\mathcal{U}\times\mathcal{V}}^n A_{\kappa_{\mathcal{U}}^{\otimes\mathcal{V}}-2}/\mathfrak{m}_{\mathcal{U}\times\mathcal{V}}^n A_{\kappa_{\mathcal{U}}^{\otimes\mathcal{V}}-2}}$ as filtered $\frac{\Lambda_{cc}}{\mathfrak{m}_{cc}^n}[\Gamma_0]$-modules, again using [10, Remark 4.1] and the description of the filtration given in [10, Sect. 4.2].

---

[12] If $(\lim_{\leftarrow n})^{(1)}$ is defined by means of [19, (1.4)], then the proof of [19, (1.6) Proposition] shows that the existence of numerable inverse limits suffices for this result being in force.

[13] Here we define $\left(\lim_{\leftarrow n}\right)^{(1)}$ using [19, (1.4)]. Then is is clear that $(C_n) \sim (D_n)$ implies $\lim_{\leftarrow n}(C_n) = \lim_{\leftarrow n}(D_n)$ and $\left(\lim_{\leftarrow n}\right)^{(1)}(C_n) = \left(\lim_{\leftarrow n}\right)^{(1)}(D_n)$.

[14] As already remarked in a previous footnote, setting $\mathfrak{m} := \mathfrak{m}_{\mathcal{U}} \otimes_{\mathcal{O}_L} \Lambda_{\mathcal{V}} + \Lambda_{\mathcal{U}} \otimes_{\mathcal{O}_L} \mathfrak{m}_{\mathcal{V}}$ one as $\mathfrak{m}^n \subset \mathfrak{m}_{\mathcal{U}\times\mathcal{V}}^n$ and $\mathfrak{m}_{\mathcal{U}\times\mathcal{V}}^{2n} \subset \mathfrak{m}^n$, implying that the $\mathfrak{m}_{\mathcal{U}\times\mathcal{V}}$-adic and the $\mathfrak{m}$-adic topologies on $A \otimes_{\mathcal{O}_L} B$ are equivalent for every $\Lambda_{\mathcal{U}}$-module A and every $\Lambda_{\mathcal{V}}$-module B. Here $A_{\kappa_{\mathcal{U}}-2} \widehat{\otimes}_{\mathcal{O}_L} \Lambda_{\mathcal{V}}$ denotes any one of these two equivalent completions. It also follows that $\frac{\Lambda_{\mathcal{U}\times\mathcal{V}}}{\mathfrak{m}_{\mathcal{U}\times\mathcal{V}}^n} \sim \frac{\Lambda_{\mathcal{U}}}{\mathfrak{m}_{\mathcal{U}}^n} \otimes_{\mathcal{O}_L} \frac{\Lambda_{\mathcal{V}}}{\mathfrak{m}_{\mathcal{V}}^n}$ (and then $\frac{\Lambda_{\mathcal{U}\times\mathcal{V}}}{\mathfrak{m}_{\mathcal{U}\times\mathcal{V}}^n}[\Gamma_0] \sim \frac{\Lambda_{\mathcal{U}}}{\mathfrak{m}_{\mathcal{U}}^n} \otimes_{\mathcal{O}_L} \frac{\Lambda_{\mathcal{V}}}{\mathfrak{m}_{\mathcal{V}}^n}[\Gamma_0]$).

In particular, we see that

$$A_{\kappa_{\mathcal{U}}-2}\widehat{\otimes}_{\mathcal{O}_L}\Lambda_{\mathcal{V}} = A_{\kappa_{\mathcal{U}}^{\otimes\mathcal{V}}-2} \text{ and } A_{\kappa_{\mathcal{U}}^{cc}-2} = \frac{A_{\kappa_{\mathcal{U}}^{\otimes\mathcal{V}}-2}}{I_{cc}A_{\kappa_{\mathcal{U}}^{\otimes\mathcal{V}}-2}} \tag{19}$$

If $\mathbf{l}_1, \mathbf{l}_2 : \mathcal{O}_{K,p}^{\times} \to \mathcal{O}(\mathcal{V})^{\times}$, then these characters again factor through the subring $\Lambda_{\mathcal{V}} \subset \mathcal{O}(\mathcal{V})$ by profiniteness of $\mathcal{O}_{K,p}^{\times}$. Hence $\psi_{E,p}^{1-\mathbf{l}_1}$ and $\overline{\psi}_{E,p}^{1-\mathbf{l}_2}$ are two characters of G_F with value in $\Lambda_{\mathcal{V}}$. Arguing as we did before (3), for every number field $F' \supset F$ and every $\Lambda_{\mathcal{V}}[G_{F'}]$-module M (resp. system of sheaves of $\Lambda_{\mathcal{V}}$-modules on $Y_1(N, p^r)_{F',\acute{e}t}$), we may consider its twist $M[1-\mathbf{l}_1, 1-\mathbf{l}_2]$ by the character $\psi_{E,p}^{1-\mathbf{l}_1}\overline{\psi}_{E,p}^{1-\mathbf{l}_2}$. Recalling that there is a canonical (iso)morphism $\pi_{cc} \circ i_{\mathcal{V}} : \Lambda_{\mathcal{V}} \to \Lambda_{cc}$, the same notation applies to $\Lambda_{cc}[G_{F'}]$-modules and systems of sheaves of Λ_{cc}-modules on $Y_1(N, p^r)_{F',\acute{e}t}$ (equivalently, use the characters $\mathbf{l}_1^{cc}$ and $\mathbf{l}_2^{cc}$ and directly work with the ring Λ_{cc}). We define $\mathcal{A}_{\kappa_{\mathcal{U}}^{\otimes\mathcal{V}}-2;1-\mathbf{l}_1,1-\mathbf{l}_2} := \mathcal{A}_{\kappa_{\mathcal{U}}^{\otimes\mathcal{V}}-2}[1-\mathbf{l}_1, 1-\mathbf{l}_2]$ and $\mathcal{A}_{\kappa_{\mathcal{U}}^{cc}-2;1-\mathbf{l}_1,1-\mathbf{l}_2} := \mathcal{A}_{\kappa_{\mathcal{U}}^{cc}-2}[1-\mathbf{l}_1, 1-\mathbf{l}_2]$. It will be convenient to introduce the following notation for systems of étale sheaves on $Y_1(N, p^r)_{F',\acute{e}t}$. If $\mathcal{F} = (\mathcal{F}_n)$ is a system of étale sheaves of $\Lambda_{\mathcal{U}}$-modules (resp. $\Lambda_{\mathcal{U}\times\mathcal{V}}$-modules), we set $\mathcal{F}\widehat{\otimes}_{\mathcal{O}_L}\Lambda_{\mathcal{V}} := \left(\mathcal{F}_n \otimes_{\mathcal{O}_L} \frac{\Lambda_{\mathcal{V}}}{\mathfrak{m}_{\mathcal{V}}^n}\right)$ (resp. $\frac{\mathcal{F}}{I_{cc}\mathcal{F}} := \left(\frac{\mathcal{F}_n}{I_{cc}\mathcal{F}_n}\right)$). Then the above discussion implies that there is an equivalence of systems of étale sheaves on $Y_1(N, p^r)_{F',\acute{e}t}$:

$$\mathcal{A}_{\kappa_{\mathcal{U}}-2}\widehat{\otimes}_{\mathcal{O}_L}\Lambda_{\mathcal{V}}[1-\mathbf{l}_1, 1-\mathbf{l}_2] \sim \mathcal{A}_{\kappa_{\mathcal{U}}^{\otimes\mathcal{V}}-2;1-\mathbf{l}_1,1-\mathbf{l}_2} \text{ and } \mathcal{A}_{\kappa_{\mathcal{U}}^{cc}-2;1-\mathbf{l}_1,1-\mathbf{l}_2} = \frac{\mathcal{A}_{\kappa_{\mathcal{U}}^{\otimes\mathcal{V}}-2;1-\mathbf{l}_1,1-\mathbf{l}_2}}{I_{cc}\mathcal{A}_{\kappa_{\mathcal{U}}^{\otimes\mathcal{V}}-2;1-\mathbf{l}_1,1-\mathbf{l}_2}}. \tag{20}$$

Let $c \in \mathbb{N}$ prime to p and $\varphi : E \to E_{\varphi}$ be as in the discussion before the statement of Theorem 1.3. Extend F by a finite extension so that it contains the field cup out by the character $\varkappa$ and take $F' = F_{\varphi}$.

Lemma 2.2 *There is a surjective Hecke equivariant morphism of $\Lambda_{cc,L}$-modules*

$$\pi_{\mathbf{f},\chi} : H^1\left(F', H^1_{\acute{e}t}\left(Y_1(N,p)_{\overline{F}'}, \mathcal{A}_{\kappa_{\mathcal{U}}^{cc}-2;1-\mathbf{l}_1,1-\mathbf{l}_2}\left(\kappa_{\mathcal{U}}^{cc}-1\right)\right)\right) \longrightarrow H^1(K, V(\mathbf{f}, \chi))$$

which is compatible with the specialization maps and π_{f_k,χ_v}.

Proof Let us consider the following two diagrams:

$$\begin{array}{ccc} H^1_{\acute{e}t}\left(Y_1(N,p^r)_{\overline{F}'}, \mathcal{A}_{\kappa_{\mathcal{U}}-2}\left(\kappa_{\mathcal{U}}^{cc}-1\right)\right)\widehat{\otimes}_L\Lambda_{\mathcal{V},L}[1-\mathbf{l}_1, 1-\mathbf{l}_2] & \longrightarrow & H^1_{\acute{e}t}\left(Y_1(N,p^r)_{\overline{F}'}, \mathcal{A}_{\kappa_{\mathcal{U}}^{\otimes\mathcal{V}}-2;1-\mathbf{l}_1,1-\mathbf{l}_2}\left(\kappa_{\mathcal{U}}^{cc}-1\right)\right) \\ \downarrow & & \downarrow \\ H^1\left(\Gamma_1(N,p^r), A_{\kappa_{\mathcal{U}}-2}\right)\widehat{\otimes}_L\Lambda_{\mathcal{V},L} & \longrightarrow & H^1\left(\Gamma_1(N,p^r), A_{\kappa_{\mathcal{U}}^{\otimes\mathcal{V}}-2}\right), \end{array}$$

$$\begin{array}{ccc} \dfrac{H^1_{\acute{e}t}\left(Y_1(N,p^r)_{\overline{F}'}, \mathcal{A}_{\kappa_{\mathcal{U}}^{\otimes\mathcal{V}}-2;1-\mathbf{l}_1,1-\mathbf{l}_2}\left(\kappa_{\mathcal{U}}^{cc}-1\right)\right)}{I_{cc,L}H^1_{\acute{e}t}\left(Y_1(N,p^r)_{\overline{F}'}, \mathcal{A}_{\kappa_{\mathcal{U}}^{\otimes\mathcal{V}}-2;1-\mathbf{l}_1,1-\mathbf{l}_2}\left(\kappa_{\mathcal{U}}^{cc}-1\right)\right)} & \longrightarrow & H^1_{\acute{e}t}\left(Y_1(N,p^r)_{\overline{F}'}, \mathcal{A}_{\kappa_{\mathcal{U}}^{cc}-2;1-\mathbf{l}_1,1-\mathbf{l}_2}\left(\kappa_{\mathcal{U}}^{cc}-1\right)\right) \\ \downarrow & & \downarrow \\ \dfrac{H^1\left(\Gamma_1(N,p^r), A_{\kappa_{\mathcal{U}}^{\otimes\mathcal{V}}-2}\right)}{I_{cc,L}H^1\left(\Gamma_1(N,p^r), A_{\kappa_{\mathcal{U}}^{\otimes\mathcal{V}}-2}\right)} & \longrightarrow & H^1\left(\Gamma_1(N,p^r), A_{\kappa_{\mathcal{U}}^{cc}-2}\right). \end{array}$$

The existence of the vertical arrows follows from [10, (79)]: they are surjective and with Galois stable kernel, thus inducing a $G_{F'}$-module structure on the $\Gamma_1(N, p^r)$ cohomology groups. The existence of the horizontal arrows follows from (19) and (20) above and the resulting squares are obviously commutative by the canonical nature of [10, (67)]. We claim that the lower horizontal arrows appearing in these commutative diagrams are isomorphisms of $\Lambda_{\mathcal{U},L}\widehat{\otimes}_L\Lambda_{\mathcal{V},L}$-modules: then their Galois equivariance follows from the surjectivity of the vertical arrows. It will follows that

$$H^1\left(\Gamma_1(N, p^r), A_{\kappa_{\mathcal{U}}^{cc}-2}\right) = \frac{H^1\left(\Gamma_1(N, p^r), A_{\kappa_{\mathcal{U}}-2}\right)\widehat{\otimes}_L\Lambda_{\mathcal{V},L}}{I_{cc,L}H^1\left(\Gamma_1(N, p^r), A_{\kappa_{\mathcal{U}}-2}\right)\widehat{\otimes}_L\Lambda_{\mathcal{V},L}}$$

as $\Lambda_{cc,L}[G_{F'}]$-modules. Then, because $V(\mathbf{f})$ appears as a quotient of $H^1\left(\Gamma_1(N, p), A_{\kappa_{\mathcal{U}}-2}\right)$ (cf. [10, (83), (84) and Sect. 5]), the existence of $\pi_{\mathbf{f},\chi}$ will follow from our definition of $V(\mathbf{f}, \chi)$ using Shapiro's lemma as in the definition of π_{f_k,χ_ν} with $(\mathcal{S}_{k-2}, f_k, l_1, l_2, \chi_\nu, \varkappa_0)$ replaced by $\left(\mathcal{A}_{\kappa_{\mathcal{U}}-2}, \mathbf{f}, \mathbf{l}_1, \mathbf{l}_2, \chi, \varkappa\right)$.

The proof of the fact that the two lower horizontal arrows that appear in the above diagrams are isomorphisms even before inverting p relies on the following sublemma, whose proof is postponed. We set $\Gamma := \Gamma_1(N, p^r)$ and let $P_\cdot$ be a resolution of the trivial Γ-module $\mathbb{Z}$ by free finitely generated $\mathbb{Z}[\Gamma]$-modules, whose existence follows from results of Borel and Serre (or using a triangulation of $Y_1(N, p^r)_{\mathbb{C}}$ as in [25, Sect. 8.1]).

Sublemma. *If* $I \subset \Lambda_{\mathcal{U}} \otimes_{\mathcal{O}_L} \Lambda_{\mathcal{V}}$ *is an ideal generated by an* $A_{\kappa_{\mathcal{U}}-2} \otimes_{\mathcal{O}_L} \Lambda_{\mathcal{V}}$*-regular sequence, then we have the vanishing* $H^2\left(\Gamma, \frac{A_{\kappa_{\mathcal{U}}-2}\otimes_{\mathcal{O}_L}\Lambda_{\mathcal{V}}}{IA_{\kappa_{\mathcal{U}}-2}\otimes_{\mathcal{O}_L}\Lambda_{\mathcal{V}}}\right) = H^2\left(\Gamma, A_{\kappa_{\mathcal{U}}^{\otimes\mathcal{V}}-2}\right) = 0.$

Let us now prove that *the lower horizontal arrow that appears in the first diagram is an isomorphism*. Now if M is any $\Lambda_{\mathcal{U}}[\Gamma]$-module, the complex $Hom_\Gamma(P_\cdot, M)$ (resp. $Hom_\Gamma(P_\cdot, M \otimes_{\mathcal{O}_L} \Lambda_{\mathcal{V}})$) computes $H^\cdot(\Gamma, M)$ (resp. $H^\cdot\left(\Gamma, M \otimes_{\mathcal{O}_L} \Lambda_{\mathcal{V}}\right)$). Because $P_\cdot$ is a free finitely generated $\mathbb{Z}[\Gamma]$-module, the natural morphism of complexes $Hom_\Gamma(P_\cdot, M) \otimes_{\mathcal{O}_L} \Lambda_{\mathcal{V}} \to Hom_\Gamma\left(P_\cdot, M \otimes_{\mathcal{O}_L} \Lambda_{\mathcal{V}}\right)$ is an isomorphism; together with the exactness of $- \otimes_{\mathcal{O}_L} \Lambda_{\mathcal{V}}$, we deduce that

$$H^i(\Gamma, M) \otimes_{\mathcal{O}_L} \Lambda_{\mathcal{V}} \xrightarrow{\sim} H^i\left(\Gamma, M \otimes_{\mathcal{O}_L} \Lambda_{\mathcal{V}}\right) \tag{21}$$

and then, for every $n \in \mathbb{N}$,

$$\frac{H^i(\Gamma, M) \otimes_{\mathcal{O}_L} \Lambda_{\mathcal{V}}}{\mathfrak{m}_{\mathcal{U}\times\mathcal{V}}^n H^i(\Gamma, M) \otimes_{\mathcal{O}_L} \Lambda_{\mathcal{V}}} \xrightarrow{\sim} \frac{H^i\left(\Gamma, M \otimes_{\mathcal{O}_L} \Lambda_{\mathcal{V}}\right)}{\mathfrak{m}_{\mathcal{U}\times\mathcal{V}}^n H^i\left(\Gamma, M \otimes_{\mathcal{O}_L} \Lambda_{\mathcal{V}}\right)}. \tag{22}$$

We now remark that, because $\mathfrak{m}_{\mathcal{U}\times\mathcal{V}}$ is generated by an $A_{\kappa_{\mathcal{U}}-2} \otimes_{\mathcal{O}_L} \Lambda_{\mathcal{V}}$-regular sequence, say $(x_1, \ldots, x_r)$, $\mathfrak{m}_{\mathcal{U}\times\mathcal{V},n} := \Lambda_{\mathcal{U}\times\mathcal{V}}\left(x_1^n, \ldots, x_r^n\right)$ is again generated by an $A_{\kappa_{\mathcal{U}}-2} \otimes_{\mathcal{O}_L} \Lambda_{\mathcal{V}}$-regular sequence for every $n \in \mathbb{N}$. Furthermore, we have $\mathfrak{m}_{\mathcal{U}\times\mathcal{V},n} \subset$

$\mathfrak{m}_{\mathcal{U}\times\mathcal{V}}^{n}$ and $\mathfrak{m}_{\mathcal{U}\times\mathcal{V}}^{rn}\subset\mathfrak{m}_{\mathcal{U}\times\mathcal{V},n}$, so that

$$\frac{A_{\kappa_{\mathcal{U}}-2}\otimes_{\mathcal{O}_L}\Lambda_{\mathcal{V}}}{\mathfrak{m}_{\mathcal{U}\times\mathcal{V},n}A_{\kappa_{\mathcal{U}}-2}\otimes_{\mathcal{O}_L}\Lambda_{\mathcal{V}}}\sim\frac{A_{\kappa_{\mathcal{U}}-2}\otimes_{\mathcal{O}_L}\Lambda_{\mathcal{V}}}{\mathfrak{m}_{\mathcal{U}\times\mathcal{V}}^{n}A_{\kappa_{\mathcal{U}}-2}\otimes_{\mathcal{O}_L}\Lambda_{\mathcal{V}}}. \tag{23}$$

Using induction on length of the regular sequence and the fact that $H^2\left(\Gamma,\frac{A_{\kappa_{\mathcal{U}}-2}\otimes_{\mathcal{O}_L}\Lambda_{\mathcal{V}}}{\mathfrak{m}_{\mathcal{U}\times\mathcal{V},n}A_{\kappa_{\mathcal{U}}-2}\otimes_{\mathcal{O}_L}\Lambda_{\mathcal{V}}}\right)=0$ thanks to the above sublemma, we deduce that

$$\frac{H^1\left(\Gamma,A_{\kappa_{\mathcal{U}}-2}\otimes_{\mathcal{O}_L}\Lambda_{\mathcal{V}}\right)}{\mathfrak{m}_{\mathcal{U}\times\mathcal{V},n}H^1\left(\Gamma,A_{\kappa_{\mathcal{U}}-2}\otimes_{\mathcal{O}_L}\Lambda_{\mathcal{V}}\right)}\overset{\sim}{\longrightarrow}H^1\left(\Gamma,\frac{A_{\kappa_{\mathcal{U}}-2}\otimes_{\mathcal{O}_L}\Lambda_{\mathcal{V}}}{\mathfrak{m}_{\mathcal{U}\times\mathcal{V},n}A_{\kappa_{\mathcal{U}}-2}\otimes_{\mathcal{O}_L}\Lambda_{\mathcal{V}}}\right). \tag{24}$$

Combining (22) , (23), (24) and then taking the inverse limit yields

$$H^1\left(\Gamma,A_{\kappa_{\mathcal{U}}-2}\right)\widehat{\otimes}_{\mathcal{O}_L}\Lambda_{\mathcal{V}}\overset{\sim}{\longrightarrow}\varprojlim_{n}H^1\left(\Gamma,\frac{A_{\kappa_{\mathcal{U}}-2}\otimes_{\mathcal{O}_L}\Lambda_{\mathcal{V}}}{\mathfrak{m}_{\mathcal{U}\times\mathcal{V}}^{n}A_{\kappa_{\mathcal{U}}-2}\otimes_{\mathcal{O}_L}\Lambda_{\mathcal{V}}}\right).$$

The claimed isomorphism now follows if we can show that $H^1(\Gamma,-)$ commutes with inverse limits by surjective maps. But using the inhomogeneous cochains one easily sees that taking the 1-boundaries preserves surjective maps, hence $\left(\frac{B^1(\Gamma,M_n)}{H^0(\Gamma,M_n)}\right)$ is Mittag-Leffler if (M_n) is an inverse system with surjective transition maps, and then the arguments of [2, Lemma 3.11] applies to show what we want.

Finally, we show that *the lower horizontal arrow that appears in the second diagram is an isomorphism*. To this end, we recall that $I_{cc,L}=\Lambda_{cc,L}\varpi_{cc}$, then we take the long exact sequence induced in cohomology by the multiplication by ϖ_{cc}-map and use the fact that $H^2\left(\Gamma,A_{\kappa_{\mathcal{U}}^{\otimes\mathcal{V}}-2}\right)=0$ by the sublemma.

It remains to *prove the sublemma*. We begin with the remark that we may assume that

$$P_{\cdot}=(P_2\to P_1\to P_0)$$

because Γ is the fundamental group of $Y_1(N,p^r)_{\mathbb{C}}$ which satisfies Poincaré duality $H^i\left(Y_1(N,p^r)_{\mathbb{C}},\mathbb{Z}\right)=H_{2-i}\left(Y_1(N,p^r)_{\mathbb{C}},\mathbb{Z}\right)$ (see also [25, Sect. 8.1]). If follows that $H^i(\Gamma,N)=0$ for every $i\geq 3$. By induction on the on length of the regular sequence, we then deduce that

$$\frac{H^2\left(\Gamma,A_{\kappa_{\mathcal{U}}-2}\otimes_{\mathcal{O}_L}\Lambda_{\mathcal{V}}\right)}{IH^2\left(\Gamma,A_{\kappa_{\mathcal{U}}-2}\otimes_{\mathcal{O}_L}\Lambda_{\mathcal{V}}\right)}\overset{\sim}{\longrightarrow}H^2\left(\Gamma,\frac{A_{\kappa_{\mathcal{U}}-2}\otimes_{\mathcal{O}_L}\Lambda_{\mathcal{V}}}{IA_{\kappa_{\mathcal{U}}-2}\otimes_{\mathcal{O}_L}\Lambda_{\mathcal{V}}}\right).$$

Hence suffices to show that $H^2\left(\Gamma,A_{\kappa_{\mathcal{U}}-2}\otimes_{\mathcal{O}_L}\Lambda_{\mathcal{V}}\right)=H^2\left(\Gamma,A_{\kappa_{\mathcal{U}}^{\otimes\mathcal{V}}-2}\right)=0$: using (21) we see that we only need to show that $H^2\left(\Gamma,A_{\kappa_{\mathcal{U}}-2}\right)=H^2\left(\Gamma,A_{\kappa_{\mathcal{U}}^{\otimes\mathcal{V}}-2}\right)=0$. We will prove that, indeed, we have $H^2(\Gamma,M)=0$ if $M=\varprojlim(M_n)$ for a Mittag-

Leffler system (M_n) of Γ-modules with the property that M_n is a direct limit of finite Γ-modules.

To this end, let us consider the category $\mathcal{A}$ of Γ-modules and let $\mathcal{A}^{\mathbb{N}}$ be the category of inverse systems of Γ-modules indexed by the natural numbers. Then the left exact functor $\varprojlim_n \circ (-)^{\Gamma}$ can be derived and, writing $H_c^i(\Gamma, (M_i))$ for the resulting cohomology groups, it follows from [19, (1.6) Proposition] that if $(M_n) \in \mathcal{A}^{\mathbb{N}}$ there is an exact sequence

$$0 \longrightarrow \left(\varprojlim_n\right)^{(1)} H^{i-1}(\Gamma, M_n) \longrightarrow H_c^i(\Gamma, (M_n)) \longrightarrow \varprojlim_n H^i(\Gamma, M_n) \longrightarrow 0. \tag{25}$$

On the other hand, setting $M := \varprojlim (M_n)$, we have $H_c^i(\Gamma, (M_n)) = H^i(\Gamma, M)$ if (M_n) is a Mittag-Leffler system: indeed, in order to compute $H_c^i(\Gamma, (M_n))$, we consider the inverse system of complexes $(Hom(P_\cdot, M_n)) = (Hom(P_\cdot, M_n))$, then apply $\varprojlim \circ (-)^{\Gamma}$ and take the cohomology groups[15]; because $\varprojlim Hom_{\Gamma}(P_\cdot, M_n) = Hom_{\Gamma}(P_\cdot, M)$ we get $H^i(\Gamma, M)$. Hence, we deduce from (25) that $H^2(\Gamma, M) = 0$ if we can show that $\left(\varprojlim_n\right)^{(1)} H^1(\Gamma, M_n) = \varprojlim H^2(\Gamma, M_n) = 0$ for (M_n) a Mittag-Leffler system such that M_n is a direct limit of finite Γ-modules. To this end, we note that the comparison with the Betti and then the étale cohomology (using the topological $K(\pi, 1)$-property of $Y_1(N, p^r)_{\mathbb{C}}$ for the first comparison) shows that $H^2(\Gamma, N) = 0$ when N is a finite Γ-module (thanks to [23, Chap. VI, Theorem 7.2], because $Y_1(N, p^r)_{\overline{F}'}$ is affine of dimension 1). Using the fact that $H^2(\Gamma, -)$ commutes with direct limits thanks to [8, Theorem 3], we deduce that $H^2(\Gamma, N) = 0$ when N is a direct limit of finite Γ-modules. This proves $\varprojlim H^2(\Gamma, M_n) = 0$ but also, taking the long exact sequence induced by

$$0 \longrightarrow N := \ker(M_{n+1} \to M_n) \longrightarrow M_{n+1} \longrightarrow M_n \longrightarrow 0$$

and using the fact that $H^2(\Gamma, N) = 0$ thanks to what we have just proved (because N is a direct limit of finite Γ-modules if M_{n+1} is a direct limit of finite Γ-modules), that

$$H^1(\Gamma, M_{n+1}) \longrightarrow H^1(\Gamma, M_n)$$

is surjective. Then $\left(\varprojlim_n\right)^{(1)} H^1(\Gamma, M_n) = 0$, as wanted.

---

[15] To see that this is the case, note that the Mittag-Leffler property allows us to replace (M_n) with an equivalent inverse system of Γ-modules with surjective transition maps. Then we see that $(Hom_{\Gamma}(P_\cdot, M_n))$ has the same property (because $Hom(P_k, -)$ is exact for every k) and $(I_n^\cdot) := (Hom(P_\cdot, M_n))$ is a resolution of $(M_n) \in \mathcal{A}^{\mathbb{N}}$. But for every fixed k, using (25) and the fact that $(H^0(\Gamma, I_n^k)) = (Hom_{\Gamma}(P_k, M_n))$ is Mittag-Leffler, it follows that $H_c^i(\Gamma, (I_n^k)) = 0$ for every $i > 0$. Hence $(I_n^\cdot)$ is an F-acyclic resolution of (M_n) for $F = \varprojlim_n \circ (-)^{\Gamma}$ and the cohomology groups of $F((I_n^\cdot)) = \varprojlim_n Hom_{\Gamma}(P_\cdot, M_n)$ compute $H_c^i(\Gamma, (M_n))$.

We can now construct the class $\mathfrak{z}_{\mathbf{f},\chi}$ as an element of $H^1(K, V(\mathbf{f}, \chi))$: indeed, the interpolation property of Theorem 1.3 will imply a posteriori that $\mathfrak{z}_{\mathbf{f},\chi} \in H^1_{geo}(K, V(\mathbf{f}, \chi))$ by a similar (and simpler) argument as the one of [10, Corollary 8.2]. The construction is inspired from the definition of the p-adic L-functions of [3], although the cohomological approach of [16] will be easier to adapt to our setting. If κ is as above, let us write A_κ^Z (resp. A_κ^X) for the space of functions defined similarly as $A_\kappa^Y := A_\kappa$ was defined but with the set Y replaced by the set Z (resp. X). Then Γ_0 (resp. $Aut_{\mathbb{Z}_p}(T)$) acts on it, $A_\kappa^X = A_\kappa^Y \oplus A_\kappa^Z$ as Γ_0-modules in view of (18) and $A_\kappa^X = \mathrm{Ind}_{Aut_{\mathbb{Z}_p}(T)}^{\Gamma_0} A_\kappa^Z$ (cf. [16, Lemma 4.3]). As in [10, Sect. 4.2] one may attach an étale sheaf $\mathcal{A}_\kappa^Z$ on $Y_1(N, p)_{F_\varphi, \acute{e}t}$ to A_κ^Z and we *define* $\mathcal{A}_\kappa^X := p_! \mathcal{A}_\kappa^Z$ on $Y_1(N)_{F_\varphi, \acute{e}t}$ (where $p_!$ is applied componentwisely to the system), which exists because $p : Y_1(N, p)_{F_\varphi} \to Y_1(N)_{F_\varphi}$ is étale (see [14, Chap. I, Sect. 8, Lemma 8.2]). Indeed, because p is finite of degree $\left[Aut_{\mathbb{Z}_p}(T) : \Gamma_0\right]$, we see that $A_\kappa^X = \mathrm{Ind}_{G_{Y_1(N)}}^{G_{Y_1(N,p)}} A_\kappa^Z$ and then, because $\mathcal{A}_{\kappa,\eta_{E,t}}^Z = A_\kappa^Z$ (as systems of filtered objects), we have $\mathcal{A}_{\kappa,\eta_{E,t,C}}^X = A_\kappa^X$ (as systems of filtered objects, by [14, Chap. I, Sect. 8, Lemma 8.2]) and the pull-back of $\mathcal{A}_\kappa^X$ to $Y_1(N, p)_{F_\varphi, \acute{e}t}$ decomposes as the direct sum $\mathcal{A}_\kappa^Y \oplus \mathcal{A}_\kappa^Z$. In particular, we have morphisms $\rho_W^X : p^* \mathcal{A}_\kappa^X \to \mathcal{A}_\kappa^W$ of sheaves on $Y_1(N, p)_{F_\varphi, \acute{e}t}$ inducing

$$\rho_W^X : H^i_{\acute{e}t}\left(Y_1(N)_{\overline{F}_\varphi}, \mathcal{A}^X_{\kappa_{\mathcal{U}}^{cc}-2;1-\mathbf{l}_1,1-\mathbf{l}_2}(\kappa_{\mathcal{U}}^{cc}-2)\right) \longrightarrow H^i_{\acute{e}t}\left(Y_1(N,p)_{\overline{F}_\varphi}, \mathcal{A}^W_{\kappa_{\mathcal{U}}^{cc}-2;1-\mathbf{l}_1,1-\mathbf{l}_2}(\kappa_{\mathcal{U}}^{cc}-2)\right) \tag{26}$$

for $W \in \{Z, Y\}$. Note that, indeed, the identification $A_\kappa^X = \mathrm{Ind}_{Aut_{\mathbb{Z}_p}(T)}^{\Gamma_0} A_\kappa^Z$ induces $\left(\frac{A_\kappa^X}{\mathfrak{m}_\mathcal{O}^n A_\kappa^X}\right)_n = \mathrm{Ind}_{Aut_{\mathbb{Z}_p}(T)}^{\Gamma_0} \left(\frac{A_\kappa^Z}{\mathfrak{m}_\mathcal{O}^n A_\kappa^Z}\right)_n$ and the exhaustive filtration by finite Γ_0-modules on $\frac{A_\kappa^Z}{\mathfrak{m}_\mathcal{O}^n A_\kappa^Z}$ induces such a filtration by finite $Aut_{\mathbb{Z}_p}(T)$-modules on $\frac{A_\kappa^X}{\mathfrak{m}_\mathcal{O}^n A_\kappa^X}$ because $\left[Aut_{\mathbb{Z}_p}(T) : \Gamma_0\right]$ is finite: since $p_!$ commutes with direct limits (by [14, Chap. I, Sect. 8, Lemma 8.2]), it follows that $\mathcal{A}_\kappa^X$ is the inverse system associated to $\left(\frac{A_\kappa^X}{\mathfrak{m}_\mathcal{O}^n A_\kappa^X}\right)_n$ as in [10, Sect. 4.2]. Fix integers $x, y, w, z \in \mathbb{Z}$ satisfying the relations $pxw - Nyz = 1$, $px \equiv 1\ mod(N)$ and $y \equiv 1\ mod(p)$. Right multiplication by the matrix $W_p = \begin{pmatrix} px & y \\ Npz & pw \end{pmatrix}$ with respect to the basis of Remark 2.1 maps Y bijectively onto Z and induces a bijection

$$W_p : H^i_{\acute{e}t}\left(Y_1(N,p)_{\overline{F}_\varphi}, \mathcal{A}^Z_{\kappa_{\mathcal{U}}^{cc}-2;1-\mathbf{l}_1,1-\mathbf{l}_2}(\kappa_{\mathcal{U}}^{cc}-2)\right) \longrightarrow H^i_{\acute{e}t}\left(Y_1(N,p)_{\overline{F}_\varphi}, \mathcal{A}^Y_{\kappa_{\mathcal{U}}^{cc}-2;1-\mathbf{l}_1,1-\mathbf{l}_2}(\kappa_{\mathcal{U}}^{cc}-2)\right). \tag{27}$$

For $W \in \{X, Y\}$, define

$$W_0 := \left\{x v_E + y \overline{v}_E \in W : x, y \in \mathcal{O}_{K,p}^\times\right\} \subset W.$$

Because $W_0 \subset W$ is a G_F-stable open and compact subset, we can define

$$\begin{aligned} Z^W_{\mathbf{l}_1,\mathbf{l}_2} &\in A^W_{\kappa^{cc}_{\mathcal{U}}-2;1-\mathbf{l}_1,1-\mathbf{l}_2}\left(\kappa^{cc}_{\mathcal{U}}-2\right)^{G_F} \subset A^W_{\kappa^{cc}_{\mathcal{U}}-2;1-\mathbf{l}_1,1-\mathbf{l}_2}\left(\kappa^{cc}_{\mathcal{U}}-2\right)^{G_{F_\varphi}} \\ &= H^0_{\text{ét}}\left(\mathfrak{Spec}\left(F_\varphi\right), \mathcal{A}^W_{\kappa^{cc}_{\mathcal{U}}-2;1-\mathbf{l}_1,1-\mathbf{l}_2}\left(\kappa^{cc}_{\mathcal{U}}-2\right)\right) \end{aligned}$$

to be the locally analytic function on W defined by the formula

$$Z^W_{\mathbf{l}_1,\mathbf{l}_2}\left(xv_E + y\overline{v}_E\right) := \begin{cases} x^{\mathbf{l}^{cc}_2-1}y^{\mathbf{l}^{cc}_1-1}, & \text{if } xv_E + y\overline{v}_E \in W_0, \\ 0, & \text{if } xv_E + y\overline{v}_E \in W - W_0. \end{cases}$$

Let us write Y^W to denote either $Y_1(N)$ or $Y_1(N, p)$ when $W = X$ or, respectively, $W = Y$. Using the pushforward $\eta_{\varphi *}$ and the morphism HS arising from the Hochschild-Serre spectral sequence yields, similarly as in [10, (156)],

$$\mathrm{AJ}_{\text{ét}} : H^0_{\text{ét}}\left(\mathfrak{Spec}\left(F_\varphi\right), \mathcal{A}^{W,\varphi}_{\kappa^{cc}_{\mathcal{U}}-2;1-\mathbf{l}_1,1-\mathbf{l}_2}\left(\kappa^{cc}_{\mathcal{U}}-2\right)\right) \longrightarrow H^1\left(F_\varphi, H^1_{\text{ét}}\left(Y^W_{\overline{F}_\varphi}, \mathcal{A}^W_{\kappa^{cc}_{\mathcal{U}}-2;1-\mathbf{l}_1,1-\mathbf{l}_2}\left(\kappa^{cc}_{\mathcal{U}}-2\right)\right)\right)$$

which is compatible with the analogue $(4)_{Y_1(N,p)}$ of (4) for the curve $Y_1(N, p)$ when $W = Y$ (here $\mathcal{A}^{W,\varphi}_{\kappa^{cc}_{\mathcal{U}}-2;1-\mathbf{l}_1,1-\mathbf{l}_2}$ denotes the pull-back of $\mathcal{A}^W_{\kappa^{cc}_{\mathcal{U}}-2;1-\mathbf{l}_1,1-\mathbf{l}_2}$ along the point $\eta_{E_\varphi,\varphi\circ t,C} : \mathfrak{Spec}\left(F_\varphi\right) \to Y_1(N)_{F_\varphi}$ determined by $\left(E_\varphi, \varphi \circ t, C\right)$). We also have

$$\varphi_* : H^0_{\text{ét}}\left(\mathfrak{Spec}\left(F_\varphi\right), \mathcal{A}^W_{\kappa^{cc}_{\mathcal{U}}-2;1-\mathbf{l}_1,1-\mathbf{l}_2}\left(\kappa^{cc}_{\mathcal{U}}-2\right)\right) \longrightarrow H^0_{\text{ét}}\left(\mathfrak{Spec}\left(F_\varphi\right), \mathcal{A}^{W,\varphi}_{\kappa^{cc}_{\mathcal{U}}-2;1-\mathbf{l}_1,1-\mathbf{l}_2}\left(\kappa^{cc}_{\mathcal{U}}-2\right)\right)$$

compatible with the analogue $(6)_{Y_1(N,p)}$ of (6) for the curve $Y_1(N, p)$ when $W = Y$. We can now define

$$\mathfrak{Z}^W_{\mathbf{l}_1,\mathbf{l}_2,\varphi} := \frac{1}{p-1}\mathrm{AJ}_{\text{ét}}\left(\varphi_* Z^W_{\mathbf{l}_1,\mathbf{l}_2}\right) \in H^1\left(F_\varphi, H^1_{\text{ét}}\left(Y^W_{\overline{F}_\varphi}, A^W_{\kappa^{cc}_{\mathcal{U}}-2;1-\mathbf{l}_1,1-\mathbf{l}_2}\left(\kappa^{cc}_{\mathcal{U}}-2\right)\right)\right)$$

and, finally,

$$\mathfrak{Z}_{\mathbf{f},\chi} := \pi_{\mathbf{f},\chi}\left(\rho^X_Y\left(\mathfrak{Z}^X_{\mathbf{l}_1,\mathbf{l}_2,\varphi}\right)\right) + \alpha_p^{-1}\pi_{\mathbf{f},\chi}\left(W_p\rho^X_Z\left(\mathfrak{Z}^X_{\mathbf{l}_1,\mathbf{l}_2,\varphi}\right)\right) \in H^1\left(K, V\left(\mathbf{f}, \chi\right)\right), \quad (28)$$

where ρ^X_W and W_p are induced by (26) and (27) respectively and α_p is the eigenvalue of $\mathbf{f}$ under the action of the W_p-operator.

In order to complete the proof of Theorem 1.3, it remains to be proved the interpolation property. We will prove it in terms of these "α-factors" that appears in (12). To this end, let us we remark that in Sect. 1.1 we defined $\mathfrak{z}^N_{l_1,l_2,\varphi} := \mathfrak{z}_{l_1,l_2,\varphi}$ working with the modular curve $Y_1(N)$ as the image of $z^N_{l_1,l_2} := z_{l_1,l_2}$, but the same construction mutatis mutandis defines a class $\mathfrak{z}^{N,p}_{l_1,l_2,\varphi}$ working with the modular curve $Y_1(N, p)$ of level $\Gamma_1(N, p)$ as the image of $z^{N,p}_{l_1,l_2}$, the polynomial function on Y defined as in (5).

Let us first assume that p *is split*. We will deduce (11) from the following two equations:

$$\mathfrak{z}_{\mathbf{f}_k,\chi_\nu} = \frac{1}{p-1}\left(1-\alpha_{p,k}^{-1}\chi_\nu\left(\mathfrak{p}\right)\right)\mathfrak{z}_{f_k,\chi_\nu}^{N,p} \text{ and } \Pi_*^\alpha\left(\mathfrak{z}_{l_1,l_2,\varphi}^{N,p}\right) = (p-1)\left(1-\alpha_{p,k}^{-1}\chi_\nu\left(\overline{\mathfrak{p}}\right)\right)\mathfrak{z}_{f_k,\chi_\nu}. \tag{29}$$

Let us set $Y^0 := Y - Y_0$. Then $Y = Y_0 \sqcup Y^0$ and, writing $\mathcal{A} := \mathcal{A}_{\kappa_{\mathcal{U}}^{cc}-2;1-\mathbf{l}_1,1-\mathbf{l}_2}$, this decomposition induces a decomposition $\mathcal{A} = \mathcal{A}_0 \oplus \mathcal{A}^0$ in which $\mathcal{A}_0$ (resp. $\mathcal{A}^0$) is associated to spaces of functions on Y_0 (resp. Y^0) and, by definition, Z_{l_1,l_2}^Y is the projection of $z_{l_1,l_2}^{N,p}$ to $\mathcal{A}_0$ if $\nu \in \Sigma_{geo}$ (cf. [10, discussion before Theorem 7.1]). When p is split, we may assume that we have $\{v_1, v_2\} = \{v_E, \overline{v}_E\}$ in Remark 2.1, $X_0 = Y_0$ and Y^0 is the set of $xv_E + y\overline{v}_E$ such that $y/x \equiv j_E \ mod(p)$ for some fixed $j_E \in \{0, ..., p-1\}$ (cf. [3, Lemma 3.7] and [16, Lemma 5.3]). Furthermore, writing $g_{\mathfrak{p}} \in G_{F_\varphi/K}$ for the element representing the image of the uniformizer at $\mathfrak{p}$ under the reciprocity law, one see that $Y^0 = g_{\mathfrak{p}}Y$ (cf. [16, (2.7) and Lemma 5.3]). First, we deduce from $X_0 = Y_0$ that we have $\mathfrak{z}_{\mathbf{f},\chi} = \mathfrak{z}_{\mathbf{f},\chi}^Y := \pi_{\mathbf{f},\chi}\left(\mathfrak{z}_{\mathbf{l}_1,\mathbf{l}_2,\varphi}^Y\right)$. Second, writing the double coset operator U_p as the union of coset operators π_j mapping Y to the set of the $xv_E + y\overline{v}_E$ such that $y/x \equiv j \ mod(p)$ for $j \in \{0, ..., p-1\}$, we deduce that the operator $U_p - g_{\mathfrak{p}}$ (resp. $g_{\mathfrak{p}}$) factors through the cohomology with coefficients in $\mathcal{A}_0$ (resp. $\mathcal{A}^0$). Then we see that

$$U_p\left(\varphi_*\left(z_{l_1,l_2}^{N,p}\right) - \varphi_*\left(Z_{l_1,l_2}^Y\right)\right) = g_{\mathfrak{p}}\varphi_*\left(z_{l_1,l_2}^{N,p}\right),$$

from which the first equality in (29) follows applying HS and projecting onto $V(\mathbf{f}_k, \chi_\nu)$. The reader should compare this discussion with the one explaining the third equality in the proof of [10, (162)]. The second equality of (29) follows once again applying HS and projecting onto $V(\mathbf{f}_k, \chi_\nu)$ from the equality $\Pi_*^\alpha\left(z_{l_1,l_2}^{p,N}\right) = (p-1)\left(1-\alpha_{p,k}^{-1}g_{\overline{\mathfrak{p}}}\right)z_{l_1,l_2}^N$ because $g_{\overline{\mathfrak{p}}}$ induces $\varepsilon(p)$ times the slash operator δ_p (cf. [16, (2.6)] and the formula for Π_*^α of the inert case discussed below).

Let us now assume that p *is inert*. Then one has $X_0 = X$ (cf. Remark 2.1 and [3, Lemma 3.7]). It follows that $Z_{l_1,l_2}^X = z_{l_1,l_2}^N$ for $\nu \in \Sigma_{geo}$ and then, unwinding (28) gives

$$\mathfrak{z}_{\mathbf{f}_k,\chi_\nu} = \frac{1}{p-1}\left(\mathfrak{z}_{f_k,\chi_\nu}^{N,p} + \alpha_{p,k}^{-1}W_p\left(\mathfrak{z}_{f_k,\chi_\nu}^{N,p}\right)\right).$$

Now the claim follows using the formula $\Pi_*^\alpha = pr_{1*} - \alpha_{p,k}^{-1}pr_{1*}W_p$ (in view of the equality $W_p = \gamma \cdot \delta_p$ for some $\gamma \in \Gamma_0(N)$ with lower right entry $\equiv p \ mod(N)$) and the fact that $W_p^2 = p \cdot \gamma$ for some $\gamma \in \Gamma_0(Np)$ with lower right entry $\equiv p \ mod(N)$ acts as multiplication by $\varepsilon(p)p^{k-2}$ on $\mathfrak{z}_{f_k,\chi_\nu}^{N,p}$ (cf. [16, proofs of Proposition 4.4 and Corollary 4.8]).

We now prove Theorem 1.4 (hence assume that p *is split*). Similarly as in the second equation of (20) above, one has $\mathcal{A}_{\kappa_{\mathcal{U}}^*-2;1-l_1^o,1-\mathbf{l}_2} = \frac{\mathcal{A}_{\kappa_{\mathcal{U}}^{cc}-2;1-\mathbf{l}_1,1-\mathbf{l}_2}}{I_{\mathbf{l}_1^{cc}=l_1^o}\mathcal{A}_{\kappa_{\mathcal{U}}^{cc}-2;1-\mathbf{l}_1,1-\mathbf{l}_2}}$ if $\kappa_{\mathcal{U}}^* = \pi_{\mathbf{l}_1^{cc}=l_1^o} \circ \kappa_{\mathcal{U}}^{cc}$ and the notation for the twists is defined as above. One has an Abel-Jacobi map

$$\mathrm{AJ}_{\acute{e}t} : H_{\acute{e}t}^0\left(\mathfrak{Spec}\left(F_\varphi\right), \mathcal{A}_{\kappa_{\mathcal{U}}^*-2;1-\mathbf{l}_1,1-\mathbf{l}_2}^{\varphi}\left(\kappa_{\mathcal{U}}^*-2\right)\right) \longrightarrow H^1\left(F_\varphi, H_{\acute{e}t}^1\left(Y_1(N,p)_{\overline{F}_\varphi}, \mathcal{A}_{\kappa_{\mathcal{U}}^*-2;1-\mathbf{l}_1,1-\mathbf{l}_2}\left(\kappa_{\mathcal{U}}^*-2\right)\right)\right).$$

Also, because $I_{\mathbf{l}_1^{cc}=l_1^o}$ is generated by a regular element, the Galois representation $V\left(\mathbf{f},\chi_{\mathbf{l}_1^{cc}=l_1^o}\right)$ appears as a quotient of $H^1_{\acute{e}t}\left(Y^W_{\overline{F}_\varphi},\mathcal{A}_{\kappa^*_{\mathcal{U}}-2;1-\mathbf{l}_1,1-\mathbf{l}_2}\left(\kappa^*_{\mathcal{U}}-2\right)\right)$ (by the same argument proving that the lower horizontal arrow that appears in the second diagram of the proof of Lemma 2.2 is an isomorphism). Let us define $\mathbf{l}_2^* := \pi_{\mathbf{l}_1^{cc}=l_1^o}\circ\mathbf{l}_2^{cc}$. Noticing that $Z^Y_{\mathbf{l}_1=l_{1,o},\mathbf{l}_2}\left(xv_E+y\overline{v}_E\right):=x^{\mathbf{l}_2^*-1}y^{l_{1,o}^{cc}-1}$, which defines an element of $H^0_{\acute{e}t}\left(\mathfrak{Spec}\left(F_\varphi\right),\mathcal{A}_{\kappa^*_{\mathcal{U}}-2;1-\mathbf{l}_1,1-\mathbf{l}_2}\left(\kappa^*_{\mathcal{U}}-2\right)\right)$, specializes to $z^{N,p}_{l_1,l_2}$ for every $\nu\in\Sigma_{geo}$ of infinity type $\left(l_1^o-1,l_2-1\right)$, the claim follows setting $\mathfrak{z}^Y_{\mathbf{l}_1=l_{1,o},\mathbf{l}_2,\varphi}:=\frac{1}{p-1}\mathrm{AJ}_{\acute{e}t}\left(\varphi_*Z^Y_{\mathbf{l}_1=l_{1,o},\mathbf{l}_2}\right)$ and then considering its image $\mathfrak{z}_{\mathbf{f},\chi_{\mathbf{l}_1^{cc}=l_1^o}}$ in $H^1\left(F_\varphi,V\left(\mathbf{f},\chi_{\mathbf{l}_1^{cc}=l_1^o}\right)\right)$ (cf. [10, Sect. 9.3]).

Remark 2.3 When p is inert, the class $\mathfrak{z}^Y_{\mathbf{f},\chi}:=\pi_{\mathbf{f},\chi}\left(\mathfrak{z}^Y_{\mathbf{l}_1,\mathbf{l}_2,\varphi}\right)$ has the property that its specialization $\mathfrak{z}^Y_{f_k,\chi_\nu}$ equals $\mathfrak{z}^{N,p}_{f_k,\chi_\nu}$. When p is split, it is the improved class $\mathfrak{z}_{\mathbf{f},\chi_{\mathbf{l}_1^{cc}=l_1^o}}$ that behaves in this way.

Remark 2.4 As explained, Theorem 1.3 generalizes to the case of Shimura curves. However, one has to modify the proof of Lemma 2.2, which uses the fact that we may work with an affine curve. To circumvent this issue a more thought analysis of the Hecke module structure of $\mathcal{A}_{\kappa-2}$ and $A_{\kappa-2}$ is required in order to define a cuspidal projector killing the cohomology groups as required. Also, in order to generalize Theorem 1.3 to Coleman families of p-power nebentype ε_p another generalization of Lemma 2.2 is required in order to realize $V\left(\mathbf{f}\right)$ as a quotient of $H^1\left(\Gamma_1\left(N,p^r\right),A_{\kappa_{\mathcal{U}}-2}\right)$. This can be done using the comparison and the control theorems proved in [15, Theorem 8.1 and Theorem 9.3].

3 Generalized Heegner Cycles and Generalized Heegner Classes

In this section we compute the étale Abel-Jacobi image of the generalized Heegner cycle Δ_φ associated to the isogeny $\varphi: E_{F_\varphi}\to E_\varphi$ and prove Lemma 1.1. To simplify notations, we base change all the objects to F_φ, but we do not stress this operation in the notation. Set $r:=k-2$ and then define

$$\mathfrak{R}^r := R^{2r}\left(E^r\times E^r\longrightarrow\mathfrak{Spec}\left(F_\varphi\right)\right)_*\left(\mathbb{Q}_p\right),\ \mathfrak{R}^r_\varphi := R^{2r}\left(E^r_\varphi\times E^r\longrightarrow\mathfrak{Spec}\left(F_\varphi\right)\right)_*\left(\mathbb{Q}_p\right)$$
$$\text{and } \mathcal{R}^r := R^{2r}\left(E_1^r\left(N\right)\times E^r\longrightarrow Y_1\left(N\right)\right)_*\left(\mathbb{Q}_p\right).$$

We write $\iota_{\eta_\varphi}: E^r_\varphi\to E^r_1\left(N\right)$ for the pull-back of $E^r_1\left(N\right)/Y_1\left(N\right)$ obtained from $\eta_\varphi:\mathfrak{Spec}\left(F_\varphi\right)\to Y_1\left(N\right)$. Note that the diagonal morphism $d_{E^r}: E^r\to E^r\times E^r$ (resp. the isogeny $\varphi\times 1_{E^r}: E^r\times E^r\to E^r_\varphi\times E^r$) induces a relative push-forward

$$d_{E^r*}:\mathbb{Q}_p=R^0\left(E^r\longrightarrow\mathfrak{Spec}\left(F_\varphi\right)\right)_*\left(\mathbb{Q}_p\right)\longrightarrow\mathfrak{R}^r\left(r\right)\ (\text{resp. } \varphi_*\times 1_{E^r*}:\mathfrak{R}^r\left(r\right)\longrightarrow\mathfrak{R}^r_\varphi\left(r\right)).$$

Similarly, because $\mathfrak{R}_{\varphi}^{r}$ is identified with the pull-back of $\mathcal{R}^{r}$ obtained from $\eta_{\varphi}:\mathfrak{Spec}\left(F_{\varphi}\right)\rightarrow Y_{1}\left(N\right)$ (since $\iota_{\eta_{\varphi}}\times 1_{E^{r}}:\left(E_{\varphi}^{r}\times E^{r}\right)/F_{\varphi}\rightarrow\left(E_{1}^{r}\left(N\right)\times E^{r}\right)/Y_{1}\left(N\right)$ is cartesian), there is a pushforward morphism $\eta_{\varphi *}:H_{\acute{e}t}^{0}\left(\mathfrak{Spec}\left(F_{\varphi}\right),\mathfrak{R}_{\varphi}^{r}\left(r\right)\right)\rightarrow H_{\acute{e}t}^{2}\left(Y_{1}\left(N\right),\mathcal{R}^{r}\left(r+1\right)\right)$. As in the proof of [10, Proposition 3.7], we may consider the following commutative diagram in which the vertical arrows comes from the Leray spectral sequence:

$$\begin{array}{ccccccc}
H_{\acute{e}t}^{0}\left(\mathfrak{Spec}\left(F_{\varphi}\right),\mathbb{Q}_{p}\right) & \overset{d_{E^{r}*}}{\longrightarrow} & H_{\acute{e}t}^{0}\left(\mathfrak{Spec}\left(F_{\varphi}\right),\mathfrak{R}^{r}\left(r\right)\right) & \overset{\varphi_{*}\times 1_{E^{r}*}}{\longrightarrow} & H_{\acute{e}t}^{0}\left(\mathfrak{Spec}\left(F_{\varphi}\right),\mathfrak{R}_{\varphi}^{r}\left(r\right)\right) & \overset{\eta_{\varphi *}}{\longrightarrow} & H_{\acute{e}t}^{2}\left(Y_{1}\left(N\right),\mathcal{R}^{r}\left(r+1\right)\right)\\
\| & & \downarrow & & \downarrow & & \downarrow\\
H_{\acute{e}t}^{0}\left(E^{r},\mathbb{Q}_{p}\right) & \overset{d_{E^{r}*}}{\longrightarrow} & H_{\acute{e}t}^{2r}\left(E^{r}\times E^{r},\mathbb{Q}_{p}\left(r\right)\right) & \overset{\varphi_{*}\times 1_{E^{r}*}}{\longrightarrow} & H_{\acute{e}t}^{2r}\left(E_{\varphi}^{r}\times E^{r},\mathbb{Q}_{p}\left(r\right)\right) & \overset{\iota_{\eta_{\varphi *}}\times 1_{E^{r}*}}{\longrightarrow} & H_{\acute{e}t}^{2r+2}\left(E_{1}^{r}\left(N\right)\times E^{r},\mathbb{Q}_{p}\left(r+1\right)\right)
\end{array}$$

This diagram is the analogue of [10, (60)]. Note that $\mathrm{AJ}_{\acute{e}t}\left(\Delta_{\varphi}\right)$ is obtained applying the morphism

$$\mathrm{HS}:H_{\acute{e}t}^{2r+2}\left(E_{1}^{r}\left(N\right)\times E^{r},\mathbb{Q}_{p}\left(r+1\right)\right)\longrightarrow H^{1}\left(F_{\varphi},H_{\acute{e}t}^{2r+1}\left(E_{1}^{r}\left(N\right)_{\overline{F}_{\varphi}}\times_{\overline{F}_{\varphi}}E_{\overline{F}_{\varphi}}^{r},\mathbb{Q}_{p}\left(r+1\right)\right)\right)$$

arising from the Hochschild-Serre spectral sequence to the image of 1 under the composition of the arrows appearing in the lower row of the above diagram. We remark that $S_{r}\times S_{r}$ acts on the modules appearing in the above diagram except $H_{\acute{e}t}^{0}\left(\mathfrak{Spec}\left(F_{\varphi}\right),\mathbb{Q}_{p}\right)$ and $H_{\acute{e}t}^{0}\left(E^{r},\mathbb{Q}_{p}\right)$ and the arrows between them are equivariant for this action. Because we further project $\mathrm{AJ}_{\acute{e}t}\left(\Delta_{\varphi}\right)$ from $H^{1}\left(F_{\varphi},H_{\acute{e}t}^{2r+1}\left(E_{1}^{r}\left(N\right)_{\overline{F}_{\varphi}}\times_{\overline{F}_{\varphi}}E_{\overline{F}_{\varphi}}^{r},\mathbb{Q}_{p}\left(r+1\right)\right)\right)$ to $X:=H^{1}\left(F_{\varphi},H_{\acute{e}t}^{1}\left(Y_{1}\left(N\right)_{\overline{F}_{\varphi}},\mathcal{S}_{r,1-l_{1},1-l_{2}}\left(r+1\right)\right)\right)$ and $S_{r}\times S_{r}$ acts trivially on $H_{\acute{e}t}^{1}\left(Y_{1}\left(N\right)_{\overline{F}_{\varphi}},\mathcal{S}_{r,1-l_{1},1-l_{2}}\left(r+1\right)\right)$, arguing as in the proof of [10, Proposition 3.7] we deduce that the image $\mathrm{AJ}_{\acute{e}t}^{[l_{1},l_{2}]}\left(\Delta_{\varphi}\right)$ of $\mathrm{AJ}_{\acute{e}t}\left(\Delta_{\varphi}\right)$ in X is obtained as follows. Let us set $V:=H^{1}\left(E_{\overline{F}_{\varphi}},\mathbb{Q}_{p}\left(1\right)\right)$ (resp. $V_{\varphi}:=H^{1}\left(E_{\varphi,\overline{F}_{\varphi}},\mathbb{Q}_{p}\left(1\right)\right)$), $V^{\vee}:=H^{1}\left(E_{\overline{F}_{\varphi}},\mathbb{Q}_{p}\right)$ and consider the projection

$$\mathrm{p}:\mathfrak{R}^{r}\left(r\right)=H_{\acute{e}t}^{2r}\left(E_{\overline{F}_{\varphi}}^{r}\times_{\overline{F}_{\varphi}}E_{\overline{F}_{\varphi}}^{r},\mathbb{Q}_{p}\left(r\right)\right)\longrightarrow\mathrm{Sym}^{r}\left(V\right)\otimes\mathrm{Sym}^{r}\left(V^{\vee}\right)\longrightarrow\mathrm{Sym}^{r}\left(V\right)\left[1-l_{1},1-l_{2}\right]$$

obtained from the Kunneth decomposition and the further projection from $V^{\otimes r}\otimes V^{\vee\otimes r}$ to $\mathrm{Sym}^{r}\left(V\right)\otimes\mathrm{Sym}^{r}\left(V^{\vee}\right)$ and then to $\mathrm{Sym}^{r}\left(V\right)\left[1-l_{1},1-l_{2}\right]$ (the latter via (2)). Here we abusively identify the sheaves on $\mathfrak{Spec}\left(F_{\varphi}\right)_{\acute{e}t}$ with the associated $G_{F_{\varphi}}$-modules. Setting $\mathfrak{S}^{r}:=\mathrm{Sym}^{r}\left(V\right)\left[1-l_{1},1-l_{2}\right](=S_{r;1-l_{1},1-l_{2}}\left(r\right))$, $\mathfrak{S}_{\varphi}^{r}:=\mathrm{Sym}^{r}\left(V_{\varphi}\right)\left[1-l_{1},1-l_{2}\right](=S_{r;1-l_{1},1-l_{2}}^{\varphi}\left(r\right))$ and $\mathcal{S}^{r}:=\mathcal{S}_{r,1-l_{1},1-l_{2}}\left(r+1\right)$, there are similar projections $\mathfrak{R}_{\varphi}^{r}\left(r\right)\rightarrow\mathfrak{S}_{\varphi}^{r}$ and $\mathcal{R}^{r}\left(r+1\right)\rightarrow\mathcal{S}^{r}$ and compatible push forward maps. Then $\mathrm{AJ}_{\acute{e}t}^{[l_{1},l_{2}]}\left(\Delta_{\varphi}\right)$ is obtained applying

$$H_{\acute{e}t}^{0}\left(\mathfrak{Spec}\left(F_{\varphi}\right),\mathfrak{S}^{r}\right)\overset{\varphi_{*}\times 1_{E^{r}*}}{\longrightarrow}H_{\acute{e}t}^{0}\left(\mathfrak{Spec}\left(F_{\varphi}\right),\mathfrak{S}_{\varphi}^{r}\right)\overset{\eta_{\varphi *}}{\longrightarrow}H_{\acute{e}t}^{2}\left(Y_{1}\left(N\right),\mathcal{S}^{r}\right)\overset{\mathrm{HS}}{\longrightarrow}H^{1}\left(F_{\varphi},H_{\acute{e}t}^{1}\left(Y_{1}\left(N\right)_{\overline{F}_{\varphi}},\mathcal{S}^{r}\right)\right)$$

to $c:=\mathrm{p}\left(d_{E^{r}*}\left(1\right)\right)$. Hence we have finished if we can verify that

$$c=\frac{1}{(l_1-1)!\,(l_2-1)!}z_{l_1,l_2}\otimes w_E^{l_1-1}\overline{w}_E^{l_2-1}. \tag{30}$$

To this end, let us fix basis $\{v_E, \overline{v}_E\} \subset V$ and $\{w_E, \overline{w}_E\} \subset V^\vee$ which are dual each other under the Poincaré pairing $P_{E_{\overline{F}}}$: $P_{E_{\overline{F}}}(v_E \otimes w_E) = P_{E_{\overline{F}}}(\overline{v}_E \otimes \overline{w}_E) = 1$ and $P_{E_{\overline{F}}}(v_E \otimes \overline{w}_E) = P_{E_{\overline{F}}}(\overline{v}_E \otimes w_E) = 0$. We may want to choose $\{v_E, \overline{v}_E\}$ and $\{w_E, \overline{w}_E\}$ so that (1) holds; this requires to extend the coefficients of the cohomology groups to K when p is not split, but we do not highlight this extension of the scalars in the notation. As usual, let $p_{r,V} : V^{\otimes r} \to \mathrm{Sym}^r(V)$ be the maximal quotient such that S_r acts trivially on $\mathrm{Sym}^r(V)$, so that the images $v_E^a \overline{v}_E^b$ of the tensors $v_E^{\otimes a} \otimes \overline{v}_E^{\otimes b}$ for $(a, b) \in \mathbb{N}\times\mathbb{N}$ such that $a + b = r$ form a basis of $\mathrm{Sym}^r(V)$. The same applies to $V^\vee$. The morphism $i_{r,V} : \mathrm{Sym}^r(V) \to V^{\otimes r}$ given by the rule sending the image $x_1 \cdot \ldots \cdot x_r \in \mathrm{Sym}^r(V)$ of $x_1 \otimes \ldots \otimes x_r \in V^{\otimes r}$ to $\sum_{\sigma\in S_r} \sigma(x_1 \otimes \ldots \otimes x_r)$ is well defined and injective and similarly for $V^\vee$. Consider the following sequence of maps

$$P^r_{E_{\overline{F}}} : \mathrm{Sym}^r(V^\vee) \otimes \mathrm{Sym}^r(V) \overset{i_r}{\hookrightarrow} V^{\vee\otimes r} \otimes V^{\otimes r} \overset{i^1_{\mathrm{K}}}{\hookrightarrow}$$

$$H^r\left(E^r_{\overline{F}}, \mathbb{Q}_p\right) \otimes H^r\left(E^r_{\overline{F}}, \mathbb{Q}_p(r)\right) \overset{i^2_{\mathrm{K}}}{\hookrightarrow} H^{2r}\left(E^r_{\overline{F}} \times_{\overline{F}} E^r_{\overline{F}}, \mathbb{Q}_p(r)\right) \overset{d^*_{E^r_{\overline{F}}}}{\longrightarrow} H^{2r}\left(E^r_{\overline{F}}, \mathbb{Q}_p(r)\right) = \mathbb{Q}_p,$$

where $i_r = i_{r,V^\vee} \otimes i_{r,V}$, i^1_{K} and i^2_{K} come from the Kunneth decomposition and $d_{E^r_{\overline{F}}}$ is the diagonal morphism. Note that $d^*_{E^r_{\overline{F}}} \circ i^2_{\mathrm{K}}$ is the Poincaré pairing $P_{E^r_{\overline{F}}}$ for $E^r_{\overline{F}}$ and $P_{E^r_{\overline{F}}} \circ i^1_{\mathrm{K}}$ is the r-tensor power $P^{\otimes r}_{E_{\overline{F}}}$ of the Poincaré pairing $P_{E_{\overline{F}}}$ for $E_{\overline{F}}$ (as it follows from the fact that $P_{E_{\overline{F}}}$ is the composition of the Kunneth morphism followed by the diagonal pull-back, together with the compatibility of the "fundamental class" with tensor products). On the other hand, a purely algebraic computation shows the following two facts.

(1) The morphisms $p_{r,V}$ and $i_{r,V^\vee}$ (resp. $p_{r,V}$ and $i_{r,V^\vee}$) are dual each other under the perfect pairing $P^{\otimes r}_{E_{\overline{F}}} \circ i_r$.
(2) We have $P^r_{E_{\overline{F}}}\left(1/a! \cdot w_E^a \overline{w}_E^b \otimes 1/d! \cdot v_E^c \overline{v}_E^d\right) = \delta_{a,c}$ (where $\delta_{a,c}$ denotes the Kronecker delta function).

Dualizing the above diagram we find

$$\mathbb{Q}_p = H^0\left(E^r_{\overline{F}}, \mathbb{Q}_p\right) \overset{d_{E^r_{\overline{F}}*}}{\longrightarrow} H^{2r}\left(E^r_{\overline{F}} \times_{\overline{F}} E^r_{\overline{F}}, \mathbb{Q}_p(r)\right) \overset{p^2_{\mathrm{K}}}{\twoheadrightarrow} H^r\left(E^r_{\overline{F}}, \mathbb{Q}_p(r)\right) \otimes H^r\left(E^r_{\overline{F}}, \mathbb{Q}_p\right)$$

$$\overset{p^1_{\mathrm{K}}}{\twoheadrightarrow} V^{\otimes r} \otimes V^{\vee\otimes r} \overset{p_r}{\twoheadrightarrow} \mathrm{Sym}^r(V) \otimes \mathrm{Sym}^r(V^\vee),$$

where the maps are the following. The first arrow is the diagonal push-forward, which is indeed the dual of $d_{E^r_{\overline{F}}*}$. Because the Kunneth decomposition is compatible with Poincaré duality p^2_{K} comes from the Kunneth decomposition and similarly for p^1_{K}. Finally, because $P_{E^r_{\overline{F}}} \circ i^1_{\mathrm{K}} = P^{\otimes r}_{E_{\overline{F}}}$, it follows from (1) that $p_r = p_{r,V} \otimes p_{r,V^\vee}$. Using

(2) and the well known expression for the dual morphism of a perfect pairing, we see that

$$p_r p_{\mathrm{K}}^1 p_{\mathrm{K}}^2 d_{E_{\overline{F}}^r *}(1) = \sum_{a+b=r} 1/a!b! \cdot v_E^a \overline{v}_E^b \otimes w_E^a \overline{w}_E^b. \tag{31}$$

The claimed (30) follows from (31) after projecting from $\mathrm{Sym}^r(V) \otimes \mathrm{Sym}^r\left(V^\vee\right)$ to $\mathrm{Sym}^r(V)[1-l_1, 1-l_2]$ and then noticing that the isomorphism $\mathrm{Sym}^r(V) = \mathrm{Sym}^r\left(Hom_{\mathbb{Q}_p}\left(V, \mathbb{Q}_p\right)\right)(r) = S_r(X)(r)$ induced by the Weil pairing (cf. the discussion before (7)) sends $v_E^{l_1-1}\overline{v}_E^{l_2-1}$ to z_{l_1,l_2}.

References

1. Andreatta, F., Iovita, A.: Katz type p-adic L-functions for primes p non-split in the CM field. Preprint (2019)
2. Andreatta, F., Iovita, A., Stevens, G.: Overconvergent Eichler-Shimura isomorphisms. J. Inst. Math. Jussieu **14**, 221–274 (2015)
3. Bertolini, M., Darmon, H.: Hida families and rational points on elliptic curves. Invent. Math. **168**, 371–431 (2007)
4. Bertolini, M., Darmon, H., Prasanna, K.: p-adic Rankin L-series and rational points on CM elliptic curves. Pac. J. Math. **260**(2), 261–303 (2012)
5. Bertolini, M., Darmon, H., Prasanna, K.: Generalised Heegner cycles and p-adic Rankin L-series. Duke Math. J. **162**(6), 1033–1148 (2013)
6. Bertolini, M., Darmon, H., Prasanna, K.: Chow-Heegner points on CM elliptic curves and values of p-adic L-functions. Int. Math. Res. Not. **2014**, 745–793 (2014)
7. Brooks, E.H.: Generalized Heegner cycles, Shimura curves, and special values of p-adic L-functions. PhD thesis
8. Brown, K.S.: Homological criteria for finiteness. Comment. Math. Helvetici **50**, 129–135 (1975)
9. Buzzard, K.: On p-adic families of automorphic forms. In: Modular Curves and Abelian Varieties, Progress in Maths, vol. 224, pp. 23–44. Birkhäuser (2004)
10. Bertolini, M., Seveso, M.A., Venerucci, R.: Reciprocity laws for balanced diagonal classes. Astérisque **434**, 77–174 (2022)
11. Bertolini, M., Seveso, M.A., Venerucci, R.: Diagonal classes and the Bloch-Kato conjecture. Münster J. Math. **13**, 317–352 (2020)
12. Castella, F., Hsieh, M.-L.: Heegner cycles and p-adic L-functions. Math. Ann. **370**, 567–628 (2018)
13. Disegni, D.: The universal p-adic Gross-Zagier formula. Invent. Math. **230**, 509–649 (2022)
14. Freitag, E., Kiehl, R.: Étale cohomology and the Weil conjectures. Springer, Berlin (1988)
15. Greenberg, M., Seveso, M.A.: p-adic families of cohomological modular forms for indefinite quaternion algebras and the Jacquet-Langlands correspondence. Canad. J. Math. **68**, 961–998 (2016)
16. Greenberg, M., Seveso, M.A., Shahabi, S.: Modular p-adic L-functions attached to real quadratic fields and arithmetic applications. J. Reine Angew. Math. **721**, 167–231 (2016)
17. Goldstein, G., Schappacher, N.: Séries d'Eisenstein et fonctions L de courbes elliptiques à multiplication complexe. J. Reine Angew. Math. **327**, 184–218 (1981)
18. Howard, B.: Variation of Heegner points in Hida families. Invent. Math. **167**, 91–128 (2007)
19. Jannsen, U.: Continuous étale cohomology. Math. Ann. **280**, 207–246 (1988)
20. Jetchev, D., Loeffler, D., Zerbes, S.L.: Heegner points in Coleman families. Proc. Lond. Math. Soc. **122**, 124–152 (2021)

21. Loeffler, D., Zerbes, S.L.: Iwasawa theory and p-adic L-functions over $\mathbb{Z}_p^2$-extensions. Int. J. Number Theory **10**, 2045–2095 (2014)
22. Loeffler, D., Zerbes, S.L.: Rankin-Eisenstein classes in Coleman families. Res. Math. Sci. **3**, 29 (2016)
23. Milne, J.S.: Étale cohomology. Princeton Mathematical Series, vol. 33. Princeton University Press, Princeton (1980)
24. Schappacher, N.: Periods of Hecke characters. Lecture Notes in Mathematics, vol. 1301. Springer, Berlin (1988)
25. Shimura, G.: Introduction to the arithmetic theory of automorphic functions. Princeton University Press, Princeton (1971)

Congruences of Modular Forms and Modularity of Tate–Shafarevich Classes

Matteo Tamiozzo

Abstract We prove, under suitable assumptions, that p-torsion Tate–Shafarevich classes for elliptic curves over the rationals are visible in quotients of Jacobians of modular curves, as predicted by a conjecture of Jetchev–Stein. The key ingredient is the non-triviality of the Bertolini–Darmon bipartite Kolyvagin system, which implies that suitable cohomology classes of the system form a basis of the Selmer group modulo p.

Keywords Tate–Shafarevich group · bipartite Kolyvagin system · Modular forms

1 Introduction

Let E be an elliptic curve over a number field K, with L-function $L(E/K, s)$; we denote by n the rank of $E(K)$, and by $|Ш(E/K)|$ the cardinality of the Tate–Shafarevich group $Ш(E/K)$. The Birch and Swinnerton-Dyer conjecture predicts that $\mathrm{ord}_{s=1} L(E/K, s) = n$, and

$$\frac{L^{(n)}(E/K, 1)}{n! P(E/K) R(E/K)} = |Ш(E/K)|, \tag{1}$$

where the regulator $R(E/K)$ and the period $P(E/K)$ are defined in [15, Definitions 1.6, 2.9].

1.1. **Bertolini–Darmon's bipartite Kolyvagin system**. Assume in the rest of this introduction that the elliptic curve E is defined over $\mathbf{Q}$, so that $L(E/\mathbf{Q}, s) = L(f, s)$ for a newform $f \in S_2(\Gamma_0(N))$, and let $K/\mathbf{Q}$ be an imaginary quadratic field. In the seminal work [4], Bertolini–Darmon introduced a strategy to study the Birch and Swinnerton-Dyer conjecture for E/K based on congruences of modular forms.

M. Tamiozzo (✉)
LAGA, Université Sorbonne Paris Nord, Villetaneuse, France
e-mail: tamiozzo@math.univ-paris13.fr

M. Longo et al. (eds.), *Elliptic Curves and Modular Forms in Arithmetic Geometry*, Springer Proceedings in Mathematics & Statistics 527,
https://doi.org/10.1007/978-3-032-13123-2_13

1.1.1. Let us first recall the basic idea to show that $E(K)$ is finite if $L(E/K, 1) \neq 0$; more details on the argument and the required assumptions can be found in the body of the text, and in [4, 30]. For a fixed prime $p > 3$, we consider newforms $f_\ell \in S_2(\Gamma_0(N\ell))$ congruent to f modulo (a prime above) p, for suitable auxiliary primes ℓ inert in K, called admissible primes (cf. Sect. 2.1.1). Heegner points on the abelian varieties A_ℓ attached to the modular forms f_ℓ yield cohomology classes $z_\ell^{BD} \in H^1(K, E[p])$ which are possibly ramified at ℓ, as A_ℓ has bad reduction at ℓ. More precisely, assuming that the left-hand side of (1) for $n = 0$ is a p-adic unit, one shows via the so-called first reciprocity law that z_ℓ^{BD} is not a Selmer class, as it fails to satisfy the relevant local condition at ℓ. Varying ℓ, this allows to construct enough classes in $H^1(K, E[p]) \smallsetminus \mathrm{Sel}(K, E[p])$ to force, by global duality, the vanishing of $\mathrm{Sel}(K, E[p])$. In particular, $E(K)$ is finite.
1.1.2. More generally, Bertolini–Darmon construct classes $z_{\mathfrak{l}}^{BD} \in H^1(K, E[p])$, indexed by suitable products of admissible primes, which are part of a bipartite Kolyvagin system Z^{BD} for $E[p]$. The reader can consult Sects. 2.1.3 and 2.3.3 for the definitions, and Remark 9 if surprised by our terminology. Using such classes–and working modulo higher powers of p–one can prove the following implication, under suitable assumptions:

$$\mathrm{ord}_{s=1} L(E/K, s) \leq 1 \Rightarrow n{=}\mathrm{ord}_{s=1} L(E/K, s), \text{ and } v_p\left(\frac{L^{(n)}(E/K, 1)}{n!P(E/K)R(E/K)}\right) \geq v_p(|\text{Ш}(E/K)[p^\infty]|).$$

We refer the reader to [3, 5], as well as to [38, §1] for a similar result for Hilbert modular forms of parallel weight two and references to further related works. In addition, as highlighted in [38], in order to conclude that the p-adic valuations of the two sides of (1) are equal one needs to know that

$$L^{alg}(E/K, 1) \equiv 0 \pmod{p} \Rightarrow \mathrm{Sel}(K, E[p]) \neq 0. \qquad (2)$$

More precisely, one needs an analogue of this implication for modular forms obtained raising the level of f modulo p, cf. [38, Theorem 1.5, Remark 6.11] and Sect. 2.3.4.

1.2. **A basis of** $\mathrm{Sel}(K, E[p])$ This note originated from our attempt to study the implication 2; our starting point is the simple observation that the first reciprocity law implies that each z_ℓ^{BD} is a Selmer class if $L^{alg}(E/K, 1)$ is not a p-adic unit. Under certain assumptions implying that Z^{BD} is non-trivial (i.e. contains at least one non-zero element) we prove in Theorem 7 that suitable classes $z_{\mathfrak{l}}^{BD}$ form a basis of $\mathrm{Sel}(K, E[p])$. A similar problem for the Heegner point Kolyvagin system was raised by Gross [14, §11] and studied by Zhang W. [41, §8.2].

Assume that $L^{alg}(E/K, 1) \equiv 0 \pmod{p}$. If Z^{BD} is non-trivial, then 2 holds (cf. Proposition 5 and Sect. 2.3.4) and, by Proposition 6, one of the classes $z_{\mathfrak{l}}^{BD}$ belongs to $\mathrm{Sel}(K, E[p]) \smallsetminus \{0\}$. However, let us emphasize that, at present, the only way we know to prove the non-triviality of Z^{BD} consists in deducing it from a generalization of 2, cf. Sect. 2.3.4.

1.3. **Modularity of Tate–Shafarevich classes**. We apply Theorem 7 to the study of visibility of Tate–Shafarevich classes; some background on the subject can be found in Sect. 3. Recall that $c \in H^1(K, E)$ is visible in an abelian variety A in which E is embedded if the image of c in $H^1(K, A)$ is trivial. Jetchev–Stein [20, Conjecture 7.1.1] conjectured that every class $c \in Ш(E/\mathbf{Q})$ is visible in a quotient of the Jacobian of a modular curve $X_0(M)$; we call a class c with this property *modular*. Concretely, this tells us that c can be realized as an E-torsor contained in a quotient of $\mathrm{Jac}(X_0(M))$.

As mentioned above, each class z_ℓ^{BD} comes from a K-point on a modular abelian variety A_ℓ with $E[p] \subset A_\ell$; it follows that the image of z_ℓ^{BD} in $H^1(K, E)$ is visible in the quotient $(E \times A_\ell)/E[p]$ (where $E[p]$ is embedded anti-diagonally; cf. Sect. 4.2.1). Generalizing this observation, we prove new instances of the conjecture of Jetchev–Stein (previously known only for classes of order at most 3). A special case of our main theorem is the following; see Theorem 5 for a more precise result.

Theorem 2 *Let $E/\mathbf{Q}$ be an elliptic curve of squarefree conductor N, without complex multiplication. Let $p > 3$ be a prime of good ordinary reduction. Assume that $\bar{\rho} : \mathrm{Gal}(\bar{\mathbf{Q}}/\mathbf{Q}) \to \mathrm{Aut}_{\mathbf{F}_p}(E[p])$ is surjective and ramified at every prime $q \mid N$ such that $q \equiv \pm 1 \pmod{p}$. Then every class in $Ш(E/\mathbf{Q})[p]$ is modular.*

1.3.2. Note that there is no restriction on the rank of $E(\mathbf{Q})$ in the theorem. The hypotheses are imposed to ensure on the one hand that the bipartite Kolyvagin system Z^{BD} can be constructed, and on the other hand that it is non-trivial. For the reader's convenience, we recall that the most important ingredients to achieve these goals are the following (cf. also the introduction of [38]).

(1) Results on the cohomology with $\mathbf{Z}/p\mathbf{Z}$-coefficients of Shimura curves and Shimura sets (multiplicity one, Ihara's lemma) used to construct Z^{BD}.
(2) The Iwasawa main conjecture–especially Skinner–Urban's divisibility–as well as a comparison of periods, employed to prove (7) (generalizing 2) and to deduce the non-triviality of Z^{BD}.

We have not attempted to make our assumptions optimal; still, they are easy to verify in practice, and they do hold true in a non-trivial example, illustrated in Sect. 4.3.

1.4. **Structure of the text**. In Sect. 2 we study bipartite Kolyvagin systems for $E[p]$ and prove Theorem 7. We formulate our arguments abstractly, hoping that this clarifies their structure and highlights the key point, i.e. the non-triviality of Z^{BD}. Along the way, we show that bipartite Kolyvagin systems for $E[p]$ are essentially unique, and their existence is equivalent to the p-parity conjecture for E/K; cf. Sect. 2.1.7. We also take the opportunity to briefly illustrate how the Birch and Swinnerton-Dyer conjecture for E/K–including the p-part of the special value formula (1)–in analytic rank at most one, as well as p-converse theorems, can be deduced from the non-triviality of Z^{BD} (without using other Kolyvagin systems). Many of the relevant arguments, collected in Sects. 2.3.5, 2.3.9 and 2.3.10, can be found, more or less disguised, scattered throughout the literature [3, 5, 16, 21, 37, 41]. The results mentioned above constitute a substantial part of the evidence currently supporting the Birch and Swinnerton-Dyer conjecture for E/K. The reader might find it of some interest that they all follow from one unifying principle (even though, of course, they have also been proved individually by other means, which we do not attempt to discuss here).

In Sect. 3, which is purely expository, we recall some some basic notions and known results in the theory of visibility of Tate–Shafarevich classes, trying to make this text accessible to readers without previous exposure to the topic. Finally, in Sect. 4 we prove our main results, namely Theorem 5 and Corollary 6.

Sections 2 and 3 are independent of each other. The reader only interested in visibility of Tate–Shafarevich classes can start from Sect. 3, then move on to Sect. 4 and refer back to Sect. 2 when needed. In order to understand the proofs in the last section, part of the general discussion in Sect. 2 is unnecessary: one only needs Theorem 7, which is a linear algebra exercise using the two reciprocity laws (cf. Sects. 2.1.3 and 2.3.4).

Acknowledgements

This note originates from my attempt to explore the way paved by Bertolini–Darmon's work (especially [4]), whose influence could hardly be overestimated. I would also like to gratefully acknowledge the influence of Christophe Cornut on this work: the germs of some of the ideas explained below were first developed in 2016 during my master thesis on [41], under Cornut's supervision. I am grateful to Christophe Cornut and Henri Darmon for independently suggesting that the results in section 2 could have applications related to visibility of Tate–Shafarevich classes. I also wish to thank Luca Dall'Ava, Fred Diamond, Michele Fornea, Aleksander Horawa, Mohamed Moakher, Naomi Sweeting and Olivier Wittenberg for helpful discussions related to the topics studied in this article, and Christophe Cornut, Luca Dall'Ava and the anonymous referee for their comments on an earlier version of this document.

The work presented in this text was supported by the National Science Foundation under Grant No. DMS-1928930, while I was in residence at the Mathematical Sciences Research Institute in Berkeley, California, during the Spring 2023 semester on Algebraic Cycles, L-values and Euler Systems. I wish to sincerely thank the staff for their support throughout my stay.

2 Generating the Mod p Selmer Group

2.1 Bipartite Kolyvagin Systems Modulo p

2.1.1. *Setting*. Let $E/\mathbf{Q}$ be an elliptic curve of conductor N, corresponding to a newform

$$f = \sum_{n=1}^{\infty} a_n q^n \in S_2(\Gamma_0(N)).$$

Let $p > 3$ be a prime number not dividing N. We assume that E has no complex multiplication, and that the Galois representation $\bar{\rho} : \mathrm{Gal}(\bar{\mathbf{Q}}/\mathbf{Q}) \to \mathrm{Aut}_{\mathbf{F}_p}(E[p])$ is surjective. Fix an imaginary quadratic field K such that the prime factors of pN are unramified in K, and write $N = N^- N^+$, where $q \mid N^-$ (resp. $q \mid N^+$) if q is inert (resp. split) in K. Assume that N^- is squarefree; the sign of the functional equation of $L(E/K, s)$ is

$$\varepsilon(E/K) = (-1)^{|\{q|N^-\}|+1}.$$

A prime number $\ell \nmid pN$ inert in K is called admissible if $\ell \not\equiv \pm 1 \pmod{p}$ and $a_\ell \equiv \pm(\ell + 1) \pmod{p}$. We denote by $\mathcal{A}$ the set of products $\mathfrak{l} = \ell_1 \ell_2 \cdots \ell_j$ of distinct admissible primes; we allow $j = 0$, in which case $\mathfrak{l} = 1$ by convention. We can write $\mathcal{A}^+ \coprod \mathcal{A}^-$, where

$$\mathcal{A}^+ = \{\mathfrak{l} = \ell_1 \ell_2 \cdots \ell_j : j \geq 0, |\{q \mid N^-\}| \not\equiv j \pmod{2}\},$$
$$\mathcal{A}^- = \{\mathfrak{l} = \ell_1 \ell_2 \cdots \ell_j : j \geq 0, |\{q \mid N^-\}| \equiv j \pmod{2}\}.$$

We will abusively denote the unique place of K above an admissible prime $\ell \in \mathcal{A}$ still by ℓ. We will need the following properties of admissible primes.

(1) The K_ℓ-module $E[p]$ splits (uniquely) as a direct sum $\mathbf{Z}/p\mathbf{Z} \oplus \mathbf{Z}/p\mathbf{Z}(1)$, and $H^1_{ur}(K_\ell, E[p]) \simeq H^1(K_\ell, \mathbf{Z}/p\mathbf{Z}) \simeq \mathbf{Z}/p\mathbf{Z}$. We set $H^1_{tr}(K_\ell, E[p]) = H^1(K_\ell, \mathbf{Z}/p\mathbf{Z}(1)) \simeq \mathbf{Z}/p\mathbf{Z}$ (cf. [4, Lemma 2.6], [16, Lemma 2.2.1]).
(2) For every $0 \neq c \in H^1(K, E[p])$ there are infinitely many $\ell \in \mathcal{A}$ such that $\mathrm{loc}_\ell(c) \neq 0$ (cf. [4, Theorem 3.2]).

2.1.2. *Selmer groups*. The local condition at $\ell \in \mathcal{A}$ for the Selmer group $\mathrm{Sel}(K, E[p])$ is the unramified one. Given $\mathfrak{l} \in \mathcal{A}$, we define $\mathrm{Sel}_{(\mathfrak{l})}(K, E[p])$ changing the local condition to $H^1_{tr}(K_\ell, E[p])$ at every $\ell \mid \mathfrak{l}$. Furthermore, for $\ell \in \mathcal{A}$ not dividing $\mathfrak{l}$, using the local condition 0 (resp. $H^1(K_\ell, E[p])$) at ℓ we obtain a Selmer group $Sel_{(\mathfrak{l})\ell}(K, E[p])$ (resp. $Sel^{\ell}_{(\mathfrak{l})}(K, E[p])$). Thus, we have four Selmer groups fitting in the following rhombus:

$$\begin{array}{ccccc} & & Sel^{\ell}_{(\mathfrak{l})}(K, E[p]) & & \\ & \nearrow & & \nwarrow & \\ Sel_{(\mathfrak{l})}(K, E[p]) & & & & Sel_{(\mathfrak{l}\ell)}(K, E[p]) \\ & \nwarrow & & \nearrow & \\ & & Sel_{(\mathfrak{l})\ell}(K, E[p]). & & \end{array}$$

By global duality we have $\dim(\mathrm{Sel}^{\ell}_{(\mathfrak{l})}(K, E[p])) - \dim(\mathrm{Sel}_{(\mathfrak{l})\ell}(K, E[p])) = 1$, and either $\mathrm{Sel}_{(\mathfrak{l}\ell)}(K, E[p]) = \mathrm{Sel}_{(\mathfrak{l})\ell}(K, E[p])$ and $\mathrm{Sel}_{(\mathfrak{l})}(K, E[p]) = \mathrm{Sel}^{\ell}_{(\mathfrak{l})}(K, E[p])$, or $\mathrm{Sel}_{(\mathfrak{l}\ell)}(K, E[p]) = \mathrm{Sel}^{\ell}_{(\mathfrak{l})}(K, E[p])$ and $\mathrm{Sel}_{(\mathfrak{l})}(K, E[p]) = \mathrm{Sel}_{(\mathfrak{l})\ell}(K, E[p])$ (cf. [13, Lemma 9], [16, Proposition 2.2.9]). It follows that

$$\dim(\mathrm{Sel}_{(\mathfrak{l}\ell)}(K, E[p])) - \dim(\mathrm{Sel}_{(\mathfrak{l})}(K, E[p])) = \pm 1, \tag{3}$$

and we have the following equivalence:

$$\begin{aligned} \mathrm{loc}_{\ell} : \mathrm{Sel}_{(\mathfrak{l})}(K, E[p]) \to H^1_{ur}(K_{\ell}, E[p]) \text{ is surjective } &\Leftrightarrow \dim(\mathrm{Sel}_{(\mathfrak{l}\ell)}(K, E[p])) \\ &= \dim(\mathrm{Sel}_{(\mathfrak{l})}(K, E[p])) - 1. \end{aligned} \tag{4}$$

The implication from left to right will be crucial for us.

Furthermore, let $r = \dim(\mathrm{Sel}(K, E[p]))$. Given $\mathfrak{l} = \ell_1 \cdots \ell_j$ in $\mathcal{A}$, the dimension of $\mathrm{Sel}_{(\mathfrak{l})}(K, E[p])$ is at least $r - j$, so

$$j < r \Rightarrow \mathrm{Sel}_{(\mathfrak{l})}(K, E[p]) \neq 0. \tag{5}$$

2.1.3. *Bipartite Kolyvagin systems for* $E[p]$. A bipartite Kolyvagin system $Z = Z^+ \coprod Z^-$ for $E[p]$ consists of two collections

$$Z^+ = \{z_{\mathfrak{l}} \in \mathbf{Z}/p\mathbf{Z}, \mathfrak{l} \in \mathcal{A}^+\}, \quad Z^- = \{z_{\mathfrak{l}} \in \mathrm{Sel}_{(\mathfrak{l})}(K, E[p]), \mathfrak{l} \in \mathcal{A}^-\}$$

satisfying the following relations (usually called first and second reciprocity law):

(RL1) for every $\mathfrak{l} \in \mathcal{A}^-$ and every $\ell \mid \mathfrak{l}$, we have

$$\mathrm{loc}_{\ell}(z_{\mathfrak{l}}) = 0 \in H^1_{tr}(K_{\ell}, E[p]) \Leftrightarrow z_{\mathfrak{l}/\ell} = 0 \in \mathbf{Z}/p\mathbf{Z};$$

(RL2) for every $\mathfrak{l} \in \mathcal{A}^-$ and every $\ell \in \mathcal{A}$, $\ell \nmid \mathfrak{l}$, we have

$$\mathrm{loc}_{\ell}(z_{\mathfrak{l}}) = 0 \in H^1_{ur}(K_{\ell}, E[p]) \Leftrightarrow z_{\mathfrak{l}\ell} = 0 \in \mathbf{Z}/p\mathbf{Z}.$$

The collection Z_0 of elements $z_{\mathfrak{l}} = 0$ for every $\mathfrak{l} \in \mathcal{A}$ satisfies the above requirements; a bipartite Kolyvagin system for $E[p]$ different from Z_0 will be called non-trivial. We will recall later how an explicit bipartite Kolyvagin system $Z \neq Z_0$ can be constructed in our situation (under suitable assumptions), relying especially on the work of Bertolini–Darmon [4] and Skinner–Urban [35]. In this section we will explore some consequences of the existence of a non-trivial Z. First of all, let us remark that either of the following assertions is equivalent to non-triviality of Z:

(i) there is $\mathfrak{l} \in \mathcal{A}^+$ such that $z_\mathfrak{l} \neq 0$;
(iI) there is $\mathfrak{l} \in \mathcal{A}^-$ such that $z_\mathfrak{l} \neq 0$.

Indeed, if (i) holds then by (RL1) for every $\ell \in \mathcal{A}$ not dividing $\mathfrak{l}$ we have $z_{\mathfrak{l}\ell} \neq 0$. Conversely, assume that $z_\mathfrak{l} \neq 0$ for some $\mathfrak{l} \in \mathcal{A}^-$. Choose an admissible prime $\ell \nmid \mathfrak{l}$ such that $\mathrm{loc}_\ell(z_\mathfrak{l}) \neq 0$ (cf. Sect. 2.1.1). By (RL2) we have $z_{\mathfrak{l}\ell} \neq 0$.

Lemma 3 *Let Z be a bipartite Kolyvagin system for $E[p]$.*

(1) If $\mathfrak{l} \in \mathcal{A}^+$ and $z_\mathfrak{l} \neq 0$ then $\mathrm{Sel}_{(\mathfrak{l})}(K, E[p]) = 0$.
(2) If $\mathfrak{l} \in \mathcal{A}^-$ and $z_\mathfrak{l} \neq 0$ then $\mathrm{Sel}_{(\mathfrak{l})}(K, E[p]) \simeq \mathbf{Z}/p\mathbf{Z}$.

Proof The first point follows from global duality, (RL1) and the existence of admissible primes as in (2) of Sect. 2.1.1 (cf. [16, Proposition 2.3.5] or [38, Corollary 7.14]). The second point follows from the first, (RL2) and (4), since by Sect. 2.1.1(2) there exists $\ell \in \mathcal{A}$ not dividing $\mathfrak{l}$ such that $\mathrm{loc}_\ell(z_\mathfrak{l}) \neq 0$.

Proposition 4 *If there exists a non-trivial bipartite Kolyvagin system for $E[p]$ then, for every $\mathfrak{l} = \ell_1 \cdots \ell_j \in \mathcal{A}$, we have*

$$\dim(Sel_{(\mathfrak{l})}(K, E[p])) \equiv |\{q \mid N^-\}| + j + 1 \pmod 2. \tag{6}$$

Proof By the previous lemma and the remark preceding it there is $\mathfrak{l}_0 \in \mathcal{A}^+$ such that $\mathrm{Sel}_{(\mathfrak{l}_0)}(K, E[p]) = 0$. Furthermore, adding a prime factor to $\mathfrak{l} \in \mathcal{A}$ changes the parity of both sides of the congruence (6) (see (3)). Hence, if $\mathfrak{l}, \mathfrak{l}' \in \mathcal{A}$ and $\mathfrak{l} \mid \mathfrak{l}'$ then (6) holds true for $\mathrm{Sel}_{(\mathfrak{l})}(K, E[p])$ if and only if it does for $\mathrm{Sel}_{(\mathfrak{l}')}(K, E[p])$. Now fix $\mathfrak{l}_0$ as above and take $\mathfrak{l} \in \mathcal{A}$. As (6) holds for $\mathrm{Sel}_{(\mathfrak{l}_0)}(K, E[p])$, it does for $\mathrm{Sel}_{(\mathrm{lcm}(\mathfrak{l}_0, \mathfrak{l}))}(K, E[p])$, hence for $\mathrm{Sel}_{(\mathfrak{l})}(K, E[p])$.

The next proposition states that non-triviality of a bipartite Kolyvagin system Z is equivalent to the converse of Lemma 3.

Proposition 5 *Let Z be a bipartite Kolyvagin system for $E[p]$. The following assertions are equivalent.*

(1) $Z \neq Z_0$.
(2) For every $\mathfrak{l} \in \mathcal{A}$,

$$\mathrm{Sel}_{(\mathfrak{l})}(K, E[p]) = 0 \Rightarrow \mathfrak{l} \in \mathcal{A}^+ \text{ and } z_\mathfrak{l} \neq 0.$$

(3) For every $\mathfrak{l} \in \mathcal{A}$,

$$\mathrm{Sel}_{(\mathfrak{l})}(K, E[p]) \simeq \mathbf{Z}/p\mathbf{Z} \Rightarrow \mathfrak{l} \in \mathcal{A}^- \text{ and } z_\mathfrak{l} \neq 0.$$

Proof

(1) $\Rightarrow$ (2) The argument below–which actually shows that (1) implies (2) and (3)–is a straightforward adaptation of the one in the proof of [37, Theorem 3.3.8]; we include the details for completeness.

Step 0. By Proposition 4 if $\mathfrak{l} \in \mathcal{A}$ and $\mathrm{Sel}_{(\mathfrak{l})}(K, E[p]) = 0$ (resp. $\mathrm{Sel}_{(\mathfrak{l})}(K, E[p]) \simeq \mathbf{Z}/p\mathbf{Z}$) then $\mathfrak{l} \in \mathcal{A}^+$ (resp. $\mathfrak{l} \in \mathcal{A}^-$). Let $Z^{\heartsuit} \subset Z$ be the set of elements indexed by $\mathfrak{l} \in \mathcal{A}$ such that $\dim(\mathrm{Sel}_{(\mathfrak{l})}(K, E[p])) \leq 1$. We will show that every $z_{\mathfrak{l}} \in Z^{\heartsuit}$ is non zero.

Step 1. Let $\mathfrak{l} \in \mathcal{A}^-$ and $\ell \nmid \mathfrak{l}$ be an admissible prime. If $z_{\mathfrak{l}}, z_{\mathfrak{l}\ell} \in Z^{\heartsuit}$ then

$$z_{\mathfrak{l}} \neq 0 \Leftrightarrow z_{\mathfrak{l}\ell} \neq 0.$$

Indeed, if $z_{\mathfrak{l}} \neq 0$ then it generates $\mathrm{Sel}_{(\mathfrak{l})}(K, E[p])$; furthermore, by (3) the dimension of $\mathrm{Sel}_{(\mathfrak{l}\ell)}(K, E[p])$ is even, hence it is zero. By (4) we must have $\mathrm{loc}_{\ell}(z_{\mathfrak{l}}) \neq 0$, hence $z_{\mathfrak{l}\ell} \neq 0$ by (RL2). Conversely, if $z_{\mathfrak{l}\ell} \neq 0$ then $z_{\mathfrak{l}} \neq 0$ by (RL2).

Step 2. Let $\mathfrak{l} \in \mathcal{A}^+$ and $\ell \nmid \mathfrak{l}$ be an admissible prime. If $z_{\mathfrak{l}}, z_{\mathfrak{l}\ell} \in Z^{\heartsuit}$ then

$$z_{\mathfrak{l}} \neq 0 \Leftrightarrow z_{\mathfrak{l}\ell} \neq 0.$$

Indeed, if $z_{\mathfrak{l}} \neq 0$ then $z_{\mathfrak{l}\ell} \neq 0$ by (RL1). Conversely, if $z_{\mathfrak{l}\ell} \neq 0$ then it generates $\mathrm{Sel}_{(\mathfrak{l}\ell)}(K, E[p])$; furthermore, $\mathrm{Sel}_{(\mathfrak{l})}(K, E[p])$ is trivial. By the discussion in Sect. 2.1.2 we have $\mathrm{loc}_{\ell}(z_{\mathfrak{l}\ell}) \neq 0$, and (RL1) implies that $z_{\mathfrak{l}} \neq 0$.

Step 3. By assumption there is $\mathfrak{l}_0 \in \mathcal{A}$ such that $z_{\mathfrak{l}_0} \neq 0$ (such a $z_{\mathfrak{l}_0}$ automatically belongs to $Z^{\heartsuit}$ by Lemma 3). Let $z_{\mathfrak{l}}$ be an element in $Z^{\heartsuit}$. We will prove that there is a "path" in $Z^{\heartsuit}$ from $z_{\mathfrak{l}_0}$ to $z_{\mathfrak{l}}$ such that the index of each element in the path is obtained from the index of the previous element adding or removing an admissible prime. By the previous steps, this will imply that $z_{\mathfrak{l}} \neq 0$, as we wanted to prove.

Note that there is $\mathfrak{l}' \in \mathcal{A}$ multiple of $\mathfrak{l}$ and $\mathfrak{l}_0$ such that $z_{\mathfrak{l}'} \in Z^{\heartsuit}$ (using Sect. 2.1.1(2) and (4), add admissible primes to $\mathrm{lcm}(\mathfrak{l}, \mathfrak{l}_0)$ if needed, until the dimension of the Selmer group drops at most to one). Therefore it suffices to show that, given any $z_{\mathfrak{l}_1}, z_{\mathfrak{l}_2} \in Z^{\heartsuit}$ with $\mathfrak{l}_1 \mid \mathfrak{l}_2$ and $\mathfrak{l}_1 \neq \mathfrak{l}_2$, we can connect $z_{\mathfrak{l}_1}$ to $z_{\mathfrak{l}_2}$ via a path as above. We prove this by induction on the number d of prime factors of $\mathfrak{l}_3 = \mathfrak{l}_2/\mathfrak{l}_1$; the statement for $d = 1$ is true by definition. Assume that $d > 1$; for every $\ell \mid \mathfrak{l}_3$ we know by induction that, if $z_{\mathfrak{l}_2/\ell} \in Z^{\heartsuit}$, then $z_{\mathfrak{l}_1}$ and $z_{\mathfrak{l}_2/\ell}$ are connected (via a path as above), hence we are done. Let us now suppose that, for every $\ell \mid \mathfrak{l}_3$, the element $z_{\mathfrak{l}_2/\ell}$ is not in $Z^{\heartsuit}$. By the discussion in Sect. 2.1.2 we must have $\dim(\mathrm{Sel}_{(\mathfrak{l}_2/\ell)}(K, E[p])) = 2$ and $\mathrm{loc}_{\ell}(\mathrm{Sel}_{(\mathfrak{l}_2)}(K, E[p])) = 0$ for every $\ell \mid \mathfrak{l}_3$. Therefore we find that $\mathrm{Sel}_{(\mathfrak{l}_2)}(K, E[p]) \subset \mathrm{Sel}_{(\mathfrak{l}_1)}(K, E[p])$. As $\mathrm{Sel}_{(\mathfrak{l}_2)}(K, E[p])$ is one-dimensional, the inclusion is an equality. Choose an admissible prime

$\ell' \nmid \mathfrak{l}_2$ such that $\mathrm{loc}_{\ell'}(\mathrm{Sel}_{(\mathfrak{l}_1)}(K, E[p])) \neq 0$; then $\mathrm{Sel}_{(\mathfrak{l}_1\ell')}(K, E[p])) = \mathrm{Sel}_{(\mathfrak{l}_2\ell')}(K, E[p])) = 0$, so $\mathrm{Sel}_{(\mathfrak{l}_2\ell'/\ell)}(K, E[p])) \simeq \mathbf{Z}/p\mathbf{Z}$ for every $\ell \mid \mathfrak{l}_3$. Now we can connect, for any $\ell \mid \mathfrak{l}_3$,

$$z_{\mathfrak{l}_1} \rightsquigarrow z_{\mathfrak{l}_1\ell'} \rightsquigarrow z_{\mathfrak{l}_2\ell'/\ell} \rightsquigarrow z_{\mathfrak{l}_2\ell'} \rightsquigarrow z_{\mathfrak{l}_2}.$$

(2) ⇒ (3) If $\mathrm{Sel}_{(\mathfrak{l})}(K, E[p]) \simeq \mathbf{Z}/p\mathbf{Z}$ then by Sect. 2.1.1(2) and (4) there is $\ell \nmid \mathfrak{l}$ admissible such that $\mathrm{Sel}_{(\mathfrak{l}\ell)}(K, E[p]) = 0$, hence $\mathfrak{l}\ell \in \mathcal{A}^+$ and $z_{\mathfrak{l}\ell} \neq 0$. Therefore $\mathfrak{l} \in \mathcal{A}^-$ and $z_{\mathfrak{l}} \neq 0$ by (RL2).

(3) ⇒ (1) Using Sect. 2.1.1(2) and (4) repeatedly we see that there is $\mathfrak{l} \in \mathcal{A}$ such that $\mathrm{Sel}_{(\mathfrak{l})}(K, E[p]) = 0$. For every $\ell \in \mathcal{A}$ not dividing $\mathfrak{l}$ we have $\mathrm{Sel}_{(\mathfrak{l}\ell)}(K, E[p]) \simeq \mathbf{Z}/p\mathbf{Z}$ (by (3)), hence $z_{\mathfrak{l}\ell} \neq 0$.

2.1.7. To sum up, the linear-algebraic gymnastics done in this section gives us the following information.

(i) The existence of a non-trivial bipartite Kolyvagin system for $E[p]$ implies that $\dim(\mathrm{Sel}(K, E[p])) \equiv |\{q \mid N^-\}| + 1 \pmod 2$ (this is the conclusion of Proposition 4 for $\mathfrak{l} = 1$).

(ii) Conversely, suppose that $\dim(\mathrm{Sel}(K, E[p])) \equiv |\{q \mid N^-\}| + 1 \pmod 2$; then the congruence (6) holds for every $\mathfrak{l} \in \mathcal{A}$ (cf. the proof of Proposition 4). For every $\mathfrak{l} \in \mathcal{A}$ such that $\mathrm{Sel}_{(\mathfrak{l})}(K, E[p]) = 0$ (resp. $\mathrm{Sel}_{(\mathfrak{l})}(K, E[p]) \simeq \mathbf{Z}/p\mathbf{Z}$) pick a non-zero element $z_{\mathfrak{l}}$ in $\mathbf{Z}/p\mathbf{Z}$ (resp. in $\mathrm{Sel}_{(\mathfrak{l})}(K, E[p])$). We will now check that these elements form a bipartite Kolyvagin system Z; by Proposition 5, any other non-trivial bipartite Kolyvagin system for $E[p]$ is obtained multiplying the elements of Z by elements of $(\mathbf{Z}/p\mathbf{Z})^\times$.
Let $\mathfrak{l} \in \mathcal{A}$ and let $\ell \nmid \mathfrak{l}$ be an admissible prime. If $\dim(\mathrm{Sel}_{(\mathfrak{l})}(K, E[p])) \geq 3$ then $\dim(\mathrm{Sel}_{(\mathfrak{l}\ell)}(K, E[p])) \geq 2$ and both $z_{\mathfrak{l}}$ and $z_{\mathfrak{l}\ell}$ are zero. If $\dim(\mathrm{Sel}_{(\mathfrak{l})}(K, E[p])) = 2$ then $z_{\mathfrak{l}} = 0$; furthermore, by (3), the dimension of $\mathrm{Sel}_{(\mathfrak{l}\ell)}(K, E[p])$ is 1 or 3. In the latter case $z_{\mathfrak{l}\ell} = 0$; in the former case, by (4) we must have $\mathrm{loc}_\ell(\mathrm{Sel}_{(\mathfrak{l}\ell)}(K, E[p])) = 0$, hence $\mathrm{loc}_\ell(z_{\mathfrak{l}\ell}) = 0$. If $\dim(\mathrm{Sel}_{(\mathfrak{l})}(K, E[p])) = 1$ then $\mathrm{Sel}_{(\mathfrak{l}\ell)}(K, E[p])$ has dimension 0 or 2, and $\mathrm{Sel}_{(\mathfrak{l}\ell)}(K, E[p]) = 0$ if and only if $\mathrm{loc}_\ell(\mathrm{Sel}_{(\mathfrak{l})}(K, E[p])) \neq 0$, hence $\mathrm{loc}_\ell(z_{\mathfrak{l}}) \neq 0 \Leftrightarrow z_{\mathfrak{l}\ell} \neq 0$. Finally, if $\mathrm{Sel}_{(\mathfrak{l})}(K, E[p]) = 0$ then $\mathrm{Sel}_{(\mathfrak{l}\ell)}(K, E[p])$ is one-dimensional and $\mathrm{loc}_\ell(\mathrm{Sel}_{(\mathfrak{l}\ell)}(K, E[p]))$ is non-zero, hence $\mathrm{loc}_\ell(z_{\mathfrak{l}\ell}) \neq 0$.

Note that $\mathrm{Sel}(K, E[p]) = \mathrm{Sel}(K, E[p^\infty])[p]$ (cf. [38, Proposition 5.8]), so the dimension of $\mathrm{Sel}(K, E[p])$ and the $\mathbf{Z}_p$-corank of $\mathrm{Sel}(K, E[p^\infty])$ have the same parity (thanks to the Cassels–Tate pairing). Therefore, the congruence in (i) is the parity conjecture for $\mathrm{Sel}(K, E[p^\infty])$. Summing up, we learn that the existence of a non-trivial Z is equivalent to the p-parity conjecture (cf. also [38, Proposition 8.8], [41, §9.2]). The conjecture is already known–in bigger generality–by work of Nekovář [28, 29], which exploits (in a different way) the non-triviality of a suitable Euler system [9].

A construction of an explicit, non-trivial Z, consisting of elements $z_{\mathfrak{l}}$ related to special values of L-functions of modular forms, yields further information on the Birch and Swinnerton-Dyer conjecture, as we will recall in Sect. 2.3 below.

2.2 **A basis of** $\mathrm{Sel}(K, E[p])$ Let Z be a non-trivial bipartite Kolyvagin system for $E[p]$. By Proposition 5, if $\mathrm{Sel}(K, E[p])$ is one-dimensional, then it is generated by the class z_1. The following theorem generalizes this observation, without imposing any restriction on the dimension of $\mathrm{Sel}(K, E[p])$.

Proposition 6 *Let Z be a non-trivial bipartite Kolyvagin system for $E[p]$, and let $r = \dim(\mathrm{Sel}(K, E[p]))$. There exists $\mathfrak{l} = \ell_1\ell_2\cdots\ell_r \in \mathcal{A}$ such that $\{z_{\mathfrak{l}/\ell_i}, 1 \le i \le r\}$ is a basis of* $\mathrm{Sel}(K, E[p])$.

Proof The case $r = 1$ was discussed before the statement of the proposition, so let us assume that $r \ge 2$. Using Sect. 2.1.1(2) and (4), choose distinct admissible primes $\ell_1, \dots, \ell_r$ such that, letting $\mathfrak{l} = \ell_1 \cdots \ell_r$, we have $\mathrm{Sel}_{(\mathfrak{l})}(K, E[p]) = 0$. By Proposition 4, for every $1 \le i \le r$ we have $\mathfrak{l}/\ell_i \in \mathcal{A}^-$, hence $z_{\mathfrak{l}/\ell_i} \in \mathrm{Sel}_{(\mathfrak{l}/\ell_i)}(K, E[p])$. Fix $1 \le i, j \le r$ with $j \ne i$. By (5) we have $\mathrm{Sel}_{(\mathfrak{l}/(\ell_i\ell_j))}(K, E[p]) \ne 0$, hence $\mathrm{loc}_{\ell_j}(z_{\mathfrak{l}/\ell_i}) = 0$ by (RL1) and Lemma 3. Therefore, $z_{\mathfrak{l}/\ell_i}$ belongs to $\mathrm{Sel}(K, E[p])$. Finally, by (RL2) and Proposition 5, the image of $(z_{\mathfrak{l}/\ell_i})_{1\le i\le r}$ via the map

$$\mathrm{loc}_{\ell_1} \oplus \cdots \oplus \mathrm{loc}_{\ell_r} : \mathrm{Sel}(K, E[p]) \to H^1_{ur}(K_{\ell_1}, E[p]) \oplus \cdots \oplus H^1_{ur}(K_{\ell_r}, E[p]) \simeq (\mathbf{Z}/p\mathbf{Z})^r$$

is the canonical basis of $(\mathbf{Z}/p\mathbf{Z})^r$ (up to multiplication by a unit on each factor).

2.3. **Bertolini–Darmon's bipartite Kolyvagin system modulo** p. We will now work with Bertolini–Darmon's bipartite Kolyvagin system Z^{BD} for $E[p]$, whose construction relies on congruences of modular forms.

2.3.1. *Level raising*. For every $\mathfrak{l} \in \mathcal{A}$, there is a cuspidal, new (normalized) eigenform $f_{\mathfrak{l}} \in S_2(\Gamma_0(N\mathfrak{l}))$ and a prime ideal $\mathfrak{p}_{\mathfrak{l}}$ of the ring $\mathcal{O}_{\mathfrak{l}} \subset \mathbf{C}$ generated by the Fourier coefficients of $f_{\mathfrak{l}}$ such that $\mathcal{O}_{\mathfrak{l}}/\mathfrak{p}_{\mathfrak{l}} \simeq \mathbf{Z}/p\mathbf{Z}$ and the residual Galois representation $\bar{\rho}_{f_{\mathfrak{l}}} : \mathrm{Gal}(\bar{\mathbf{Q}}/\mathbf{Q}) \to \mathrm{GL}_2(\mathcal{O}_{\mathfrak{l}}/\mathfrak{p}_{\mathfrak{l}})$ is isomorphic to $\bar{\rho}$. This follows from [11, Theorem 1], as explained in the proof of [41, Theorem 2.1] (note that the isomorphism $\bar{\rho}_{f_{\mathfrak{l}}} \simeq \bar{\rho}$ forces $\mathcal{O}_{\mathfrak{l}}/\mathfrak{p}_{\mathfrak{l}} \simeq \mathbf{Z}/p\mathbf{Z}$). Let $\bar{\mathcal{O}}_{\mathfrak{l}}$ be the integral closure of $\mathcal{O}_{\mathfrak{l}}$, and fix a prime ideal $\bar{\mathfrak{p}}_{\mathfrak{l}} \subset \bar{\mathcal{O}}_{\mathfrak{l}}$ such that $\bar{\mathfrak{p}}_{\mathfrak{l}} \cap \mathcal{O}_{\mathfrak{l}} = \mathfrak{p}_{\mathfrak{l}}$.

2.3.2. *Assumptions*. In addition to the hypotheses in Sect. 2.1.1, let us assume throughout the rest of this section that N is squarefree and, if either $q \mid N^-$ and $q \equiv \pm 1 \pmod{p}$, or $q \mid N^+$, then $\bar{\rho}$ is ramified at q. In particular, on the one hand hypothesis CR in [33] holds; on the other hand, the local conditions defining $\mathrm{Sel}(K, E[p])$ can be expressed purely in terms of the $\mathrm{Gal}(\bar{K}/K)$-module $E[p]$; in particular, the mod $\mathfrak{p}_{\mathfrak{l}}$ Selmer group of $f_{\mathfrak{l}}$ is isomorphic to $\mathrm{Sel}_{(\mathfrak{l})}(K, E[p])$ (cf. the proof of [41, Theorem 5.2]).

2.3.3. *Construction of* Z^{BD} Under the above assumptions, recall that the bipartite Kolyvagin system $Z^{BD} = \{z_{\mathfrak{l}}^{BD}, \mathfrak{l} \in \mathcal{A}\}$ is constructed as follows. For every $\mathfrak{l} \in \mathcal{A}^+$

(resp. $\mathfrak{l} \in \mathcal{A}^-$) let $B_\mathfrak{l}$ be the quaternion algebra over $\mathbf{Q}$ ramified at $N^-\mathfrak{l}\infty$ (resp. $N^-\mathfrak{l}$), and let $\mathrm{Sh}_{N^+}(B_\mathfrak{l}^\times)$ be the Shimura set (resp. Shimura curve) of level $\Gamma_0(N^+)$ attached to $B_\mathfrak{l}^\times$. Let $Cl(\mathcal{O}_K)$ be the ideal class group of $\mathcal{O}_K$, which acts on points of $\mathrm{Sh}_{N^+}(B_\mathfrak{l}^\times)$ with CM by $\mathcal{O}_K$ (cf. [9, §1.2, §3.8, §5.3]).

(1) For $\mathfrak{l} \in \mathcal{A}^+$, the vector space of functions $\mathrm{Sh}_{N^+}(B_\mathfrak{l}^\times) \to \bar{\mathcal{O}}_\mathfrak{l}/\bar{\mathfrak{p}}_\mathfrak{l}$ on which the Hecke algebra acts as on $f_\mathfrak{l}$ is non-zero (cf. [41, §6.1]), hence so is the space of $\mathbf{F}_p$-valued functions on $\mathrm{Sh}_{N^+}(B_\mathfrak{l}^\times)$ with the same property. Furthermore, this $\mathbf{F}_p$-vector space is one-dimensional (cf. [33, Proposition 6.5, Theorem 6.2]). Fixing a basis $\bar{f}_\mathfrak{l}^{B_\mathfrak{l}}$, the element $z_\mathfrak{l}^{BD} \in \mathbf{Z}/p\mathbf{Z}$ is the sum $\sum_{\sigma \in Cl(\mathcal{O}_K)} \bar{f}_\mathfrak{l}^{B_\mathfrak{l}}(Q^\sigma)$, for a fixed point Q with CM by O_K (cf. [3, §2.2.1]).
(2) For $\mathfrak{l} \in \mathcal{A}^-$, the class $z_\mathfrak{l}^{BD} \in \mathrm{Sel}_{(\mathfrak{l})}(K, E[p])$ is obtained from the divisor $\sum_{\sigma \in Cl(\mathcal{O}_K)} Q^\sigma$ on $\mathrm{Sh}_{N^+}(B_\mathfrak{l}^\times)$, for a fixed point Q with CM by O_K (cf. [3, §2.3.2]). We will explain this in more detail in Sect. 4.2.2, where the precise construction will be needed.

The proof that the above classes satisfy the relations yielding a bipartite Kolyvagin system is non-trivial; it rests on the structure of special fibers of Shimura curves–allowing to realize level raising and Jacquet–Langlands functoriality geometrically–as well as on Ihara's lemma and multiplicity one modulo p. See [4, §8, 9] and [33].

2.3.4. *Non triviality of* Z^{BD} For $\mathfrak{l} \in \mathcal{A}^+$, the element $z_\mathfrak{l}^{BD}$ is non-zero if and only if the algebraic part of the special value $L^{alg}(f_\mathfrak{l}/K, 1)$, as defined in [41, (6.5)], is not congruent to zero modulo $\bar{\mathfrak{p}}_\mathfrak{l}$ (this follows from Gross's formula, cf. [41, Corollary 6.2]). Now, we have the following results (in increasing order of difficulty):

(1) the ($\bar{\mathfrak{p}}_\mathfrak{l}$-adic) valuation of the Tamagawa numbers $t_q(f_\mathfrak{l})$ at primes $q \mid N^+$ is zero by Sect. 2.3.2;
(2) the valuation of the ratio between the Gross period used to define $L^{alg}(f_\mathfrak{l}/K, 1)$ and the canonical period $\Omega_{f_\mathfrak{l}}^{can}$ equals the sum of the valuations of the Tamagawa numbers at primes dividing $N^-\mathfrak{l}$ [33, Theorem 6.8];
(3) assume that E is ordinary at p. Then the following implication holds (Tamagawa numbers at places dividing N^+ are irrelevant in view of (1)):

$$\mathrm{Sel}_{(\mathfrak{l})}(K, E[p]) = 0 \Rightarrow \frac{L(f_\mathfrak{l}/K, 1)}{\Omega_{f_\mathfrak{l}}^{can} \prod_{q \mid N^-\mathfrak{l}} t_q(f_\mathfrak{l})} \not\equiv 0 \pmod{\bar{\mathfrak{p}}_\mathfrak{l}}. \tag{7}$$

This follows from work of Kato and Skinner–Urban (cf. [41, Theorem 7.1], whose third hypothesis is satisfied under our assumptions by [34, Theorem 1.1]). As a consequence of (1), (2) and (3), the equivalent conditions of Proposition 5 are satisfied, hence $Z^{BD} \neq Z_0$. Indeed, if $\mathfrak{l} \in \mathcal{A}$ and $\mathrm{Sel}_{(\mathfrak{l})}(K, E[p]) = 0$ then $\mathfrak{l} \in \mathcal{A}^+$ (by the parity conjecture [31, Theorem B], or by [41, Theorem 7.1] and [Corollary 1.9], and by the previous discussion we have $L^{alg}(f_\mathfrak{l}/K, 1) \not\equiv 0 \pmod{\bar{\mathfrak{p}}_\mathfrak{l}}$).

2.3.5. As we have seen above, for $\mathfrak{l} \in \mathcal{A}^+$ the element $z_\mathfrak{l}^{BD}$ is related to the central value $L(f_\mathfrak{l}/K, 1)$; similarly, for $\mathfrak{l} \in \mathcal{A}^-$ the class $z_\mathfrak{l}^{BD}$ is related to $L'(f_\mathfrak{l}/K, 1)$ (by the Gross–Zagier formula for Shimura curves [39]). Thanks to these relations, the abstract results in the previous sections acquire the following arithmetic meaning.

(1) Lemma 3(1) tells us that, if $\mathfrak{l} \in \mathcal{A}^+$ and $L^{alg}(f_\mathfrak{l}/K, 1) \not\equiv 0 \pmod{\bar{\mathfrak{p}}_\mathfrak{l}}$, then $\mathrm{Sel}_{(\mathfrak{l})}(K, E[p]) = 0$. In particular, if $L(E/K, 1) \neq 0$ then $E(K)$ is finite. A proof of this implication in much greater generality, using this circle of ideas, can be found in [30].
Similarly, it follows from Lemma 3(2) that, if $\mathrm{ord}_{s=1} L(E/K, s) = 1$, then $E(K)$ has rank one, and $Ш(E/K)[p] = 0$ if $z_1^{BD} \neq 0$.
(2) Proposition 5 relates the non-triviality of Z^{BD} to converses of the statements in the previous point, which in turn yield various "converse theorems". For instance, Proposition 5(3) implies that, if $E(K)$ has rank one and $Ш(E/K)$ is finite, then $\mathrm{ord}_{s=1} L(E/K, s) = 1$, a converse of the theorem of Gross–Zagier and Kolyvagin.
Of course, it would be of interest if one could prove that $Z^{BD} \neq Z_0$ directly, without going through the "mod p converse" (7).
(3) The proof of the implication (2) $\Rightarrow$ (3) in Proposition 5 is essentially an abstract formulation of the argument in [41, §7.2], which is at the heart of Zhang's proof of the non-triviality of the Heegner point Kolyvagin system. In a nutshell, Zhang reduces the latter to the non-triviality of Z^{BD}.
(4) Finally, Proposition 6 yields Theorem 7 below.

Theorem 7 *Let $E/\mathbf{Q}$ be an elliptic curve of squarefree conductor N, without complex multiplication. Let K be an imaginary quadratic field such that the prime factors $q \mid N$ are unramified in K, and write $N = N^- N^+$, where $q \mid N^-$ (resp. $q \mid N^+$) if q is inert (resp. split) in K. Let $p > 3$ be a prime not dividing $N\mathrm{disc}(K/\mathbf{Q})$, and let $r = \dim(\mathrm{Sel}(K, E[p]))$. Assume that E is ordinary at p, and that $\bar{\rho} : \mathrm{Gal}(\bar{\mathbf{Q}}/\mathbf{Q}) \to \mathrm{Aut}_{\mathbf{F}_p}(E[p])$ has the following properties:*

(1) $\bar{\rho}$ is surjective;
(2) if $q \mid N^-$ and $q \equiv \pm 1 \pmod{p}$ then $\bar{\rho}$ is ramified at q;
(3) if $q \mid N^+$ then $\bar{\rho}$ is ramified at q;

Then there exists $\mathfrak{l} = \ell_1 \ell_2 \cdots \ell_r \in \mathcal{A}$ such that $\{z_{\mathfrak{l}/\ell_i}^{BD}, 1 \leq i \leq r\}$ is a basis of $\mathrm{Sel}(K, E[p])$.

Proof The assumptions ensure that Z^{BD} is defined and non trivial, cf. Sects. 2.3.3 and 2.3.4; hence, the theorem follows from Proposition 6.

2.3.7. A weaker version of the theorem was proved in [41, Theorem 11.1].

Note that Theorem 7 is valid without any restriction on the rank of $E(K)$. If the rank is bigger than one, no general construction of a set of (topological) generators of $\mathrm{Sel}(K, T_p(E))$ is known; on the other hand, Theorem 7 tells us that $\mathrm{Sel}(K, E[p])$ is always generated by classes coming from CM points on suitable Shimura curves. However, for $r > 1$ the bases $\{z_{\mathfrak{l}/\ell_i}^{BD}, 1 \leq i \leq r\}$ of $\mathrm{Sel}(K, E[p])$ obtained through the theorem are very non-canonical, and their relation with the arithmetic properties of E is mysterious. The Birch and Swinnerton-Dyer conjecture predicts that

$\text{ord}_{s=1} L(E/K, s) \leq r$, and equality holds if and only if all the Selmer classes $z^{BD}_{\mathfrak{l}/\ell_i}$ in the statement of the theorem come from K-points of E. However, our approach offers no way to understand which classes have this property, and it is unclear how to relate the elements $z^{BD}_{\mathfrak{l}/\ell_i}$ to (higher derivatives of) $L(E/K, s)$. For these reasons, the above theorem does not seem to have direct applications to the Birch and Swinnerton-Dyer conjecture. We will instead apply it in the next section to the study of visibility of Tate–Shafarevich classes.

Remark 8 In view of the aforementioned relations between the elements $z^{BD}_{\mathfrak{l}}$ and L-values, the reciprocity laws (RL1), (RL2) can be roughly interpreted as algebraic incarnations of the idea, suggested in [25], that congruences of modular forms should correspond to congruences between values of their L-functions. We will not attempt to list here the myriad of works related to this topic, but we would like to remind the reader of [19], where Jochnowitz suggests to look at "examples where the forms are congruent, but the root numbers are opposite [...] This should happen, for example, any time you had a newform of square-free level that was congruent mod p to a form of lower level. It would be interesting to work out some concrete examples, and see if the derivatives come in as we predicted." [19, p. 255]. The reciprocity laws for Z^{BD} are a manifestation of the fact that Jochnowitz's visionary prediction was essentially correct.

2.3.9. *Primitivity of Bertolini–Darmon's bipartite Kolyvagin system.* Let us collect further consequences of the non-triviality of Z^{BD}, which will not be needed in the sequel of this article but might have independent interest. The construction of Z^{BD} recalled above can be extended (working with modular forms modulo p^k congruent to f) to obtain elements in $\mathbf{Z}/p^k\mathbf{Z}$ and $\text{Sel}_{(\mathfrak{l})}(K, E[p^k])$ for $k \geq 1$. See [4, 33]; the relevant abstract theory was developed in [16]. The condition in Proposition 5(1) can be seen as an analogue in this setting of the notion of primitivity of usual Kolyvagin systems [27, Definition 4.5.5]. Analogously to [27, Theorem 4.5.6], this property implies that the inequalities of [38, Theorem 1.5] are equalities (cf. also [16, Theorem 2.5.1]). For E/K and p as in Theorem 7, this yields the p-part of the Birch and Swinnerton-Dyer special value formula in analytic rank zero and one (in the rank one case, the formula was obtained in [41, Theorem 10.2] as a byproduct of the primitivity of the Heegner point Kolyvagin system); this idea was developed in [3, 5].

Note that to ensure the non-triviality of Z^{BD}–which guarantees that the resulting bound on the size of the Selmer group is optimal–we had to impose assumptions implying the vanishing of the valuation of suitable Tamagawa numbers, cf. §2.3.4(1). For classical Kolyvagin systems, more precise relations between the failure of primitivity and the p-adic valuation of Tamagawa numbers have been established, cf. [6, 18].

2.3.10. *Converse theorems.* Non-triviality of Z^{BD} also implies more precise converse theorems than the one mentioned in §2.3.5(2). For instance, let us sketch an argument proving the following implication, assuming that Z^{BD} is non trivial (which is the case under the assumptions of Theorem 7):

$$\mathrm{Sel}(K, E[p^\infty]) \text{ has } \mathbf{Z}_p\text{-corank one} \Rightarrow \mathrm{ord}_{s=1} L(E/K, s) = 1. \tag{8}$$

Write $\mathrm{Sel}(K, E[p^\infty]) = \mathbf{Q}_p/\mathbf{Z}_p \oplus S$ with $S = T \oplus T$ finite. Let us fix an integer k larger than twice the length $l(T)$ of T. In the rest of this subsection, we will work with Bertolini–Darmon's bipartite Kolyvagin system modulo p^k, whose elements will be denoted by $z^{BD}_{\mathfrak{l},k}$. As in [16] and [38], the elements $z^{BD}_{\mathfrak{l},k}$ are obtained as reductions modulo p^k of the analogous elements attached to $\mathfrak{l} \in \mathcal{A}_{2k}$ squarefree product of primes $\ell \nmid pN$ inert in K such that $\ell \not\equiv \pm 1 \pmod{p}$ and $a_\ell \equiv \pm(\ell + 1) \pmod{p^{2k}}$.

By the Gross–Zagier formula for Shimura curves, it suffices to prove that the class $z^{BD}_{1,k} \in \mathrm{Sel}(K, E[p^k])$ (coming from the divisor considered in §2.3.3(2) for $\mathfrak{l} = 1$) is non-zero. For every $\mathfrak{l} \in \mathcal{A}_{2k}$ with an even number of prime factors we have an isomorphism $\mathrm{Sel}_{(\mathfrak{l})}(K, E[p^k]) \simeq \mathbf{Z}/p^k\mathbf{Z} \oplus T_\mathfrak{l} \oplus T_\mathfrak{l}$ (cf. [16, Proposition 2.2.7, Corollary 2.2.10]). We will prove by induction on $l(T_\mathfrak{l})$ that, for such $\mathfrak{l}$, if $l(T_\mathfrak{l}) \leq l(T)$ then $z^{BD}_{\mathfrak{l},k} \neq 0$.

Choose (using [4, Theorem 3.2]) $\ell_1 \in \mathcal{A}_{2k}$ not dividing $\mathfrak{l}$ such that $\mathrm{loc}_{\ell_1}(\mathrm{Sel}_{(\mathfrak{l})}(K, E[p^k])) = \mathbf{Z}/p^k\mathbf{Z}$, so that $\mathrm{Sel}_{(\mathfrak{l}\ell_1)}(K, E[p^k]) \simeq T_\mathfrak{l} \oplus T_\mathfrak{l}$ by global duality. If $T_\mathfrak{l} = 0$ then $z^{BD}_{\mathfrak{l}\ell_1,k} \neq 0$, so $z^{BD}_{\mathfrak{l},k} \neq 0$ by the second reciprocity law. Otherwise, let p^t be the exponent of $T_\mathfrak{l}$ and write $T_\mathfrak{l} = \mathbf{Z}/p^t\mathbf{Z} \oplus T'$. Choose $\ell_2 \in \mathcal{A}_{2k}$ not dividing $\mathfrak{l}\ell_1$ such that loc_{ℓ_2} is injective on $\mathbf{Z}/p^t\mathbf{Z} \subset T_\mathfrak{l}$. The group $\mathrm{loc}_{\ell_2}(\mathrm{Sel}_{(\mathfrak{l}\ell_1)}(K, E[p^k]))$ is contained in $H^1_{ur}(K_{\ell_2}, E[p^k])[p^t] \simeq \mathbf{Z}/p^t\mathbf{Z}$ (because $p^t T_\mathfrak{l} = 0$), therefore $\mathrm{loc}_{\ell_2}(\mathrm{Sel}_{(\mathfrak{l}\ell_1)}(K, E[p^k])) \simeq \mathbf{Z}/p^t\mathbf{Z}$. The analogue modulo p^k of the diagram in §2.1.2 (cf. [16, Proposition 2.2.9]) implies that $\mathrm{Sel}_{(\mathfrak{l}\ell_1\ell_2)}(K, E[p^k])$ is isomorphic to $\mathbf{Z}/p^k\mathbf{Z} \oplus T' \oplus T'$. Hence $T' \simeq T_{\mathfrak{l}\ell_1\ell_2}$, and $z^{BD}_{\mathfrak{l}\ell_1\ell_2,k} \neq 0$ by the inductive hypothesis.

We will show that $\mathrm{loc}_{\ell_2}(z^{BD}_{\mathfrak{l}\ell_1\ell_2,k}) \neq 0$, which implies that $z^{BD}_{\mathfrak{l}\ell_1,k} \neq 0$ and $z^{BD}_{\mathfrak{l},k} \neq 0$ by the reciprocity laws, concluding the proof. We argue by contradiction. Let $e \geq 0$ be the greatest integer such that $z^{BD}_{\mathfrak{l}\ell_1\ell_2,k} \in p^e\mathrm{Sel}_{(\mathfrak{l}\ell_1\ell_2)}(K, E[p^k])$. The proof of [38, Theorem 8.3] shows (under our assumption that $Z^{BD} \neq Z_0$) that e equals $l(T')$; furthermore, $z^{BD}_{\mathfrak{l}\ell_1\ell_2,k}$ is contained in a submodule of $\mathrm{Sel}_{(\mathfrak{l}\ell_1\ell_2)}(K, E[p^k])$ isomorphic to $\mathbf{Z}/p^k\mathbf{Z}$ (cf. [16, Lemma 3.3.6], [38, Proposition 7.9]). It follows that, if $\mathrm{loc}_{\ell_2}(z^{BD}_{\mathfrak{l}\ell_1\ell_2,k}) = 0$, then $\mathrm{loc}_{\ell_2}(\mathrm{Sel}_{(\mathfrak{l}\ell_1\ell_2)}(K, E[p^k]))$ is killed by $p^{l(T')}$, hence by $p^{l(T_\mathfrak{l})}$. This contradicts global duality, because $\mathrm{loc}_{\ell_2}(\mathrm{Sel}_{(\mathfrak{l}\ell_1)}(K, E[p^k]))$ is also killed by $p^{l(T_\mathfrak{l})}$ and $2l(T_\mathfrak{l}) < k$.

In the ordinary case, the implication (8) was obtained, more indirectly, in [41, Theorem 1.3] as a consequence of the non-triviality of the Heegner point Kolyvagin system and of Kolyvagin's structure theorem for the Selmer group [22]. In the supersingular case, (8) was proved (under suitable assumptions) in [10, Theorem 6.10].

2.3.11. *Addendum to* [38]. In this paragraph, let K/F be a quadratic CM extension of a totally real field. As mentioned above, in [38] a version of Bertolini–Darmon's bipartite Kolyvagin system is used to give upper bounds on the size of Selmer groups of Hilbert modular forms f of parallel weight two and analytic rank at most one over K. For modular elliptic curves E over F, these bounds are a step towards the

p-part of the Birch and Swinnerton-Dyer special value formula (1) for E/K. For instance, suppose that $L(E/K, 1) \neq 0$. As discussed in [38, Remarks 5.5, 5.6], if one knows that the relevant inequality in [38, Theorem 1.5] is an equality, a comparison between the period $P(E/K)$ and the one appearing in the special value formula of [40, Theorem 7.1] is required to deduce the p-part of (1). This comparison is known under suitable assumptions for $F = \mathbf{Q}$ [33, Theorem 6.8], but we would like to clarify that it appears to be much deeper for general F. For instance, for F of even degree and E with good reduction everywhere one needs to show that the quotient of the period in [40, Theorem 7.1] by the product of the archimedean periods of E/K is a p-adic unit. This is closely related to conjectures of Ichino–Prasanna [17, Conjectures A, C] (whose rational, Hodge-theoretic counterpart is Oda's conjecture [32, p. xii]).

Remark 9 We conclude this section with a brief remark on our choice of notation and terminology. We denoted bipartite Kolyvagin systems by Z because of the relation with zeta elements which is apparent in the example of Z^{BD}. Furthermore, we chose the name "bipartite *Kolyvagin* system" instead of the more common "bipartite *Euler* system" because Z^{BD} resembles more a classical Kolyvagin system than an Euler system. First of all, as already pointed out in the introduction of [4], there are no Kolyvagin derivatives in the definition of Z^{BD}: ramification of the classes $z_{\mathfrak{l}}^{BD} \in H^1(K, E[p])$ at places dividing $\mathfrak{l}$ comes from bad reduction of the Shimura curves used to construct $z_{\mathfrak{l}}^{BD}$, rather than from ramification of suitable field extensions of K as in Kolyvagin's construction of the Euler system of Heegner points. In addition, let us point out that the structure of $\mathrm{Sel}(K, E[p^\infty])$ can be described in terms of invariants attached to systems of elements (modulo p^k for arbitrary k) satisfying generalizations of (RL1) and (RL2), as in [21, §4]. This is akin to the information one can obtain from Kolyvagin systems, cf. for instance [27, Proposition 4.5.8].

Note that the fact that Z^{BD} is not constructed as a "Kolyvagin derivative of an Euler system" appears to be crucial for several purposes. Firstly, it ensures that all the classes in Z^{BD} are directly related to special values of L-functions, which in turn allows to establish the non-triviality of Z^{BD} (a similar remark can be found in [16, §1]). Secondly, the fact that all the cohomology classes in Z^{BD} directly come from points will be essential for our applications to visibility below.

The above discussion leads us to wonder whether there are other ways to construct directly objects behaving like Kolyvagin systems.

3 Background on Visibility

3.1. **Visibility**. Let K be a number field, E/K an elliptic curve and $\iota : E \to A$ an embedding of E into an abelian variety A (over K). A class $c \in H^1(K, E)$ is visible in A (via ι) if $\iota(c) = 0 \in H^1(K, A)$. In other words, denoting by $\gamma : A \to B = A/E$ the quotient map, the class c is visible in A if it is the image of a point $P \in B(K)$ via the connecting map $B(K) \to H^1(K, E)$. Concretely, this means that c corresponds to the E-torsor $\gamma^{-1}(P) \subset A$.

Let us start by recalling some basic facts about visibility.

Lemma 4 *For every $c \in H^1(K, E)$, the following assertions hold true.*

(1) Let L/K be a finite extension such that $\mathrm{res}_K^L(c) = 0 \in H^1(L, E)$. The class c is visible in an abelian variety of dimension $[L : K]$.
(2) Let $m \geq 1$ be an integer. If $c \in Ш(E/K)[m]$ then c is visible in an abelian variety of dimension at most m.

Proof (1) The class c is visible in $A = \mathrm{Res}_{L/K} E_L$, as explained in the proof of [2, Proposition 1.3].
(2) This is [2, Proposition 2.4]. The point is that c is in the image of $\mathrm{Sel}(K, E[m])$, whose elements give rise to E-torsors with an effective divisor of degree m, as observed by Cassels (cf. the proof of [7, Theorem 1.3]).

3.2. **Visibility and Congruences**. Let $\iota : E \to A$ be an embedding. By Poincaré's reducibility theorem, we can choose an abelian subvariety $E^\perp \subset A$ such that there is a short exact sequence

$$0 \to E \cap E^\perp \to E \times E^\perp \to A \to 0$$

where $E \cap E^\perp$ is finite over K and is embedded anti-diagonally in $E \times E^\perp$. It follows that, if $c \in H^1(K, E)$ is visible in A, then $(c, 0)$ is in the image of $H^1(K, E \cap E^\perp)$ (in particular, c is killed by the exponent of $E \cap E^\perp$). Reversing the argument yields a classical construction which allows to produce visible classes (cf. [8, §3]); we will make use of a variation of it in the next section. Let us illustrate the construction first in the simplest example: suppose that F is an elliptic curve, p is a prime and $E[p] \simeq F[p]$ is an isomorphism. Let $A = (E \times F)/E[p]$, where the embedding is anti-diagonal; the following diagram is commutative:

$$\begin{array}{ccc} E[p] & \longrightarrow & F \\ \downarrow & & \downarrow \\ E & \longrightarrow & A. \end{array}$$

Let $P \in F(K)$, and let $c \in H^1(K, F[p])$ be its image via the connecting map $F(K)/pF(K) \to H^1(K, F[p])$. Then the image of c in $H^1(K, E)$ is visible in A. In particular, if c belongs to $\mathrm{Sel}(K, E[p])$, then it yields a visible class in $Ш(E/K)[p]$.

3.2.1. *Visibility of* $Ш(E/K)[3]$ *in Abelian Surfaces*. A first non-trivial application of the above construction is Mazur's theorem [26] asserting that every class in $Ш(E/K)[3]$ is visible in an abelian surface (note that Lemma 4 only guarantees that such a class is visible in an abelian threefold). Mazur shows that, for every $c \in \mathrm{Sel}(K, E[3])$, there is an elliptic curve F with an isomorphism $F[3] \simeq E[3]$

such that c goes to zero in $H^1(K, F)$. To do so, Mazur's idea in a nutshell is to consider the moduli space $\mathcal{S}(E, c)$ of elliptic curves F together with an isomorphism $E[3] \simeq F[3]$, and a section of an F-torsor representing the image of c in $H^1(K, F)$. As $c \in \mathrm{Sel}(K, E[3])$, the variety $\mathcal{S}(E, c)$ has a point over every completion of K, and one needs to show that it has a K-point. The key input is the fact that $\mathcal{S}(E, c)/\bar{K}$ is birationally equivalent to $\mathbf{P}^2$, which allows to apply the local-global principle (to a suitable Brauer–Severi variety constructed from $\mathcal{S}(E, c)$).

3.2.2. Note that only for small n we can have many elliptic curves F/K such that $F[n] \simeq E[n]$ (the moduli space of such F is a form of the modular curve of full level n, whose genus increases); this suggests that classes in $\text{Ш}(E/K)[n]$ should not, in general, be visible in abelian surfaces for n large. Explicit examples for $n = 6, 7$ are given in [12].

3.3. **Visibility in Modular Abelian Varieties.** Let us now restrict to $K = \mathbf{Q}$. Let $E/\mathbf{Q}$ be an elliptic curve of conductor N; fix a modular parametrization $\pi : X_0(N) \to E$ and assume that the induced map $E \to J_0(N) = \mathrm{Jac}(X_0(N))$ is injective (such an (E, π) is called optimal). With the notation of §3.2, we can take $E^\perp$ to be the kernel of the map $J_0(N) \to E$ induced by π, and we deduce that classes $c \in H^1(K, E)$ which are visible in $J_0(N)$ are killed by the modular degree, i.e. the degree of π (cf. [8, p. 20]). It is not always the case that $\text{Ш}(E/\mathbf{Q})$ is killed by the modular degree, hence Tate–Shafarevich classes are not visible in $J_0(N)$ in general (cf. [8, §4]).

3.3.1. *Visibility at Higher Level.* It is natural to wonder whether classes in $\text{Ш}(E/\mathbf{Q})$ which are not visible in $J_0(N)$ become visible in $J_0(M)$ for some multiple M of N (the analogous phenomenon of capitulation of ideal classes of abelian number fields in cyclotomic fields is studied in [36]). We will be interested in a variation of this question, explained in the next paragraph. An example of visibility of Tate–Shafarevich classes at higher level is given in [2, §4.2], where the authors exhibit an elliptic curve of conductor 5389 with Tate–Shafarevich classes invisible in $J_0(5389)$, but visible in $J_0(7 \cdot 5389)$; as explained before Proposition 4.2 of *loc. cit.*, the construction is related to level raising.

3.3.2. *Modularity.* As in [20, §2], by a J_0-modular abelian variety we mean an abelian variety over $\mathbf{Q}$ which is a quotient of $J_0(M)$ for some $M \geq 1$. We say that a class $c \in \text{Ш}(E/\mathbf{Q})$ is modular if there is an embedding $\iota : E \to A$ of E into a J_0-modular abelian variety A such that c is visible in A. The following is a special case of [20, Conjecture 7.1.1].

Conjecture 5 (cf. [20, Conjecture 7.1.1]) Every class in $\text{Ш}(E/\mathbf{Q})$ is modular.

Remark 6 As pointed out in [20, Remarks 7.1.2, 7.1.3], the following more precise problems would also be worth investigating; they might offer a way to relate the previous conjecture to the conjectural finiteness of $\text{Ш}(E/\mathbf{Q})$ (note that classes in $\text{Ш}(E/\mathbf{Q})$ which are visible via a fixed $\iota : E \to A$ form a finite group by the Mordell–Weil theorem).

(1) Study the (minimal) levels of the J_0-modular abelian varieties where the classes in $\text{Ш}(E/\mathbf{Q})$ become visible.
(2) Given a prime $\ell \nmid N$, the two degeneracy maps $X_0(N\ell) \to X_0(N)$ induce maps $J_0(N) \to J_0(N\ell)$; repeating the argument we get maps $J_0(N) \to J_0(N\ell_1 \cdots \ell_i)$ for every product of distinct primes not dividing N. For $E \subset J_0(N)$, determine which classes in $\text{Ш}(E/\mathbf{Q})$ are visible in $J_0(N\ell_1 \cdots \ell_i)$ via (linear combinations of) such maps.

3.3.5. *Known Results*. The class $0 \in \text{Ш}(E/\mathbf{Q})$ is modular, because E is modular. Besides this, in [20] the authors give the following theoretical evidence supporting Conjecture 5.

(1) Classes in $\text{Ш}(E/\mathbf{Q})$ of order 2 or 3 are modular [20, Proposition 7.2.1], because they are visible in an abelian surface (by Lemma 4 and §12.1.1).
(2) By [20, Proposition 7.2.2], classes in $\text{Ш}(E/\mathbf{Q})$ which split over an abelian extension of $\mathbf{Q}$ are visible in quotients of $\text{Jac}(X_1(M))$, $M \geq 1$.

To conclude this section, we would like to mention that the relations between visibility, the Birch and Swinnerton-Dyer special value formula, and congruences of modular forms, at the heart of the results presented below, were explored in a series of articles by Agashe, cf. for instance [1].

4 Modularity of $\text{Ш}(E/\mathbf{Q})[p]$

4.1. **Statements**. The aim of this section is to prove new instances of Conjecture 5, as in Corollary 6 below. Our main result is the following.

Theorem 5 *Let $E/\mathbf{Q}$ be an elliptic curve of squarefree conductor N, without complex multiplication. Let K be an imaginary quadratic field such that the prime factors $q \mid N$ are unramified in K, and write $N = N^-N^+$, where $q \mid N^-$ (resp. $q \mid N^+$) if q is inert (resp. split) in K. Let $p > 3$ be a prime not dividing $N\text{disc}(K/\mathbf{Q})$. Assume that E is ordinary at p, and that $\bar{\rho} : \text{Gal}(\bar{\mathbf{Q}}/\mathbf{Q}) \to \text{Aut}_{\mathbf{F}_p}(E[p])$ has the following properties:*

(1) $\bar{\rho}$ is surjective;
(2) if $q \mid N^-$ and $q \equiv \pm 1 \pmod{p}$ then $\bar{\rho}$ is ramified at q;
(3) if $q \mid N^+$ then $\bar{\rho}$ is ramified at q;

Then there is a J_0-modular abelian variety $A/\mathbf{Q}$ and an embedding $\iota : E \to A$ such that every $c \in \text{Ш}(E/K)[p]$ is visible in $A \times_{\mathbf{Q}} K$ via $\iota \times_{\mathbf{Q}} K$.

Before discussing the proof of the theorem, let us deduce a similar result over $\mathbf{Q}$.

Corollary 6 *Let $E/\mathbf{Q}$ be an elliptic curve of squarefree conductor N, without complex multiplication. Let $p > 3$ be a prime not dividing N. Assume that E is ordinary at p, and that $\bar{\rho} : \text{Gal}(\bar{\mathbf{Q}}/\mathbf{Q}) \to \text{Aut}_{\mathbf{F}_p}(E[p])$ has the following properties:*

(1) $\bar{\rho}$ *is surjective;*
(2) if $q \mid N$ *and* $q \equiv \pm 1 \pmod{p}$ *then* $\bar{\rho}$ *is ramified at* q*;*

Then every class in $\text{Ш}(E/\mathbf{Q})[p]$ *is modular.*

Proof Choose an imaginary quadratic field K of discriminant coprime with Np, and such that $q \mid N$ is split (resp. inert) in K if $\bar{\rho}$ is ramified (resp. unramified) at q. Then E/K satisfies the assumptions of the previous theorem, so there is a J_0-modular abelian variety $A \supset E$ such that every class $c \in \text{Ш}(E/K)[p]$ is visible in $A \times_{\mathbf{Q}} K$. As the restriction map $\text{Ш}(A/\mathbf{Q})[p] \to \text{Ш}(A/K)[p]$ is injective, every class in $\text{Ш}(E/\mathbf{Q})[p] \subset \text{Ш}(E/K)[p]$ is visible in A.

4.2. **Visibility of the Classes** $z_{\mathfrak{l}}^{BD} \in H^1(K, E[p])$ Our aim is to prove Theorem 5; the assumptions and notation of the theorem are in force from now on. We will work with Bertolini–Darmon's bipartite Kolyvagin system $\{z_{\mathfrak{l}}^{BD}, \mathfrak{l} \in \mathcal{A}\}$ for $E[p]$, introduced in §2.3. The key ingredient is the following proposition.

Proposition 7 *For every* $\mathfrak{l} \in \mathcal{A}^-$*, the image of* $z_{\mathfrak{l}}^{BD} \in H^1(K, E[p])$ *in* $H^1(K, E)$ *is visible in a quotient of* $J_0(N\mathfrak{l})$*.*

4.2.2. *Construction of the Classes* $z_{\mathfrak{l}}^{BD} \in H^1(K, E[p])$. Fix $\mathfrak{l} \in \mathcal{A}^-$. Let us first describe precisely the construction of the class $z_{\mathfrak{l}}^{BD} \in H^1(K, E[p])$. As in §2.3.1 we have a newform $f_{\mathfrak{l}} \in S_2(\Gamma_0(N\mathfrak{l}))$, a prime ideal $\mathfrak{p}_{\mathfrak{l}}$ of the ring $\mathcal{O}_{\mathfrak{l}} \subset \mathbf{C}$ generated by its Fourier coefficients, and a prime ideal $\bar{\mathfrak{p}}_{\mathfrak{l}} \subset \bar{\mathcal{O}}_{\mathfrak{l}}$ such that $\bar{\mathfrak{p}}_{\mathfrak{l}} \cap \mathcal{O}_{\mathfrak{l}} = \mathfrak{p}_{\mathfrak{l}}$. Let $\mathfrak{m}_{\mathfrak{l}}$ be the kernel of the $\mathcal{O}_{\mathfrak{l}}/\mathfrak{p}_{\mathfrak{l}}$-valued morphism of the Hecke algebra attached to $f_{\mathfrak{l}}$. Let $J_{\mathfrak{l}}$ be the Jacobian of the Shimura curve $\text{Sh}_{N^+}(B_{\mathfrak{l}}^{\times})$ (compactified if $N^-\mathfrak{l} = 1$) of level $\Gamma_0(N^+)$ attached to the quaternion algebra ramified at $N^-\mathfrak{l}$. By [33, Proposition 4.4] we have an isomorphism of $\text{Gal}(\bar{\mathbf{Q}}/\mathbf{Q})$-modules $J_{\mathfrak{l}}[p]/\mathfrak{m}_{\mathfrak{l}} \simeq E[p]$ (unique up to multiplication by a unit in $\mathbf{Z}/p\mathbf{Z}$). Furthermore, $J_{\mathfrak{l}}(K)/\mathfrak{m}_{\mathfrak{l}} \simeq \text{Pic}(\text{Sh}_{N^+}(B_{\mathfrak{l}}^{\times}))(K)/\mathfrak{m}_{\mathfrak{l}}$ since $\mathfrak{m}_{\mathfrak{l}}$ is non-Eisenstein; hence, quotienting by $\mathfrak{m}_{\mathfrak{l}}$ the p-descent exact sequence for $J_{\mathfrak{l}}(K)$, we obtain maps

$$\text{Pic}(\text{Sh}_{N^+}(B_{\mathfrak{l}}^{\times}))(K)/\mathfrak{m}_{\mathfrak{l}} \to H^1(K, J_{\mathfrak{l}}[p])/\mathfrak{m}_{\mathfrak{l}} \to H^1(K, J_{\mathfrak{l}})[p]/\mathfrak{m}_{\mathfrak{l}} \tag{9}$$

whose composite is zero. Therefore, we obtain a map

$$\text{Pic}(\text{Sh}_{N^+}(B_{\mathfrak{l}}^{\times}))(K)/\mathfrak{m}_{\mathfrak{l}} \to H^1(K, J_{\mathfrak{l}}[p]/\mathfrak{m}_{\mathfrak{l}}) \simeq H^1(K, E[p]).$$

The class $z_{\mathfrak{l}}^{BD}$ is defined to be the image of the divisor $\sum_{\sigma \in Cl(\mathcal{O}_K)} Q^{\sigma}$.

4.2.3. *Proof of Proposition* 7 Let $I_{\mathfrak{l}}$ be the kernel of the $\mathcal{O}_{\mathfrak{l}}$-valued morphism of the Hecke algebra attached to $f_{\mathfrak{l}}$. If $\mathfrak{l} = 1$, choose a surjection $\varphi : J_1 \to E$ factoring through I_1. If the induced map $\varphi : J_1[p] \to E[p]$ is zero, we can write $\varphi = p\varphi'$ for some $\varphi' : J_1 \to E$. Repeating the process, we may (and will) assume that $\varphi : J_1[p] \to E[p]$ is non-zero, hence it induces an isomorphism $J_1[p]/\mathfrak{m}_1 \simeq E[p]$ (since the source is an irreducible $\text{Gal}(\bar{K}/K)$-module). The image of $\sum_{\sigma \in Cl(\mathcal{O}_K)} Q^{\sigma}$ in $H^1(K, J_1)[p]/\mathfrak{m}_1$ via the composite of the maps in (9) is zero, hence the image of z_1^{BD} in $H^1(K, E)$ is zero.

From now on, let us suppose that $\mathfrak{l} \neq 1$. If $f_{\mathfrak{l}}$ has rational Fourier coefficients, so that via the Eichler–Shimura construction one can attach to it an elliptic curve $E_{\mathfrak{l}}$ quotient of $J_{\mathfrak{l}}$, then the visibility of $z_{\mathfrak{l}}^{BD}$ in $(E \times E_{\mathfrak{l}})/E[p]$ follows from the argument in §3.2. The general case will be established via a variation of that argument.

We will construct an abelian variety $A_{\mathfrak{l}}$ with an action of $\bar{\mathcal{O}}_{\mathfrak{l}}$, isogenous to the quotient of $J_{\mathfrak{l}}$ attached to (the Jacquet–Langlands transfer of) $f_{\mathfrak{l}}$. More precisely, we will construct a surjection $J_{\mathfrak{l}} \to A_{\mathfrak{l}}$ which factors through $I_{\mathfrak{l}}$, giving rise to a non-zero map $J_{\mathfrak{l}}[p]/\mathfrak{m}_{\mathfrak{l}} \to A_{\mathfrak{l}}[\bar{\mathfrak{p}}_{\mathfrak{l}}]$ and yielding a commutative diagram:

$$\begin{array}{ccc} H^1(K, J_{\mathfrak{l}}[p]) & \longrightarrow & H^1(K, J_{\mathfrak{l}})[p] \\ \downarrow & & \Big| \\ H^1(K, J_{\mathfrak{l}}[p]/\mathfrak{m}_{\mathfrak{l}}) & & \Big| \\ \downarrow & & \downarrow \\ H^1(K, A_{\mathfrak{l}}[\bar{\mathfrak{p}}_{\mathfrak{l}}]) & \longrightarrow & H^1(K, A_{\mathfrak{l}})[\bar{\mathfrak{p}}_{\mathfrak{l}}]. \end{array} \tag{10}$$

Admitting the existence of such a diagram for the moment, let us conclude the proof of Proposition 7. The map $J_{\mathfrak{l}}[p]/\mathfrak{m}_{\mathfrak{l}} \to A_{\mathfrak{l}}[\bar{\mathfrak{p}}_{\mathfrak{l}}]$ is injective (as it is non-zero and the source is an irreducible $\mathrm{Gal}(\bar{K}/K)$-module). Calling Φ the image, we have $\Phi \simeq E[p]$, and the class $z_{\mathfrak{l}}^{BD}$ is visible in $(E \times A_{\mathfrak{l}})/\Phi$ (as usual, the action of Φ is anti-diagonal). Indeed, since $H^1(K, A_{\mathfrak{l}})[\bar{\mathfrak{p}}_{\mathfrak{l}}]$ is killed by $\bar{\mathfrak{p}}_{\mathfrak{l}}$, the right vertical map in (10) factors through $\mathfrak{m}_{\mathfrak{l}}$. Therefore, we obtain a version of (10) with the first row replaced by $H^1(K, J_{\mathfrak{l}}[p])/\mathfrak{m}_{\mathfrak{l}} \to H^1(K, J_{\mathfrak{l}})[p]/\mathfrak{m}_{\mathfrak{l}}$.

By (9) the image of $\sum_{\sigma \in Cl(\mathcal{O}_K)} Q^{\sigma}$ in $H^1(K, J_{\mathfrak{l}}[p])/\mathfrak{m}_{\mathfrak{l}}$ goes to zero in $H^1(K, J_{\mathfrak{l}})[p]/\mathfrak{m}_{\mathfrak{l}}$, hence the image of $z_{\mathfrak{l}}^{BD}$ in $H^1(K, A_{\mathfrak{l}}[\bar{\mathfrak{p}}_{\mathfrak{l}}])$ goes to zero in $H^1(K, A_{\mathfrak{l}})$. Visibility of $z_{\mathfrak{l}}^{BD}$ in $(E \times A_{\mathfrak{l}})/\Phi$ follows from the commutative diagram

$$\begin{array}{ccc} \Phi & \longrightarrow & A_{\mathfrak{l}} \\ \downarrow & & \downarrow \\ E & \longrightarrow & (E \times A_{\mathfrak{l}})/\Phi. \end{array}$$

Finally, the abelian variety $A_{\mathfrak{l}}$ admits a surjective morphism from the $\mathfrak{l}$-new quotient of $J_0(N\mathfrak{l})$; as E is a quotient of the $\mathfrak{l}$-old part of $J_0(N\mathfrak{l})$, we conclude that there is a surjection $J_0(N\mathfrak{l}) \to (E \times A_{\mathfrak{l}})/\Phi$.

4.2.4. *Construction of the Diagram* (10). As in [41, §3.7], we take $A_{\mathfrak{l}}$ to be an abelian variety in the isogeny class of $J_{\mathfrak{l}}/I_{\mathfrak{l}}$ with a surjection $\varphi : J_{\mathfrak{l}} \to A_{\mathfrak{l}}$ factoring through $I_{\mathfrak{l}}$, and such that the induced $\mathcal{O}_{\mathfrak{l}}$-action on $A_{\mathfrak{l}}$ extends to an action of $\bar{\mathcal{O}}_{\mathfrak{l}}$. Let e be the largest integer such that $\bar{\mathfrak{p}}_{\mathfrak{l}}^{e}$ divides $p\bar{\mathcal{O}}_{\mathfrak{l}}$, and write $A_{\mathfrak{l}}[p] = A_{\mathfrak{l}}[\bar{\mathfrak{p}}_{\mathfrak{l}}^{e}] \times T$. Let $\alpha \in \bar{\mathcal{O}}_{\mathfrak{l}}$ be an element with $\bar{\mathfrak{p}}_{\mathfrak{l}}$-adic valuation 1, and with valuation 0 at the other p-adic places of $\bar{\mathcal{O}}_{\mathfrak{l}}$. The idea is that the bottom map in (10) should arise from the long exact sequence coming from "multiplication by $\bar{\mathfrak{p}}_{\mathfrak{l}}$". Even though the ideal $\bar{\mathfrak{p}}_{\mathfrak{l}}$ might not be principal, the desired map can be constructed starting from the map

$H^1(K, A_{\mathfrak{l}}[p]) \to H^1(K, A_{\mathfrak{l}})[p]$, multiplying by α^{e-1} and projecting onto the $\bar{\mathfrak{p}}_{\mathfrak{l}}^e$-torsion part.

Let us make this idea precise. As in [41, §3.7], we take φ such that the image of the induced map between p-adic Tate modules is not contained in $\bar{\mathfrak{p}}_{\mathfrak{l}} T_p(A_{\mathfrak{l}})$. Then the image of the composite $J_{\mathfrak{l}}[p] \to A_{\mathfrak{l}}[p] \to A_{\mathfrak{l}}[\bar{\mathfrak{p}}_{\mathfrak{l}}^e]$ is not contained in $\bar{\mathfrak{p}}_{\mathfrak{l}} A_{\mathfrak{l}}[\bar{\mathfrak{p}}_{\mathfrak{l}}^e]$, so it is not killed by α^{e-1}. We have the following commutative diagram, where the bottom vertical maps are induced by multiplication by α^{e-1} and projection onto the $\bar{\mathfrak{p}}_{\mathfrak{l}}^e$-torsion part:

$$\begin{array}{ccc} H^1(K, J_{\mathfrak{l}}[p]) & \longrightarrow & H^1(K, J_{\mathfrak{l}})[p] \\ \downarrow{\scriptstyle\varphi} & & \downarrow{\scriptstyle\varphi} \\ H^1(K, A_{\mathfrak{l}}[p]) & \longrightarrow & H^1(K, A_{\mathfrak{l}})[p] \\ \downarrow{\scriptstyle\alpha^{e-1}} & & \downarrow{\scriptstyle\alpha^{e-1}} \\ H^1(K, A_{\mathfrak{l}}[\bar{\mathfrak{p}}_{\mathfrak{l}}]) & \longrightarrow & H^1(K, A_{\mathfrak{l}})[\bar{\mathfrak{p}}_{\mathfrak{l}}]. \end{array}$$

Let ψ be the composite of the maps $J_{\mathfrak{l}}[p] \xrightarrow{\varphi} A_{\mathfrak{l}}[p] \xrightarrow{\alpha^{e-1}} A_{\mathfrak{l}}[\bar{\mathfrak{p}}_{\mathfrak{l}}]$. Our choice of φ ensures that ψ is non-zero. Furthermore, it factors through $I_{\mathfrak{l}}$ because φ does. The target is killed by $\bar{\mathfrak{p}}_{\mathfrak{l}}$, hence by $\bar{\mathfrak{p}}_{\mathfrak{l}} \cap \mathcal{O}_{\mathfrak{l}} = \mathfrak{p}_{\mathfrak{l}}$. Therefore, the map ψ factors through $\mathfrak{m}_{\mathfrak{l}}$, yielding (10).

4.2.5. *Proof of Theorem* 5 We may, and do, assume that $r = \dim(\mathrm{Sel}(K, E[p])) \geq 2$ (if $r = 0$ there is nothing to prove; for $r = 1$, Theorem 7 ensures that $\mathrm{Sel}(K, E[p])$ is generated by z_1^{BD}, which comes from a K-point, hence $\text{Ш}(E/K)[p] = 0$). Take $\mathfrak{l} = \ell_1 \ell_2 \cdots \ell_r \in \mathcal{A}$ as in Theorem 7. For $1 \leq i \leq r$, let $\mathfrak{l}^{(i)} = \mathfrak{l}/\ell_i \in \mathcal{A}^-$, and let $A_{\mathfrak{l}^{(i)}}$ be the abelian variety constructed in §4.2.4. Let

$$A = (E \times A_{\mathfrak{l}^{(1)}} \times \cdots \times A_{\mathfrak{l}^{(r)}})/E[p]^r,$$

where the embedding $E[p]^r \to E \times A_{\mathfrak{l}^{(1)}} \times \cdots \times A_{\mathfrak{l}^{(r)}}$ sends $(x_1, \ldots, x_r)$ to $(-x_1 - x_2 - \cdots - x_r, x_1, x_2, \ldots, x_r)$. The map $E \to A$ sending x to $(x, 0, \ldots, 0)$ is an embedding, and for every $1 \leq i \leq r$ we obtain a diagram

$$\begin{array}{ccccc} E[p] & \longrightarrow & A_{\mathfrak{l}^{(i)}} & & \\ \downarrow & & \downarrow & & \\ E & \longrightarrow & (E \times A_{\mathfrak{l}^{(i)}})/E[p] & \longrightarrow & A. \end{array}$$

By the proof of Proposition 7 we know that every $z_{\mathfrak{l}^{(i)}}^{BD} \in \mathrm{Sel}(K, E[p])$ yields a class in $\text{Ш}(E/K)[p]$ which is visible in $(E \times A_{\mathfrak{l}^{(i)}})/E[p]$, hence in A. Therefore, all linear combinations of the images of the classes $z_{\mathfrak{l}^{(i)}}^{BD}$ in $\text{Ш}(E/K)[p]$ are visible in A. Theorem 7 implies that every class in $\text{Ш}(E/K)[p]$ is visible in A.

Finally, fix $1 \leq j \leq r$. The abelian variety $A_{\mathfrak{l}^{(j)}}$ (isogenous to the quotient of $J_0(N\mathfrak{l}^{(j)})$ attached to $f_{\mathfrak{l}^{(j)}}$) is the only one among the $A_{\mathfrak{l}^{(i)}}$ which is a quotient of the

$\mathfrak{l}^{(j)}$-new part of $J_0(N\mathfrak{l})$. On the other hand, E is a quotient of the $\mathfrak{l}$-old part of $J_0(N\mathfrak{l})$. Therefore, there is a surjection $J_0(N\mathfrak{l}) \to A$.

Remark 8 From the proof of Theorem 5 we obtain the following additional information: if $r = \dim(\mathrm{Sel}(K, E[p]))$, then there is $\mathfrak{l} = \ell_1 \cdots \ell_r \in \mathcal{A}$ such that $\text{Ш}(E/K)[p]$ is visible in a quotient of $J_0(N\mathfrak{l})$. It would of course be more satisfying to be able to replace r by (a number only depending on) $\dim(\text{Ш}(E/K)[p])$, but we are unable do this as our argument does not distinguish $E(K)/pE(K)$ from $\text{Ш}(E/K)[p]$ (cf. §2.3.7).

4.3. **Example**. Let us consider the elliptic curve (with Cremona label 307010b1)

$$E : Y^2 + XY + Y = X^3 + 511495021X + 14412635791156$$

of conductor $N = 2 \cdot 5 \cdot 11 \cdot 2791$. According to the LMFDB database [23, Elliptic curve 307010.b2], this is an optimal quotient of $J_0(N)$, with modular degree $m_E = 435265488$, analytic rank 0 and analytic order of $\text{Ш}(E/\mathbf{Q})$ equal to 169 (note that the p-part of the Birch and Swinnerton-Dyer special value formula for $E/\mathbf{Q}$ holds for $p = 13$ by [35, Theorem 2]). As $m_E \not\equiv 0 \pmod{13}$, by the discussion in §3.3 the classes of order 13 in $\text{Ш}(E/\mathbf{Q})$ are not visible in $J_0(N)$. On the other hand, for $p = 13$ the assumptions of Corollary 6 are satisfied: (1) follows from [24, Theorem 4], (2) holds as no prime $q \mid N$ is congruent to $\pm 1 \pmod{13}$, and E is ordinary at 13. Therefore $\text{Ш}(E/\mathbf{Q})[13]$ is visible in a quotient of $J_0(N\mathfrak{l})$ for some $1 \neq \mathfrak{l} \in \mathcal{A}$ (which we did not attempt to find explicitly).

References

1. Agashe, A.: A visible factor of the special L-value. J. Reine Angew. Math. **644**, 159–187 (2010)
2. Agashe, A., Stein, W.: Visibility of Shafarevich-Tate groups of abelian varieties. J. Number Theory **97**(1), 171–185 (2002)
3. Berti, A., Bertolini, M., Venerucci, R.: Congruences between modular forms and the Birch and Swinnerton-Dyer conjecture. In: Elliptic curves, modular forms and Iwasawa theory. Springer, Cham, pp. 1–31 (2016)
4. Bertolini, M., Darmon, H.: Iwasawa's main conjecture for elliptic curves over anticyclotonic $\mathbb{Z}_p$ -extensions. Ann. Math. (2) **162**(1), 1–64 (2005)
5. Bertolini, M., Longo, M., Venerucci, R.: The anticyclotomic main conjectures for elliptic curves. arXiv preprint, arXiv:2306.17784
6. Büyükboduk, K.: Tamagawa defect of Euler systems. J. Number Theory **129**(2), 402–417 (2009)
7. Cassels, J.W.S.: Arithmetic on curves of genus 1. IV: Proof of the Hauptvermutung. J. Reine Angew. Math. **211**, 95–112 (1962)
8. Cremona, J., Mazur, B.: Visualizing elements in the Shafarevich-Tate group. Exp. Math. **9**(1), 13–28 (2000)
9. Cornut, C., Vatsal, V.: Nontriviality of Rankin-Selberg L -functions and CM points. In: L-functions and Galois representations. Cambridge University Press, Cambridge, pp. 121–186 (2007)

10. Castella, F., Wan, X.: Perrin-Riou's main conjecture for elliptic curves at supersingular primes. Math. Ann. **389**(3), 2595–2636 (2024)
11. Diamond, F., Taylor, R.: Lifting modular mod l representations. Duke Math. J. **74**(2), 253–269 (1994)
12. Fisher, T.: Invisibility of Tate-Shafarevich groups in abelian surfaces. Int. Math. Res. Not. **2014**(15), 4085–4099 (2014)
13. Gross, B.H., Parson, J.A.: On the local divisibility of Heegner points. In: Number theory, analysis and geometry. Springer, Berlin, pp. 215–241 (2012)
14. Gross, B.H.: Kolyvagin's work on modular elliptic curves. In: L-functions and arithmetic. Lond. Math. Soc. Lect. Note Ser. 153, 235–256 (1991)
15. Gross, B.H.: Lectures on the conjecture of Birch and Swinnerton-Dyer. In: Arithmetic of L-functions. American Mathematical Society, Providence, pp. 169–209 (2011)
16. Howard, B.: Bipartite Euler systems. J. Reine Angew. Math. **597**, 1–25 (2006)
17. Ichino, A., Prasanna, K.: Periods of quaternionic Shimura varieties. I. Contemp. Math. 762, American Mathematical Society, Providence (2021)
18. Jetchev, D.: Global divisibility of Heegner points and Tamagawa numbers. Compos. Math. **144**(4), 811–826 (2008)
19. Jochnowitz, N.A.: p-adic conjecture about derivatives of L-series attached to modular forms. In: p-adic monodromy and the Birch and Swinnerton-Dyer conjecture. American Mathematical Society, Providence, pp. 239–263 (1994)
20. Jetchev, D., Stein, W.: Visibility of the Shafarevich-Tate group at higher level. Doc. Math. **12**, 673–696 (2007)
21. Kim, C.-H.: A higher Gross-Zagier formula and the structure of Selmer groups. Trans. Am. Math. Soc. **377**(5), 3691–3725 (2024)
22. Kolyvagin, V.A.: On the structure of Selmer groups. Math. Ann. **291**(2), 253–259 (1991)
23. The LMFDB Collaboration: The L-functions and modular forms database. https://www.lmfdb.org
24. Mazur, B.: Rational isogenies of prime degree. Invent. Math. **44**, 129–162 (1978)
25. Mazur, B.: On the arithmetic of special values of L functions. Invent. Math. **55**, 207–240 (1979)
26. Mazur, B.: Visualizing elements of order three in the Shafarevich-Tate group. Asian J. Math. **3**(1), 221–232 (1999)
27. Mazur, B., Rubin, K.: Kolyvagin systems. Mem. Am. Math. Soc. 799, American Mathematical Society, Providence (2004)
28. Nekovář, J.: On the parity of ranks of Selmer groups. II. C. R. Acad. Sci. Paris, Sér. I, Math. **332**(2), 99–104 (2001)
29. Nekovář, J.: Selmer complexes. Astérisque 310, Société Mathématique de France, Paris (2006)
30. Nekovář, J.: Level raising and anticyclotomic Selmer groups for Hilbert modular forms of weight two. Can. J. Math. **64**(3), 588–649 (2012)
31. Nekovář, J.: Some consequences of a formula of Mazur and Rubin for arithmetic local constants. Algebra Number Theory **7**(5), 1101–1120 (2013)
32. Oda, T.: Periods of Hilbert modular surfaces. Prog. Math. 19, Birkhäuser, Basel (1982)
33. Pollack, R., Weston, T.: On anticyclotomic μ-invariants of modular forms. Compos. Math. **147**(5), 1353–1381 (2011)
34. Ribet, K.A.: On modular representations of $\mathrm{Gal}(\overline{\mathbb{Q}}/\mathbb{Q})$ arising from modular forms. Invent. Math. **100**(2), 431–476 (1990)
35. Skinner, C., Urban, E.: The Iwasawa Main Conjectures for GL_2. Invent. Math. **195**(1), 1–277 (2014)
36. Schoof, R., Washington, L.C.: Visibility of ideal classes. J. Number Theory **130**(12), 2715–2731 (2010)
37. Sweeting, N.: Kolyvagin's conjecture and patched Euler systems in anticyclotomic Iwasawa theory. arXiv preprint, arXiv:2012.11771
38. Tamiozzo, M.: On the Bloch-Kato conjecture for Hilbert modular forms. Math. Z. **299**(1–2), 427–458 (2021)

39. Yuan, X., Zhang, S.-W., Zhang, W.: The Gross-Zagier formula on Shimura curves. Ann. Math. Stud. 184, Princeton University Press, Princeton (2013)
40. Zhang, S.-W.: Gross-Zagier formula for GL_2. II. In: Heegner points and Rankin L-series. Cambridge University Press, Cambridge, pp. 191–241 (2004)
41. Zhang, W.: Selmer groups and the indivisibility of Heegner points. Camb. J. Math. **2**(2), 191–253 (2014)

Matteo Longo · Marco Adamo Seveso ·
Rodolfo Venerucci · Stefano Vigni
Editors

Elliptic Curves and Modular Forms in Arithmetic Geometry

Celebrating Massimo Bertolini's 60th Birthday, Milano, Italy, September 12–16, 2022

Springer

Editors
Matteo Longo
Dipartimento di Matematica "Tullio Levi-Civita"
Università degli Studi di Padova
Padova, Italy

Marco Adamo Seveso
Dipartimento di Matematica "Federigo Enriques"
Università degli Studi di Milano Statale
Milan, Italy

Rodolfo Venerucci
Dipartimento di Matematica "Federigo Enriques"
Università degli Studi di Milano Statale
Milan, Italy

Stefano Vigni
Dipartimento di Matematica
Università degli Studi di Genova
Genova, Italy

ISSN 2194-1009 ISSN 2194-1017 (electronic)
Springer Proceedings in Mathematics & Statistics
ISBN 978-3-032-13122-5 ISBN 978-3-032-13123-2 (eBook)
https://doi.org/10.1007/978-3-032-13123-2

Mathematics Subject Classification: 11G05, 11F11, 11G35, 11G25, 11G15, 11G40

This Springer imprint is published by the registered company Springer Nature Switzerland AG
The registered company address is: Gewerbestrasse 11, 6330 Cham, Switzerland

The manufacturer's authorised representative in the EU is Springer Nature Customer Service Centre GmbH, Europaplatz 3, 69115 Heidelberg, Germany. If you have any concerns regarding our products, please contact ProductSafety@springernature.com

Printed and bound by CPI Group (UK) Ltd, Croydon, CR0 4YY
07/10/2026
02229354-0001